Statistics for Biology and Health

More information about this series at http://www.springer.com/series/2848

Statistics for Biology and Health

Lanju Zhang
Editor

Max Kuhn, Ian Peers, Stan Altan
Section Editors

Nonclinical Statistics for Pharmaceutical and Biotechnology Industries

Springer

Editor
Lanju Zhang
Nonclinical Statistics, Abbvie Inc
North Chicago, IL, USA

ISSN 1431-8776 ISSN 2197-5671 (electronic)
Statistics for Biology and Health
ISBN 978-3-319-23557-8 ISBN 978-3-319-23558-5 (eBook)
DOI 10.1007/978-3-319-23558-5

Library of Congress Control Number: 2015958677

Springer Cham Heidelberg New York Dordrecht London

Printed on acid-free paper

Springer International Publishing AG Switzerland is part of Springer Science+Business Media (www.springer.com)

Preface

Drug discovery and development is a long, costly, and high-risk process. It starts with the identification and validation of a disease target and the generation and optimization of compounds that potentially have some important effect on the target. This is followed by a battery of in vitro and in vivo (animal) studies to characterize the safety profile and finally clinical testing on human beings. Only then can a new drug filing be made to regulatory agencies such as the FDA and the EMA to secure the legal right to market a drug product. In parallel, a manufacturing process and associated analytical methods are developed and validated to produce drug product in the appropriate dosage form of consistent quality in the appropriate dosage form for the many required clinical studies and then post-approval marketing. This is a concerted collaboration involving many scientific disciplines and organizations against the backdrop of a heavily regulated industry.

Moving from the early discovery phase to the marketing of a drug is a high-risk proposition. Less than 1/5,000 compounds make it through the development pipeline to successful marketing. So it is no surprise that pharmaceutical companies strive to establish an ample pipeline of therapeutically important compounds with significant activity and a good safety profile. These compounds can only come from basic discovery experiments, preclinical safety studies, and formulation and manufacturing process studies employing good quality control methods. In these nonclinical areas, discovery, preclinical development, and chemistry, manufacturing, and controls (CMC), variability is the norm and most decisions are data driven. Statistics plays a significant role in moving drug candidates through the drug development pipeline, in decision making and assessing risks, and in directly improving the efficiency of the drug development process.

The novice statistician just entering this exciting and rich field of nonclinical statistical applications, however, lacks good reference books describing important statistical tools necessary for their work and professional development. In the literature, the authors are only aware of two reference books covering selected topics in nonclinical statistics. One is *Pharmaceutical Statistics Using SAS: A Practical Guide* (Dmitrienko, Chuang-Stein, and D'Agostino 2007), with four out of 14 chapters discussing nonclinical topics. The other is *Statistics in Drug*

Research: Methodologies and Recent Developments (Chow and Shao 2002), with three out of ten chapters discussing topics in the CMC area. Many important references to nonclinical statistical applications are scattered throughout numerous scientific and statistical journals, industry standards, and regulatory documents. So clearly there is a need for a reference book that collects and summarizes the wide diversity of nonclinical topics into a single volume. With this in mind, we are proud to present the first book that is completely dedicated to the diverse range of nonclinical topics, bringing together relevant discussions of statistical methods in all three nonclinical areas: discovery, nonclinical development, and CMC.

The book is intended to be a reference book for scientists and statisticians in the pharmaceutical and biotechnology industries, as well as in regulatory agencies. It can also serve as a textbook for a statistical consulting course for a statistics graduate program. Our aim is to also provide an excellent resource for academic researchers to understand the current challenges within nonclinical statistics and direct research in this area. It is also the authors' hope that this book will inspire professional statisticians to write additional reference books so as to build a library of reference materials for nonclinical statistics, matching the level of reference materials available to clinical statisticians.

This book is divided into four parts. Part I introduces the book with three chapters. Chapter 1 defines nonclinical statistics, discusses the nonclinical statistical profession, and points out pathways to making it a discipline. Chapter 2 surveys current nonclinical statistical contributions in regulatory agencies with a focus on the US FDA. Chapter 3 gives a roadmap on how to become a good nonclinical statistician.

Part II of the book, including Chaps. 4–8, focuses on the statistical and scientific problems in early drug discovery. This is a broad field; Chap. 4 is a broad overview that introduces this section of the discovery pipeline. The other chapters in the second part focus on target discovery using genetic markers, compound screening via high-throughput screening (HTS), compound optimization, early safety screening, and computational chemistry models.

Part III, including Chaps. 9–14, addresses the statistical challenges associated with working in nonclinical safety assessment and drug development. It is designed to be a clear and accessible presentation of up-to-date technical material combined with practical insight into its application. Experts working in the field share their insights covering six related areas: an overview of the statistician's role and contribution within nonclinical drug safety assessment; statistical aspects of key regulatory safety studies; clinical assay development for biological macromolecules; regulatory perspectives on design and interpretation of carcinogenicity studies, including new research on statistical methods; design and evaluation of drug combination studies; and the increasingly important role of biomarkers in pharmaceutical discovery and development.

Part IV, including Chaps. 15–26, is dedicated to important topics concerning the chemistry, manufacturing, and controls (CMC) aspects of drug development, covering both large molecules (biologics) and small molecules. These follow on the heels of the discovery phase of drug development touched on above. Numerous formulation studies are carried out that combine the API with other substances, known as excipients, to produce the drug product that possesses desirable properties of appearance, shelf stability, and bioavailability. The scientific studies required to produce the final drug product also involve analytical method development studies. These are comprised of studies that develop analytical methods to characterize the physical and chemical properties of the drug product. All of these activities in the CMC area are governed by considerable regulatory oversight in the form of US Code of Federal Regulations sections, or their European and international requirements as well as numerous guidance documents pertaining to pharmaceutical product development. Given this brief description, Part IV covers a broad array of topics that expand and elucidate the regulatory and statistical aspects of these broad areas of CMC studies briefly described above. The topics covered are analytical method validation, lifecycle approach to bioassay, quality by design, process validation, process capability and statistical process control, stability modeling, in vitro dissolution, content uniformity, acceptance sampling, chemometrics and predictive modeling, and comparability studies. Each chapter focuses on a key aspect of analytical method development, formulation development, or associated statistical considerations in manufacturing.

While these 26 chapters include cross-referenced material, they can be read as individual contributions. Collectively they represent a rich source of current information, good practice, practical advice, and guidance with examples.

The editors are grateful to all the contributors who took time to write the chapters of this book. They are leaders and top experts in the field. Their broad view on the topic, in-depth discussion, and stimulating advice for future directions make this book an invaluable reference for practicing statisticians and scientists, academic researchers, and regulatory reviewers. We are indebted to numerous reviewers who helped to edit and improve the chapters. We thank Jonathan Gurstelle who helped to initiate this project before he left Springer. We are indebted to Matthew Amboy and Christine Crigler of Springer Sciences for their professional help in guiding us from the preparation of manuscripts through the production of the book. Our sincere gratitude goes to our families for their patience and support.

North Chicago, IL, USA — Lanju Zhang
Groton, CT, USA — Max Kuhn
Alderley Park, UK — Ian Peers
Raritan, NJ, USA — Stan Altan

Contents

Contributors

Stan Altan is Senior Director and Research Fellow of Nonclinical Statistics at Janssen Research & Development, LLC. He received his Ph.D. from Temple University in Biometrics. Stan is a Fellow of the American Statistical Association. Over the past 30+ years, he has supported drug discovery, toxicology, pharmaceutical and chemical development, biologics, and all phases of clinical studies. Stan is a founding member of the Nonclinical Biostatistics Leaders' Forum, the IQ Consortium Statistical Leadership Group and the AAPS CMC Statistics Focus Group. He is on the editorial board of the Statistics in Biopharmaceutical Research journal, an ASA publication.

Steven A. Bailey joined the Pfizer Drug Safety R&D Statistics Team from Wyeth in 2010. Prior to that, he had 26 years of experience in Drug Safety support at Wyeth. Steven has particular expertise in the support of reproductive toxicology and juvenile animal studies, as well as experience in general toxicology, safety pharmacology, carcinogenicity and safety biomarker development. Steven earned his B.S. in mathematics from Carnegie-Mellon in 1982, his M.S. in statistics from the same school in 1983, and is currently an American Statistical Association Accredited Professional Statistician (PStat).

Luc Bijnens holds master's degrees in zoology and biostatistics and a Ph.D. degree in biology. He spent the earlier part of his career in academia in Europe and Africa. After having worked in the pharmaceutical field for some time, he joined Johnson and Johnson (J&J) to support clinical oncology. After that, he built the European nonclinical biostatistics team. He is a visiting professor at the University of Hasselt and he played a major role in statistical societies in Belgium and Europe. He has mentored many master and Ph.D. level students leading to publication of their work in peer reviewed journals.

Bruno Boulanger holds a Ph.D. in experimental psychology from the University of Liège. After a post-doctorate at the Université Catholique de Louvain and the University of Minnesota in Statistics, he joined Eli Lilly in Belgium in 1990. Since then, he gathered 20 years of experience in several areas of pharmaceutical research and industry. He holds various positions in Europe and in the USA. Bruno is the chief scientific officer of Arlenda and Senior Lecturer at the Université of Liège. He is also a USP Expert, member of the Committee of Experts in Statistics. Bruno has authored more than 100 publications in applied statistics.

Richard K. Burdick is Emeritus Professor of Statistics in the W. P. Carey School of Business, Arizona State University (ASU). He also recently worked for 10 years as Quality Engineering Director for Amgen, Inc. His research and consulting interests are in CMC statistical applications and measurement systems analysis. He has written over 50 journal articles and 3 books. Dr. Burdick is a Fellow of the American Statistical Association and a member of the American Society for Quality. He received his Bachelor's Degree in statistics from the University of Wyoming. He received his Master's and doctorate degrees in statistics from Texas A&M University.

Rita M. Cantor is a Professor of Human Genetics in the David Geffen School of Medicine at UCLA. She conducts statistical genetics research in genetics and genomics and has found genetic associations with complex traits such as autism and coronary artery disease. This includes developing methods and conducting analyses for quantitative traits and gene by environment interactions.

Xiaoyu (Cassie) Dong, Ph.D. is a Statistics Reviewer at Office of Biostatistics in CDER, FDA. She received Ph.D. in statistics from the UMBC in 2011. After a half year job at MedImmune, she joined FDA in 2012. Her primary review work concentrates on the product quality assurance, including product specification setting, stability, sampling plan, dissolution, process characterization/validation, assay analysis and biosimilar assessment. In addition, she actively conducts research on product quality. Cassie is the author of 16 scientific journal articles on various aspects of pharmaceutical statistics and science. She is also an invited speaker for a number of international/national conferences.

Gary Gintant is a Research Fellow in the Dept. of Integrative Pharmacology, Integrated Science and Technology, at AbbVie. He is involved in multiple drug discovery and safety activities and initiatives internally;external activities include various cardiac safety initiatives (such as ILSI/HESI Proarrhythmia Models Project, theCardiac Safety Research Consortium, and the Comprehensive in Vitro Proarrhythmia Assay Initiative) while serving on various journal editorial boards, NIH study sections, and Safety Pharmacology Society committees. Research interests include cardiovascular pharmacology, cellular electrophysiology/ion channels, arrhythmias, application of stem-cell derived cells and tissues to drug discovery efforts, and translational medicine. He gained his MA, M.Phil. and PhD. degrees from the College of Physicians and Surgeons of Columbia University, NY, and was on faculty at Wayne State Univ. School of Medicine in Detroit MI prior to joining Abbott/AbbVie.

Hanspeter Gubler, Ph.D. is Senior Scientific Business Analyst/Business Consultant and Novartis Leading Scientist in the Informatics Department of the Novartis Institutes for BioMedical Research (NIBR). He has broad experience in statistical data analysis and statistical modeling of in-vitro assay data in drug discovery. He is lead designer of all calculation components of the global NIBR in-house High-Throughput Screening, Profiling and Compound Combination data analysis software systems, as well as several more specialized systems for the analysis of biophysical experiments. He was previously Director and Group Leader of the Informatics and Automation Technology group in the Novartis Lead Discovery Center.

Chris Harbron has over 20 years of industrial experience working as a statistician. He has worked within Pfizer, AstraZeneca and now Roche in positions which have covered the complete discovery and development pipeline. Chris has a strong interest in biomarkers and their application in personalized healthcare, understanding biological processes and more general decision making and has published and presented externally and widely on these subjects.

Lynne Hare is a consulting statistician with 45 years of experience emphasizing business process improvement in R&D, manufacturing and other strategic functions. Serving a large client base, he has helped bring about culture change in research by accelerating speed to the successful launch of new products and processes and in manufacturing through the reduction of process variation. He has led statistics organizations at Kraft Foods, the National Institute of Standards and Technology, Unilever and Hunt-Wesson Foods. He is a Fellow of the American Statistical Association and of the American Society for Quality.

Athula Herath is Statistical Director of MedImmune (AstraZeneca Biologics Unit) based at the Translational Sciences Department of MedImmune in Cambridge (UK). He has been with AZ for the past 9 years and prior to that worked for Nestle (Switzerland) as the Group Manager of the Nestle's Research Centre in Lausanne. Before joining Nestle, Athula was the Director of Biostatistics and Bioinformatics of Oxford GlycoSciences in Oxford, UK. He obtained his Ph.D. in computer science (statistical computing) from the University of Keele (UK) and was a Lecturer of Computing Sciences at the University of Keele before joining Industry in 1998. Athula currently chairs the Biomarker Special Interest Group of the British Pharmaceutical Statisticians Association (PSI).

Buffy Hudson-Curtis is a nonclinical statistician working at GlaxoSmithKline. She has spent the majority of her career in research and development, and is currently working in drug manufacturing. She has experience with oral solid dosage forms, topical formulations, inhalation products and biopharmaceutical products. Buffy, who received her Ph.D. in statistics from North Carolina State University in 2001, is thankful for the opportunity to author this chapter with her former classmate and coworker, Dr. Steven Novick.

Craig Hyde has worked in the field of statistics and genetics since 2001 and has been at Pfizer since 2004, now as Director of Research Statistics supporting human genetics. His work spans analysis of epigenetics, nextgen sequencing, candidate gene studies, and GWAS for target discovery, as well as pharmacogenomics studies in clinical settings. His current focus is on Mendelian randomization, metabolomics and loss of function variants. He completed his formal education at the University of Arizona, where he received a Ph.D. in applied mathematics in 1998.

Matthew T. Jackson grew up in Palmerston North, New Zealand, and attended the University of Canterbury where he studied mathematics and performance bassoon. He moved to Pittsburgh, PA, for his graduate studies, attending both

Carnegie Mellon University and the University of Pittsburgh. Matthew completed his Ph.D. in category theory and logic in 2006, and after a short spell of teaching, was employed from 2008 until 2014 as a nonclinical biostatistics reviewer at the US Food and Drug Administration. He is currently working on software development for Amgen Inc.

Robert J. Kubiak heads a research group responsible for validation of bioanalytical methods for pharmacokinetic measurements and immunogenicity assessments at MedImmune, Gaithersburg, Maryland. He worked on development and validation of immunoassays at Tandem Labs, PPD and Meso Scale Discovery. He holds a doctorate in medicinal chemistry from University of Illinois at Chicago and a master's degree in biotechnology from Johns Hopkins University.

Max Kuhn is a Senior Director in Research Statistics in Pfizer R&D in Groton, CT. He has over 15 years of experience in the pharmaceutical and diagnostic industries. His interests are in the application of machine learning models and estimation problems in general. He is the co-author of the best selling text Applied Predictive Modeling and the author of eight R packages.

David LeBlond is a statistical consultant with 35 years of experience in the pharmaceutical and medical device industries. He is active on statistical expert committees and panels for PhRMA, PQRI, AAPS and USP, holds degrees in biochemistry (M.S., Ph.D.) and statistics (M.S.), and has co-authored more than 50 peer-reviewed publications in these fields.

Pierre Lebrun completed a master's degree in computer sciences and economy, followed by a master's in statistics at the University of Louvain-la-Neuve in Belgium. Next, he completed his Ph.D. in statistics in 2012 at the University of Liege (Belgium), in the topic of Bayesian models and design space applied to pharmaceutical industry. Pierre is now working for Arlenda, a company specialized in biostatistics for pharmaceutical development. Topics of interest include Bayesian statistics, Quality by Design, (bio)assay development and validation, process validation, formulation, manufacturing and release of drugs and vaccines, using effective Design Space strategies.

Dingzhou Li joined Nonclinical Statistics at Pfizer, Inc. in 2008 and is currently on the Drug Safety R&D Statistics Team. Dingzhou has supported multiple areas in Discovery and Drug Safety R&D, including assay development and validation, pharmacokinetics and metabolism, safety pharmacology, genetic toxicology and safety biomarkers. Prior to joining Pfizer, Dingzhou was in the Discovery Statistics Function of Eli Lilly, Inc. from 2004 to 2008. Dingzhou holds a B.S. in physics from Wuhan University in China, an M.S. in biostatistics and a Ph.D. in physics from the University of Michigan-Ann Arbor.

Jason J.Z. Liao received his Ph.D. in statistics in 1997 from the Department of Statistics, The University of Michigan. Since then, he has been working in the biopharmaceutical industry with increasing responsibilities. He has excellent statistical knowledge and practical experience in whole drug development for chemical drugs, biologics, vaccines and biosimilars/follow-on biologics. He is actively promoting more appropriate statistics and more powerful statistics in solving real pharmaceutical applications and has published about 40 peer-reviewed articles.

Karl K. Lin is an Expert Mathematical Statistician/Team Leader (Applications in Pharmacology and Toxicology), CDER/FDA. He received his B.A. and M.A. degrees in economics from National Taiwan University, and Master's, and Ph.D. degrees in statistics from the University of California, Berkeley, and the University of Michigan, respectively. He has been with the FDA since 1987. His main responsibilities at FDA include statistical reviews and regulatory research in animal toxicology/pharmacology studies, stability studies and bioavailability/bioequivalency studies of human new and generic drugs. He has had extensive publications in his areas of responsibilities by himself or in collaboration with professionals in USA and abroad.

Yang Lu is an Assistant Professor in Residence at the Harbor-UCLA Medical Center, David Geffen School of Medicine at UCLA and the Los Angeles Biomedical Research Institute. With a background in health policy and health economics, her research focuses on improving adherence to treatment regimen and outcomes of chronic conditions. One of her research interests is to apply findings in precision medicine to pharmacogenetic and pharmacoeconomic studies.

Min Min received her Ph.D. in statistics from University of Maryland, College Park, in December 2007. She joined the FDA in June 2008 and worked as a statistical reviewer with The Pharmacology and Toxicology team for six and a half years. Currently, she is a statistical reviewer for Division of Gastroenterology and Inborn Errors Products (DGIEP) team. Her research areas include mixed models, meta-analysis, and longitudinal and categorical data analysis.

Steven Novick is a seasoned pharmaceutical industry statistician. Having spent the majority of his career with GlaxoSmithKline, he is currently a senior statistician with Arlenda. Steven authored and co-authored about 1 dozen CMC and drug-discovery statistics articles, including a seminal paper on DDU testing via the parametric tolerance interval test. He received his Ph.D. from North Carolina State University in 2000. Steven is gratified to co-author this chapter with his long-time friend and former NC State classmate, Dr. Buffy Hudson-Curtis.

Ian Peers is a senior Leader in Global Medicines Development at AstraZeneca. He received his B.Sc. from the University of Wales, Bangor and his Master's and Ph.D. degrees from the University of Manchester. He has over 26 years of experience working as an academic and industrial statistician in Research and Development with 14 years of experience in biopharmaceuticals research including 11 years as Global Head of Statistics at AstraZeneca. Ian has collaborated with academic researchers globally, has supervised doctoral and post-doctoral students, worked in several therapeutic areas and has published and presented on preclinical, translational and clinical research. He is a professional Charted Statistician (CStat) of the Royal Statistical Society and holds an Honorary Professorship in the Faculty of Natural Sciences at the University of Stirling.

Katherine Perez-Morera is a writer and a pharmacist who graduated with a Pharm.D. from the University of Florida in 2012. She spent her formative years in genetics and pharmacodynamics labs, and now focuses her professional life on promoting outstanding patient care and studying neuroscience; she's also passionate about spreading the knowledge she's acquired by writing expository essays as well as fictional stories from a scientific perspective. She can be contacted at perezmorera@gmail.com and she welcomes questions.

John Peterson received his B.S. degree from the Stony Brook University (double major: applied mathematics and computer science). He received his Ph.D. in statistics from the Pennsylvania State University. John is a Fellow of the American Statistical Association, an Associate Editor of the Journal of Quality Technology and a co-founding member of the Industrial Statistics Section of the International Society for Bayesian Analysis. He has published extensively on response surface methodology and its role in process optimization. John is a Senior Director in the Statistical Sciences group at GlaxoSmithKline Pharmaceuticals. His current research involves using Bayesian statistical methodology for process optimization.

Bill Pikounis is Head of Nonclinical Statistics & Computing at Janssen Pharmaceuticals, the pharmaceutical division of Johnson & Johnson. He oversees worldwide statistical services and products to serve areas of discovery, safety, and manufacturing of biotech (biologic, large) molecule, vaccine, and small molecule (pill, tablet) products. He received his Ph.D. from the University of Florida. Bill is an American Statistical Association (ASA) Fellow. His professional technical interests lie in data graphs, resampling methods, resistance and robustness, longitudinal data analysis, statistical computing and software solutions. For more details, see http://billpikounis.net/.

David M. Potter currently leads the Drug Safety R&D Statistics Team at Pfizer Inc., and has served in both technical and global leadership positions in Nonclinical Statistics since joining Pfizer in 1998. David's focus is on how quantitative methods can help organizations make high-quality decisions in the face of uncertainty. He holds a B.S. in economics from the University of Pennsylvania's Wharton School, and M.S. and Ph.D. degrees in statistics from the University of Wisconsin-Madison.

Michelle Quinlan is an Associate Director in Clinical Pharmacology Biostatistics at Novartis Oncology. She received her Ph.D. in statistics from the University of Nebraska-Lincoln in 2010. She was the 2008 Honorable Mention Gertrude Cox Scholarship recipient. She joined the PQRI Stability Shelf Life Working Group in 2007 as a graduate student and her dissertation research focused on evaluating current methodology and developing enhanced methodologies for shelf life estimation. Her research interests include quantile regression with random effects, baseline correction methods for TQT studies and PK/PD modeling.

Mohammad Atiar Rahman is Acting Division Deputy Director and Expert Mathematical Statistician, CDER/FDA. He received his B.Sc. and M.Sc. in statistics from Karachi University, Pakistan, and his M.S. and Ph.D. in statistics from Monash University, Australia and State University of New York at Albany. He has worked for the FDA from 2000 to present, and during 1990–1996. He worked for Berlex Laboratories and Otsuka America Pharmaceuticals between 1996 and 2000. He has published ten research papers and three book chapters along with numerous presentations in various meetings and seminars on topics related to FDA statistical review of carcinogenicity studies of new pharmaceuticals.

Eric Rozet is Senior Statistician at Arlenda in Liège (Belgium). He has more than 12 years of experience in nonclinical statistics and particularly in statistical aspects related to (bio)assays and processes: QbD and robust optimization, validation, transfer, inter-laboratories studies, uncertainty assessment, etc. Eric trains analysts of the Pharmaceutical Industry on topics such as optimization, validation, robustness and transfer of assays and processes. He is also author of numerous articles in applied statistics and is regularly giving presentations on these subjects. Eric has a B.Sc. degree in bio-engineering, a Master's degree in biostatistics and a Ph.D. degree in pharmaceutical sciences.

Timothy Schofield is a Senior Fellow in the Analytical Biotechnology department at MedImmune. Prior to joining MedImmune, Tim worked at Arlenda, Inc. as Managing Director and Head of Nonclinical Statistics, at GSK in US Regulatory Affairs, and at the Merck Research Laboratories heading the Nonclinical Statistics unit. Tim is a member of the USP Statistics Expert Committee and has participated in industry initiatives related to Quality by Design, analytical method development and validation, stability and specifications. Tim received a Master of Arts degree in Statistics and Operations Research in 1976 from the University of Pennsylvania in Philadelphia.

Meiyu Shen is a team leader and a senior statistical reviewer at the Center of Drug Evaluation and Research, Food and Drug Administration (FDA). She obtained her Ph.D. in statistics from Statistics Program of Department of Mathematics at University of Maryland at College Park in 2015. She obtained her Ph.D. in chemical engineering from the Department of Chemical Engineering at Iowa State University in 1999. She received four FDA Outstanding Service Awards for significant efforts and contributions in statistics related to reviews and research in chemistry and manufacturing control and biosimilar evaluation. She has published more than 35 papers in the statistical, medical and engineering journals.

Lei Shu received her Ph.D. in statistics from Purdue University in 2008. After graduation, she joined Abbott/AbbVie as a nonclinical statistician in supporting early phase drug safety studies: in-vitro/in-vivo cardiac safety screening, genotoxicity, preclinical toxicology and animal carcinogenicity studies. She was the major statistician who supported the developing of the novel QTiSA-HT cardiac screening test in Abbott/AbbVie. She is now a Senior Biostatistics manager at Astellas Pharma.

Ronald D. Snee, Ph.D. is Founder of Snee Associates, LLC, a firm that provides guidance to pharmaceutical and biotech executives in their pursuit of improved business performance using Quality by Design and other improvement approaches. He worked at DuPont for 24 years prior to starting his consulting career. He has authored several articles on how to successfully implement QbD, co-authored two books on QbD and speaks regularly at pharmaceutical and biotech conferences. He has published 5 books and more than 270 papers in the fields of quality, performance improvement, management and statistics. He has received numerous awards and honours for his contributions in these fields. Ron received his B.A. from Washington and Jefferson College and M.S. and Ph.D. degrees from Rutgers University.

Perceval Sondag holds a Bachelor's degree in physical therapy and a master's degree in statistics from the University of Louvain-la-Neuve (Belgium). After working in Quality Control and Operational Research in Hospital setting, he joined Arlenda in 2013 and specialized in Bayesian modeling and nonclinical statistics.

Marie C. South obtained a first class honours degree in mathematics from Cambridge University, followed by an M.Sc. in statistics with applications in medicine from Southampton University. She subsequently worked for the Medical Research Council Biostatistics Unit, before gaining a Ph.D. from the University of Newcastle upon Tyne. Since 1994, Marie has worked for AstraZeneca (formerly Zeneca) supporting manufacturing and subsequently drug discovery. In addition to roles supporting oncology and neuroscience, she has spent the greater part of her recent career providing global leadership for a team providing expert statistical input into nonclinical studies for drug safety assessment.

Helen Strickland is the Senior Statistical Consultant in the Product Lifecycle Management Competency Center in GSK. She received a B.S. in chemistry in 1984 and an M.S. in statistics in 1996 from North Carolina State University. Helen worked as a Quality Control Chemist for Novozymes from 1984 to 1991. In 1992, she joined GSK as an analytical chemist. She has over 23 years of experience working with orally inhaled products and oral solid dosage products. She has extensive experience providing statistical support to quality control evaluations, stability protocols, process validation studies and in vitro equivalence comparisons.

Walter W. Stroup is a Professor of Statistics at the University of Nebraska, Lincoln. He received his Ph.D. in statistics from the University of Kentucky in 1979. His areas of expertise are statistical modeling and design of experiments. He joined the PQRI Stability Shelf Life Working Group in 2007 and received PQRI's 2009 Excellence in Research award. He is author or co-author of four mixed model textbooks and has published extensively in this area. He has taught numerous workshops, directed several graduate student research projects, and has consulted with government and industry organizations on pharmaceutical issues in statistics.

Cheng Su has a Ph.D. in statistics from North Carolina State University and 18 years of experience working in nonclinical statistics areas (9 years with Roche Palo Alto, 9 years with Amgen Washington). He led groups of statisticians supporting drug discovery and translational sciences areas. He is interested in applying statistical thinking and solutions to make a difference in drug research and development. Specific areas of expertise include high throughput screening, drug combination, predictive modeling, genomics and genetics, biomarkers, and automated analysis system. He is currently a Biostatistics Director with Amgen, working in early development clinical trials after recent company reorganization.

Steven F. Thomson received his M.S. in mathematics and his M.S. in statistics and an ABD (Master's degree + 2 years) from the University of Kentucky in 1972 and 1977, respectively. Prior to joining the FDA in August of 1994, he was a statistical research associate for several years at the Fred Hutchinson Cancer Research Institute in Seattle, Washington and a statistical/statistical computing consultant at the University of Kentucky. His statistical interests include Bayesian statistics, statistical computing, linear models and latent variable models. He has several co-authored book chapters and papers in the above areas of interest.

Yi Tsong is Director of Division of Biometrics VI, in the Office of Biostatistics of Center for Drug Evaluation and Research. He leads a statistical group to support research and review of studies of CMC, analytical biosimilar, new drug bioequivalence, Pharm-tox, Thorough QT, drug abuse potential and abuse deterrent. He joined FDA for 28 years. He is active in research in all aspects of regulatory statistics. He also serves as Associate Editors of SIM and JBS. He graduated in 1979 with a Ph.D. in statistics from University of North Carolina at Chapel Hill.

Wen Wu has 15 years of industrial experience in statistics at GSK, Roche and AstraZeneca, and has 20 years of hands on experience in data science, with 48 peer-reviewed publications. He holds dual Ph.Ds in chemometrics and analytical chemistry, an M.Sc. in pharmacology and a B.Sc. in pharmacy. He has been a member of the editorial advisory board of Chemometrics and Intelligent Laboratory System since 2006, and is also a Chartered Chemist (CChem), Chartered Scientist (CSci) and Fellow of the Royal Society of Chemistry (FRSC). He was included in the 2008 Edition of Who's Who in the world and the 2007 Edition of Who's Who in Science and Engineering.

Hyuna Yang is a trained statistician, has worked on statistical genetics/bioinformatics, and recently works on CMC area.

Phillip Yates is a Manager in Pfizer's BioTherapeutics Statistics group in Groton, CT. As a nonclinical statistician, he primarily supports the Pharmacodynamics and Drug Metabolism organization in most areas of small and large molecule assay development. He also has experience in the area of discovery target biology and the analysis of –omics data.

Frank Ye, Ph.D., is currently Director of Quality and has responsibility for Quality operations in China including Quality of distributed products, local manufacturing and joint ventures. Previously, Dr. Ye was Director of Quality Engineering in Amgen with responsibility of providing analytical and statistical techniques aimed at maximizing the effectiveness and efficiency of Amgen processes to deliver quality products. Before joining Amgen, Dr. Ye worked in GlaxoSmithKline and Schering-Plough as Principal Statistician supporting pharmaceutical development and manufacturing. Dr. Ye holds a B.S. in computer sciences and an M.S. in statistics from the University of Oregon, and a Ph.D. in biostatistics from the University of North Carolina at Chapel Hill.

Jianchun Zhang is a principal statistician at MedImmune. As a nonclinical biostatistician, he provides statistical support to scientists during the entire biologics drug development cycle from preclinical R&D to post-marketing commitments, including preclinical studies, assay and process development, translational science, manufacture, etc. In particular, he works on immunogenicity assay development and validation and has several publications in this regard. He received his Ph.D. in statistics in 2010 from Purdue University.

Lanju Zhang is Director in Statistics and Head of Nonclinical Statistics Group in the department of Data and Statistical Sciences at AbbVie. He leads a group providing statistical support to preclinical studies and CMC areas. Prior to moving to AbbVie, he was in MedImmune with increasing responsibilities to support all nonclinical areas. He is active in research and has published many papers and book chapters in nonclinical and clinical areas. He received his Ph.D. in statistics in 2005 from University of Maryland Baltimore County.

Part I
Introduction

Chapter 1
Introduction to Nonclinical Statistics for Pharmaceutical and Biotechnology Industries

Lanju Zhang and Cheng Su

Abstract Drug discovery and development is a long, complicated, risky, and costly process. Most decisions in the process are made based on data with uncertainties, which provides a natural field for statistics. Clinical statistics is a shining example of how an industry has embraced statistics as an equal partner. However, it is not the case of nonclinical statistics. In this chapter, we give a brief introduction to drug discovery and development process, including chemistry, manufacturing and controls (CMC), statistical applications in these nonclinical areas, and the current landscape of clinical and nonclinical statistics in pharmaceutical and biotechnology companies. Then we try to define nonclinical statistics, discuss the nonclinical statistical profession, identify possible causes of predicaments in nonclinical statistics, and point out potential directions to change the status quo.

Keywords CMC • Drug research and development • Nonclinical statistics • Target identification

1.1 Introduction to Drug Research and Development (R&D)

The mission of drug research and development (R&D) is to bring new safe and efficacious medicines to patients. Many have reported that the process of developing a new medicine is long and costly. According to one report by Innovation.org ("Drug discovery and Development: understanding the R&D process"), the entire R&D process of developing a new medicine takes about 10–15 years and costs about 800 million to 1 billion dollars. These numbers include the cost of thousands of failures; for every 5000–10,000 compounds that enter into the development process, only one receives marketing approval, underlying the high risk of this business.

L. Zhang (✉)
Data and Statistical Sciences, AbbVie Inc., North Chicago, IL, USA
e-mail: lanju.zhang@abbvie.com

C. Su
Research and Translational Sciences Biostatistics, Amgen Inc., Thousand Oaks, CA, USA

L. Zhang (ed.), *Nonclinical Statistics for Pharmaceutical and Biotechnology Industries*, Statistics for Biology and Health, DOI 10.1007/978-3-319-23558-5_1

Recent reports indicated that the cost is increasingly higher. The lengthy process and high cost highlight the difficulty of developing a new medicine, as well as the opportunities for pharmaceutical companies to improve R&D productivity.

The importance of statistics has long been recognized in clinical trials and reinforced by many regulatory and industry guidelines. Drug companies dedicate most of their statistical resources to these stages of drug development in comparison to earlier exploratory and preclinical studies. This is mostly driven by the necessity of meeting stringent regulatory requirements. In contrast, for the earlier stage studies for which guidelines are not available or limited, there is often a need to demonstrate the benefits of statistical rigor before they are accepted by relevant stakeholders. Many of these earlier studies fall into the realm of nonclinical statistics, which we aim to define in this chapter and discuss throughout the book. It is imperative for us, as nonclinical statisticians, to help our scientists and business leaders understand the promise of a more efficient R&D process through a systematic application of statistical thinking and techniques in experimental design, data analysis and interpretation to improve the success rate in each stage of the R&D process.

In this book, we aim to define nonclinical statistics as an interdisciplinary subject and highlight significant statistical applications in the nonclinical arena to increase the visibility of our profession and provide guidance for new statisticians who are entering this field.

1.1.1 Drug Discovery and Development Process

Sufficient understanding of the subject matter of science is critical to a statistician's ability to solve the right problems using the *right* approaches. The *right* approaches should not only be statistically sound but also meet the level of practical needs. Hence, successful nonclinical statisticians need to have an adequate understanding of the scientific area he/she works on. In what follows we provide a brief review of the drug R&D process to enhance this understanding.

1.1.1.1 Target Identification

A target is generally a single molecule such as a gene or protein that is involved in disease pathway and upon intervention leads to changes in disease progression. Target discovery consists of target identification and validation. Common methods of target identification include genetic association, genomic association and phenotypic screening. As a simple example for genomic association study, comparison of gene expressions between disease and normal subjects could lead to a list of differentially expressed genes which provide a starting point for target discovery. Once potential targets are identified, they need to be fully validated. A multi-validation approach would increase confidence on the involvement of identified targets in disease through study of various functional aspects. One class of such approaches is to verify the association between disease relevant phenotypes and modification of target related genes. Examples of this approach include transgenic animal model when a target gene is knocked out, or in vitro system that gene expressions are silenced through interfering RNA (siRNA), or use of CRISPR technology to modify the genome. It is common that same methods are used in both target identification and validation.

1.1.1.2 Lead Generation and Optimization

The next step is to find drug candidates that interact with the chosen target. The process includes screening, lead identification and lead optimization. In screening, the strategy is to search through a library of compounds (often in the scale of tens of thousands) to identify a shorter list with strong interactions with the target in binding or cell based assays. One such approach is high-throughput screening and the resulting list is called hits. In the lead identification stage, a hit is considered a lead if it possesses many "drug-like" chemical and biological properties in absorption, distribution, metabolism, excretion and toxicity (ADME/Tox), and pharmacokinetics (PK). In the lead optimization stage, structure modifications are made to the leads to make them more effective and safer, e.g., to increase the selectivity, the ability to bind to the target instead of other proteins. The optimized leads become drug candidates and move to the next development stage.

1.1.1.3 Nonclinical Development

A drug candidate is tested extensively for pharmacological and toxicological properties through in vitro and in vivo studies. According to ICH guideline M3(R2), nonclinical safety assessment usually includes pharmacology studies, general toxicity studies, toxicokinetic and nonclinical PK studies, reproduction toxicity studies, genotoxicity studies and in some case carcinogenicity studies. Other nonclinical studies to assess phototoxicity, immunotoxicity, juvenile animal toxicity and abuse liability should be conducted on a case-by-case basis. The need for nonclinical safety studies and their relation to the conduct of human clinical trials is delineated in the guidance ICH M3(R2).

In this stage, researchers also start working on drug manufacturing process and formulation methods. After adequate information of the safety profile is gained from animal studies, the company will file an investigational new drug (IND) application to regulatory agencies such as FDA for initiating clinical trials. All studies prior to IND constitute preclinical development. Some animal studies, for example carcinogenicity studies and long term toxicity studies, will be conducted during the clinical development stage. It is for this reason that we call all these animal studies together as nonclinical development.

1.1.1.4 Translational Science

Here we refer to the studies that aid the extrapolation of drug use from animal to human. It is not a completely new concept; however its recent emphasis in the Critical Path Initiatives of FDA 2004 signaled a paradigm shift in drug research and development. Translational studies include late discovery, nonclinical development and early clinical studies, so it is not an independent phase of R&D per se. The importance of translational science has been recognized by many drug companies that have formed organizations specialized in translation research, though the exact scope could differ from company to company. Some even add a new stage, namely "Exploratory" to their R&D process diagram to emphasize the translational science studies that are not represented in either nonclinical development or clinical development.

A key focus is the use of biomarkers and human tissues. Through biomarker studies in human tissues (e.g., blood), researchers could gain insight on drug's PD effects and PK/PD relationship, which could lead to a better decision on target population selection and dosing decision for clinical trials. An important part of the effort, which starts at earlier stages, is to identify and verify promising biomarkers and to develop and validate bioassays that can reliably measure the biomarkers. Translational studies may use human samples from healthy donors, previous or ongoing clinical trials. This area of research has seen substantial growth and the use of innovative technologies and statistical methods.

1.1.1.5 Clinical Development

A drug candidate is studied in human subjects to evaluate its clinical effectiveness and safety through three phases of clinical trials. Although nonclinical statisticians do not usually support clinical trial design, execution, analysis and final report, it is beneficial for them to understand the clinical development milestones as many nonclinical development and CMC decisions are dependent on clinical development.

1.1.1.5.1 Phase I

Phase I clinical trials typically (except oncology) enroll healthy volunteers to investigate the absorption, distribution, and excretion of an investigational compound in the human body to find its most tolerable dose (MTD). The trials tend to be small and quick. The PK profile is also studied to gain insight on dose, exposure, and safety relationship.

1.1.1.5.2 Phase II

Efficacy of the drug candidate is explored in trials (phase IIa) with up to several hundred target patients treated for usually several weeks or a few months (Proof of Concept). If the drug candidate shows certain level of effect, a larger dose ranging study (phase IIb) involving 3–5 dose levels is conducted to establish the drug candidate's dose response (efficacy and safety) profile, so that an optimal dose may be selected for the next phase III testing.

1.1.1.5.3 Phase III

This phase of trials are also called confirmatory trials, which often involve thousands of patients, adequately long follow-up, and proper statistical design and analysis with stringent type I error control to confirm the drug candidate is efficacious and safe at the selected dose in the targeted patient population. The results of these phase III trials (typically two or more) form the foundation of a new drug application (NDA) for regulatory approval of marketing authorization.

1.1.1.5.4 Post-approval Studies

These studies are sometimes called phase IV trials. These include trials are conducted after a drug is marketed to gather information of its effect in various populations and any side effect of long term use. They also include trials that compare the effectiveness or safety of a drug to other treatment options.

Figure 1.1 depicts a typical drug discovery and development process.

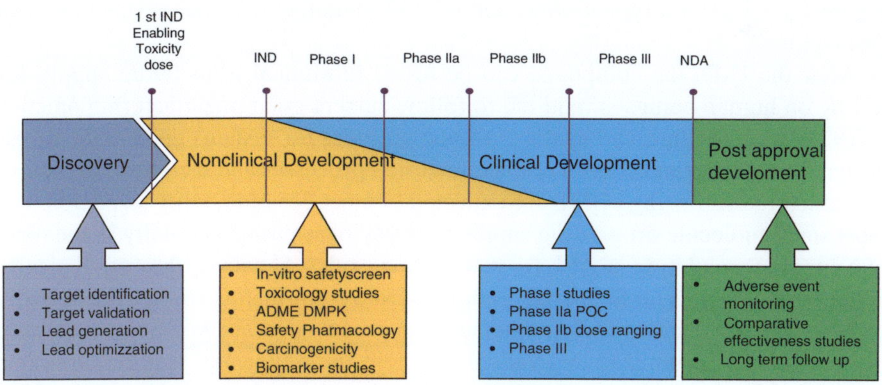

Fig. 1.1 Drug discovery and development process

1.1.2 Chemistry, Manufacturing and Control (CMC)

After a lead compound is discovered and moves through different development stages, a supply of the compound is required for a sequence of testing in vitro, in vivo and in humans to establish its safety and efficacy profiles. The scale and quality standard of drug supply varies at different stages, and accordingly, the manufacturing processes and analytical methods also evolve. In the following, we will discuss drug manufacturing and quality testing at different stages from discovery to submission and post approval, using small molecules as an example.

At the discovery stage, after a disease target is identified, high throughput screening assays are used to generate hits— molecules that interact with the target in the desirable way. The physiochemical properties of the best "hit" compounds, sometimes called leads, are then tested with different assays. Cardiotoxicity and hepatotoxicity will also be tested before a compound moves to the preclinical stage. In these assays at the discovery stage, the amount of each compound required is usually at a milligram level for small molecules. Drug supply at this level can be synthesized by discovery scientists in a flask. Purity of the synthesized drug is characterized but impurities are not usually identified or quantified at this stage.

At the preclinical stage, toxicity and other safety studies require a significantly larger amount of drug, from a scale of grams to hundreds of grams. The production of such drug supply is transferred to a department in the Research and Development (R&D) organization, responsible for manufacturing active pharmaceutical ingredients (APIs). A manufacturing process is developed, along with analytical methods to characterize the API and to identify and quantify impurities. Raw materials are sourced with appropriate quality control. Sometimes outcome of safety studies and drug metabolism tests may indicate a heightened concern for safety or bioavailability, leading to a modification to the synthesization, an optimization of manufacture process and analytical methods, or a re-sourcing of some raw materials. Figure 1.2 includes a typical flowchart of API manufacture process (the part with dashed lines).

After the IND, the compound can be tested in clinical trials. Drug supply for testing on human beings is required to follow current good manufacturing practice (cGMP), which is enforced by the FDA and provides for systems that assure proper design, monitoring and control of manufacturing processes and facilities. Medicines manufactured following cGMP are expected to meet desired quality standards. For most small molecule drugs, drug supply for early phase trials is usually in the form of a simple formulation of API, for example, capsules. However, they are typically in the form of tablets for final drug product. Before confirmatory trials, a formulation of API has been decided and a manufacturing process and analytical methods have been developed, optimized and validated. These trials typically require kilograms or more of the drug products, at least three batches to monitor the robustness of the process and analytical methods. Some of the drugs will be put on stability study at the proposed storage condition (typically room temperature). The data will be used to project a shelf life based on critical quality attributes (for example, potency and

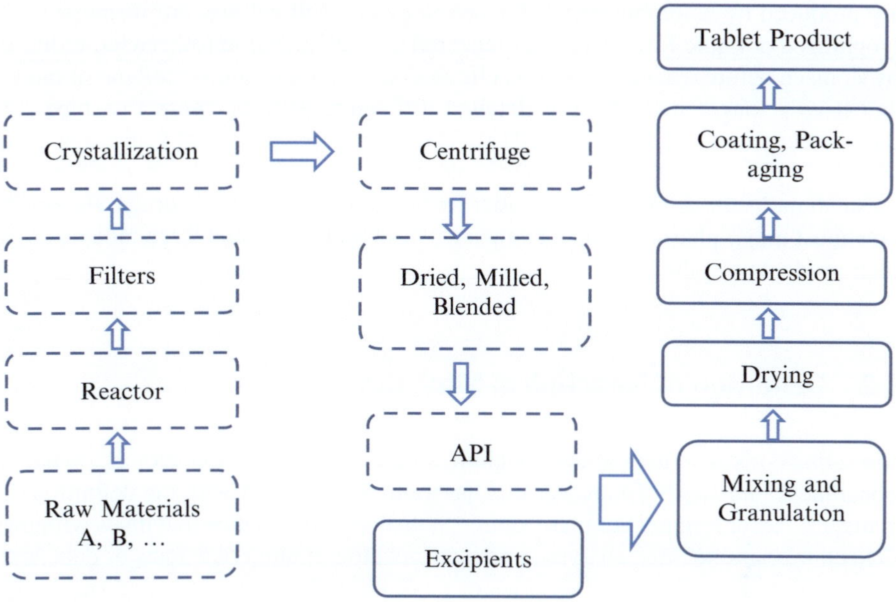

Fig. 1.2 Drug manufacturing and tableting process

impurity) and corresponding specifications (ICH Q1). Figure 1.2 includes a typical tableting process (the part with solid lines).

Since the efficacy and safety of the drug confirmed in these trials is conditional on the quality of drug used, which in turn is conditional on the manufacturing processes and analytical methods, the commercial scale manufacturing after the drug is approved for marketing should use the same manufacturing process and analytical methods as for the confirmatory clinical trials. Any change to the process or analytical methods needs to show improvement or comparability to those used for registration trials.

1.1.3 Differences Between Small and Large Molecules

The discussion in Sects. 1.1.1 and 1.1.2 focuses on small molecules. Another type of therapeutics is large molecules or biologics that requires some differences in terms of discovery, development and manufacture. Small molecules are usually chemical compounds that have molecular weight less than 500 Da, are produced by chemical synthesization and often administered orally in the formulation of tablets, are metabolized by liver and gut, enter systemic circulation through gut wall, enter all sites of the body by penetrating cell membranes, and have a short half-life. Large molecules are peptides or proteins that have molecular weight more than 500 Da, are produced by recombinant DNA technology and cell culture, are formulated in lyophilized or liquid format and administered through injection (otherwise, reduced by stomach before reaching systemic circulation), cannot penetrate cell membranes, and have a longer half-life. For detailed difference between these two types of therapeutics, refer to Samanen (2013). The differences historically distinguished traditional pharmaceutical and biotechnology companies in that the former focused on small molecule drugs while the latter focused on large molecule drugs. However, now most major pharmaceutical companies have both small and large molecules in their product portfolios.

1.2 Definition of Nonclinical Statistics

Since this book is written about nonclinical statistics, the first question is, "what is nonclinical statistics?" To answer this question, we will start with the definition of statistics. According to Dictionary.com, "Statistics is the science that deals with the collection, classification, analysis, and interpretation of numerical facts or data, and that, by use of mathematical theories of probability, imposes order and regularity on aggregates of more or less disparate elements." Statistical science has found applications in many fields, leading to branches with distinct names. For example, ecomometrics is statistics applied to economics, psychometrics is statistics applied to psychology, and actuary is statistics applied to life insurance, and so on.

Biostatistics is the application of statistics in fields related to biology, including medicine, pharmacy, agriculture, and fishery. Obviously, clinical biostatistics, statistics that supports clinical trials, is part of biostatistics. In fact, to many people in pharma/biotech industries, clinical biostatistics and biostatistics are interchangeable, due to the fact that in most universities departments of Biostatistics are located in medical schools and the emphasis of training is on statistical methods for clinical trials.

Clinical biostatistics seems to be a mature discipline. There is an academic department of this discipline in many universities. There are numerous journals dedicated to the subject, such as, statistics in medicine, journal of biopharmaceutical statistics, statistics in biopharmaceutical research, biometrics, biometrical journal, etc. Online searches reveal a huge collection of books dedicated to clinical biostatistics. In pharmaceutical or biotechnology companies, there is usually a department of biostatistics, organizationally located in clinical development in most companies, comprising mostly clinical statisticians. Clinical statisticians, as regulated in ICH E6 and E9, participate in design, monitoring, data analysis and reporting, and final result interpretation of clinical trials. They are a key member in clinical project teams, along with clinicians or physicians. As Lendrem (Lendrem 2002) commented, this is a "shining example of how an industry has embraced statistics and statisticians as equal partners in a joint venture."

In contrast, the story for nonclinical statistics is quite different. First, there is not such a department in many companies. Instead, there is usually just a functional group. Second, the group name is not universal. According to a 2008 nonclinical statistics leadership survey (referred to as Survey 2008 in the following), the group name includes "pharmaceutical sciences statistics", "Preclinical and Development Statistics", "Biometrics Research", "Statistical Services", "Research&Translational Sciences Statistics", "Preclinical and Research Biostatistics", with "Nonclinical Statistics" as the most common one. These names reveal what scientific functions the nonclinical statistics group supports. For example, Preclinical and Research Biostatistics may support preclinical development and discovery, whereas Preclinical and Development Statistics could support preclinical development and CMC. We will discuss why we prefer "Nonclinical Statistics" shortly. Third, there are a much smaller number of nonclinical statisticians compared to clinical statisticians. More details on this come later. Lastly, nonclinical statisticians usually do not have a formal role in any project team. More often, they serve as internal consultants.

In spite of the divergence, we try to give a definition for "Nonclinical Statistics." We define "Nonclinical Statistics" as statistics applied to areas other than clinical trials in pharmaceutical/biotechnology industries; these areas primarily comprise discovery, nonclinical development/translational science and CMC. Note that we use nonclinical development instead of preclinical development because the former concerns all safety studies after a compound's discovery, whereas the latter only includes safety studies prior to clinical trials. After the IND and first in human milestones, and parallel to clinical development, many safety studies are still conducted, such as carcinogenicity studies, long term toxicity studies. Similarly CMC includes API and drug product development as well as corresponding analytical method development and validation in parallel to clinical development. The drug supply from CMC needs to nonclinical development, clinical trial materials and commercial marketing. We also extend this definition to include translational sciences where many biomarker studies take place in clinical trials.

The relationship between Statistics, Biostatistics, Clinical Biostatistics and Nonclinical Statistics is depicted in Fig. 1.3. As is mentioned earlier, in pharma/biotech industries, biostatistics and clinical Biostatistics are often used interchangeably, but historically biostatistics should encompass statistics related to biology, including those used in discovery and nonclinical development. However, we don't use "Nonclinical Biostatistics" because much statistics in CMC has little to do with biology, but is more related to chemistry.

This definition is by no means the most appropriate one. However, it is based on the authors' extensive discussion with nonclinical statistics group leaders across the industries. Many nonclinical statistics conferences are often organized in these three areas. We will use this as our formal working definition and as our basis to structure the whole book.

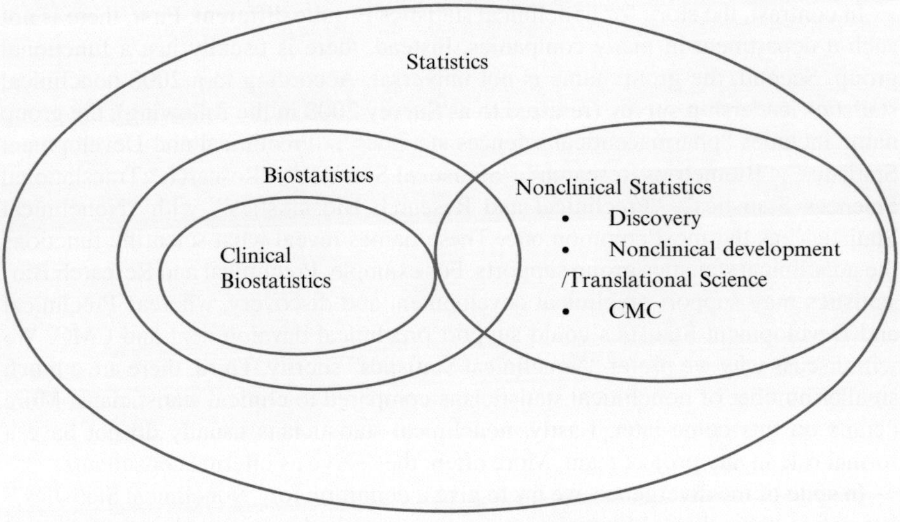

Fig. 1.3 Definition of nonclinical statistics and relationship between biostatistics and nonclinical statistics (*Circle areas* are not in scale)

1.3 Some Statistical Applications in Nonclinical Areas

The collection of statistical applications in nonclinical areas is enormous and rapidly growing, a result of the vastly diverse research and development objectives in nonclinical areas and the evolving strategies and technologies. In this section, we highlight some common statistical applications in nonclinical studies, with details to follow in the forthcoming chapters.

Design of Experiment (DOE) principles should be applied to any and all experiments. For any studies aiming to draw conclusions from comparisons, it is critical to consider the source of variability, bias, the balance of design and adequate sample size. The principle of randomization and blocking applies to CMC experiments or in vivo animal models that generate "small" data sets, and also to next generation sequencing experiments that generate "large" data set. In situations where the focus is to find conditions that produce desirable outcomes in areas such as drug manufacturing, lab automation and assay development, methods of screening and optimization designs are used routinely.

Assay Development and Validation is important throughout the R&D process as assays are the basic research tools for quantification of biological and chemical outcomes. Assay cut point determination, linearity, limits setting (detection, quantification), assay quality control criteria, precision, robustness, and assay transfer require many statistical methodologies, such as design of experiment, variance component analysis, predictive intervals, equivalence testing, parallelism testing, data transformation, outlier detection and nonlinear curve fitting.

Predictive Modeling (also called classification, pattern recognition, data mining) has applications in numerous problems wherever there is a desire to learn about outcomes without actually conducting large experiments. To name a few examples, prediction of compound activity using chemical descriptors in high throughput screening or in low throughput quantitative structure activity relationship (QSAR) studies, prediction of preclinical/clinical toxicity based on gene expression profile, prediction of clinical efficacy from gene expression and protein biomarkers, or prediction of sample concentration using spectrum data (chemometrics). This is a fast growing area as Big Data analytics are gaining popularity in many industries.

Statistical Quality Control and Process Control applies to nonclinical areas in high throughput screening, instrument automation, laboratory operation, process development and drug manufacturing. The methodologies include statistical process control charts, acceptance testing, equivalence testing, and process capability.

Many nonclinical research areas are not regulated by industrial or regulatory guidelines and sample sizes are often limited due to resource constraint. These aspects provide exciting opportunities for nonclinical statisticians to develop innovative and fit-for-purpose solutions to make an impact in the fields. It also calls for research and collaborations with scientists that would lead to industry standards.

1.4 Nonclinical Statistics Profession

1.4.1 Current Status in Pharma/Biotech Industry

Grieve (2002) asked "Do statisticians count?" In the paper, he acknowledged that "statisticians have been forced onto companies by regulation." Of course, he is referring to clinical statisticians. Because of regulation, clinical statisticians have integrated successfully into the clinical development process. They are an important project team member. They may not be counted adequately as desired, but they are in a good position and in large number to exert their influence if they follow Grieve's strategy: be more entrepreneurial, more communicative, and more persistent.

However, only a few nonclinical areas are regulated. For example, there is statistical guidance on carcinogenicity studies and stability studies, so they are the nonclinical areas to justify for statistical support most easily. However, other areas are left in a vacuum. The correlation between regulation and statistical support can also be seen by the number of clinical and nonclinical statisticians. Although it is hard to get an accurate number, thanks to Linkedin.com, we can have a rough estimate. We searched in LinkedIn, on June 13 2014, using key word statistician with constraint "pharmaceutical and biotechnology industry", and found 2362 statisticians in pharma and 611 in biotech. In total there are 2973 statisticians in pharma/biotech industries. Then we used key word "nonclinical statisticians", "discovery statisticians", "CMC statisticians", "preclinical statisticians", and found

41, 248, 46, 69, respectively. We believe in Linkedin's nomenclature nonclinical statistician is the same as CMC Statistician, so there are approximately 363 nonclinical statisticians, according to our definition. The percentage is about 12 %.

Of course, these numbers are not exact, because many statisticians may not have a Linkedin account. On the other hand, some statisticians may be counted more than once in nonclinical statisticians, for example, myself in both CMC and preclinical statisticians. However, we believe these numbers are reasonably close to truth. To validate, we searched again using key word statistician with a constraint on current company equal to Abbvie and Amgen. The returned number was almost the same as the number of statisticians in these two companies. Also according to nonclinical leadership forum 2008 survey, there were 243 nonclinical statisticians from 11 major pharma/biotech companies. Since most small companies do not have nonclinical statisticians, we believe 363 is a reasonable estimate of number of nonclinical statisticians in the industries now.

What does 12 % imply? For every eight to nine clinical biostatistics positions there is only one nonclinical statistics position. Accordingly, there are much more high ranking positions in clinical than nonclinical statistics. This seems to reinforce a perception among statisticians, according to Lendrem, that "venturing into non-clinical is nothing less than professional suicide, and anyone who does so will be consigned to a career wilderness." However, we argue that this is a misconception. In fact, professionally clinical and nonclinical statistics provide similar professional opportunities. For clinical statisticians, more positions mean more competing candidates. Usually we have seen similar speed of career advancement for most clinical and nonclinical statisticians.

1.4.2 Number of Nonclinical Statisticians

This is a fundamental problem, because as statisticians we all know that a smaller sample size means lower power. We estimated that there are about 363 nonclinical statisticians, accounting for about 12 % of all statisticians in pharma/biotech industries. A question arises. "Is this size of nonclinical statisticians right for the industries?" Maybe we don't need more nonclinical statisticians. To answer this question, we must recall that statistics is the science of decision making under uncertainty. Almost all decisions in the drug discovery and development, no matter clinical or nonclinical, are made under uncertainty and data driven. "Why is it then considered appropriate to conduct preclinical research without insisting on the same level of statistical rigor and quality?" asked Peers et al. (2012). They pointed an important factor contributing to the situation, which is that these nonclinical decisions are at producers' risk. In other words, companies are comfortable that because nonclinical decisions will not induce any regulatory compliance issue, they can afford to make wrong decisions and swallow the cost. This is surprising with the consideration of skyrocketing research and development cost in the industries and poor quality of research studies in discovery and nonclinical development (Begley and Ellis 2012).

It may be helpful to consider the idea of quality by design for CMC. It is recognized that quality cannot be tested into the final product. Quality has to be built into product through an optimal design of manufacturing process and good control strategy during the process. Similarly, the effectiveness (efficacy and safety) cannot be tested into a compound through clinical development. It has to be built into a compound through optimal decision making and good control strategy during the discovery, nonclinical development, clinical development, and CMC processes. This calls for statistical support from the beginning to the end; in particular, equal quality statistical supports in discovery and nonclinical developments and CMC.

In this sense, obviously nonclinical statistics groups are under-sized. To change the situation, nonclinical statisticians must be more entrepreneurial, more communicative and more persistent to convince R&D management of the value of statistics in nonclinical decision making, with data of successes from using statistics and failures from not using statistics. This is also a fertile research area. We should clearly and collectively identify specific nonclinical areas where statisticians can add most value, how to determine resources, how to align objectives, how to influence regulatory counterparts, and share success stories or lessons learned.

1.4.3 Interaction with Academia and Regulatory Bodies

The most important factor contributing to the success of clinical biostatistics is regulatory enforcement on statisticians involved in design, monitoring, analysis, and report of clinical trials (Grieve 2002). However, recognizing the value of statistics for clinical trials did not happen within the regulatory bodies. Rather, it was the result of the joint forces of medical academia and government. Take the randomized controlled trial as an example. Fisher pioneered randomized, controlled experiments in agriculture as early as 1926 (Fisher 1926). However, it was Hill (1937) who advocated and used randomization in medical studies later. Not until 1970s was it mandated in the government statute and became the gold standard of modern medical research method. Requirement of a statistician in such trials is a byproduct of this change. Notably, the pharmaceutical industry protested against the legislation initially and only the pressure of compliance with the law led to the boom of clinical statistics in pharmaceutical industries. For an excellent review, refer to Meldrum (2000).

The nonclinical statistics professional community will need to take a similar route to integrate into drug development process. Unfortunately, the value of statistical supports is also under-appreciated in academia. Very often for clinical research in universities supported by a grant, a statistician is included in the grant proposal. Rarely this is the case for biological research studies, discovery studies, biochemistry studies. Science students coming out of school environment with little appreciation for statistics when designing, analyzing and reporting studies in their academic labs will bring the same mentality to the industry and the regulatory agencies.

There is also a disconnection between nonclinical statistical tools and biostatistics departments in academia. For example, one important statistical tool for CMC area is design of experiment (DOE). However, in department of statistics or biostatistics of many universities, very few faculty members research this area. Even fewer students will focus on their dissertations in this area. There are some statistical professors in engineering schools whose research areas are on DOE, but they produce a limited number to none of students with statistics degree. It is vital for pharmaceutical industry to engage more professors on nonclinical statistical research and to produce more students trained in nonclinical areas.

An interdisciplinary program that brings together researchers from chemistry, biology, and statisticians in universities, may be helpful to recognize the value of statistics in nonclinical studies. The research results will translate into regulatory guidance that calls for formal involvement of nonclinical statisticians in nonclinical areas.

1.4.4 Interaction with Internal Partners

The success of a nonclinical statistician and a nonclinical statistical group relies highly on how we interact with our collaborators within the company. The readers are referred to Chap. 3 for guidance on how to become a successful nonclinical statistician.

1.5 Conclusions

Drug discovery and development is a long and costly process. The failure rate of drug submissions has been increasing. All these contribute to the skyrocketing cost of health care. To improve the probability of success, innovation has been initiated in clinical development through FDA's critical path initiative. Adaptive design, for example, has been encouraged as one of the enablers. However, it should be recognized that effectiveness cannot be tested into a compound. Effectiveness should be built into a compound by design through discovery and nonclinical development. In other words, improving efficiency in discovery and early development can supply clinical development with better drug candidates, and thus a higher chance of clinical testing success. Like the contribution of clinical biostatistics to clinical trials, nonclinical statistics can contribute equally to decision making in discovery, nonclinical development and CMC areas.

There is great potential for nonclinical statistics to grow. However, Grieve (2002) pointed out that it is statisticians that matter, not statistics. It is nonclinical statisticians' responsibility and mission to unlock the value of nonclinical statistics, and disseminate it to industry partners, academic peers and regulatory agencies. Only by adding value to the business and having that value recognized, and by introducing quality standard and regulation, can the nonclinical statistical profession grow to its full potential.

References

Begley GC, Ellis LM (2012) Raise standards for preclinical cancer research. Nature 483: 531–533

Fisher RA (1926) The arrangement of field experiments. J Ministry Agric (G B) 33:503–513

Food and Drug Administration (2004) Innovation or Stagnation: Challenge and opportunity on the critical path to new medical products. Washington DC, USA

Grieve AP (2002) Do statisticians count? A personal view. Pharm Stat 1:35–43

Hill AB (1937) Principles of medical statistics. I. The aim of the statistical method. Lancet 1:41–43

Lendrem D (2002) Statistical support to non-clinical. Pharm Stat 1:71–73

Meldrum ML (2000) A brief history of the randomized controlled trial: from oranges to lemons to the gold standard. Hematol Oncol Clin North Am 14:745–760

Peers IS, Ceuppens PR, Harbron C (2012) In search of preclinical robustness. Nat Rev 11:733–774

Samanen J (2013) Similarities and differences in the discovery and use of biopharmaceuticals and small-molecule chemotherapeutics. In: Ganellin R, Roberts SM, Jefferies R (eds) Introduction to biological and small molecule drug research and development, theory and case studies, pp 161–204

Chapter 2
Regulatory Nonclinical Statistics

Mohammad Atiar Rahman, Meiyu Shen, Xiaoyu (Cassie) Dong, Karl K. Lin, and Yi Tsong

Abstract The nonclinical statistics teams in the Center of Drug Review and Research of the Food and Drug Administration (FDA) conduct regulatory reviews, statistical consultation, and statistical methodology development in nonclinical regulations. In this chapter, we provide a brief description of the two teams and provide two examples in statistical research development. In the first example, we describe the historical background and evolution of statistical methodology development in the last 20 years for the acceptance sampling and lot evaluation procedures on dose content uniformity involved with FDA Chemistry Manufacturing, and Control (CMC) Statistics Team. In the second example, we illustrate the research activities of Pharmacological/Toxicological (Pharm-Tox) Statistics Team at FDA with the background and evaluation of multiple pairwise comparisons in animal carcinogenetic studies.

Keywords Chemistry manufacturing, and control • Acceptance sampling • Content uniformity • Pharmacological/toxicological studies

2.1 Background

Regulatory Nonclinical Statistics in the Center for Drug Evaluation and Research (CDER) of the Food and Drug Administration (FDA) consists of two teams: Chemistry, Manufacturing, and Control (CMC) and Pharmacological/Toxicological study teams. The two nonclinical statistics teams are located within the Division of Biometrics VI of the Office of Biostatistics.

This presentation reflects the views of the authors and should not be construed to represent FDA's views or policies.

M.A. Rahman • M. Shen • X. Dong • K.K. Lin • Y. Tsong (✉)
Division of Biometrics VI, Office of Biostatistics, Office of Translational Science, Center for Drug Evaluation and Research, Food and Drug Administration, 10903 New Hampshire Ave., Silver Spring, MD 20993, USA
e-mail: yi.tsong@fda.hhs.gov

L. Zhang (ed.), *Nonclinical Statistics for Pharmaceutical and Biotechnology Industries*, Statistics for Biology and Health, DOI 10.1007/978-3-319-23558-5_2

2.1.1 CMC Statistics Team

The CMC Statistics team provides statistical expertise to FDA to ensure product quality through review, regulation, and research for new drugs, biological products, biosimilar products, and generic drugs. Currently, this team consists of seven Ph.D. level statisticians including one team leader and one technical leader.

There are two types of reviews: consultation review and regulatory review. For the consultation review, the CMC Statistics team responds to the consultation requests from CDER/FDA on a broad range of CMC issues, including new drug CMC reviews, scale-up and post-approval changes (SUPAC), new drug bioequivalence including in-vivo and in-vitro, generic drug in-vitro bioequivalence product stability, product specification, botanical drug product consistency, post-marketing quality surveillance, and other quality issues. These consultation reviews provide important statistical support to many CMC related offices in CDER/FDA. These consults may be requested by Office of Policy for Pharmaceutical Quality (OPPQ), Office of New Drugs (OND), Office of Generic Drugs (OGD), Office of Product Quality (OPQ), Office of Biotechnology Products (OBP), Office of New Drug Products (ONDP), Office of Lifecycle Drug Products (OLDP), Office of Testing and Research (OTR), Office of Process and Facilities (OPF), and Office of Surveillance (OS).

The aim of a statistical CMC consultation review is to provide statistical expertise to address specific issues in a regulatory submission, e.g., the shelf life determination for a product. It may also provide evaluation on statistical approaches proposed by sponsors, e.g., to evaluate the suitability of a stability model used in a submission. Under the consultation setup, review comments from the CMC Statistics team may not be conveyed directly to the sponsor nor is a part of the regulatory decision making.

The typical process of a CMC statistical consultation is outlined as follows. Chemist reviewers or biologist reviewers send a CMC statistical consultation request to the CMC Statistics team through the project manager. Upon receiving the request, the CMC Statistics team leader assigns the work to one CMC team member. The assigned primary statistical reviewer and the team leader or technical leader will meet with the chemist or biologist to discuss the review issues. Once the work is completed, the statistical consultation review will be put into the FDA's filing system and then will be signed off by the reviewer, the team leader, and the division director.

In addition to the consultation review, the CMC Statistics team conducts regulatory reviews for biosimilar submissions. As part of the review team since 2013, the CMC statistical review team has been playing a key role in reviewing the statistical assessment of analytical similarity for Investigational New Drug (IND) and Biological License Application (BLA) submissions for biosimilar biological products. Our review comments will be conveyed to the sponsor directly and are critical to the regulatory decision making. Our statistical findings and evaluation may be presented at Advisory Committee meetings.

Besides the consultation and biosimilar reviews introduced earlier, the CMC Statistics team also develops CMC statistical methodologies for product quality assurance. We will illustrate this with one example in Sect. 2.2.

2.1.2 Pharmacological/Toxicological Review Team

The risk assessment of a new drug exposure in humans usually begins with an assessment of the risk of the drug in animals. It is required by law that the sponsor of a new drug conducts nonclinical studies in animals to assess the pharmacological actions, the toxicological effects, and the pharmacokinetic properties of the drug in relation to its proposed therapeutic indications or clinical uses. Studies in animals, designed for assessment of toxicological effects of the drug, include acute, subacute, subchronic, chronic toxicity studies, carcinogenicity studies, reprotoxicology studies, and pharmacokinetic studies.

The statistical reviews and evaluations of toxicology studies of new drugs is an integrated part of FDA drug review and approval process. The Pharm/Tox Statistics Team in the Office of Biostatistics in the Center for Drug Evaluation and Research of FDA is responsible for this area of the review and approval process. An assessment of the risk for carcinogenicity includes life-time tests in mice and rats. The primary purpose of a long-term animal carcinogenicity experiment is to determine the oncogenic potential of a new drug when it is administered to animals for the majority of their normal lifespan.

Regular long-term (chronic) carcinogenicity studies of a new drug are usually planned for 2 years (104 weeks) in rats and mice. However, in the 1990s ICH started allowing a drug sponsor to conduct a 26-week transgenic mouse study to replace the regular 2-year mouse study in its new drug application submission. At least three dose groups and a negative control and a positive control (treated with a known carcinogen, e.g., *p*-cresidine, *N*-methyl-*N*-nitrosourea, benzene, or 12-*O*-tetradecanoyl-phorbol-13-acetate (TPA)) are used in the transgenic mouse study with 25–30 mice sex/group. The histopathology endpoints of the transgenic mouse study are the same as those used in the 2-year studies except the 26-week study using Tg.AC transgenic mice. The interest in detecting oncogenic potential of a new drug is to test if there are statistically significant positive linear trends (or dose–response relationships) induced by the new drug. Based on the interest, a typical statistical review and evaluation of the carcinogenicity studies of a new drug performed by FDA nonclinical statisticians includes the essential parts described below. The statistical methods used in the FDA review and evaluation are based on results of FDA internal research and guidances or guidelines of regulatory agencies and research institutions such as WHO, ICH, NIH inside and outside U.S.

In the analysis of tumor data, it is essential to identify and adjust for the possible differences in intercurrent mortality (or longevity) among treatment groups to eliminate or reduce biases caused by these differences. Intercurrent mortality refers to all deaths not related to the development of a particular type or class of tumors to be analyzed for evidence of carcinogenicity. Like human beings, older animals have many times higher probability of developing or dying of tumors than those of younger age. The Cox's Test, the generalized Wilcoxon or Kruskal–Wallis test, and the Tarone trend tests are routinely used to test the heterogeneity in survival distributions and significant dose–response relationship (linear trend) in mortality. The choice of a survival-adjusted method to analyze tumor data depends on the role which a tumor plays in causing the animal's death. Tumors can be classified as "incidental", "fatal", and "mortality-independent (or observable)" according to the contexts of observation described in the WHO monograph by Peto et al. (1980). Tumors which are directly or indirectly responsible for the animal's death, but are merely observed at the autopsy of the animal after it has died of some unrelated causes, are said to have been observed in an incidental context. Tumors which kill the animal either directly or indirectly are said to have been observed in a fatal context. Tumors, such as skin tumors, whose times of criterion attainment (that is, detection of the tumor at a standard point of their development, other than the times or causes of death, are of primary interest in analyses, are said to have been observed in a mortality-independent (or observable) context. To apply a survival-adjusted method correctly, it is essential that the context of observation of a tumor be determined as accurately as possible.

Different statistical techniques have been proposed for analyzing data of tumors observed in different contexts of observation. For example, the prevalence method, the death-rate method, and the onset-rate method are recommended for analyzing data of tumors observed in incidental, fatal and mortality-independent contexts of observation, respectively, in Peto et al. (1980). Misclassifications of incidental tumors as fatal tumors, or of fatal tumors as incidental tumors, will produce biased results. When a tumor is observed in a fatal context for a set of animals and is observed in an incidental context for the other animals in the experiment, data should be analyzed separately by the death-rate and the prevalence methods. Results from the different methods can then be combined to yield an overall result. The combined overall result can be obtained by simply adding together the separate observed frequencies, the expected frequencies, and the variances, or the separate T statistics and their variances.

The prevalence method, the death-rate method, and the onset-rate method use a normal approximation in the test for the positive linear trend or difference in tumor incidence rates. It is also well known that the approximation results will not be stable and reliable, and mostly tends to underestimate the exact p-values when the total numbers of tumor occurrence across treatment groups are small. In this situation, it is advisable to use the exact permutation trend test to test for the positive linear trend. The exact permutation trend test is a generalization of the Fisher's exact test to a sequence of $2x(r + I)$ tables. The widely used prevalence method, the death rate method, and the onset rate methods for analyzing incidental, fatal, and mortality independent tumors, respectively, described in previous sections rely on good information about cause of death of tumors. There are situations in which investigators have not included cause of death information in their statistical analyses and electronic data sets.

To avoid the use of the cause-of-death information needed in the above Peto methods described in the WHO monograph, the Bailer–Portier poly-3 (in general poly-k) tests have been proposed for testing linear trends in tumor rates. These tests are basically modifications of the survival unadjusted Cochran–Armitage test for linear trend in tumor rate. The Cochran–Armitage linear trend test is based on a binomial assumption that all animals in the same treatment group have the same risk of developing the tumor over the duration of the study. This assumption is thus no longer valid if some animals die earlier than others.

The Bailer–Portier poly-3 test adjusts for differences in mortality among treatment groups by modifying the number of animals at risk in the denominators in the calculations of overall tumor rates in the Cochran–Armitage test to reflect "less-than-whole-animal contributions for decreased survival". The modification is made by defining a new number of animals at risk for each treatment group. After weighting the pros and cons of the Peto methods and of the poly-k method, the FDA nonclinical statisticians have recently switched from the Peto methods to the poly-k method in their statistical reviews and evaluations of carcinogenicity studies of new drugs.

Interpreting results of carcinogenicity experiments is a complex process. Because of inherent limitations, such as the small number of animals used, low tumor incidence rates, and biological variation, a carcinogenic drug may not be detected (i.e. a false negative error is committed). Also, because of a large number of statistical tests performed on the data (usually 2 species, 2 sexes, 20–30 tissues examined, and 4 dose levels), there is a large probability that statistically significant positive linear trends or differences in some tumor types are purely due to chance alone (i.e. a false positive error is committed). Therefore, it is important that an overall evaluation of the carcinogenic potential of a drug should be made based on the knowledge of multiplicity of statistical significance of positive linear trends and differences, historical information, and other information of biological relevance.

In order to reduce the false positive rate, statistical reviewers in CDER use data of the concurrent control group(s) and historical control data to classify common and rare tumors, and adopt the following decision rule in their evaluation: A positive linear trend (dose–response relationship) is considered not to occur by chance of variation alone if the p-value is less than 0.005 for a common tumor, and 0.025 for a rare tumor. For the test of a pairwise increase in incidence rate, the significance levels of 0.01 and 0.05 are used, respectively.

To ensure that the committed false negative rate is not excessive, statistical reviewers collaborate with the reviewing pharmacologists, pathologists, and medical officers to evaluate the adequacy of the gross and histological examination of both control and treated groups, the adequacy of the dose selection, and the durations of the experiment in relation to the normal life span of the tested animals.

In negative studies, the statistical reviewers will perform a further evaluation of the validity of the design of the experiment, to see if there were sufficient numbers of animals living long enough to get adequate exposure to the chemical, and to be at risk of forming late-developing tumors, and to see if the doses used were adequate to present a reasonable tumor challenge to the tested animals. In Sect. 2.3, we provide one example in animal study design to illustration the regulatory nonclinical statisticians' contribution in evaluating and developing more advance design of animal studies. A summary of the chapter is given in Sect. 2.4.

2.2 Example of CMC Methodology Development

As an example of regulatory methodology development and evaluation conducted by the CMC Statistics team at CDER/FDA, we will describe the statistical methodology development of dose content uniformity assessment for both small and large sample sizes during the last 10 years.

As one of the most important quality attributes for drugs, dose content measures the amount of the active ingredient of the product relative to the label claim (LC). For a therapeutic product, most of the dose contents should be within (85,115)%LC to ensure the homogeneity of the product. We can evaluate the content uniformity through the acceptance sampling, which is required by FDA to meet the quality standards for ensuring the consistency of the dose content with the label claim. The U.S. Pharmacopoeia (USP) publishes the dose content uniformity (DCU) sampling acceptance procedure and its revision for applicable to products seeking licensure in the US market every 5 years. The Europe Pharmacopeia (EP) and Japan Pharmacopeia (JP) frequently publish the DCU testing procedure used in Europe and Japan, respectively. The DCU procedure used in these three regions may be different due to the differences in quality requirements and statistical considerations. In the following, we outline the USP, EU, and JP testing procedures.

USPXXIV that was published by USP in 2005 recommended a two-tier sampling acceptance procedure as follows.

1st tier: A sample of 10 units is collected. The lot complies with the USP DCU requirement if the dose content of each unit is within (85 %, 115 %) LC and RSD (i.e. the sample standard deviation divided by the sample mean) is less than 6 %. If it fails to comply, we move to the 2nd tier.

2nd tier: Additional 20 units are randomly sampled and measured. The lot complies with the DCU requirement if the dose content of each of 30 units is within (75 %, 125 %) LC, no more than 1 unit has the dose content outside (85 %, 115 %) LC, and RSD $\leq$ 7.8 %. It fails the DCU requirement otherwise. The requirement of none of the content of the 30 units is allowed to be outside (75 %, 125 %) LC is often referred to as zero tolerance condition.

EPIII defines the DCU test as a two-tier procedure similar to USPXXIV procedure but without requirement on RSD.

1st tier: 10 units are sampled. The lot complies with the DCU requirement if the dose content of each of all 10 units is within (85 %, 115 %) LC. It fails to comply if more than 1 unit has dose content outside (85 %, 115 %) LC or at least 1 unit has dose content outside (75 %, 125 %) LC. Otherwise, we move to the 2nd tier.

2nd tier: Additional 20 units are randomly sampled and measured. The lot complies with the DCU requirement if dose content of each of 30 units is within (75 %, 125 %) LC, and no more than 1 unit has dose content outside (85 %, 115 %) LC. It fails the DCU test otherwise.

JPXIV testing procedure is based on the tolerance limit and also has a zero tolerance requirement. JPXIV consists of two tiers as follows. 1st tier: 10 units are randomly sampled and tested. The lot complies if all 10 units' dose contents are within (75 %, 125 %) of LC and 85 % < ($\overline{x}-2.2$*s) <115 %, where $\overline{x}$ and s respectively are the sample mean and the sample standard deviation of the 10 units. It fails to comply if at least 1 unit has the dose content outside (75 %, 125 %) of LC. Otherwise, randomly sample 20 units more tested and move on to the 2nd tier.

2nd tier: The lot complies if all 30 units' dose contents are within (75 %, 125 %) of LC and 85 % < ($\overline{x}-1.9$s) <115 %, where $\overline{x}$and s are the sample mean and the sample standard deviation of the 30 units, respectively. Otherwise, the lot does not comply with the DCU requirement.

In 2006, FDA CMC Statistics team evaluated the statistical properties of the USP, EU, and JP procedures outlined above with the operating characteristics curves. We proposed that a lot is accepted if only a small proportion of units in the lot have the dose contents below a lower specification or above an upper specification limit. Our proposed procedure is a two-tier sampling plan based on the two one-sided tolerance intervals. We further adapt Pocock's group sequential boundaries to control the confidence levels at the two tiers. Our proposed DCU testing procedure (Tsong and Shen 2007) is described as follows.

1st Tier: Sample 10 units, the lot is accepted if

$$\overline{x}_{11}-115\% < A^{U_1} \text{ and } \overline{x}_{11}-85\% > A^{L_1},$$

where $A^{U_1} = -\,C(\alpha_1)(s_1/\surd\,10)$ - $s_1 Z_{0.9375}$. $A^{L_1} = C(\alpha_1)(s_1/\surd\,10)$ - $s_1 Z_{0.9375}$; $\bar{x}_1$ and s_1 are the sample mean and the sample standard deviation of the 10 units, respectively. Otherwise, move to the 2nd tier and sample an additional 20 units.

2nd Tier: We accept the lot if

$$\overline{x}_2-115\% < A^{U_2} \text{ and } \overline{x}_2-85\% > A^{L_2},$$

where $A^{U_2} = -\,C(\alpha_2)(s_2/\surd\,30)$ - $s_2 Z_{0.9375}$. $A^{L_2} = C(\alpha_2)(s_2/\surd\,30)$ - $s_2 Z_{0.9375}$; $\bar{x}_2$ and s_2 are the sample mean and the sample standard deviation of the 30 units, respectively. Otherwise, we conclude that the lot fails to comply with the DCU requirement. We remove the zero tolerance requirement in the procedure by allowing a small probability of any sample falling outside of the zero tolerance limits (75 %, 125 %) LC under normality assumption.

In order to harmonize the acceptance sampling plans across the United States, Europe, and Japan regions, a harmonization procedure (U.S. Pharmacopoeia XXV

2010) was developed to replace USP XXIV and EP III plans. The two-stage harmonized procedure is derived based on a sequential procedure using the two-sided tolerance interval combined with an indifference zone for the sample mean and zero tolerance criteria for the observed dose content of each unit.

The two-stage harmonized procedure for DCU is described in (U.S. Pharmacopoeia XXV 2010).

For the first stage, the sample mean ($\bar{x}$) and the sample standard deviation (s) of the 10 units are calculated. The indifference zone (M) at the first stage is defined as $M = 98.5$ % if $\bar{x} <98.5$ %; $M = 101.5$ % if $\bar{x} >101.5$ %; and $M = \bar{x}$ if $98.5\% \leq \bar{x} \leq 101.5\%$. The two-sided tolerance interval is calculated as ($\bar{x} - 2.4s$, $\bar{x} + 2.4s$), where the constant 2.4 can be interpreted as the tolerance coefficient with approximately 87.5 % coverage and a confidence level of 90.85 % for a sample size of 10. The dose content uniformity is accepted if all 10 samples are within (75 %, 125 %)M and ($\bar{x} - 2.4s$, $\bar{x} + 2.4s$) is covered by (M-15 %, M+15 %). If the lot fails to be accepted, go to the second stage.

In Stage 2, additional 20 samples are randomly collected. With a total of 30 samples, a tolerance coefficient of 2.0 is used for calculation. The constant 2.0 can be interpreted as the tolerance coefficient with 87.5 % coverage and 95.14 % confidence level for a sample size of 30. The lot is accepted if all 10 samples are within (75 %, 125 %)M and ($\bar{x} - 2s$, $\bar{x} + 2s$) is covered by (M-15 %, M+15 %). Otherwise, the lot fails the DCU test.

FDA CMC Statistics team evaluated this harmonized USP procedure. Based on our evaluation, it is biased toward the lot with the true mean deviating from 100 % label claim. In other words, the probability of passing the lot with an off-target mean is higher than that of the lot with an on-target mean (100 % LC) based on simulations (Shen and Tsong 2011).

Since 2007, the pharmaceutical industry has expressed an interest in conducting large sample testing for dose content uniformity due to the availability of near-infrared spectroscopy in the manufacturing process. With this NIRS technology, continuously testing of the dose content without destroying the units becomes possible. Thus, the pharmacopeia acceptance sampling procedure for small samples should be extended to large samples. Many statistical testing procedures have been proposed for this purpose. The approach proposed by the FDA CMC Statistics team is a large sample DCU testing procedure based on the two one-sided tolerance intervals (TOSTI). Our proposed approach maintains a high probability to pass the USP compendia by restricting the TOSTI OC curves for any given sample size to intersect with the USP OC curve for a sample size of 30 at the acceptance probability of 90 % when the individual unit is assumed to be normally distributed with an on-target mean of 100 %LC (Shen et al. 2014). The derivation of the tolerance coefficient K for any large sample size n was provided in Shen et al. (2014), Dong et al. (2015), and Tsong et al. (2015). We denote this extension as PTIT_matchUSP90 method in the remaining section.

The large sample dose uniformity tests with two options were published in European Pharmacopeia (Council of Europe 2012). EU option 1 is a parametricapproach based on a two-sided tolerance interval approach with an indifference zone and a counting limit for the number of dosage units outside 75 %M −125 %M, with M defined the same way as that in the USP Harmonized procedure. EU option 2 is a non-parametric approach developed for non-normally distributed contents of dosage units. The EU option 2 is actually a counting procedure with two acceptance criteria. The number of dosage units outside of $[1 - L1, 1 + L1]\,M$ and $[1 - L2, 1 + L2]\,M$ are required to be no more than $C1$ and $C2$, respectively, where $L1$, $L2$, $C1$, and $C2$ are defined in Table 2.9.47.-2 of European Pharmacopeia (Council of Europe 2012).

FDA CMC Statistics team compared PTIT_matchUSP90 method with the two options of European Compendia under normality and mixture of two normal variables (Shen et al. 2014). In this work, we found that the two options of European Compendia give very different acceptance probabilities. In addition, the acceptance probabilities of both parametric and non-parametric options are higher than that obtained from our proposed PTIT_matchUSP90 procedure. Furthermore, these two EU options in European Pharmacopoeia 7.7 still have the same bias property as in the harmonized procedure recommended in USP XXV.

Here FDA CMC Statistics team compare the acceptance probability of the EU option 1 in European Pharmacopoeia Supplement 8.1 with the PTIT_matchUSP90 against the coverage within 85–115 %LC under the normality assumption. The results are shown in Fig. 2.1. As can be seen, the acceptance probability of EU

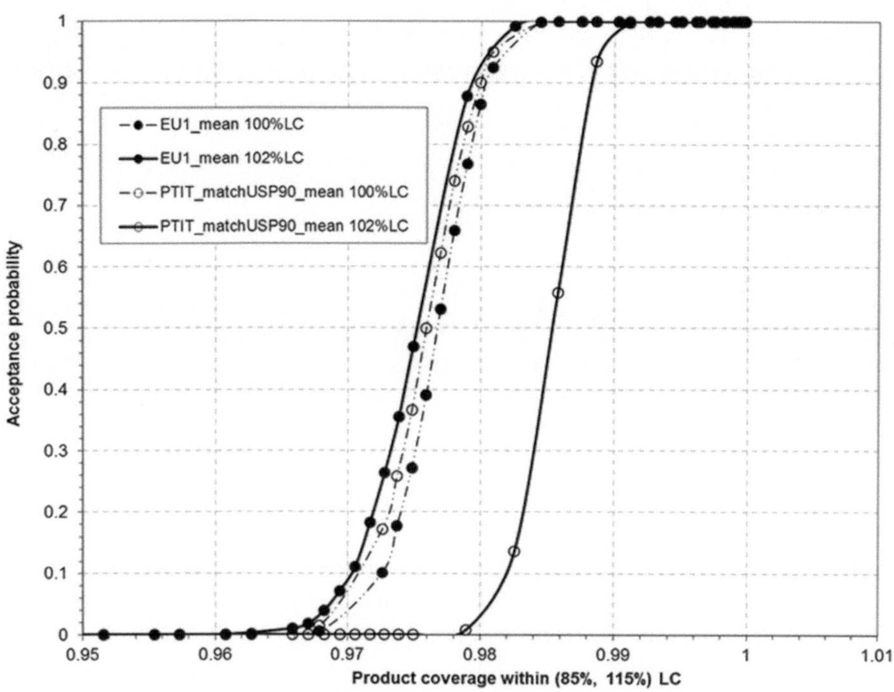

Fig. 2.1 Bias of European Union option 1 for individual dose content distributed as independent and identical normal variable with sample size $n = 1000$

Table 2.1 Comparison of the acceptance probability between the PTIT_USPmatch90 and EU option 2 method

	Acceptance probability	
Sample size, *n*	EU option 2	PTIT_USPmatch90
100	0.6458	0
150	0.5276	0
200	0.6047	0
300	0.455	0
500	0.3509	0
1000	0.2075	0

option 1 approach with a mean content of 102 %LC is higher than that with a mean of 100 %LC. We also compare the acceptance probability of the EU option 2 (nonparametric method) with those of the PTIT_matchUSP90 procedure when the individual unit dose content follows a uniform distribution in the range from 85 % to 115 % with 97 % probability and a value 84 % with 3 % probability in Table 2.1. For this particular distribution, the acceptance probability of USP harmonized DCU procedure is 3.72 % for a sample size of 30 units. Table 2.1 shows that the EU option 2 has acceptance probabilities higher than 50 % for sample sizes up to 200. The acceptance probability only reduces when the sample size is significantly larger than 1000. On the other hand, such a lot would have almost 0 % probability of passing USP DCU harmonized procedure and the PTIT_macthUSP90. The EU option 2 does not take the content variability into consideration and misses the purpose of dose content uniformity test. Further research on appropriate comparison is in progress.

2.3 Examples of Carcinologicity Study Methods Development

One of the responsibilities of the Pharmacological—Toxicological Statistical Review Team is to keep track of new statistical methodologies developed in the area of animal carcinogenicity studies and works on the development of new or modified methodologies better suited for carcinogenicity data analysis. Following are two examples of the team's research efforts.

Pairwise Comparisons of Treated and Control Group In the carcinogenicity data analysis routinely the treated groups are compared to the control group as primary or additional tests. For these pairwise comparisons, by convention only data from the selected two groups are used by ignoring data from the other dose groups. The test is termed as the unconditional test. Members of the Pharm–Tox Statistics team proposed two modifications to this conventional test. In the first modification, we proposed to use the data from all dose groups in variance calculation, in the spirit of test for contrasts of the general ANOVA analysis. The test is termed as the conditional test. It is shown that the asymptotic relative efficiency (ARE) of conditional modification versus unconditional test is greater than 1. The second modification is to use the variance estimation method proposed by Hothorn and Bretz (2000). It is shown through simulation study that the second modification provides more power in both exact and asymptotic situations and has higher power than the unconditional test. Furthermore, the simulation results showed that asymptotically the conditional test always has more power than the second modified test. The detailed results were published in Rahman and Tiwari (2012).

Multiple Contrast Type Tests A typical animal carcinogenicity study involves the comparison of several dose groups to a negative control group for dose response relationship. The typical shape of the outcome (proportion of tumor bearing animals vs. dose levels) is assumed to be linear or approximately linear. However, in practice the shape of the outcome may turn out to be concave, convex or some other non-linear curve. The Cochran–Armitage (CA) test is most frequently used to test the positive dose response relationship. This test is based on a weighted linear regression on proportions. It is well known that the CA test lacks power for the nonlinear shape of the outcomes. For general shape of outcomes, Hothorn and Bretz (2000) proposed the multiple contrasts test (MC). This test suggests the use of the maximum over several single contrasts, where each of them is chosen appropriately to cover a specific dose response shape. In mathematical form the MC test statistic T^{MC} is defined as

$$\mathrm{T}^{\mathrm{MC}} = \max\left\{\mathrm{T}_1{}^{\mathrm{SC}},\ \mathrm{T}_2{}^{\mathrm{SC}},\ \mathrm{T}_3{}^{\mathrm{SC}}\right\},$$

where, MC = Multiple Contrast, SC = Single Contrast,
T^{MC} = Multiple Contrast test proposed by Hothorn and Bretz,
$\mathrm{T}_1{}^{\mathrm{SC}}$ = Test with Helmert contrast $\underset{\sim}{c}^{(1)} = (-1, -1, \ldots, -1, k)$, powerful for convex profiles,
$\mathrm{T}_2{}^{\mathrm{SC}}$ = Test with Linear contrast $\underset{\sim}{c}^{(2)} = (-k, -k+2, \ldots, k-2, k)$, powerful for linear profiles,
$\mathrm{T}_3{}^{\mathrm{SC}}$ = Test with step contrast $\underset{\sim}{c}^{(3)} = (-1, -1, \cdots - 1,\ 1, \ldots, 1)$ for k odd, $\underset{\sim}{c}^{(3)} = (-1, -1, \cdots - 1,\ 0,\ 1, \ldots, 1)$ for k even, powerful for sub-linear profiles, and,

$$T_a^{SC} = \frac{\sum_i \frac{r_i}{n_i} c_i^{(a)}}{\sqrt{p(1-p)\sum_i \frac{\left(c_i^{(a)}\right)^2}{n_i}}},$$

for the ith dose group, r_i is the number of tumor bearing animals, n_i is the number of animals at risk, $c_i^{(a)}$ are the elements of $\underset{\sim}{c}^{(a)}$, p_i is the proportion of tumor bearing animals and p is the common value of p_1, $p_2, \ldots, p_k$ under null hypothesis.

Hothorn and Bretz compared their test with the CA test and concluded that the MC test on the average is more powerful than the CA test. A team member took interest in investigating this topic and found that the MC method performs well for convex outcome, but not as good for concave outcome. He also proposed a new test method based on the maximum of sequential Cochran–Armitage (SCA) test over dose groups. In mathematical form SCA test statistic T^{SCA} is defined as

$$T^{SCA} = \max\left\{T_1{}^{CA},\ T_2{}^{CA},\ T_3{}^{CA}\right\},$$

where, $T_1{}^{CA}$ = CA test with dose groups $\underset{\sim}{d}^{(1)} = (0,1,2,3)$; $T_2{}^{CA}$ = CA test with dose groups $\underset{\sim}{d}^{(2)} = (0,1,2)$; and $T_3{}^{CA}$ = CA test with dose groups $\underset{\sim}{d}^{(3)} = (0,1)$,

and $T_a^{CA} = \sqrt{\frac{N}{r(N-r)}} \dfrac{\sum_i \left(r_i - \frac{n_i}{N} r\right) d_i^{(a)}}{\sqrt{\sum_i \frac{n_i}{N}\left(d_i^{(a)}\right)^2 - \left(\sum_i \frac{n_i}{N} d_i^{(a)}\right)^2}}.$

This new test has similar power as CA and MC tests for linear dose response, and has higher power for concave outcome. The MC test still has higher power for convex dose response.

In 2014, members of Pharm–Tox Statistics team evaluated the approaches with their interpretation of the results of a carcinogenicity experiment. If a significant linear dose response is found, the interpretation is straight forward. However, they found that the interpretation of other shapes of the dose response is difficult and somewhat subjective. Finding a highly statistically significant U-shaped or sign-curved dose response may not have any practical value. Therefore, finding a simple dose response like linear or at most quadratic is very important for practical purposes. The MC method has relatively simple models. However, it is a combination of several linear and non-linear models. If a significant dose response is found using the MC model, it is difficult to know what kind of dose response it is. On the other hand since the SCA methods are based on linear model, the interpretation of any findings from these tests is easy. Also, since the SCA test is based on the maximum of all possible CA tests, one additional advantage of SCA test is that it can capture a positive dose response, which may be present in a part of the data. For example in a concave data like 6, 10, 15, and 6 for control, low, medium and high groups, where a clear positive linear dose response is evident in the first three dose levels, the CA test shows a non-significant dose response ($p = 0.346$); however, SCA test still captures such dose response ($p = 0.035$).

The major criticism of both the HB and SCA methods is that the dose response shape is not known prior to the completion of the experiment. This is of serious concern which hinders the practical use of these methods. For this reason the agency has not yet adopted these methods as a part of their regular tumor data analysis. However, a post hoc analysis can still be performed, in relation to the observed the outcome pattern after the completion of the experiment.

A manuscript of the evaluation of SCA method has been submitted to the Journal of Biopharmaceutical Statistics for publication.

2.4 Summary

In summary, FDA regulatory nonclinical statisticians played many critical roles in review, regulation, and research for drug development and evaluation. Each of the two teams consists of five to seven members. They make significant contribution to the public health through assessing the carcinogenicity potential, product quality and product manufacturing control of each biopharmaceutical product seeking licensure in the US market.

References

Council of Europe (2012) Uniformity of dose units (UDU) using large sample sizes, 7th edn. Chapter 2.9.47. of European Pharmacopeia 7.7. Report Pub Co Ltd, pp 5142–5245

Dong X, Tsong Y, Shen M (2015) Equivalence tests for interchangeability based on two one-sided probabilities. J Biopharm Stat 24(6):1332–1348

Hothorn LA, Bretz F (2000) Evaluation of animal carcinogenicity studies: Cochran–Armitage trend test vs. multiple contrast test. Biometrical J 42:553–567

U.S. Pharmacopoeia XXV (2010) USP 905 uniformity of dosage units, Mack Printing Company, Easton

Peto R, Pike MC, Day NE, Gray RG, Lee PN, Parish S, Peto J, Richards S, Wahrendorf J (1980) Guidelines for sample sensitive significance test for carcinogenic effects in long-term animal experiments. In: Long term and short term screening assays for carcinogens: a critical appraisal, international agency for research against cancer monographs, Annex to supplement. World Health Organization, Geneva, pp 311–426

Rahman MA, Tiwari RC (2012) Pairwise comparisons in the analysis of carcinogenicity data. Health 4(10):910–918

Shen M, Tsong Y (2011) Bias of the USP Harmonized test for dose content uniformity, Stimuli to the revision process, vol 37. Mack Printing Company, Easton

Shen M, Tsong Y, Dong X (2014) Statistical properties of large sample tests for dose content uniformity. Therapeutic Innovation Regulatory Sci 48(5):613–622

Tsong Y, Shen M (2007) Parametric two-stage sequential quality assurance test of dose content uniformity. J Biopharm Stat 17:143–157

Tsong Y, Dong X, Shen M, Lostritto RT (2015) Quality assurance test of delivery dose uniformity of multiple-dose inhaler and dry powder inhaler drug products. J Biopharm Stat. doi:10.1080/10543406.2014.972510

Wald A, Wolfowitz J (1946) Tolerance limits for a normal distribution. Ann Math Stat 19:208–215

Chapter 3
How To Be a Good Nonclinical Statistician

Bill Pikounis and Luc Bijnens

Abstract All fields profess commonly expressed criteria for its individual professionals to be successful. For the pharmaceutical/biotechnology industry that is the scope of this book, there are many accounts in the field of statistics of what it takes to be a good statistician. The goal of this chapter is to focus on specific characteristics for nonclinical statisticians which we believe are essential to be viewed as "good" professionals, either as individual contributors or managers.

Keywords Adaptability • Collaboration • Consultation • Enterprise Perspective • Negotiation • Resourcing • Statistical Software

3.1 Introduction

Many books, articles, and web content have been published that at least partially consider the concept of how to be a good statistician. See for example, Hahn and Doganaksoy (2011) and references therein. Our goal for this chapter is to focus on behaviors that we believe define good nonclinical statisticians.

Our content will be broad and lack concreteness in places due to the overview nature of the topic. We do hope to provide at least high-level practical advice and references for further study from our personal experiences, and therefore use a mix of first, second, and third person perspectives to the reader.

Our views are based on a combination of over 35 years as nonclinical statisticians in the pharmaceutical industry. We lead groups in the United States and Europe, and our scopes of responsibilities and accountabilities are global. There is no official definition of "nonclinical", and please see Chap. 1 and other chapters in this book for examples of the span of nonclinical. For this chapter, we presume three general areas outside of clinical trials: discovery, manufacturing, and safety. Each of these

B. Pikounis, Ph.D. (✉)
Janssen Research and Development, Johnson & Johnson, Spring House, PA
e-mail: bpikouni@its.jnj.com

L. Bijnens, Ph.D.
Janssen Research and Development, Johnson & Johnson, Beerse, Belgium

L. Zhang (ed.), *Nonclinical Statistics for Pharmaceutical and Biotechnology Industries*, Statistics for Biology and Health, DOI 10.1007/978-3-319-23558-5_3

three general areas also contains multiple other broad areas, of course, and we will touch on some of these in more specific descriptions. Whatever your definition of nonclinical statistics is and its areas of application, we believe that we have had similar experiences within our company.

There will be two main sections in addressing what makes a good nonclinical statistician. The first is common to all who identify themselves, or are identified, as a nonclinical statistician. We refer to this role as an *individual contributor*. The second section covers the additional duties of managing and leading a nonclinical statistics group, which we will refer to as *manager*. There is overlap to the principles in both roles, but we feel there are strategic factors of the manager role which are specific for success of nonclinical groups within their larger organization.

While we both feel our years of experiences and the people networks we have built as nonclinical statisticians provide us with meaningful perspectives to write this chapter, we do know we could have missed important points, or provided points where other nonclinical statisticians disagree. Our chosen criteria of what is "good" as a nonclinical statistician may be incomplete or disagreeable to you as a reader. We certainly invite you to contact us with any thoughts to help us clarify and learn about other factors you feel produce a good nonclinical statistician.

3.2 Individual Contributor

3.2.1 End-to-End Responsibility for Projects

In contrast to clinical development studies and trials within the pharmaceutical/biotechnology industry, there are generally no requirements for nonclinical statisticians to be consulted or to be part of the team for design or data evaluation of studies or experiments. When a scientist, engineer, or researcher wishes or is asked to enlist the help of a statistician for a nonclinical need, there is an initial contact.

3.2.1.1 The Initial Request

These initial, "request for help" contacts are variable in their frequency, like an inhomogeneous Poisson process. They are also variable in urgency, importance, complexity, and familiarity. The requestor may be someone you have never worked with before. Or it may be someone you have been actively collaborating with.

It is critical for a good nonclinical statistician to make no assumptions about the request or the requestor before a first meeting is held. It is certainly tempting to worry about yet another unplanned challenge that will add stress to an already heavy workload. But there is always the opportunity to start or continue building a relationship, valuably contribute to an important scientific or business endeavor, or expand your knowledge and skills.

No requests for help should ever be turned down. After a request comes in, it needs to be dealt with in a timely manner. Our policy experience is to respond within the next business day. This results in a maximum of 2 full business days, given the modern nature of global time zones and 24/7 connectivity. The initial response may be no more than an acknowledgment that the request has been received, reviewed, and will be dealt with soon.

The minimum obligation is to provide an initial consultation, which invariably means one hour or less. Within our large company, requests occasionally come to us from outside our official scope of nonclinical pharmaceutical research and development, due to network and reputation. We owe it to our company, and as statistics professionals, to properly perceive and understand the need of the requestor without pre-judgment. Then a healthy dialogue can take place to come up with options which may or may not involve continued collaboration with a nonclinical statistician from the staff.

3.2.1.2 The Initial Consultation

A request as previously described will typically turn into a project. Our definition of "project" here is that at least one more action needs to be taken beyond the initial consultation to fulfill this request and declare it as completed (Allen 2002). This determination of a project comes at the end of the initial consultation. One can optionally declare the actual discussion of the initial consultation as a project, if no further tasks are defined and agreed to for the nonclinical statistician to execute. If tasks for further collaboration are agreed to, the nonclinical statistician should proceed with defining and managing a project or projects within their organization system, stemming from the request. Projects can range to days to weeks to months to years.

We recommend an initial request be followed-up with a sincere desire to meet with a person and understand the entire context of the statistical needs. As mentioned earlier, this necessary level of understanding is impossible from an email or other communication channels. Except for rare, straightforward cases, such as explaining the difference between a standard deviation and standard error, replying to an email with a solution that you believe answers the question must be avoided. (Another category of exception is document reviews.) The prospective collaborator may believe their request is a simple question or questions, but virtually always it is not. All this can be clarified with an initial consultation. If the initial consultation can feasibly take place face-to-face, it should, especially for new or early relationships. In our modern world, this is not always possible due to geographical or time constraints.

The professionalism of a nonclinical statistician will start to reveal itself immediately to a requestor in how they handle the request and proceed to the initial consultation. As the saying goes, "first impressions count", whether fair or not. Use phrases on behalf of yourself or your organization such as:

- "Thank you for reaching out."
- "I/We will be glad to help."
- "May I suggest we set up a day/time for an initial consultation so I can fully understand your data, objectives, and scientific and business contexts?"
- "If you have any data or background materials you can provide beforehand, please email them to me."

None of these phrases commit you to any obligations beyond the initial consultation. Even if they do provide background materials, you may not have the time to do more than glance at them, which is fine since they will not provide all the information you need. Your genuine interest will also be clearer to the requestor if you take the initiative to set up the appointment time and location that makes it convenient for them.

When you meet, allow your requestor and (potential) collaborator to provide all the necessary information in order to understand what is needed. Strive for a good conversation and healthy dialogue at all times. There are boundless references on how to do this, especially for the "listen first" and "seek first to understand" (Covey 1989) stages. It is easier said than done, of course, but this constant attunement is critical to the quality of the relationship.

The researcher requestor may not have a clear idea of how statistics can help. A key charge for the nonclinical statistician is to translate the scientific questions into statistical frameworks such as (but not only!) hypothesis testing or interval estimation, whether the request involves data at hand or involves the planning of a study. Key technical pieces involve understanding of experimental units, factors, endpoints, design, missing data, excluded data, etc. Appreciation and understanding of the underlying science is needed as well, since later evaluation and the quality of the interpretation of the data depends on it. The rest of this book offers a comprehensive canvas of the diversity of research & development areas where a nonclinical statistician can help. Good general traits for the consultation aspects of statistician can be found for example in Boen and Zahn (1982) or within Hahn and Doganaksoy (2011).

The researcher may have already completed some analysis of the data and seeks your review or your assistance with misunderstanding or limitations of the software. This is increasingly common due to the wide availability of software for storing and analysis of data. In projects that involve big data, collaboration with other quantitatively trained colleagues is essential. For now, we mention that open receptiveness to ideas about statistical methods and data analysis methods is essential. If the initial consultation reveals that the project will be a more 1-to-1 or 1-to-small-group collaboration, the nonclinical statistician will have dominant control of the choice of statistical methods. If the nonclinical statistician is part of an interdisciplinary team with other quantitative colleagues, then confidence, but not arrogance, is needed to mutually choose appropriate data analytic techniques. Depending on experience and the complexity of the problem, one might find themselves unsure at all what statistical methods will be needed. It is perfectly all right to say "I don't know" or that "I don't have experience," and also assure your requestor that you have the ability to search for accurate and sensitive approaches from literature, colleagues, etc..

Completion of the initial consultation requires a clear definition of *what is needed* and *when it is needed*, particularly if an urgent business milestone is approaching. Here is where good negotiation skills on part of the nonclinical statistician are a must. Besides the people present in the initial consultation meeting, there are other considerations of stakeholders that are not present that will be taken into account, and these must be disclosed. It may be that a presentation to management is coming up next week and one or a few slides of statistical content are needed to cover evaluations and interpretations of a key endpoint. A more comprehensive report can wait until later, or eventually may not be needed at all. Everyone is busy, and we have found universal recognition by researcher colleagues that the ideal situation of fast as possible delivery of results *and* perfect work is not attainable. One solution may turn out to be for the nonclinical statistician to serve as an advisor for the researcher to continue doing their own analysis with their own software package. As Allen (2002) describes, every project can always be done better with more time and/or information. Allen furthermore advises, and we paraphrase here: "Lack of time is not the major issue. The real problem is a lack of clarity and definition about the project and what is required."

One more topic we wish to mention here is the "curse of knowledge," and its effect on interactions, starting with the initial consultation. As Heath and Heath (2007) discuss, "once we know something, we find it hard to imagine what it was like not to know it." As with all professional fields, constant self-checking of this natural human tendency is needed in order to prevent obstacles in communication by a nonclinical statistician to a non-statistician. Statistical jargon will be needed when a statistical expert verbally discusses, writes, or presents statistical concepts related to problems and solutions. Awareness to simplify as much as possible with the listener's perspective in mind will be beneficial to effectively communicate complexity and to build credibility. Practice of this behavior is needed all the way through and to the end of the project in order to continue reception of its benefit.

3.2.1.3 Lead Responsibility to Complete the Project

The initial consultation also unequivocally identifies the lead nonclinical statistician contact for the project. Usually this is the statistician who assumes the lead for the initial consultation in the first place. Exceptions to this would include professionals in the early stage of their nonclinical statistics career, where a more senior colleague will assist. In more complicated, longer term projects, this lead statistician may also need to coordinate with other nonclinical statisticians to perform portions of the work. We feel the taking on of full professional responsibility and accountability for a project is a vital component of the nonclinical statistician who is identified as the lead. It is an end-to-end endeavor.

3.2.1.3.1 Prioritization

The demand of choices on how to spend time on one project in relation to all others requires daily assessment. Every collaborator believes their project is important and wishes for their project to be attended to as soon as possible. Regulatory and business critical needs related to your company's portfolio of medicines should also take precedence. If this is not clear, your supervisor or line management should be able to clarify for you. In addition to these external forces, it can be helpful to ensure you do not only do work as it shows up, but also to dedicate time to defining your work, and to dedicate time to doing that predefined work (Allen 2002). Self and time management are universal problems in today's fast-paced world, so seeking out a personal organization system that you can execute will be advantageous.

3.2.1.3.2 Data Transfer

The lead nonclinical statistician will need to work with the data directly in whatever form it comes in. Excel spreadsheets and other tabular formats specific to software remain most common across discovery and manufacturing areas. The transfer of data via structured queries (SQL) from standard relational database systems remains infrequent. It is reasonable to request data of certain format from a requestor, keeping in mind the tradeoffs of the time it requires, the building of the relationship, and the business needs. If you can find a format where additional processing by your statistical software can reshape or transform the original data into formats for the needed statistical functions or procedures, proceed with that.

One example of this is the wide format for longitudinal data, where each experimental unit is a separate row, into the long and narrow format of one observation per row for standard modeling syntax of SAS or R. Another is the combination of different endpoints or factor levels across different worksheets or tables, to potentially merge into larger tabular data sets in preparation for exploration and modeling.

It is always best to ask if a machine readable format can be produced, especially if a PDF, or word-processing document, or even paper hardcopy of the data is initially provided. The delay and risk of a bad outcome from re-entering data always outweighs the additional work to avoid such intractable formats. If data transfer and data management remain an issue after reasonable negotiation, be clear that delays due to needed manual entry or copy and pasting, and careful verification, will be substantial. This can be acutely painful when compliance and regulatory procedures and expectations are added, and reviews, approvals, and signoffs are needed.

3.2.1.3.3 Statistical Methodology and Software

Accurate and sensitive assessment of relationships, trends, structures, and patterns will continue to grow in importance as economic pressures increase in the pharmaceutical/biotechnology industry. Proper application of statistical tools assists efficiencies needed in the pharmaceutical research and development process.

A great feature of being a nonclinical statistician is the liberty to choose methods for solving a problem of design or data analysis. Modern methods and computing power have never been more accessible to apply to data sets of small to moderately large size. In later subsections we discuss the fundamental need for good nonclinical statisticians to continuously learn and apply methods that are new to them.

Approaches need to include exploration of the data, fitting models, formal analyses to gauge the magnitude of effects, and checking of model assumptions. The exploration part should heavily rely on data graphs and visualization.

SAS and R appear to be the key essential programming environments for nonclinical statisticians to use these days. Proficiency in both is ideal to allow for the availability of each system's qualities, though we have seen good nonclinical statisticians rely essentially on one or the other of R or SAS exclusively. Another important category of statistical software for a nonclinical statistician is those used by research colleagues, namely for instance SAS JMP, Graph Prism, and Minitab. Proficiency in these provides the option for research colleagues to perform analyses accurately themselves through advice or formal education and training from nonclinical statisticians.

One of John Tukey's most famous quotes (1962) is: "Far better an approximate answer to the right question, which is often vague, than an exact answer to the wrong question, which can always be made precise." This seems to us particularly insightful for nonclinical statisticians. Often nonclinical statisticians work with data sets that come from inadequate design. There is always a tension between quality and timeliness in today's tumultuous business and scientific environments. There is never one single right analysis approach to any data set, since that implies that any other approach is incorrect. As described earlier, a good nonclinical statistician takes advantages of all their earned professional liberties to clarify the right questions and come up with approaches, results, and interpretations that are meaningful to their collaborators, and to their organization and its medicines. Then the patients receiving those medicines ultimately benefit.

3.2.1.3.4 Delivery of Results

It is critical for nonclinical statisticians to document their project work, in order to effectively curate and communicate results, interpretations, and conclusions to their research partners. Researchers often use results from such documents in their reports. These may be incorporated into laboratory notebooks, or may become part of official company R&D reports. The form of a "report" document should be flexible, as long as the content is sharp. A report write-up could be an email, or a short section of summary and details, or a slide deck, for example. Fulfillment of simpler requests such as sample size calculations, review of a manuscript, or explanation of a statistical concept, are common candidates for such quick communications.

Formal standalone reports are needed on occasion. These would include a Statistical Methodology section, Appendices, References, etc. One example of these are "source" documents for Chemistry, Manufacturing, and Control (CMC) sections of regulatory filings. Such statistical source documents are submitted and approved in a company's technical document repository.

The aforementioned structure of two primary sections, Summary and Details, is one we have found useful for the category of informal reports. The Summary is presented first to provide the bottom-line-up-front content of most interest to the reader. This summary section should not exceed a page. For the following Details section, the focus is on presentation of results, with judicious use of graphs, tables, and text. Statistical methodology descriptions are minimized unless the researcher requests it. The main principle is to present results and interpretations as soon as possible and in a clear, concise manner.

Whatever the type, documents are critical products for nonclinical statisticians to produce for their collaborators. They sharpen the thoughts, actions, and decisions of the nonclinical statistician. They also help provide an essential record of the work for future reference as knowledge sharing amongst peers, or for knowledge recollection by the original author when a need to reopen, refer, or build upon a long ago completed project arises.

3.2.1.3.5 Management of Project Materials and Closure

In prior sections we touched on aspects of autonomy for application of methods and communication of results for nonclinical statisticians. Each individual nonclinical statistician will develop a personal system for dealing with their projects and those should be primarily driven by their autonomy. Outcomes of those projects, however, require a responsibility and accountability to the larger organization. Electronic folders which can be accessed or zipped up to contain and pass along all materials related to the project—particularly the key components of input data, program code, and communication of results—should be clear enough to be reconstructed at a later date and shared upon request with colleagues who may have a similar project and may actually have to cover for the originally responsible nonclinical statistician.

The recently very visible topic of "Reproducible Research" and its connections with statistical practice are important to consider. (See for example Begley and Ellis 2012.) Shotwell (2013) presented a survey of recent investigations and recommendations that should be heeded and practiced by nonclinical statisticians in their own controllable sphere of influence. Good nonclinical statisticians who practice this continuously contribute to organization knowledge, and benefit themselves and everyone by freely sharing experiences and expertise. Paraphrasing the Clarebout principle on the specific computational dimension of reproducible research (de Leeuw 2001), it represents true scholarship that goes beyond just an *advertisement* of the scholarship.

3.2.2 *Service and Partnerships*

Peers of ours have expressed their dismay at nonclinical statistics groups being regarded as a "service." The related concept of being a server to a customer, or synonymously, client, is even more troubling to these peers. We can understand this, if the definition of service centers on the illusion of "push button" statistical analyses transactions, or just doing such statistical analyses "fast." Following Pink (2012), we suggest starting with definition of service as "improving other lives, and in turn, improving the world." For nonclinical statisticians in the pharmaceutical/biotechnology industry, this has a natural connection with our contribution toward the effectiveness, safety, and availability of medicines for patients.

Greenleaf (1982) introduced the idea of "servant leadership," and this has shown to be effective for hierarchical relationships as well as for peer-to-peer relationships. Certainly the practice of service needs to be balanced with a respect amongst all parties. It does not require the nonclinical statistician to be deferential to demands. With proper mutual understanding and negotiation, service helps build the longer term relationship of scientific collaboration, partnership, and transformation that nonclinical statisticians aspire to achieve with researchers. We can move from labels such as "clients" and "customers" to "partners" and "collaborators" and back again without any contradiction.

3.2.3 *People Networks*

The number of professionals identifying themselves as nonclinical statisticians is small relative to clinical statisticians in the pharmaceutical/biotechnology industry. It is therefore most important for nonclinical statisticians to build their people network of nonclinical statisticians within their organization and with nonclinical statisticians outside of it. Given the diversity of nonclinical statistics projects and collaborations, along with the growing interest of many professional quantitative fields for medical data, and the volatility of the business environment, the strength of contributions and careers depends on relationships built through individual networks. Technical advice from your network will help you get your work done, and cultural advice from your network will help you navigate ongoing and acute challenges in managing your career.

3.2.4 *Research, Presentations, and Publications*

Nonclinical statisticians are applied scientists. The main difference between statisticians in pharmaceutical research and statisticians in academic settings is the fact that the scientific problems come from pharmaceutical industry R&D.

The methodological questions at hand are usually generated by research projects that need to be resolved. The underlying reasons for the research are not uniquely methodological since there is a pharmaceutical problem at the basis. For that reason we often call the work of statisticians in the pharmaceutical/biotechnology industry "applied statistics" work, as the application drives the statistical research.

With good researcher collaborator relationships will come co-author invitations for your statistical input into a paper, poster, or presentation. This will as least involve statistical content for tables or text, and possibly graphs. When summarization of the statistical methods is requested, it needs to be provided in a clear and concise manner.

First-author presentations and publications are also essential to career development and being a good nonclinical statistician. They provide credentials of scientific rigor that are universally recognized.

3.2.5 *Adaptability*

Also called flexibility or versatility, the trait of adaptability is especially vital to being a good nonclinical statistician. It is more true that roles and responsibilities stretch across multiple boundaries. Such elasticity has been shown to be valuable for organizations to cope with tumultuous business conditions (Pink 2012). Organizations, especially larger ones, where fixed skills or specialties could be established and grow, are punished when inevitable upheaval occurs in business or organizational cycles. A good nonclinical statistician will effectively respond to new challenges even when first unfamiliar with the context of the problem, and later when considering and deciding amongst appropriate methods for a solution. Projects range from a one-hour consultation to weeks to months and even years. There will be a mix of these in the nonclinical statistician's project workload at any time. There is no way to predict what the number and variety of projects will be in the foreseeable future of say, more than one month out.

A good nonclinical statistician readily accepts opportunities to enhance skills, work with new areas, and expand impact on their organization, even when initially taken out of their comfort zone.

3.2.6 *Stretching: An Example of Translational Research*

Earlier in this chapter we mentioned three primary categories of focus for nonclinical: discovery, manufacturing, and safety. The discovery area is very broad, with examples in Chaps. 4–8 of this book and further discussion on Chap. 1. If we broaden the concept of discovery to include the bridge area of preclinical-to-clinical translation, or animal-to-human, then this is an area that seems to be generally under represented by nonclinical statisticians. Translational research, as introduced in Chap. 1, also covers research more broadly in the translational phase, such as with biomarkers. Building on the theme of the previous section on adaptability, those good nonclinical statisticians working in this underserved area are an example of stretching out of classically defined nonclinical areas.

This translational area is an important point where nonclinical statisticians, clinical statisticians, and other quantitative scientists such as clinical pharmacologists and modelers should collaborate. The first clinical trials of a potential therapy rely heavily on findings from pre-clinical experiments. The information does not only need to go from pre-clinical to clinical but information collected in the clinical phases also needs to be fed back to the pre-clinical researchers. This feedback in the field of translational medicine is becoming more and more important. Pre-clinical researchers, data analysts and statisticians use that information to build their system biology and pharmacokinetic/pharmacodynamics (PK/PD) models (Bijnens et al. 2012; Jacobs et al. 2008). Modelling is then used to simulated real trials in silico based on the information coming from pre-clinical and updated information coming from clinical trials. That way it is hoped that there is less failure in the later phases of clinical research because safety, efficacy and the variability of the measurements can be anticipated in a more reliable manner.

3.2.7 *Professional Service*

Following our advice earlier on networking, good avenues to meet with colleagues and other professionals and have an impact is through service to the statistics profession. Judging science fairs and reaching out to secondary schools and universities to teach, or provide one-off career-oriented presentation and discussion, offer general avenues where the specifics of valuable nonclinical statistician impact can be highlighted as well. Not only is joining your local professional chapter and its national organization important, but participating in events and taking on leadership roles such as an officer or serving on committees will help you identify and reduce or close skill gaps in communications and teamwork.

3.3 Manager

For the purposes of this section, a manager is someone who leads a team of people within a nonclinical statistics function, with official designation of the team as individual direct reports to the manager. We will use the shorter term of "individuals" to denote such individual direct reports. The term manager here also includes the person who officially heads the entire nonclinical statistics function within the organization.

There is a vast literature on leadership and management that contain principles that apply to nonclinical statistician managers as they do to managers across other professions. In the following subsections we concentrate on characteristics for nonclinical statistics managers. These characteristics focus on how to be a good nonclinical statistician in a managerial role. The terms of manager and leaders are synonymous for our purposes here, as we implicitly presume that a good nonclinical statistics manager must be a good leader.

3.3.1 Enterprise Perspective

In the introduction we presumed a simplified partition of discovery, safety, and manufacturing as parts of nonclinical statistician functional scope for the purposes of this chapter. A good nonclinical statistics manager may only be officially responsible for their scope of responsibility and accountability within one of these areas. Nonetheless, as a manager, you and your team will have many interdependencies with other functions that make up the large picture of the company. Beyond technical competence, the success of your team and yourself depends on a perspective beyond "what is in it for me and my team." Missions, visions, and values from the overall company and other functions in your company need sufficient attention above your personal views. Ask yourself: How does your perspective serve your organization, your company and the world? Why should anyone care about your perspective? What assumptions underlie your perspective? It is equally fair and important to seek out and understand the answers to these questions from the perspectives of the functions you work with as well. Strategic company initiatives that do not involve statistical expertise are valuable opportunities for you or individual team members to strengthen and broaden your reputation and your function's reputation to be recognized as willing and able to accomplish difficult projects that are needed by the enterprise.

3.3.2 Business Critical

Earlier in this chapter, prioritization was discussed at the individual level. It is particularly challenging for the nonclinical statistics manager since requests for help tend to come on an ad hoc basis. Unlike our clinical statistician counterparts, there is no standard requirement for a statistician to be part of the official team for an new molecular entity (NME)—compound, biologic, vaccine, etc.—under development or already approved.

While the pace of the world to some degree presents critical responsibilities and accountabilities every day, there is a periodic frequency of truly business critical requests that arise from time to time with true importance and urgency. As mentioned previously, regulatory and business critical needs related to your company's portfolio of medicines from early stage decision stage research gates through late stage development through commercial should be obvious and take precedence. Such requests can come from teams or individual executives. It is important to diagnose such questions quickly and ensure you and your team are clear and have consensus on the prioritization and the reasoning behind it. As a manager it is your obligation to communicate the reasoning.

3.3.3 Engagement and Movement

Persuasion and influence are vital skills for a nonclinical statistics manager. There are individual direct report employees to engage, peers to build coalitions with, and upper level management to move to decide and act.

Two particular intrinsic motivator categories we have found to be successful for nonclinical statistics individuals are autonomy and mastery (Pink 2011). We mentioned earlier the liberty to choose statistical and computing methodologies in data evaluation. This also provides unlimited opportunities to learn and improve skills by studying and doing. The end-to-end full professional responsibility detailed earlier in the Individual Contributor section is a skill we expect individuals to develop by the end of their first year as a member of the nonclinical statistics group.

Navigation between personal and work lives is also something good managers support for their individual direct reports. Once a project has been defined well enough, today's portability of personal computer laptops and devices allows the completion of that work to be done anywhere, anytime. Established one-to-one meeting appointments between manager and individual on a bi-weekly, monthly, or even weekly basis are simple and effective to ensure issued potential problems—technical and human—are detected, discussed, and worked through. Everyone seeks trust, and as a manager these sessions provide opportunities to gain and display confidence in individuals so they can build their confidence in themselves, and also confidence in you.

Watkins (2003) discusses the need to build coalitions with peers in the horizontal dimension. For a nonclinical statistics manager, the range of peers should include collaborators and their managers. Are there people you should meet or better connect since they are important to your success? Invest in the relationship to build capital before there is a need to ask for something later. The investment will also help clarify the landscape of supporters, opponents, and those in-between who will need to be further understood with respect to what can move them to the supporter side. Sending an email or calling "cold" to request an introduction has invariably worked in our experiences, as the standard response from recipients we contact has been "thanks for reaching out" and "I would be glad to meet."

Persuasion and movement of upper management typically involves the goal of convincing them to provide resources they have in exchange for something the function of nonclinical statistics can provide. The better we can persuade that those resources will make their world better, the better chance we have to succeed from moving along the spectrum of "not value-added" to "beneficial" to "critical." Skills in formal business planning must be developed by nonclinical statistics managers in order to effectively propose resource solutions to meet demand. In the next subsection we describe some resourcing approaches we have undertaken.

It has been helpful in our experience to compile monthly highlights from all staff to use for updates to management and to also have on hand for later reference when examples of breadth and depth are needed to illustrate our valuable contributions in a pitch for, or defense of, resources. We also have a tracking system so that each request from researchers is logged and volume metrics with some demographics such as therapeutic or functional area can be compiled and reported on a quarterly basis or as needed.

3.3.4 Resourcing

Clinical statisticians generally have a regulatory mandate at their companies to be involved end-to-end for the planning and execution of a clinical trial. At these same companies, there is no analogous general mandate for nonclinical statisticians to be involved in studies and experiments in any of the three general areas of discovery, safety, or manufacturing postulated in this chapter.

At our company a model of "lead statistician" has evolved for every new molecular entity (NME) in development to be the first and single point of contact for manufacturing needs related to the product. This typically starts just before or soon after the commencement of Phase 3 clinical trials. The lead statistician commitment stays through the health authority filing, approval, and post-approval follow-up needs, such as different formulations, potential modification of specifications, and extension of shelf-life. While the NME program is an obvious focus and prioritization of the individual lead statistician's time, it does not exclude other nonclinical statistician team members to assist to complete assignments related to the NME program. Nor does it preclude the nonclinical statistician from lead professional responsibility on other projects, including those in the discovery or safety areas. More experienced nonclinical statisticians at our company handle lead statistician roles for manufacturing support of two or more products.

At this time of writing, our reputable statistical support for manufacturing within the company has produced invitations to collaborate on programs such as Process Validation strategies (FDA Guidance 2011), Design of Experiments (DOE) for earlier formulations, Design Spaces, and predictive scale-down models.

To have resource supply to meet these demands for statistical support, we must work within the formal company project management system to quantify and forecast the demand. This is a mix of internal Full Time Equivalent (FTE) units to use for permanent internal hiring and contractors. The supply to meet demand is also supplemented by deliverables from external partners, which typically involves special capabilities, or else capacities, that cannot be efficiently met by internal resources. (See the later subsection on Education and Software Strategies.) Even when FTE formulas exist, quantification of the demand is an approximate endeavor due to the dynamics of product development. For nonclinical statistics, we try to compensate for these unknown errors based on subjective prior experience. Here again, reputation counts for a lot in order for the company to trust the demand estimates from nonclinical statistics so that resource requests to minimize the gap between supply and demand are fairly considered.

The nonclinical statistician support resourcing model for manufacturing therefore relies on (1) the assumption of statistical support for every NME in the portfolio, and (2) meaningful, defensible estimates of FTE demand for each of those NMEs. The sum of these individual NME FTE demand estimates provide a basis for the number of nonclinical statisticians in the group to support manufacturing demand.

Ideally the forecast demand model for nonclinical statistician support of manufacturing should similarly apply to areas of statistician support for safety and discovery. At our company we are working formally with initiatives to this end at the time of the writing of this chapter. For safety, the requirements of small molecule animal toxicology studies and large molecule anti-drug antibody/immune response assays present natural anchors to quantify demand of nonclinical statistician support. For discovery, the initial focus has been on programs entering lead optimization stage, when a lead candidate or small number of lead candidates have emerged from screening and are on the critical path toward declaration of an NME and beyond to a milestone of First in Human (FIH). This opportunity at our company has emerged primarily due to recognition of statistics as one critical piece of a top-down focus to improve procedures and practices for robust, reproducible research in discovery. See for example, Begley and Ellis (2012) on issues highlighted for scientific reproducibility.

Once there is the approval to hire, where do good nonclinical statisticians come from? It is virtually impossible to get many (if any) candidates with direct nonclinical statistics experience for an opening. There needs to be an openness to a range of qualifications, most likely in fields like clinical statistics, or medical devices, medical sciences, computer/information/data/quantitative sciences, etc. Candidates without direct pharmaceutical or even health care experiences could be considered. Our interviewing process places high importance on attitude, motivation, and the potential of the candidate to fit with our culture, even if they lack the direct technical experience at the hiring stage. Of course, a sufficient baseline of technical skills needs to be demonstrated. Adler (2013) and Murphy (2011) are good references to draw upon in order to develop a systematic approach to attracting and evaluating candidates that will be valuable performers beyond the first year after their potential hire. This perspective has allowed us to confidently invest in hires that are at the beginning or very early in their career out of school, along with more senior statisticians who do not have direct nonclinical statistics experience. Experienced individual contributors are counted upon to mentor and assist newer hires for their technical and communication development, in combination with coaching by the team manager.

3.3.5 *Education and Software Strategies*

The previous Resources subsection touched on education and software as part of the strategy to meet demand for statistical knowledge and capabilities for nonclinical studies and experiments.

3.3.5.1 Education

Our researcher customers have always had access to data analysis software, and nowadays that is more true than ever. At our company, GraphPad Prism, Minitab, SAS JMP, and Excel are licensed, available, and supported on company issued personal computers. Earlier we mentioned that virtually all researchers will do some form of data analysis for a given study themselves. Some will enlist the help of a nonclinical statistician for primary analysis goals. While this typically involves the core relationship of providing the statistician the data and objectives, an alternative request might be a review of an analysis already performed by their own software. Another alternative request would be along these lines, for example: How do I do this in GraphPad Prism, or SAS JMP? Have I done this the right way? Could you show me how? Would you write up a guide for me?

Nonclinical statisticians need not feel threatened by researchers doing their own data analysis. Our experiences assure us that critically important studies, from a business and/or regulatory perspective, will come to our attention so that we can provide our expert services as appropriate. With the number of studies and experiments and researchers always guaranteed to outnumber statisticians with a high ratio, there is no practical way for nonclinical statisticians to have a hand in all of these through a one-to-one consulting model. It is a win for nonclinical statisticians to guide researchers to appropriate analyses, interpretation, and presentation they can do themselves, and it is a win for researchers in terms of efficiency.

Two interdependent approaches to guide researchers to do analyses themselves come from education and software when requested by departments, teams, and committees, etc., nonclinical statisticians should present on fundamental statistical issues and concepts such as the use of the appropriate statistical methods, logarithmic transformations, sample size calculations, presentation of results, interpretation of p-values, etc.

The modern day-to-day pressures of work for everyone renders the implementation of short courses that span multiple days or even one full day highly unlikely. Researchers and their management have consistently indicated to us that prepared presentations of 1–2 h are welcome. With the help of external partners we have developed a series of such partial day modules on statistical concepts for discovery researchers that include fundamentals, comparisons of groups, dose-response models, and experimental designs. Each module consists of the classic lecture-based presentation, and then followed by a hands-on portion where a group of up to 20 researchers bring their laptops with software such as GraphPad Prism or SAS JMP installed, and work through examples with the instructors to reinforce the concepts of the earlier presentation. All slide deck materials and data sets are then made available at our actively maintained internal website for later references. Some of the module content has been recorded for on-demand streaming.

The internal website also provides our nonclinical statistical group a platform to provide information about the group as well as educational resources. In addition to the aforementioned course module materials, we provide short two to three page articles on topics such as outliers to deal with commonly recurring questions. The outlier article in particular serves as official guidance from our group to the research community.

3.3.5.2 Software

Computer technologies continue to hold great promise for leveraging statistical methodology by nonclinical statisticians for researchers. The last subsection described the combination of off-the-shelf software with education to provide benefit. The development and sustained support of customized software tools are also a valuable investment to provide statistical methods that have one or more characteristics of widespread applicability, incorporation of modern techniques, high repetition, or automated handling of data that is otherwise prone to errors from manual handling.

We use a mix of SAS and R engines to build such customized tools. It is preferable to maintain the code centrally on a server to ease updates when they are required. R scripts are made available via an open source enterprise service bus that can be called in many ways, e.g. Web interface, Excel, or other software that has network connectivity application programming interfaces. This way the R source code does not have to be copied to the computer of each individual scientist. At the same time scientists do not need to install R on their computer. Open source computer languages also facilitate straightforward exchange of new methods with colleagues in the industry, and academic and governmental institutes. That way the code can be tested and improved when required. These days there is a trend for pharmaceutical companies to encourage precompetitive collaboration since they are usually not interested in purchasing patents for statistical methods. This new climate encourages a fast turnover of methodological research. Often the new statistical methods are published in applied statistical journals and the corresponding open-source computer code is made available for open source licensed use.

Good customized statistical tools require understanding and application of good software principles and lifecycle management. Metrics on the usage of the software are important to collect and report to assess and justify continued value. As mentioned previously with education, we have found these software development capabilities require external partnerships, where clear deliverables can be negotiated.

3.3.6 Internal and External Partnerships

Many quantitative and data analytic groups of professionals beyond statisticians have emerged in pharmaceutical companies in recent years, correlated with the massive collection of data by internal and external sources. The internet facilitates the exchanging of such big data and the exchange of ideas amongst all quantitative scientists to combine and transform such data to information and ultimately, knowledge. Moreover, pre-competitive collaborations in international networks amongst industrial, contract and academic quantitative scientists are essential to tackle the questions and problems of these large and complex data sets. Good nonclinical statisticians, as characterized in this chapter, are well-suited to embrace these challenges and helpfully respond to the opportunities.

References

Adler L (2013) The essential guide for hiring & getting hired. Workbench Media, Atlanta

Allen D (2002) Getting things done: the art of stress-free productivity. Penguin, New York

Begley CG, Ellis LM (2012) Drug development: raise standards for preclinical cancer research. Nature 483(7391):531–533

Bijnens L, Van den Bergh A, Sinha V, Geys H, Molenberghs G, Verbeke T, Straetemans R, Kasim A, De Ridder F, Balmain-Mackie C (2012) A meta-analytical framework to include historical data in allometric scaling. Stat Biopharm Res 4(2):205–215

Boen J, Zahn D (1982) The human side of statistical consulting. Lifetime Learning Publications, Belmont

Covey S (1989) The seven habits of highly effective people, New York:Sirnon & Schuster

de Leeuw J (2001) Reproducible research: the bottom line. Technical Report, Department of Statistics, UCLA

Greenleaf R (1982) The servant as leader. Robert K. Greenleaf Center for Servant-Leadership, Indianapolis

FDA (2011) Guidance for industry process validation: general principles and practices. Rockville, MD

Hahn G, Doganaksoy N (2011) A career in statistics: beyond the numbers. Wiley, Hoboken

Heath C, Heath D (2007) Made to stick: why some ideas survive and others die. Random House LLC, New York

Jacobs T, De Ridder F, Rusch S, Van Peer A, Molenberghs G, Bijnens L (2008) Including information on the therapeutic window in bioequivalence acceptance. Pharm Res 25(11): 2628–2638

Murphy M (2011) Hiring for attitude. McGraw-Hill, New York

Pink DH (2011) Drive: the surprising truth about what motivates us. Penguin, New York

Pink DH (2012) To sell is human: the surprising truth about moving others. Penguin, New York

Shotwell M (2013) Barriers to reproducible research and a web-based solution. Presented at the 36th annual Midwest Biopharmaceutical Statistics Workshop, Muncie, Indiana

Tukey JW (1962) The future of data analysis. Ann Math Stat 33:1–67

Watkins M (2003) The first 90 days: critical success strategies for new leaders at all levels. Harvard Business Press, Boston

Part II
Statistical Methods for Drug Discovery

Chapter 4
Statistical Methods for Drug Discovery

Max Kuhn, Phillip Yates, and Craig Hyde

Abstract This chapter is a broad overview of the drug discovery process and areas where statistical input can have a key impact. The focus is primarily in a few key areas: target discovery, compound screening/optimization, and the characterization of important properties. Special attention is paid to working with assay data and phenotypic screens. A discussion of important skills for a nonclinical statistician supporting drug discovery concludes the chapter.

Keywords Assay Development • Assay Validation • GWAS • Bioassay • Target • ADME • Dose Response

4.1 Introduction

Drug discovery, as defined in this chapter, includes the earliest portion of the pharmaceutical pipeline and starts at the inception of a new project to the delivery of a viable drug candidate (i.e., "a CAN"). The CAN should have:

- acceptable potency for the biological target,
- good pharmacokinetic properties, such as its absorption, distribution, metabolism and elimination (ADME),
- no known toxicities issues, and
- good manufacturability.

Once these characteristics are demonstrated, the development process can concentrate on a single molecule for formulation, further pre-clinical evaluations, and eventually more focused clinical testing. The road to a drug candidate can be a

M. Kuhn (✉) • C. Hyde
Pfizer Global R&D, Groton, CT, USA
e-mail: max.kuhn@pfizer.com

P. Yates
Pfizer's BioTherapeutics Statistics Group, Groton, CT, USA

Pfizer Global R&D, Groton, CT, USA

L. Zhang (ed.), *Nonclinical Statistics for Pharmaceutical and Biotechnology Industries*, Statistics for Biology and Health, DOI 10.1007/978-3-319-23558-5_4

long and arduous process that balances a number of different, and often competing, goals. In this chapter, we provide an overview of the drug discovery process and highlight general concepts as well as areas where nonclinical statistics can have a significant impact on the quality of drugs.

It is no secret that a high percentage of drug candidates fail during clinical testing. Arrowsmith (2011a,b) estimates that a large majority of failures are either related to efficacy (approximately 51–66 %) or safety (about 20 %). It can be assumed that these failures could be prevented by either improving the disease target/rational or by designing better molecules. However, once the drug candidate is finalized there is little that can be done to mitigate these issues since any problematic aspects are "baked in" to the compound. As the cost of obtaining regulatory approval for a new molecular entity continues to increase, failing to reduce the likelihood of a late-stage failure for a CAN during clinical testing poses a serious risk to the continued viability of the pharmaceutical industry.

Given this, the early discovery phase is the primary opportunity for statistics to add value to the project. This can be accomplished using basic statistical techniques, such as:

- good experimental design (including sample randomization and blinding),
- characterization and reduction of assay variability,
- elimination of subjectivity,
- clarification and estimation of uncertainty, and
- suitable statistical analysis.

It is important that the medicinal chemist, target biologist, bioanalyst, project (management) lead, etc., understand and appreciate the value of each of these items. In contrast to the number of statisticians involved in clinical trials and other late-stage regulatory matters, the number of discovery scientists far outnumbers the number of statisticians in drug discovery. The remainder of this chapter discusses different aspects of the early discovery process from both a statistical and non-statistical perspective. In each section, areas of opportunity where statistics can have a positive impact on the drug candidate are also discussed.

For the purposes of discussion, we emphasize the discovery and development of small chemical compounds here. Large molecule therapies, e.g., biologics, proteins or polypeptides, vaccines, gene therapy, or oligonucleotides, can present a different set of challenges and concerns. For example, most small molecules are orally administered and processed by the gastrointestinal tract whereas large molecules are often delivered via injection or infusion. Emerging technologies that make use of stem cells, a combination of small and large molecules, e.g., an antibody-drug conjugate, or harness the host's immune system to treat a complex disorder such as cancer are therapeutic examples that represent the leading edge of drug discovery. Also, in this chapter, we use the terms "compound" and "molecule" interchangeably.

The remainder of this chapter is divided into discussions of two distinct stages of early drug discovery: target selection and compound optimization. The former is focused on finding and exploring the appropriate biological rational for a molecule while the latter is concerned with making the best possible medicine. These stages have little in the way of commonality; one is primarily focused on biology whereas the other is very chemistry-centric (at least for small molecules). The statistical skills needed for each are similar only in the most general sense.

4.2 Targets and Phenotypes

There are several paths to a drug candidate. The most common strategy in the modern era is *target-based* discovery. Here, a specific biological entity, such as a protein or enzyme, is targeted for modification (Landry and Gies 2008). For example, the Janus kinase (JAK) family of enzymes is associated with different aspects of oncology and inflammatory diseases, such as psoriasis, arthritis and Crohn's disease (Clark et al. 2014; Luo and Laaja 2004; Wilks 2008). The four proteins in this family are important to many diseases because they can interfere with signaling pathways by interacting with certain cell membrane receptors. The "JAK-STAT pathway," which also involves signal transducers and activators of transcription (STAT) genes, blocks signals and thus impedes transcriptional regulation.

Target-based strategies hinge on the assumption that enough is known about the disease biology to be confident that this highly reductionist approach will be effective. This type of project judges the activity or potency of the molecule by measuring the effect of the drug on the specific molecular target at the site or tissue of action. This may or may not be an effective strategy. The next section demonstrates that biological pathways, even when well-characterized or understood, are complex. The interplay between biological components, such as proteins, enzymes and signaling factors, is convoluted. There are often regulatory or compensatory sub-systems that can neutralize the effect of inhibiting a single target.

Alternatively, phenotypic screens (Eggert 2013; Swinney 2013; Zheng et al. 2013), a former mainstay of drug development, do not focus on a specific molecular target. Instead, this approach focuses on a phenotype: some observable characteristic of an organism that is correlated to the disease state in humans. In comparison to target-based discovery, there may be little or no assumed understanding of what target(s) are being affected by compounds that demonstrate a potent effect on the phenotype. The phenotype may also be more relevant to the disease than the molecular target. For example, a compound may have dynamic effects on various signaling pathways and these types of effects cannot be evaluated in a simple biochemical screen for a single target. Similarly, a cell-based or *in vivo* phenotypic screen can evaluate the compound's ability to penetrate the cell. Swinney and Anthony (2011) reviewed a cohort of approved drugs and found that, for first-in-class small molecules, more approved drugs were a result of programs driven by the phenotype than specific targets. They also note that most biologics were driven by target-based approaches. While enjoying a bit of a resurgence in current drug discovery efforts, phenotypic screens are not a panacea. Between concerns over model organism choice and translation worries, challenges in creating cell-based models predictive of disease that are amenable to practical concerns such as cost and throughput, a need for a basic understanding of the biochemical underpinning linking a compound to a phenotype, etc., these screens present both scientific and practical challenges for their effective use.

To illustrate the potential of a more phenotype-driven approach, cystic fibrosis (CF) is an autosomal recessive disorder caused by mutations in a gene that affects epithelial membranes (Ratjen and Doring 2003). The disease can affect several tissues, including the pancreas, liver, sinuses and lungs. The airway surface liquid (ASL) model mimics the liquid meniscus that can be found in human airways (Verkman et al. 2003). When the airways are impacted by the disease, the phenotype is a reduced amount of liquid. A phenotypic screen might use a membrane of appropriate cells with the CF mutation and measure the impact of a compound on the amount of meniscus observed (Harvey et al. 2011). Potent compounds found using this type of screen are known to affect the observed physiological response and may hit more than one target. While not discussed here, target promiscuity (or the possibility for off-target effects) is part of the rationale for using a phenotypic screen in early discovery toxicology studies.

The difference between a target, biomarker and phenotype can be difficult to conceptualize and may be somewhat ill-defined in practice. The notion of a target is often used broadly to include well-know causal indicators of poor health, such as:

- "bad cholesterol" (i.e., LDL) for heart disease, or
- glycated hemoglobin (HbA1c) as a surrogate for plasma glucose concentration in patients with diabetes mellitus.

While these are, in a sense, legitimate "targets", the compounds used to moderate them typically do so by focusing on a more specific protein. A more fitting use of the word "target" is when it is used to describe a direct gene product (referred to as the *gene* or *genetic target*). Similar ambiguity exists when describing phenotypes. For example, one might consider LDL as a phenotype since it is an "observable characteristic." However, this particular protein would probably not be considered a proper phenotype since atherosclerosis is a consequence of high circulating LDL levels. In other words, LDL is the means to the end (i.e., vascular disease) and not the end itself. In addition to the target and phenotype, there is often the need for a clinical endpoint that will be used to gauge a compound's efficacy when used in human subjects. The causal order or interrelationships between a target, biomarker, phenotype, clinical endpoint, and human disease is often complex and may defy a simple reductionist model.

4.3 Target Identification and Characterization

Before embarking on a detailed discussion of target discovery strategies, a short primer is needed on the *central dogma of molecular biology* (Crick 1970) which, when simplified, amounts to:

DNA makes RNA and RNA makes protein.

DNA (and the rare variations within) lies at the start of the fundamental biological process for creating proteins. Traditional estimates ranging from 34 % (Gregory 2005) to newer estimates of up to 75 % (Djebali et al. 2012) of the genome are *transcribed* into functional RNA, and a contiguous region involved in the formation of a single functional RNA molecule after transcription (and ultimately a single protein where applicable) is typically what defines a "gene", and hence this transcription process is also known as *gene expression*. Most of these RNA molecules contain regions called *exons* which form a protein by a process known as translation (or sometimes simply "coding"), whereby consecutive triplets of RNA nucleotides form *codons* that each become any of 20 amino acids (in humans) as determined by the codon sequence, beginning with the first "start" codon (ATG) and ending with the occurrence of any of several stop codons. The resulting chain of amino acids then comprises the final peptide or protein after extensive bending and folding. The remaining transcribed segments of the original DNA sequence that are cut away during the RNA maturation process and hence occur between the exons are known as introns, while the final RNA contains an untranslated region (UTR) on either side of the protein coding (exonic) region (Watson 1992). Only 1.5 % of the human genome lies within exons.

There are many paths to a viable drug target. Lindsay (2003) provides an overview of different approaches. Some targets are found by perturbing normal biological processes at some point along the path of the central dogma. For example, RNA transcription can be disrupted in a relevant cell type using short, interfering RNA (siRNA) assays (Jones et al. 2004). The effect of *knocking down* a gene in a pathway can be measured on some relevant functional endpoint. siRNA screens can be used to investigate entire genomes using high throughput assay systems similar to those described in this chapter to optimize compounds. Zhang (2011) and the following chapter in this book describe statistical methods for analyzing siRNA screens. Many other methods for studying and/or perturbing biological pathways exist. However, this section will focus on the investigation of differences in *human DNA* to find good drug targets.

For heritable diseases, the target discovery is increasingly done through the use of genetic association studies, and with the advent of microarrays, these are often Genome Wide Association Studies (GWAS) (Burton et al. 2007) intended to cover all the common variation in the human genome. This is actually an appropriate level at which to begin the discovery process because the human genetic (DNA) sequence is static throughout a person's lifetime and is unambiguously either neutral or at the beginning of any causal chain initiating a biological process in which it is involved. Chapter 7 of this text describes the nuances of GWAS studies in more detail.

There are in total about 3.2 billion DNA base pairs along the total length of all 23 chromosomal pairs. However, only about 0.1 % (Jorde and Wooding 2004) of these vary across individuals, while the other 99.9 % of the DNA sequence is identical for all humans. Genetic association studies revolve around the aforementioned 0.1 % of DNA sequence that varies across individuals. Most of these variations are isolated and are called SNPs (Single Nucleotide Polymorphisms), other types of variation including indels (insertions/deletions) and copy number variations (repeated stretches of sequence with variable multiplicity). GWAS chips in particular are primarily designed to genotype bi-allelic SNPs intended to "tag" the entire genome by virtue of having at least one genotyped SNP minimally correlated with every un-typed SNP in the genome. This correlation, called Linkage Disequilibrium (LD), is a consequence of localized contiguous regions of the genome being inherited intact by the reproduction process.

4.3.1 Statistical Aspects of GWAS Studies

By genotyping bi-allelic SNPs, GWAS chips specify which of two possible alleles (from among A, C, T, or G) is present in each of the two chromosomal copies. Because the error rate for such calls is very low, this eliminates nearly all the variability in the tested feature. In fact, the result is essentially a three-level categorical variable: the subject either has two copies of the first allele, two copies of the second allele, or one of each. In practice, the genotype is usually modeled as having an additive effect on the phenotype in question, meaning the phenotypic estimate for subjects with one copy of each allele is constrained to lie precisely halfway between the estimates for subjects having two copies of the first allele and subjects having two copies of the second allele. Under this assumption, genotype is simply coded as a numerical variable equal to the number of minor (rare) alleles: hence, 0, 1, or 2. Note that because the SNPs on a GWAS chips are simply meant to tag their correlated (by LD) neighbors, this approach works well even when the underlying effect from some un-typed functional SNP is dominant or recessive (meaning, fully born by only one copy of the rare allele, or absent except two copies of the rare allele), the effect estimate for a correlated tagging SNP will still manifest itself as more additive.

Under either the additive or categorical model, the statistical method for a genetic association analysis is simply either a linear model with a quantitative phenotype as the dependent variable, or a logistic regression model for case/control phenotypes, with genotype as the tested independent variable. Obviously, risk factors for the phenotype should be adjusted for covariates where they are felt to be independent of the anticipated therapeutic strategy or genetic influence. Such simple analyses are entirely valid, but under certain conditions:

1. The subjects must be unrelated.
2. The subjects should be from a single specific ethnic heritage. Otherwise Simpson's paradox (Simpson 1951) can cause incorrect results due to allele frequencies differing by ethnic groups.
3. For similar reasons as the previous condition, each analyzed SNP should be in Hardy Weinberg equilibrium, meaning the allelic probabilities in each chromosomal copy are independent and a consequence of random mating with respect to the alleles.
4. When the endpoint is quantitative, it should be normally distributed (after transformation where necessary).

These basic analyses are supported in the GWAS setting by the ubiquitously used PLINK software[1] (Purcell et al. 2007). This software provides valuable QC and data management tools in addition to executing basic analyses. Other packages exist to handle more complex situations, such as Merlin (Abecasis et al. 2001) or EMMAX (Kang et al. 2010) to handle GWAS in cohorts of related individuals.

Two important concepts apply to the statistical interpretation of GWAS results. The first is obvious—there is a huge multiple testing burden to overcome, given that some types of chips contain one million SNPs. This is exacerbated by the increasing number of SNPs which can be imputed by algorithms such as IMPUTE (Howie et al. 2009) and MACH (Li et al. 2010) from reference datasets. The 1000 genomes sequencing project (McVean et al. 2012) has recently revealed many more rare variants, increasing the number of imputable SNPs from the 2.5 million traditionally imputed well over nine million. Fortunately, the method of Nyholt (2004), later modified for GWAS by Gao (2011) along with a second method by Li and Ji (2005), allow estimation of the effective number of independent signals in these GWAS. Since many GWAS SNPs have correlated partners due to LD, and the more SNPs that are imputed, the more LD is present, it turns out that eve from nine million SNPs, the estimated number of independent signals does not exceed four million, and so a significance threshold of $p < 1.0 \times 10^{-08}$ can safely be considered to correspond to a Bonferroni-adjusted cutoff of $p < 0.05$, and this is likely a bit conservative. Importantly, where there is a pre-specified list of likely candidate genes, this strict threshold can be avoided under a hierarchical multiple testing paradigm, where the top tier candidates only need be adjusted for their own multiplicity.

The second important concept that applies to statistically interpreting GWAS results is less intuitive but true nonetheless: for discovering novel targets, effect size does not matter as long as there is a reproducibly significant association between a SNP and disease. The reason for this is that any SNP, particularly the tagging SNPs typed for a GWAS, likely have at best a modest effect on the gene itself. Even if the SNP changes an amino acid, this is only one change in a very large chain of molecules and cannot in general be expected to alter the function of the gene by

[1] `http://www.pngu.mgh.harvard.edu/purcell/plink/`.

much. Indeed, the strongest GWAS associations with disease or macrophenotypes still typically explain less than 5 % of the variation in the endpoint. However, as long as there is no doubt that a small change in the gene corresponds to a change in disease, then one might expect that a complete chemically-induced knock-out or amplification of the gene might still have a clinically relevant effect.

4.3.2 Challenges with Biological Interpretation

Despite seemingly needing only a good p-value, there are still a number of challenges to identifying a novel target from even the most significant GWAS association. Chief among these is the assignment of the SNP to a gene. Although only SNPs in the exonic region of a gene can affect the coding sequence of the final protein, RNA expression is activated or amplified by SNPs outside these exonic regions. Further, even a SNP in the exonic region of a gene may actually be promoting expression of another nearby gene, possibly on the opposite strand (each chromosomal copy consists of paired strands of complementary sequence in opposing directions, with each gene expressing on only one of these strands), although the role of exonic SNPs on the gene in which they are located can be looked up bioinformatically and usually indicates whether it causes a functional or consequential change for the encoded protein. Nevertheless, in both gene-dense regions and for SNPs far from any gene, untangling which gene a SNP might be affecting can be particularly challenging. While an obvious remedy to this problem is to leverage knowledge of the genes in the region, this is often limited or unavailable. Another approach is to fine-map the gene, though recently this has been replaced by simply using more advanced technologies to sequence the region around the gene. The hope is that more functional, but likely rare (due to negative evolutionary selection) variants will be discovered in the gene. However, these techniques are also quite expensive, and obtaining sufficient subjects to verify association between a rare functional variant and disease is challenging.

At this point, RNA gene expression and proteomics become valuable and necessary. Note that the same hybridization technology is used for both GWAS chips and RNA gene expression arrays, but whereas the former simply detects and quantifies proportions of sequence mismatches and so reads out as a fairly unambiguous genotype call, the latter is based on quantifying a signal, and so yields a continuous value proportional to the amount of gene expressed. Since expression can vary by tissue, this provides an important tool for prosecuting GWAS variants (or putative functional variants in those genes) and determining the relevant associated gene: genes whose expression levels most associate with the SNP in the tissues most relevant for the disease are the most likely candidates among those in the region of a SNP. When a SNP is associated with gene expression levels in one or more tissues, it is called an expression quantitative trait loci, or eQTL (Rockman and Kruglyak 2006). A number of public databases, including Genevar (Grundberg et al. 2012) and more recently GTex from the Broad Institute

(Lonsdale et al. 2013), are now available for looking up eQTLs by tissue. Also recently, the ENCODE project (Dunham et al. 2012) has attempted to catalog the functional elements of all DNA including regions between genes, identifying, for example, chromatin regions thought to be more likely to regulate or turn on gene expression. Ideally, the most compelling evidence relating a SNP to a gene would be associations between the SNP and the direct protein product of the gene. However, because of the complex folding that proteins undergo, they cannot be detected en masse by sequence hybridization. Rather, mass spectrometry is often used, and many proteins are not directly detectable. Likewise, the assays for detecting proteins can be considerably more expensive and less reliable than genotyping or even gene expression assays.

A second complication with turning GWAS hits into a therapeutic strategy is that even when the SNP is well-mapped to a gene, determining the direction of effect on the protein is not always clear. Association of a SNP with a decrease in expression (which comes from detected RNA) may actually mean more protein is being synthesized from the RNA, and some SNPs may affect the rate of translation rather than expression. Also, a SNP may affect protein activity in opposition to its effect on protein or RNA abundance.

A commonly used tool for verifying directionality (and the applicability) of a gene is transgenic mice. Although not available for all genes, studying the changes or conditions of mouse knock-outs (i.e. mice with a specific gene suppressed) under induced stress or of various phenotypic statuses can provide strong evidence both as to the directionality a treatment should take, as well as simply verifying the relevance of the suspected gene to the disease.

Transgenic mice are also useful for verifying the potential causality of the gene, which addresses a third complication with turning a GWAS result into therapeutic strategy: despite genetics being only a possible cause and not an effect, it is not guaranteed that a strong association of a SNP with disease is not caused by a latent environmental or other unmeasured risk factor. Moreover, indirect targeting of biomarkers of poor health (usually through direct targeting of a gene) will not be effective unless the biomarker is causal, which may not be the case even where the biomarker is prospectively associated with incident disease. However, the question of causality can be addressed by the so-called "Mendelian Randomization" approach (Smith and Shah 2003), which tests the causality of a measurable quantitative biomarker thought to be on the causal path between a gene and disease. The concept is fairly straightforward: since genotype can only be a cause and not an effect, then if the SNP is significantly (and specifically) associated with an intermediate biomarker, and that biomarker is associated with the disease, then under the presumption of causality, one can predict what effect the SNP should have on disease.[2] A SNP-to-disease association that is both significant and consistent with the prediction from the constituent arms (SNP-to-biomarker, and

[2]The term "Mendelian Randomization" refers to the notion that we are randomized at birth to the genetic "treatment" of the SNP.

biomarker-to-disease) would indicate causality of both the SNP and the biomarker as an intermediate. Historically, such studies have not usually been possible because large sample sizes are needed to be powered to detect what are often subtle genetic effects on disease, and also because of a lack of intermediate biomarkers concurrent with genetic data. However, recently increased efforts have gone into profiling intermediate molecular biomarkers and microphenotypes in subjects with GWAS. A number of high-throughput metabolomics panels and lipidomics panels have been developed by companies such as Metabolon and Liposcience, and in fact recently the largest GWAS of a full metabolomics panel was published, meta-analyzing the UKTwins and KORA cohorts (Shin et al. 2014). Likewise, many GWAS of protein panels have also been recently published (Kim et al. 2013).

Where a SNP is already of interest due to a disease association, Mendelian Randomization mainly serves to identify or verify the intermediate molecule leading to the disease. For verifying a gene, this could simply be levels or activity of the direct protein product. However, many therapeutic strategies are designed to indirectly target (via some direct genetic target) previously observed biomarkers of prospective risk. Perhaps the most successful such therapy is statins, designed to reduce LDL cholesterol, which have also been subsequently shown to reduce coronary heart disease (CHD). Indeed, Mendelian Randomization can easily demonstrate that genetic associations in the HMGCR gene targeted by statins are commensurately associated with both LDL and CHD in a way consistent with causality, given the known LDL to CHD associations.

The Mendelian Randomization technique is particularly well-suited, however, for demonstrating the *non-causality* of an intermediate trait. Recent publications have shown that Mendelian Randomization can be used to show that prospective epidemiological associations of HDL (Voight et al. 2012), CRP (The C Reactive Protein Coronary Heart Disease Genetics Collaboration 2011), and sPLA2-IIA (Holmes et al. 2013) with incident coronary heart disease are all not likely to be causal.

4.3.3 A Summary of GWAS Limitations

Despite these tangible results and the many published GWAS findings, there are inherent limitations to GWAS beyond simply the limitations in interpretation and conversion to a therapeutic strategy that we have already discussed. First, the likelihood of detecting an association via GWAS depends on the heritability of the disease. Likewise, while Mendelian diseases (with a single genetic cause) are easy to detect and analyze via genetics, most common heritable diseases are complex in that they are heterogeneous in nature, and so likely have a multitude of genetic contributors, each corresponding to a different "strata" of the disease. Unfortunately, the nature of the heterogeneity is typically hidden without already knowing the genetics, so complex diseases continue to be a challenge, overcome only by increasingly large sample sizes.[3] In addition to heterogeneity, some diseases have a high degree of unmeasured environmental influence, and this can also wash out a GWAS signal. Finally, even if a genetic target is unambiguously identified as causal toward disease, there are still at least two things that can go wrong.

- A gene discovered by GWAS may be causal of the disease, but that does not necessarily mean that targeting it can reverse the disease once it has occurred.
- The gene may not be "druggable". This is a notion referring both to its likely safety (genes involved in too many pathways are likely dangerous to knock down), and the ability of the target to bind with high affinity to a small molecule/drug. Fortunately, "druggability" has been cataloged or predicted for many of the 20,000 or so known genes (Griffith et al. 2013).

Despite these limitations, emphasis in the pharmaceutical industry is increasingly leaning toward only pursuing targets with human genetic evidence, particularly targets initially discovered in humans. This is partly a reaction to the high historical attrition rate of compounds based solely on discovery in animal models but which ultimately did not translate back into humans. Note that animal models and *in vitro* cell line assays are still a highly used and important tool for validating putative genetic targets. However, the additional assurance from discovery in humans comes from the idea that it would take almost a conspiratorial level of bad luck for a discovery to be made solely in humans, generate a hypothesis that is verified in animal models and cell line assays, but then have that hypothesis fail to translate back into humans. Hence, requiring human evidence at the forefront of target selection is a sound strategy, but traditional animal models and *in vitro* methods are still necessary and valuable tools for following up these discoveries. In fact, the ability to directly follow up human genetic evidence in these settings has recently increased dramatically with the development of the CRISPR gene-editing system (Hwang et al. 2013). New technologies such as this, coupled with the increasing emphasis and availability of human genetic data measured against increasingly deep levels of molecular measurements, should facilitate much advancement in the understanding of causal pathways leading to disease and thereby provide many new sound therapeutic strategies far into the future.

4.4 An Overview of the Chemistry Optimization Process

Assuming that an appropriate target or phenotype has been identified, the basic drug screening process is similar for either type of project. Hughes et al. (2011) provide an excellent overview of the stages of the early discovery process. Swinney and Anthony (2011) and Swinney (2013) contrast target- and phenotypic-based screens.

[3]Thankfully, the academic community has been highly co-operative with one another in creating large consortia to produce meta-analyses from many smaller GWAS studies that total to hundreds of thousands of subjects.

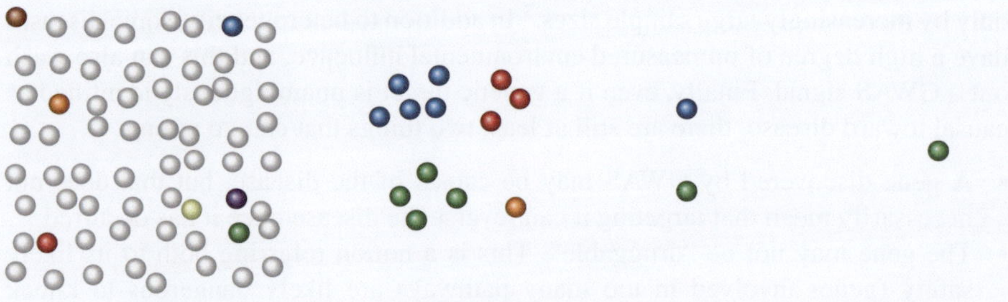

Fig. 4.1 A graphical representation of the compound optimization process

Figure 4.1 shows a diagram that illustrates the winnowing process of compounds over the course of a discovery project.

- A large number of diverse compounds are screened for activity using *in vitro* assays. Depending on the circumstances, there may only be a small percentage of biologically active "hits." In Fig. 4.1, these are represented as colored balls.
- Hits in primary screening assay(s) are followed up to find molecules that have the potential for further optimization (i.e., 'leads'). Hopefully, one or more chemical *series* of compounds with common sub-structures are generated by the initial screen. These are shown in Fig. 4.1 as clusters of balls with the same color. In other cases, the screening hits may result in "singleton" molecules. The initial set of hits are re-screened to exclude promiscuously active compounds and logistical artifacts.
- The *hit-to-lead* or *lead discovery* process refines the molecules and considers characteristics beyond potency, such as pharmacokinetic quantities and basic physicochemical properties.
- *Lead optimization* focuses on a few compounds and/or chemical series to optimize and evaluate.
- After lead optimization, there may be a few promising compounds (or members of a series) and one must be nominated to be the drug *candidate* for the project. This choice may be driven be the initial *in vivo* assessments of safety, intellectual property issues, and so on.

The next chapter is a thorough summary of practical and statistical aspects of analyzing data from large-scale screening campaigns.

Phenotypic screens may lend themselves to a more streamlined process. The measurement of the phenotype is likely made using cell-based assays or possibly an *in vivo* assay. In these cases, higher quality hits may be found. For example, if a compound has poor properties (e.g., permeability) it will not be able to enter the cell and therefore show little efficacy. Similarly, potential toxicities are more likely to be demonstrated in these screens than in simple ligand-binding biochemical assays. For example, cell-based assays can usually be supplemented with cell viability assays to obtain rudimentary estimates of toxicity.

4.5 Assay Data in Drug Discovery

As one might expect, there are usually a large number of assays used to optimize molecules and these vary in complexity and type. The *primary pharmacology* assays are usually related to potency against the target or phenotype. In traditional target-based projects, initial potency measurements are from simple biochemical ligand-binding assays. Compounds that show sufficient activity against the target are then run against some type of functional assay to verify the relevance of the hit.

The initial screen for potency is often conducted at a single concentration; molecules that prove interesting might then be followed by a dilution series of the concentration to verify that a dose-response relationship exists. To do this, the assay value is determined over a range of concentrations and a statistical dose-response model is fit to the data. Figure 4.2 shows an example of such a dataset from a phenotypic screen where the assay outcome is larger for more potent compounds. One common method for quantifying the potency of a compound is to calculate the *effective concentration* or *EC*. For these data, we fit a four parameter logistic model to the assay data using standard nonlinear regression methods that assume normally distributed errors:

$$Y_i = \beta_1 + \frac{\beta_2 - \beta_1}{1 + \exp\{\beta_4(\log x_i - \beta_3)\}} + \epsilon_i$$

where β_1 is the minimum value of the outcome, β_2 is the maximum, β_3 is the concentration that corresponds to a 50 % increase in the outcome, β_4 is a parameter for the slope of the curve and $\epsilon_i \stackrel{iid}{\sim} N\left(0, \sigma_\epsilon^2\right)$, $i = 1, \ldots, n$. It is common to estimate the concentration that delivers 50 % effectiveness, known as the EC_{50}. For these data, there is a good dose-response and an acceptable model fit that yields $EC_{50} = \hat{\beta}_3 = 1.037$ with a 95 % confidence interval of (0.898, 1.198). Had the data for this compound not shown a systematic increase in the assay result, we would be more likely to consider the hit an aberration. Single concentration assays are easily amenable to high throughput screens where thousands or a million-plus compounds may be tested. Failure to show a dose-response for a suitable concentration range can be a cause for a compound's attrition despite initial results.

Secondary pharmacology assays are usually associated with toxicity endpoints and perhaps ADME properties. The phrase "secondary pharmacology" is most associated with the former but can also include assays to determine the drug-likeness of the initial hits. For example, lipophilicity, permeability and other factors are important considerations when developing drugs. Other factors are also evaluated such as the ease of synthesis, if the molecule is amenable to parallel/combinatorial chemistry, etc. Additionally, when the target is part of a gene family selectivity assays can be instrumental in ensuring that an appropriate and specific potency is achieved. In the previously mentioned JAK-STAT pathway, four important genes are JAK1, JAK2, JAK3 and Tyk2. When optimizing the chemistry, there may be interest in determining the potency of one target to another and this may be measured using the ratios of EC_{50} values for a pair of these genes.

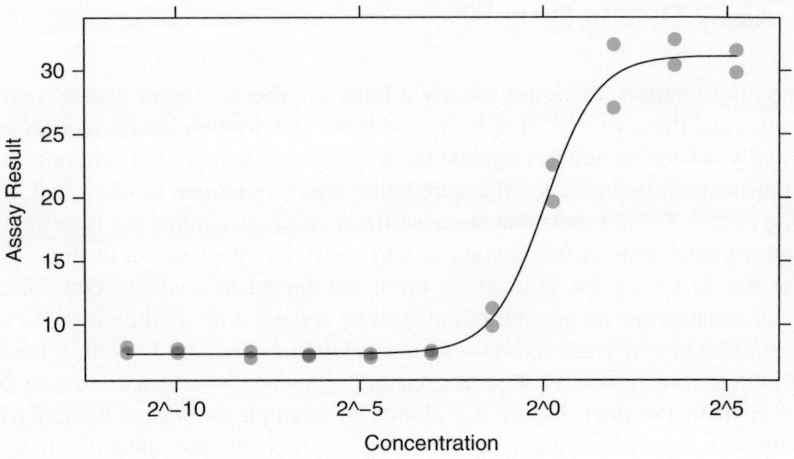

Fig. 4.2 A typical dose-response experiment with a nonlinear regression line to estimate the effective concentration

The following sub-sections in this chapter discuss aspects of assay data and how statistics can be used to increase assay quality. Chapter 8 describes the design and analysis of an *in vivo* assay to characterize cardiac safety liabilities.

When considering the particulars of assay data, there is more than one perspective to contemplate. One might consider the delineation between the *producers* (i.e., the screening groups) and the *consumers* (including medicinal chemists, biologists and project teams) of the assay. Each group has their own set of conflicting objectives for assay data and it is often the case that a successful drug discovery project is one that considers both viewpoints. Statisticians may be in a position to mediate between these groups and the last subsection below discusses a method for facilitating such a discussion.

4.5.1 Important Aspects of Assay Data

During the assay development phase it is advantageous to characterize the nature of the assay outcome so that this information can be used to properly analyze the data. If the assay result is appreciably skewed, transformations of the data are likely to be needed. It is important to discuss these findings with the scientists; often naturally skewed data are mistaken for symmetric data with outliers. Outlier identification techniques tend to have poor properties when dealing with very small samples sizes and can be counter-productive in this context. Also, when working with EC_{50} data, there is the potential for *off-scale* results when 50 % inhibition was not achieved at the largest dose. In this case, censored data arises (Kalbfleisch and Prentice 1980) and the data are usually reported as "$> C_{max}$" where C_{max} is the largest dose used in the experiment (left-censoring can also occur but is less common). Many scientists are unaware of the effects of censoring and the available data analysis methods that can be used. As such, it is not uncommon for a censored data point to be replaced with C_{max}. This can lead to serious bias in mean estimates and severely underestimate the variance of the mean since multiple censored values would result in the same value repeated under the guise of being known.

Scientists often work with parameters that have biological significance but whose sampling distribution may be unknown. Permeability is commonly measured utilizing Caco-2 or MDCK cells and is typically defined via a first order differential equation. The efflux ratio (Wang et al. 2005), defined here as the apparent permeability out of the cell divided by the apparent permeability into the cell,

$$ER = \frac{P_{app}(B \to A)}{P_{app}(A \to B)},$$

is an example of an interpretable parameter comprised of empirical estimates whose sampling distribution may not be readily apparent. Ranking compounds in terms of their efflux potential is an area where a statistician can contribute expertise. Measuring transporter activity via various assays invites parallels to inter-observer agreement problems in statistics. For example, transporter activity can be measured using various (transfected) cell lines, sandwich culture human hepatocytes (a cell-based system that can better mimic the dynamics of intact liver tissue via the formation of bile canaliculi), primary cells (intact tissue, e.g., liver slices), or via mutant *in vivo* animal models. Comparing in-house assay performance to that obtained using one or more contract research organizations is also possible. Apart from comparing the accuracy and precision/reproducibility of various assay estimates, cost, donor availability, amenability to automated screening systems, etc., can dictate the acceptable parameters for an assay.

Another challenge in discovery is the general lack of absolute standards. Not surprisingly, a shortage of absolute standards facilitates the widespread use of relative measures of comparison. For example, whereas a statistician considers a standard deviation (or variance) as a meaningful summary measure for dispersion many scientists prefer to use the coefficient of variation to measure uncertainty. Relative comparisons, e.g., an assay's relative error to a known target or another estimated quantity, are commonly used to interpret data. Fold changes are used extensively, e.g., comparing the potency of two or more compounds. Unfortunately, our experience is that the default level of evidence for equality required by many scientists is that two data points be "within twofold" of each other. This comparison may not include a careful discussion of what sources of variability were involved in forming the comparison. Also, since parameters such as bioavailability and plasma protein binding are defined as fractions, relative comparisons involving fractions occur. Propagation of error techniques such as Fieller's theorem (Fieller 1954) and other basic results from mathematical statistics are useful. While simulation or Monte Carlo techniques can prove beneficial, the ability to determine or approximate closed form solutions should not be overlooked. Closed form solutions prove useful in approximating confidence intervals, can often be included in basic computational settings or tools, or assist with sample size calculations. Unlike later stage confirmatory clinical studies subject to regulator scrutiny, a first order approximation to an early discovery problem may provide a suitable level of rigor. Unlike some large clinical studies that recruit tens of thousands of patients, discovery efforts often involve making decisions with imperfect or limited amounts of data.

4.5.2 *Improving and Characterizing Assay Quality*

There are two basic phases of assay development where statistics can play an important role: the characterization and optimization of an assay. Characterization studies are important as they are an aid to understanding the assay's operating characteristics as well as help identify areas for improvement. Improving the assay via statistical methods, such as sequential experimental design, can have a profound positive impact. The statistician would work with the scientist to understand the important experimental factors that can be explored and efficient experimental designs can be used to optimize these factors. Haaland (1989) and Hendriks et al. (1996) provide examples of this type of data analysis.

Many of the key questions for an assay are related to variation and reproducibility. For example, a characterization of the sources and magnitudes of various noise effects is very important. If some significant sources of noise cannot be reduced, how can they be managed? For plate-based assays, there may be substantial between-well variation (i.e., "plate effects") and so on. A general term for experiments used for the purpose of understanding and quantifying assay noise is *measurement system analysis*. These methods are often the same as those used in industrial statistics to measure repeatability and reproducibility (Burdick et al. 2003, 2005). Two examples of such experiments are discussed in Chap. 6. Initial conversations with the assay scientists can help understand which aspects of the assay have the highest risk of contributing unwanted systematic variation and these discussions can inform subsequent experiments.

Investigations into possible sources of unwanted bias and variation can help inform both the replication strategy as well as the design of future experiments. If the sources of variation are related to within-plate effects, it may be advisable to replicate the samples in different areas of the plate and average over these data points. Also, it is common for the final assay result to be an average of several measurements (where the replicates are likely technical/subsamples and not experimental replicates). Robust summary statistics, such as the median, can mitigate the effect of outliers with these replicates. In cases where systematic problems with the assay cannot be eliminated, "Block what you can; randomize what you cannot" (Box et al. 2005). Blocking and randomization will help minimize the effect of these issues on experiments comparing different compounds. For example, some assays require donors for biological materials, such as liver cells, and the donor-to-donor variation can be significant. Unless there are enough biological materials to last through the entirety of the project, blocking on the donor can help isolate this unwanted effect. Other statistical methods, such as repeated measures analysis, can be used to appropriately handle the donor-to-donor variation but are useless if the experimental layout is not conducive to the analysis. As another example, light/dark cycles or cage layout can have significant effects on *in vivo* experiments and can be dealt with using blocking or randomization. These unwanted sources of noise and/or bias in experiments can be severe enough to substantially increase the likelihood that a project will fail.

Statistical process control (Montgomery 2012) can be useful for monitoring assay performance. While most compounds screened for activity are measured a limited number of times, control compounds are typically used to track assay performance across time. Multiple control compounds may be used to gauge a particular assay's behavior over a diverse compound space. For an assay consumer, how their compound fares relative to the known behavior of one or more control compounds invites statistical comparisons. In addition to using control compound data to pass/fail individual assay results, these data are also used on occasion to adjust or scale the assay value for a compound under consideration. This can again suggest the need for applying basic concepts such as Fieller's theorem. Comparable to the development of normalization methods used in the analysis of microarray data, methods that attempt to mitigate for extraneous sources of variability may be applied. For example, standard curves are often used where tested compounds are reported in comparison to an estimated standard profile. In part, these normalization or adjustment schemes are often used because the biologist or chemist lack engineering-like specifications for interpreting data. Use of external vendors in the early discovery process can necessitate the need for monitoring assay quality or performing inter-laboratory comparisons.

The elimination of subjectivity is also important. In *in vitro* experiments subjectivity can surface when assembling the data. For example, a policy or standard for handling potential outliers can be critical as well as a simple definition of an outlier. In *in vivo* experiments, there is a higher potential for subjectivity. Some measurements may be difficult to observe reliably, such as the number of eye-blinks. In these cases, a suitable strategy to mitigate these factors should be part of the assay protocol.

4.5.3 Conceptualizing Experimental Robustness

As previously mentioned, both the assay producer and the assay consumer can have different, perhaps opposing, views on the definition of a "fit for purpose" assay. In the end, both parties desire to have a successful experiment where a definitive resolution of an experimental question is achieved, which we term a *robust experiment*. Neither of these groups may have a high degree of statistical literacy and may not realize what factors (besides more common project-management related timelines and costs) influence experimental robustness. First and foremost, the consumers of the assay should have some *a priori* sense of what is expected from an assay. Comparable to sample size estimation procedures, some notion of the level of signal that the assay should be able to detect is critical. Two examples that we routinely encounter include:

- For a single dose potency assay, the requirement might be initially stated as "reliably detect a percent inhibition greater than 50 %."
- For selectivity assays, the target might be phrased as the "ability to differentiate between a selectivity ratio of 10-fold and 50-fold."

In the first case, more work may be needed to express the required signal in terms of a comparison, such as "be able to differentiate a 10 % difference in inhibition." Without this information, the probability that the assay will meet the expectations of the consumers decreases. Our experience is that this requirement is rarely discussed prior to the development of the assay and is a common source of conflict that surfaces after the assay protocol has been finalized and the assay is in regular use.

The replication strategy of the assay should be informed by the required signal, the levels of noise, cost and other factors. It is important that this strategy be the result of an informed discussion between the assay producers and consumers. To this end, we frame the discussion in terms of increasing experimental robustness using a modification of a formula from Sackett (2001),

$$\text{Experimental Robustness} = \frac{\text{required signal}}{\text{assay noise}} \times \sqrt{\text{true replicates}}.$$

Here, the signal is related to the expectations of the experimenter. To resolve small, subtle differences between conditions would correspond to a small signal and, all other things being equal, reduce the robustness. Similarly, the magnitude of the assay noise will have a profound effect on the robustness. The third item, the number of replicates, is a proxy for the overall experimental design but this level of generality effectively focuses the discussion. This equation can help the teams understand that, in order to achieve their goal, trade-offs between these quantities may be needed.

4.6 Compound Screening

During initial screening, a large number of compounds are evaluated, typically using high-throughput assays. While compound screening may be synonymous with "high throughput screening" there are several variations on how the corpus of compounds is chosen and/or assayed.

- One simple approach is to screen the entire "file" of compounds (those that have been previously synthesized and have sufficient chemical matter for testing).

Given the cost and logistical issues of screening such a large number of compounds, many of which may not be viable drugs, this is viewed as a sub-optimal approach.
- Focused screens evaluate a subset of molecules based on some *a priori* criteria. For example, a project targeting the central nervous system might constrain their screen to compounds that are known or predicted to cross the blood-brain barrier. As another example, a focused screen may only include compounds that are believed to have activity against a certain target class (e.g., kinases, G protein-coupled receptors, etc.).
- Fragment-based screening (Murray and Rees 2009) takes small chemical structures and screens them for activity. Here, hits are likely to have low biological activity and optimization techniques may combine two or more active fragments into a more potent molecule. Alternatively, the fragment may be used as the starting point to design new molecules.
- Yet another method for high throughput screening pools multiple compounds into a single well for screening (Kainkaryam and Woolf 2009; Remlinger et al. 2006). Bona fide hits are determined by deconvoluting the results across multiple wells containing the active compound.

Often, especially with target-based projects, the initial screen may be conducted with each compound tested at a single concentration. In basic biochemical screens the assay results may be the percent inhibition of the target. For phenotypic screens the experimental design and the nature of the assay endpoint can vary.

For simple initial screening designs for binding assays, there is usually an expectation that most molecules have negligible biological activity and basic statistical analysis of such data focus on finding the "outlier" signal in a sea of inactivity. This analysis has largely been commoditized in scientific software. However, there are a number of aspects of the analysis that should be reviewed based on what is known about the assay. For example, the activity values can be normalized based on the positive and negative controls but this depends on the controls being appropriate and stable over time. Also, compensating for row, column or edge effects may be required for a plate-based assay. Malo et al. (2006) is an excellent overview of statistical techniques for primary screening data.

In target-based screens, once a molecule is estimated to have sufficient potency a secondary battery of assays are used to further characterize the molecule. First, steps are taken to ensure that the hit was not an aberration by re-synthesizing the compound for repeated testing. The compounds are also evaluated for impurities (Hermann et al. 2013). A dose-response experiment is usually performed to get a more refined estimate of potency via an EC_{50}. If the primary pharmacological assay is biochemical, such as a ligand binding assay, a functional assay may also be run to confirm that the hit is relevant. As mentioned previously, other considerations are made regarding the molecule: the ease of synthesis, if the molecule is amenable to parallel/combinatorial chemistry, solubility, permeability, etc.

If the potency and ancillary characteristics of the compound are acceptable, the compound is usually termed a *lead compound* and further optimized. Depending on the project and the success rate, screening may continue to find more hits in case there are complications or unexpected issues with the current set of leads. It is common for project teams to develop one or more backup compounds in parallel with the lead compound in case an issue is found with the lead molecule.

4.7 Compound Optimization

Lead optimization is usually accomplished by taking one or more initial lead compounds and sequentially making changes to the molecules to improve its characteristics. Often, this process assumes a good structure activity relationship (SAR), meaning that changes to the molecule can yield reliable results (apart from measurement error). For example, Ganesh et al. (2014) describe the types of structural substitutions and changes that can be made to increase selectivity and other characteristics from a starting set of molecules. In our experience, many of these types of changes are based on the intuition and experience of the medicinal chemists. However, the source of structural modifications can also be suggested by quantitative, model-based methods. Chapter 6 describes techniques such as activity cliffs, Free-Wilson models and other data-driven approaches that can help predict the change in the characteristics of a molecule for a specific structural modification. Virtual modifications can be attempted by using QSAR (Quantitative SAR) models to predict specific characteristics prior to synthesizing the compound. In any case, lead optimization can be difficult due to the large number of characteristics to optimize.

A typical lead optimization experiment would consist of several similar molecules and a panel of outcomes. The medicinal chemist is usually interested in determining whether or not any differences in the assay results are "real" (i.e., above the experimental noise in the system) or not. In some cases, if there is no statistical difference between molecules for one characteristic, a structural modification might be accepted if it produces a significant improvement in some other quality. For this reason, there is an advantage to using confidence intervals instead of formal statistical hypothesis tests. Chemists are usually interested in knowing *how much of an effect* is produced by a structural change. Confidence intervals on differences in means are effective at answering this type of question since they are in the original units and directly show the uncertainty in the parameter of interest. Hypothesis tests, on the other hand, answer the question *is there an effect*. The corresponding p-value obfuscates both the direction and magnitude of the change. See Evans and Dawson (1988), Matthews and Altman (1996) and Sterne (2001) for a broader discussion of the issue.

4.8 Characterization of Important Properties

Despite the investigative nature of screening and optimization activities, a rich set of concerns are present in early discovery that may reach well into the late stages of development. While dated, Kola and Landis (2004) document that real strides have been made in reducing compound attrition due to ADME issues. Still, a compound that appears effective and safe in initial screens can fail for ADME-related concerns. If the compound presents a potential risk for adverse interactions with other concomitant medications, i.e., a drug-drug interaction, that poses a serious obstacle to further development and may invite subsequent discussions with regulators. Statisticians in early discovery can benefit, both directly and indirectly, from understanding aspects of various regulatory guidances and how they may apply to characterizing a compound's properties. For example, guidances pertaining to drug interaction studies, bioequivalence, and stability studies have proven useful in our early discovery experience. Dose, an important human parameter related to a compound's potency that is subject to intense scrutiny during clinical testing, will be considered in early discovery despite the associated difficulties of accurately predicting human dose. If the half-life of a compound is either too short or too long that can impact a desired dosing regimen or suggest toxicity concerns. Regardless of the amount or extent of the characterization work performed in pre-clinical discovery, the ability to extrapolate from *in vitro* or *in vivo* assay data to predict human pharmacokinetics (PK), human dose, or pharmacodynamics (PD) is the ultimate goal. While published nearly two decades ago, Lin and Lu (1997) discuss aspects of PK and metabolism that are relevant to drug discovery today and link a compound's properties to late stage development concerns.

ADME experts study the interplay between a chemical moiety and a biological system. These experts work alongside biologists, pharmacologists, medicinal chemists, and other project members to advance an asset towards a regulatory filing. How the organism affects the compound and how the compound impacts the organism is their domain of expertise. How and where in the body is the compound metabolized? Prior to elimination, will any of the intermediate metabolites formed *in vivo* affect the safety profile for the compound? Can *in vitro* data be used to predict *in vivo* pharmacology? Yang et al. (2010) lists several assays used at various stages in development. Some examples include: hepatic clearance, permeability, transporter activity, plasma protein binding, and both CYP inhibition and induction assays. Each assay, either individually or collectively, presents opportunities to apply the concepts of statistical thinking.

As suggested earlier, predicting human PK and dose is critical. The methods associated with *in vitro-in vivo* extrapolation (IVIVE) or *in vitro-in vivo* correlations (IVIVC) could comprise a separate chapter devoted to PK modeling. Curry et al. (2002) provide an overview here and relate these efforts to Phase I studies. Apart from statistical considerations, a wide range of scientific assumptions may apply to make effective use of these models. Allometry, a method that uses scaling equations, is still used to predict human PK based on non-human *in vitro* and *in vivo* data. Apart from the obvious use of allometry to predict human PK from rat PK, it may be needed to predict human PK for limited subpopulations such as those involving pediatric subjects (Anderson and Holford 2008). Since these models use, for example, weight or body surface area and clearance values obtained via different assays (e.g., microsomal or hepatocyte) propagation of error techniques are useful here. Additional empirical scaling factors may be introduced to the model to improve prediction quality. The number of parameters in a given PK or PD model can grow quickly. For example, clearance may consist of separate renal, biliary, and hepatic components and hepatic clearance may be distributed across multiple liver enzymes. For some assays, an *in silico* tool may provide a prediction that can be contrasted with a gene reporter assay, a biochemical assay, a stem cell assay or a cell line assay. Estimating the therapeutic index for a compound, a measure typically defined as the ratio of the highest non-toxic drug exposure that achieves the desired efficacy, is a critical attribute of a compound. Muller and Milton (2012) illustrate a translational therapeutic index grid that contains data obtained from several forms of cell- or tissue-based assays, three different animal species, and multiple doses for a COX2 inhibitor and document challenges that are inherently statistical in nature.

Increasingly, as ADME-related sciences advance the promise and complexity of 'personalized medicine' can appear in discovery. Various ADME characteristics may differ between genders or across preclinical animal species. Genotyping platforms may be used to explore subject-level ADME differences. Since genetic traits can create subgroups of poor or extensive metabolizers or affect transporter activity, these data may be necessary for developing a suitable pharmacokinetic model. Rendic and Di Carlo (1997) list several environmental factors, e.g., smoking, nutrition, alcohol or drug usage, thought to influence the metabolism of xenobiotics by liver enzymes. Apart from the potential these characteristics may have in the later stages of clinical development for a compound, attempts to understand metabolic activity in terms of physicochemical properties, e.g., molecular weight, lipophilicity, or solubility, can be performed.

Statisticians in drug discovery will also be exposed to statistical tools commonly used in later stages of clinical development. Population PK-PD models, or PopPK-PD, are used to model PK or PD data using software platforms such as NONMEM[4] or Phoenix winNonlin.[5] These tools might use mixed effect models to relate PK or PD data to a variety of covariates, e.g., age, weight, dose, metabolic or genotype parameters. One can encounter these models in translational animal studies and they are often touted for their ability to model sparse or incompletely sampled data. Refer to Bonate (2011) for a discussion of nonlinear mixed effect models and their application in modeling PopPK-PD data. Similar to PopPK-PD models, physiologically-based pharmacokinetic models (PBPK) models are used to model PK data. Here, one attempts to create a physiological model, via a set of interconnected tissue compartments, that can reflect key ADME processes

[4]`http://www.iconplc.com`.

[5]`http://www.certara.com`.

in an organism. Espie et al. (2009) review PBPK models and their use in drug development. In contrast to PopPK-PD models, which borrow heavily from mixed effect modeling techniques, PBPK models are defined via a set of differential equations. Some obvious examples of ADME compartments in a PBPK model could include the stomach, gut, liver, kidney, and adipose tissue that are connected by arterial and venous blood flow. These models, despite their mechanistic basis, can also contain empirical parameters that affect their use and interpretation.

As the use of quantitative methods and models continues to expand in the ADME sciences, the development and use of *in silico* tools has increased rapidly. These tools are used to assess the risk of drug-drug interactions and impact clinical study design, to explore hard-to-evaluate pediatric subpopulations, or inform the choice of which compounds to synthesize and investigate further. These tools, given their use in predicting human PK, are also used to study dosing regimens or to simulate food effect studies. Simcyp[6] and GastroPlus[7] are examples of two tools in use in industry. While these tools are largely used by ADME experts, they present novel opportunities for statisticians since they allow for computer experiments to be performed. While sophisticated design layouts and models are possible, our expertise in understanding estimate variability, how it propagates through an integrated model, concerns about model robustness or parameter sensitivity, etc., may be our most valuable contribution. These tools often employ algorithms or models from the published literature and may require expert judgment for effective use.

4.9 Important Skills for Nonclinical Statisticians

The preceding sections identified several technical areas of analysis that can be useful for statisticians working in drug discovery, including: regression analysis (e.g., linear and nonlinear regression models), survival analysis, multiple comparisons, Bayesian analysis, (non)linear mixed models, experimental design, censored data models and process monitoring. Outside of target discovery, the majority of the statistical knowledge required to make a substantive impact on the discovery process can be found in Box et al. (2005) and Cochran and Cox (1950). Basic experimental design and statistical analysis, while not tantalizing or groundbreaking, are the most appropriate tools for solving issues with assays and experiments used to contrast molecules. One cannot overemphasize the need for a solid understanding of basic experimental statistics which, sadly, is continually deprioritized in academic statistics programs.

However, drug discovery research is continually investigating new technologies and methodologies and these may present opportunities that require more

[6]`http://www.simcyp.com/`.

[7]`http://www.simulations-plus.com/`.

sophisticated technical skills. Assays can take many forms and their output can be complex. Our discussion here regarding characterization work has not been comprehensive in scope. For example, the use of imaging modalities in discovery can be diverse. Preclinical imaging studies may examine whole body tissue distribution of a radiolabeled ligand in rodents, MRI methods are used to profile drug candidates in suitable animal models, and flow cytometry is routinely used in cell-based assays. Just as the proliferation of omics platforms (e.g., genome, transcriptome, metabolome, etc.) has created a sea of quantitative data in the biological sciences, the need and use of statistical methods is diverse in drug discovery. Statistical thinking, a concept long touted in the quality engineering field that links processes and variation, is a unique perspective that can complement the skill set and training of the average discovery biologist or chemist.

Statistical support for high content screening (HCS) assays provides a good case study to illustrate the need for more sophisticated technical skills. HCS is a platform for generating multivariate data for individual cells using fluorescence microscopy and image analysis (Bickle 2010; Haney et al. 2006). Using different fluorescent wavelengths, HCS assays can simultaneously measure multiple cell characteristics in individual cells such as the cell membrane, nuclear membrane, cytoskeleton, etc. Cell images are generated and these can be analyzed to segment specific cells and quantify a variety of different characteristics for each cell. For example, Hill et al. (2007) describe an assay that quantitates a number of cell characteristics such as the total fluorescent intensity of the cell nucleus and aspects pertaining to cell geometry (via the ratio of the cell area to the area of a convex hull). These *cellular* metrics can be aggregated over the population within a cell into *wellular* statistics that can be used for analysis. Using these data, specific types of biological activity can be quantitatively measured. For example, the project team might be interested in changes in cell morphology, nuclear translocation of a specific protein, mitotic arrest or other phenotypes (Korn and Krausz 2007).

There are several important areas where statistics can contribute to an HCS assay. First and foremost, a high quality assay that is reproducible and stable is key. Again, basic experimental design or quality control methods can greatly contribute here. Once the images are obtained, cell segmentation and feature generation (Shariff et al. 2010; Soille 2003) are also critical. Also, when converting cellular data to wellular results, there are a number of opportunities to use first principles of statistics to minimize variance using within-cell normalization techniques. The content of the screen can then be used in multivariate analyses to better understand the nature of the changes elicited by compounds. This may include unsupervised methods, such cluster or principle component analysis (Johnson and Wichern 2007), to discover new phenotypes that were not initially identified or help identify previously unknown cell toxicity issues. Also, these data can be used as inputs into machine learning models (Kuhn and Johnson 2013) that could be used to predict the biological activity of new compounds. This type of methodology is also discussed in Chap. 6. In summary, HCS is an example of a discovery activity that requires simple statistical tools as well as more advanced quantitative methodologies. There are myriad other occasions that can be found across drug discovery.

In addition to analytical skills, there are certain computational skills and tools which will be helpful, if not necessary, for a statistician entering the field of drug discovery. Primarily, these include familiarity with both Linux and R (R Core Team 2014). Due to its open source status, many valuable packages for analyzing or pre-processing data from new technologies are available in R, and in particular the Bioconductor (Gentleman et al. 2004) suite of R packages contains many valuable packages for pre-processing and analyzing microarray data for both expression and genetic analysis. Also, since microarray analyses can be extremely high-throughput in nature, use of grid computing is highly valuable, and most such computing environments are native to Linux, while the license-free nature of R makes it ideal to run across a grid. In addition, many of the aforementioned staple tools in genetics—Plink, IMPUTE, MACH, and EMMAX—are primarily available as command-line executables for Linux (or the syntactically identical MacOS X), while use on a Windows machine, where available, is typically confined to the much clunkier MS-DOS environment. Obviously, in addition to R and Linux, familiarity with the specific software packages just mentioned here and in previous sections of this chapter will be valuable also.

For a statistician who is interested in working in drug discovery there are several ways to get started. First, try to develop a good understanding of the relevant sciences. For example, basic knowledge of cell biology is crucial. Texts, such as Alberts et al. (2007) or Alberts et al. (2013), can significantly increase the effectiveness of a statistically trained individual. From the preceding section, a good knowledge of genetics is crucial for target discovery activities. For compound optimization, a basic knowledge of pharmacology is important. For example, the first few chapters of Rang et al. (2007) are sufficient.

Also, there are several conferences that have "nonclinical" or "preclinical" content that is directly applicable, such as the Midwest Biopharmaceutical Statistics Workshop[8] or the recurring Nonclinical Biostatistics Conference.[9]

4.10 Conclusions

Early drug discovery is fertile with interesting and important opportunities for statistical contributions. For those who are interested in science this can be an extremely exciting area in which to work. The projects tend to be diverse and, in our experience, fun. Across the entire drug pipeline, this phase has the most potential to add value to a drug by improving the molecule before it is progressed to later phases of extensive (and expensive) development.

[8] `http://www.mbswonline.com`.

[9] `http://bit.ly/1qilzvh`.

Acknowledgements We would like to thank David Potter and Bill Pikounis for providing feedback on a draft of this chapter.

References

Abecasis G, Cherny S, Cookson W, Cardon L (2001) Merlin-rapid analysis of dense genetic maps using sparse gene flow trees. Nat Genet 30(1):97–101

Alberts B, Johnson A, Lewis J, Raff M, Roberts K, Walter P (2007) Molecular biology of the cell. Garland Publishing, New York

Alberts B, Bray D, Hopkin K, Johnson A, Lewis J, Raff M, Roberts K, Walter P (2013) Essential cell biology. Garland Publishing, New York

Anderson B, Holford N (2008) Mechanism-based concepts of size and maturity in pharmacokinetics. Ann Rev Pharmacol Toxicol 48(1):303–332

Arrowsmith J (2011a) Trial watch: phase III and submission failures: 2007–2010. Nat Rev Drug Discov 10(2):87–87

Arrowsmith J (2011b) Trial watch: phase II failures: 2008–2010. Nat Rev Drug Discov 10(5): 328–329

Bickle M (2010) The beautiful cell: high-content screening in drug discovery. Anal Bioanal Chem 398(1):219–226

Bonate P (2011) Pharmacokinetic-pharmacodynamic modeling and simulation. Springer, Berlin

Box GEP, Hunter S, Hunter W (2005) Statistics for experimenters: design, innovation, and discovery. Wiley, Hoboken

Burdick R, Borror C, Montgomery D (2003) A review of methods for measurement systems capability analysis. J Qual Technol 35(4):342–354

Burdick R, Borror C, Montgomery D (2005) Design and analysis of gauge R&R studies: making decisions with confidence intervals in random and mixed ANOVA models, vol 17. SIAM, Philadelphia

Burton P, Clayton D, Cardon L, Craddock N et al (2007) Genome-wide association study of 14,000 cases of seven common diseases and 3,000 shared controls. Nature 447(7145):661–678

Clark J, Flanagan M, Telliez J-B (2014) Discovery and development of janus kinase (JAK) inhibitors for inflammatory diseases. J Med Chem 57(12):5023–5038

Cochran W, Cox G (1950) Experimental designs. Wiley, New York

Crick F (1970) Central dogma of molecular biology. Nature 227(5258):561–563

Curry S, McCarthy D, DeCory H, Marler M, Gabrielsson J (2002) Phase I: the first ppportunity for extrapolation from animal data to human exposure. Wiley, New York, pp 95–115

Djebali S, Davis C, Merkel A, Dobin A et al (2012) Landscape of transcription in human cells. Nature 489(7414):101–108

Dunham I, Kundaje A, Aldred S, Collins P et al (2012) An integrated encyclopedia of DNA elements in the human genome. Nature 489(7414):57–74

Eggert U (2013) The why and how of phenotypic small-molecule screens. Nat Chem Biol 9(4):206–209

Espie P, Tytgat D, Sargentini-Maier M, Poggesi I, Watelet J (2009) Physiologically based pharmacokinetics (PBPK). Drug Metab Rev 41(3):391–407

Evans S, Dawson P (1988) The end of the p value? Br Heart J 60(3):177

Fieller E (1954) Some problems in interval estimation. J R Stat Soc Ser B (Methodological) 16(2):175–185

Ganesh T, Jiang J, Yang M, Dingledine R (2014) Lead optimization studies of cinnamic amide EP2 antagonists. J Med Chem 57(10):4173–4184

Gao X (2011) Multiple testing corrections for imputed SNPs. Genet Epidemiol 35(3):154–158

Gentleman R, Carey VJ, Bates D et al (2004) Bioconductor: open software development for computational biology and bioinformatics. Genome Biol 5:R80

Gregory R (2005) Synergy between sequence and size in large-scale genomics. Nat Rev Genet 6(9):699–708

Griffith M, Griffith O, Coffman A, Weible J, McMichael J, Spies N, Koval J, Das I, Callaway M, Eldred J, Miller C, Subramanian J, Govindan R, Kumar R, Bose R, Ding L, Walker J, Larson D, Dooling D, Smith S, Ley T, Mardis E, Wilson R (2013) DGIdb: mining the druggable genome. Nat Methods 10(12):1209–1210

Grundberg E, Small K, Hedman A, Nica A et al (2012) Mapping cis- and trans-regulatory effects across multiple tissues in twins. Nat Genet 44(10):1084–1089

Haaland P (1989) Experimental design in biotechnology, vol 105. CRC Press, Boca Raton

Haney S, Lapan P, Pan J, Zhang J (2006) High-content screening moves to the front of the line. Drug Discov Today 11(19–20):889–894

Harvey P, Tarran R, Garoff S, Myerburg M (2011) Measurement of the airway surface liquid volume with simple light refraction microscopy. Am J Respir Cell Mol Biol 45(3):592–599

Hendriks M, de Boer J, Smilde A (1996) Robustness of analytical chemical methods and pharmaceutical technological products. Elsevier, Amsterdam

Hermann J, Chen Y, Wartchow C, Menke J, Gao L, Gleason S, Haynes N, Scott N, Petersen A, Gabriel S, Vu B, George K, Narayanan A, Li S, Qian H, Beatini N, Niu L, Gan Q (2013) Metal impurities cause false positives in high-throughput screening campaigns. ACS Med Chem Lett 4(2):197–200

Hill A, LaPan P, Li Y, Haney S (2007) Impact of image segmentation on high-content screening data quality for SK-BR-3 cells. BMC Bioinf 8(1):340–353

Holmes M, Simon T, Exeter H, Folkersen L et al (2013) Secretory phospholipase A2-IIA and cardiovascular disease. J Am Coll Cardiol 62(21):1966–1976

Howie B, Donnelly P, Marchini J (2009) A flexible and accurate genotype imputation method for the next generation of genome-wide association studies. PLoS Genet 5:e1000529

Hughes J, Rees S, Kalindjian S, Philpott K (2011) Principles of early drug discovery. Br J Pharmacol 162(6):1239–1249

Hwang W, Fu Y, Reyon D, Maeder M, Tsai S, Sander J, Peterson R, Yeh J-R, Joung J (2013) Efficient genome editing in zebrafish using a CRISPR-Cas system. Nat Biotechnol 31(3): 227–229

Johnson R, Wichern D (2007) Applied multivariate statistical analysis, 6th edn. Prentice Hall, New York

Jones S, de Souza P, Lindsay M (2004) siRNA for gene silencing: a route to drug target discovery. Curr Opin Pharmacol 4(5):522–527

Jorde L, Wooding S (2004) Genetic variation, classification and 'race'. Nat Genet 36:S28–S33

Kainkaryam R, Woolf P (2009) Pooling in high-throughput drug screening. Curr Opin Drug Discov Dev 12(3):339–350

Kalbfleisch J, Prentice R (1980) The statistical analysis of failure time data. Wiley, New York

Kang H, Sul J, Service S, Zaitlen N, Kong S, Freimer N, Sabatti C, Eskin E (2010) Variance component model to account for sample structure in genome-wide association studies. Nat Genet 42(4):348–354

Kim S, Swaminathan S, Inlow M, Risacher S, The Alzheimer's Disease Neuroimaging Initiative (ADNI) (2013) Influence of genetic variation on plasma protein levels in older adults using a multi-analyte panel. PLoS ONE 8(7):e70269

Kola I, Landis J (2004) Can the pharmaceutical industry reduce attrition rates? Nat Rev Drug Discov 3(8):711–716

Korn K, Krausz E (2007) Cell-based high-content screening of small-molecule libraries. Curr Opin Chem Biol 11(5):503–510

Kuhn M, Johnson K (2013) Applied predictive modeling. Springer, Berlin

Landry Y, Gies J-P (2008) Drugs and their molecular targets: an updated overview. Fundam Clin Pharmacol 22(1):1–18

Li J, Ji L (2005) Adjusting multiple testing in multilocus analyses using the eigenvalues of a correlation matrix. Heredity 95(3):221–227

Li Y, Willer C, Ding J, Scheet P, Abecasis G (2010) MaCH: using sequence and genotype data to estimate haplotypes and unobserved genotypes. Genet Epidemiol 34(8):816–834
Lin J, Lu A (1997) Role of pharmacokinetics and metabolism in drug discovery and development. Pharmacol Rev 49(4):403–449
Lindsay M (2003) Target discovery. Nat Rev Drug Discov 2(10):831–838
Lonsdale J, Thomas J, Salvatore M, Phillips R et al (2013) The genotype-tissue expression (GTEx) project. Nat Genet 45(6):580–585
Luo C, Laaja P (2004) Inhibitors of JAKs/STATs and the kinases: a possible new cluster of drugs. Drug Discov Today 9(6):268–275
Malo N, Hanley J, Cerquozzi S, Pelletier J, Nadon R (2006) Statistical practice in high-throughput screening data analysis. Nat Biotechnol 24(2):167–175
Matthews J, Altman D (1996) Statistics notes: interaction 2: compare effect sizes not P values. Br Med J 313(7060):808–808
McVean G, Altshuler D, Durbin R, Abecasis G et al (2012) An integrated map of genetic variation from 1,092 human genomes. Nature 491(7422):56–65
Montgomery D (2012) Introduction to statistical quality control. Wiley, New York
Muller P, Milton M (2012) The determination and interpretation of the therapeutic index in drug development. Nat Rev Drug Discov 11(10):751–761
Murray C, Rees D (2009) The rise of fragment-based drug discovery. Nat Chem 1(3):187–192
Nyholt D (2004) A simple correction for multiple testing for single-nucleotide polymorphisms in linkage disequilibrium with each other. Am J Hum Genet 74(4):765–769
Purcell S, Neale B, Todd-Brown K, Thomas L, Ferreira M, Bender D, Maller J, Sklar P, de Bakker P, Daly M, Sham P (2007) PLINK: a tool set for whole-genome association and population-based linkage analyses. Am J Hum Genet 81(3):559–575
R Core Team (2014) R: a language and environment for statistical computing. R Foundation for Statistical Computing, Vienna. http://www.R-project.org
Rang H, Dale M, Ritter J, Moore P (2007) Pharmacology. Churchill Livingstone, Edinburgh
Ratjen F, Doring D (2003) Cystic fibrosis. Lancet 361(9358):681–689
Remlinger K, Hughes-Oliver J, Young S, Lam R (2006) Statistical design of pools using optimal coverage and minimal collision. Technometrics 48(1):133–143
Rendic S, Di Carlo F (1997) Human cytochrome P450 enzymes: a status report summarizing their reactions, substrates, inducers, and inhibitors. Drug Metab Rev 29(1–2):413–580
Rockman M, Kruglyak L (2006) Genetics of global gene expression. Nat Rev Genet 7(11):862–872
Sackett D (2001) Why randomized controlled trials fail but needn't: 2. Failure to employ physiological statistics, or the only formula a clinician-trialist is ever likely to need (or understand!). Can Med Assoc J 165(9):1226–1237
Shariff A, Kangas J, Coelho L, Quinn S, Murphy R (2010) Automated image analysis for high-content screening and analysis. J Biomol Screen 15(7):726–734
Shin S, Fauman E, Petersen A, Krumsiek J et al (2014) An atlas of genetic influences on human blood metabolites. Nat Genet 46(6):543–550
Simpson E (1951) The interpretation of interaction in contingency tables. J R Stat Soc Ser B (Methodological) 13:238–241
Smith G, Shah E (2003) Mendelian randomization: can genetic epidemiology contribute to understanding environmental determinants of disease? Int J Epidemiol 32(1):1–22
Soille P (2003) Morphological image analysis: principles and applications. Springer, Berlin
Sterne J (2001) Sifting the evidence—what's wrong with significance tests? Another comment on the role of statistical methods. Br Med J 322(7280):226–231
Swinney D (2013) Phenotypic vs. target-based drug discovery for first-in-class medicines. Clin Pharmacol Ther 93(4):299–301
Swinney D, Anthony J (2011) How were new medicines discovered? Nat Rev Drug Discov 10(7):507–519
The C Reactive Protein Coronary Heart Disease Genetics Collaboration (2011) Association between c reactive protein and coronary heart disease: mendelian randomisation analysis based on individual participant data. Br Med J 342:d548

Verkman A, Song Y, Thiagarajah J (2003) Role of airway surface liquid and submucosal glands in cystic fibrosis lung disease. Am J Physiol Cell Physiol 284(1):C2–C15

Voight B, Peloso G, Orho-Melander M, Frikke-Schmidt R, Barbalic M, Jensen M, Hindy G, Holm H, Ding E, Johnson T et al (2012) Plasma HDL cholesterol and risk of myocardial infarction: a mendelian randomisation study. Lancet 380(9841):572–580

Wang Q, Rager J, Weinstein K, Kardos P, Dobson G, Li J, Hidalgo I (2005) Evaluation of the MDR-MDCK cell line as a permeability screen for the blood-brain barrier. Int J Pharm 288(2): 349–359

Watson J (1992) Recombinant DNA. Macmillan, New York

Wilks A (2008) The JAK kinases: not just another kinase drug discovery target. Semin Cell Dev Biol 19(4):319–328

Yang H, Liu X, Chimalakonda A, Lu Z, Chen C, Lee F, Shyu W (2010) Applied pharmacokinetics in drug discovery and development. Wiley, Hoboken, pp 177–239

Zhang X (2011) Optimal high-throughput screening: practical experimental design and data analysis for genome-scale RNAi research. Cambridge University Press, Cambridge

Zheng W, Thorne N, McKew J (2013) Phenotypic screens as a renewed approach for drug discovery. Drug Discov Today 18(21–22):1067–1073

Chapter 5
High-Throughput Screening Data Analysis

Hanspeter Gubler

Abstract An overview over the role and past evolution of High Throughput Screening (HTS) in early drug discovery is given and the different screening phases which are sequentially executed to progressively filter out the samples with undesired activities and properties and identify the ones of interest are outlined. The goal of a complete HTS campaign is to identify a validated set of chemical probes from a larger library of small molecules, antibodies, siRNA, etc. which lead to a desired specific modulating effect on a biological target or pathway. The main focus of this chapter is on the description and illustration of practical assay and screening data quality assurance steps and on the diverse statistical data analysis aspects which need to be considered in every screening campaign to ensure best possible data quality and best quality of extracted information in the hit selection process. The most important data processing steps in this respect are the elimination of systematic response errors (pattern detection, categorization and correction), the detailed analysis of the assay response distribution (mixture distribution modeling) in order to limit the number of false negatives and false discoveries (false discovery rate and *p*-value analysis), as well as selecting appropriate models and efficient estimation methods for concentration-response curve analysis.

Keywords Compound and RNAi screening processes • Data quality control • Data normalization • Correction of systematic response errors • Hit identification and ranking • Dose-response curve analysis

5.1 Introduction

5.1.1 HTS in Drug Discovery

The beginnings of High-Throughput Screening (HTS) in the pharmaceutical and biotech industry go back to the early 1990s when more and more compounds needed

H. Gubler (✉)
Novartis Institutes for BioMedical Research, NIBR Informatics, Basel, Switzerland
e-mail: hanspeter.gubler@novartis.com

L. Zhang (ed.), *Nonclinical Statistics for Pharmaceutical and Biotechnology Industries*, Statistics for Biology and Health, DOI 10.1007/978-3-319-23558-5_5

to be tested for a broader series of targets in an increasing number of biological assay systems. In some companies the investigation of larger compound series for activity in a biochemical or cell-based assay systems had its origins in natural product screening, but was extended to look for modulating effects of compounds from the historical and growing in-house collections and added libraries of compounds from combinatorial synthesis. The goal of HTS is the identification of a subset of molecules (small molecular compounds, siRNA, antibodies, antibody conjugates, etc.) from a larger library which have a modulating effect on a given biological target. A large part of HTS work was, and still is, centered on investigating the effects of small molecules against various intra- and extracellular molecular targets and identifying compounds and compound series with a desired mode of action. In the past 20 years these compound collections have been strongly and continually enhanced by complementing the initially available sets with further sets from in-house syntheses, carefully selected additions from commercial sources, various classes of natural products, and known drugs. Stored libraries of compounds covering a broad chemical space are thus available for repeated screening or picking for special purpose investigations and have reached several 100,000 in many biotech and screening service companies, as well as academic facilities, and >1 million (up to 2 million) in most major pharmaceutical companies. Automated compound stores and retrieval systems are in use in most of the companies and allow fast replication of compound library copies into assay plates for regular screening and fast picking of sub-libraries of smaller sets of compounds for more focused screens and follow-up confirmation and verification of any activity found in the broader initial (primary) screening rounds (Fox et al. 2006; Pereira and Williams 2007; Mayr and Fuerst 2008; Mayr and Bojanic 2009; Macarron et al. 2011).

On the experimental side essentially all High-Throughput Screening experiments are performed in a highly automated, standardized and controlled fashion using microtiter plates with 96, 384 or 1536 wells, i.e. grids of 'reaction containers' embedded in rectangular plastic plates following an industry standard format with typically between 50–300 μL (96), 10–100 μL (384) and 1–10 μL (1536) working volumes containing the biological material (proteins, cells, cell fragments), assay buffer, reagents and the compound (or other) sample solutions whose activities will be determined. Nowadays the industrial HTS labs essentially only use 384- and 1536 well plates, whereas lower throughput labs may still perform a part of their experiments and measurements in 96-well plates. Most of the quality control diagnostics and data analysis aspects discussed later in this chapter can—and should—be applied irrespective of actual plate formats and throughput (ultrahigh, high, mid or low).

In essentially the same time period the sequencing of the human (and other) genomes has allowed to identify several thousand potential molecular targets for pharmaceutical intervention, some (but by far not all) of them coming with an understanding of the function and, thus, allowing the pursuit of drug discovery efforts. Large efforts in functional genomics, i.e. the determination of the function of genes, RNA transcripts and the resulting protein products as well as their regulation are needed on top of generating the pure sequence information to identify potentially druggable targets (Sakharkar et al. 2007; Bakheet and Doig 2009). The methods of identification and validation of disease-relevant molecular targets are wide and diverse (Kramer and Cohen 2004; Hughes et al. 2011). Information from DNA, RNA and protein expression profiling, proteomics experiments, phenotypic observations, RNAi (RNA interference) screens and extensive data mining and bioinformatics approaches is used to link information on diseases, gene functions and biological interaction networks with molecular properties. In RNAi screening larger libraries of siRNA or shRNA samples are used to investigate the modulation of gene function, the subsequent modification of protein expression levels and resulting loss of function in large scale high-throughput experiments (genome-scale RNAi research, genome-wide screens seeking to identify all possible regulators of a biological process, or screens limited to a subset of target genes related to a specific biological pathway) using phenotypic readouts (Matson 2004; Root et al. 2006). Several of the target identification and validation approaches thus share a series of technological elements (automation, measurement equipment and data analysis methods) with the 'classical' plate-based small molecule HTS, and most of the data quality assessment methods and hit selection steps described below can be directly applied in these areas. Some of the differences in specific analysis steps for particular types of probes, e.g. for RNAi screens will be mentioned.

A high degree of automation and standardized assay execution processes are key ingredients for effective HTS and quite large investments into the development and deployment of robotics and automated laboratory equipment have been made by the industry in the past two decades. Many vendors have also developed smaller automation workstations or work cells which can be installed and used in smaller laboratories. As with most initially specialized technologies we are also seeing here a movement from the centralized industry labs into low- and mid-throughput laboratories which are more distributed in drug discovery organizations, and also into academic institutions (Kaiser 2008; Baker 2010). The large body of experience gained over more than two decades in the pharmaceutical industry and larger government laboratories about optimized sample management and screening processes, possible screening artifacts and suitable data analysis techniques can beneficially be applied in the institutions which have more recently adopted these technologies. On the other hand, some of the statistical error correction methods described further below were elaborated within the last decade in academic institutions and are now benefitting the whole screening community.

Robotic screening systems come either as fully integrated setups where all assay steps (plate movement, reagent addition, compound addition, incubation, readouts, discarding of used plates) are scheduled and executed in automated fashion, or as a series of separate independent workstation cells which are used sequentially, often with a 'manual' transfer of plate batches between them. The large fully integrated HTS systems can process and measure up to 100,000 compounds or even more per day (ultra-high throughput screening, uHTS), depending on assay technology, details of the assay protocol and processing times for individual steps. Workstation based mid-throughput screening (MTS) typically reaches throughputs of 10,000–20,000 samples/day (e.g. in batches of 20–50 384 well plates per day). Throughput will naturally be smaller if lengthy process steps like e.g. incubation phases are needed or if the measurement time is prolonged, either because of limitations of the measurement instrument, or because the course of a biochemical reaction needs to be followed for a certain amount of time. The aspect of performing the complete set of experiments of a particular screen in *separate batches* of *individual plates* is clearly very important and will have an effect on the necessary data analysis steps.

The optimization of the high throughput screening processes has over time gone through different phases which have initially focused on obtaining higher throughput and larger numbers of screened entities, then on higher sophistication in assay optimization, standardization of processes, miniaturization, added pilot-, counter- and orthogonal screening experiments, as well as improved analysis techniques, i.e. a focus on higher efficiency and better data quality, and in the last 8–10 years the screening and decision making processes were set up in much more flexible ways to better allow diversified and case-by-case handling for each campaign—either full deck screens or adapted focused screens including best possible validation of results with parallel specificity and selectivity screens to obtain higher quality and better characterized hits than resulting from the 'raw' hit list of the main screen (Mayr and Fuerst 2008; Macarron et al. 2011).

Counter screens are used to identify compound samples which don't show an activity directed towards the intended biological target, but which nonetheless give positive readouts ('false positive responses' in an assay) by interfering with the readout mechanism or act otherwise nonspecifically. Some compounds will even do this in concentration dependent manner, thus mimicking a desired activity. This can occur due to aggregation, colored components of assay 'cocktail', fluorescence, inhibition of detection enzymes (reporter mechanism), cytotoxicity, etc. (Thorne et al. 2010; Hughes et al. 2012).

An orthogonal screen is based on an assay which uses a different format or a different readout mechanism to measure the same phenomenon and confirm that the activity is really directed towards the target of interest. Compounds which are active in both the original and orthogonal assay are usually prioritized for follow-up work.

Selectivity screens are used to determine whether a particular compound is acting solely on the target of interest or also on other targets of the same family (e.g. enzymes, protease inhibitors, related ion channels, receptor families, etc.).

Counter-, orthogonal and selectivity screens are thus used to stratify the hit list of the putative actives on the target of interest. The selection of the main screening assay and the setup of suitable filter-experiments coupled with *optimal data analysis approaches* to extract the cleanest and most complete information possible are important ingredients for the success of HTS-based hit discovery projects. Secondary follow-up assays for further confirmation and quantification of the desired modulation of the target and possibly determination of mechanisms of action will need similar care in processing of the underlying plate-based data and best possible characterization of concentration-dependent responses. Hit compounds or compound series possessing suitable pharmacological or biological and physicochemical properties and characterized in such a complete fashion, including a final structural verification, can then be considered possible starting points for further chemical optimization, i.e. become a 'lead compound' or a 'lead series' in a given drug discovery project (Hughes et al. 2011).

High-Throughput Screening methods and tools in their diverse forms (using biochemical, cell- or gene-based assay systems) with small molecular compounds, siRNA, shRNA, antibodies, antibody drug conjugates, or other probes to modulate the intended target or biological process of interest using a wide variety of assay and readout technologies have thus become an *essential research tool for drug discovery*, i.e. screening for hits and leads, functional genomics (target discovery), biomarker detection and identification in proteomics using mass spectrometry readouts , automated large scale characterization of samples (e.g. for sample quality assessment), as well as the detailed characterization of hit series with biophysical measurements, often using label-free assay technologies, and ADMET (absorption, distribution, metabolism, excretion, toxicity) screens (Macarron et al. 2011). In all cases *the choice of the most suitable statistical methods* for the different data analysis steps forms an important part of the usefulness and success of this toolset.

5.1.2 HTS Campaign Phases

Compound Screening Typically several different rounds of screens are run for each project in compound based drug discovery. An initial primary screen is applied to assess the activity of a collection of compounds or other samples and to identify hits against a biological target of interest, usually employing single measurements ($n = 1$) due to the sheer number of samples which need to be processed. The *primary* screen identifies actives from a large diverse library of chemical probes, or alternatively, from a more focused library depending on pre-existing knowledge on a particular target. Selected hits and putative actives are then processed in second stage *confirmation* screen, either employing replicates at a particular single concentration, or a concentration response curve with just a few concentration points. In Table 5.1 we show the typical experimental characteristics (numbers of replicates, numbers of concentrations) of various screening phases for a large compound based HTS campaign in the author's organization as an illustration for the overall experimental effort needed and the related data volumes to be expected. The numbers are also representative for other larger screening organizations. The addition of counter- or selectivity measurements will of course lead to a correspondingly higher effort in a particular phase.

If primary hit rates are very high and cannot be reduced by other means then counter-screens may need to be run in parallel to the primary screen (i.e. one has to run two large screens in parallel!) in order to reduce the number of candidates in the follow phase to a reasonable level and to be able to more quickly focus on the more promising hit compounds. If primary hit rates are manageable in number then such filter experiments (whether counter- or selectivity-screens) can also be run in the confirmation phase. Selectivity measurements are often delayed to the

Table 5.1 Typical phases in compound-based high-throughput screening campaigns

Screening phase	# of replicates	# concentrations/ sample	Total # of different test samples	Total # of wells
Pilot	2–3	1	10^3–10^4	$2{\cdot}10^4$
Primary	1	1	10^5–$1.5{\cdot}10^6$	10^5–$1.5{\cdot}10^6$
Confirmation (experiment with replicates)	2–4	1	10^3–$5{\cdot}10^4$	10^4–$2{\cdot}10^5$
Confirmation (experiment with concentration dependent measurements)	1	2–4	10^3–$5{\cdot}10^4$	10^4–$2{\cdot}10^5$
Validation (detailed concentration dependence of response, potency determination)	1–4	8–12	10^3–$5{\cdot}10^4$	10^4–10^6

validation screening phase in order to be able to compare sufficient details of the concentration-response characteristics of the compounds which were progressed to this stage on the different targets, and not just have to rely on single-point or only very restricted concentration dependent measurements in the previous phases. Thus, the actual details of the makeup of the screening stages and the progression criteria are highly project dependent and need to be adapted to the specific requirements.

A *validation* screen, i.e. a detailed measurement of full concentration-response curves with replicate data points and adequately extended concentration range, is then finally done for the confirmed actives, as mentioned, possibly in parallel to selectivity measurements.

The successful identification of interesting small-molecule compounds or other type of samples exhibiting genuine activity on the biological target of interest is dependent on selecting suitable assay setups, but often also on adequate 'filter' technologies and measurements, and definitely, as we will see, also on employing the most appropriate statistical methods for data analysis. These also often need to be adapted to the characteristics of the types of experiments and types of responses observed. All these aspects together play an important role for the success of a given HTS project (Sittampalam et al. 2004).

Besides the large full deck screening methods to explore the influence of the of the particular accessible chemical space on the target of interest in a largely unbiased fashion several other more 'knowledge based' approaches are also used in a project specific manner: (a) focused screening with smaller sample sets known to be active on particular target classes (Sun et al. 2010), (b) using sets of drug-like structures or structures based on pharmacophore matching and in-silico docking experiments when structural information on the target of interest is available (Davies et al. 2006), (c) iterative screening approaches which use repeated cycles of subset screening and predictive modeling to classify the available remaining sample set into 'likely active' and 'likely inactive' subsets and including compounds predicted to likely show activity in the next round of screening (Sun et al. 2010), and (d) fragment based screening employing small chemical fragments which may bind weakly to a biological target and then iteratively combining such fragments into larger molecules with potentially higher binding affinities (Murray and Rees 2008).

In some instances the HTS campaign is not executed in a purely sequential fashion where the hit analysis and the confirmation screen is only done *after* the complete primary run has been finished, but several alternate processes are possible and also being used in practice: Depending on established practices in a given organization or depending on preferences of a particular drug discovery project group one or more intermediate hit analysis steps can be made before the primary screening step has been completed. The two main reasons for such a partitioned campaign are: (a) To get early insight into the compound classes of the hit population and possible partial refocusing of the primary screening runs by sequentially modifying the screening sample set or the sequence of screened plates, a process which is brought to an 'extreme' in the previously mentioned iterative screening approach, (b) to get a head start on the preparation of the screening plates for the confirmation run so that it can start *immediately* after finishing primary screening, without any 'loss' of time. There is an obvious advantage in keeping a given configuration of the automated screening equipment, including specifically programmed execution sequences of the robotics and liquid handling equipment, reader configurations and assay reagent reservoirs completely intact for an immediately following screening step. The careful analysis of the primary screening hit candidates, including possible predictive modeling, structure-based chemo-informatics and false-discovery rate analyses can take several days, also depending on assay quality and specificity. The preparation of the newly assembled compound plates for the confirmation screen will also take several days or weeks, depending on the existing compound handling processes, equipment, degree of automation and overall priorities. Thus, interleaving of the mentioned hit finding investigations with the continued actual physical primary screening will allow the completion of a screen in a shorter elapsed time.

RNAi Screening RNAi 'gene silencing' screens employing small ribonucleic acid (RNA) molecules which can interfere with the messenger RNA (mRNA) production in cells and the subsequent expression of gene products and modulation of cellular signaling come in different possible forms. The related experiments and progression steps are varying accordingly. Because of such differences in experimental design aspects and corresponding screening setups also the related data analysis stages and some of the statistical methods will then differ somewhat from the main methods typically used for (plate-based) single-compound small molecule screens. The latter are described in much more depth in the following sections on statistical analysis methods.

Various types of samples are used in RNAi (siRNA, shRNA) screening: (a) non-pooled standard siRNA duplexes and siRNAs with synthetically modified structures, (b) low-complexity pools (3–6 siRNAs with non-overlapping sequences) targeting the same gene, (c) larger pools of structurally similar silencing molecules, for which measurements of the loss of function is assessed through a variety of possible phenotypic readouts made in plate array format in essentially the same way as for small molecule compounds, and also—very different—(d) using large scale pools of shRNA molecules where an entire library is delivered to a single population of cells and identification of the 'interesting' molecules is based on a selection process.

RNAi molecule structure (sequence) is essentially known when using approaches (a) to (c) and, thus, the coding sequence of the gene target (or targets) responsible for a certain phenotypic response change can be inferred in a relatively straightforward way (Echeverri and Perrimon 2006; Sharma and Rao 2009). RNAi hit finding progresses then in similar way as for compound based screening: An initial *primary* screening run with singles or pools of related RNAi molecule samples, followed by a *confirmation* screen using (at least 3) replicates of single or pooled samples, and if possible also the individual single constituents of the initial pools. In both stages of screening statistical scoring models to rank distinct RNAi molecules targeting the *same* gene can be employed when using pools and replicates of related RNAi samples (redundant siRNA activity analysis, RSA, König et al. 2007), thus minimizing the influence of strong off-target activities on the hit-selection process. In any case candidate genes which cannot be confirmed with more than one distinct silencing molecule will be eliminated from further investigation (false positives). Birmingham et al. (2009) give an overview over statistical methods used for RNAi screening analysis. A final set of *validation* experiments will be based on assays measuring the sought phenotypic effect and other secondary assays to verify the effect on the biological process of interest on one hand, and a direct parallel observation of the target gene silencing by DNA quantitation, e.g. by using PCR (polymerase chain reaction) amplification techniques). siRNA samples and libraries are available for individual or small sets of target genes and can also be produced for large 'genome scale' screens targeting typically between 20,000 and 30,000 genes.

When using a large-scale pooled screening approach, as in d) above, the selection of the molecules of interest and identification of the related RNA sequences progresses in a very different fashion: In some instances the first step necessary is to sort and collect the cells which exhibit the relevant phenotypic effect (e.g. the changed expression of a protein). This can for example be done by fluorescence activated cell sorting (FACS), a well-established flow-cytometric technology. DNA is extracted from the collected cells and enrichment or depletion of shRNA related sequences is quantified by PCR amplification and microarray analysis (Ngo et al. 2006). Identification of genes which are associated with changed expression levels using 'barcoded' sets of shRNAs which are cultured in cells both under neutral reference conditions and test conditions where an effect-inducing agent, e.g. a known pathway activating molecule is added (Brummelkamp et al. 2006) can also be done through PCR quantification, or alternatively through massive parallel sequencing (Sims et al. 2011) and comparison of the results of samples cultured under different conditions. These are completely 'non-plate based' screening methods and will not be further detailed here.

5.2 Statistical Methods in HTS Data Analysis

5.2.1 General Aspects

In the next sections we will look at the typical process steps of preparing and running a complete HTS campaign, and we will see that statistical considerations play and important role in essentially all of them because optimal use of resources and optimal use of information from the experimental data are key for a successful overall process. In the subsequent sections of this chapter we will not touch on all of these statistical analysis aspects with the same depth and breadth, not least because some of these topics are actually treated in more detail elsewhere in this book, or because they are not very specific to HTS data analysis. We will instead concentrate more on the topics which are more directly related to plate-based high-throughput bioassay experiments and related efficient large scale screening data analysis.

5.2.2 Basic Bioassay Design and Validation

Given a particular relevant biological target the scientists will need to select the assay method, detection technology and the biochemical parameters. This is sometimes done independently from HTS groups, or even outside the particular organization. Primarily the sensitivity (ability to detect and accurately distinguish and rank order the potency of active samples over a wide range of potencies, e.g. based on measurements with available reference compounds) and reproducibility questions need to be investigated in this stage. Aspects of specificity in terms of the potential of a particular assay design and readout technology to produce signals from compounds acting through unwanted mechanisms as compared to the pharmacologically relevant mechanism also need to be assessed.

The exploration and optimization of assay conditions is usually done in various iterations with different series of experiments employing *(fractional) factorial design, replication and response surface optimization methods* to determine robust response regions. The potentially large series of assay parameters (selection of reagents, concentrations, pH, times of addition, reaction time courses, incubation times, cell numbers, etc.) need to be investigated at multiple levels in order to be able to detect possible nonlinearities in the responses (Macarron and Hertzberg 2011). The practical experimentation involving the exploration of many experimental factors and including replication often already takes advantage of the existing screening automation and laboratory robotics setups to control the various liquid handling steps, which otherwise would be rather cumbersome and error-prone when performed manually (Taylor et al. 2000). Standard methods for the analysis of designed experiments are used to optimize dynamic range and stability of the assay readout while maintaining sensitivity (Box et al. 1978; Dean and Lewis 2006). Most often the signal-to-noise ratio *SNR*, signal window *SW* and Z'-factor are used as optimization criteria. See the section on assay quality measures and Table 5.2 below for the definition of these quantities.

Table 5.2 Selected assay quality metrics

	Quality control measure	Estimation expression	Comments
a	Coefficient of Variation for Controls and References, CV	$\frac{s_x}{\bar{x}}$	Simple statistic giving rough indication of signal variability, without consideration of actual range of obtainable responses
b	High/Low Ratio, HLR Signal/Background Ratio, SBR	$\frac{\bar{x}_N}{\bar{x}_P}, \frac{\bar{x}_P}{\bar{x}_N}$	HLR and SBR quantify the ratio of the maximal to minimal response values but without considering the variability of the signals
c	Signal/Noise Ratio, SNR	$\frac{\lvert\bar{x}_N - \bar{x}_P\rvert}{s_N}$	'Dynamic range' of signal in relation to standard deviation of neutral controls. Variation of positive controls not taken into account
d	Signal Window Coefficient, SW	$\frac{\lvert\bar{x}_N - \bar{x}_P\rvert - 3\,(s_P + s_N)}{s_N}$	Akin to SNR, but only considering signal range 'outside' of 3*s* limits of the controls, i.e. the 'usable' signal window to quantify responses differing from the controls (Sittampalam et al. 1997)
e	Z'-factor *RZ*′-factor (robust) Assay Window Coefficient	$1 - \frac{3\ (s_P + s_N)}{\lvert\bar{x}_P - \bar{x}_N\rvert}$	Use of mean $\bar{x}$ and standard deviation s_x for Z' or median $\tilde{x}$ and mad estimators $\tilde{s}_x$ for RZ', respectively. Ratio of 'usable' signal window size outside of 3*s* limits in relation to the full signal range (relative signal window size between controls) (Zhang et al. 1999)
f	*V*-factor Conceptually related to Z′-factor, but also considering the variation of a set of intermediate response values, not just the variation at the response range limits	$1 - \frac{6\,\bar{s}}{\lvert\bar{x}_P - \bar{x}_N\rvert}$	$\bar{x}$ is average standard deviation of replicate measurements at multiple concentrations with effects between $\bar{x}_N$ and $\bar{x}_P$, or the average standard deviation of the residuals of a fitted model (Ravkin 2012; Bray and Carpenter 2012)
g	*SSMD*, strictly standardized mean difference	$\frac{\bar{x}_P - \bar{x}_N}{\sqrt{s_P^2 + s_N^2}}$	*SSMD* expressions for unequal variance and unequal sample sizes can be found in Zhang et al. (2007). *SSMD* estimates tends to be closer to population values than the equivalent *t*-statistics. In practice *SSMD* statistics are predominantly used to assess the significance of effect sizes in RNAi screening, see Zhang (2008, 2011a, b)

Instead of mean $\bar{x}$ and standard deviation s estimators the median $\tilde{x}$ and mad $\tilde{s}$ can be used as robust plug-in replacements

5.2.3 Assay Adaptation to HTS Requirements and Pilot Screening

The adaptation of assay parameters and plate designs with respect to the positioning and numbers of control samples, types of plates, plate densities and thus, liquid volumes in the individual plate wells, to fulfill possible constraints of the automated HTS robotics systems needs to follow, if this was not considered in the initial assay design, e.g. when the original bioassay was designed independently of the consideration to run it as a HTS. The initially selected measurement technology often has an influence on the obtainable screening throughput. If these design steps were not done in initial collaboration with the HTS groups, then some of the assays parameters may have to be further changed and optimized with respect to relevant assay readout quality measures (see Table 5.2) for the assay to be able to run under screening conditions. Again, experimental design techniques will be employed, albeit with a now much reduced set of factors. These assay quality metrics are all based on the average response levels $\bar{x}$ of different types of controls, corresponding to readout for an inactive probe (*N*, neutral control) or the readout for a 'fully active' probe (*P*, positive control) and the respective estimates of their variability (standard deviation) s.

The quality control estimators in Table 5.2 are shown as being based on the mean $\bar{x}$ and standard deviation s, but in practice the corresponding outlier resistant 'plug-in' equivalents using the median $\tilde{x}$ and median absolute deviation (mad) $\tilde{s}$ estimators are often used, where $\tilde{s}$ includes the factor 1.4826 to ensure consistency with s, so that $E\left(\tilde{s}(x)\right) = E\left(s(x)\right) = \sigma$ if $x \sim N(\mu, \sigma^2)$ (Rousseeuw and Croux 1993).

The quality measures which incorporate information on variability of the controls often much more useful than simple ratios of average values because probability based decision making in the later hit selection process needs to take into account the distributions of the activity values. In order to gain a quick overview and also *see* aspects of the distribution of the measured values (e.g. presence of outliers, skewness) which are not represented and detected by the simple statistical summary values the data should always be *visualized* using e.g. categorized scatterplots, boxplots, strip plots and normal quantile-quantile plots.

Possible modifications of the original assay design may entail gaining higher stability of response values over longer time scales or in smaller volumes to be able to run larger batches of plates, shortening some biochemical reaction or incubation phases to gain higher throughput, or to have smaller influence of temperature variations on responses, etc. Such modifications may sometimes even have to be done at the cost of reduced readout response levels. As long as the assay quality as measured by a suitable metric stays above the acceptance criteria defined by the project group such 'compromises' can usually be done without large consequences for the scientific objectives of the HTS project.

Besides the optimization of the assay *signal range* and *signal stability* an important part of the quality determination and validation in the assay adaptation phase is the initial determination of assay *reproducibility* for samples with varying degrees of activity using correlation- and analysis of agreement measures, as well providing rough *estimates on expected false positive and false negative rates* which are based on the evaluation of single point %-activity data at the planned screening concentration and the determination of the corresponding concentration-response data as an activity reference for the complete standard 'pilot screening library' of diverse compounds with known mechanisms of action and a wide range of biological and physicochemical properties for use in all HTS assays in the pilot screening phase (Coma et al. 2009a, b). See Table 5.3 for an overview of typical analyses performed in this phase.

These investigations with the standard pilot screening library will sometimes also allow one to gain early information on possible selectivity (to what extent are compounds acting at the target vs. other targets of the same family) and specificity (is the compound acting through the expected mechanism or through an unwanted one) of the assay for an already broader class of samples available in the pilot screening library than are usually investigated in the initial bioassay design.

While the detailed evaluation of factors and response optimization using experimental design approaches with replicated data are feasible (and necessary) at the assay development and adaptation stages, this is no longer possible in the actual large primary HTS runs. They will usually need to be executed with $n = 1$ simply because of the large scale of these experiments with respect to reagent consumption, overall cost and time considerations. Because of this bulk- and batch execution nature of the set of measurements collected over many days or weeks it is not possible to control all the factors affecting the assay response. Given these practical experimental (and resource) restrictions, the *observed random and systematic variations of the measured responses need to be accommodated and accounted for in the data analysis steps* in order to extract the highest quality activity- and activity-rank order information possible. The use of optimal data normalization and hit selection approaches are key in this respect.

5.2.4 Assay Readouts, Raw and Initial Derived Values

Raw readouts from plate readers are generated in some instrument-specific units (often on an 'arbitrary' scale) which can be directly used for data quality and activity assessment, but in other assay setups there may be a need for an initial data transformation step. This can be needed when performing time- or temperature-dependent measurements which need a regression analysis step to deliver derived readouts in meaningful and interpretable physical units (assay readout endpoints), e.g. using inverse estimation on data of a calibration curve, determining kinetic parameters of a time-course measurement or protein melting transition temperatures in a thermal shift assay, or extracting the key time course signal characteristics in a kinetic fluorescence intensity measurement. Sometimes such derived information can be obtained directly from the instrument software as an alternate or additional 'readout', but when developing new assays and measurement methods

Table 5.3 Quality metrics and methods in assay adaptation and pilot screening phase

	Type of analysis, metric	Methods, tools	Comments
a	Correlation analysis	Pearson correlation coefficient ρ ($n = 2$), Spearman rank correlation ($n = 2$), intraclass correlation coefficient ICC ($n > 2$)	Reproducibility of n replicate measurements, reliability analysis (Coma et al. 2009a, b)
b	Analysis of agreement	Bland–Altman plot, scale-location plots: $s_{x,i} \sim f(\overline{x}_i)$ or $\lvert\Delta_{x,i}\rvert \sim f(\overline{x}_i)$, where s is the standard deviation and Δ is the range of data at a particular $\overline{x}$ value	Reproducibility of replicate measurements, reliability analysis, assessment of heteroscedasticity (Bland and Altman 1986; Sun et al. 2005)
c	Normality, deviation from normality	Normal probability plot, Kolmogorov–Smirnov Test, Anderson–Darling Test	Comparison of control and reference sample distributions with normal distributions (Wu and Liu 2008)
d	Initial estimate of expected false positive (FP) and false negative (FN) rates	Comparison of primary screening results with results of concentration response experiments	FP and FN rate estimation under pilot screening conditions (Zhang et al. 2005; Coma et al. 2009a, b; Ilouga and Hesterkamp 2012)
e	MSR, minimum significant ratio (smallest potency ratio between compounds which is statistically significant at level α, usually $\alpha = 0.05$)	$10^{Z_{\alpha/2}\sqrt{2}\hat{s}}$ where $\hat{s}$ is the sample standard deviation of a set of independent repeated $\log_{10}(IC_{50})$ potency determinations and $Z_{\alpha/2}$ the standard normal distribution quantile	Smallest statistically significant potency ratio (IC_{50} ratio) between any pair of compounds (Eastwood et al. 2005, 2006)

it is sometimes necessary to be able to add different, more sophisticated or more robust types of analyses as ad-hoc data preprocessing steps. It is a definite advantage if a standard software framework is available where such additional analysis steps and related methods can be easily explored, developed and readily plugged into the automated assay data processing path, e.g. by using R scripts (R Development Core Team 2013) in the Pipeline Pilot® (http://accelrys.com/products/pipeline-pilot/), Knime (https://www.knime.org/) or similar analytics platform included upstream of an organization's standard screening data processing system, where data thus transformed can then easily be processed using the available standard HTS data analysis methods in the same way as any other type of screening reader output.

Assay and readout technologies have gone through many changes and advancements in the past years. Whereas initially in HTS the measurement of only one or very few (2–3) readout parameters per well (e.g. fluorescence intensities at two different wavelengths) was customary—and still is for many practical applications—the advent of automated microscopy and cellular imaging coupled with automated image analysis (image based High Content Analysis or Screening, HCA, HCS) which can detect changes in the morphology of cells or of separately labeled cell compartments (nucleus, membrane, organelles, etc.), thus resulting in a large number of parameters for a given well or even for each individual cell, has led to the need for the exploration and evaluation of suitable multivariate statistical data analysis methods (Hill et al. 2007). Intensities, textures, morphological and other parameters from the segmented images are captured at several different wavelengths and corresponding feature vectors are associated with each identified object or well (Abraham et al. 2004; Carpenter 2007; Duerr et al. 2007; Nichols 2007). Cell level analysis enables the analysis of the various cell-cycles and the separation of the effects of the probes on cells in a particular state (Loo et al. 2007; Singh et al. 2014). Besides the now quite broadly used image based HCS approaches there are several other assay technologies which produce multivariate readouts of high dimensions, Cytof (Qiu et al. 2011), Luminex gene expression profiling (Wunderlich et al. 2011), RPA (van Oostrum et al. 2009), laser cytometry (Perlman et al. 2004), and others, with medium throughput . For most of these technologies the *most suitable* optimal data analysis methods are still being explored. Questions of normalization, correction of systematic errors, discrimination and classification are under active investigation in many labs (Reisen et al. 2013; Kümmel et al. 2012; Abraham et al. 2014; Singh et al. 2014; Smith and Horvath 2014; Haney 2014). It is clear that all these different types of assay technologies can benefit from a common informatics infrastructure *for large scale multivariate data analysis*, which includes a large set of dimension reduction, feature selection, clustering, classification and other statistical data analysis methods, as well as a standardized informatics systems for data storage and metadata handling, coupled to high performance computing resources (compute clusters) and large volume file stores and databases (Millard et al. 2011).

The high numbers of readout parameters (300–600) (Yin et al. 2008; Reisen et al. 2013) which must be simultaneously analyzed and the much higher data volumes which need to be processed introduce new aspects into high-throughput screening data analysis which are usually not covered by the available features in the established standard screening informatics systems (Heyse 2002; Gunter et al. 2003; Kevorkov and Makarenkov 2005; Gubler 2006; Boutros et al. 2006; Zhang and Zhang 2013). This makes these data much more challenging to analyze from the point of view of methodology, complexity of assay signals and the sheer amounts of data. But it is clear that these types of screening technologies and efficient methods to analyze the large data volumes will become even more important and widespread in future. While one can say that the analysis methods for standard HTS data have been *largely* settled—at least from the point of view of the *main recommended data processing and quality assurance* steps as outlined in this chapter—this is definitely not yet the case for the high dimensional multivariate screening data analysis, especially when going to the single cell level. Note that the screening literature occasionally refers to *multi-parametric* analysis in this context. Systematic investigations on advantages and disadvantages of particular methods and the preferred approaches for determining assay and screening quality metrics, correction of systematic response errors, classification of actives, etc. with such types of data are ongoing and are naturally more complex than for the cases where just a few readout parameters can be processed in a largely independent manner up to the point where the final values need to be correlated to each other (Kümmel et al. 2012).

5.2.5 Assay Quality Measures

The overall error which accumulates over the many different chemical, biological and instrumental processing steps to obtain the final readout in a screening assay needs to be kept as small as possible so that there is high confidence in the set of compounds identified as active in a screening campaign. The assay quality metrics to measure and monitor this error are based on simple location and scale estimates derived from raw readout data from the different types of wells on a microtiter plate (zero-effect and full inhibition of full activation controls for normalization of the data, reference controls exhibiting responses in the middle of the expected response range, background wells, and test sample wells). Different quality indicators have been proposed to measure the degree of separability between positive and zero-effect (neutral) assay controls: Signal to background ratio or high-low ratio, coefficient of variation, signal to noise ratio, Z- and Z'-factor (not to be confused with a Z-score) (Zhang et al. 1999), strictly standardized mean difference (*SSMD*) (Zhang et al. 2007) and others are in routine use to optimize and measure assay response quality (see Table 5.2).

The Z'-factor has become an accepted and widely used quality metric to assess the discriminatory power of a screening assay. It is a relative measure and quantifies the 'usable window' for responses between the upper and lower controls outside of their respective $3s$ limits. Z' can be between $-\infty$ (if the control averages which define the response limits are identical), 0 when the two $3s$ limits 'touch' each other, and 1 if the standard deviation of the controls becomes vanishingly small. Z' is an empirical point measure and the derivation of its large sample interval estimator was only recently published (Majumdar and Stock 2011). The sampling uncertainty of Z' should be considered when setting acceptance thresholds, especially for lower density plates with small numbers of control wells. Small sample intervals can be estimated by bootstrap resampling (Iversen et al. 2006). Other quality indicators than those listed were proposed and described in the literature (e.g. assay variability ratio, signal window and others), but are not so widely used in standard practice because they are related to the Z'-factor and don't represent independent information (Sui and Wu 2007; Iversen et al. 2006). The V-factor is a generalization of the Z'-factor to multiple response values between $\overline{x}_N$ and $\overline{x}_P$ (Ravkin 2004).

Some *screening quality problems* can occur for actual sample wells which are *not captured by control well data* and the measures listed in Table 5.2, e.g. higher variability for sample wells than for control wells, additional liquid handling errors due to additional process steps for sample pipetting, non-uniform responses across the plates, etc. Such effects and tools for their diagnosis are described in more detail further below in the screening data quality and process monitoring section.

In Fig. 5.1 we show an example of the behavior of the High/Low control ratio (*HLR*) and the Z' factor for a set of 692 1536-well plates from a biochemical screen exhibiting several peculiarities: (a) clear batch boundary effects in the ratio of the *HLR* values for batch sizes varying between 80 and 200 plates, (b) 'smooth' time dependence of the *HLR* (fluorescence intensity) ratio due to the use of continuous assay product formation reaction and related detection method, (c) no 'strongly visible' influence of the varying *HLR* on the Z'-factor, i.e. a negligible influence of the varying *HLR* on the relative 'assay window', (d) an interleaved staggering pattern of *HLR* which is due to the use of a robotic system with two separate processing lanes with different liquid handling and reader instruments. This latter aspect may be important to take into account when analyzing the data because any systematic response errors, if they occur at a detectable and significant level, are likely to be different between the two subsets of plates, hence a partially separate analysis may need to be envisaged. We also see that for assays of this nature setting a tight range limit on *HLR* will not make sense; only a lower threshold could be useful as a potential measurement failure criterion.

5.2.6 Screening Data Quality and Process Monitoring

Automated screening is executed in largely unattended mode and suitable procedures to ensure that relevant quality measures are staying within adequate acceptance limits need to be set up. Some aspects of statistical process control (SPC) methodology (Shewhart 1931; Oakland 2002) can directly be transferred to HTS as an 'industrial' data production process (Coma et al. 2009b; Shun et al. 2011).

Data quality monitoring of larger screens using suitably selected assay quality measures mentioned above and preferably also for some of the additional screening

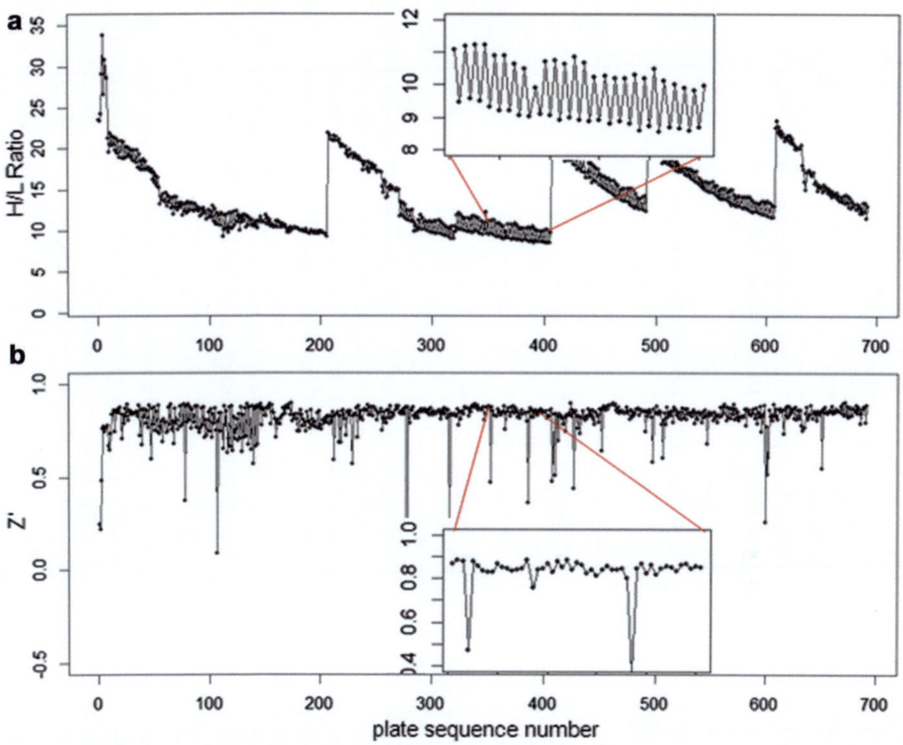

Fig. 5.1 (**a**) Screening quality control metrics: High/Low ratio *HLR* for complete screening run showing batch start and end effects, signal changes over time and alternating robot lane staggering effects (*inset*). (**b**)Z'-factor (assay window coefficient) for the same plate set with occasional low values for this metric, indicating larger variance of control values for individual plates, but generally negligible influence of robot lane alternation on assay data quality (*inset*, saw tooth pattern barely visible)

quality measures listed below in Table 5.4, can be done online with special software tools which analyze the data in an automated way directly after the readout is available from the detection instruments (Coma et al. 2009b), or at least very soon after completing the running of a plate batch with the standard data analysis tools which are in use, so that losses, potential waste and the need for unnecessarily repetitions of larger sets of experiments are minimized. As in every large scale process the potential material and time-losses and the related financial aspects cannot be neglected and both plate level and overall batch level quality must be maintained to ensure a meaningful completion of a screening campaign.

Systematic response errors in the data due to uncontrolled (and uncontrollable) factors will most likely also affect some of the general and easily calculated quality measures shown in Table 5.4, and they can thus be indirect indicators of potential problems in the screening or robotic automation setup. When using the standard HTS data analysis software tools to signal the presence of systematic response

Table 5.4 Screening quality metrics

	Metric	Estimation expression	Comments
a	*Z*-factor *RZ*-factor (robust) Screening Window Coefficient	$1-\frac{3\ (s_P+s_C)}{\lvert\overline{x}_P-\overline{x}_C\rvert}$	Mean, standard deviation or median, mad use in the same way as for *Z′*/*RZ′* factor (Zhang et al. 1999)
b	Maximum, minimum, or range of systematic error $\hat{S}$ on plate *p*	$\max_p\left(\hat{S}_{ip}\right)$, $\min_p\left(\hat{S}_{ip}\right)$, $\max_p\left(\hat{S}_{ip}\right)-\min_p\left(\hat{S}_{ip}\right)$	H. Gubler, 2014, unpublished work
c	*VEP*, Fraction of response variance 'explained' by the estimated systematic error components (pattern) on plate *p*	$\tilde{s}_p^2\left(\hat{S}_{ip}\right)\ /\ \tilde{s}_p^2\left(x_{ip}\right)$	Coma et al. (2009b)
d	Moran's *I*, spatial autocorrelation coefficient	$\frac{n\sum_{i=1}^{n}\sum_{j=1}^{n}w_{ij}\left(x_i-\overline{x}\right)\left(x_j-\overline{x}\right)}{\sum_{i=1}^{n}\sum_{j=1}^{n}w_{ij}\sum_{i=1}^{n}\left(x_i-\overline{x}\right)^2}$ with neighbor weights w_{ij} (e.g. $w_{ij}=1$ if *i* and *j* are neighbor wells within a given neighborhood distance δ, e.g. $\delta=2$, sharing boundaries and corner points, and $w_{ij}=0$ otherwise, and also $w_{ii}=0$). Also distance based weights $w_{ij}=1/d_{ij}^{\alpha}$ can be used	Use for statistical test of null hypothesis that no significant spatial correlation is present on a plate, i.e. no systematic location dependent sample response deviation from the average null-effect response exists (Moran 1950; Murie et al. 2013) Other spatial autocorrelation coefficients can also be used for this test, e.g. Geary's *C* ratio (Geary 1954)

These measures are focused on sample-well area of plates

Instead of mean $\overline{x}$ and standard deviation *s* estimators the median $\tilde{x}$ and mad $\tilde{s}$ are often used as robust plug-in replacements. In order to roughly determine systematic error estimates $\hat{S}_{ip}$ in the *initial* HTS *monitoring* stage a model which is very quickly calculated can be chosen (e.g. spatial polish, polynomial or principal pattern models, see Table 5.6). For the *final analysis* of the screening production runs the most suitable error modeling method for the determination of the systematic plate effects will be used and the screening quality measures before and after correcting the data for the systematic errors can be compared errors or a general degradation of the readout quality, then the broad series of available diagnostic tools can efficiently flag and 'annotate' any such plate for further detailed inspection and assessment by the scientists. This step can also be used to automatically categorize them for inclusion or exclusion in a subsequent response correction step.

A note on indexing in mathematical expressions: In order to simplify notation as far as possible and to avoid overloading the quoted mathematical expressions we will use capital index letters to indicate a particular subset of measured values (e.g. P and N as previously mentioned, C compound samples), and we will use single array indexing of measured or calculated values of particular wells i whenever possible. In those situations where the exact row and column location of the well in the two-dimensional grid is important we will use double array indexing ij. The explicit identification of the subset of values on plate p is almost always required, so this index will appear often.

These additional metrics are relying on data from the sample areas of the plates and will naturally provide additional important insight into the screening performance as compared to the control sample based metrics listed in Table 5.2. As for the previously mentioned control sample based assay quality metrics it is even more important to visualize such additional key screening data quality metrics which are based on the (compound) sample wells in order to get a quick detailed overview on the behavior of the response data on single plates, as well as its variation over time to obtain indications of data quality deteriorations due to instrumental (e.g. liquid handling failures), environmental (e.g. evaporation effects, temperature variations) and biochemical (e.g. reagent aging) factors, or due to experimental batch effects, e.g. when using different reagent or cell batches. Direct displays of the plate data and visualizations of the various assay quality summaries as a function of measurement time or sequence will immediately reveal potentially problematic data and suitable threshold settings can trigger automatic alerts when quality control metrics are calculated online.

Some of the listed screening quality metrics are based on direct estimation of systematic plate and well-location specific experimental response errors S_{ijp}, or are indicators for the presence of spatial autocorrelation due to localized 'background response' distortions, e.g. Moran's I coefficient which also allows the derivation of an associated p-value for the 'no autocorrelation' null hypothesis (Moran 1950). Similar visualizations as for the HLR and Z'-factor shown in Fig. 5.1 can also be generated for the listed screening quality metrics, like the Z-factor (screening window), Moran coefficient I, or the VEP measure.

In Fig. 5.2 we show examples of useful data displays for response visualizations of individual plates. In this case both the heatmap and the separate platewise scatterplot of all data, or boxplots of summary row- and column effects of a 384-well plate clearly show previously mentioned systematic deviations of the normalized response values which will need to be analyzed further. In the section on correction of systematic errors further below we also show an illustration of the behavior of the Moran coefficient in presence of systematic response errors, and after their removal (Fig. 5.6).

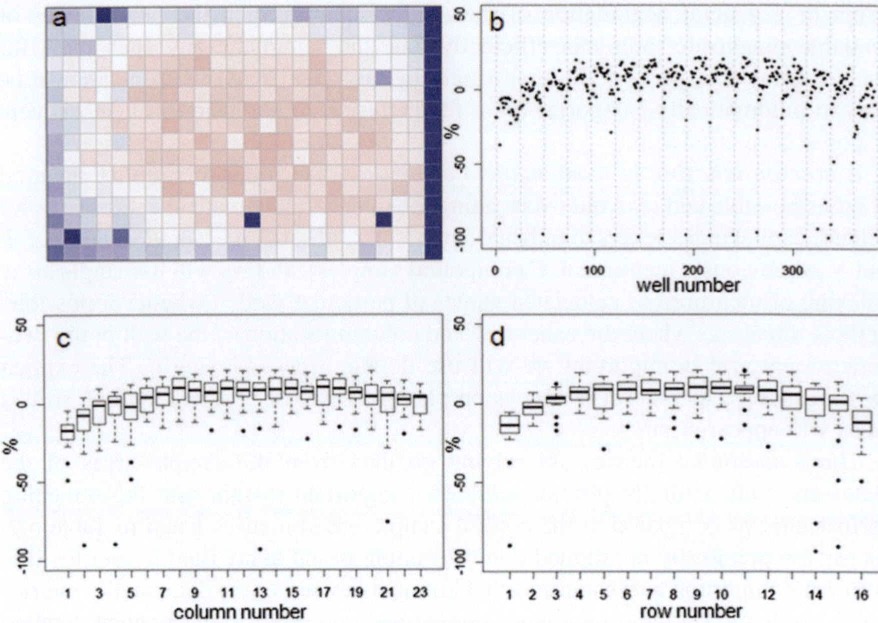

Fig. 5.2 (**a**) Example plate data heatmap showing visibly lower signal in border areas due to evaporation effects. (**b**) Scatterplot of normalized response values of all individual wells with grid lines separating different plate rows. (**c**) Boxplot of column summaries, and (**d**) Boxplot of row summaries, all showing location dependent response differences across the plate and offsets of row- and column medians from 0

It is also recommended to regularly intersperse the screening plate batches with sets of quality control plates without compounds (providing a control of liquid handling performance) and plates containing only inhibitor or activator controls and, if relevant, further reference compound samples exhibiting intermediate activity to provide additional data on assay sensitivity. 'QC plates' without compounds or other types of samples (just containing assay reagents and solvents) are also very helpful for use in correcting the responses on the compound plates in cases where it is not possible to reliably estimate the systematics errors from an individual plate or a set of sample plates themselves. This is e.g. the case for plates with large numbers of active samples and very high 'hit rates' where it is not possible to reliably determine the effective 'null' or 'background' response, as well as for other cases where a large number of wells on a plate will exhibit nonzero activity, especially also all concentration- response experiments with the previously selected active samples from the primary screens. For these situations we have two possibilities to detect and correct response values: (a) Using the mentioned QC control plates (Murie et al. 2014), and (b) using plate designs with 'uniform' or 'uniform random' placement of neutral control wells across the whole plate, instead of placing the controls in a particular column or row close to the plate edges (Zhang 2008), as is often done in practice. Such an arrangement of control well locations will allow the estimation of systematic spatial response deviations with the methods which only rely on a small number of free parameters to capture the main characteristics and magnitude of the background response errors, as e.g. polynomial response models.

5.2.7 *Integration of Diagnostic Information from Automated Equipment*

Modern liquid handlers (e.g. acoustic dispensers) often have the capability to deliver success/failure information on the liquid transfer for each processed well and this information can be automatically joined with the reader data through plate barcode and well matching procedures, and then considered for automated valid/invalid flagging of the result data, provided suitable data management processes and systems are available. Also other types of equipment may be equally able to signal failures or performance deterioration at the plate or well level which then can also be correlated with reader data and be integrated in suitable alerting and flagging mechanisms for consideration during data analysis.

5.2.8 *Plate Data Normalization*

Uncontrolled experimental factors will influence the raw assay readouts on each individual plate. This can be likened to the existence of multiplicative and possibly also additive systematic errors between individual plates p which can be represented by the measurement model $x_{ip} = \lambda_p (m + \kappa_i) + b_p$, where x_{ip} is the measured raw value in well i on plate p , m is the 'intrinsic' null-effect response value of the assay, κ_i is a sample (compound) effect, with $\kappa_i = 0$ if inactive, and λ_p and b_p are plate-, instrument- and concentration dependent gain factors and offsets, respectively. x_{ip} can then also equivalently be represented as

$$x_{ip} = m_p + \lambda_p \kappa_i + \epsilon_{ip} \approx \overline{x}_p + \gamma_{ip} + \epsilon_{ip} \tag{5.1}$$

when also including an error term $\epsilon_{ip} \sim N\left(0, \sigma_p^2\right)$ and setting $m_p = \lambda_p m + b_p$, $\gamma_{ip} = \lambda_p \kappa_i$. We make $E\left(\epsilon_{ip}\right) = \sigma_p$ explicitly depend on plate index p because this often corresponds to experimental reality, especially between different plate batches. Reagent aging and evaporation as well as time shifts in some process steps will usually lead to smooth trends in plate averages within batches, whereas e.g. the effect of cells plated at very different times may show up as a more discrete batch effect of responses. Plate-level normalization procedures use the response values of specific wells to bring the response values into a standardized numerical range which can be easily interpreted by scientists, usually a 0–100 % scale with respect to the no-effect and the 'maximal' effect obtainable with the given assay parameters. In this particular context 'plate normalization' is simply understood to adjust the *average responses* between the different plates: *Control based normalization* via specifically selected control wells, *sample based normalization* via the ensemble of all sample wells *if* the number of active wells is 'small enough'. The use of robust estimation procedures with high breakdown limits is critical for successful sample based normalization because of the likely presence of 'outliers', i.e. active compounds with larger response deviations in usually one and sometimes both possible response directions.

Normalization of Compound Screening Data In the experiment- and plate design for small molecule compound screening one usually has two types of controls to calibrate and normalize the readout signals to a 0–100 % effect scale: A set of wells (neutral or negative controls) corresponding to the zero effect level (i.e. no compound effect) and a set of wells (positive controls) corresponding to the assay's signal level for the maximal inhibitory effect, or for activation- or agonist assays, corresponding to the signal level of an effect-inducing reference compound. In the latter case the normalized %-scale has to be understood to be relative to the chosen reference which will vary between different reference compounds, if multiple are known and available, and hence has a 'less absolute' meaning compared to inhibition assays whose response scale is naturally bounded by the signal level corresponding to the complete absence of the measured biological effect. In some cases an effect-inducing reference agonist agent may not be known and then the normalization has to be done with respect to neutral controls only, i.e. using Fold- or any of the *Z*- or *R*-score normalization variants shown in Table 5.5.

Normalization of RNAi Screening Data In RNAi screens which use functional readouts any gene silencing event can in principle lead to an inhibition or to an enhancement of the observed phenotypic readout (e.g. a simple cell viability measurement). It is thus advisable to use several different types of controls to be able to assess screen quality and calibrate the response scale. In any case, siRNA controls which have no specific silencing effect on the target gene(s) need to be designed and used as negative controls, and siRNAs which target some genes having a previously known association with the biological process under study and leading to a modulation of the desired readout can be used as positive controls. In a particular screen it is desirable to employ positive controls of different strengths (e.g. weak inhibition, strong inhibition) to compare the often strongly varying observed siRNA effects to the effect sizes of known influencers of a complex biological pathway, and also to use controls exhibiting different effect directions (inhibitors, enhancers) to be able to assess the reliability of the assay response scale in *either* direction. Besides the natural use of the positive controls as screening QC indicators to assess screen stability, derive assay window quality Z'-factors, etc. the effect sizes of the positive controls need to be considered as almost 'arbitrary' reference response levels allowing to classify individual siRNA responses as 'weak' to 'very strong', which is similar to the use of different types of agonist assay reference wells in compound screening providing different relative normalization scales as described

Table 5.5 Data normalization and score calculations

	Normalization or scoring measure	Estimation expression	Comment
a	Percent of Control	$100\,\frac{x_i}{\overline{x}_{N,p}},\ 100\,\frac{x_i}{\overline{x}_{C,p}}$	Compensates for plate to plate differences in N = neutral control/negative control average (no compound effect) or C = sample average
b	(Normalized) Percent Inhibition (NPI) or Normalized Percent Activation/Percent Activity (NPA) with appropriately chosen neutral N and positive P controls	$\pm 100\,\frac{x_i - \overline{x}_{N,p}}{\overline{x}_{P,p} - \overline{x}_{N,p}},\ \pm 100\,\frac{x_i - \overline{x}_{C,p}}{\overline{x}_{P,p} - \overline{x}_{C,p}}$	N = neutral control/negative control (=no compound effect), P = positive control (=full inhibition or full activation control) Can also be based on sample (compound) average $\overline{x}_{C,p}$ of each plate instead of neutral control average $\overline{x}_{N,p}$
c	Percent Change	$\pm 100\,\frac{x_i - \overline{x}_{N,p}}{\overline{x}_{N,p}},\ \pm 100\,\frac{x_i - \overline{x}_{C,p}}{\overline{x}_{C,p}}$	(see comment for row b)
d	Fold (control based, sample based)	$\frac{x_i}{\overline{x}_{N,p}},\ \frac{x_i}{\overline{x}_{C,p}}$	Based on neutral control average $\overline{x}_{N,p}$ or sample average $\overline{x}_{C,p}$ of each plate
e	Z-score, using mean and standard deviation estimators for $\overline{x}$ and s_x, respectively	$\frac{x_i - \overline{x}_p}{s_{x,p}}$	Compensates for plate to plate differences in average value *and* variance s_x can be determined from negative controls N or from samples C: $s_x = s_N$ or $s_x = s_S$
f	R-score, similar to Z-score, but using median and mad estimators $\tilde{x}$ and $\tilde{s}_x$, respectively	$\frac{x_i - \tilde{x}_p}{\tilde{s}_{x,p}}$	Often preferred over Z-scores because of relatively frequent presence of outliers and/or asymmetries in the distributions
g	One-sided Z- or R-scores. Scoring only for samples which exhibit an effect in the desired direction	Separate Z- or R-scores for $\{i : x_i \geq \overline{x}\}$, or $\{i : x_i \leq \overline{x}\}$ with calculation of s_x for corresponding subset only	Compensates for plate to plate differences in average and in variance for asymmetric distributions

(continued)

Table 5.5 (continued)

	Normalization or scoring measure	Estimation expression	Comment
h	Generalized *R*-score calculation of platewise or batchwise corrected response values. Scoring after elimination of estimated systematic error *S* at each plate/well location.	$\frac{x_{ijp} - \tilde{x}'_{ijp}}{\tilde{s}_{x',p}}$ where $\tilde{x}'_{ijp} = \tilde{x}_p + \hat{S}_{ijp}$ is the 'true' null-effect reference response which is corrected for the estimated spatial and temporal systematic errors $\hat{S}$, and the scale $\tilde{s}_{x'} = mad\,(\boldsymbol{x} - \boldsymbol{x}')$ is calculated from the data after this location-dependent data re-centering step Any of the estimation procedures listed in Table 5.6 can be used to determine $\hat{S}_{ijp}$, provided they are at all applicable for the particular systematic patterns which are observed	Scoring after correction of systematic errors: Implicitly includes the *B*-score methodology as a special case See Table 5.6 for various modeling and estimation approaches to obtain $\hat{S}_{ijp}$
i	*B*-score (original method) *B*-score (simplified method), is a special case of the generalized *R*-score calculation (see row h)	$\frac{x_{ijp} - \left(medpolish\left(r_{ip}, c_{jp}\right) + smooth_p\left(r_i, c_j\right)\right)}{\tilde{s}_{x,p}}$ $\frac{x_{ijp} - medpolish\left(r_{ip}, c_{jp}\right)}{\tilde{s}_{x,p}}$	Most papers use the *B*-score values without the use of the edge-preserving $smooth_p$ correction term which is part of the original implementation (Brideau et al. 2003). For details of median polish procedure see Mosteller and Tukey (1977)

Instead of mean $\bar{x}$ and standard deviation s estimators the median $\tilde{x}$ and mad $\tilde{s}$ are often used as robust plug-in replacements in the normalization expressions in the last paragraph. Plate-based RNAi screening data are usually normalized to the Fold scale based on the neutral controls (see Table 5.5) and actual relative effect 'sizes' of individual single siRNA samples in either direction are either assessed by using *R*-scores, and in the case of replicates by using a *t*-statistic, or preferably the *SSMD*-statistic for unequal sample sizes and unequal variances between siRNA and neutral controls (Zhang 2011a, b). This type of *SSMD*- based normalization, and in a similar way the consideration of magnitudes of *R*-score values, already touches on the hit selection process as described further below.

Table 5.5 lists the most frequently used data normalization expressions and it can easily be seen that the values from the various expressions are centered at 0, 1, or 100 and linearly related to each other. The normalized values are proportional to the biological modulation effect κ_i, provided the assay and readout system response is linear as assumed in Eq. (5.1), and when disregarding ϵ_{ip} and systematic response errors. For nonlinear (but monotone) system responses the rank orders of the normalized activity values are maintained. With the presence of experimental random and systematic measurement errors this strict proportionality and/or concordance in rank ordering of the sample activities κ_i within and across plates is of course no longer guaranteed. Especially in cell-based assay system the assay response variability can be relatively large and also the relative rank ordering of sample activities from primary screening and the follow-up IC_{50} potency determinations can differ considerably, mostly due to concentration errors in the different experiments (Gubler et al. 2013).

Besides these *average plate effects* ('*gain factor differences*') other response differences at the well, row or column level occur frequently in actual experimental practice. They often occur in a similar fashion on multiple plates within a batch and it is important to take these *location- and batch dependent errors* into account to be able to identify hits in an efficient way. We deal with the detection and correction of these effects separately in the next section, but of course they belong to the same general topic of assay response normalization.

If we observe plate-to-plate or batch-to-batch variations of σ_p then the use of *Z*- or *R*-score-based normalization is advised to allow the application of a meaningful experimentwise hit selection threshold for the entire screen. If systematic response errors are detected on the plates then the final scoring for hit selection needs to be done with the *corrected* activity values, otherwise the hit list will be clearly biased. The scoring expressions are based on estimated center $\tilde{x}$ and scale $\tilde{s}$ values and correspond to a *t*-statistic, but the actual null-distribution of the inactive samples in a primary screen will usually contain a very large number of values, so that we can assume $f_0 \sim N(0, 1)$ for the *R*-scores of this null-effect subset after correction. We will revisit this aspect in the section on hit-selection strategies.

5.2.9 Well Data Normalization, Detection and Correction of Systematic Errors in Activity Data

In this section we deal with the detection, estimation and correction of and assay response artifacts. The series of different process steps in assay execution and the different types of equipment and environmental factors in a typical plate based screen often lead to assay response differences across the plate surfaces (i.e. non-random *location effects*), both as smooth trends (e.g. due to time drifts for measurements of different wells in a plate), bowl shapes (often due to environmental factors like evaporation, temperature gradients), or step-like discrete striping patterns (most often due to liquid handling equipment imperfections (dispensing head, needle or pin) leading to consistently lower or higher than average readings, and combinations of some or all of these types of effects. Also gradients in the step-like artifacts can sometimes be observed due to the time ordering of dispensing steps. Often, but not always, these effects are rather similar on a whole series of plates within a measurement batch which obviously will help in estimation and subsequent correction procedures. Individual well patterns can obviously only be estimated if they repeat on all or a subset (batch) of plates, otherwise they are confounded with the effect of the individual compound samples. Automatic partitioning of the sets of readouts to reflect common patterns and identification of those respective different systematic patterns is an important aspect for efficient and effective response correction steps. Temporal changes of non-random response patterns are related to batch-wise assay execution, reagent aging effects, detection sensitivity changes or changes in environmental factors and may appear gradual or sudden.

It is obvious that systematic spatial or temporal response artifacts will introduce bias and negatively affect the effectiveness of hit finding especially for samples with weak and moderate size effects and will influence the respective false decision rates in hit selection. Such effects should thus be corrected before attempting hit selection or using these data for other calculation steps. Especially when considering fragment based screens with low-molecular samples of relatively low potency, natural product (extract) screens with low amounts of particular active ingredients, or RNA interference screens where small to moderate size effects can be of interest (corresponding to full knockdown of a gene with a small effect, or partial knockdown of a gene with strong effects), or if one simply is interested in detecting active samples in the whole range of statistically significant modulating effects then these response correction methods become crucial to allow optimized and meaningful analyses. The positive influence of the response correction on the hit confirmation rate, the reproducibility of the activity in follow-up screens or secondary assays can be clearly demonstrated (Wu et al. 2008).

An important prerequisite for successful estimation of response corrections using the actual screening sample data is the assumption that the majority of these samples are inactive and that active samples are randomly placed on the various plates. For screens with high rates of non-zero response wells it is advised to place neutral control wells (showing no effect on the assay readout) spread 'uniformly' across the plates, and not, as is often the case, in the first or last columns of the plates, and use them to check for the occurrence, estimation and correction of systematic response errors. Sometimes the plate layout designs which can be produced by compound management and logistics groups in an automated way are limited due to restrictions of the robotic liquid handling equipment, but in order to produce reliable screening results an optimal design of control well placement is important and liquid handling procedures should be adapted to be able to produce assay plates in such a way. Such specially designed plates with a suitable spatial spread and location of control wells can be used to derive 'smooth' (e.g. polynomial) average response models for each plate (or for a set of plates) which do not rely on the assumption that the majority of the test samples are inactive. For example in RNAi screening many of the samples or sample pools have usually some effect in the assay, leading to a large range of responses and related problems to efficiently use correction methods which rely on estimations of the null response based on the sample wells themselves. The response level of 'truly inactive' samples is difficult to determine in this case and consequently an efficient plate designs with well-chosen controls in the described sense or the use of interspersed control plates for error correction in a measurement batch can become important (Zhang 2008). Cell-based screens, including RNAi screens, often need additional and prolonged incubation periods which often exacerbate assay noise, response bias and artifacts in the border regions of the plates.

In actual practice it also happens that structurally and bioactivity-wise similar compounds are placed near each other because of the way the stored compound master plates were historically constructed (often from groups of similar compounds delivered to the central compound archives in a batch) even for 'random' HTS libraries, or simply due to the makeup of plates in focused libraries which exhibit larger rates of activity on selected target classes (e.g. enzyme inhibitors). Modern compound management setups today allow a more flexible creation of screening plates, but the presence of spatial clusters of activity or of known subsets of very likely active samples of on particular plates need to be considered for the decision to include or exclude selected plate subsets from the standard way of background response estimation and related processing.

Well-level assay response normalization and correction of the spatial and temporal patterns is in essence just a 'more sophisticated' form of the sample based normalization mentioned in the previous paragraph. Because of the expected presence of at least some 'active' wells (i.e. outliers for the purpose of background response estimation) it is highly advisable to use *robust* (*outlier resistant*) *estimation methods* when relying on the actual screening sample wells to derive the response models. The robustness breakdown limits for different methods are of course quite variable and need to be considered separately for each. The breakdown bounds for the median polish procedure were elaborated by Hoaglin et al. (1983).

As mentioned in the process monitoring section graphical displays are important to visually detect and often also quickly diagnose the potential sources of the error patterns. Also the visualizations of the error patterns and of the subsequently corrected data, including suitable graphics of the corresponding (now improved) quality metrics is an important practical element of screening quality assurance (Brideau et al. 2003; Coma et al. 2009b).

It is also important to note that data correction steps should only be applied if there is evidence for actual systematic errors, otherwise their application can result in variance bias, albeit with a magnitude strongly dependent on the actual correction method used. Such variance bias can have an influence on the hit identification step because the corresponding activity threshold choices can be affected in an unfavorable fashion. Suitably constructed quality metrics which are based e.g. on Moran's I spatial autocorrelation coefficient (see Table 5.4 item d) can be used to determine whether a systematic error is present and whether corresponding response correction should be applied or not. 'Suitably' means that the weight matrix needs to minimally cover the neighborhood region which is expected to exhibit strong correlations when systematic errors of a particular type are present, e.g. by using a neighborhood range $\delta = 2$ around grid point $\{i_0, j_0\}$ for setting the weights $w_{ij} = 1$ for all $\{i, j : (0 \leq |i - i_0| \leq \delta) \wedge (0 \leq |j - j_0| \leq \delta)\}$ with $w_{i_0 j_0} = 0$ and $w_{ij} = 0$ otherwise for the situations where discrete response 'striping' effects in every second row or column can occur due to some liquid handling errors besides the possible smoother spatial responds trends across the plate surface. Different δ extents in row and column directions or use of different weighting functions altogether may be more optimal for other types of expected patterns.

We have separated the assay response normalization into *plate-level normalization* as outlined in the previous section, including the calculation of various assay quality metrics according to Tables 5.1 and 5.2, and the possible subsequent *row, col and well-level effect response adjustment* procedures. In essence the latter can be considered as a *location-dependent sample-based data normalization step*. In Table 5.6 we show several such modeling and correction methods for location dependent systematic errors of well-level data in plate based screening which have been developed and described within the past 10–15 years (Heyse 2002; Heuer et al. 2002; Brideau et al. 2003; Kevorkov and Makarenkov 2005; Gubler 2006; Malo et al. 2006; Makarenkov et al. 2007; Birmingham et al. 2009; Bushway et al. 2010; Dragiev et al. 2011; Zhang 2011a; Mangat et al. 2014)

In Fig. 5.2 we have already seen data containing typical systematic response errors. As an illustration of model performance obtained with some of the approaches listed in Table 5.6 we show the same data in Fig. 5.3a with three different error model representations in Fig. 5.3b, c and the resulting corrected data, after applying the *loess* error model based correction in Fig. 5.3e. The corresponding row-wise boxplot of the corrected data in Fig. 5.3e can be compared to uncorrected case in Fig. 5.2d and the resulting smaller variances as well as better centering on zero are immediately evident.

For further illustration of various diagnostics and response modeling methods we will here use a simulated screening plate data of limited size (50 384-well plates) with normalized percent inhibition data scaled between 0 (null effect) and −100 (full inhibition) exhibiting several features which are typically found in real HTS data sets: Edge effects due to evaporation, response trends due to temperature gradients,

Table 5.6 Estimation methods for modeling of systematic plate response errors S_{ijp}

	Model category	Response model Estimation method	Comment
a	Spatial array response polishing	$x_{ijp} = M_p + R_{ip} + C_{jp} + \epsilon_{ijp}$ $\hat{S}_{ijp} = stats::medpolish_p\left(r_{ip}, c_{jp}\right)$ $\hat{S}_{ijp} = stats::medpolish_p\left(r_{ip}, c_{jp}\right) + smooth_p\left(r_i, c_j\right)$ $\hat{S}_{ijp} = trimmed.meanpolish_p\left(r_{ip}, c_{jp}, \alpha\right)$	Iterative plate response polishing (Brideau et al. 2003; H. Gubler, 2014, unpublished work)
b	Linear models, Polynomial models	$x_{ijp} = M_p + \left(R_{ip} + C_{jp}\right)^{1..n} + \epsilon_{ijp}$ $\hat{S}_{ijp} = MASS::rlm\left(\sim 1 + r_{ip} * c_{jp}\right)$ $\hat{S}_{ijp} = MASS::rlm\left(\sim poly_p\left(r_{ip}, c_{jp}, n\right)\right)$	Multiple Linear regression models (incl. possible higher order and interaction terms), nth degree polynomial models, ANOVA models with interaction terms (Kevorkov and Makarenkov 2005; Dragiev et al. 2011)
c	Linear mixed effects (LME) models	$x_{ijp} = M_p + \left(R_{ip} + C_{jp}\right)^{1..n} + \epsilon_{ijp}$ $\hat{S}_{ijp} = lme4::lmer_p\left(r_{ip}, c_{jp}, n\right),$ $\hat{S}_{ijp} = robustlmm::rlmer_p\left(r_{ip}, c_{jp}, n\right)$ e.g. using a simple 1st order LME model ~1 + r*c + (1|r) + (1|c), which allows more flexibility than *medpolish* (which does not incorporate any interaction terms)	Mixed linear fixed and random effects model (H. Gubler, 2014, unpublished work)
d	Nonparametric (NP) smoothing, local polynomial regression model	$x_{ijp} = M_p + f^{smooth}\left(R_{ip}, C_{jp}\right) + \epsilon_{ijp}$ $\hat{S}_{ijp} = f_p^{NP}\left(r_{ip}, c_{jp}, \alpha, deg\right)$ e.g. $\hat{S}_{ijp} = stats::loess_p\left(r_{ip}, c_{jp}, \alpha, deg\right)$ or $\hat{S}_{ijp} = locfit::locfit_p\left(r_{ip}, c_{jp}, \alpha, deg\right)$	Robust local regression models (Gubler 2006; Zhang 2011a)
e	Nonparametric Array Filter	$x_{ijp} = M_p + f^{filter}\left(R_{ip}, C_{jp}\right) + \epsilon_{ijp}$ $\hat{S}_{ijp} = f_p^{HMF}\left(r_{ip}, c_{jp}, HMF\right)$ A 1st pass HM filter step can also be combined with e.g. a 2nd polynomial *rlm*, local polynomial *loess* regression, or even a principal pattern modeling step	Hybrid Median Filters (HMF) (Bushway et al. 2010)

(continued)

Table 5.6 (continued)

	Model category	Response model Estimation method	Comment
f	Process specific explicit modeling of combined *smooth* response error and liquid handling (*lh*) error (which typically occurs as a 'striping' pattern)	$x_{ijp} = M_p + f_p^{smooth}(R_{ip}, C_{jp}) + f_p^{lh}(tip(R_i, C_j), order(R_i, C_j)) + \epsilon_{ijp}$ $\hat{S}_{ijp} = f_p^{smooth}(r_{ip}, c_{jp}) + f_p^{lh}(tip_{ij}, order_{ij})$ e.g. with $f^{smooth} = MASS::rlm(\sim poly(r_{ip}, c_{jp}, \ldots))$ or with $f^{smooth} = stats::loess(r_{ip}, c_{jp}, \ldots)$ and the separate regression term $f^{lh}(tip, order) = rlm(\ldots)$ considering the appropriate subsets of wells and their respective (dispensing-) *order* for *tip* specific sub-terms of f^{lh}. Estimation of f^{lh} is usually based on considering data from multiple plates p for reliable parameter estimation of a linear or polynomial function, possibly with a cutoff at some upper *order* value	Model relies on availability of sufficient data for simultaneous estimation of f^{sm} and f^{lh} parameters, as well as detailed knowledge of plate processing steps. Robust regression methods should preferably be used H. Gubler, 2014, unpublished work
g	Principal pattern model	$x_{ijp} = M_p + f(W_{ijp}) + \epsilon_{ijp}$ $\hat{S}_{ijp} = svd^{-1}(base::svd(x_{ijp}), k) = \boldsymbol{U}_{1..k}\boldsymbol{D}_{1..k}\boldsymbol{V}'_{1..k}$ With back-transformation svd^{-1} (i.e. signal reconstruction) using only the leading $k \ll n$ of the $n = \min(r_{max}c_{max}, p_{max})$ components The *svd* (*pca*, *eof*) method used should preferably be an outlier resistant version(as e.g. *robpca* from R package *rrcov*, or equivalent). Matrices $\boldsymbol{U}$, $\boldsymbol{D}$ and $\boldsymbol{V}$ are the left singular vectors, singular values and right singular vectors of $\boldsymbol{X}$	Principal Pattern representations: Can use singular value decomposition (SVD), principal component analysis (PCA), empirical orthogonal function (EOF) calculation methods. Cannot be used for single plates (Gubler 2006)
h	Principal pattern model	$x_{ijp} = M_p + f(W_{ijp}) + \epsilon_{ijp}$ $\hat{S}_{ijp} = ICA^{-1}(ICA(x_{ijp}), k) = \boldsymbol{A}_{1..k}\boldsymbol{S}_{1..k}$ with back-transformation ICA^{-1} (i.e. signal reconstruction) using only the leading $k \ll n$ components, where $\boldsymbol{A}$ and $\boldsymbol{S}$ are the independent component and mixing matrices for the original responses $\boldsymbol{X}$. The *ICA* method is e.g. implemented in R function *fastICA::fastICA*, but is also available for other languages and systems	Principal Pattern representations: Independent Component Analysis, ICA. Cannot be used for single plates (Gubler 2006; Hyvarinen et al. 2001)

i	Two-step IQM (interquartile mean) normalization	$x_{ijp} = M_p + W_{ijp} + \epsilon_{ijp}$ Plate-wise interquartile mean normalization $x'_{ip} = IQM_p\left(x_{ip}\right)$ *followed* by well-wise interquartile mean normalization $x''_{ip} = IQM_i\left(x'_{ip}\right)$ $\hat{S}_{ijp} = \hat{W}_{ijp} = \hat{x}''_{ip}$	2nd step may need to be done on a per batch basis if distinctly different error patterns are present on subsets of plates, but 2nd step is optional (Mangat et al. 2014)
j	Spatial polish and well normalization (SPAWN)	$x_{ijp} = M_p + R_{ip} + C_{jp} + W_{ijp} + \epsilon_{ijp}$ Plate-wise B-score $B_{ijp} = \frac{x_{ijp} - stats::medpolish\left(r_{ip}, c_{jp}\right)}{\tilde{s}_{x,p}}$ or generalized R-score calculation $R_{ijp} = \frac{x_{ijp} - x'_{ijp}}{\tilde{s}_{x',p}}$ with $x'_{ijp} = \tilde{x}_p + \hat{S}_{ijp}$ (see also Table 5.5 h and i) using any of the methods in the previous table rows, B-scores being a special case of the generalized R-scores, *followed* by calculation of a well-average or of a linear model of the scores $\hat{W}_{ij}(p) = rlm\left(B_{ijp} \sim p\right)$, subsequent subtraction of the well-effects $\hat{B}'_{ijp} = B_{ijp} - \hat{W}_{ij}(p)$ and renewed scoring calculation $B'_{ijp}/\tilde{s}_{B',p}$, or in the equivalent way with the R_{ijp} values.	Two-step plate polish and well normalization of time-ordered plates is illustrated here for B-scores (~median polish, Makarenkov et al. 2007; Murie et al. 2013), but can be handled in a similar way for all other $\hat{S}_{ijp}$ estimation methods when using generalized R-scores 2nd step may need to be done on a per batch basis if distinctly different error patterns are present on subsets of plates

Modeling approaches differ by underlying response model structure with M (mean effect), R (row effects), C (column effects), W (well effects) and their assumed interactions. Modeling approaches also need to be selected for their ability to represent certain types of patterns. Liquid handling stripes can e.g. only be adequately represented by models f, g, h (and only in a limited way with model a which often leads to response bias in some wells along rows or columns if only partial striping occurs). For conciseness of notation we are using R package and function names, when available, to represent the main calculation steps for the estimation of the systematic error patterns. Please refer to the literature references for details of modeling functions which are not prefixed with R package names

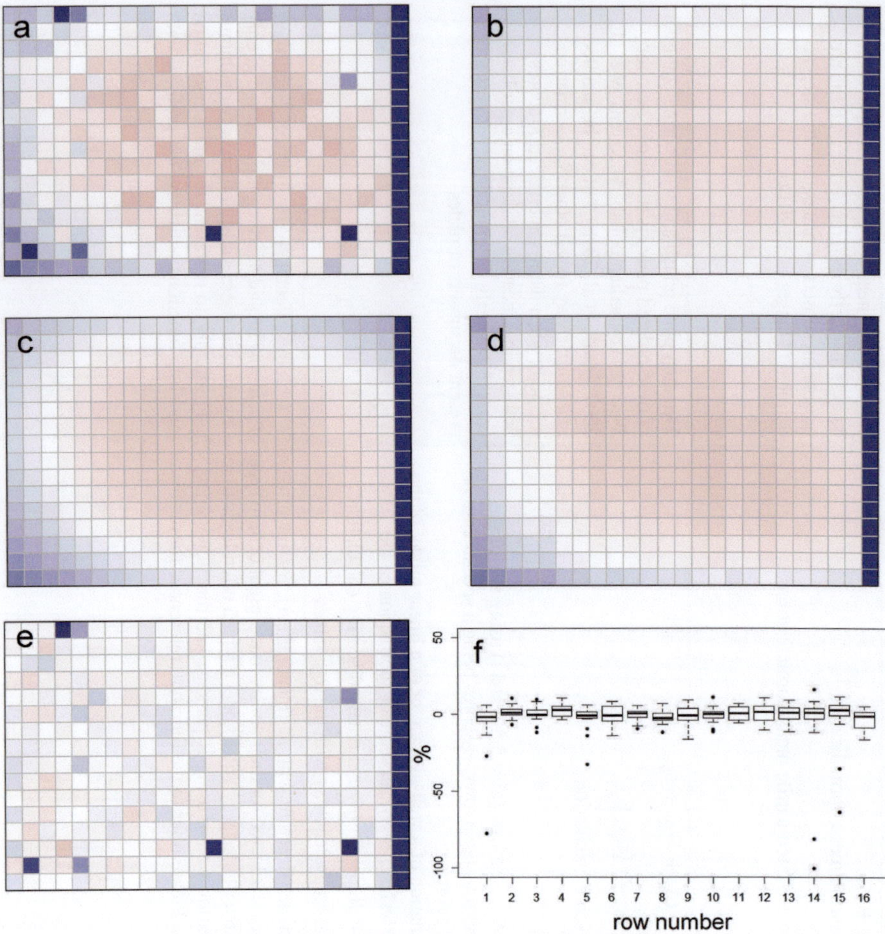

Fig. 5.3 (**a**) Same normalized plate data as in Fig. 5.2a. (**b**) Median polish representation of systematic error, which is inadequate in this case because a R*C interaction term is not included. (**c**) *loess* model representation of systematic error using 2nd degree polynomial local regression. (**d**) Robust mixed linear effects model representation of systematic error using *rlmer* function with R*C interaction terms. (**e**) Corrected data after *loess* error pattern subtraction. (**f**) Boxplot of row summaries corresponding to corrected data of Fig. 5.3e (compare with Fig. 5.2d, without correction)

liquid handling response stripes which vanish after a series of dispensing steps in interleaved row-wise dispensing, single dispensing needle malfunction, plate batch-effects with different dominant error patterns between batches, assay response noise $\sim N\left(0, \sigma^2\right)$ with $\sigma = 5$, and a response distribution of the randomly placed hits with an overall hit rate of 5 % which is approximated as ~*Gamma*(1, 0.025), leading to a median inhibition of hits around −30 %, i.e. obtaining a smaller number of hits with strong inhibition and a larger number with moderate and small values which is the usual situation screening.

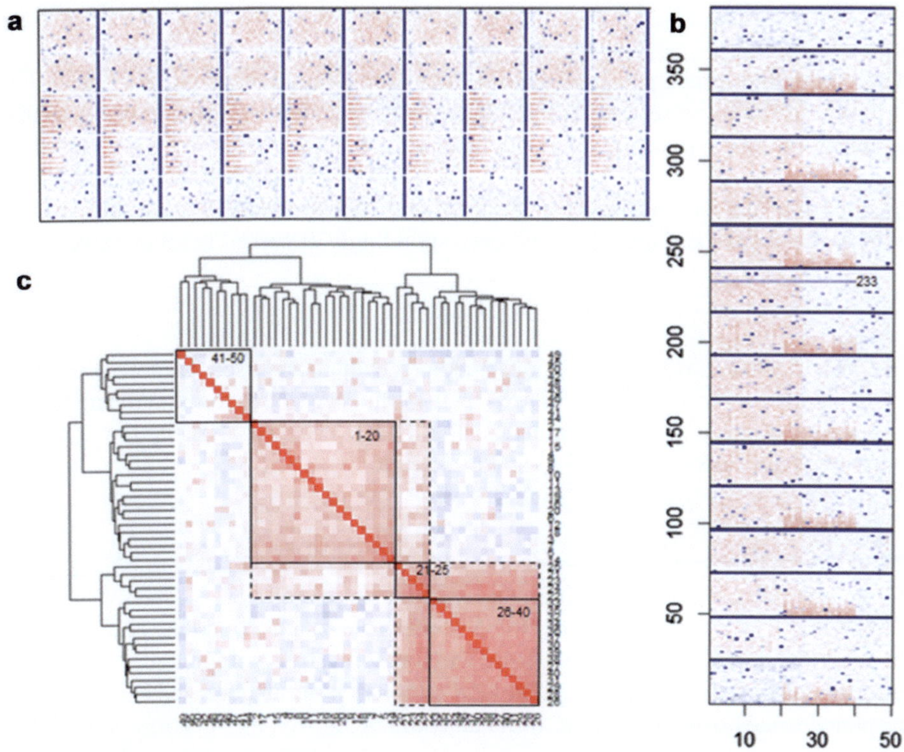

Fig. 5.4 (**a**) Heatmaps of simulated data set of 50 plates. Different types of systematic errors are visible, see text (**b**) Assay-map of the same data set, in this arrangement and ordering of the well-index allowing a quick identification of row- and individual well-effects (e.g. response error for well 233 in a subset of plates). (**c**) Plate-by-plate correlation matrix of well values for the complete data set, allowing the identification and grouping of plates with similar error patterns either visually or using hierarchical clustering

This simulated data set is represented as a heat map series in Fig. 5.4a and as the corresponding 'assay map' (Gubler 2006) in Fig. 5.4b. The assay map is a single heat map of the complete data set where the 2D row- and col- location is transformed into a linear well-index and data displayed as a contiguous (well-index, plate measurement sequence index) map. Response errors affecting particular wells or well-series in a set of sequentially processed plates are then immediately visible, as e.g. in our case a consistently lower response of well 233 in plates 1 to 40 due to a "defective" dispensing tip. Other batch effects are visible in both types of heatmaps.

Our principal goal in performing HT screens is the identification of interesting compounds which perturb the response of the biological test system. As per Eq. (5.1) and extending the indexes to two dimensions i,j we can model the observed response x_{ijp} in row i, column j and plate p as $x_{ijp} = \overline{x}_p + \gamma_{ijp} + \epsilon_{ijp}$, allowing for a simple estimation of the relative compound effect size as

$$\hat{\gamma}_{ijp} = x_{ijp} - \overline{x}_p. \tag{5.2}$$

The compound effect γ_{ijp} is of course fully confounded with ϵ_{ijp} for the $n = 1$ primary screening case. Separate estimation of γ_{ijp} and ϵ_{ijp} can only be made when measuring replicates, as is often done in RNAi screens, or when running confirmation screens with replicates. As previously discussed, in actual screening practice we almost always have systematic response errors S_{ijp} present in the experimental data, and thus now find correspondingly disturbed readout values

$$x_{ijp} = \overline{x}_p + \gamma'_{ijp} + S_{ijp} + \epsilon_{ijp}. \tag{5.3}$$

In Table 5.6 we have listed various methods which are used to determine the systematic error terms S and focus now on the influence of this term on the apparent compound effect γ'_{ijp} as determined in the data normalization step. For a given measured data value x_{ijp} Eqs. (5.1) and (5.3) give

$$\gamma'_{ijp} = \gamma_{ijp} - S_{ijp} \tag{5.4}$$

for an otherwise unchanged null effect readout level $\overline{x}_p$ and observed activity value x_{ijp}. Without a suitable estimate for S_{ijp} we have again complete confounding of compound effect, systematic error and random error components. By using information based on spatial and temporal correlation of the observed experimental response values from a series of wells an estimate $\hat{S}_{ijp}$ for the location- and plate-sequence related systematic error can be obtained. Using this estimated value the term $\left(\overline{x}_p + \hat{S}_{ijp}\right)$ is now in essence the *shifted plate- and location-dependent actual null-effect reference level* which needs to be used in the different normalization expression if we want to consider the influence of $\hat{S}_{ijp}$ on the relative compound responses on the % scale. Equation (3) describes the behavior of assay response values around the average null-effect (neutral control) response $\overline{x}_p = \overline{x}_{N,p}$, but when e.g. calculating normalized percent inhibition values a 'similar' systematic error $\hat{S}'_{ijp}$ is also assumed to be present at the level of the full effect controls $\overline{x}_{P,p}$. This value which participates in defining the %-effect scaling at location *i,j*, cannot be directly estimated from the data. While it is usually possible to determine $\hat{S}_{ijp}$ near the $\overline{x}_{N,p}$ response level quite reliably because in small molecule compound HTS most samples are inactive, and we thus possess a lot of data for its estimation, or we are able to base the estimation on designed control well positions and their responses, we can only make *assumptions* about the influence of the systematic error factors on response values around the $\overline{x}_{P,p}$ level, unless detailed experiments about this aspect would be made. A purely additive response shift as per Eq. (5.3) for *all* possible magnitudes of the assay responses is an unreasonable assumption, especially if all values derived from a particular readout technology are limited to values ≥ 0. In the case of an inhibition assay the positive control $\overline{x}_P$ corresponds usually to small readout signal values $\left(\overline{x}_{P,p} < \overline{x}_{N,p}\right)$ and we can either assume that $S'_{ijp} = 0$ at small signal levels, or we can assume a fractional reduction of the amplitude of the systematic error which is scaled linearly with the corresponding "High/Low" ratio as

$$\hat{S}'_{ijp} = \hat{S}_{ijp}\frac{\overline{x}_{P,p}}{\overline{x}_{N,p}}, \tag{5.5}$$

i.e. a *multiplicative influence of the systematic errors on the assay signal* magnitude. A typical normalization expression (e.g. for *NPI*) with explicit consideration of the systematic error contributions at the $\overline{x}_N$ and—if relevant—at the $\overline{x}_P$ level forms the basis for integrating the response value correction into a reformulated expression for *corrected NPI* values:

$$\begin{aligned} NPI_{i,p\ corrected} &= 100\frac{x_i - \left(\overline{x}_{N,p} + \hat{S}_{i,p}\right)}{\left(\overline{x}_{P,p} + \hat{S}'_{i,p}\right) - \left(\overline{x}_{N,p} + \hat{S}_{i,p}\right)} \\ &= 100\frac{x_i - \left(\overline{x}_{N,p} + \hat{S}_{i,p}\right)}{\overline{x}_{P,p}\left(1 + \hat{S}_{i,p}/\overline{x}_{N,p}\right) - \left(\overline{x}_{N,p} + \hat{S}_{i,p}\right)}, \end{aligned} \tag{5.6}$$

noting that this is a now a *plate- and well-location dependent normalization of the assay response data* x_i which *eliminates* the systematic batch-, plate- and well-level response distortions. As mentioned previously the corresponding modification for other simple normalization expressions can be derived in an analogous way. The difference between varying assumptions for the behavior of S' with changing absolute response leads to slightly different values for $NPI_{corrected}$, the difference between the two being larger for smaller High/Low ratios of the assay controls (dynamic range of the assay response). Using Eq. (5.5) and the *NPI* expression from Table 5.5 we obtain:

$$NPI_{i,p\ corrected} = \frac{NPI_{i,p}\left(\overline{x}_{N,p} - \overline{x}_{P,p}\right) + 100\hat{S}_{i,p}}{\left(\overline{x}_{N,p} - \overline{x}_{P,p}\right)\left(1 + \hat{S}_{i,p}/\overline{x}_{N,p}\right)}, \tag{5.7}$$

or for the $S'_{i,p} = 0$ case:

$$NPI_{i,p\ corrected} = \frac{NPI_{i,p}\left(\overline{x}_{N,p} - \overline{x}_{P,p}\right) + 100\hat{S}_{i,p}}{\overline{x}_{N,p} + \hat{S}_{i,p} - \overline{x}_{P,p}}. \tag{5.8}$$

These are useful relationships because in practice the simple $NPI_{i,p}$ which are based on plate-level controls without consideration of location-dependent effects are already available from the preceding data analysis step which converts the 'arbitrary' raw data values to a common % or fold scale. The *NPI* values are then used for various diagnostic graphics (heat maps, scatterplots, boxplots, etc.) and provide a basis for comparing the uncorrected and the subsequently corrected response values in a quick overview fashion (and to help in the visual detection of systematic response artifacts). The described $\hat{S}'_p$ correction ambiguity does of course not have any influence on the *scoring* methods which only rely on data centering with respect to the null-effect levels $\bar{x}_N$, or better, with respect to the estimated *actual* null-effect response levels $\left(\bar{x}_N + \hat{S}_{ij}\right)$.

It is clear that the modeling approaches have to be chosen according to the actual types of patterns occurring in the data set, hence visual inspection, possible partitioning of data sets, and choice of an optimal model have to go hand in hand. Automatic partitioning of the plate data set can be done very efficiently by clustering the $p \times p$ correlation matrix of the pairwise inter-plate well values for all p plates. Response values for the samples or for samples together with neutral controls can be included in the correlations. Larger correlation values will be found for the plates which exhibit similar spatial response distortions, while we will have $E\left(corr\left(\boldsymbol{x}_k, \boldsymbol{x}_l\right)\right) = 0$ for independent random distributions of the responses on plates k and l. The correlation matrix for our example data set with added (hierarchical) clustering information is shown in Fig. 5.4c. The main 4 sub-clusters can clearly be associated with the 4 discernable groups of plates with distinct error patterns (edge effects: 1- 20, edge effects + striping: 21–25, striping only 26–40, no systematic pattern: 41–50), in complete agreement with the structure of the data set.

Another method for the grouping of 'similar' plates which can be used for the purpose of partitioning is changepoint analysis (Hinkley 1970; Horvath 1993; Barry and Hartigan 1993) in a suitable set of QC indicators which are sensitive to specific differences in the systematic error pattern in the ordered sequence of plates. The first two components $\boldsymbol{U}_i\boldsymbol{D}_i,\ i = 1..2$ of a robust principal pattern analysis using the *robpca* method (Hubert et al. 2005) for the entire simulated plate data set are shown in Fig. 5.5a, b.

The two represented principal patterns clearly correspond to the main visible systematic error features of particular subsets of the plates with evaporation edge effects, liquid handling stripes in alternate rows which taper off at higher column numbers, as well as the single-well pipetting failure at ($row = 10$, $col = 17$). The corresponding principal component loadings $\boldsymbol{V}'_i$ are shown in Fig 5.5c, d, respectively. Now we can use these *PCA* loadings, or similarly, the *ICA* mixture weights, from such an exploratory diagnostic analysis of the complete data set for changepoint (i.e. pattern- or plate batch-boundary) detection as indicated in these figures. The red horizontal lines indicate the extent of the data series with common mean and variance properties according to an analysis using the *PELT* method (Killick et al. 2012) which is also implemented in the R *changepoint* package (Killick and Eckley 2014). The superset of the changepoint locations from these 2 principal component loadings is in complete agreement with the pattern- and plate 'batch' boundaries ($k = 20, 25, 40$) which we had identified before and correspond to the properties of the generated data set. For plates 41 to 50 the average contribution from either of these two pattern components is close to 0, as indicated by the position of the red (mean) lines. The information from such an overview analysis, jointly together with the indicators of the significance of the

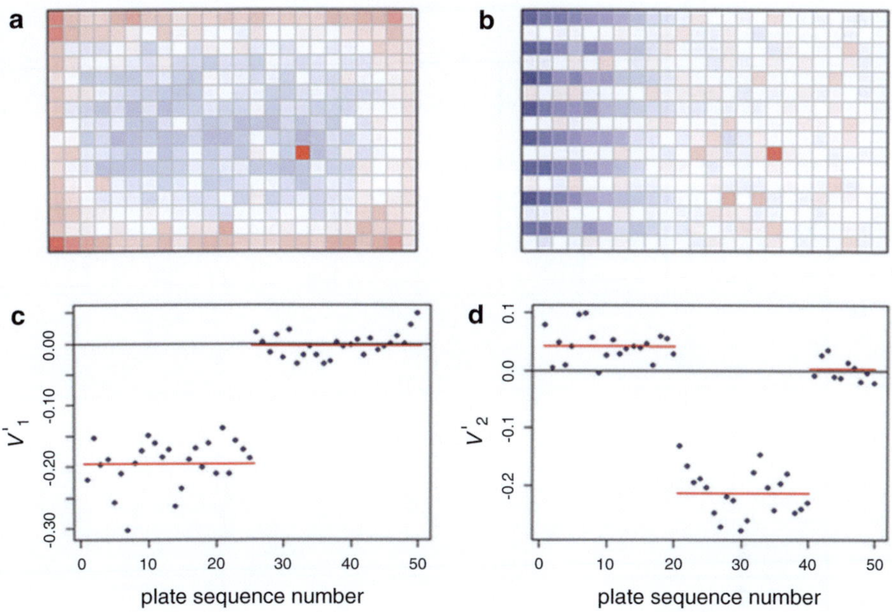

Fig. 5.5 (**a**) First principal pattern, and (**b**) second principal pattern extracted from the example plate data set with a robust principal component analysis, reflecting the dominant error patterns in the data: evaporation edge effects, tapered striping in alternate rows and defective pipetting tip in well 233. (**c**) loading factors of first principal pattern, and (**d**) loading factors for 2nd principal pattern, both with overlaid changepoint detection segments (*red lines*)

Moran autocorrelation coefficients *I* from each plate allow us to quickly judge which plate sets we should include or exclude from any pattern modeling steps.

The principal pattern methods can be used both as quick *diagnostic* pattern detection tool, as just described, and of course also as basis for the *data correction* step. In Fig. 5.6 we demonstrate the application of this 2-component robust principal component model to the whole data set, with normalized data shown in a, error pattern in b, and resulting corrected data set in c. We also illustrate the behavior of the spatial autocorrelation coefficient before and after data correction and clearly see that the autocorrelated response contributions are removed to such an extent that Moran's *I* values (Fig. 5.6d, e) are significantly reduced and most of the corresponding *p*-values (Fig. 5.6f, g) are now below the $\alpha = 0.05$ level which is indicated by the dashed line.

Based on these 'cleaned' screening plate data sets where now all or most of the discernable systematic response errors have been eliminated using one of the described error-modeling and correction approaches the further screening data analysis steps are either (a) hit sample selection in primary screening, or (b) calculation of further quantities of interest based on data from a whole series of individual wells originating from one or several plates (e.g. concentration-response curve characterizations, compound potency determinations).

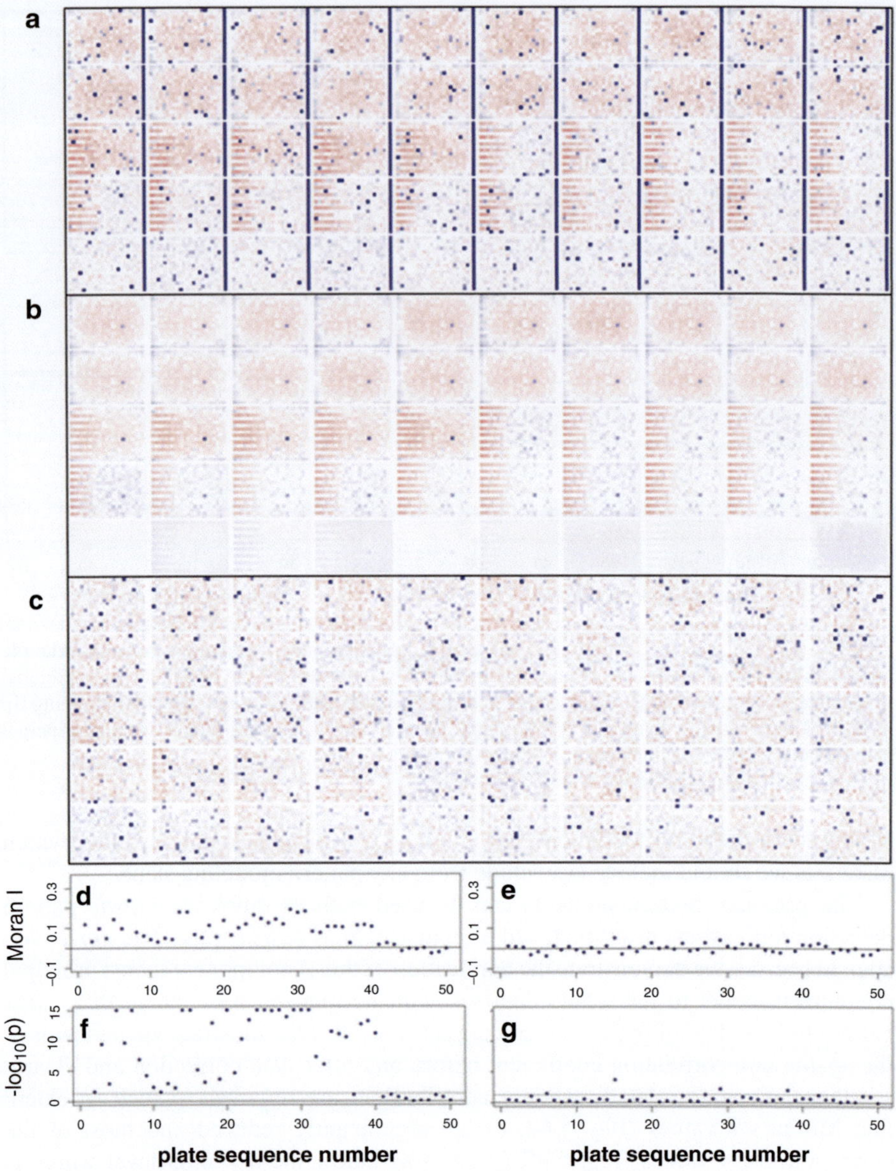

Fig. 5.6 (**a**) Example plate set, normalized data. (**b**) Systematic errors of plate data as estimated with a two-component principal pattern model. (**c**) Corrected plate data set after elimination of estimated systematic error values. (**d**) Moran autocorrelation coefficient for example plate data set before correction, and (**e**) after correction. (**f**) *p*-values of Moran coefficients before correction, and (**g**) their *p*-values after correction

The ultimate optimality criterion for the error modeling and data correction steps is of course a maximized confirmation rate, i.e. a minimized rate of false discoveries in the list of selected hits, possibly under the constraint of an upper limit of the number of samples which can be selected and progressed into a next screening stage because of economic and resource constraints, and as small as possible false negative rate. Minimizing the false discovery rate in the primary screening hit list is expected to result in maximized confirmation rate of the putative hits in follow-up experiment, which is either a screen measuring the activity of the hits with independent replicates or a screen with concentration dependent measurements (Zhang et al. 2000; Wu and Liu 2008; Prummer 2012).

5.2.10 Sample Activity Ranking, Scoring and Tests of Significance

Since the hit identification from a single-concentration primary HTS with many expected actives is a large scale multiple hypothesis testing problem we need to consider practical methods for controlling the false-positive rate. The basic goal is to identify as many significant actives as possible which are not part of the null distribution, while at the same time incurring only a low proportion of false positives. The 'false discovery rate' statistic introduced two decades ago (Benjamini and Hochberg 1995) and related follow-up work by many authors (Efron et al. 2001; Efron 2004, 2010; Storey 2002; Storey and Tibshirani 2003; Dalmasso et al. 2005; Strimmer 2008a) has led to further development of the methodology and to its adaptation to the practical needs of large scale biological experimentation in genomics and screening. It is an ideal statistic for this purpose, because a straightforward normal p-value based threshold is not very informative in terms of presence of the interesting non-null component, because it is only related to the proportion of the samples with null-activity above this threshold. Methods like the q-value analysis and others (Storey and Tibshirani 2003; Strimmer 2008a) were shown to maintain high statistical power in many situations of biological experimentation and convey the necessary information about the proportion of the significant actives which are statistical false positives. These methods are very useful in practice because the 'success rate' of the subsequent follow-up verification of the detected activity from the primary screen is directly related to the q-value threshold, and the follow-up experimental confirmation and verification work (and related 'costs') can then be optimized in this respect.

In the screening data analysis context the mentioned false discovery rate FDR (proportion of false positives from the potentially interesting number of samples above the selection threshold, $FDR = FP/(TP + FP)$) and the false negative rate FNR (proportion of missed actives from the total number of real actives, $FNR = FN/(TP + FN)$) are the most informative and useful criteria for decision making at the primary hit selection stage.

From the explanations in the previous sections it is clear that the overall activity ranking of the probes to select the most 'interesting' should be done based on data which are *corrected* for any systematic response artifacts on a particular plate and also across the various plates measured in a screen. Activity ranking in primary screening can either be done using the normalized responses or by considering a related score value. Whereas the well–level data correction approaches have adjusted for systematic bias in the average responses, there may still be remaining differences in terms of systematic shifts of the assay variance, either plate-to-plate, batch-to-batch, or systematic differences related to the sample classes exposed to the screen (e.g. natural product fractions or extracts, single small molecular compounds, compound-mixtures, combinatorial library pools, siRNA pools etc.). Performing the analysis of the screening results based on scoring approaches can account for such differences in the assay "noise". Its calculation can be based either on the (robust) variance estimation for the test sample area of the plates, or on the neutral control samples. In all analyses of corrected %-scale normalized or scored data we are working with scaled residuals $r_{ijp} \propto x_{ijp} - \overline{x}_p$ for hit ranking and selection.

If the estimation of the scale value $\hat{s}_p$, which is used for determining *Z*- or *R*-scores is based on the control samples, then the often limited number of them on a single plate can result in a large uncertainty of this estimate, especially for the lower plate densities of 96 and 384. For smaller n it may be advisable to use shrinkage estimation methods for the determination of the variance (Cui et al. 2005; Tong and Wang 2007; Murie et al. 2009) and obtain a regularized and more efficient estimate $\tilde{s}_p$ by borrowing information from other ('neighboring') plates to prevent individual increases in the rates of false positives and/or false negatives when the corresponding platewise score values are over- or underestimated. But also for higher density plates variance shrinkage can be advantageous if the control well areas on some individual plates contain systematic errors which may not have been eliminated by the previously described response corrections if the 'process-history' of sample wells and control wells differ, which can be the case in some types of assays. The decomposition of the plate set into subsets belonging to particular measurement batches with similar response properties and separate estimation within these batch boundaries as well as adaptive local adjustment of the shrinkage parameters λ and w_p in

$$\tilde{s}_p^2 = \lambda \hat{s}_p^2 + (1 - \lambda)\, \overline{s}_{p,w_p}^2 \tag{5.9}$$

on the time-ordered set of plates p will likely lead to a more efficient scale estimate $\tilde{s}_p$, but the resulting score values may not be bias-free (see further below). The shrinkage parameters can be optimized by e.g. minimizing the calculated false discovery rate for a given *Z*- or *R*-score threshold, or maximizing the number of putative hits for a given preset *FDR* value, i.e. the fraction of identified hits which are "not interesting" based on purely statistical considerations. λ describes the mixing between the value $\hat{s}_p$ of a particular individual plate and a component $\overline{s}_{p,w_p}$ which has higher bias and lower variance, and which itself depends on a 'smoothing' parameter w_p. The calculation of $\bar{s}_{p,w_p}$ can e.g. be based on local averaging or on kernel smoothing of the values from 'neighboring' plates (Prummer 2012).

A scatterplot of (normal distribution) *p*-values from *Z*- or *R*-scores on the *y*-axis and the NPI, percent change, or fold change values (Cui and Churchill 2003) on the *x*-axis can be very useful to identify samples which exhibit a certain minimal %-effect change, and at the same time assess their statistical 'significance' (probability that the null-hypothesis of 'no biological' activity is true). Similar types of visualizations are also used in gene expression analysis, genome scale RNAi screens, or genome-wide association studies (GWAS).

As an illustration of the hit analysis we now return to the actual biochemical example screen which we have used in the section on assay- and screening QC metrics to show the typical behavior of selected measures for a complete screen with around 1 Mio pure compound samples in 692 1536-well plates. In this particular screen the plate data set was corrected with a 'robust' *SVD* modeling procedure which was composed of an initial median polish run, trimming of those data points which exhibit large residual values $r_{ijp} > 4\ mad_p\left(x_{ijp}\right)$ by replacing their values with the model values, and finally calculating a *SVD* model across the whole screen using this modified data set where the set of wells with large activity (*NPI*) values were thus effectively omitted from the final modeling step (method not listed in Table 5.6). This is the REOF procedure which is available, among several others previously listed in the standard Novartis NIBR in-house HTS data analysis software system (Gubler 2006). The scatterplot of normalized activity values (negative *NPI* values) and the corresponding normal distribution *p*-values calculated from the *R*-scores are shown in Fig. 5.7. This (half-) volcano plot allows a good simultaneous assessment of activity values and their statistical significance. A similar plot can of course be generated from %-activity and *R*-score values for situations where proper *p*-values cannot be obtained.

A threshold along the %-activity or along the *p*-value or score axis can e.g. be chosen so that the total number of hits is below some maximal number of samples which can be progressed to the next screening stage, while simultaneously considering the false discovery rate as outlined below. A %-activity threshold can also include a consideration of the minimal potency of potential hit samples, i.e. an estimate of the maximal acceptable IC_{50} value which is of interest in further analysis by transforming the *NPI* values to such an IC_{50} by assuming an 'ideal' concentration response relationship (see Eq. (5.11) below). For example when setting a threshold at 50 % inhibition we would expect to be able to detect hit compounds with IC_{50} values which are smaller than the concentration used in the screening run (Gubler et al. 2013). For samples without known concentration in screening (e.g. siRNA, shRNA) such a direct translation can of course not be done.

A two component mixture model for the overall distribution function of the normalized activity values of all results from a particular screen can be defined as

$$f(x) = \pi_0 f_0(x) + (1 - \pi_0) f_A(x) \tag{5.10}$$

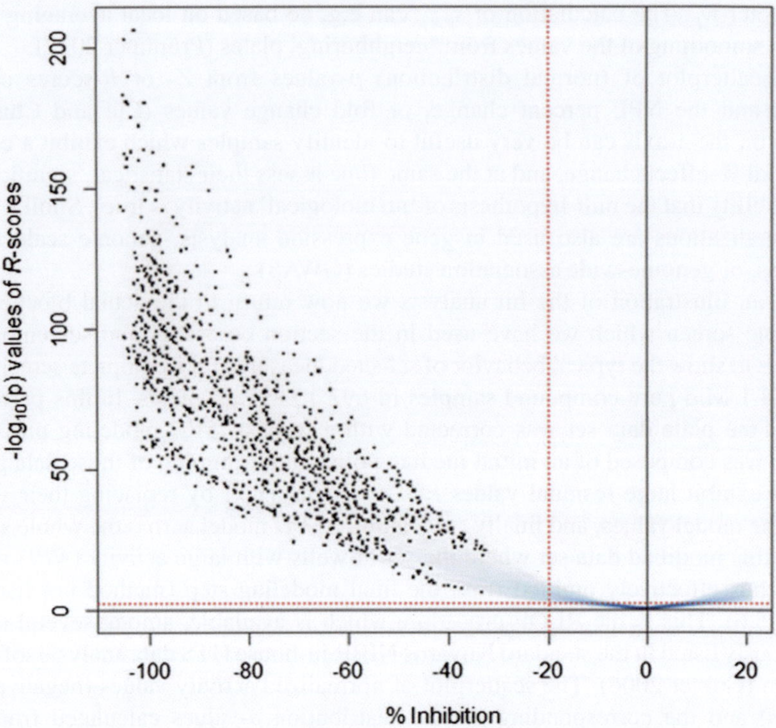

Fig. 5.7 (Half-) volcano plot of an actual 1Mio sample HTS with %-inhibition values on *x*-axis and -$\log_{10}(p)$ values of *R*-scores on *y*-axis. *R*-scores were corrected for average bias as outlined in text. %-inhibition values and initial *R*-scores calculated after removal of systematic response errors

with proportion π_0 of the null-distribution function $f_0(x)$ for the inactive samples and proportion $(1 - \pi_0)$ for the alternative distribution $f_A(x)$ of the active samples. When applying this approach in practice to a *large number* of *Z*-, *R*- or equivalent score values, then it is reasonable to assume $f_0(x) \sim N(0, 1)$, provided that centering, possible error correction and the scaling values were estimated correctly for the calculation of *x*. Especially the unbiased estimation of $\tilde{s}$ is crucial for transforming scores into normal probability *p*-values, as $f_0(p) \sim U(0, 1)$ only when this prerequisite is fulfilled. In this sense the shape of the distribution for 'larger' *p*-values, where the influence of f_A becomes negligible, can be used as a practical consistency check for the appropriateness of the scale estimates $\tilde{s}$ used in the score calculations.

In practice we may also encounter situations where $f_0(x)$ is not normally distributed as is the case e.g. after the application of the simple median polish procedure. Given an initial normal distribution of values the residuals will in this case usually appear leptokurtic with a central narrower peak and a related density depression in its immediate neighborhood due to the various 'median' operations, whereas the wider tail areas follow relatively closely a normal distribution density with a slight excess at larger x. In such situations it is also very likely that the scale $\tilde{s}$ will be underestimated and score values come out too large when compared to the original residual scale. Due to both possible reasons it does also no longer make sense to assign normal distribution p-values to the resulting corrected score values x. If there is interest to explicitly extract $f_A(x)$ from such an activity- or score distribution then a Bayesian mixture modeling approach with appropriate empirical priors may be used, now with $f_0(x)$ itself already being a mixture of a narrow and a broad component. If this is not done, or cannot be done, then simple activity rankings will have to suffice for hit identification in routine screening data analysis, but the additional highly useful information on e.g. estimated false discovery rates cannot be obtained from the data. In any case, a *lower limit* for the hit-threshold, in terms of %-activity or score values, can be derived from $\tilde{s}_N$ of the N controls, because in practice we always have $\tilde{s}_N \leq \tilde{s}_C$ due to additional experimental variability for the compound samples C as compared to the more homogeneous set of neutral controls N.

In our particular illustration of screening hit analysis the R-scores were initially calculated using the platewise $\tilde{s}_N$ values based on a *loess* based variance shrinkage procedure as described above. The corresponding score values were then used to estimate the mixture components and perform a false discovery rate analysis using the *fdrtool* R package (Strimmer 2008b). In order to compensate for any remaining bias in the score estimates x the following procedure was applied: An average R-score bias correction factor was derived from comparing the null distributions $f_0(x_N) \sim N(0, s_N^2)$ for $\{x_N : x \in N\}$ and $f_0(x_C) \sim N(0, s_C^2)$ for $\{x_C : x \in C\}$, then rescaling the scores x as $x^{'} = x/s_N$, resulting in $f_0(x_N^{'}) \sim N(0, 1)$ and $f_0(x_C^{'}) \sim N(0, s_C^2/s_N^2)$ where s_C/s_N should now be close to 1 if we assume that the same *average variance estimation bias* is present in the control and compound data samples. In this case this is borne out by the actual data from this screen and can be seen in Fig. 5.8a where we obtain $f_0(x_C^{'}) \sim N(0, 1.02^2)$ after the described bias correction with the normal null-density scaling factor of $s_N = 0.93$.

Using the same set of scaled R-score values $x^{'}$ we can also see that the corresponding normal distribution p-values in Fig. 5.8b show the 'expected' behavior, i.e. they are essentially flat in the large p region corresponding to a normal f_0 density with an average value of the proportion of samples following the null distribution of $\hat{\pi}_0 = 0.84$ and with a peak related to the alternative f_A distribution of the non-null samples at small p-values. Incidentally, we obtain consistent π_0 values of 0.86 also from *fdrtool* and 0.85 from the *qvalue* R packages (Storey and Tibshirani 2003). This consistency allows us to have confidence in the related tail area '*Fdr*' (Efron 2004) values as reported by *qvalue* (see Fig. 5.8c) where we can see that $Fdr \leq 0.1$ for up to a total number of identified hits (significant tests) of $n_{hit} \cong 49{,}000$, also in agreement with the direct p-value histogram analysis of the *Fdr* fraction. This q-value (≤ 0.1) corresponds in this particular case to a one-sided p-value of 0.006 and a Z (R-score) threshold of around 2.5. When limiting n_{hit} to 16,000 by selecting a R-score cutoff of around 4, then we obtain an expected *Fdr* close to 0. This means that we can expect a very high hit confirmation rate—close to 100 %—in a follow-up verification experiment.

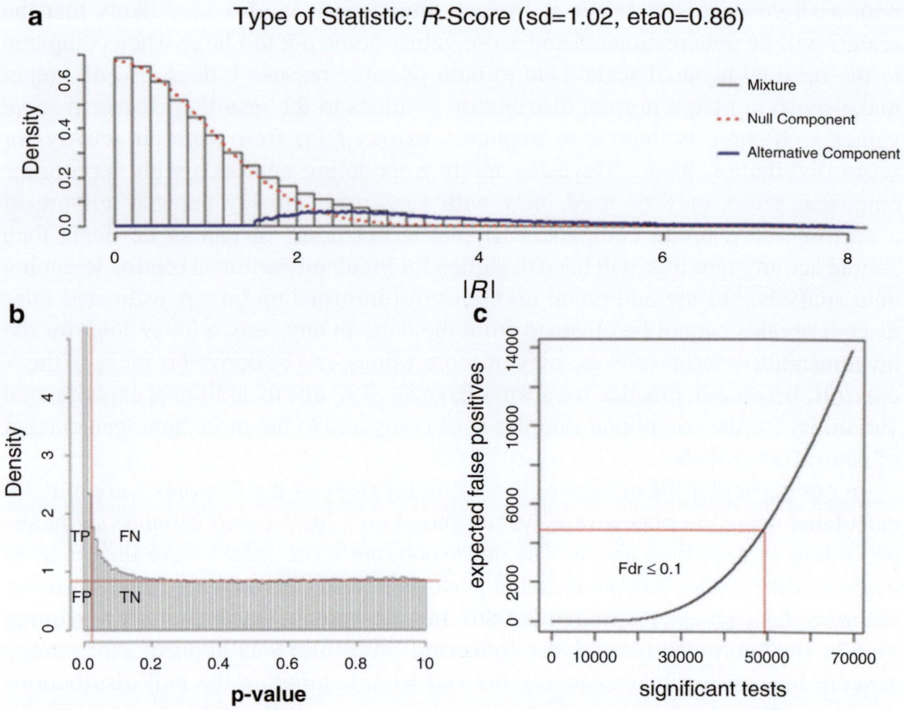

Fig. 5.8 (**a**) Decomposition of bias-corrected *R*-score values into null- and alternative mixture components using R package fdrtool. (**b**) *p*-value distribution of the same dataset with indication of hit selection threshold of $p = 0.006$ (*red vertical line*) and estimated percentage of null samples $\hat{\pi}_0 = 0.84$ (*red horizontal line*). (**c**) Expected false positives as function of total number of significant tests following from tail area false discovery rate analysis using R package qvalue, with red marker lines for $Fdr = 0.1$

With an average $\tilde{s}_N$ value of 4.5 % (rescaled as per the described average bias correction procedure), this score threshold can be converted into a %-inhibition (*NPI*) threshold of $4{\cdot}4.5\ \% = 18.0\ \%$. Disregarding the different plate-to-plate variances which are taken into account in the *R*-scores, but not in the *NPI* values, we can then loosely say that the actually chosen threshold value of 20 % inhibition for this screen is thus quite 'conservative' with respect to the false discovery rate criterion. From such a mixture distribution- and *p*-value histogram analysis we can also get estimates for the proportion of the non-detected active samples for the threshold setting corresponding to $Fdr = 0.1$ example value and obtain a false negative $FNR = 0.39$ in this case, or $FNR = 0.68$ for the threshold at *R*-score $= 4$. A set of *p*-value distribution and *FDR* analyses for different screening assays done by Prummer (2012) shows quite good agreement between estimated confirmation rates and the actually observed ones for a series of screens where follow-up experiments were performed as complete concentration-response measurements.

When analyzing screening data with replicates the (compound-) sample specific data variability can be estimated, resulting in the possibility to apply more powerful for hit-selection methods than for the $n = 1$ case where we are essentially assuming that every sample has the same variability. Of course it is always possible to simply average the normalized responses from the replicates and analyze them in the same way as the unreplicated primary data. This is not a real problem in small molecule compound screening, but e.g. for RNAi screening where we expect differences of variability among different siRNA samples it is essential to apply statistical methods which consider these possible sample specific differences. When comparing different RNAi treatment effects to each other the use of good estimates of the variability and use of efficient statistics becomes crucial. Many different approaches for analyzing and identifying hits from replicated data have been developed for gene expression array data (Kerr and Churchill 2001; McDavid et al. 2013; Baldi and Long 2001; Smyth 2004) and for RNAi screening (Birmingham et al. 2009), including Bayesian modeling and hypothesis testing approaches (Zhang et al. 2008). *SSMD*-based analysis (see Table 5.2) and hit selection rules for RNAi screening data with and without replication, as well as statistical methods for comparison of gene effects between groups are discussed in depth in a book by Zhang (2011a).

Gene expression experiments are usually using both biological and technical replication, i.e. measuring readout variability of responses from different sample sources and from different measurements of the same sample in a given experiment. Also in RNAi screening we can have similar situations: replicates from different sample preparations or simply multiple measurements of the same sample. Estimation and modeling of the error components from the replicated sample, the replicated measurements from of a single sample and combining these with the estimates from the neutral control samples can be done in different ways to provide more efficient variance estimation for subsequent hypothesis testing. Regularized variance estimations for use in hit selection methods often use shrinkage and Bayesian approaches and will lead to more powerful ‘regularized’ hypothesis testing (e.g. regularized t-statistics, Baldi and Long 2001; Tusher et al. 2001; Murie et al. 2009). The advantage of Bayesian methods is to allow incorporating balanced information from sample wells and control wells in the posterior distributions and then assigning probabilities to the test samples of belonging to either of the no-effect, activation or inhibition groups. The wide area of statistical methods for gene expression analysis and identification of differentially expressed genes (equivalent to effect-size assessment and hit selection for replicated screening data) was worked on extensively in the past 10–15 years and it is impossible to cover in this short chapter and we refer the reader to the extensive base of published literature on (primarily) microarray, but also RNAi screening data analysis (Birmingham et al. 2009, and references therein).

As mentioned previously in the section on data normalization for RNAi screens the siRNA effects are often better quantified by using the *SSMD*-statistic than by using *R*-scores, especially when working with replicates. In a similar fashion than shown in the previous volcano plot example for a compound-based ‘%-inhibition screen' we can also plot p-values based on *SSMD* analysis and the Fold-change values, instead of using the Z- or R-score based p-values and the %-inhibition values, to arrive at an equivalent visualization (even including the various types of controls in the same plot for additional information content). The derivation of *FP*, *FN* and related *FDR* rates based on the *SSMD*-statistic is outlined in much detail in Zhang (2011a).

In contrast to the RNAi screening and gene expression analysis areas publications on the topic of replication for small-molecule HTS data in full primary screens are practically non-existent because replicate measurements are essentially only done in the initial pilot phase where the focus is primarily on *assessing the degree of agreement*, and then again in the confirmation stage, i.e. *after* the primary screening hit selection where the analysis is again more centered around *verifying* the initially detected primary screening effect of the putative hits. Determining responses from replicated measurements using samples from the same or from different formulations of the same compound at a single concentration and/or the determination of a verifiable and plausible concentration dependence of the sought effect is the prime focus in this phase, but this is no longer a large-scale multiple hypothesis testing process with the associated need to control false positive (false discovery) and false negative rates.

We note again that the final hit list is often additionally filtered by considering counter screening results besides the results from the main screen, and we also need to mention that the HTS community often uses the term 'false positives' in the sense that a chemical test sample may well be verifiably active in a given assay through a target-independent effect, but inactive towards the actual biological target, as e.g. detected by a counter screen. This is obviously not the same use of the term 'false positive' as in statistics.

In compound screening additional elimination and selection steps are often made to modify the primary hit list, either by applying filters which identify unwanted chemical structure elements and compounds possessing specific chemical property values, or also augmenting the hit list based on structure-activity considerations ('hit list enrichment') with related similar compounds for subsequent confirmation or other follow-up experiments (Sun et al. 2010; Varin et al. 2010; Mulrooney et al. 2013).

5.2.11 Calculation of Derived Quantities from Normalized Plate and Well %Activity Data

The estimated standard errors and the unbiasedness of derived summary results or of parameters of calculations of derived quantities which are themselves *based on a whole series of %-activity values in different wells in one or multiple plates* also depend on best possible plate data quality in a very similar way as the data used for primary hit selection. Thus, every effort should be made to apply the response error correction methods outlined above also in these types of experiments. Because most of these types of experiments rely on a high percentage of non-null data the previously described modeling approaches can often not be used directly. Useful models of the systematic errors can nonetheless be established by using control data from suitable plate designs ('uniform' placement of control wells across the plates), or often better, by a separate set of *control plates measured in parallel to the sample plates*. The advantage of the latter approach is that we are then also able to model and correct for possible striping, edge effects and more 'complicated' patterns than e.g. linear or 2nd degree polynomial response surface dependencies which are essentially the only practical possibilities when using a limited number of control wells as pattern model anchor points on the sample plates. An example of such control plate correction approaches (CPR, control plate regression) and assessment of their success in concentration-response screens was described by Murie et al. (2014).

The most frequent case of calculations of such derived results from High-Throughput Screening plate experiments is the determination of the concentration-response curve (CRC) characteristics of the compound samples and the estimation of the parameters of the simple phenomenological sigmoidal 4-parameter-logistic (4PL) Hill curve function

$$y = f^{4PL}(x) = A_{inf} + \frac{\left(A_0 - A_{inf}\right)}{1 + 10^{\alpha\left(\log_{10}(x) - \log_{10}(AC_{50})\right)}} \tag{5.11}$$

if a significant concentration dependence exists in the experimental range employed in such a screen (Ritz and Streibig 2005; Formenko et al. 2006). In Eq. (5.11) we have concentration x, response y, and the 4 function parameters A_0, A_{inf} (lower and upper plateau values), AC_{50} (inflection point) and α (Hill slope parameter), and $f^{4PL}(AC_{50}) = \left(A_{inf} + A_0\right)/2$. Note our use of the generic term AC_{50} instead of the specific terminology IC_{50} for inhibition (*IC*: inhibitory concentration) and EC_{50} for activation experiments (*EC*: effective concentration).

The original Hill function (Hill 1910) has a theoretical basis in simple ligand binding theory, albeit with some unrealistic mechanistic assumptions (Goutelle et al. 2008). For all practical purposes in screening, especially for more complex cellular assays where potentially a whole cascade of biochemical reaction steps is involved, it should be viewed more as a convenient empirical model. Rankings of the 'strength' of the validated hit compound effects are usually based on the AC_{50} parameter, but for some types of assays A_{inf}, $\left(A_{inf} - A_0\right)$ or an 'area under the curve' effect measure which combines potency and efficacy into a single number are also considered (Huang and Pang 2012).

Even if the compounds for concentration dependent measurements are pre-selected to have shown activity in the previous concentration-independent screening primary runs the number of cases of 'incomplete' curves is usually still quite high, of course dependent on chosen hit threshold, previously used primary screening concentration and actual concentration range of the CRC experiments. In the HTS confirmation and validation screens the concentration range is held constant and not adapted on a compound by compound basis as might be done in low throughput compound profiling experiments. An accurate and independent determination of all regression function parameters is then not always possible and the analysis software needs to produce sensible results also in the initial presence of high degrees of parameter correlations, e.g. by automatic identification of ill-determined parameters and sensible application of constraints, or by switching to simpler fallback models, if needed.

HTS production-quality software for automated nonlinear regression fitting of large series (i.e. thousands or tens of thousands) of concentration-response curves needs to work in a highly robust fashion, use autostarting methods for estimation of initial parameters, and be able to handle outliers, e.g. by using resistant regression algorithms (*M*-estimation, *IRLS* iterative reweighted least squares) (Normolle 1993; Street et al. 1988). Also the likely presence of outliers needs to be signaled as diagnostic information. The analysis methods implemented in the software also need to be able to deal flexibly with other frequent 'unusual' situations like completely inactive compounds which lead to ~constant activity values over the full concentration range making it impossible to derive the 4 parameters of Eq. (5.11) in a meaningful way, or with the mentioned 'incomplete curves' where e.g. the response plateau at higher concentrations is not reached, and generally with unusual non-sigmoidal curve shapes (e.g. bell-shaped, inversely bell-shaped, multiphasic).

If experimental curve shapes do not follow the sigmoidal shape of Eq. (5.11) then derived parameters can be strongly biased, up to the level of being *completely* misleading to the analyst, e.g. when trying to represent a bell-shaped curve with f^{APL}. Thus, the detection of lack of fit and the selection of alternate models for the regression analysis is an important topic for large scale concentration response curve fitting in screening. A pragmatic modeling approach is to move away from a parametric representation of a given curve and choose a nonparametric model f^{NP} as per Eq. (5.12)

$$y = f^{NP}(x, \Theta). \tag{5.12}$$

if this is necessary. This model switching decision can e.g. be based on predefined lack of fit criteria together with the occurrence of large parameter dependency values. Depending on the actual model choice for f^{NP}, a preparatory optimization of one or more model parameters Θ for use in the actual curve analysis may be needed to reflect the experimental design and the typical assay variability. This can e.g. be done by assessing penalized likelihood measures, e.g. the Akaika information criterion (Akaike 1974), or using generalized cross validation (Craven and Wahba 1979) on the complete curve data set by varying Θ within appropriate ranges. A practical choice for f^{NP} is e.g. a smoothing spline function (Frommolt and Thomas 2008), possibly including monotonicity constraints (Kelly and Rice 1990). It is understood that the x values in these functions are the logarithmic concentrations $\log_{10}(x)$. For simplicity of notation we are not explicitly mentioning the logarithms in the remainder of this section.

Instead of the previous 4 model parameters in Eq. (5.11) we then use surrogate values which provide an equivalent characterization of the concentration-response relationship, e.g. the set of values $f(x_{min})$, $f(x_{max})$, $\min(f(x))$, $\max(f(x))$, $\arg\min_x f(x)$, $\arg\max_x f(x)$, and also the 'absolute' AC_{50} (concentration of the intersection of the fitted curve with the 50 % level as given by the N and P controls) abs $AC_{50} = \{x : f(x) = 50\}$, as well as an approximate 'equivalent Hill slope' $\alpha \approx 4(df/dx)_{x=absAC_{50}} / (\ln(10)\,(f\,(x_{max}) - f\,(x_{min})))$, together with proper consideration of non-existent or multiple intersections in the set $\{x : f(x) = 50\}$.

Such a nonparametric approach is also useful for confirmation CRC experiments where only 2–4 different concentrations may be used to verify the existence of reasonable concentration dependences and to obtain a rough estimate of abs AC_{50} for the putative hits, making it essentially impossible to use parametric nonlinear regression methods. For the analysis of such experiments the choice of interpolation splines for f^{NP} are preferred over smoothing spline functions.

For some types of assays (e.g. cell proliferation inhibition) the abs AC_{50} value—which can of course also be directly calculated from $f^{4PL}(x)$ when using the model represented by Eq. (5.11)—can be biologically more meaningfully interpreted than the AC_{50} function parameter (Sebaugh et al. 2011). In such experiments the concentration where *50 % response inhibition with respect to the control values* is reached will have an easily interpretable meaning, whereas the position of the AC_{50} concentration (i.e. the inflection point of the curve) can e.g. be influenced by the fact that we can obtain cell count readings at high concentrations which lie below the 100 % inhibition baseline value (when cells are killed) or that the 100 % inhibition level is not reached (cell growth cannot be completely inhibited), i.e. we can have $\Delta A = |A_{inf} - A_0| > 100\ \%$ or $\Delta A < 100\ \%$. As a consequence the position of the inflection point is less informative than the position of the intersection of the response curve with a particular prescribed inhibition level.

When considering AC_{50} or abs AC_{50} potency rank orders for the final selection of hits in the confirmation or validation screens the MSR (minimum significant ratio) which was optimally already derived in the assay adaptation stage (see Table 5.3) can be used as an indicator for assessing the 'significance' of potency differences (Eastwood et al. 2005, 2006). Other relevant measures in this context, when comparing the potency values from the target specific screen and one or several parallel selectivity screens, are the MSSR (minimum significant selectivity ratio) and MSRSR (minimum significant ratio of selectivity ratios) values. They are calculated in a similar way as MSR (Goedken et al. 2012) and give information on the confidence limits of the selectivity ratio $SR = AC_{50}$(off-target assay)/AC_{50}(on-target assay) and of ratios of SR values for different compounds.

5.3 Open Questions and Ongoing Investigations in the HTS and HCS Data Analysis Methods Field

Many aspects of small-molecule and RNAi High-Throughput Screening data analysis were explored in the past 10–15 years and several publications describe the details of the most relevant statistical aspects for the analysis of HTS data:

- Detection, modeling and correction of systematic error patterns of the plate-based screening readouts and to minimize selection bias, and thus allowing
- Optimal activity scoring and ranking of active features in the hit selection process to minimize the false discovery rate, minimize the number of false negatives and maximize the success rate of follow-up screening stages.

Nonetheless, in the author's opinion several specific areas merit to be explored and reported in more depth for the benefit of the whole HTS and HCS community:

- Which are the most optimal plate designs for replicate single concentration and concentration-response experiments? Which are the most optimal locations for replicate data points distributed on a single plate, or on multiple 'replicate' plates, given certain types of systematic errors?
- Is there an overall 'best' modeling and correction method for certain types of systematic error patterns? Which methods are overall most efficient and have the least amount of bias?
- Standard testing datasets and platform for comparing different analysis methods, including test and performance results.
- Response error modeling methods involving many *median* operations in their calculation (like e.g. the median polish and the simplified B-score methods) often lead to strongly non-normal distributions of the residuals. What is the best and *most practical* way to perform false discovery rate analyses in such cases?

The following are questions, active research topics and future directions of necessary statistical work in the area of high-dimensional multivariate screening data analysis. The 'classical' HTS assays, readout technologies and associated data analysis methods as outlined in this chapter will keep a lot of their present importance, and even become more pervasive in academia and smaller biological research laboratories for target identification, target validation and screening for active chemical features. But active research and development of statistical data analysis methods in the HTS/HCS field now center much more on these general questions:

- What are the most optimal normalization, feature selection, dimension reduction, error correction, scoring and classification methods for high-dimensional multivariate data from phenotypic image based High-Content Screening and other similar sources of such data? Under which conditions are the particular methods applicable? What are their advantages and disadvantages?

- What is the influence of imaging parameters and noise on the localization of subcellular features and what is the influence of different types of image analysis artifacts on HCS data analysis? How do these affect the derived (usually lower dimensional) final measures and classification results?
- What are the most suitable and informative analysis methods, including normalization and possible systematic error correction questions, for single cell data and multivariate time-course data?

References

Abraham VC, Taylor DL, Haskins JRL (2004) High content screening applied to large-scale cell biology. Trends Biotechnol 22(1):15–22

Abraham Y, Zhang X, Parker CN (2014) Multiparametric Analysis of Screening Data: Growing Beyond the Single Dimension to Infinity and Beyond. J Biomol Screen 19(5):628–639

Akaike H (1974) A new look at the statistical model identification. IEEE Trans Autom Contr 19(6):716–723. doi:10.1109/TAC.1974.1100705

Baker M (2010) Academic screening goes high-throughput. Nat Methods 7:787–792. doi:10.1038/nmeth1010-787

Bakheet TM, Doig AJ (2009) Properties and identification of human protein drug targets. Bioinformatics 25(4):451–457

Baldi P, Long AD (2001) A Bayesian framework for the analysis of microarray expression data: regularized t-test and statistical inferences of gene changes. Bioinformatics 17:509–519

Barry D, Hartigan JA (1993) A Bayesian analysis of change point problems. J Am Stat Assoc 88:309–319

Benjamini Y, Hochberg Y (1995) Controlling the false discovery rate: a practical and powerful approach to multiple testing. J R Stat Soc Ser B 57(1):289–300

Birmingham A, Selfors LM, Forster T, Wrobel D, Kennedy CJ, Shanks E, Santoyo-Lopez J, Dunican DJ, Long A, Kelleher D, Smith Q, Beijersbergen RL, Ghazal P, Shamu CE (2009) Statistical methods for analysis of high throughput RNA interference screens. Nat Methods 6(8):569. doi:10.1038/nmeth.1351

Bland JM, Altman DG (1986) Statistical methods for assessing agreement between two methods of clinical measurements. Lancet 1:307–310

Boutros M, Bras LP, Huber W (2006) Analysis of cell-based RNAi screens. Genome Biol 7:R66. doi:10.1186/gb-2006-7-7-r66

Box GEP, Hunter WG, Hunter JS (1978) Statistics for experimeers: an introduction to design, data analysis and model building. Wiley, New York. ISBN:0-471-09315-7

Bray MA, Carpenter AE (2012) Advanced assay development guidelines for image-based high content screening and analysis. In: Sittampalam GS (ed) Assay guidance manual. http://www.ncbi.nlm.nih.gov/books/NBK53196/. Accessed 15 Oct 2014

Brideau C, Gunter B, Pikounis B, Liaw A (2003) Improved statistical methods for hit selection in high-throughput screening. J Biomol Screen 8(6):634–647

Brummelkamp TR, Fabius AWM, Mullenders J, Madiredjo M, Velds A, Kerkhoven RM, Bernards R, Beijersbergen RL (2006) An shRNA barcode scrren provides insight into cancer cell vulnerability to MDM2 inhibitors. Nat Chem Biol 2(4):202–206

Bushway PJ, Azimi B, Heynen-Genel S, Price JH, Mercola M (2010) Hybrid median filter background estimator for correcting distortions in microtiter plate data. Assay Drug Dev Technol 8(2):238–250. doi:10.1089/adt.2009.0242

Carpenter AE (2007) Image-based chemical screening. Nat Chem Biol 3(8):461–465. doi:10.1038/nchembio.2007.15

Coma I, Clark L, Diez E, Harper G, Herranz J, Hofmann G, Lennon M, Richmond N, Valmaseda M, Macarron R (2009a) Process validation and screen reproducibility in high-throughput screening. J Biomol Screen 14:66–76

Coma I, Herranz, J, Martin J (2009) Statistics and decision making in high-throughput screening. In: William P, Janzen WP, Bernasconi P (eds) High-throughput screening. Methods in molecular biology, vol 565. Humana, Totowa. ISBN:978-1-60327-257-5

Craven P, Wahba G (1979) Smoothing noisy data with spline functions. Numer Math 31:377–403

Cui X, Churchill GA (2003) Statistical tests for differential expression in cDNA microarray experiments. Genome Biol 4(4):210

Cui X, Hwang JTG, Qiu J, Blades NJ, Churchill GA (2005) Improved statistical tests for differential gene expression by shrinking variance components estimates. Biostatistics 6(1): 59–75

Dalmasso C, Broet P, Moreau T (2005) A simple procedure for estimating the false discovery rate. Bioinformatics 21(5):660–668. doi:10.1093/bioinformatics/bti063

Davies JW, Glick M, Jenkins JL (2006) Streamlining lead discovery buy aligning in silico and high-thtoughput screening. Curr Op Chem Bio 10:343–351

Dean A, Lewis S (eds) (2006) Screening: methods for experimentation in industry, drug discovery, and genetics. Springer, New York. ISBN 978-1-4419-2098-0

R Development Core Team (2013) R: a language and environment for statistical computing. R Foundation for Statistical Computing, Vienna. http://www.Rproject.org. ISBN:3900051070

Dragiev P, Nadon R, Makarenkov V (2011) Systematic error detection in experimental high-throughput screening. BMC Bioinformatics 12:25

Duerr O, Duval F, Nichols A, Lang P, Brodte A, Heyse S, Besson D (2007) Robust Hit identification by quality assurance and multivariate data analysis of a high content, cell based assay. J Biomol Screen 12(8):1042–1049

Eastwood BJ, Chesterfield AK, Wolff MC, Felder CC (2005) Methods for the design and analysis of replicate-experiment studies to establish assay reproducibility and the equivalence of two potency assays. In: Gad S (ed) Drug discovery handbook. Wiley, New York

Eastwood BJ, Farmen MW, Iversen PW, Craft TJ, Smallwood JK, Garbison KE, Delapp NW, Smith GF (2006) The minimum significant ratio: a statistical parameter to characterize the reproducibility of potency estimates from concentration-response assays and estimation by replicate-experiment studies. J Biomol Screen 3:253–261

Echeverri CJ, Perrimon N (2006) High-throughput RNAi screening in cultured cells – a user's guide. Nat Rev Genet 7:373–384

Efron B (2004) Large-scale simultaneous hypothesis testing: the choice of a null hypothesis. J Am Stat Assoc 99(465):96–104

Efron B (2010) Large-scale inference. Cambridge University Press, Cambridge. ISBN 978-0-521-19249-1

Efron B, Tibshirani R, Storey J, Tusher V (2001) Empirical Bayes analysis of a microarray experiment. J Am Stat Assoc 96:1151–1160

Formenko I, Durst M, Balaban D (2006) Robust regression for high-throughput screening. Comput Methods Prog Biomed 82:31–37

Fox S, Farr-Jones S, Sopchak L, Boggs A, Nicely HW, Khoury R, Biros M (2006) High-throughput screening: update on practices and success. J Biomol Screen 11(7):864–869

Fox SJ (ed) (2002) High throughput screening 2002: new strategies and technologies. High Tech Business Decisions, Moraga

Frommolt P, Thomas RK (2008) Standardized high-throughput evaluation of cell-based compound screens. BMC Bioinformatics 9:475

Geary RC (1954) The contiguity ratio and statistical mapping. Inc Stat 5(3):15–145. doi:10.2307/2986645

Goedken ER, Devanarayan V, Harris CM, Dowding LA, Jakway JP, Voss JW, Wishart N, Jordan DC, Talanian RV (2012) Minimum significant ratio of selectivity ratios (MSRSR) and confidence in ratio of selectivity ratios (CRSR): quantitative measures for selectivity ratios obtained by screening assays. J Biomol Screen 17(7):857–867. doi:10.1177/1087057112447108

Goutelle S, Maurin M, Rougier F, Barbaut X, Bourguignon L, Ducher M, Maire P (2008) The Hill equation: a review of its capabilities in pharmacological modelling. Fundam Clin Pharmacol 22(6):633–648. doi:10.1111/j.1472-8206.2008.00633.x

Gubler H (2006) Methods for statistical analysis, quality assurance and management of primary high-throughput screening data. In: Hüser J (ed) High-throughput screening in drug discovery. Methods and principles in medicinal chemistry, vol 35. Wiley-VCH GmbH, Weinheim, pp 151–205. doi:10.1002/9783527609321.ch7

Gubler H, Schopfer U, Jacoby E (2013) Theoretical and experimental relationships between percent inhibition and IC50 data observed in high-throughput screening. J Biomol Screen 18(1):1–13. doi:10.1177/1087057112455219

Gunter B, Brideau C, Pikounis B, Liaw A (2003) Statistical and graphical methods for quality control determination of high-throughput screening data. J Biomol Screen 8(6):624–633

Haney SA (2014) Rapid assessment and visualization of normality in high-content and other cell-level data and its impact on the interpretation of experimental results. J Biomol Screen 19(5):672–684

Heuer C, Haenel T, Prause B (2002) A novel approach for quality control and correction of HTS based on artificial intelligence. Pharmaceutical Discovery and Development 2002/03, PharmaVentures Ltd., Oxford

Heyse S (2002) Comprehensive analysis of high-throughput screening data. In: Bornhop DJ et al (eds) Proceedings of the SPIE, Biomedical Nanotechnology Architectures and Applications 4626, pp 535–547

Hill AV (1910) The possible effects of the aggregation of the molecules of haemoglobin on its dissociation curves. J Physiol 40:iv–vii

Hill AA, LaPan P, Li Y, Haney SA (2007) Analysis of multiparametric HCS data. In: Haney SA (ed) High content screening: science, techniques and applications. Wiley, New York. doi:10.1002/9780470229866.ch15

Hinkley DV (1970) Inference about the change-point in a sequence of random variables. Biometrika 57(1):1–17

Hoaglin DC, Mosteller F, Tukey JW (1983) Understanding robust and exploratory data analysis. Wiley, New York. ISBN 0-471-09777-2

Horvath L (1993) The maximum likelihood method for testing changes in the paramaters of normal observations. Ann Stat 21(2):671–680

Huang S, Pang L (2012) Comparing statistical methods for quantifying drug sensitivity based on in vitro dose–response assays. Assay Drug Dev Technol 10(1):88–96. doi:10.1089/adt.2011.0388

Hubert M, Rousseuw PJ, Vanden Branden K (2005) ROBPCA: a new approach to robust principal component analysis. Technometrics 47:64–79

Hughes JP, Rees S, Kalindjian SB, Philpott KL (2011) Principles of early drug discovery. Br J Pharmacol 162:1239–1249

Hughes M, Inglese J, Kurtz A, Andalibi A, Patton L, Austin C, Baltezor M, Beckloff M, Sittampalam S, Weingarten M, Weir S (2012) Early drug discovery and development guidelines: for academic researchers, collaborators, and start-up companies. In: Sittampalam S et al (eds) Assay guidance manual. Eli Lilly & Company and the National Center for Advancing Translational Sciences, Bethesda. http://www.ncbi.nlm.nih.gov/books/NBK92015/

Hyvarinen A, Karhunen J, Oja E (2001) Independent component analysis. Wiley, New York. ISBN 978-0471405405

Ilouga PE, Hesterkamp T (2012) On the prediction of statistical parameters in high-throughput screening using resampling techniques. J Biomol Screen 17(6):705–712. doi:10.1177/1087057112441623

Iversen PW, Eastwood BJ, Sittampalam GS (2006) A comparison of assay performance measures in screening assays: signal window. Z′-factor and assay variability ratio. J Biomol Screen 11(3):247–252

Kaiser J (2008) Industrial-style screening meets academic biology. Science 321(5890):764–766. doi:10.1126/science.321.5890.764

Kelly C, Rice J (1990) Monotone smoothing with application to dose-response curves and the assessment of synergism. Biometrics 46(4):1071–1085

Kerr MK, Churchill GA (2001) Experimental design for gene expression microarrays. Biostatistics 2(2):183–201

Kevorkov D, Makarenkov V (2005) Statistical analysis of systematic errors in high-throughput screening. J Biomol Screen 10(6):557–567

Killick R, Eckley IA (2014) Changepoint: an R package for changepoint analysis. J Stat Software 58(3):1–19

Killick R, Fearnhead P, Eckley IA (2012) Optimal detection of changepoints with a linear computational cost. J Am Stat Assoc 107(500):1590–1598

König R, Chiang CY, Tu BP, Yan SF, DeJesus PD, Romero A, Bergauer T, Orth A, Krueger U, Zhou Y, Chanda SK (2007) A probability-based approach for the analysis large scale RNAi screens. Nat Methods 4(10):847–849

Kramer R, Cohen D (2004) Functional genomics to new drug targets. Nat Rev Drug Discov 3(11):965–972

Kümmel A, Selzer P, Siebert D, Schmidt I, Reinhardt J, Götte M, Ibig-Rehm Y, Parker CN, Gabriel D (2012) Differentiation and visualization of diverse cellular phenotypic responses in primary high-content screening. J Biomol Screen 17(6):843–849. doi:10.1177/1087057112439324

Loo LH, Wu LF, Altschuler SJ (2007) Image-based multivariate profiling of drug responses from single cells. Nat Methods 4(5):445. doi:10.1038/NMETH1032

Macarron R, Hertzberg RP (2011) Design and implementation of high-throughput screening assays. Mol Biotechnol 47(3):270–285. doi:10.1007/s12033-010-9335-9

Macarron R, Banks MN, Bojanic D, Burns DJ, Cirovic DA, Garyantes T, Green DVS, Hertzberg RP, Janzen WP, Paslay JW, Schopfer U, Sittampalam GS (2011) Impact of high-throughput screening in biomedical research. Nat Rev Drug Discov 10:188–195. doi:10.1038/nrd3368

Majumdar A, Stock D (2011) Large sample inference for an assay quality measure used in high-throughput screening. Pharm Stat 1:227–231

Makarenkov V, Zentilli P, Kevorkov D, Gagarin A, Malo N, Nadon R (2007) An efficient method for the detection and elimination of systematic error in high-throughput screening. Bioinformatics 23:1648–1657

Malo N, Hanley JA, Cerquozzi S, Pelletier J, Nadon R (2006) Statistical practice in high-throughput screening data analysis. Nat Biotechnol 24:167–175

Mangat CS, Bharat A, Gehrke SS, Brown ED (2014) Rank ordering plate data facilitates data visualization and normalization in high-throughput screening. Biomol Screen 19(9): 1314–1320. doi:10.1177/1087057114534298

Matson RS (2004) Applying genomic and proteomic microarray technology in drug discovery. CRC, Boca Raton. ISBN 978-0849314698

Mayr LM, Bojanic D (2009) Novel trends in high-throughput screening. Curr Opin Pharmacol 9(5):580–588

Mayr LM, Fuerst P (2008) The future of high-throughput screening. J Biomol Screen 13:443–448

McDavid A, Finak G, Chattopadyay PK, Dominguez M, Lamoreaux L, Ma SS, Roederer M, Gottardo R (2013) Data exploration, quality control and testing in single-cell qPCR-based gene expression experiments. Bioinformatics 29(4):461–467. doi:10.1093/bioinformatics/bts714

Millard BL, Niepel M, Menden MP, Muhlich JL, Sorger PK (2011) Adaptive informatics for multifactorial and high-content biological data. Nat Methods 8(6):487

Moran PAP (1950) Notes on continuous stochastic phenomena. Biometrika 37(1):17–23. doi:10.2307/2332142

Mosteller F, Tukey J (1977) Data analysis and regression. Addison-Wesley, Reading. ISBN 0-201-04854-X

Mulrooney CA, Lahr DL, Quintin MJ, Youngsaye W, Moccia D, Asiedu JK, Mulligan EL, Akella LB, Marcaurelle LA, Montgomery P, Bittker JA, Clemons PA, Brudz S, Dandapani S, Duvall JR, Tolliday NJ, De Souza A (2013) An informatic pipeline for managing high-throughput screening experiments and analyzing data from stereochemically diverse libraries. J Comput Aided Mol Des 27(5):455–468. doi:10.1007/s10822-013-9641-y

Murie C, Woody O, Lee AY, Nadon R (2009) Comparison of small n statistical tests of differential expression applied to microarrays. BMC Bioinformatics 10:45. doi:10.1186/1471-2105-10-45

Murie C, Barette C, Lafanechere L, Nadon R (2013) Single assay-wide variance experimental (SAVE) design for high-throughput screening. Bioinformatics 29(23):3067–3072. doi:10.1093/bioinformatics/btt538

Murie C, Barette C, Lafanechere L, Nadon R (2014) Control-plate regression (CPR) normalization for high-throughput screens with many active features. J Biomol Screen 19(5):661–671. doi:10.1177/1087057113516003

Murray CW, Rees DC (2008) The rise of fragment-based drug discovery. In: Edward Zartler E, Shapiro M (eds) Fragment-based drug discovery: a practical approach. Wiley, Hoboken. ISBN 978-0-470-05813-8

Ngo VN, Davis RE, Lamy L, Yu X, Zhao H, Lenz G, Lam LT, Dave S, Yang L, Powell J, Staudt LM (2006) A loss-of-function RNA interference screen for molecular targets in cancer. Nature 441:106–110

Nichols A (2007) High content screening as a screening tool in drug discovery. Methods Mol Biol 356:379–387

Normolle DP (1993) An algorithm for robust non-linear analysis of radioimmunoassay and other bioassays. Stat Med 12:2025–2042

Oakland J (2002) Statistical process control. Routledge, Milton Park. ISBN 0-7506-5766-9

Pereira DA, Williams JA (2007) Origin and evolution of high throughput screening. Br J Pharmacol 152:53–61

Perlman ZE, Slack MD, Feng Y, Mitchison TJ, Wu LF, Altschuler SJ (2004) Multidimensional drug profiling by automated microscopy. Science 306(5699):1194–1198

Prummer M (2012) Hypothesis testing in high-throughput screening for drug discovery. J Biomol Screen 17(4):519–529. doi:10.1177/1087057111431278

Qiu P, Simonds EF, Bendall SC, Gibbs KD Jr, Bruggner RV, Linderman MD, Sachs S, Nolan GP, Plevritis SK (2011) Extracting a cellular hierarchy from high-dimensional cytometry data with SPADE. Nat Biotechnol 29(10):886–891. doi:10.1038/nbt.1991

Ravkin I (2004) Quality measures for imaging-based cellular assays. Poster #P12024, Society for Biomolecular Screening. Annual Meeting Abstracts. http://www.ravkin.net/posters/P12024-Quality%20Measures%20for%20Imaging-based%20Cellular%20Assays.pdf. Accessed 15 Oct 2014

Reisen F, Zhang X, Gabriel D, Selzer P (2013) Benchmarking of multivariate similarity measures for high-content screening fingerprints in phenotypic drug discovery. J Biomol Screen 18(10):1284–1297. doi:10.1177/1087057113501390

Ritz C, Streibig JC (2005) Bioassay analysis using R. J Stat Softw 12(5):1–22. http://www.jstatsoft.org/

Root DE, Hacohen N, Hahn WC, Lander ES, Sabatini DM (2006) Genome-scale loss-of-function screening with a lentiviral RNAi library. Nat Methods 3(9):71. doi:10.1038/NMETH92

Rousseeuw PJ, Croux C (1993) Alternatives to the median absolute deviation. J Am Stat Assoc 88(424):1273–1283. doi:10.2307/2291267

Sakharkar MK, Sakharkar KR, Pervaiz S (2007) Druggability of human disease genes. Int J Biochem Cell Biol 39:1156–1164

Sebaugh JL (2011) Guidelines for accurate EC50/IC50 estimation. Pharm Stat 10:128–134

Sharma S, Rao A (2009) RNAi screening – tips and techniques. Nat Immunol 10(8):799–804

Shewhart WA (1931) Economic control of quality of manufactured product. Van Nostrand, New York. ISBN 0-87389-076-0

Shun TY, Lazo JS, Sharlow ER, Johnston PA (2011) Identifying actives from HTS data sets: practical approaches for the selection of an appropriate HTS data-processing method and quality control review. J Biomol Screen 16(1):1–14. doi:10.1177/1087057110389039

Sims D, Mendes-Pereira AM, Frankum J, Burgess D, Cerone MA, Lombardelli C, Mitsopoulos C, Hakas J, Murugaesu N, Isacke CM, Fenwick K, Assiotis I, Kozarewa I, Zvelebil M, Ashworth A, Lord CJ (2011) High-throughput RNA interference screening using pooled shRNA libraries and next generation sequencing. Genome Biol 12(R104):1–13

Singh S, Carpenter AE, Genovesio A (2014) Increasing the content of high-content screening: an overview. J Biomol Screen 19(5):640–650. doi:10.1177/1087057114528537

Sittampalam GS, Iversen PW, Boadt JA, Kahl SD, Bright S, Zock JM, Janzen WP, Lister MD (1997) Design of signal windows in high-throughput screening assays for drug discovery. J Biomol Screen 2:159

Sittampalam GS, Gal-Edd N, Arkin M, Auld D, Austin C, Bejcek B, Glicksman M, Inglese J, Lemmon V, Li Z, McGee J, McManus O, Minor L, Napper A, Riss T, Trask OJ, Weidner J (eds) (2004) Assay guidance manual. Eli Lilly & Company and the National Center for Advancing Translational Sciences, Bethesda. http://www.ncbi.nlm.nih.gov/books/NBK53196/

Smith K, Horvath P (2014) Active learning strategies for phenotypic profiling of high-content screens. J Biomol Screen 19(5):685–695

Smyth GK (2004) Linear models and empirical bayes methods for assessing differential expression in microarray experiments. Stat Appl Genet Mol Biol 3(1):1–26

Storey JD (2002) A direct approach to false discovery rates. J R Stat Soc Ser B 64:479–498

Storey JD, Tibshirani R (2003) Statistical significance for genome-wide experiments. Proc Natl Acad Sci 100:9440–9445

Street JO, Carrol RJ, Ruppert D (1988) A note on computing robust regression estimates via iteratively reweighted least squares. Am Stat 42:152–154

Strimmer K (2008a) A unified approach to false discovery rate estimation. BMC Bioinformatics 9:303. doi:10.1186/1471-2105-9-303

Strimmer K (2008b) fdrtool: a versatile R package for estimating local and tail area-based false discovery rates. Bioinformatics 24(12):1461–1462

Sui Y, Wu Z (2007) Alternative statistical parameter for high-throughput screening assay quality assessment. J Biomol Screen 12(2):229–234

Sun D, Whitty A, Papadatos J, Newman M, Donnelly J, Bowes S, Josiah S (2005) Adopting a practical statistical approach for evaluating assay agreement in drug discovery. J Biomol Screen 10(5):508–516

Sun D, Jung J, Rush TS, Xu Z, Weber MJ, Bobkova E, Northrup A, Kariv I (2010) Efficient identification of novel leads by dynamic focused screening: PDK1 case study. Comb Chem High Throughput Screen 13(1):16–26

Taylor PB, Stewart FP, Dunnington DJ, Quinn ST, Schulz CK, Vaidya KS, Kurali E, Lane TR, Xiong WC, Sherrill TP, Snider JS, Terpstra ND, Hertzberg RP (2000) Automated assay optimization with integrated statistics and smart robotics. J Biomol Screen 5(4):213–226

Thorne N, Auld DS, Inglese J (2010) Apparent activity in high-throughput screening: origins of compound-dependent assay interference. Curr Opin Chem Biol 14(3):315–324. doi:10.1016/j.cbpa.2010.03.020

Tong T, Wang Y (2007) Optimal shrinkage estimation of variances with applications to microarray data analysis. J Am Stat Assoc 102(477):113–122. doi:10.1198/016214506000000126

Tusher VG, Tibshirani R, Chu G (2001) Significance analysis of microarrays applied to the ionizing radiation response. Proc Natl Acad Sci 98(9):5116–5121

van Oostrum J, Calonder C, Rechsteiner D, Ehrat M, Mestan J, Fabbro D, Voshol H (2009) Tracing pathway activities with kinase inhibitors and reverse phase protein arrays. Proteomics Clin Appl 3(4):412–422

Varin T, Gubler H, Parker CN, Zhang JH, Raman P, Ertl P, Schuffenhauer A (2010) Compound set enrichment: a novel approach to analysis of primary HTS data. J Chem Inf Model 50(12): 2067–2078. doi:10.1021/ci100203e

Wu Z, Liu D, Sui Y (2008) Quantitative assessment of hit detection and confirmation in single and duplicate high-throughput screenings. J Biomol Screen 13(2):159–167. doi:10.1177/1087057107312628

Wunderlich ML, Dodge ME, Dhawan RK, Shek WR (2011) Multiplexed fluorometric immunoassay testing methodology and troubleshooting. J Vis Exp 12(58):pii:3715

Yin Z, Zhou X, Bakal C, Li F, Sun Y, Perrimon N, Wong ST (2008) Using iterative cluster merging with improved gap statistics to perform online phenotype discovery in the context of high-throughput RNAi screens. BMC Bioinformatics 9:264. doi:10.1186/1471-2105-9-264

Zhang XD (2008) Novel analytic criteria and effective plate designs for quality control in genome-scale RNAi screens. J Biomol Screen 13:363–377

Zhang XD (2011a) Optimal high-throughput screening: practical experimental design and data analysis for genome-scale RNAi research. Cambridge University Press, Cambridge. ISBN 978-0-521-73444-8

Zhang XD (2011b) Illustration of SSMD, Z Score, SSMD*, Z* score and t statistic for hit selection in RNAi high-throughput screening. J Biomol Sreen 16(7):775–785

Zhang XD, Zhang Z (2013) displayHTS: a R package for displaying data and results from high-throughput screening experiments. Bioinformatics 29(6):794–796. doi:10.1093/bioinformatics/btt060

Zhang JH, Chung TD, Oldenburg KR (1999) A simple statistical parameter for use in evaluation and validation of high throughput screening assays. J Biomol Screen 4:67–73

Zhang JH, Chung TD, Oldenburg KR (2000) Confirmation of primary active substances from high throughput screening of chemical and biological populations: a statistical approach and practical considerations. J Comb Chem 2(3):258–265

Zhang JH, Wu X, Sills MA (2005) Probing the primary screening efficiency by multiple replicate testing: a quantitative analysis of hit confirmation and false screening results of a biochemical assay. J Biomol Screen 10:695. doi:10.1177/1087057105279149

Zhang XD, Ferrer M, Espeseth AS, Marine SD, Stec EM, Crackower MA, Holder DJ, Heyse JF, Strulovici B (2007) The use of strictly standardized mean difference for hit selection in primary RNA interference high-throughput screening experiments. J Biomol Screen 12(4):497–509. doi:10.1177/1087057107300646

Zhang XD, Kuan PF, Ferrer M, Shu X, Liu YC, Gates AT, Kunapuli P, Stec EM, Xu M, Marine SD, Holder DJ, Strulovici B, Heyse JF, Espeseth AS (2008) Hit selection with false discovery rate control in genome-scale RNAi screens. Nucleic Acids Res 36(14):4667–4679. doi:10.1093/nar/gkn435

Chapter 6
Quantitative-Structure Activity Relationship Modeling and Cheminformatics

Max Kuhn

Abstract This chapter describes quantitative tools for analyzing chemical structures and relating them to assay results using statistical models. The focus is on prediction of new compounds as well as the exploratory analysis and data mining of large compound databases. Other issues related to how these analytical methods are used are discussed.

Keywords Machine learning • Molecular descriptors • Applicability domain

6.1 Overview

Given one or more assays to measure the effectiveness of a molecule against at target or phenotype, a great deal of time and effort in the early phases of the drug discovery process is used to modify early hits to create molecules that have attractive properties (see Chap. 4 for an overview). The structure-activity relationship (SAR) assumption is that similar molecules have similar properties. Based on this hypothesis, medicinal chemists use their experience and intuition to change one or more atoms to make improvements. Once a set of virtual compounds are designed, they are synthesized and tested using a battery of relevant assays. For a sense of scale, a typical project will design and synthesize hundreds to thousands of compounds.

This chapter describes to general classes of analytical methods, quantitative structure-activity relationship (QSAR) and Chemoinformatics, that are used to analyze chemical data. These techniques can assist the chemist using empirical data on existing compounds to make predictions about new compounds. These predictions can help inform the medicinal chemist prior to synthesizing a molecule or to predict secondary properties prior to the generation of assay results.

QSAR models utilize chemical structures of molecules to predict the activity or potency. However, the term QSAR can refer to models used to predict a wide

M. Kuhn (✉)
Pfizer Global R&D, Groton, CT, USA
e-mail: max.kuhn@pfizer.com

L. Zhang (ed.), *Nonclinical Statistics for Pharmaceutical and Biotechnology Industries*, Statistics for Biology and Health, DOI 10.1007/978-3-319-23558-5_6

variety of characteristics, such as permeability, greasiness or safety endpoints. The term quantitative structure-property relationships (QSPR) may be more appropriate in many cases but is less commonly used.

One definition of Cheminformatics comes from Brown (1998):

> "The use of information technology and management has become a critical part of the drug discovery process. Cheminformatics is the mixing of those information resources to transform data into information and information into knowledge for the intended purpose of making better decisions faster in the area of drug lead identification and organization."

There are obvious similarities between Cheminformatics and Data Mining (Han et al. 2006; Tan et al. 2005) and Data Science (Schutt and O'Neil 2013). In each case, potentially large and complex data are used to answer questions and predict future outcomes.

In additional to the large number of compounds created and stored in pharmaceutical companies, large amounts of publicly available chemical and biological data (such as PubChem) are also utilized. Cheminformatics uses data mining techniques with retrospective data to inform and catalyze the process of creating high quality molecules. As we will see later in this chapter, cheminformatics encompasses standard data mining approaches as well as specialized techniques informed by chemical principals.

In this chapter, we introduce QSAR and Cheminformatics with several examples. Other practical considerations are also illustrated. Statistical methods (and opportunities) are also described in each section.

6.2 Quantifying Molecular Structure

The structure of a molecule can be represented in several ways. One of the most common is the simplified molecular-input line-entry system (Weininger 1988) or SMILES for short. SMILES strings are basically chemical formulas. For example, acetaminophen contains eight carbon atoms, nine hydrogens, a nitrogen and two oxygen atoms (i.e. $C_8H_9NO_2$) with a the SMILES string:

$$\texttt{CC(NC1} = \texttt{CC} = \texttt{C(O)C} = \texttt{C1)} = \texttt{O}$$

Although SMILES strings have their limitations they have proven to be an effective representation of a compound.

To create many types of models, some numeric representation of the structure is required. There are a wide variety of such *chemical descriptors* (Leach and Gillet 2007). Some examples:

- Simple count of specific atoms or bonds (e.g. the number of covalent bonds or the number of carbons) are commonly used.
- Fingerprint descriptors are binary indicators for specific sub-structures or fragments. There are potentially millions of possible fingerprint descriptors. In many data sets, fingerprint descriptors tend to be very sparse, meaning that they are mostly zero across different compounds.

- There are numerous descriptors that are represented by continuous values such as molecular weight, surface area, volume and positive charge. Other, more esoteric descriptors exist such as the flexibility index of the longest chain in the molecule.

Some descriptors used are *estimated* based on pre-defined models. For example, lipophilicity (or "greasiness") is an important property. The most common representations of this property is log*P*, where *P* is the partition coefficient that measures the ratio in concentrations between water and octanol. For example, polar compounds will tend to concentrate in water and yield lower log*P* value. There are assay systems that can measure log*P*. However, when used as inputs into QSAR models, this descriptor is usually quantified by an existing model. See Machatha and Yalkowsky (2005) for examples.

For continuous descriptors, it is common to observe high degrees of between-predictor correlations. This can occur for several reasons. First, there are many different methods for calculating certain descriptors. For example, different versions of surface area include or exclude some atoms (e.g., nitrogen or oxygen) which can lead to extremely high correlations. Secondly, there are many different descriptors that quantify the same underlying characteristic of the molecule. The number of bonds is likely correlated with the number of atoms in the molecule and these in-turn have a relationship to the size and weight of a compound. This characteristic of molecular descriptors can induce severe multicollinearity (Myers 1990) in the data which can have a significant effect on some statistical models.

For example, using the data from Karthikeyan et al. (2005) where 4173 compounds were used to predict the melting point of a molecule. Using a set of 202 molecular descriptors, the average absolute correlation between pairs of descriptors was 0.26. However, a principal component analysis (Abdi and Williams 2010) of the data indicated that the first three components of the data accounted for 96 %, 3.2 % and 0.4 % of the variance, respectively. This implies that the vast majority of the descriptors used in the analysis were capturing the same (linear) information and are redundant.

6.3 Structure Based Models

One simplistic model that relates structure to potency was created by Free and Wilson (1964). In some situations, portions of a molecule can be treated as a single unit that can be substituted in different ways. An *R group* is a place holder for a structure that can be attached to an end of a molecule (i.e. a side chain). A set of compounds might be represented as one or more core molecules that are constant along with several possible R groups. For example, Free and Wilson describe a single core molecule with two locations for substitutions (i.e. R_1 and R_2). In their data, the possible values of R_1 were either H or CH_3 while R_2 could have been $N(CH_3)_2$ or $N(C_2H_5)_2$. Based on this, there are four possible molecules that could be represented in this way. In practice, the number of "levels" in the *R* groups is typically large.

The *Free-Wilson model* is a simple ANOVA model that is additive in the R groups. A possible design matrix for these data might be

Molecule	Intercept	R_1	R_2
Core + H + $N(CH_3)_2$	1	0	0
Core + CH_3 + $N(CH_3)_2$	1	1	0
Core + H + $N(C_2H_5)_2$	1	0	1
Core + CH_3 + $N(C_2H_5)_2$	1	1	1

where R_1 is an indicator for compounds that contain CH_3 and R_2 is an indicator for $N(C_2H_5)_2$. To understand the relationship between potency and molecular structure, a linear model could be used:

$$y_i = \mu + \alpha R_{i1} + \beta R_{i2} + \epsilon_i$$

where y_i are the potency values for compound i, μ is the grand mean and the ϵ_i are the model residuals that might be assumed to be normally distributed under the standard assumptions for ordinary linear regression. The utility of this model is twofold. First, it may be possible to get accurate potency predictions for new molecules whose combinations of R groups have not been synthesized or assayed. This depends on how effective the additive structure of the Free-Wilson model is at describing the data. The second use of the model is for the chemist to understand how changes in structure might increase or decrease potency for the current set of molecules currently under consideration. In practice, a Free-Wilson model would usually have effects for different R groups that have many different substructures (instead of only two distinct values for R_1 and R_2 shown above). This is case, there will be many regression parameter estimates that the chemist can use to understand the effect of adding each R-group structure.

The stereotypical QSAR model tends to be more sophisticated and uses a wider variety of compounds/descriptors and has a stronger focus on prediction. Instead of modeling specific R groups, they might include more general predictors such as atom counts or descriptors of size and charge. More general descriptors (i.e. not based on R groups) would allow more direct prediction of new molecules whereas the Free-Wilson model is confined to predictor compounds within the range of the R groups observed in the data set.

QSAR models can generally be grouped into classification or regression models. While these are imperfect labels, we use classification to denote the prediction of a discrete outcome (e.g. toxic or non-toxic) while regression is used to denote models that predict some continuous value (e.g. EC_{50}, solubility, etc.). The type of predictive model used can vary greatly. In some cases, simple linear regression models are used while others might use more complex machine learning models to fit the data. Most statistical prediction models can be grouped in terms of their bias and variance (Friedman 1997). Models with low variance tend of have high bias.

Examples of these models are linear regression and linear discriminant analysis. They are numerically stable models (i.e. low variance) that lack the ability to model complex trends in the data (i.e. high bias). In the other extreme are models that are very flexible and can fit to most any pattern in the data (hence low bias) but may the propensity to over-fit the model to patterns that may or may not generalize to new data. These models also tend to be unstable (i.e. high variance), meaning that changes to the data can have considerable effects on the model. An example of one such model is the artificial neural network (Bishop 2007). The choice between these two class of models tends to depend on the QSAR modeler and their prior education and experiences. For descriptions and discussions of different types of predictive models, see Hastie et al. (2008) or Kuhn and Johnson (2013).

An an example, Kauffman and Jurs (2001) use predictive models to estimate the potency of cyclooxygenase-2 (COX-2) inhibitors. They modeled the $log(EC_{50})$ values of compounds that were created by four chemical series. Their focus was on topological descriptors, which are derived by converting the SMILES string to a 2D network diagram of atoms then calculating summary metrics on this graph. These descriptors convey information related to the size and shape of the molecule (among other characteristics). Their models used such 74 descriptors. They evaluated two different models to predict the $log(EC_{50})$ values: ordinary linear regression and artificial neural networks.

Given an initial pool of 273 compound, they split the data into three partitions:

- a *training set* of 220 compounds used to estimate (or "train") the model parameters
- a *validation set* of 26 compounds used primarily to tune the neural network meta-parameters and
- a *test set* of 27 compounds that are utilized to obtain an unbiased estimate of model performance.

The test set root mean square error value for the ordinary linear regression was 0.655 log units and the neural networks was able to obtain a value of 0.625 log units. There was no appreciable difference between the neural network and linear regression models that were above and beyond the experimental variation. Given a new set of compound structures, the potency can be predicted and these values can be used to rank or prioritize which compounds should be synthesized or given more attention.

Note that the model building process has an emphasis on empirical validation using a set of samples specifically reserved for this purpose. While is an extremely important characteristic of predictive modeling in general, there is an added emphasis using QSAR modeling. This is a stark contrast to most classical statistical methods where statistical hypothesis tests (e.g. lack of fit) are calculated from the training set statics and used to validate the model. The emphasis on a separate test set of samples is not often taught in most regression modeling textbooks. Also, for many classic statistical models, the appropriateness of the model is might be judged by a statistical criterion that is not related to model accuracy (e.g. the binomial likelihood). Here, the focus is on creating the most accurate model rather than the most statistically legitimate model. One would hope that a statistically sound model would be the most accurate but this is not always the case. Friedman (2001) describes an example where "[...] degrading the likelihood by overfitting actually *improves* misclassification error rates. Although perhaps counterintuitive, this is not a contradiction; likelihood and error rate measure different aspects of fit quality."

There are abundant examples of QSAR models in journals such as the *Journal of Chemical Information and Modeling*, the *Journal of Cheminformatics*, the *Journal of Chemometrics*, *Molecular Informatics* and *Chemometrics and Intelligent Laboratory Systems*. Many of the methodologies developed in these sources have applications outside of QSAR modeling. Additionally, it is very common for articles on these journals to contain the sample data in supplementary files, which enables the reader to reproduce and extend the techniques discussed in the manuscripts.

6.4 Non-structure Models

The intent of QSAR models is to use existing data to predict important characteristics to increase the efficiency and effectiveness of drug design. While structural descriptors are often used, there are occasions where assay data exists that can be used instead.

For example, compound "de-risking" is the process of understanding potential toxicological liabilities based on existing data. Maglich et al. (2014) created models to assess the potential for reproductive toxicity in males by measuring a compound's effect on steroidogenic pathways. They used 83 compounds known to be reproductive toxicants and 79 "clean" compounds and ran assays to measure a number hormone levels as well as the RNA expression of several important genes. These models can then be used to screen new molecules for reproductive toxicity issues.

In another safety-related model, Sedykh et al. (2010) use the biological *in vitro* assay outcomes from dose-response curves as predictors of *in vivo* toxicity. Their analysis showed that a model using the assay data and molecular descriptors improved the predictive power of the model.

6.5 Other Aspects of QSAR Models

The field of QSAR modeling has matured to the point where the current literature has been exploring the more subtle issues related to predictive modeling.

6.5.1 Applicability Domains and Model Confidence

One key consideration for project teams is choosing between *global* and *local* models. Local models are built using the compounds that have been generated to-date for the current project. Global models are built using a much larger, broader set of compounds. In some cases, there may be global and local QSAR models for the same characteristic. In theory, global models should be better than local models since they have typically use more diverse data to train the models. However, in practice, local models tend to do better since they are built with the most *relevant data*. The earlier example from Kauffman and Jurs (2001) was a local model that used compounds from four chemical series (as opposed to a highly diverse compound set).

This phenomenon has lead computational chemists to create *applicability domain* models (Jaworska et al. 2005; Netzeva et al. 2005; Weaver and Gleeson 2008) to supplement the predictive QSAR model. When predicting a new compound, the applicability domain model attempts to quantify how similar the compound is to those used to train the QSAR model. In this way, the degree of extrapolation can be assessed when a chemist is presented with a model prediction.

Another approach to evaluating the quality of a model prediction is to include some measure of uncertainty with the prediction. For example, prediction intervals (Myers 1990) for a new sample can be calculated for many models. These intervals reflect the noise in the individual prediction and include the noise in the model as well as the how much extrapolation is involved. Prediction intervals are usually wider than the more commonly used confidence intervals (which correspond to the model mean value rather than the individual prediction). One issue with prediction intervals is that they can only be analytically calculated for a small set of simple models (e.g. linear regression). More computationally intensive methods for creating these intervals exist using bootstrapping (Mojirsheibani 1998; Mojirsheibani and Tibshirani 1996) but have not been evaluated with complex machine learning models. Another approach to conveying uncertainty in the predictions is to create a surrogate measure than has a correlation with the quality of the prediction. For example, Keefer et al. (2013) describe a confidence metric for regression models. This metric is based on a measure of model error for compounds that are near the sample being predicted. They also show that this error rate is correlated with the local test set error and thus reflects the quality of prediction.

6.5.2 Training and Test Set Selection

As knowledge and experience is accrued by medicinal chemistry community, the types of compounds being currently developed can be very different then those created in prior years. Additionally, long-run assays may drift over time or the assay format may change. As structures and assays trends change over time, the modeler should be reassessing what data should be used to train and validate QSAR models. A good training set should be structurally diverse but should also contain relevant data.

It is not uncommon to update models over time as new compound data are generated. Additionally, simple random sampling may not be the best approach for splitting data into training and test sets. For example, a strong argument can be made to use the most recent data for the test set. Also, since the underlying SAR assumption is that structurally similar compounds have similar characteristics, it would make sense to ensure that two compounds that have a high degree of structural similarity are included in either the training or test set but not both. There are several ways to accomplish this, as described in Martin et al. (2012). One method, maximum dissimilarity sampling (Clark 1997; Snarey et al. 1997), adds compounds to the training or test sets that are most dissimilar to those already included in those collections.

6.5.3 Reliability of Assay Data and the Effect on QSAR Models

When developing QSAR models using assay results, it is critical to understand the nature of the experimental data. For example, it is not uncommon for assays to be run at multiple locations, including outsourcing companies. It is critical to understand the potential differences and biases in the data caused by inconsistent results across sites. Unwanted systematic differences in the experimental results will directly propagate noise into model predictions.

Similarly, the amount of experimental noise should be quantified prior to QSAR modeling. One way of accomplishing this is the use variance component analysis (Brown and Prescott 2006; Burdick et al. 2005) to quantify different sources of variation within an assay. For example, a basic experimental design might be to measure 30 chemically diverse compounds across 3 weeks. Within each week, each compound is replicated twice. The results of this experiment would allow the statistician to compute components of variation related to the compound-to-compound noise, the week-to-week noise and the within-week noise. The last term actually measures all other noise sources that are unrelated to compound or week. If the assay is run using compounds on plates, this term might reflect any "plate effects" where the assay results are partially determined based on their location on the plate.

In the case where the assay is perfect, all of the variation would be attributed to the differences between compounds. Here, the week-to-week and within-week variance components would be zero. In the worse case where the assay is noisy to the point of being uninformative, the compound-to-compound variance would be the smallest. In practice, the results are somewhere in-between. Regardless, these variance components can give direction to the scientists on where the assay could be improved (e.g. within plate effects).

The QSAR modeler can also benefit from these results. Suppose that 50 % of the variation in the data is not related to the compound but caused by the process of conducting the assay. This limits the ability of any model to accurately predict compounds. One metric used to measure the performance of a predictive model is the R^2 statistic, a.k.a. the coefficient of determination. This value measures how much of the information in the data can be explained by a model. If R^2 is near one, the model is able to explain most of the patterns in the data while a value near zero implies that the model is ineffective. If 50 % of the variation in the data is irreducible measurement noise, the modeler should not expect the R^2 statistic to be greater than 0.5. Knowing this prior to modeling prevents the modeler from "chasing noise" and helps the project team understand the limitations of the assay data and, by extension, the QSAR model.

As an example, a lab scientist requested an assay characterization analysis of an ADME permeability assay. The assay produces a ratio of two experimental measurements. At the time of the analysis, 52 compounds had been generated with at least two replicates across different weeks (overall there were between two and six replicates). On average, there were 2.4 replicates per compound. A visualization of the per-week averages are shown in Fig. 6.1. The *x*-axis is the ratio data on the log scale and the compounds have been sorted by their assay value. The colors indicate different weeks. A variance component analysis of these data contained model terms for compound-to-compound, week-to-week and within-week (which is effectively the final residual term). Once the variance components were estimated using the log of the ratios, the greatest source of variation was compound-to-compound (95.0 % of the total variance), followed by week-to-week (0.5 %) and within-week (4.5 %). Although there is some variation in Fig. 6.1, which appears to increase at the very

Fig. 6.1 Experimental results for a measurement systems analysis of an ADME endpoint

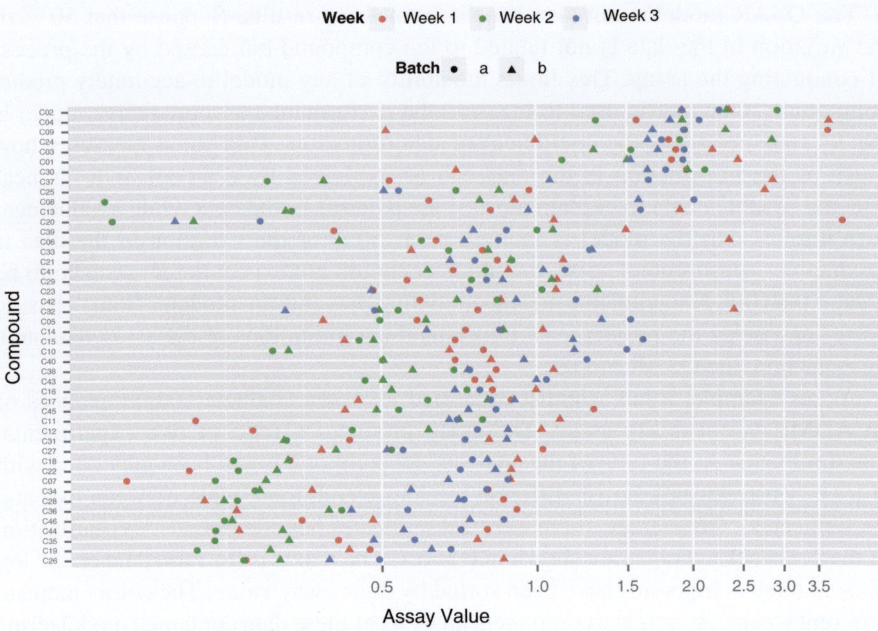

Fig. 6.2 Experimental results for a RNA expression assay

high-end of the assay dynamic range, the overall effect is overshadowed by the natural variation between compounds. These data are a good potential substrate for a QSAR model.

Another example is a characterization done for a cell-based assay that measures RNA expression. In this case, an experiment was designed using 46 compounds. Each was assayed across 3 weeks and, within each week, two separate preparations of the assay materials were created. Within each preparation, three technical replicates were generated. The assay uses a plate-based format and each plate contained all 46 compounds and the technical replicates were generated within the same plate. Figure 6.2 shows the data averaged over the technical replicates. Clearly, there is considerable variation in the data and the noise generated by the measurement system is likely to be larger than the compound-to-compound variation. A variance component analysis here had terms for compound-to-compound, week-to-week, batch-within-week and within-batch. As in the previous example, the last term is a catch-all for other sources of variation not accounted for by the other terms. The percentages of total variation in the data were estimated as follows: batch-within-week (49.9 %), compound-to-compound (36.8 %), week-to-week (11.4 %) and within-batch (1.9 %). Approximately two thirds of the variation in the data is attributable to unwanted systematic factors caused by the assay. This should help set expectations of the QSAR modeler and the project team regarding the utility of the assay as well as motivating the need to conduct experiments to improve the assay robustness.

6.5.4 Multiparameter Optimization

A project team working to develop new chemical matter are tasked with balancing a multitude of different properties. For example, the compound must be potent enough to show efficacy but cannot cause toxicities. Making sure that all the properties are taken into account can be difficult when predictions from numerous models are presented to the chemist. One approach to mitigating this issue is multiparameter optimization (MPO) where each compound is given a composite score that factors in multiple characteristics. One example is to use desirability functions, first created by Harrington (1965) and popularized by Derringer and Suich (1980) in the context of statistical experimental design. In the context of QSAR modeling, a scientist would assign functions that translates each characteristic of a compound to a measure of desirability between zero (i.e. completely undesirable) to one (i.e. most desirable). For example, Fig. 6.3 shows a set of scores where the target selectivity should be maximized while the active permeability is optimal when the efflux ratio is minimized. These scores are then combined into an overall desirability score (e.g. using a geometric average) that is used to rank compounds. Figure 6.3 shows a contour plot of the geometric average of the two desirability scores across a range of the two individual outcomes. When a new compound is predicted, this surface is used to translate both QSAR predictions to an overall metric.

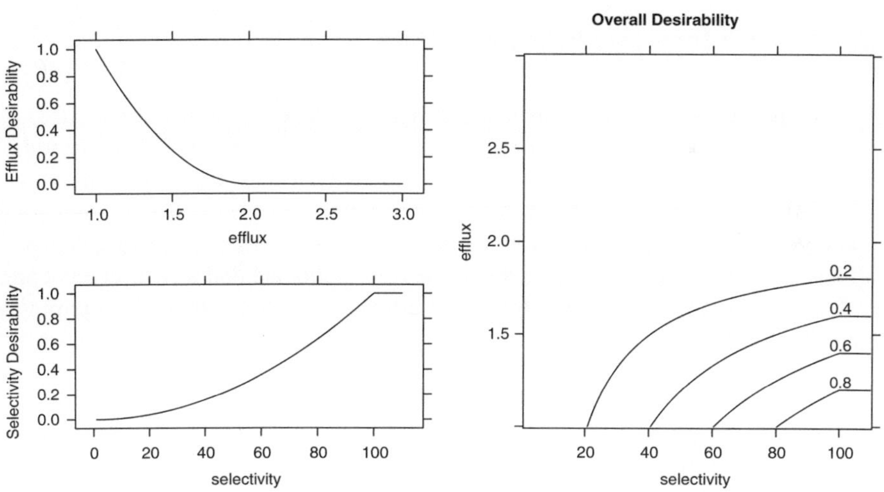

Fig. 6.3 Individual and overall desirability scores

6.5.5 The Effect of Binning

One pathological behavior that is common to QSAR and Chemometric analysis is the binning of continuous data into discrete groups. While this approach to data analysis makes the interpretation of data supposedly easier (e.g. 25 % of compounds have "high activity") the overall effect of binning is overwhelmingly negative. For example, Kenny and Montanari (2013) shows that the discretation of continuous data can significantly inflate correlations my artificially reducing variation. In cases where the continuous data are unimodal, visualization of binned data can over-accentuate patterns in the data that may not be reproducible in future data sets, especially if there is no objective rational for the cut points that define the bins. We have found significant propensity of binning to over-fit the model to the current data set, especially if the data set is not large. As an example, Austin and Brunner (2004) show that binning of data can lead to appreciable increase in the rate of false discoveries of statistically significant relationships. Also, Sect. 20.4 of Kuhn and Johnson (2013) illustrate the degradation of model performance when the model outcome (e.g. potency) is converted from a number to a discrete group. In our experience, scientists tend to bin continuous data not because it is the best approach for the analysis, but because the visualization software with which they are most familiar lack better analytical and graphical tools.

6.6 Cheminformatic Mining of Data

One common question from management might be "can we reduce the number of compounds that are synthesized and still effectively optimize drugs?" To answer this question, a Cheminformatics group might retrospectively analyze the compounds submitted for synthesis and their results. From this analysis, a set of rules may be analytically derived that can reduce synthesis costs. Using a separate data set, perhaps from a different project's data, these rules validated and tested. In any case, the hypothesis can be loosely defined and a heuristic process may be required to arrive at a solution.

Another example of Cheminformatics is to develop tools and analyses that might help medicinal chemists construct new molecules or visualize complex data sets. In large pharmaceutical companies, databases may exist that contain large numbers of previously synthesized compounds and their corresponding assay results. As mentioned in a previous section, medicinal chemists might try to improve a molecule by adding one or more atoms to an end of an existing core structure. A chemoinformatician might interrogate these historical data to find matched molecular pairs (Griffen et al. 2011; Leach et al. 2006). These are pairs of compounds that differ only by a specific, small structural change. For a specific change (or *transformation*), a set of pairs can be tabulated and the resulting effect on the potency (or other characteristics of interest) can be quantified. The changes caused by a specific transformation can help guide the chemists to prioritize compounds for synthesis.

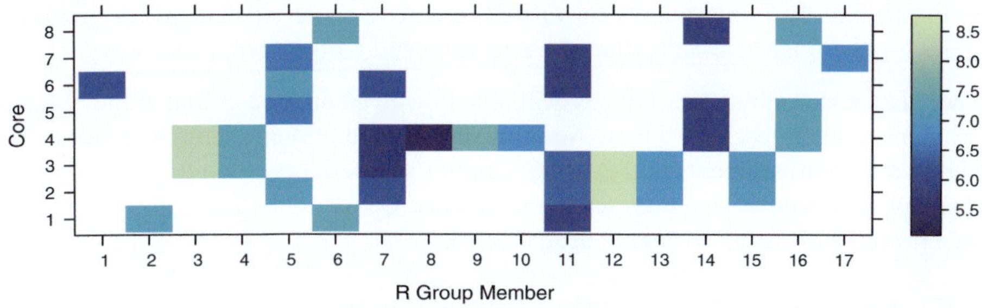

Fig. 6.4 A SAR matrix showing the level of activity for different structural cores and R groups in an existing data set

A related approach is the creation of SAR maps (Agrafiotis et al. 2007) or SAR matrices (Wassermann et al. 2012). An example is shown in Fig. 6.4. From an on-going project, a pre-existing set of compounds with measured potency results were assembled. Using the SAR matrix method of Wassermann et al. (2012), the molecules are broken down into fragments and a subset of structurally similar compounds is created and organized into a two-dimensional matrix structure. The actual fragment structures in Fig. 6.4 are anonymized and, for simplicity, are labeled as the molecule *core* and *R group*. In this figure, smaller potency values are better. From this, the 11th core was consistently potent. Also, structure #4 in the *R* group was, for some compounds, able to achieve good potency. However, the combination of the two was not synthesized or assayed and this visualization might lead a chemist to investigate such a compound. For each data set, many SAR matrices are generated and there are many open questions regarding them, such as

- What makes a good matrix? Are there some that we should ignore? For example, does the sparsity matter?
- How chemically diverse are the compounds in the average SAR matrix?
- Are the visual trends shown in each matrix quantitatively predictive for the missing compounds?

These, and other questions, provide a fertile ground to make a strong statistical impact on compound optimization.

6.7 Summary

The analysis of chemical data for the purpose of drug discovery sets has many advantages:

- frequently large sample sizes
- the predictive nature of the problems allow for immediate feedback loops that help the statistician understand the effectiveness of their work

- in machine learning models for QSAR, creativity is an advantage for creating high-performance models (that are well validated with external data sets).

There are many public data sources available to gain an understanding of the typical problems and to test new ideas. Additionally, computational chemists tend to be highly quantitative scientists that, in this author's experience, tend to be collaborative and data-driven. Given these characteristics, QSAR and Cheminformatics can be a rewarding area to support in drug discovery.

Acknowledgements Thanks to Scot Mente and David Potter for providing feedback on draft versions of this manuscript.

References

Abdi H, Williams L (2010) Principal component analysis. Wiley Interdiscip Rev Comput Stat 2(4):433–459

Agrafiotis DK, Shemanarev M, Connolly PJ, Farnum M, Lobanov VS (2007) SAR maps: a mew SAR visualization technique for medicinal chemists. J Med Chem 50(24):5926–5937

Austin P, Brunner L (2004) Inflation of the type I error rate when a continuous confounding variable is categorized in logistic regression analyses. Stat Med 23(7):1159–1178

Bishop C (2007) Pattern recognition and machine learning. Springer, New York

Brown F (1998) Chemoinformatics: what is it and how does it impact drug discovery? In: Bristol J (ed) Annual reports in medicinal chemistry vol 33. Academic, New York, pp 375–384

Brown H, Prescott R (2006) Applied mixed models in medicine. Wiley, New York

Burdick R, Borror C, Montgomery D (2005) Design and analysis of gauge R&R studies. SIAM, Philadelphia

Clark R (1997) OptiSim: an extended dissimilarity selection method for finding diverse representative subsets'. J Chem Inf Comput Sci 37(6):1181–1188

Derringer G, Suich R (1980) Simultaneous optimization of several response variables. J Qual Technol 12(4):214–219

Free S, Wilson J (1964) A mathematical contribution to structure-activity studies. J Med Chem 7(4):395–399

Friedman J (1997) On bias, variance, 0/1-loss, and the curse-of-dimensionality. Data Min Knowl Disc 1(1):55–77

Friedman J (2001) Greedy function approximation: a gradient boosting machine. Ann Stat 1189–1232

Griffen E, Leach A, Robb G, Warner D (2011) Matched molecular pairs as a medicinal chemistry tool. J Med Chem 54(22):7739–7750

Han J, Kamber M, Pei J (2006) Data mining: concepts and techniques. Morgan Kaufmann, San Francisco

Harrington E (1965) The desirability function. Ind Qual Control 21(10):494–498

Hastie T, Tibshirani R, Friedman J (2008) The elements of statistical learning: data mining, inference and prediction. Springer, Berlin

Jaworska J, Nikolova-Jeliazkova N, Aldenberg T (2005) QSAR applicability domain estimation by projection of the training set descriptor space: a review. Altern Lab Anim 33(5):445–459

Karthikeyan M, Glen R, Bender A (2005) General melting point prediction based on a diverse compound data set and artificial neural networks. J Chem Inf Model 45(3):581–590

Kauffman G, Jurs P (2001) QSAR and *k*-nearest neighbor classification analysis of selective cyclooxygenase-2 inhibitors using topologically-based numerical descriptors. J Chem Inf Comput Sci 41(6):1553–1560

Keefer C, Kauffman G, Gupta R (2013) Interpretable, probability-based confidence metric for continuous quantitative structure-activity relationship models. J Chem Inf Model 53(2): 368–383

Kenny P, Montanari C (2013) Inflation of correlation in the pursuit of drug-likeness. J Comput Aided Mol Des 27(1):1–13

Kuhn M, Johnson K (2013) Applied predictive modeling. Springer, Berlin

Leach A, Gillet V (2007) An introduction to chemoinformatics. Springer, Berlin

Leach A, Jones H, Cosgrove D, Kenny P, Ruston L, MacFaul P, Wood J, Colclough N, Law B (2006) Matched molecular pairs as a guide in the optimization of pharmaceutical properties; a study of aqueous solubility, plasma protein binding and oral exposure. J Med Chem 49(23):6672–6682

Machatha S, Yalkowsky S (2005) Comparison of the octanol/water partition coefficients calculated by ClogP, ACDlogP and KowWin to experimentally determined values. Int J Pharm 294(1–2):185–192

Maglich J, Kuhn M, Chapin R, Pletcher M (2014) More than just hormones: H295R cells as predictors of reproductive toxicity. Reprod Toxicol 45:77–86

Martin T, Harten P, Young D, Muratov E, Golbraikh A, Zhu H, Tropsha A (2012) Does rational selection of training and test sets improve the outcome of QSAR modeling? J Chem Inf Model 52(10):2570–2578

Mojirsheibani M (1998) Iterated bootstrap prediction intervals. Stat Sin 8:489–504

Mojirsheibani M, Tibshirani R (1996) Some results on bootstrap prediction intervals. Can J Stat 24(4):549–568

Myers R (1990) Classical and modern regression with applications, vol 2. Duxbury Press, Belmont, CA

Netzeva T, Worth T, Aldenberg A, Benigni R, Cronin M, Gramatica P, Jaworska JS, Kahn S, Klopman G, Marchant C (2005) Current status of methods for defining the applicability domain of (quantitative) structure-activity relationships. Altern Lab Anim 33:155–173

Schutt R, O'Neil C (2013) Doing data science. O'Reilly, Sebastopol, CA

Sedykh A, Zhu H, Tang H, Zhang L, Richard A, Rusyn I, Tropsha A (2010) Use of *in vitro* HTS-derived concentration-response data as biological descriptors improves the accuracy of QSAR models of *in vivo* toxicity. Environ Health Perspect 119(3):364–370

Snarey M, Terrett N, Willett P, Wilton DJ (1997) Comparison of algorithms for dissimilarity-based compound selection. J Mol Graph Model 15(6):372–385

Tan P, Steinbach M, Kumar V (2005) Introduction to data mining. Addison-Wesley, New York

Wassermann A, Haebel P, Weskamp N, Bajorath J (2012) SAR matrices: automated extraction of information-rich SAR tables from large compound data sets. J Chem Inf Model 52(7): 1769–1776

Weaver S, Gleeson P (2008) The importance of the domain of applicability in QSAR modeling. J Mol Graph Model 26(8):1315–1326

Weininger D (1988) SMILES, a chemical language and information system. 1. Introduction to methodology and encoding rules. J Chem Inf Comput Sci 28(1):31–36

Chapter 7
GWAS for Drug Discovery

Yang Lu, Katherine Perez-Morera and Rita M. Cantor

Abstract Genome-wide association studies (GWAS) are playing a major role in identifying genetic associations with complex traits and disorders. It is anticipated this basic work will lead to a better understanding of the etiologies of such traits and disorders. In pharmacogenetics, application of GWAS methods is evolving. Here, the traits and disorders of interest derive from the diversity in patient response to medications. It is anticipated that GWAS will be successful in identifying the genotypes of individuals who respond well to a medication with correct dosages and highlighting those who will exhibit particular side effects. The GWAS process for a medication includes: (1) an appropriate study sample with variation in the medication response; (2) genotypes measured by arrays in members of the study sample; (3) quality control, imputation and correction for population stratification on the GWAS genotypes; (4) SNP association tested with the drug response using appropriate statistical methods for categorical and/or quantitative traits and appropriately corrected levels of statistical significance. In this chapter, GWAS is motivated by a brief summary of known pharmacogenetics findings. We then follow with the appropriate background information to understand the genetics concepts that support GWAS and a clear delineation of the statistical methods used to conduct and interpret a successful GWAS. We conclude with the future of this evolving methodology.

Keywords Pharmacogenetics • Gene mapping • SNPs • Association testing

Y. Lu
Harbor-UCLA Medical Center, Los Angeles Biomedical Research Institute, Torrance, CA, USA

K. Perez-Morera • R.M. Cantor (✉)
University of California–Los Angeles, Los Angeles, CA, USA
e-mail: RCantor@mednet.ucla.edu

L. Zhang (ed.), *Nonclinical Statistics for Pharmaceutical and Biotechnology Industries*, Statistics for Biology and Health, DOI 10.1007/978-3-319-23558-5_7

7.1 Emerging Pharmacogenetics and the Anticipated Role of GWAS

Genome-wide association studies (GWAS) are poised to make major contributions to the emerging field of personalized medicine. This field is focused on identifying the genetic differences among individuals, which can successfully lead to very specific risk assessments and subsequent treatments based on particular genetic profiles. Pharmacogenetics is the application of this concept to the inter-individual differences in responses to medications. Over the last 50 years, it has become increasingly clear that there are substantial differences in individual responses to the drugs used to treat symptoms and illnesses. These differences include the overall response to the particular drug, the response to specific doses, and to its adverse effects. A substantial portion of these differences are likely to result from the genetic variations among the patients. However, most of those genetic contributors have not yet been identified. It is anticipated that once appropriate genetic drug response profiles are identified, they will provide this critical information for successful applications to drug indications, dosage, warnings, and ultimately indicate precautions on drug interactions and adverse effects. This form of personalized medicine is becoming available and applied to a growing number of drug molecules, in medical fields ranging from infectious diseases to oncology.

In this first section, we review some of the developments in pharmacogenetics in order to motivate the need for GWAS studies and illustrate feasibility. We follow this broad introduction with a review of the essential basic genetics, statistics and design principles that will promote for an understanding of the implementation and interpretation of GWAS in the subsequent sections.

Personalized medicine has been a factor in drug treatment since at least the 1960s. During that decade the genetics of anesthesia use, allergic reactions, and *warfarin* dosing were explored (Kalow 1964; West and Harris 1964; Solomon 1968). We illustrate the evolution of this preliminary recognition to by describing the history of *warfarin*, an anticoagulant that was initially used to reduce the rat population in the early 1950s (Davis 1951). Later in that decade, reports of *warfarin*'s use as a human anticoagulant became increasingly common, and by 1956, it was referred to as the "anticoagulant of choice" (Shapiro 1953). As its use grew, clinicians began reporting an unusual bleeding response in some patients. Initially, the response to *warfarin* was measured by estimating the prothrombin time (PT). The International Normalized Ratio (INR) supplanted the PT, because it allowed standardization of clotting measurements. As this approach was being developed and applied, clinicians began observing widely disparate INRs among subjects who were being given the same dosage of *warfarin* (Bentley et al. 1986; Morrison et al. 1989). This was attributed to differences in the metabolism of warfarin (Solomon 1968). Identifying the source of the discrepancies was critical for *warfarin*'s usage since the drug has a narrow therapeutic range, and maintaining INR values within the required range can result in patients having a healthy life, a potentially lethal bleeding episode, or a crippling coagulation episode.

In recent years, successful GWAS have confirmed the role of two genetic alleles as the sources of the differences in the response to *warfarin*. These alleles reside in the VKORC1 and CYP2C9 genes (Takeuchi et al. 2009). Knowledge of whether an individual carries these polymorphisms is essential for predicting the patient response. However, clinical use of VKORC1 and CYP2C9 allele testing for *warfarin* use is currently somewhat controversial, because the cost of testing everyone on whom the drug is to be used exceeds that of the gain in savings that results from testing patients for the alleles (Patrick et al. 2009). On the other hand, there is also support for its application in certain subpopulations (Patrick et al. 2009). Fortunately, as personalized medicine and pharmacogenetics become applied in a wide range of situations, the cost of testing will be reduced overall, and will likely cease to be a long-term concern.

The recent addition of a Black Box Warning to *abacavir*'s label by the FDA is another indication of the increasing importance of genotyping patients for risk alleles before medication usage. *Abacavir* is an effective medication in the treatment of HIV. However, this drug is a compound that can lead to severe multi-organ hypersensitivity reaction among individuals who are genetically at risk due to the presence of the HLA-B*5701 allele. Prior information regarding this allele can be used effectively to prevent fatal events as well as ease fears before the medication is taken. In addition, the *abacavir* FDA warning explains that even in an HLA-B*5701 negative individual, the drug should be discontinued and the patient not re-challenged, upon any sign of a rash (U.S. Food and Drug Administration 2014). This is because HLA-B*5701 is not likely to be the only hypersensitivity allele for *abacavir*, and other risk alleles at the same or different genes remain unknown. It is anticipated that GWAS can be used to identify additional alleles at other genes that predispose to this negative response. Hypersensitivity to *abacavir* is not the only immunopathogenic response that has been studied through pharmacogenetics. Other HLA class I and II alleles have been linked to hypersensitivity reactions precipitated by other medications (Karlin and Phillips 2014). A prominent example is HLA-B*1502 positivity among patients of Southeast Asian (specifically Chinese, Thai, Malaysian and Indian) ancestry and potential hypersensitivity to both *carbamazepine* and *phenytoin* (Chung et al. 2004; Locharernkul et al. 2008; Mehta et al. 2009; Chang et al. 2011) This hypersensitivity could be expressed as the T-cell mediated Stevens-Johnson syndrome/toxic epidermalnecrolysis (SJS/TEN), which is a painful sheet-like mucosal and skin loss, sometimes resulting in death (Chung et al. 2004; Locharernkul et al. 2008; Mehta et al. 2009; Chang et al. 2011).

Pharmacogenetics has come to play an important role in cancer treatment. As one example, *irinotecan*, a topoisomerase 1 inhibitor mostly used for treating colon cancer, may lead to potentially lethal neutropenia, if given to patients that are homozygous at the UGT1A1*28 allele (Innocenti et al. 2004). This particular allele is carried by about 10% of the U.S. population and is more common among individuals of African and European descent (Beutler et al. 1998; Chen et al. 2014).

Variations in genetic alleles also influence the treatment of cardiovascular conditions. Angiotensin-Converting-Enzyme Inhibitors (ACEIs) and HMG-CoA reductase inhibitors are groups of drug entities that have significant clinical use in hypertension and dyslipidemia, respectively. In the 1960s and 1970s, the venom of the Bothrops Jararaca was determined to contain peptidases that lowered blood pressure through multiple mechanisms. The synthetic analogues of these peptidases are ACE Inhibitors and they are well established in the treatment of hypertension to the point that the newly published JNC-8 guidelines emphasize them as one of the four first-line alternatives for hypertension care. Although ACE Inhibitors have successfully been used in hypertension treatment for years, patients of African descent appear to receive less of a tension-lowering benefit from them (Peck et al. 2013). In addition, angioedema, a potentially life-threatening side effect of ACE Inhibitors, has been observed more often among African Americans than among Americans with a European background (Brown et al. 1996). These differences among populations highlight the likely genetic contribution to the responses. Knowing the genotypic differences that result in both the decrease in responsiveness and more severe side effect profile among African Americans could be used to predict whether a specific admixed individual would benefit from an ACE Inhibitor or whether the use of that drug could lead to angioedema. An additional question that emerges in the clinical setting arises with HMG-CoA reductase inhibitors, commonly known as statins. These medications present with an adverse effect profile that often includes myopathy, which in some instances develops into rhabdomyolysis, a potentially lethal condition that can deteriorate into kidney failure. Only a small fraction of the individuals on statins develop the condition and there are GWAS studies underway to identify their risk profiles.

Table 7.1 lists some known pharmacogenetic factors that illustrate the importance of expanding and applying this knowledge to achieve a more precise form of drug treatment. For each drug we provide the disorders for which it has been deemed an effective treatment, the drug response and predisposing alleles and the method by which this knowledge has been achieved. Most findings have been identified with candidate genes, where known biology suggests the best genes to test. GWAS are conducted from a completely unbiased point of view where no gene is prioritized over another. This is preferable for complete gene discovery. As GWAS become an important approach to improve efficacy and safety in pharmacotherapy, we are likely to move into an era when policy makers and public and private insurance payers will analyze whether the cost for genotyping to achieve the benefits of personalized medicine warrants the cost. Currently, however, the essential information to achieve this goal remains sparse. The limited available literature suggests that the cost effectiveness of genotyping is sensitive to its actual per capita cost, which is diminishing due to technological advances. For example, in the case of *warfarin* dosing for patients with atrial fibrillation, the cost effectiveness ratio of genotyping is less than $20,000 per quality-adjusted life year (QALY), far below the commonly accepted threshold of $100,000/QALY, if genotyping costs $200 per person (Patrick et al. 2009). Moreover, with current higher life expectancies, patients will often bear the burden of multiple chronic conditions. Drug-related predictive genetic information will therefore be essential for good health, making genotyping even more cost effective.

Table 7.1 Drugs showing pharmacogenetics effects of genetic alleles

Drug name	Drug indication(s)[a]	Effect allele(s) (Reference)	Method of allele identification
Abacavir	HIV-1 (U.S. Food and Drug Administration et al. 2013)	Multiorgan clinical syndrome **HLA-B*5701** (Mallal et al. 2002)	Candidate gene
ACE Inhibitors	Hypertension, acute myocardial infarction, heart failure, myocardial infarction prophylaxis, postmyocardial infarction, reduction of cardiovascular mortality, stroke prophylaxis[b,c] (Clinical Pharmacology Database 2010a, b, 2012, 2013a; Byrd et al. 2008; Pare et al. 2013)	Angiodema risk-**MME** **rs989692**[d] (Pare et al. 2013)	GWAS,[e] Candidate gene
Allopurinol	Gout, hyperuricemia, nephrolithiasis[b] (Clinical Pharmacology Database 2011)	Serious dermatological reaction **HLA-B*5801** (Hung et al. 2005)	Candidate gene
Atomoxetine	ADHD (U.S. Food and Drug Administration, Center for Drug Evaluation and Research 2014b)	Increased response (poor metabolizers) **CYP2D6**[f] (Ring et al. 2002) Response **NET/SLC6A2**[g] (Ramoz et al. 2009)	NET/SLC6A2: candidate gene[g]
Carbamazepine	Epilepsy, trigeminal neuralgia	Serious dermatological reaction **HLA-B*1502** (Man et al. 2007)	Candidate gene

(continued)

Table 7.1 (continued)

Drug name	Drug indication(s)[a]	Effect allele(s) (Reference)	Method of allele identification
HMG-CoA reductase Inhibitor	Hypercholesterolemia, hyperlipoproteinemia, hypertriglyceridemia, myocardial infarction prophylaxis and stroke prophylaxis[b,h] (Clinical Pharmacology Database 2013b)	Myalgia risk **SLCO1B1/rs4363657** (Gelissen and McLachlan 2014; The SEARCH Collaborative Group 2008)	GWAS
Irinotecan	Colorectal metastatic carcinoma (U.S. Food and Drug Administration, Center for Drug Evaluation and Research 2006)	Neutropenia risk- **UGT1A1*28** (Ando et al. 1998)	Candidate gene
Phenytoin	Seizures (U.S. Food and Drug Administration, Center for Drug Evaluation and Research 2014a)	Serious dermatological reaction **HLA-B*1502** (Chung et al. 2004; Man et al. 2007)	Candidate gene
Warfarin	Coagulation prevention	Bleeding risk- **CYP2C9*2 and *3** and **VKORC1 1639** (Aithal et al. 1999; D'Andrea et al. 2005)	CYP2C9: PCR genotyping[i] VKORC1: pos cloning, candidate gene, GWAS

[a]Indication refers to FDA label except for drug classes HMG-CoA reductase inhibitor and ACE Inhibitors
[b]Information unavailable from FDA website. Obtained from Clinical Pharmacology Database
[c]Based on *lisinopril, enalapril, fosinopril,* and *ramipril*. These ACE Inhibitors were used by subjects in the Nashville Tennessee study. Also according to Pare et al. (2013), *ramipril* was used in the ONTARGET trial
[d]MME polymorphism was identified via candidate gene analysis
[e]No association of genome-wide significance has been found
[f]Might not be clinically significant because low-affinity enzymes might take over metabolization when 2D6 activity is compromised
[g]Needs further assessment
[h]Indications specific for *simvastatin* since SLCO1B1-simvastatin association is currently the most robust
[i]By the time genotyping efforts were undertaken for CYP2C9, warfarin had been in use for decades. Once alleles were identified for patients in the clinical setting, researchers matched individuals' alleles to the warfarin dose were the respective patient had been stabilized

7.2 GWAS Background: Overview of Gene Mapping

Our genes are located in a linear fashion along strands of deoxyribonucleic acid (DNA) that are divided into 23 pairs of chromosomes within the cell nucleus. This linear architecture provides the opportunity to map and identify the genes that contribute to traits such as drug response. To begin, we assess whether a trait such as drug response has a genetic component and we estimate the trait's heritability. A significant heritability means that a fraction of the inter-individual variation in that trait is the result of variation due to genes (The 1000 Genomes Project Consortium 2010). These heritable traits are excellent candidates for gene mapping, which is designed to identify the specific genes contributing to the trait using information on gene marker locations along the chromosomes. Unfortunately, it is usually difficult to assess whether a particular response to a drug is heritable. This information would require the analysis of panels of related individuals, such as monozygotic and dizygotic twin pairs, who both receive the same pharmacologic agent. Concordance of response in those who share a greater proportion of their genes, the monozygotic twins, should be statistically greater than concordance in the dizygotic twin pairs. Such information is rarely available in an observational setting.

Linkage analysis is a well-established analytic tool that was used very extensively during the 1960s through the middle of the 2000s to map the genes for heritable traits to their chromosome locations. Linkage is based on a process that has been referred to as "reverse genetics". That is, the approach works in the reverse order than the model describing how genes operate, biologically. While genes act in a forward fashion to produce and regulate the proteins that lead to a trait, reverse genetics starts with individuals having been measured for the trait, and uses linkage analysis, GWAS or other approaches to identify the predisposing genes. Here, the genes are identified last.

Reverse genetics became fully feasible in the 1990s, when a very extensive panel of multi-allelic markers or genetic variations spanning the human genome was identified. Linkage analysis is a statistical method that follows genotyped marker alleles and measured trait values within family pedigrees to identify chromosome regions where the marker alleles and trait values are aligned. Alignment is assessed statistically and helps us infer that alignment is seen in a certain chromosome region more than one would expect by chance alone. That is, we infer that the gene leading to some aspect of the trait value is 'linked' to a marker with a known location along the chromosomes. If the whole genome is analyzed for linkage with the trait, the approach is referred to as a full genome linkage scan. If specific genes are tested, they are referred to as "candidate genes". In summary, in the regions exhibiting significant linkage, we infer that the gene affecting the trait is close to the linked marker, and reverse genetics is achieved. With linkage, the resolution at the locus is usually quite poor, as many genes can reside within a linked region. Nevertheless, a statistically significant linkage result limits the search for the predisposing gene to those in the linked region. Using the advances made by the Human Genome Project, reverse genetics has been very effective in identifying genes causing rare single gene traits that only result from mutations in a single or a few different genes.

A major change in gene mapping occurred during the last 10 years, with the development of the essential tools for successful GWAS. "Reverse genetics" is also used to conduct GWAS, however, the genotyped markers are bi-allelic, having only two versions, and referred to as single nucleotide polymorphisms (SNPs). These markers occur more frequently and are spaced much more closely than linkage markers. The collection of SNPs on our chromosomes developed over evolution. SNPs are random changes to the DNA that are passed down among humans over time. Most SNPs have no detectable effects on those who inherit them, but their proximity on the chromosomes to the changes that do predispose to traits of interest makes them valuable. This chapter is devoted to presenting the current study designs and methods for testing the GWAS SNPs to identify predisposing genes through association, which can ultimately inform pharmacogenetics.

7.3 Concepts, Designs and Statistical Methods for GWAS

We begin with a brief overview of the GWAS approach, assuming the trait under analysis is binary. For example, an individual either exhibits a particular side effect when given the drug, or they do not. Individuals that exhibit this trait form the sample of cases, and a group matched for age, sex, ethnicity, and other relevant factors, including ethnicity, form the control group. Both samples are genotyped using very dense SNP marker panels that have become available on commercial arrays. These genotyping arrays have evolved over time to contain an increasing number of SNPs, and most often we see studies with one million SNPs, each having a known location along the chromosomes. Genes that contribute to the trait are identified by statistically testing the individual SNPs for an association with the trait, in this case comparing the cases and controls. It should be emphasized that the SNPs themselves are just landmarks along the chromosomes and are not likely to predispose to developing the trait. They just mark alleles that contribute to the trait by their proximity. That is, identifying that a SNP is associated with the trait indicates there is likely to be an allele of a gene that contributes to the risk for developing the trait within a small chromosomal region close to that SNP. This is because the commercially available GWAS SNP panels have been designed to capitalize on an important genetic feature referred to as linkage disequilibrium (LD).

7.3.1 Linkage Disequilibrium Among SNPs that Are in Close Proximity

LD is a genetic factor that allows us to conduct GWAS arrays that provide the genotypes of one million SNPs when there are an estimated 30 million over the whole population. It is ubiquitous along the chromosomes and reflects the genetic

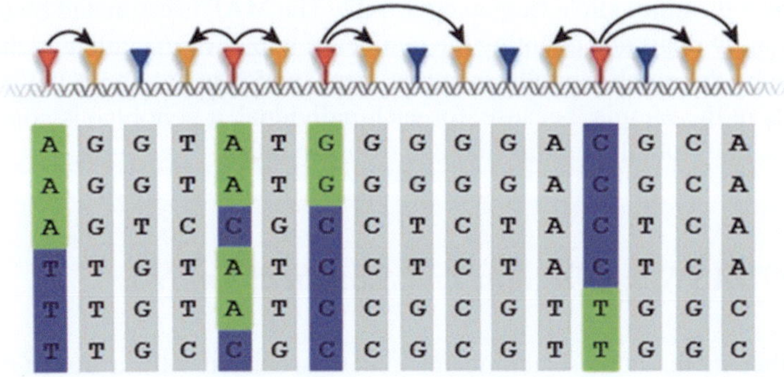

Fig. 7.1 Illustration of tagging SNPs

history of a population. Thus, we expect LD to be found between the trait allele and a nearby SNP allele if the trait and SNP alleles have traveled together in close proximity on the same chromosome throughout time and there have been only a few crossovers between them. Crossovers occur when gametes are formed and the probability of a crossover between two SNPs is correlated with the distance between the SNPs. We can use the genotype of one allele to predict the genotype of the other if LD exists between the two. The mechanism by which fewer genotyped SNPs can capture the variation of other SNPs in close proximity is referred to as "tagging." Tagging is illustrated Fig. 7.1, where the SNP genotype sequence for 16 SNPs is given for 6 people, with each person's DNA sequence represented by a row. It is clear that the SNPs are not distributed independently within individuals. There are patterns that reflect LD. For example, the first and second SNPs are in LD and they 'tag' each other. Therefore, only one of them has to be genotyped. If a person has an "A" at the first SNP, they will have a "G" at the second. Likewise, a "T" at the first indicates there is a "T" at the second. Similarly, SNPs 4, 5 and 6 tag each other.

In the next sections we describe the biological concepts applied in conducting GWAS and explain the appropriate statistical methods to identify important associations leading to enhancement of our pharmacogenetics knowledge.

7.3.2 Alleles and Minor Allele Frequencies

Alleles are the key elements in genetic variation. They differ from each other in a small genetic element and result in alternate versions of the same gene. Some alleles will change the protein product of a gene, some will affect the amount of protein product produced, and some will be neutral and have no obvious effect. Population geneticists study the frequency of alleles and make inferences about population history from them. Those interested in personalized medicine use the frequencies to identify the genes contributing to traits. One way to describe an allele in a population is to estimate its minor allele frequency (MAF). The MAF is estimated by drawing a sample of individuals and counting the number of copies of that allele in the sample and divide by the number of alleles at that locus, which is two times the number of individuals in the sample, because, for each individual, a genotype is composed of two alleles, one from the mother and one from the father.

7.3.3 GWAS Study Designs and Statistical Analyses

Unlike other methods of searching for genes associated with a trait, GWAS are noteworthy for being genome-wide, producing an unbiased search that does not rely on candidate genes. To begin, the study sample is collected. Then, a microarray of SNPs is genotyped for each individual. The microarray is purchased from a company that specializes in genotyping chips and the genotyping is usually done at a facility that specializes in that work. Most genetic centers have core facilities with the technology and expertise to generate genotypes. The current commercial genotyping chips have about one million SNPs on them. Sometimes there are specialized chips that have been designed for specific classes of disorders. Regardless of the microarray used, the methodology to clean and process the data will be the same. This is described in Sect. 7.5. Currently, there is preliminary work generating SNP genotypes using exome or whole genome sequencing, but the high cost and other considerations associated with data processing have not yet made it fully feasible.

7.3.3.1 Case and Control Designs for GWAS

A case and control study design is used when the outcome of interest is dichotomous. For example, a dichotomous trait in pharmacogenetics might be whether an individual shows response to a drug. In the study sample of those who take the drug, those who have a positive response might be categorized as "cases", and those who do not might be categorized as "controls". All individuals are genotyped using the same microarray platform. There will be one million genotypes per individual, so special software to accommodate this volume of data is required. The PLINK software accommodates extensive genotype data (Plink 2009). The files are compressed into a binary format, and data cleaning and statistical analyses programmed into PLINK are conducted on the files that have been compressed. The case and control samples are tested for differences in MAF over the one million SNPs, each undergoing a separate statistical test. Multiple testing is a major concern, and the accepted GWAS approach is discussed in Sect. 7.4.1. The cases and controls should all be from the same ethnic group, as different ethnicities have different allele frequencies, which could lead to a significant number of type 1 statistical errors.

7.3.3.1.1 Case Control Studies: Tests of Equality of Proportions of Allele Frequencies

The key analytic element of a GWAS is the MAF of a SNP. In a case control study, the MAFs of the genotyped SNPs are compared using a test of equality of proportions. The null hypothesis is that there is no difference in MAF between the cases and controls. The alternative can have three forms, depending upon how the SNP is expected to act. The first is that there is a difference in the MAF between the cases and controls. If the alternative is that the MAF predisposes to the trait, there will be a one-sided test that the case MAF exceeds the control MAF. If the opposite is the alternative, the allele is expected to be protective. The data on allele counts, two for each individual, populate a 2×2 contingency table with the case and control labels across the top and the two SNP alleles down the side. A χ^2 test of difference in proportions is then conducted (Agresti 2013). If the MAF is very small, Fisher's Exact Test is used. Each SNP is tested separately and those exceeding a threshold p-value are considered associated with the trait. Replication of the association in an independent study sample is essential.

7.3.3.1.2 Case Control Studies: Cochran–Armitage Trend Test of Genotype Frequencies

If we believe that the trait of interest varies according to the genotype of the individual (that is, how many minor alleles are present), the Cochran–Armitage Trend Test is used (Agresti 2013). It compares the genotype distributions of the case and control samples using a 2×3 contingency table, where each individual contributes their genotype to the count. Again, each SNP is tested separately and those exceeding the p-value threshold are considered associated with the trait. However, here we assume the genotypes follow an additive genetic model, where the number of minor alleles a person carries for the SNP has an impact on the trait.

7.3.3.1.3 Case Control Studies with Covariates: Logistic Regression

If there are known or expected covariates that influence the trait, their inclusion in the analyses is accomplished using logistic regression (Agresti 2013). The genotypes for each SNP are coded by counting the number of minor alleles, which are 0, 1 or 2. Each SNP is then tested using a logistic regression model having the significant covariates. The regression coefficient for the SNP is tested for significance to identify association.

7.3.3.2 Family Based Designs for GWAS

An alternative study design to those collecting samples of cases and controls uses family data. This approach to conducting a SNP association analysis is based on allele transmission rather than on a comparison of MAFs. The study sample is a collection of trios composed of the person with the trait and his or her parents. All three members of the trios are genotyped and the GWAS is conducted using the statistical analyses described below. The SNP association test has an expected value of 0.5. That is, at a particular SNP, each parent will transmit one allele of their genotype to their child; their other allele is not transmitted. There is, thus, a comparison of two SNP alleles for each parent: the transmitted versus the un-transmitted. The test conducted is referred to as the Transmission Disequilibrium Test (TDT). It identified those alleles that are preferentially transmitted to those who exhibit the trait of interest. An important feature of this approach is that the test protects a study from population stratification, where the difference in allele frequencies among populations can lead to type 1 statistical errors, that can occur when the cases and controls are from different ethnic groups. The TDT test can be performed using the PLINK software. For the TDT, McNemar's Test for Matched Pairs (Agresti 2013) is used. The expectation of the null hypothesis is such: over the sample of trios, the two alleles of the SNP will be transmitted from the parents to the child with the trait in a 50:50 fashion. For each SNP, only the data from parents that carry two different alleles is used. McNemar's test does not allow the inclusion of covariates or the analysis of quantitative traits. Using other statistical approaches, numerous extensions of the TDT that include covariates sibling pairs and quantitative traits have been published (Lu and Cantor 2007; Lange et al. 2002).

7.3.3.3 GWAS with Quantitative Traits

In the field of pharmacogenetics, precision is critical. For example, we may need to determine the optimal dosing for patients based on their genetic background, or the response to a drug may be may be best measured by a quantitative trait such as the severity of a negative response. Fortunately, GWAS can also be conducted on such traits. Here, the analysis is conducted on the sample of individuals receiving an effective dose or a sample of individuals that exhibit the negative response to the drug. The GWAS will identify those SNPs that are associated with the most effective dose or the degree of negative response in individuals. If the trait is normally distributed in the sample, a one-way Analysis of Variance can be used (Scheffé 1959). For a given SNP, the categories will be its genotypes. For examples, for alleles "a" and "b", the genotypic categories are "aa", "ab" and "bb". Each individual is categorized by their genotype for that SNP and that SNP then is tested. For quantitative traits that are not normally distributed, a non-parametric test, such as the Wilcoxon Rank Sum Test (Lehmann 2006), can be used.

7.3.3.3.1 Population Based Tests of Quantitative Traits with Covariates: Linear Regression

As with qualitative traits, if adjustment for covariates is appropriate, linear regression can be conducted on a single sample with quantitative values, testing each SNP with a factor for the number of minor alleles in the individual's genotype (Mosteller and Tukey 1977).

7.4 Interpreting GWAS Tests and Statistics

GWAS generate a large number of p-values because so many individual tests are conducted on the individual SNPs. In addition, these tests are not independent because the SNPs on the genotyping arrays exhibit some level of LD. The greatest challenge has been to design studies that achieve adequate power to detect the likely effect sizes of the individual SNPs. That is, the power a GWAS provides is a function of the size of the study sample, the statistical test used, and the genetic model of inheritance of the trait under analysis, including the effect size. In reference to the model, for traits where variation is solely the result of a single allele, effect sizes are very large and smaller sample sizes are adequate. However, for traits that have a more complex pattern of inheritance, which happens more often than not, much larger study samples are required to detect the small effects of each of the genes involved. Heritability of the trait is necessary, but not sufficient, as the heritability of a complex trait may be divided among a large number of loci, each with a small effect. Locus-specific effect sizes are not easily predicted, so that an appropriate choice of a pharmacogenetics trait for analysis may require some luck. Below we present the accepted approaches to setting appropriate levels of significance, then viewing and prioritizing the results.

7.4.1 Identifying Associations: Statistical Significance

The genome-wide level of significance for a GWAS has been set at 5×10^{-8} by the GWAS research community. It is based on a modified Bonferroni correction that takes into account the level of LD among the SNPs on the genotyping platforms. However, many genes have already been identified for responses to pharmacologic agents, and the investigator may want to assess these prior to or instead of testing a full GWAS panel. These genes may include known potential sources of pharmacogentic variability that include the genetic target, HLA, and ADME/PK genes. The methods of analysis are the same as those used for a full GWAS, and only the level of significance to identify SNP associations will be impacted. That is, for targeted studies, those focused on a particular set of genes, chromosome region or genes in a particular genetic pathway, the number of SNPs tested will be fewer and the level of significance will be less stringent than 5×10^{-8}. In most instances of a targeted study, the SNPs are pruned to remove LD and a Bonferroni correction for the number of SNPs tested is then usually applied. Following a full GWAS or a targeted study with a new finding, replication of an association is tested in an independent study sample assessing only the significant SNPs from the first sample. A successful replication is predicated on the same trait and the same SNP being associated in the same direction as observed in the original sample.

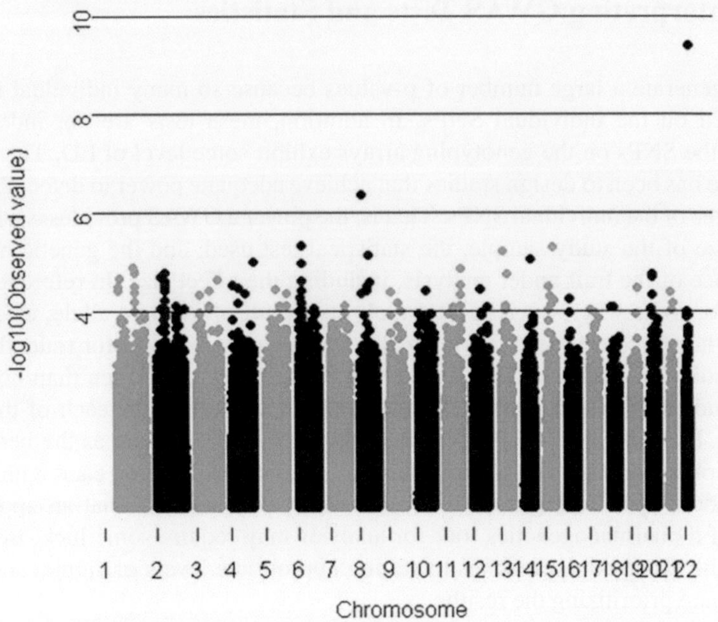

Fig. 7.2 Manhattan plot

7.4.2 *Visualizing Associations with a Manhattan Plot*

With such a large number of analyses, it is can be very helpful to visualize the results. Figure 7.2 is referred to as a Manhattan plot of the p-values of the association tests. It is organized by chromosome along the horizontal axis, going from 1 to 22. Along the horizontal axis, the SNPs are represented in their base pair order for each chromosome. The vertical axis is the negative of the log of the p-value ($-\log_{10}$ (p)), so that the higher along that axis, the more significant the p-value. Chromosome 8 shows a signal higher than most of the others, but it does not cross the 5×10^{-8} threshold for a GWAS. On the other hand, there is a SNP on chromosome 22 that is more significant than the threshold, and is prioritized for further studies.

7.4.3 *Assessing Bias with a Quantile–Quantile Plot*

It may be that there is bias in the sample that has not been corrected. Examples are undetected genetic relationships among the members of the study sample or undetected population stratification due to multiple ethnicities within the study

Fig. 7.3 Q–Q plot

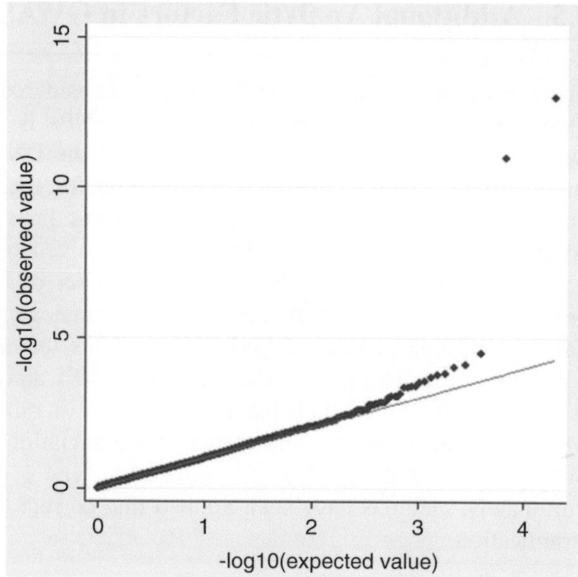

sample. These lead to an increased level of significance that can be detected with a quantile-quantile plot (Q–Q plot). Figure 7.3 is an example of a Q–Q plot. The vertical axis indicates the observed values of the test statistic. Each value is plotted against its expected value under the null distribution. Commonly used test statistics include p-value, χ^2, or t statistic (Pearson and Manolio 2008). When p-values are used as the test statistic, they are usually transformed to $-\log_{10}$ (p) to facilitate plotting very small values. The expected and observed test statistics are ranked from smallest to largest and sorted accordingly on the horizontal and vertical axes. Each dot in the Q–Q plot represents a SNP included in the GWAS analysis. As per the usual practice, an X = Y reference line has been included in the figure. Here, as is also usual for a study that does not exhibit bias, the vast majority of the SNPs are very close to this line.

When there is no association found in a GWAS, all the dots will be on or very close to the line. When there is some association found in a GWAS, one will observe that there is a small set of SNPs (dots) deviating from the reference line. If (1) the SNPs deviating from the line cluster at the end, and (2) the rest of the SNPs closely follow the reference line. that suggests the GWAS associations are true positives. In contrast, there may be potential data issues if SNPs with very high observed values do not also have very high expected values, and a large number of SNPs deviate from the reference line (Ehret 2010). Here our Q–Q plot indicates there are two associated SNPs with highly significant p-values and no detectable bias.

7.5 Additional Analytic Factors in GWAS

There are additional factors which must be considered when conducting a GWAS. First, preliminary work on the genotyped SNPs is critical. This work includes quality control (QC) to assess the quality of the DNA for each individual in the study sample and the quality of genotyping for each SNP. The QC criteria are outlined below. Second, SNP imputation allows us to cover the genome in a more comprehensive way. That is, SNP arrays evolved by increasing the number of genotypes covered. Given the extensive number of GWAS conducted, it is now possible to impute large numbers of additional genotypes for individuals with only one million genotyped SNPs. This allows for more extensive analyses with a minimal cost. Third, the differences in MAFs among the members of a study sample may be due to their having a different or mixed ethnic background. Such MAF differences can be interpreted as true associations when they only derive from differences in ethnicity. This can inflate the type 1 statistical error substantially. Fortunately, methods have been applied that correct for this unwanted population stratification.

7.5.1 Genotype Quality Control for SNP Arrays

QC of the array-produced SNP genotypes is an essential component of a successful GWAS. Analyses are conducted to assess both the DNA quality of individuals (based on summary statistics of their array-based genotypes) and the quality of the platform used to genotype each of the individual SNPs. Specific filtering criteria are used to eliminate individuals and SNPs that do not meet QC standards. One such criterion used to maintain a person in the data set is 'missingness'. The threshold usually used is that if greater than 1% of a person's genotypes are missing, that person is omitted from the analysis. SNPs are omitted if the rate of their 'missingness' among the individuals in the data set exceeds a threshold such as 1 %. One can also evaluate genotyping accuracy by comparing the estimated MAF in the study sample with its value in an online SNP database, such as the Thousand Genomes Project. Hardy-Weinberg equilibrium is also used to assess the quality of SNP genotypes (Gomes et al. 1999). This criterion uses the observed proportions of the three genotypes "aa", "ab", and "bb" in the study sample of controls for each SNP to test whether these observed proportions match a theoretical expectation of $p^2 + 2pq + q^2 = 1$, where p is the estimated MAF and $q = 1 - p$. The PLINK software will conduct this test for each of the one million genotyped SNPs, and thus, multiple testing is an issue for this QC criterion. We usually set 0.0001 as a threshold for this test, and omit those SNPs with a more significant p-value. It should be noted that 100 SNPs are expected to qualify for omission by chance alone if 1,000,000 are tested.

7.5.2 *Genotype Imputation*

Current arrays measure around 1 to 1.5 million SNPs, resulting in some chromosomal regions with minimal coverage. In addition, the SNPs on the different microarray platforms are not the same. Thus, there is not consistent coverage, although there is usually considerable overlap. Fortunately, there is substantial LD in most chromosome regions which allow genotype imputation. Large scale online data sets have provided information on LD and have been used by software to impute SNP genotypes. IMPUTE version 2 is currently the most commonly used software for genotype imputations. It is available at https://mathgen.stats.ox.ac.uk/impute/impute_v2.html.

7.5.3 *Population Stratification*

Population stratification results from an unrecognized mixture of populations within a sample and within individuals. Differences in MAF are inherent in different populations but are also the hallmark of a SNP association. Thus, population stratification can lead to type 1 statistical errors when using GWAS to identify associations, and it is important to recognize and correct for it population. The accepted correction method is to conduct analyses identifying the principal components of the genotype data. Different populations will have different coefficients for the individual SNPs that comprise the components. Stratification is then corrected by including the principal components as covariates in the association analysis. These analyses are easily conducted using the PLINK software which incorporates the Eigenstrat software into the association analysis (Price et al. 2006).

7.6 The Future of GWAS

The impact of GWAS on Pharmacogenomics is currently small, but growing steadily. GWAS allow us to identify small chromosomal regions that are likely to carry the genes whose variants have a substantive impact on a patient's response to medications. In addition to informing the prescription of currently available drugs, the variety of prescribed drugs is likely to continue to grow, thus making it critical to identify the relevant genetic variants for their successful prescription. As we undergo this process, the unbiased genome-wide approach will become even more vital. However, the technology and statistical approaches that support GWAS are evolving (Cantor et al. 2010). We are beginning to develop relatively inexpensive methods to genotype at the most dense level—whole genome sequencing. The approaches to mine the raw sequence data are likely to be very complex, but once accomplished, the statistical methods described here will continue to allow GWAS to provide the foundation for personalized medicine.

References

Agresti A (2013) An introduction to categorical data analysis, 3rd edn. Wiley, Hoboken, New Jersey

Aithal GP, Day C, Kesteven PJ, Daly AK (1999) Association of polymorphisms in the cytochrome P450 CYP2C9 with warfarin dose requirement and risk of bleeding complications. Lancet 353(9154):717–719

Ando Y, Saka H, Asai G, Sugiura S, Shimokata K, Kamataki T (1998) UGT1A1 genotypes and glucuronidation of SN-38, the active metabolite of irinotecan. Ann Oncol 9(8):845–847

Bentley DP, Backhouse G, Hutchings A, Haddon RL, Spragg B, Routledge PA (1986) Investigation of patients with abnormal response to warfarin. Br J Clin Pharmacol 22(1):37–41

Beutler E, Gelbart T, Demina A (1998) Racial variability in the UDP-glucuronosyltransferase 1 (UGT1A1) promoter: a balanced polymorphism for regulation of bilirubin metabolism? Proc Natl Acad Sci 95(14):8170–8174

Brown NJ, Ray WA, Snowden M, Griffin MR (1996) Black Americans have an increased rate of angiotensin converting enzyme inhibitor-associated angioedema. Clin Pharmacol Ther 60(1): 8–13

Byrd JB, Touzin K, Sile S et al (2008) Dipeptidyl peptidase IV in angiotensin-converting enzyme inhibitor associated angioedema. Hypertension 51(1):141–147

Cantor RM, Lange K, Sinsheimer JS (2010) Prioritizing GWAS results: a review of statistical methods and recommendations for their application. Am J Hum Genet 86(1):6–22

Chang C-C, Too C-L, Murad S, Hussein SH (2011) Association of HLA-B*1502 allele with carbamazepine-induced toxic epidermal necrolysis and Stevens–Johnson syndrome in the multi-ethnic Malaysian population. Int J Dermatol 50(2):221–224

Chen YJ, Hu F, Li CY et al (2014) The association of UGT1A1*6 and UGT1A1*28 with irinotecan-induced neutropenia in Asians: a meta-analysis. Biomarkers 19(1):56–62

Chung WH, Hung SI, Hong HS et al (2004) Medical genetics: a marker for Stevens-Johnson syndrome. Nature 428(6982):486

Clinical Pharmacology Database (2010a) Fosinopril monograph. Gold Standard

Clinical Pharmacology Database (2010b) Lisinopril monograph. Gold Standard

Clinical Pharmacology Database (2011) Allopurinol monograph. Gold Standard

Clinical Pharmacology Database (2012) Ramipril monograph. Gold Standard

Clinical Pharmacology Database (2013a) Enalapril monograph. Gold Standard

Clinical Pharmacology Database (2013b) Simvastatin monograph. Gold Standard

D'Andrea G, D'Ambrosio RL, Di Perna P et al (2005) A polymorphism in the VKORC1 gene is associated with an interindividual variability in the dose-anticoagulant effect of warfarin. Blood 105(2):645–649

Davis DE (1951) Observations on rat ectoparasites and typhus fever in San Antonio, Texas. Public Health Rep 66(52):1717–1726

Ehret GB (2010) Genome-wide association studies: contribution of genomics to understanding blood pressure and essential hypertension. Curr Hypertens Rep 12(1):17–25

Gelissen IC, McLachlan AJ (2014) The pharmacogenomics of statins. Pharmacol Res 88:99–106

Gomes I, Collins A, Lonjou C et al (1999) Hardy–Weinberg quality control. Ann Hum Genet 63(6):535–538

Hung S-I, Chung W-H, Liou L-B et al (2005) HLA-B*5801 allele as a genetic marker for severe cutaneous adverse reactions caused by allopurinol. Proc Natl Acad Sci U S A. 102(11): 4134–4139

Innocenti F, Undevia SD, Iyer L et al (2004) Genetic variants in the UDP-glucuronosyltransferase 1A1 gene predict the risk of severe neutropenia of irinotecan. J Clin Oncol 22(8):1382–1388

Kalow W (1964) Pharmacogenetics and anesthesia. Anesthesiology 25:377–387

Karlin E, Phillips E (2014) Genotyping for severe drug hypersensitivity. Curr Allergy Asthma Rep 14(3):418

Lange C, DeMeo DL, Laird NM (2002) Power and design considerations for a general class of family-based association tests: quantitative traits. Am J Hum Genet 71(6):1330–1341
Lehmann EL (2006) Nonparametric: statistical methods based on ranks. Springer, New York, NY
Locharernkul C, Loplumlert J, Limotai C et al (2008) Carbamazepine and phenytoin induced Stevens-Johnson syndrome is associated with HLA-B*1502 allele in Thai population. Epilepsia 49(12):2087–2091
Lu AT, Cantor RM (2007) Weighted variance FBAT: a powerful method for including covariates in FBAT analyses. Genet Epidemiol 31(4):327–337
Mallal SND, Witt C, Masel G, Martin AM, Moore C, Sayer D, Castley A, Mamotte C, Maxwell D, James I, Christiansen FT (2002) Association between presence of HLA-B*5701, HLA-DR7, and HLA-DQ3 and hypersensitivity to HIV-1 reverse-transcriptase inhibitor abacavir. Lancet 359(9308):727–732
Man CBL, Kwan P, Baum L et al (2007) Association between HLA-B*1502 allele and antiepileptic drug-induced cutaneous reactions in Han Chinese. Epilepsia 48(5):1015–1018
Mehta TY, Prajapati LM, Mittal B et al (2009) Association of HLA-B*1502 allele and carbamazepine-induced Stevens-Johnson syndrome among Indians. Indian J Dermatol Venereol Leprol 75(6):579–582
Morrison M, Caldwell A, McQuaker G, Fitzsimons EJ (1989) Discrepant INR values: a comparison between Manchester and Thrombotest reagents using capillary and venous samples. Clin Lab Haematol 11(4):393–398
Mosteller F, Tukey JW (1977) Data analysis and regression. Addison-Wesley, Reading
Pare G, Kubo M, Byrd JB et al (2013) Genetic variants associated with angiotensin-converting enzyme inhibitor-associated angioedema. Pharmacogenet Genomics 23(9):470–478
Patrick AR, Avorn J, Choudhry NK (2009) Cost-effectiveness of genotype-guided warfarin dosing for patients with atrial fibrillation. Circ Cardiovasc Qual Outcomes 2(5):429–436
Pearson TA, Manolio TA (2008) How to interpret a genome-wide association study. JAMA 299(11):1335–1344
Peck RN, Smart LR, Beier R et al (2013) Difference in blood pressure response to ACE-Inhibitor monotherapy between black and white adults with arterial hypertension: a meta-analysis of 13 clinical trials. BMC Nephrol 14:201
Plink (2009) Whole genome association analysis toolset. http://pngu.mgh.harvard.edu/~purcell/plink/
Price AL, Patterson NJ, Plenge RM, Weinblatt ME, Shadick NA, Reich D (2006) Principal components analysis corrects for stratification in genome-wide association studies. Nat Genet 38(8):904–909
Ramoz N, Boni C, Downing AM et al (2009) A haplotype of the norepinephrine transporter (Net) gene Slc6a2 is associated with clinical response to atomoxetine in attention-deficit hyperactivity disorder (ADHD). Neuropsychopharmacology 34(9):2135–2142
Ring BJ, Gillespie JS, Eckstein JA, Wrighton SA (2002) Identification of the human cytochromes P450 responsible for atomoxetine metabolism. Drug Metab Dispos 30(3):319–323
Scheffé H (1959) The analysis of variance. Wiley, New York
Shapiro S (1953) Warfarin sodium derivative: (coumadin sodium); an intravenous hypoprothrombinemia-inducing agent. Angiology 4(4):380–390
Solomon HM (1968) Variations in metabolism of coumarin anticoagulant drugs. Ann N Y Acad Sci 151(2):932–935
Takeuchi F, McGinnis R, Bourgeois S et al (2009) A genome-wide association study confirms VKORC1, CYP2C9, and CYP4F2 as principal genetic determinants of warfarin dose. PLoS Genet 5(3), e1000433
The 1000 Genomes Project Consortium (2010) A map of human genome variation from population-scale sequencing. Nature 467(7319):1061–1073
The SEARCH Collaborative Group (2008) SLCO1B1 variants and statin-induced myopathy—a genomewide study. New England J Med 359(8):789–799

U.S. Food and Drug Administration (2014) Information for healthcare professionals: Abacavir (marketed as Ziagen) and Abacavir-containing medications. http://www.fda.gov/Drugs/DrugSafety/PostmarketDrugSafetyInformationforPatientsandProviders/ucm123927.htm. Accessed 4 July 2014

U.S. Food and Drug Administration, Center for Drug Evaluation and Research (2006) Camptosar NDA 20-571/S-030 literature review update. U.S. Food and Drug Administration, Center for Drug Evaluation and Research http://www.fda.gov/Drugs/DrugSafety/PostmarketDrugSafetyInformationforPatientsandProviders/ucm123927.htm

U.S. Food and Drug Administration, Center for Drug Evaluation and Research (2013) Ziagen highlights of prescribing information, indications. U.S. Food and Drug Administration, Center for Drug Evaluation and Research http://www.fda.gov/Drugs/DrugSafety/PostmarketDrugSafetyInformationforPatientsandProviders/ucm123927.htm

U.S. Food and Drug Administration, Center for Drug Evaluation and Research (2014a) Phenytoin label, indications, FDA approved labeling. U.S. Food and Drug Administration, Center for Drug Evaluation and Research http://www.fda.gov/Drugs/DrugSafety/PostmarketDrugSafetyInformationforPatientsandProviders/ucm123927.htm

U.S. Food and Drug Administration, Center for Drug Evaluation and Research (2014b) Strattera highlights of prescribing information, indications. U.S. Food and Drug Administration, Center for Drug Evaluation and Research http://www.fda.gov/Drugs/DrugSafety/PostmarketDrugSafetyInformationforPatientsandProviders/ucm123927.htm

West GB, Harris JM (1964) Pharmacogenetics–a fresh approach to the problem of allergy. Ann N Y Acad Sci 118:441–452

Chapter 8
Statistical Applications in Design and Analysis of In Vitro Safety Screening Assays

Lei Shu, Gary Gintant and Lanju Zhang

Abstract In this chapter, we introduce statistical applications used in the design and analysis of a high throughput in vitro screening assay, QTiSA-HT (an acronym for QT-inotropy-Screening Assay-High Throughput), a proprietary in vitro platform used to characterize concentration-dependent effects of drugs that affect cardiac repolarization and contractility. Specifically, we discuss the design and analysis of cumulative, ascending dose concentration response studies, calculation of appropriate sample sizes, and the use of statistical significance tests and equivalence margins to provide robust estimates of true drug effects based on both concurrent and historical vehicle-control data. The goal of this chapter is to showcase how we search for solutions to real scientific problems arising in early phases of drug safety screening using statistical methods and tools.

Keywords Cardiac safety screening • Cumulative dose response design • Equivalence test • False positive control • Repeated measure analysis • Re-sampling method

8.1 Introduction

During assay development and testing, we often face such questions as "how reliable and sensitive is the assay?", "what is the appropriate sample size to ascertain drug effects in the assay?" and "how to best control false positive rate?" etc. The application of appropriate statistical methods is essential to answer these types of questions and to efficiently guide early drug discovery efforts. Such guidance can profoundly influence assay development through considerations related to experimental design,

L. Shu (✉)
Data Science, Astellas US LLC., Northbrook, IL 60062
e-mail: lei.shu@astellas.com

L. Zhang
Data and Statistical Sciences, AbbVie Inc., North Chicago, IL, USA

G. Gintant
Safety Pharmacology, AbbVie Inc., North Chicago, IL, USA

L. Zhang (ed.), *Nonclinical Statistics for Pharmaceutical and Biotechnology Industries*, Statistics for Biology and Health, DOI 10.1007/978-3-319-23558-5_8

data evaluation and interpretation, and meaningful translation of results to inform early drug discovery efforts and compound selection. Unfortunately, it is not very common to seek guidance from statisticians when developing novel assays.

In early stage of drug discovery, usually a large number of compounds are evaluated, typically using high-throughput assays as discussed in Chap. 4. In this chapter, we discuss statistical applications used to support the development of an in vitro screening assay to detect cardiac safety liabilities during candidate selection. This proprietary automated assay, called QTiSA-HT (QT-inotropy-Screening-Assay- High-Throughput), is used to characterize concentration dependent effects of evolving drug candidates on (a) cardiac repolarization (thus affecting the QT interval on the ECG) and (b) cardiac contractility (thus affecting cardiac inotropy, or strength of contractions), based on responses to sequential increasing drug concentrations measured using isolated rabbit ventricular myocytes. In early stages of lead optimization, there may be thousands of compounds to be screened quickly for decision making. Time- and cost-effectiveness, as well as compound availability, are all important when designing such early screening assays. Equally important is the need to understand and minimize the incidence of false-positive results that could lead to the unwarranted elimination of potential novel therapeutics in early screening efforts. While recognizing that the methods introduced here may represent only one of multiple solutions, our goal is to present how we searched for practical, statistically-based solutions while highlighting how statistics beneficially impacted early drug discovery efforts.

This chapter is organized as follows. In Sect. 8.2, we introduce background material regarding the utility and goals of the QTiSA-HT screening assay, along with a description of the data generated and problems with interpretation. Three statistical applications are subsequently described in the next three sections. In Sect. 8.3, we introduce a repeated measure cumulative dose response design for QTiSA-HT experiment and discuss how to calculate the sample size for this design using SAS procedure PROC GLIMMIX. In Sect. 8.4, we briefly discuss the statistical methods used for analyzing repeated measure data and apply the general mixed model to analyze data collected from such experiments. In Sect. 8.5, we propose a re-sampling method to establish an equivalence margin based on historical control data. This equivalence margin defines a realistic threshold used to identify potential meaningful responses (consistent with safety liabilities) in the early screening funnel. In Sect. 8.6, we introduce a two-step procedure which combines the significance test and equivalence margin for reporting a scientifically meaningful positive signal and discuss its advantage in a high throughput screening setting. Finally, we present overall conclusions in Sect. 8.7.

8.2 Introduction of QTiSA-HT and Statistical Issues

Cardiac safety remains of highest priority when developing novel drug candidates. Traditional safety pharmacology studies are largely focused on in vivo effects related to physiologic systems essential for sustaining life (cardiovascular, respiratory, CNS effects), and drug effects on the heart represent a primary focus of overall safety. With regards to cardiac electrophysiology, a rare but potentially lethal arrhythmia known as Torsades de Pointes (TdP) is associated with drugs that delay ventricular repolarization. On a surface ECG, delayed ventricular repolarization is manifest as prolongation of the QT interval, a recognized surrogate marker for proarrhythmia that can be measured after careful analysis of ECG recordings from preclinical and clinical studies. Even moderate concentration-dependent prolongation of the QT interval observed clinically may lead to discontinuation of compound, after significant time, effort, and other resources have been expended on the compound.

To avoid a finding of clinical QT prolongation in later development phases, efforts have focused on the early in vitro detection of compounds that delay cardiac repolarization. Studies evaluating hERG current are often conducted during early drug discovery as part of "frontloading" safety pharmacology studies (Cavero 2009), and this assay is included as part of the required ICH S7B guidelines (ICH S7B). hERG current is a repolarizing potassium current in human ventricular myocytes that plays a prominent role in defining repolarization. Multiple drugs have been shown to reduce hERG current, leading to delayed repolarization and predisposition to an arrhythmia known as Torsades-de-Pointes. However, it is now recognized that hERG represents one of multiple ionic currents that influences cardiac repolarization, and that hERG studies represent a convenient but incomplete assessment of either delayed repolarization or proarrhythmic risk (Gintant et al. 2006; Sager et al. 2014). Further, the lack of specificity of the hERG functional current assay leads to the premature and unwarranted attrition of promising drug candidates and contributes to declining pharma productivity.

QTiSA-HT evaluates drug effects on the integrated responses of acutely isolated ventricular myocytes, which reflects the summation of effects on all ionic currents involved with each heart beat (including hERG current). This alternative approach to evaluating delayed repolarization relies on evaluating changes in the duration of the cardiac action potential (an electrical signal), providing a more realistic assessment of an integrated effect on multiple ionic currents (not just the hERG current). QTiSA-HT also evaluates effects on cardiac contractions (a mechanical event that occur with each heartbeat) based on contractions of stimulated myocytes (see Fig. 8.1).

Using QTiSA-HT, it is possible to characterize concentration-dependent drug effects on ventricular repolarization and contractility based on responses of electrically stimulated myocytes. One challenge of this approach involves applying a so called cumulative dose-response design for data collection. This cumulative design, as compared to individually testing each concentration on a separate set of myocytes, is advantageous for the efficient screening of more compounds in a shorter period of time with fewer myocyte preparations. However due to the nature of this design, observations collected from the same set of the myocytes in a cell chamber are highly correlated because the same myocytes were measured multiple times. This provides a challenge in regards to incorporating within-subject (in this case, the same set of the myocytes in the cell chamber is considered as subject) correlation when analyzing data. Further, since the correlation will be considered

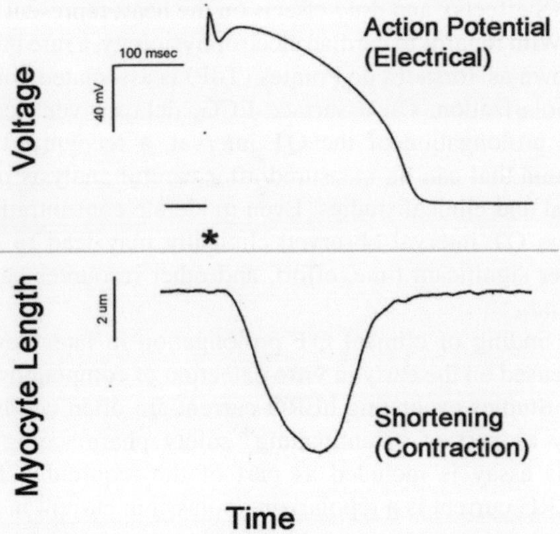

Fig. 8.1 Basis for QTiSA-HT cardiac liability assay. The *upper panel* depicts a ventricular action potential, an electrical response of a ventricular myocyte plotted as the change in transmembrane potential (in millivolts [mV]) across a ventricular membrane in response to an electrical stimulus (at *asterisk*)). The lower panel depicts the evoked mechanical response of the myocyte, a contraction wave (or twitch), represented as a transient shortening of the myocyte length (measured in micrometers [μm]) that occurs with each action potential). Prolongation of the action potential duration may lead to proarrhythmia, and changes in the extent of myocyte shortening may adversely affect cardiac contractility leading to reducing pump function and hemodynamic effects. QTiSA-HT provides screening information regarding a drugs' electrophysiologic (proarrhythmic) and contractility liabilities using optical techniques to assess contractions of multiple isolated cardiac myocytes in vitro triggered by electrical stimulation

in the analysis, it should also be considered when calculating the sample size for experiments. In addition to the within-subject correlation, another challenge involved in the power or sample size evaluation for this experimental design is the day-to-day variability of the assay. Due to differences in myocyte preparations made daily, there is substantial day-to-day variability of measurements even for the control data. The sample size calculation is aiming to detect signal for all future testing compounds which are likely tested in different days. To accommodate this day-to-day variation, a more robust and precise estimation on the assay variance is required for sample size calculations. Finally, to be more effective in guiding the selection of drug candidates early in drug discovery, the biggest challenge for this screening test is that higher throughput experiments, compared to more traditional, manually-based experiments, require automated techniques to provide an early, quick estimate of a safety signal for potential cardiac risk. Statistical approaches to determine the safety signal need to be more robust to control false positive rate in such early stage.

8.3 Cumulative Dose Response Design and Sample Size Calculation

The QTiSA-HT assay follows a cumulative dose-response design (Fig. 8.2) consisting of (a) two recording periods termed dispense 0 [D0] and dispense 1 [D1] which provide baseline parameters, (b) four recording periods (dispenses 2, 3, 4, 5) measuring responses to sequential ascending concentrations of drug (also referred to as time points or dose levels), (c) two recording periods (dispenses 6, 7) assessing drug washout and reversibility of drug effects. The ascending drug concentrations typically used are 3, 10, 30 100 μM concentrations. We call the entire process from dispense 0 to dispense 7 a test-run. Test-runs may be conducted with either drugs or vehicle-control. Each test-run is performed independently in each cell chamber over time. To account for day-to-day variability of the myocyte preparations, the same drug is tested on at least two different days, with a new myocyte preparation made each experimental day.

For QTiSA-HT experiment, data collected represent repeated measurements at different dispenses (dose levels) from the same experimental preparation (same myocytes cells at the same cell chamber). For example, dispense 2, 3, 4, and 5 is corresponding to dose level 3, 10, 30 and 100 μM, respectively as shown in Fig. 8.2. This cumulative dose-response design saves time (no need to establish baseline values for each concentration tested), saves test compound (using incremental addition to next target concentrations), reduces the number of experimental preparations needed (multiple concentration responses per one preparation), and reduces experimental variability (since same preparation is used to collect the data

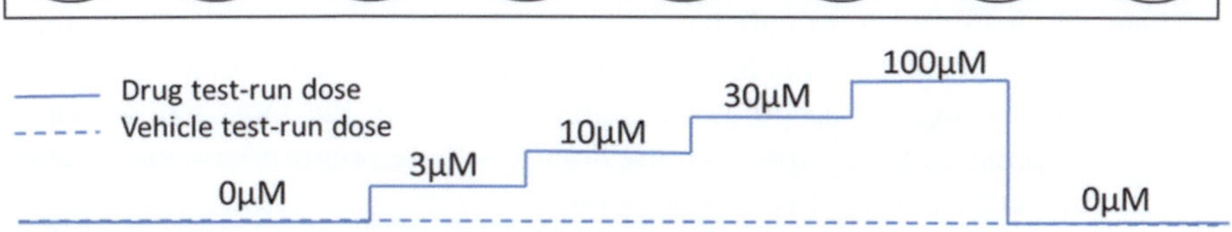

Fig. 8.2 Cumulative dose response design for QTiSA-HT experiment. A typical "test-run" consisting of two baseline "dispenses" (D0–D1)", followed by 4 periods of ascending (cumulative) concentrations of test compound (D2–D5) followed by two washout periods (D6–D7). Dispense numbers from D0 to D7 in above diagram represent the time of events during which experimental parameters (representing changes in repolarization or contractility of isolated myocytes) are measured. The total duration of each test-run is 7.5 min, enabling its use in a higher throughput experimental format

from all concentrations). Such a design is also useful for biological responses that may not demonstrate stable responses with multiple periods of drug exposures and washout.

On each day, four to eight test-runs recording effects of vehicle-control experiments are performed to supply a concurrent control for all drug test-runs conducted that day. Thus, a typical dataset for screening one drug consists of data collected from 2 days with 6 test-runs per day, involving 2 test-runs for drug and 4 test-runs for vehicle-control. The data collected from vehicle-control are shared amongst all drugs tested that day.

As we discussed in Sect. 8.2, this cumulative design is advantageous for the efficient screening of more compounds in a shorter period of time. However due to the nature of this design, it is quite a challenge when calculating the sample size. The goal of sample size calculation for this particular design is to answer: how many test-runs should be conducted for each drug to have sufficient power to detect a meaningful drug effect? As we discussed above, the current QTiSA platform typically allows at least 4 shared test-runs for control experiments and 2 test-runs for drug per day. Therefore, this sample size calculation is essentially to answer how many days do we need to perform the drug test-runs for screening a same drug? For example, if the answer is 3 days, then in total we will perform 6 test-runs for a screening drug with 2 test-runs performed per day. If a design allows us to conduct 54 runs per day, we can complete the screenings for 25 compounds in 3 days since 50 test-runs for drug can be conducted in each day (the rest 4 test-runs will be reserved for vehicle control). The less the number of days we need, the more compounds we are able to screen and make decision in a fixed time frame.

Statistical test power is the probability of rejecting the null hypothesis when the null hypothesis is false. For example, in our QTiSA experiment, one of the null hypotheses is that there is no concentration-dependent drug effect on cardiac repolarization. Performing power analysis and sample size estimation is an important aspect of any experimental design. Usually, larger sample size provides greater statistical power for a given experimental setting. A good experimental design should provide sufficient power with minimal sample size for the purpose of efficiency. For screening assay, sample size is critical not only to defining statistical power, but also affects assay cost, throughput, and timelines.

There are many ways to calculate the sample size or statistical power for a repeated measurement experimental design. A simple way is to calculate the sample size based on a single time point measurement, usually the most interesting or critical time point. A better approach uses multiple time points after consideration of underlying correlation between time points.

In this screening experiment, the time point or dispense is corresponding to dose concentration level. We are not only interested in testing the drug effect at each dose level, but also interested in testing the dose response trend. Therefore all dose levels are equally important. Sample size calculation based on a single time point, i.e. single dose level, may not provide sufficient power when testing dose response trend. More importantly, because responses to ascending sequential drug concentrations are measured from the same myocytes in this cumulative dose response design, it is obvious that measurements collected from the set of the myocytes at the same cell chamber are correlated over time. From an initial analysis of a large set of control data, the estimated within-subject (here subject is corresponding to the set of the myocytes at the same chamber) correlation is about 0.72 and 0.83 for cardiac repolarization and cardiac contractility respectively. Details about this initial analysis will be discussed later in this section. The high correlations observed from the control data suggest that the within-subject correlation should be considered when calculating the sample sizes as well as analyzing the experimental data. The appropriate statistical method to analyze the experimental data will be discussion in Sect. 8.4.

There are many ways to calculate the sample size when correlation is considered in the setting of repeated measure design. Here we introduce a simple method that can be performed either by theoretic formula or by simulation (Hedeker et al. 1999; Basagana and Spiegelman 2010; Comulada and Weiss 2010). The following four steps describe how to incorporate the correlation when calculating sample size using SAS PROC GLIMMIX procedure (Stroup 2010).

The first step is to estimate the parameters including control means of each variable at each dispense (dispense D2-D5), the overall variance and the correlation coefficient between different dispenses for the same subject, from analyzing the historical control data. This step is a must-done step when performing sample size calculation using any statistical method. A repeated measurement ANOVA model was fitted on a historical control dataset which consists of data collected from a large set of the most recently done control experiments to estimate these parameters. In this model, since the control data includes many days, day is treated as random effect, so the variance component of day can also be estimated from this model. This model is very similar to the one used for testing the drug effect which will be described in Sect. 8.4, but it only applies to the control data and day effect is treated as random.

The second step is to generate "means" data based on the estimated parameters from the first step. According to the experimental design, we typically perform 2 test-runs per day for the same screening compound. Assuming we conduct the experiments for the same compound in m days, then the total number of drug test-runs is 2 m and the total number of the corresponding concurrent control test-runs is 4 m. When using PROC GLIMMIX to do the calculation, the only data we need are the means at each dispense for each group (drug or control) on each day. The overall variance and the correlation coefficient will be provided and held fixed when performing the analysis.

The following example demonstrates how to calculate sample size for testing dose response trend on cardiac repolarization when within-subject correlation is considered. In this example, we assume the day-to-day variation $\mu_{day} \sim N\ (0, 49)$, and the estimated vehicle-control means at dispenses 2, 3, 4 and 5 is 24, 37, 45 and 50 respectively. These parameter estimations are from step 1. To test dose response trend, we assume the effect size for dispense 5 (high dose) is Δ, the effect size for dispense 2 (low dose) is $\Delta/4$, and the effect sizes for dispense 3 (mid-low dose) and dispense 4 (mid-high dose) are $\Delta/2$. Different dose response relationship can be assumed through different specification on the effect size at each dispense. The SAS code below can generate a series of testing datasets with different Δ range from 0 to 60 ms and different number of days (m = 2, 3 and 4) for the purpose of producing a power curve. A similar procedure can be used to calculate sample size for pair-wise comparison. Notice that the day-to-day variation is also incorporate in the data generation.

The following code only generates one realization of day effect. To account for the randomness of day effect, multiple "means" datasets corresponding to different realizations of day effect should be generated and their resulting sample sizes should be compared. Usually to be more conservative, the largest sample size corresponding to the worst scenario will be used as the final sample size.

```
data means;
do m=2 to 4 by 1;
  do delta=0 to 60 by 1;
   do day=1 to m;
        dayvar=7*rannor(-1);
      do dispense=2 to 5;
        z=24*(dispense=2)+37*(dispense=3)+45
            *(dispense=4)+50*(dispense=5);
        do exp=1 to 6;
        id=day*100+exp;
        if exp < 5 then trt=0; else trt=1;
        y=z+dayvar+delta*(trt=1)*(dispense=2)/4
                   +delta*(trt=1)*(dispense=3)/2
                   +delta*(trt=1)*(dispense=4)/2
                   +delta*(trt=1)*(dispense=5);

        output;
       end;
     end;
    end;
   end;
 end;
drop exp dayvar z;
run;
proc sort; by m delta day id dispense; run;
```

In step 3, the generated "means" datasets from step 2 are used as the input datasets for the PROC GLIMMIX procedure to output a contrast dataset "contrast" which will be used to calculate power in step 4. The estimated variances and correlation coefficient are specified in the PARMS statement. For both cardiac contractility and repolarization parameters collected from QTiSA-HT assay, we observed that the variances are not homogeneous across dispenses, so we use a first order heterogeneous autoregressive (ARH(1)) covariance structure to analyze the data. This ARH(1) structure will be discussed in Sect. 8.4. For example, the variance at each dispense (dispense 2-5) and the correlation coefficient are held as 89, 315, 232, 239 and 0.72, respectively for testing cardiac repolarization. The PARMS statement with hold option tells the procedure to use the values provided in the statement as the corresponding parameter estimates in the current program. This is why we only need generate "means" data in step 2 and do not need consider the covariance matrix, since it is provided in this PARMS statement. The SAS code for PROC GLIMMIX is provided in the following.

```
PROC GLIMMIX data=means;
    BY m delta;
    CLASS day trt dispense id;
    MODEL y=day trt dispense trt*day trt*dispense
                day*dispense day*dispense*trt;
    RANDOM dispense/sub=id type=arh(1) residual;
    PARMS 89 315 232 239 0.72 / hold=1,2,3,4,5;
CONTRAST 'Dose Response' trt -1 1
   trt*dispense -0.1 -0.2 -0.2 -0.5
   0.1 0.2 0.2 0.5 ;
ODS OUTPUT contrasts=contrast;run;
```

Notice that we use the exact same analysis model (Eqs. (8.1)–(8.3) in Sect. 8.4) to output the contract dataset that is subsequently used for power calculation. This is a nice feature of using this procedure to perform power analysis, because the exact same analysis model can be used to perform the power analysis, which is often not feasible using other software. In the CONTRAST statement, we specify the linear coefficients for testing the dose response trend. Different coefficients can be specified through this statement. To keep consistent, the same coefficients should always be used when analyzing the real experimental data.

The last step is to calculate the power based on the "contrast" dataset outputted from step 3. In previous PROC GLIMMIX procedure, a contrast output dataset with number of degree of freedom and F values are generated. This dataset can be used to computer non-centrality parameter, and then we can use probability statements for F-distribution to determine critical value and compute the power. SAS code for the power calculation is provided in the following:

```
data power;
    set contrast;
    alpha=0.05;
    ncparm=numdf*fvalue;
    fcrit=finv(1-alpha,numdf,dendf,0);
    power=1-probf(fcrit,numdf,dendf,ncparm);
 run;
```

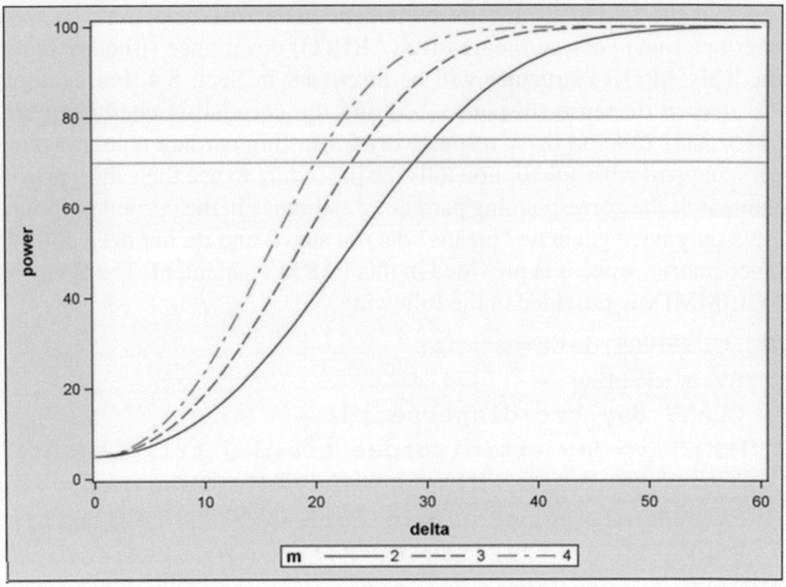

Fig. 8.3 Power vs. delta for cardiac repolarization. m is the number of days to perform drug test-runs

The results based on the above calculation are presented in several power curves (Fig. 8.3). Figure 8.3 shows 3 power curves corresponding to three sample size setting (m = 2, 3 and 4). Figure 8.3 shows that our current 2-day (blue curve with m = 2) experimental design with 2 test-runs of drug experiments and 4 test-runs of control experiments at each day has sufficient power (>=75 %) to detect reasonable dose response trend (Δ = 7.5, 15, 15, 30 msec at corresponding dose levels) on testing the cardiac repolarization. Similar results on testing cardiac contractility are obtained using the similar procedure. In addition, we also calculated the power for pair-wise comparison at each dose level. The results also support that the two-day design is sufficient. To perform the power analysis for pair-wise comparison at each dose, the similar procedure can be used, except that the contrast statement should be modified to reflect the pair-wise comparison. For example, to test pair-wise comparison between high dose and control, just specify the contrast statement as follows:

CONTRAST 'Drug vs. Control at High dose' trt

− 1 1 trt ∗ dispense 0 0 0 − 1 0 0 0 1;

8.4 Repeated Measure Analysis

An important feature of repeated measures data is the correlations across observations for the same subject. For example, annual weight measurements on a person are more similar to one another than to the measurements on other person. For QTiSA-HT experiment discussed in Sect. 8.3, the experimental unit is one test-run (consisting of seven dispenses over time). Measurements collected in one test-run are highly correlated and cannot be treated as independent measurements. The objectives of repeated measurement analysis often are to examine and compare response trends over time. This can involve comparisons of treatments at specific times, or averaged over time. It can also involve comparisons of times within the same treatment. Since observations on the same subject are not independent, appropriate estimates of variability for hypothesis test must be considered.

There are several statistical methods used for analyzing repeated measures data. The simplest one is the separate analysis by time point. Although the time correlation holds true, this simplest approach ignores the correlation between different time points within the same subject, and does not directly address the objectives of examining and comparing trends over time. Univariate analysis of variance or multivariate analyses of time contrast variables using PROC GLM procedure have also been used to analyze the overall trend or average effect over time. This approach either ignores the covariance issues (which may result in incorrect conclusions from the statistical analysis) or avoid the issues, resulting in inefficient analyses or wasting data (Littell et al. 1998). Here we adopt the general linear mixed model because it allows for directly modeling the covariance structure. The mixed model analysis can be implemented in the MIXED or GLIMMIX procedures of the SAS System (Little et al. 2000).

For our data, the repeated measurements are the values derived at each concentration level in one test-run, e.g., cardiac contractility or cardiac repolarization, at different dispenses. The experiment is a cumulative dose response design, so the different dispenses correspond to different dose levels. The average value of the same set of myocytes in the same cell chamber is used for analysis, as we wish to consider the average response of multiple myocytes within each cell chamber to aggregate random noise at the level of the myocyte. An example dataset is shown in Fig. 8.4. To test for a dose response trend, the collected data are treated as repeated measure data and analyzed using general linear mixed model. There are three fixed effects in our model:

- Dispense (2, 3, 4 and 5) or dose level (low, mid-low, mid-high and high),
- Treatment group (drug and vehicle)
- Day to perform the test-runs (day 1 and 2).

In our analysis, we not only include the main effects of the above three fixed effects, but also include their two-way and three-way interactions. The reason to include these interaction terms in the model is that it is expected that some interactions may be significant for some drugs. For example, there is significant dispense by

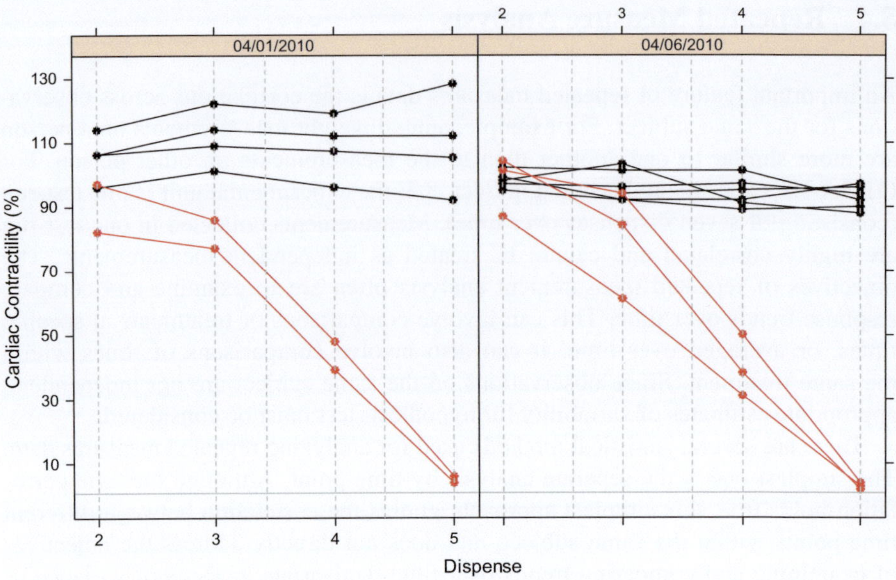

Fig. 8.4 An example analysis ready dataset for testing cardiac contractility. This example dataset is from two separate days of experiments (*left* and *right* panels). Data points plotted are the average values of cardiac contractility of the myocytes collected from the corresponding cell chamber at each dispense for each test-run. *Black dots and lines* represent vehicle control responses; *red dots and lines* represent concentration-dependent decreases in contractility with ascending drug concentrations; each line represents one test-run. In this example, contractility shows concentration-dependent declines with increasing drug concentration in dispenses 2, 3, 4 and 5

treatment interaction for some drugs with a nonlinear dose response curve, or there may be a significant day by treatment interaction for some experiments because day-to-day differences affect the drug effect differently at different dose level. It is still affordable to include all these interaction terms in the model since we have at least 48 data points collected for each compound. It is a little cumbersome to test the interactions for each compound and decide to remove some interaction terms from the model. Keep in mind that there will be over a hundred compounds that need to be analyzed within a short period. A future work is to create an automatic procedure that can test these interactions and include the most appropriate terms in the analysis for each compound.

The linear mixed model is described in Eq. (8.1). As mentioned above, since the responses from the same set of the myocytes in the same cell chamber are measured at multiple times at different dispenses, the values are correlated over time for the same test-run. As measurements from the same test-run but from different dispense ($j = 2, 3, 4,$ *and* 5) are correlated, we model this within-subject covariance by ARH(1) structure, in which the errors follow a first order heterogeneous autoregressive process [Eq. (8.2)] and the measurements from different test-runs are independent of each other, so their covariance is zero [Eq. (8.3)]. The ARH (1) correlation structure (results not shown) was shown to be appropriate based on an initial analysis on a big set of historical control data and several sets of testing data on standard drugs.

$$\begin{aligned} Y_{ijkd} &= \textit{Value from run i at dispense j in group k at day d} \\ &= \mu + dispense_j + group_k + day_d \\ &\quad + dispense_j \times group_k + dispense_j \times day_d + group_k \times day_d \\ &\quad + dispense_j \times group_k \times day_d + \epsilon_{ijkd} \end{aligned} \tag{8.1}$$

Within Subject Covariance

$$= Var\left(\begin{pmatrix} \epsilon_{i2kd} \\ \epsilon_{i3kd} \\ \epsilon_{i4kd} \\ \epsilon_{i5kd} \end{pmatrix}\right) = \begin{bmatrix} \sigma_2^2 & \rho\sigma_2\sigma_3 & \rho^2\sigma_2\sigma_4 & \rho^3\sigma_2\sigma_5 \\ \rho\sigma_2\sigma_3 & \sigma_3^2 & \rho\sigma_3\sigma_4 & \rho^2\sigma_3\sigma_5 \\ \rho^2\sigma_2\sigma_4 & \rho\sigma_3\sigma_4 & \sigma_4^2 & \rho\sigma_4\sigma_5 \\ \rho^3\sigma_2\sigma_5 & \rho^2\sigma_3\sigma_5 & \rho\sigma_4\sigma_5 & \sigma_5^2 \end{bmatrix} \tag{8.2}$$

Between Subjects Covariance

$$= Cov\left(\epsilon_{ijkd}, \epsilon_{i'j'k'd'}\right) = 0 \textit{ when } i \neq i' \tag{8.3}$$

The following SAS code is used to perform the repeated measure analysis on the data collected from our QTiSA-HT experiments. In model statement, three fixed effects and their two-way and three-way interactions are specified. In random statement, the ARH(1) correlation structure is specified for the data collected from each test-run indicated by "id", which is the unique name of the test-run performed for each drug. Again, straight speaking, for different drug, a different correlation structure may be better than the pre-specified ARH(1) structure, however, for the same reason as we mentioned for including all the interaction terms in our model, it is not efficient to test the covariance structure for every compound using manual process unless there is an automatic process available to use. Based on the results from analyzing hundreds of the historical compounds including many sets of the standard drugs (drugs are well-known for their cardiac risk), this correlation structure in general works. Notice that in addition to test the dose response trend as we discussed in Sect. 8.3, four "estimate" statements are added in the PROC GLIMMIX procedure to perform the pair-wise comparison test at each dose level. The results of pair-wise comparisons are important information on determining the safe dose range.

```
PROC GLIMMIX data=analysis;
by compound;
class day group dispense id;
model CTT=day group dispense group*day group*dispense
```

```
    day*dispense day*dispense*group/ddfm=kr;
  random dispense/sub=id(group) type=ARH (1) residual;
  estimate 'Dose Response' trt -1 1 trt*dispense
     -0.1 -0.2 -0.2 -0.5 0.1 0.2 0.2 0.5 ;
  estimate  'Drug vs. Control  at High dose' trt
     -1 1 trt*dispense 0 0 0 -1 0 0 0 1 ;
  estimate  'Drug vs. Control  at Mid-high dose' trt
     -1 1 trt*dispense 0 0 -1 0 0 0 1 0 ;
  estimate  'Drug vs. Control  at Mid-low dose' trt
     -1 1 trt*dispense 0 -1 0 0 0 1 0 0 ;
  estimate  'Drug vs. Control  at Low dose' trt
     -1 1 trt*dispense -1 0 0 0 1 0 0 0 ;
  run;
```

Figures 8.5a, b present results on testing cardiac repolarization and contractility from analyzing data collected for four standard drugs. Bars plotted in each panel are mean differences between drug and control (white bars) and their corresponding standard deviations (grey bars) at the four dose levels for changes in repolarization (Fig. 8.5a, upper panels) and contractility (Fig. 8.5b, lower panels). A star indicates the p-value from the corresponding pair-wise comparison is less or equal to 0.05 and the p-value appears in the strip of each panel is from the dose response trend test. Both the trend test and pair-wise tests are important for decision making. As we can see from these two figures, some compounds may show positive when looking at the trend test p-value, but the pair-wise comparison may show no effect at low dose level, for example, Diltiazem shows positive dose response on cardiac repolarization ($P < 0.001$), but there appears no drug effect at low and mid-low doses. Additionally, some drugs may affect only repolarization or contractility. For example, Moxifloxacin may affect cardiac repolarization, but not affect cardiac contractility. Decisions should be made by evaluating both parameters.

8.5 Establish a Scientifically Meaningful Threshold

It is very important to control the false positive rate when screening compounds for safety liabilities in early stages of drug discovery. If a safe compound is mistakenly categorized as a positive compound, the company may lose the opportunity to development a safe and efficacious successful compound. In contrast, false negative compounds may be tolerated at this early stage if subsequent screening efforts are in place to detect safety liabilities before progressing the compound substantially farther in the development pipeline.

In the case of QTiSA-HT experiments, as we discussed in Sect. 8.3, we use a set of historical control data to calculate the sample size for our repeated measure

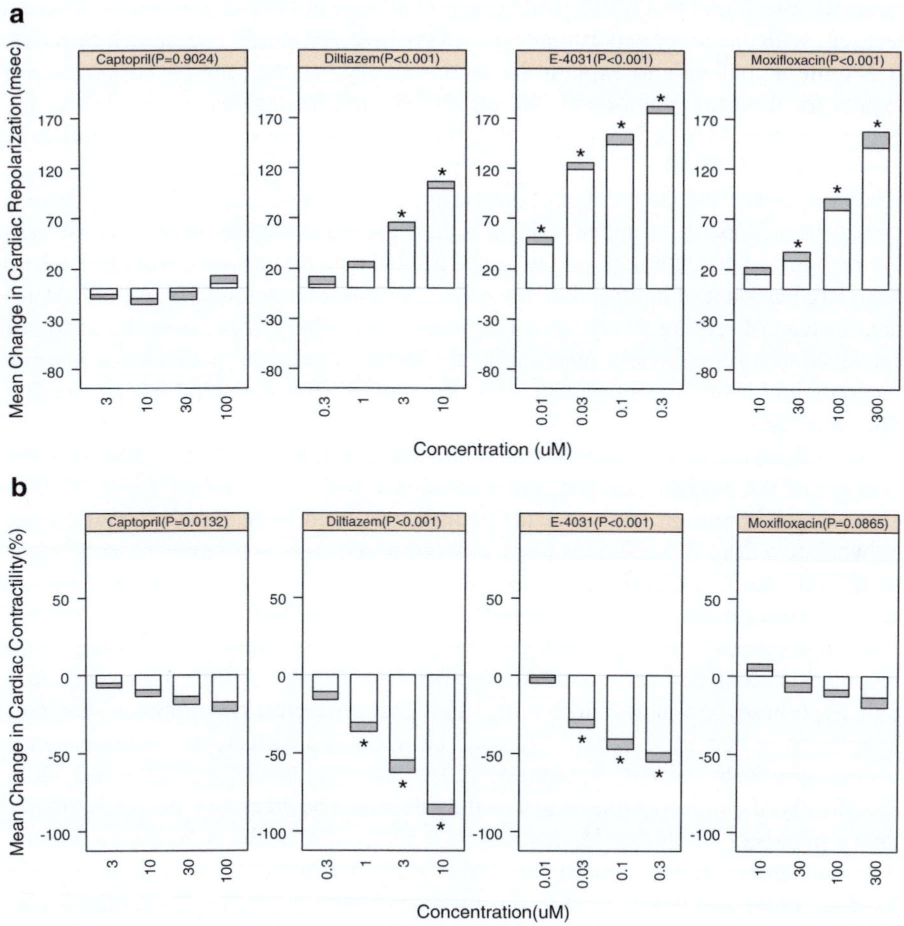

Fig. 8.5 (**a**) Drug effect on cardiac repolarization. (**b**) Drug effect on cardiac contractility

design. The variability of the assay is estimated more precisely in this way because control data from multiple days are used. However, when performing the statistical analysis described in Sect. 8.4 for each compound, only the concurrent control data are used as the reference in testing. If we conduct the experiment in two days, only 2 days data are used. From pure experimental perspective, using the concurrent control is the correct approach because it controls many factors which may bias the test result (for example, day-to-day variation arising due to differences in myocyte preparation each day). However, since we only have limited days to conduct the experiments for testing one compound, there is still a chance that the true assay variation is underestimated or overestimated with such limited concurrent data. As a matter of fact, we observed many small p-values from pair-wise comparisons but with negligible treatment differences, even after multiplicity adjustment. In some cases we also found that results for a compound were not consistent across different test sets, with one set of experiments shown positive, but another set shown negative. It is quite a challenge to explain the inconsistency between the two opposite test results for the same compound. We attributed such inconsistencies to day-to-day assay variation. A simple way to solve this problem is to increase the number of experimental days, allowing for a more precise and robust estimate of variability. However, increasing the number of days requires more time and resources. If there is no way to increase the number of days, is there any other way to solve this problem? We recognized that although we only have limited data for concurrent control, we do have large amount of control data over time. Would it be possible to use all control data instead of relying solely on concurrent control data? This raises the utility of establishing a scientifically meaningful threshold, namely an equivalence margin, based on all historical control data, and then applying this threshold when reporting drug test results.

To understand how to establish an equivalence margin, we briefly introduce the concept of the equivalence test. An equivalence test (Blackwelder 1998; Phillips 1990) is an important statistical tool commonly used to test the bioequivalence between two drug formulations (e.g., bioequivalence test in pharmacokinetic studies). Equivalence test and significance test are both used to comparing two means and are quite similar in terms of overall testing strategy. Both approaches assume a "Null Hypothesis", testing whether there is sufficient evidence to reject the "Null Hypothesis" and conclude the "Alternative Hypothesis" is true. However, their null hypotheses are stated differently. In significance testing approach, the null hypothesis is that there is no difference between two means, but in equivalence testing approach, the null hypothesis is that the two means are different. More specifically, the null hypothesis is that the difference between two means is greater than a pre-specified threshold, denoted by δ, referred to as the equivalence margin. The equivalence margin usually represents a difference that is not large enough to have any biological implication, so δ sometimes is also called biological meaningful difference. Often prior knowledge or expertise is used to define an appropriate δ.

To further explain the differences between equivalence test and significance test, we look at an example dataset from our QTiSA-HT experiment on a well-known negative control anti-inflammatory drug Indomethacin. Using the model described in Sect. 48. to test cardiac repolarization effect at high dose (dispense 5), the p-value based on a two-sided T test (significance testing approach) is less than 0.05. Hence, we reject the null hypothesis and conclude that the drug effect on cardiac repolarization at the high dose is significant. In contrast, suppose we specify in advance that a difference in cardiac repolarization of 45 ms is not considered to be biological meaningful ($\delta = 45$ using the above notation). The null hypothesis for equivalence testing, $|\mu_T - \mu_C| > \delta$, can be written as

$$\mu_T - \mu_C > \delta \quad or \quad \mu_T - \mu_C < -\delta$$

From the data we collected, the mean difference between drug and control is -23 ($\mu_T - \mu_C = -23$) and the 95 % one-sided lower and upper confidence limits is −34 and −12 respectively. Therefore, in this case, we reject the null hypothesis $|\mu_T - \mu_C| > \delta$ and conclude that the drug and control are equivalent. This example shows that sometimes the significant testing approach may conclude statistically significant drug effect, but the effect may not be deemed biologically meaningful.

From the above example, one method to test equivalence is to test the joint null hypothesis that the mean difference is not as large as the upper value of a specified range and not below the lower bound of the specified range of equivalence. By rejecting both hypotheses, we can conclude that $|\mu_T - \mu_C| \leq \delta$, or that our difference falls within the range specified. This method is called a two one-sided tests (TOST) (Schuirmann 1987). Another method is called CI (confidence interval) approach. This method specifies a range of values (say $-\delta$ to δ) that would constitute equivalency among groups and then determine the appropriate confidence interval for the mean difference between the groups to see if the CI for the difference between means falls entirely within the range of equivalency (if either lower or upper end falls beyond one does not claim equivalence). This is actually equivalent to the TOST outcome. For both approaches, we need to specify the equivalence margin δ. The following steps are used to establish the equivalence margin for our QTiSA-HT experimental data:

1. Generate sham datasets by a re-sampling method: Assume we have a large dataset of the historical control data. For example, the most recent 3 months data, consisting of 100 test-runs of vehicle-control experiments. In this step, the control dataset is used to create many sets of sham experimental data by randomly assigning some control test-runs as drug test-runs. The randomization procedure in this data re-sampling mimics the real experiment. For example, for a set of control data collected from 30 days, in each day, there are about 6-8 control test-runs. Simulated datasets can be created based on the procedure described below: Shuffle the 30 days, randomly select 2 days (according to real experiment design), in each day, randomly select 4 test-runs as control test-runs, and from the rest of the test-runs, pick up 2 test-runs as drug test-runs. The newly created 6 test-runs are treated as one sham dataset. Repeating this procedure 1000 times, 1000 sham datasets are created. The difference between these 1000 sham datasets and the real experimental dataset is that we know the truth that there should not be any drug effect for all these 1000 datasets since they are all from control data. However, due to random error, especially due to day-to-day variability of the assay, there are still some datasets claimed to have statistically significant drug effect. The proportion of these miss-claimed datasets is corresponding to the false positive error rate. In this step, we suggest including the most recently collected control data for sham datasets generation to estimate the assay variability as precisely as possible.
2. Analyze the sham datasets: In step (1), a set of sham datasets are created. In this step, we use the repeated measure ANOVA approach introduced in Sect. 8.4 to analyze these 1000 sham datasets. For each analysis, the 95 % one-sided upper and lower limits are created for each dataset.

3. Calculate false positive error rate for a pre-specified δ: From step (2), we have the upper and lower confidence limits for each sham dataset. These confidence limits are compared to the pre-specified δ. For example, for a value of δ = 30 % (a 30 % change of the cardiac contractility), the upper and lower limits will be checked to see if the results are within the interval of [−30 %, 30 %]. If they are within the interval, we conclude there is no drug effect for this set; otherwise, we conclude there is drug effect. In this way, we can easily calculate the proportion of the simulated datasets which are rejected among 1000 datasets for the fixed δ, which is in essence the Type I error rate. Repeating this calculation for a range of δ values provides a curve comparing the Type I error rate versus δ.
4. Select final equivalence margin: In step (3), we obtained the false positive error rate versus δ curve (Fig. 8.6). This curve can be used to determine a reasonable equivalence margin based on different criterion on controlling the Type I error rate (Table 8.1). For example, using a cutoff value of 1 % for the type I error rate (horizontal dashed lines in Fig. 8.6), the corresponding equivalence margins are presented in Table 8.1. Individual equivalence margin is obtained for each dose level, because data suggest the noises at different dose levels are quite different, especially for cardiac repolarization. We can also use a similar method to construct the margin for trend test statistics. However, it is more useful for researchers to have the equivalence margin for each dose level. The equivalence margin should be dynamically updated to provide more quick and accurate information for decision making.

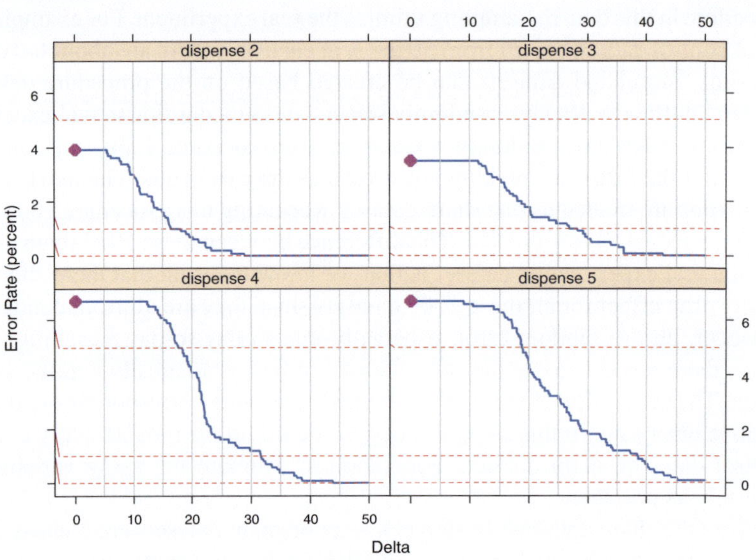

Fig. 8.6 Type I Error Rate vs. Effect Size (δ) for cardiac contractility for ascending drug concentrations (drug dispenses). The results were based on 1000 sham datasets for each δ

Table 8.1 Equivalence margins used for each dose level (false positive error rate controlled at 1 %)

Dose level	Change of cardiac contractility (%)	Change of cardiac repolarization (ms)
Low (dispense 2)	20	35
Mid-low (dispense 3)	30	40
Mid-high (dispense 4)	35	40
High (dispense 5)	40	45

8.6 A Two-Step Approach to Report Testing Results

In Sect. 8.4, we discussed a repeated measure model to analyze the experimental data using concurrent control data. This model can be used to test dose response trend as well as pair-wise comparisons for dose selection. Significant testing approaches are used to report the testing results. In Sect. 8.5, we introduced a re-sampling method to establish equivalence margins based on historical control data for all dose levels. The reason to establish equivalence margins is to evaluate whether the observed drug effect is large enough to be scientifically meaningful. So an equivalence testing approach is applied.

However, we realize that using either of these approaches alone can be problematic as shown in Fig. 8.7. If only performing significant test, there are chances that false positive results are reported with small drug effect, when the true variance is underestimated by concurrent control data. This can be illustrated by scenario A in Fig. 8.7, in which the confidence interval is well within the estimated equivalence interval, i.e. $\left[-\hat{\delta}, +\hat{\delta}\right]$, so the equivalence test concludes no drug effect while significant test shows positive since the confidence interval excludes zero. In this scenario, because the confidence interval is also within the true equivalence interval, i.e. $[-\delta, +\delta]$, clearly the significant test is a false positive result and the equivalence test result is preferred. If only performing the equivalence test, there are chances that equivalence is rejected when there is no drug effect. This can be illustrated by scenario D, in which the confidence interval includes zero but the upper limit exceeds the estimated equivalence margin, i.e. $+\hat{\delta}$. In this scenario, since the upper limit is still within the true equivalence margin, i.e. $+\delta$, the significance test captures the truth and the equivalence test claims a false signal. Scenarios B and C in Fig. 8.7 are consistent for both test approaches and both tests are correct.

To best control false positive results, which is very important at such early stage, we applied a two-step approach in reporting the test results from our QTiSA-HT experimental data. First, we conduct the significance test as described in Sect. 8.4 for each dispense/dose level: if the p-value is not significant ($P > 0.05$), we conclude there is no signal; otherwise, we proceed to the second step. In step 2, we perform an equivalence test using the equivalence margin established in Sect. 8.5. If the equivalence test is also rejected (i.e., consistent with the result from significance test), the compound will be reported as positive; otherwise, it will be still reported as negative regardless of its small p-value obtained from the significance test.

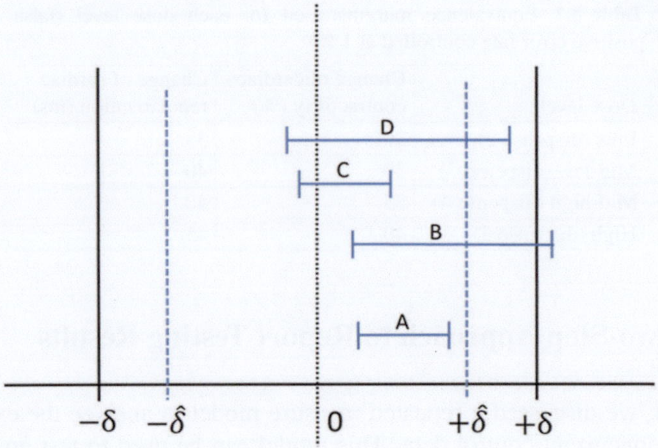

Fig. 8.7 Significant test vs. equivalence test in different test scenarios. The *solid black vertical lines* are corresponding to the true equivalence intervals $[-\delta, +\delta]$. The *blue dotted vertical lines* are corresponding to the estimated equivalence intervals $[-\hat{\delta}, +\hat{\delta}]$ based on historical control data. The *blue solid horizontal lines* with the segments are corresponding to the 95 % confidence intervals based on significant test. When $\delta < \hat{\delta}$, there are test scenarios corresponding to false negative results, which is not discussed here

The advantage of using this two-step approach is that it allows the use of historical vehicle-control data to minimize false positive results. For example, in Fig. 8.7, scenario A and B are concluded as positive if using significant test alone and scenario B and D are concluded as non-equivalent if using equivalence test alone. Both 1-step tests have one false positive result (scenario A for significant test and scenario B for equivalence test). The two-step approach only concludes scenario B as positive and makes no mistake. We recognize that this two-step approach may increase the false negative rate since more positive results are eliminated. However, as we mentioned previously that at such early stage of drug screening, false negative compounds may be tolerated if subsequent screening efforts are in place to detect safety liabilities before moving forward the compound substantially farther in the development process. For example, in vivo measures of QT prolongation from a follow-up assay can help to control false negative compounds.

8.7 Conclusions

In this chapter, we introduced several statistical applications in the design and analysis of an in vitro safety screen assay (QTiSA-HT) used in early drug discovery. To speed up the screening process, a cumulative dose response design is used for collecting repeated measures data at different dose levels. Although for simplicity, it is very common to calculate the sample size based on single dose testing, testing dose response trend as well as drug effect at each dose are equally important for this assay. A simple method using SAS PROC GLIMMIX procedure was introduced to compute the sample size for this repeated measure design that not only considers the within-subject correlation, but also can test the dose response trend. This method only requires generating mean data and is very flexible for the purpose of conducting different tests (e.g., trend test or pair-wise comparisons).

In the context of high throughput screening, we often have limited material for early screening and limited time to complete and analyze experiments. To help control false positive compounds, we incorporated historical control data by establishing an equivalence margin and then compared the effect between drug and concurrent control to the equivalence margin to confirm the statistical test results using the two-step approach.

A re-sampling method was introduced to establish the equivalence margins, a value specifically tied to the experimental design. Re-sampling can be easily performed to update the equivalence margins when necessary. This way of incorporating historical data is very helpful in the context of either safety or efficacy screening in early drug development. The same idea was also used to find a meaningful threshold for testing concentration levels. The advantage of using the two-step approach we described is that it doesn't require any change on the way to analyze the data, but rather modifies the reference values provided by the historical vehicle-control data. So it is simple to implement in practice, especially for non-statisticians to conduct the tests. This method can be used in any context that there is an abundance of historical control data and very few concurrent control data in real practice. In addition, this method also allows for comparing the concurrent control with the historical control, which is also important in monitoring experimental quality and reliability.

There are many ways to incorporate the historical (vehicle-control) data. For example, a hierarchal Bayesian model is another way to utilize both the historical and current control. However, how to appropriately weight the relatively larger historical data in the final results can be difficult, and the interpretation of the statistical model and results to non-statisticians can also be challenging.

Finally, to move the compounds quickly in the pipeline as we discussed in Sect. 8.2, the early screening assays like QTiSA, require automated techniques to provide a quick estimate of a safety signal for potential cardiac safety liabilities. The statistical methods that we discussed in Sects. 8.3 to 8.6 are all built in a series of standard computer programs developed in SAS and R. Scientists are able to deliver a standard statistical report within 24 h after completion of data collection.

References

Cavero I (2009) Exploratory safety pharmacology: a new safety paradigm to de-risk drug candidates prior to selection for regulatory science investigations. Expert Opin Drug Saf 8(6):627–647

Gintant GA, Su Z, Martin RL, Cox BF (2006) Utility of hERG assays as surrogate markers of delayed cardiac repolarization and QT safety. Toxicol Pathol 34(1):81–90

Sager PT, Gintant G, Turner JR, Pettit S, Stockbridge N (2014) Rechanneling the cardiac proarrhythmia safety paradigm: a meeting report from the Cardiac Safety Research Consortium. Am Heart J 167(3):292–300

Littell RC, Henry PR, Ammerman CB (1998) Statistical analysis of repeated measures data using SAS procedures. J Anim Sci 76:1216–123

Hedeker D, Gibbons RD, Waternaux C (1999) Sample size estimation for longitudinal designs with attrition. J Educ Behav Stat 24:70–93

Basagana X, Spiegelman D (2010) Power and sample size calculations for longitudinal studies comparing rates of change with a time-varying exposure. Stat Med 29(2):181–192

Comulada WS, Weiss RE (2010) Sample size and power calculations for correlations between bivariate longitudinal data. Stat Med 29(27):2811–2824

Phillips KF (1990) Power of the Two One-Sided Tests Procedure in Bioequivalence. J Pharmacokinet Biopharm 18(2):137–144

Blackwelder WC (1998) Equivalence trials. In: Encyclopedia of biostatistics, vol 2. Wiley, New York, pp 1367–1372

Schuirmann D (1987) A comparison of the two one-sided tests procedure and the power approach for assessing the equivalence of average bioavailability. J Pharmacokinet Biopharm 15(6): 657–680

Stroup W (2010) Introduction to generalized linear mixed models using SAS® PROC GLIMMIX, ASA GLMM workshop

Part III
Statistical Methods for Nonclinical Development

Chapter 9
Nonclinical Safety Assessment: An Introduction for Statisticians

Ian S. Peers and Marie C. South

Abstract This chapter provides an overview of the nonclinical drug safety testing process, and the statistical challenges associated with work in this area. Whilst other chapters in this book focus on specific types of designs and analyses which you may encounter during nonclinical drug development, we provide here the context for a statistician working in nonclinical safety assessment, and seek to prepare them for some of the practical issues and decisions they are likely to face. We will describe the scope and framework for the studies run within a nonclinical safety assessment programme, and recommend when and how a statistician needs to engage to add value. We will look generally at design and analysis considerations for safety studies which, whilst not unique to this context, are particularly pertinent to this area of work. Finally we will highlight some practical considerations and industry trends. If this chapter provides you with insight that complements rather than replicates what is taught in conventional statistics courses, and inspires you to believe that statistical work in this area can be both valuable and rewarding, we will have achieved our goal.

Keywords Drug safety • Toxicity • Safety pharmacology • Experimental design • Monitoring • Validation • Analysis • Continuous improvement

---

I.S. Peers (✉)
AstraZeneca, London, UK

University of Stirling, Stirling, UK

Inferstats Consulting Ltd, Cheshire, UK
e-mail: ian.peers@astrazeneca.com; ian.peers@stir.ac.uk; ian.peers@inferstats.com

M.C. South
AstraZeneca, London, UK

Inferstats Consulting Ltd, Cheshire, UK
e-mail: drmcsouth-statistics@outlook.com; marie.south@inferstats.com

L. Zhang (ed.), *Nonclinical Statistics for Pharmaceutical and Biotechnology Industries*, Statistics for Biology and Health, DOI 10.1007/978-3-319-23558-5_9

9.1 An Overview of Nonclinical Drug Safety Assessment

> Concern for man and his fate must always form the chief interest of all technical endeavours. (Albert Einstein).

In order for a substance to progress into becoming a new medicine, it is critical that it has the right safety characteristics, and that these have been appropriately taken into consideration throughout the drug's development. The following sections provide an introduction to the importance of nonclinical safety assessment, and an overview of the studies typically conducted to support the development of a new drug.

9.1.1 The Importance of Nonclinical Drug Safety Testing

An analysis of the causes of drug candidate attrition for AstraZeneca's small molecule portfolio over the period 2005 to 2010 (Cook et al. 2014) showed that out of 33 projects reviewed which closed during the pre-clinical development phase, 27 (82 %) of these projects gave unacceptable safety as the primary reason for closure. The pre-clinical phase was defined as the first Good Laboratory Practice (GLP, see Sect. 9.1.3) dose of a candidate drug in a safety study through to the investigational new drug (IND) submission. Cook et al. reviewed a similar number of projects which were closed during each of Phase I and Phase II clinical trials, finding 62 % and 30 % of these projects respectively were stopped for safety-related reasons. Whilst the learning from this review led to many steps being taken to address safety-related attrition, it remains true that the statistician who finds himself or herself working in nonclinical safety testing is supporting a critical part of the drug discovery and development process.

To put AstraZeneca's findings into context, a cross industry analysis of Phase II and Phase III drug candidate failures between 2011 and 2012 (Arrowsmith and Miller 2013) reported that 29 (28 %) out of 105 drug projects with reasons for failure evaluated were stopped due to safety concerns. This included insufficient therapeutic index, that is, the ratio between the blood concentration at which the drug becomes toxic and the concentration at which the drug is effective was too small. So safety issues continue to play a major role in project closure right through to late phase clinical testing.

Of course, the later a project fails in the discovery or development process, the more expensive the failure is, and hence a great deal of time, effort and money is being expended in trying to detect safety issues as early as possible. This means that the safety testing cascade encompasses a wide range of in vitro and in vivo studies run from early discovery through to late development, providing many opportunities for the statistician to make a real, beneficial difference.

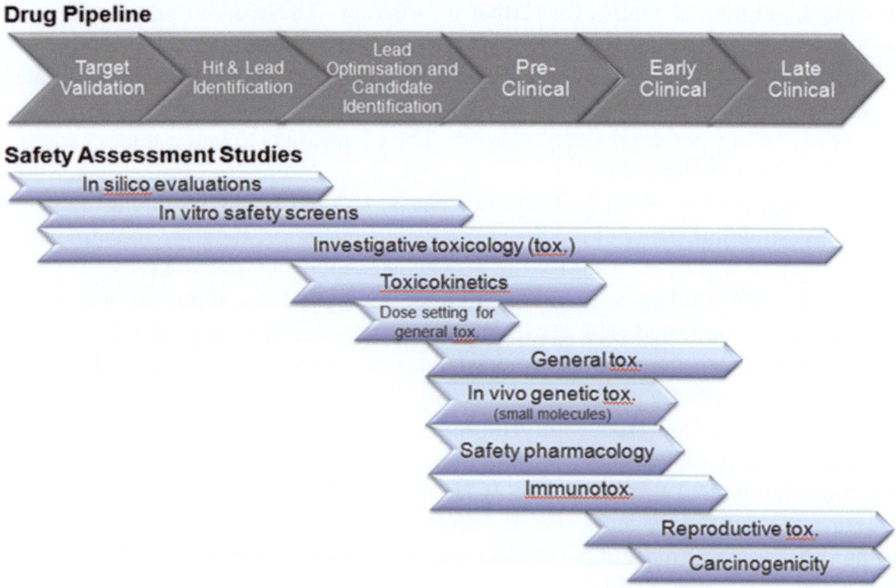

Fig. 9.1 Nonclinical safety studies which may be conducted for a typical drug project

9.1.2 *Types and Timings of Drug Safety Studies*

9.1.2.1 Overview of Nonclinical Safety Studies

Figure 9.1 shows the different types of safety studies which may be conducted for a typical drug project. It illustrates the general sequence and approximate timing of the studies required in order to take a drug into humans, and then to support more advanced clinical testing. The timing of the studies is designed to support the scientific decision-making sequence; in addition, the type and nature of clinical trials permitted at any point in drug development is limited by which nonclinical safety studies have been completed. Other safety studies may be required e.g. tissue cross reactivity studies for species selection for biotechnology-derived pharmaceuticals.

Even before any 'wet work' is conducted to support the safety aspects of a drug project, in silico evaluations may be performed e.g. to identify key moieties within a molecule that are associated with toxicity. These in silico predictions, however, are only ever as good as the data that supports them from previous studies. Other early safety work includes in vitro safety screening studies, such as those used in cardiovascular risk screening (see Chap. 8). Such studies provide an early indication of which substances have the potential to cause specific types of adverse effects or harm to humans. By their nature, these studies flag up a potential hazard rather than quantifying the risk to humans, as the translation from in vitro screen to clinical effect may not be fully established. However, the early warnings allow scientists to either control or eliminate the hazard (e.g. by moving to another chemical series), or to conduct additional studies for further information. These additional studies may be a well-defined cascade, or a set of specific investigative studies.

Investigative toxicology studies may be in vitro or in vivo and can be initiated at any point during the drug discovery and development process in response to the identification of a potential safety concern. Their purpose is to better understand the hazard, and the likelihood that this will become an unacceptable clinical risk. Since these studies are by definition bespoke, that is, tailored to the particular question under investigation, many of the general principles mentioned during the remainder of this chapter will be relevant for the statistician supporting such studies.

Whilst early studies tend to be high or medium throughput so that many substances can be tested (low cost), most safety studies of a more standard nature are run later when the number of substances or series being considered for any drug project is greatly reduced (high cost). The statistician should be aware of the International Conference on Harmonisation (ICH) guidelines which cover the remaining types of safety studies shown in Fig. 9.1, with specific guidance for biotechnological products. The ICH meets at least twice a year and brings together the regulatory authorities of Europe, Japan and the USA with the aim of harmonising pharmaceutical testing requirements for all pharmaceutical agents under four categories: Safety, Quality, Efficacy and Multi-disciplinary.

The following sections provide a brief introduction to each family of studies covered by ICH Safety Guidelines S1 through S8. We summarise briefly the purpose of the studies and highlight for the statistician some relevant points and sources of further information. Readers should, in addition, be aware of the multi-disciplinary guideline ICH M3(R2) (ICH 2009). This represents the consensus that exists regarding the type and duration of nonclinical safety studies and their timing to support the conduct of human clinical trials and marketing authorisation for pharmaceuticals. This includes the duration of repeated dose studies, specialist toxicology areas e.g. phototoxicity, combination studies and the use of biomarkers in exploratory clinical studies. For a good overall introduction to statistical methods used in toxicology studies see (Jarvis et al. 2011).

9.1.2.2 Carcinogenicity Studies

The aim of a carcinogenicity study is to identify a test substance's tumorigenic potential in animals and to assess the relevant risk in humans. Typically 2 year carcinogenicity studies will support the registration of medicines with an intended patient use of more than 6 months. A carcinogenicity study is usually required before an application can be made for marketing approval of a pharmaceutical, but not necessarily prior to the conduct of large scale clinical trials as indicated in ICH MR3(R2) (ICH 2009). ICH S1A (ICH 1995), S1B (ICH 1997) and S1C(R2) (ICH 2008) provide guidelines for rodent carcinogenicity studies, but at the time of writing a change to the S1 guidelines is proposed: to introduce a more comprehensive and integrated approach to addressing the risk of human carcinogenicity of pharmaceuticals, and to clarify and update the criteria for deciding whether the conduct of a 2-year rodent carcinogenicity study of a given pharmaceutical would add value to this risk assessment. Chapter 12 provides results of recent research exploring some aspects of the experimental design and analysis of carcinogenicity studies.

9.1.2.3 Genotoxicity Studies

ICH S2(R1) (ICH 2011a) describes the standard studies used in genetic toxicology and provides guidance on interpretation of results. In vitro and in vivo tests are used to detect substances that induce genetic damage to DNA, the goal being to characterise the risk for carcinogenic effects originating from changes in genetic material. These studies are typically only needed for small molecules. The regulatory genetic toxicology screening cascade will generally include: a bacterial gene mutation assay (e.g. Ames), an in vitro mammalian cell assay for gene mutation and/or chromosome aberrations (e.g. in vitro micronucleus assay, in vitro mouse lymphoma Tk gene mutation assay), and an in vivo assay for chromosomal effects (e.g. rodent micronucleus). For an introduction to the study design and statistical power considerations for the widely used rodent micronucleus test, statisticians are referred to (Hayes et al. 2009).

9.1.2.4 Toxicokinetics and Pharmacokinetics

Pharmacokinetics (PK) is the study of the fate of substances within the body over time and of how the body absorbs, distributes, metabolises, and eliminates a substance. PK parameters (e.g. clearance) are calculated using the concentrations of substances measured in biological matrices, usually plasma. Toxicokinetics (TK) is the description of the systemic exposure of a substance in toxicity studies using PK parameters. The overall aim of TK studies is to relate the exposure achieved in animals to the substance dose level and the time course of a toxicity study. TK data also play a role in the clinical arena, assisting the setting of limits for human exposure and the calculation of safety margins. ICH S3A (ICH 1994a) and S3B (ICH 1994b) provide guidance on TK and PK studies. Chapter 11 contains further information on PK measurement.

Toxicity studies which may be usefully supported by toxicokinetic information include studies of single and repeated dose toxicity, reproductive toxicity, genotoxicity, carcinogenicity and safety pharmacology. TK data may be obtained from all animals on a toxicity study, from representative subgroups, from satellite groups or in separate studies. A separate assessment of the effect of repeated dosing on the accumulation of the substance and/or metabolites within tissues may be needed in some situations.

Statisticians can provide important input, for example in advising when to transform data and in ensuring that data summaries include appropriate estimates of variability: the inter-individual variation of kinetic parameters is often large. Small numbers of animals are usually involved in generating TK data and understanding individual animal responses may be of more value than a refined statistical analysis of group data.

9.1.2.5 Toxicity Testing

A key part of the nonclinical safety evaluation of a pharmaceutical product is repeated dose and chronic toxicity testing in rodents and non-rodents. In general, clinical development trials of up to 2 weeks in duration will be supported by repeated dose toxicity studies in 2 species (1 non-rodent) for a minimum duration of 2 weeks, conducted to Good Laboratory Practice standards (see Sect. 9.1.3). Clinical trials which last longer than this, up to 6 months, should be supported by repeated dose toxicology studies of at least equivalent duration. ICH S4 (ICH 1998) documents the consensus view on the required duration of nonclinical chronic toxicity testing required: a 6 month rodent and a 9 month non-rodent study will generally support dosing for longer than 6 months in clinical trials. The statistician needs to be aware that these studies have many endpoints and relatively small numbers of animals per group. Descriptive statistics, and expert scientific judgment, together with knowledge of "normal" ranges for parameters, are critical to interpretation of study outcomes; the statistical significance of effects should be interpreted in this broader context. For a good introduction to the design and size of general toxicity studies of various durations see (Sparrow et al. 2011). Chapter 10 provides further information.

9.1.2.6 Reproductive Toxicology

The purpose of reproductive toxicology tests is to assess any impact of the substance tested on mammalian reproduction. This includes male and female fertility, embryo-foetal development and pre- and post-natal development. These animal studies are important in providing information in product labels for men or women wishing to have children and especially for pregnant and lactating women. Although literature, in vitro assays and studies in non-pregnant animals may provide early indications of the potential for reproductive toxicity during drug discovery, the main reproductive toxicity studies will usually come later, after 1 month general toxicity studies in rodents and non-rodents. Reproductive toxicology tests are described in ICH S5(R2) (ICH 2000a). A key statistical consideration for these studies is randomisation of animals to groups, with the pregnant female, the dam, often being the experimental unit, and taking into account the need to spread sibling animals or animals pregnant by the same stud male across groups. Care also needs to be taken with statistical analysis to allow for the dam/litter being the experimental unit. It should be noted that descriptive statistics are important, as is biological plausibility, when evaluating data from these studies, whilst inferential statistics may be used only as support for interpretation of results. Sizing of studies is also an important consideration, as the study must give rise to a sufficient number of litters; hence allowance must be made for some females failing to become pregnant. Other factors to consider in setting group size are the prevalence of events in control populations and the nature of the endpoint(s) being considered (continuous measure or otherwise). Chapter 10 provides further information.

9.1.2.7 Biotechnological Products

Biotechnology-derived pharmaceuticals (biopharmaceuticals) include products derived from characterised cells through the use of a variety of expression systems such as bacteria, yeast, insect, plant, and mammalian cells. These were initially developed in the early 1980s, with the first marketing authorisations granted later in the decade. The guidance for these products in ICH S6(R1) (ICH 2011b) is based on a critical review of experience with submission of applications for biopharmaceuticals. It sets out to provide general principles for designing scientifically acceptable nonclinical safety evaluation programs for these products. Whilst the details are specific to biotechnological products, the main study types or groupings covered are those which form the titles of the companion ICH Safety guidelines. The guidance highlights the criticality of group size in affecting ability to detect toxic events which may be associated with these products; hence a statistician is likely to have a key input into the determination of number of animals per dose.

9.1.2.8 Safety Pharmacology

Safety pharmacology studies investigate the potential undesirable pharmacodynamic effects of a substance on physiological functions in relation to exposure in the therapeutic range and above. The core battery of tests explores the potential for harm to the central nervous system, the cardiovascular system and the respiratory system. Supplementary studies exploring e.g. renal or gastrointestinal effects may be required. These studies are covered by the guidelines ICH S7A (ICH 2000b) and S7B (ICH 2005a) which highlight a key area for statistical contribution. The guidance states that the size of the groups should be sufficient to allow meaningful scientific interpretation of the data generated. Thus, the number of animals or isolated preparations should be adequate to demonstrate or rule out the presence of a biologically significant effect of the test substance, taking into account the size of the biological effect that is of concern for humans. The value added through appropriate powering of safety pharmacology studies is thus widely recognised. It is also worth noting that these studies have a wide variety of endpoints representing continuous, ordinal and nominal data types, which accordingly ensures that a wide variety of statistical methods are applicable to the data generated. Chapter 10 provides additional information and see also (Pugsley et al. 2008).

9.1.2.9 Immunotoxicology

ICH S8 (ICH 2005b) provides recommendations on the nonclinical testing of non-biologicals for immunotoxicity. The term immunotoxicity in this guideline primarily refers to immunosuppression, i.e. a state of increased susceptibility to infections or the development of tumours. Initially, a number of factors are taken into account and a weight of evidence approach used to determine if further immonotoxicology testing is required. Examples of the factors considered are the structural properties of the compound, and observations from standard general toxicity studies in animals, e.g. certain haematological changes. Chapter 11 provides further information on immunogenicity evaluation.

9.1.3 A Word About Good Laboratory Practice

The regulatory authorities in many parts of the world, including the European Union and the United States of America, require studies designed to evaluate the safety of a new chemical or biological substance to be completed in accordance with Good Laboratory Practice (GLP) principles. Some countries, such as the United Kingdom, have implemented their own statutory regulations (based on the GLP principles), which means that this work is required by law to comply with the regulations. The GLP principles provide a framework to ensure that the planning, execution, monitoring, recording, reporting and archiving of a study are conducted in a suitably rigorous manner. The purpose of GLP is to provide assurance to regulatory authorities that the results presented for a given study reflect accurately what happened in the study and are therefore reliable for assessing the safety or risk associated with the substance under test. The statistician working on GLP studies will require GLP training, and periodic refresher training, and will need to follow defined standard operating procedures (SOPs) as well as adhering to a variety of other formal requirements such as the maintenance of records of qualifications, training, and ongoing continuous professional development. Analytical or computer systems that are used to analyse data and produce statistical results to be included in the study report for a GLP study must be validated.

9.2 Starting the Dialogue: The Statistician and the Safety Scientist

'The time has come,' the Walrus said 'to talk of many things.' (Lewis Carroll, in Through the Looking-Glass and What Alice Found There, 1872)

9.2.1 The Right Time to Talk

"We need your help to analyse some extra data collected on one of our studies please." said the voice on the end of the phone. "We just need to see whether levels of 'A' in plasma change significantly with dose-level and time post-dose when this species is treated with compound 'Y'." "Was a statistician involved in the design of the additional data collection?" came the reply. "Oh no, this extra investigation was added on to a standard study, so the number of animals and the design was already set. We just need help with the data analysis."

This sort of scenario is by no means rare. 'Minor' changes to an otherwise standard study to collect additional information is often referred to as 'signal searching', and it may not be immediately obvious to the experimentalist that the chances of detecting interpretable signals are increased if the study design is discussed with a statistician before the study starts, and before resources e.g. animals are ordered, which is often some time ahead of study execution. At the point of data analysis, however, it may be too late for the statistician to help. To quote Ronald Fisher: "To call in the statistician after the experiment is done may be no more than asking him to perform a post-mortem examination: he may be able to say what the experiment died of."

When the statistician and scientist met following the phone call described above, it became apparent that the effects of dose-level and time on 'A' couldn't be accurately interpreted. Samples for measurement of 'A' had only been collected after the animals had been dosed with compound, and due to the relatively small study size, and relatively large variation in 'A' between animals and across time, effects were unclear. Had a small number of pre-dose samples been taken to indicate each animal's baseline level and variation in 'A', the outcome might have been different. Had the statistician been engaged earlier, pros and cons of taking pre-dose samples in this situation might have been discussed, even though these were not required for the 'standard' part of the study's purpose.

Discovering difficulties at the analysis stage that were not considered at the design stage is by no means limited to the in vivo setting. An example of a common shortcoming that hinders the evaluation of in vitro study data is insufficient characterisation of the assay's variation on different testing occasions; that is, different occasions when the entire assay process is replicated from start to finish, including the sample preparation for the substance(s) under test. A conversation with a statistician when setting up, evaluating or routinely running in vitro assays can help to draw out the different potential sources of variation for the assay, and ensure these are appropriately understood and allowed for during the validation and routine running of the assay.

Although the examples given are not unique to nonclinical safety testing, it is worth being particularly alert in this environment, where many studies are run and analysed in a fairly standard manner, and the value of contacting the statistician about a minor change to objective or design may not be so obvious as for a bespoke study. Even when a specifically designed investigative study is required, the investigator may by default propose to run the study using a traditional design with which they are experienced, and in such a situation the statistician may be able to suggest a more efficient and effective design or analysis, or simply recommend minor adjustments to ensure that the objectives of the study can be met.

Of course, if the statistician is named as playing an official role in a regulated safety study, they will, as part of routine procedures, have the opportunity to review and comment upon the study plan. However, when this is not the case, how can early dialogue between study personnel and statisticians be encouraged? How can the message be shared that it's never too early to contact a statistician, but often too late?

Any individual scientist probably learns this most effectively through their own experience and examples of working with statisticians, hence the relationship between the statistician and scientist is a key factor. Anything which facilitates this relationship e.g. co-located working is therefore likely to help. However, partly due to frequent personnel changes, this is sometimes a slow and not necessarily sufficient or enduring way to spread the message widely in an organisation. The statistician could proceed by working to integrate statistical considerations into the existing relatively formal procedures in nonclinical safety departments. For example, the key personnel involved in safety studies will require specific documented training, so it makes sense to try and ensure that an informal meeting with a statistician and some illustrations of when and where and how to engage a statistician form part of this training, in addition to any formal statistical training the scientist receives. Additionally, it may help to document in standard departmental procedures when and how a statistician needs to be engaged to review individual study proposals and protocols. Peers et al. (2014) describe the benefits this can have when implemented systematically within an organisation.

Another way to encourage such dialogue and interaction early is to discuss regularly with managers current examples of the positive impact of early statistician/scientist collaboration, using these opportunities to increase awareness of common pitfalls to be avoided. This will help to secure managers as advocates who will encourage their staff to engage early with statisticians.

9.2.2 The Right Things to Talk About

Whenever the statistician is involved in the advance plans for a study, there will usually be a lot more to talk about than may initially be apparent. We focus here on three questions which may be particularly pertinent for nonclinical safety studies, especially in vivo.

What Do We Already Know about this compound/drug/target from our studies or from the literature? If a safety study, investigative or otherwise, is being run, there's likely to be relevant prior information of some sort. Understanding where the study sits in a program of work, and what prior knowledge exists, and with what level of confidence, is important when setting up designs. As an example, one investigative pharmaceutical safety study involved studying two different compounds. More relevant nonclinical information was available for one of these compounds than the other. Although the initial design proposed had all groups in the study being the same size, after statistical consultation, it was agreed that the study goals could be achieved using fewer animals for those groups to be dosed with the more fully characterised compound.

What Is Fixed and What Is Flexible? If asked to comment on a study design, the statistician will do well to establish: which aspects of the study are fixed e.g. due to regulatory or other requirement; which aspects are established practice and would cause difficulties if altered; and where there is scope to be flexible. The inclusion of positive controls on studies serves as an example here. For safety pharmacology studies, the ICH guideline S7A (ICH 2000b) states: "In well-characterised in vivo test systems, positive controls may not be necessary. The exclusion of controls from studies should be justified." So for an individual study, the agreed regulatory view determines that positive controls should either be included or their exclusion justified. It is likely that each organisation will have established practice relating to this which needs to be taken into account too: if it is established practice to include positive controls on all studies, then all study design, analysis and reporting systems and protocols will be set up to support this. However, there may still remain some flexibility e.g. it may sometimes be appropriate to propose a smaller group size for the positive control group than for the other groups.

Why Are Things Done This Way? If things are done traditionally in your organisation in a certain way, do try to understand why before questioning or challenging based on your experience from working in a different area, or your theoretical knowledge. An example here is the determination of cage layout plans, and whether cages belonging to the same treatment group are kept together, or spread in a balanced way across rows and columns of the cage racking system. There are competing considerations here. On the one hand, it is desirable to reduce to an absolute minimum any opportunity for cross contamination between cages of animals receiving different treatments, and in many animal room settings, written procedures may require that this is done by keeping all cages for any treatment group underneath one another on the same rack. This also tends to facilitate speedier dosing, group by group, as any formulation can be quickly administered to relevant cages easily identifiable through their co-location. The competing consideration is the possibility that bias could creep in unintentionally when groups are experiencing very slightly different processes during the study such as the timing of receiving their dose and marginally different environmental conditions associated with different locations in the room. As caging and air-handling systems change over time, the balance of risks and benefits associated with different practices will change, so it's worth asking the "why" question periodically, to see if the balance has changed favouring a change in practice.

9.2.3 *The Right People to Talk To*

The statistician may find that an overwhelming number of people are involved in any given safety study, particularly for GLP in vivo studies. Whilst the Study Director represents the single point of study control with ultimate responsibility for the overall scientific conduct of the study, any study is likely to require the input of multiple specialists, including experts in pathology, clinical pathology, formulation, toxicokinetics and toxicology, together with those who ensure excellence in the running of the study through co-ordination and delivery of in-life animal procedures. Some, maybe all of these individuals, will have important opinions on study design, including how many animals might be needed, how best to allocate animals to groups, and other scientific and practical considerations.

If you work in a contract research organisation or a scenario where the statistician's input is limited primarily to performing power calculations and providing a statistical analysis plan to match a given design, then interaction with the Study Director may suffice. However, wherever possible and appropriate, it is far more effective and satisfying if the statistician can be included as an expert in meetings alongside other experts who together form the study design team. For example, if a specific haemodynamic marker is a critical endpoint in a particular study, the statistician may venture to suggest that randomisation of animals to groups should take into account a baseline measurement of this parameter to ensure balanced groups at the outset. It is much easier and more efficient to discuss the pros and cons of such a suggestion if the study in-life co-ordinator and the clinical pathology representative are sat around the same table as the Study Director and statistician, bringing together perspectives on design questions such as these.

Section 9.2 has focused in on the nuts and bolts of ensuring that statistical considerations are raised at the right time, in the right way, and with the right people. Section 9.3 will outline some of the typical characteristics of safety studies which may be unfamiliar to a statistician coming into this area.

9.3 Design Considerations for Safety Studies

> At first we were amused at the dramatic (over) design of the hotel, but it seemed like sometimes design overshadowed function. There was no hot water in the bathroom. (TripAdvisor 2009)

It is possible to fall into traps at opposite ends of the spectrum when it comes to designing nonclinical drug safety studies: to accept current practice with insufficient questioning, or to get so overwhelmed thinking through all the considerations that study functionality gets overshadowed. In order to help the statistician new to this domain avoid each of these extremes, we outline here a few characteristics of studies and their related design issues which may be encountered more frequently in this setting than in other areas.

9.3.1 Multiple Critical Study Endpoints

The individual new to toxicology studies can be forgiven if they are surprised by just how many endpoints producing different types of data are measured in a single study. For example, a typical non-rodent general toxicity study consists of at least 3 animals per sex in each of 4 treatment groups: vehicle control, low dose, intermediate dose, high dose, possibly with additional animals to assess recovery (see Sect. 9.3.3). Meanwhile, the number of endpoints to be evaluated will far exceed the number of animals used, with a typical list of study outcomes including: clinical behavioural observations (e.g. reduced motor activity); bodyweight and food consumption; organ weights, plus gross and microscopic pathology observations on multiple tissue types (kidneys, liver, lungs etc.); in excess of ten different haematology measures; in excess of ten different plasma chemistry measures; and other parameters of interest. What is noticeable is not just the number of measures, but the variety of types of measures. Clinical observations of animal behaviour are tabulated by animal and frequency. Microscopic pathology involves taking samples from many tissues from each animal at necropsy, and evaluating these using an ordinal scale. Other measures are continuous, but with varying distributions including normal and log-normal.

For studies with this characteristic, the role of the conventional statistical power calculation is limited. Whilst data are generated for each individual endpoint, the outcome of the study is an expert toxicological interpretation, taking into account the various data types, whether or not evaluated via a formal statistical test. The diligent reader of the ICH guidelines will already have encountered this, since they emphasise the importance of biological interpretation alongside or above statistical significance in a number of places. Group sizes for such studies tend to be based on cross-industry consensus, as described for example in (Sparrow et al. 2011). Whilst keeping these things in mind, the statistician should be aware that they may still be able to help at the design stage. As soon as a study designed for general purposes is adapted to address, in addition, a specific question relating to a specific endpoint, a review of the study's properties (e.g. power) relative to that specific endpoint is likely to be relevant.

9.3.2 Toxicity: Looking for Absence, Presence, or a Rare Event?

Paracelsus, the founder of the discipline of toxicology, is attributed with the observation that all things are toxic, it is only the dose that distinguishes a remedy from a poison. Those working on general toxicity studies will recognise this need to characterise the dose-related toxicity response of a substance: part of the purpose of such studies will be to identify the dose/exposure level at which there are no observed adverse effects, that is, the NOAEL (see Chap. 10); at the same time, some evidence of a toxic effect, such as clinical observations or bodyweight loss, is generally to be expected in association with treatment at the highest dose. A consequence of the latter is that positive control animals are not considered relevant for these studies.

In contrast, as mentioned previously, for safety pharmacology studies we need to demonstrate that real effects can be reliably found on an ongoing basis, especially for endpoints where effects may only be seen infrequently. This is achieved by using a positive control group on each study, or by running regular studies to confirm that the dose-response effect for compounds with known toxicity can be detected.

Another way to build confidence in study results, particularly when toxic effects are absent for a specific study, is through monitoring of the level and variation seen in key endpoints for vehicle control groups over time. This is appropriate for the situation where a standard type of study is run often and repeatedly in the same place. Monitoring using control charts (Wheeler 2000) helps to confirm whether or not a study is performing similarly to past studies and results are stable over time. An appropriate indicator of the level and variation in key responses should be plotted. The level could be the mean of one or more key responses in the vehicle control group, assuming this is not particularly sensitive to the specific vehicle or dosing route used. Figure 9.2 shows an example control chart. Similar charts can be used to monitor a relevant within-study measure of variation, for example the standard error of the effect size estimate, that is, of the difference between means for a compound dosed group and the vehicle group. If both the vehicle group level and

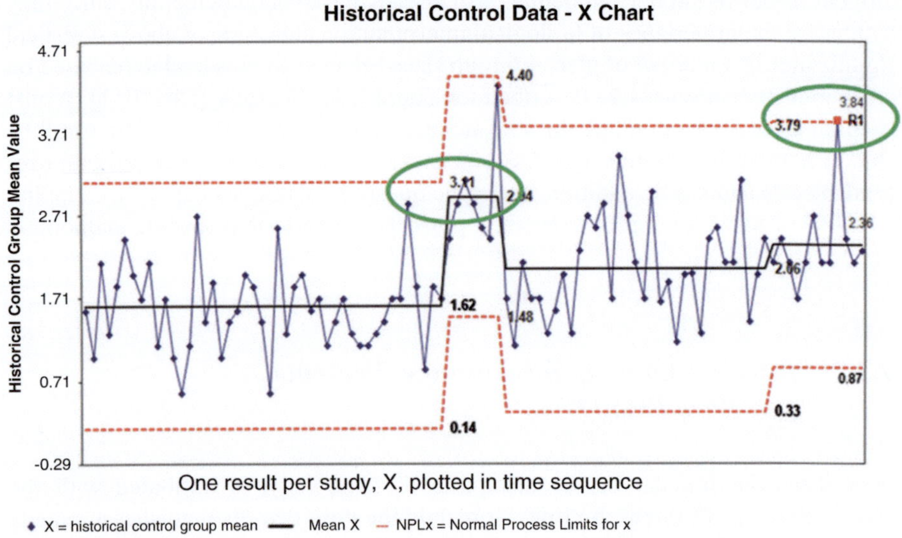

Fig. 9.2 Example of a statistical process control chart being used to monitor vehicle control group mean values over time. Step changes in the mean, or a single point outside the normal process limits (examples *circled*) indicate a process change, and may warrant investigation

the within-study variation lie within normal process limits on such a control chart, this adds confidence in the study outcomes, even if no significant effects associated with compound-dosing are found.

It should be noted that some specific drug-related toxicity outcomes, such as certain tumours or malformations in reproductive toxicity studies, are rare. Studies will generally have little power to detect these kinds of low frequency events, and a dose response is hard to demonstrate. In such situations, both statistician and scientist need to be aware that lack of statistical significance of an effect or observation does not necessarily rule out biological relevance. Confidence intervals for relevant quantities can be especially useful in such situations to indicate a range of effect sizes which are consistent with the observation made in the specific study. In addition, an analysis of the frequency of occurrence in the control group will be relevant for some rare events such as lesion incidence. The best historical control data are in-house data of the last 5 years. External data from the same strain and breeder are good substitutes when in house data are unavailable (Ettlin et al. 2010).

9.3.3 Use of Recovery Groups

This is an important topic that statisticians new to safety assessment may not have encountered previously. A recovery group is a group of animals that experience a non-dosing period that follows the main dosing phase of a study. The purpose is to understand whether toxicities observed at the end of the dosing phase are partially or completely reversible. Particular consideration should be given to inclusion of recovery arms in study designs including biopharmaceuticals with long half-lives. The reader is referred to (Pandher et al. 2012) for discussion of this topic. The article includes a helpful summary of various ICH recommendations that refer to inclusion of recovery arms in nonclinical safety studies. There is ongoing discussion about the need for and timing of recovery assessment.

The statistician may be asked to advise on the number of animals for use in a recovery group, and should take care to understand both existing practice, and information relevant to the specific substance and study in question. Good questions to ask relate to the likely incidence of any anticipated toxic effects, and whether there is any likelihood, particularly for longer studies, of animal losses in the recovery group. The statistical analysis of recovery phase data should be agreed at the planning stage. Summary statistics only may be appropriate for the recovery phase, especially if there is no vehicle recovery group for reference or if group sizes are small. If statistical comparisons are planned, it is important to remember that studies are not typically powered to allow lack of statistical significance between groups or timepoints to be interpreted as equivalence.

9.3.4 Are Very Small In Vivo Studies Worth Running?

In the authors' experience there are occasions when only small amounts of compound are available for very early safety studies. The question a statistician may be required to address is whether some early safety information from a small study is likely to be better than no information at all. A couple of points are worth making here.

The first is that for any study to be worthwhile, it has to be well run, however small. A possible temptation when compound is in short supply is to set up experiments with the potential to be biased, for example, by deliberately using lighter animals in high-dosed groups, since treatments are usually administered in proportion to bodyweight. Whilst it may not be possible to quantify the effects of such non-random animal allocation procedures, any deliberate introduction of bias such as this cannot be recommended as a reliable or valid scientific method. So, the ability to apply good practice is a key consideration, which will then increase the chance that the results will reproduce and generalise.

The second point to make is that where a reliable body of prior evidence and a reproducible animal model exists, very small studies can be incredibly useful. To illustrate this, consider the following example. A candidate drug compound, CDX, in nonclinical development was found to decrease the amplitude of the electrocardiogram (ECG) trace in all compound-dosed animals in a non-rodent safety study. An alteration of this nature is unusual, but was extremely reproducible across all animals in the study. After some research and investigation, an anaesthetised rat protocol was established which replicated this effect every time CDX was administered (see Fig. 9.3). Additional endpoints were also measured that provided more mechanistic understanding of the effects of CDX on the ECG.

This model was then used to assess very small quantities of other compounds generated by the project team with the goal of eliminating this safety concern.

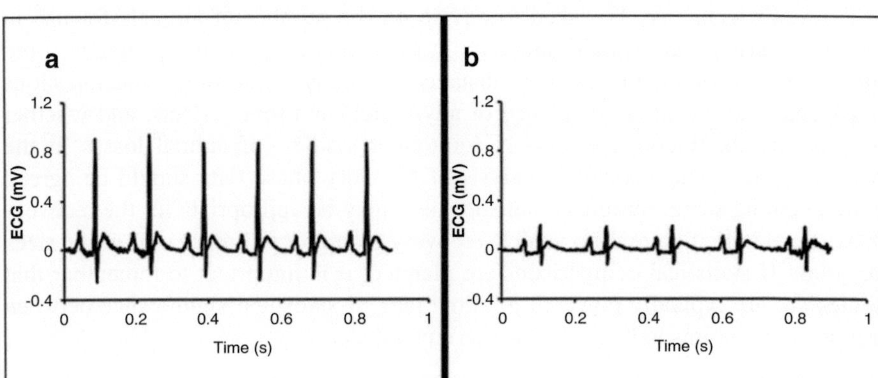

Fig. 9.3 Typical electrocardiogram (**a**) pre- and (**b**) post-dosing animals with CDX

Initially, each compound was tested in a single animal, with positive (CDX-treated) and negative (vehicle-dosed) control animals being included at regular intervals. A positive result for a test-compound led to no further synthesis, whilst a negative result was a trigger for more extensive testing to add confidence that the undesirable effect was absent.

Whilst this screening process worked well, a key point for the statistician to remember whenever a very small study is proposed is to gain as much information as possible about the position of the study in the overall programme of work. It can often be the case that a larger single study or use of an adaptive study design will offer more information and more confidence for decision making and will ultimately use fewer animals than a sequence of smaller studies.

9.3.5 Design of In Vitro Screen Validation Studies: A Brief Word

The desire to predict safety liabilities as early as possible in the drug discovery and development process, together with the desire to reduce the extent to which animals are used in scientific research, increases the drive to find reliable in vitro screening assays which predict toxicities that would otherwise be identified later via in vivo or clinical studies. Usually in these later stages the properties of the molecule are fixed and subsequent efforts focus on managing rather than eliminating the risk.

The construction of higher volume in vitro toxicity profiling assays is a balance between biological complexity, robustness, reliability, throughput and associated costs. This reductionist approach requires an appreciation of the limitations of the biology of the in vitro assay system and its corresponding predictivity. For example, there are multiple mechanisms that lead to hepatotoxic pathology, only some of which may be captured and reported appropriately by an in vitro assay under assessment.

The validation of a new in vitro assay designed to predict a particular safety liability, for example the potential to give rise to liver toxicity in the clinic, is usually accomplished by selecting a set of compounds known to be associated with such toxicity in the clinic (or in vivo), together with a set of compounds known to be clean of this toxicity, and then seeing how well the new assay predicts the known outcome. The choice of compounds to use to validate the assay may not be straightforward. The frequency and severity of e.g. clinical liver toxicity associated with two different compounds can vary widely, so even defining a validation compound's true status as "toxic" or "non-toxic" may require some cross-scientist consensus. Similarly, validation compounds may be associated with multiple toxicities in the clinic further adding complexity to their selection for a defined assay characterisation.

A further consideration (see Sect. 9.2.1) is the need to characterise sufficiently the occasion to occasion variation associated with a new assay. Without this, we cannot be sure how consistent any predictions will be in practice for the same compound if it were analysed on multiple occasions over time. Variation from occasion to occasion, when the whole assay process is replicated from start to finish, can greatly exceed the variation seen when samples from the same source are tested in duplicate within the same assay run. So, if a validation experiment is carried out on a single occasion without additional understanding of sources of variation, an incomplete picture of likely assay performance in routine use will be generated.

In addition, the precision of each measure (e.g. sensitivity, specificity etc.) used to characterise the quality of the new assay for prediction purposes (see Sect. 9.4.3) will be limited by how many positive and negative compounds are used in the validation, and this needs to be defined to balance both the required precision of the statistical estimates, and the practical constraints. As an example, if 40 compounds known to give rise to the toxicity under test are used to establish the sensitivity of the assay, and 35 of these are detected as positive by the assay, the sensitivity estimate is 35/40 = 88 %. This may be an acceptable value, but the precision of the estimate is not extremely high even with 40 compounds: the one-sided lower 95 % confidence limit is calculated to be 76 %, and should be taken into account in interpreting the result. Section 9.4.3 contains further comments on the analysis of such data. Section 9.4 will provide an overview of some aspects of statistical analysis particularly relevant for nonclinical safety studies.

9.4 Analysis Considerations for Safety Studies

> Take some more tea," the March Hare said to Alice, very earnestly. "I've had nothing yet," Alice replied in an offended tone, "so I can't take more." "You mean you can't take less," said the Hatter: "it's very easy to take more than nothing. (Lewis Carroll, Alice's Adventures in Wonderland, 1865)

It may be surprising just how often the same data viewed from a different angle may lead to alternative interpretations. As the overview (see Sect. 9.1) made clear, the analysis and interpretation of data from drug safety studies requires careful scientific and statistical input. Ideally, the statistical analysis plan for any study will be written alongside the study design, since the two are inextricably related. However, in the same way that safety studies tend to generate some specific design-related challenges, there are similar considerations when it comes to choosing which way to go with the analysis. Specific types of analysis are discussed in other chapters, so we restrict ourselves here to some general comments.

9.4.1 Graphical Analyses Versus Formal Statistics

It is worth emphasising the value of plotting the raw data for safety studies, particularly since graphical output is not always a strength of automated systems which may be employed in the analysis. The formal reporting of many safety studies

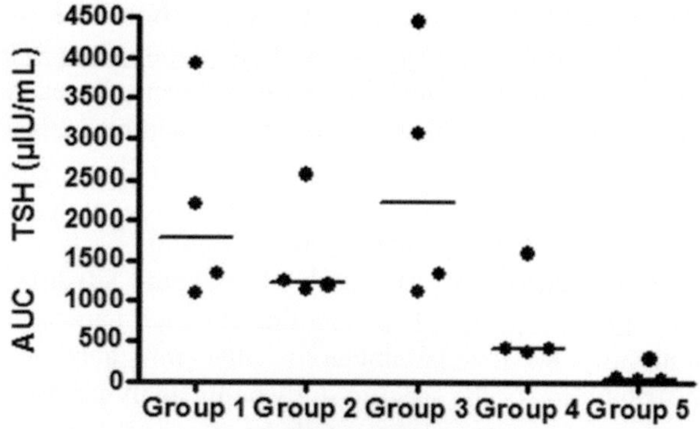

Fig. 9.4 Thyroid-stimulating hormone (TSH) response represented by area under the curve (AUC) for five groups receiving different treatments. *Each dot* represents an individual animal, and *each line* the group median

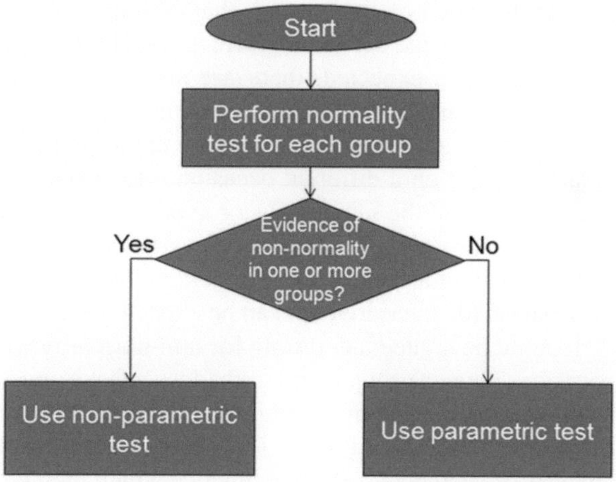

Fig. 9.5 Example of a simple analysis decision tree

will include tables of values and of summary statistics, but these may be insufficient to reveal the full nature of the patterns in the data. In particular, the default summary may include the mean and standard deviation for each group in the study, when in fact alternatives, e.g. the median, may better represent the data. For in vivo studies, it is informative to plot individual animal results not just summaries. Figure 9.4 illustrates one type of graph which is typically useful, giving far more insight than a simple table of group means and standard deviations.

9.4.2 Decision Tree Approach Versus Fixed Analysis Approach

Figure 9.5 illustrates a very simple first step which might be used within an automated system applying a decision tree approach. When a study is to be analysed using a formal statistical test, a decision tree is one way of specifying how the data will be analysed. We won't discuss for the example in Fig. 9.5 which test for normality or which size of p-value should be applied if using this approach. Our purpose here is to consider the implications of using such an approach versus a specific alternative: that of defining in advance for each study outcome how it will be analysed i.e. in this example, whether it will be analysed using a parametric or non-parametric test. In considering these alternatives, we need to remember that in the context of nonclinical safety assessment, a great deal of historical data is generally available for the majority of study endpoints.

The decision tree approach offers some advantages. Theoretically, using our example in Fig. 9.5, it ensures that for each parameter analysed, evidence of non-normality for that parameter is taken into account, based on the within-group distributions for the specific study being analysed. It requires no prior knowledge of how any parameter might usually be distributed, and the resulting analysis approach may be defended on the grounds that an impartial and consistent process has been used to arrive at the decision on which tests to use to analyse the data.

As might be expected, there are also some drawbacks to such an approach. One consequence is that a particular endpoint e.g. bodyweight may be compared across treatment groups using a non-parametric test on one occasion, and using a parametric test on a different occasion: the different occasion could be a different timepoint within the same study, or another study of the identical design but testing a different substance. It is open to question as to whether this is appropriate for an endpoint where a great deal of historical data is available indicating what kind of distribution the measurement can be expected to follow.

It could be argued that testing for non-normality within each study will allow for occasions when a group in the study does behave unusually, for example with one or more extreme results in a group. The counter argument is that proper examination of the raw data should allow for discussion and evaluation of the impact on the analysis of such occurrences, a practice which may be more easily overlooked if an automated procedure is deemed to have taken care of it.

Proponents of fixing the analysis for a given endpoint may argue that testing e.g. for deviation from normality, will not be powerful anyway in studies with small group sizes. Using an existing larger body of historical data to specify in advance the analysis to be performed will ensure that the analysis approach has no dependence on study size, and that power calculations based on the significance test to be applied can be used reliably to confirm the appropriateness of any group size over time for a given study type and endpoint.

Given the vast array of different types of safety studies, and the many practical and scientific considerations which influence the analysis and interpretation of the data, it is not possible to do more here than compare and contrast these two different approaches, which are likely to be encountered in the nonclinical drug safety testing environment. The intention here, as with the following section, is to provide the statistician with some preliminary awareness of the relevant issues, to help them prepare to engage on these matters within their own organisation.

9.4.3 Analysis of In Vitro Screen Validation Studies

The purpose of an in vitro safety screen was described in Sect. 9.1.2.1. For an overview of how to characterise data from a screening test designed to predict an outcome, readers are advised to study the sequence of three short but informative articles on diagnostic tests (Altman and Bland 1994a, b, c). These describe the key summary measures commonly used to characterise screening tests, which are: sensitivity, specificity, positive and negative predictive values (PPV and NPV) and positive and negative likelihood ratios (PLR and NLR). In addition, for a test which gives a continuous readout, Altman and Bland describe the use of Receiver Operating Characteristic (ROC) curves to establish a threshold differentiating a positive from a negative test outcome.

A few additional points with respect to the analysis of such data are worth making here. Firstly, as mentioned in Sect. 9.3.5, the precision of the estimates of sensitivity, specificity, PPV, NPV, PLR and NLR depends on the number of positive and negative compounds included in the study. It is good practice when quoting these values to include, for example, a 95 % confidence interval. Sensitivity, specificity, PLR and NLR do not depend on the prevalence of the toxicity and hence are useful for gaining an initial understanding of how well the new test detects compounds according to whether they are positive or negative, and how a positive (or negative) test result changes the odds of a compound being positive rather than negative for toxicity. However, to understand in practice how many substances which test positive are likely to be true positives, an estimate of the prevalence of the toxicity in question within the population of substances being tested is needed, and this cannot be obtained from the validation study, and may be difficult to estimate.

All of the above assume a simple "yes/no" type outcome and prediction, but in vitro assays may be used in other ways e.g. to rank compounds according to their potential hazard, or to contribute to a risk score. However the data will be used, the general principles underlying the above comments on analysis, and comments on design in Sect. 9.3.5 are worthy of consideration.

9.4.4 Other Key Analysis Questions

Here, for awareness, are some of the other key analysis questions which arise with some frequency in the nonclinical drug safety setting.

Most General or Most Powerful Testing Approach? A commonly used design for in vivo studies is one in which four separate groups of animals are dosed with either vehicle, low, intermediate or high doses of a compound; such a design can be analysed in more than one way, e.g. it could be analysed using a test such as those proposed by Williams (1971) or Shirley (1977), applicable when the data are consistent with a monotonic dose-related trend; or it could be analysed using a more general ANOVA approach followed by pairwise testing between each drug-dosed group and the vehicle group. The latter is a more general approach, applicable to studies which use a design similar to the one described, but where the groups may not represent sequential doses of the same compound; the former approach will be more powerful for the specific situation where a monotonic dose-related response is to be expected and evaluated. An organisation may be limited as to which different analysis approaches it is willing to apply, depending on the validated statistical systems and level of statistical support available to them. The statistician should be aware of the implications, and wherever possible engaged in decision-making related to this.

Repeated Measures or Timepoint by Timepoint? Many safety studies involve measuring a particular endpoint, e.g. heart rate, at multiple timepoints within a period. When this is the case, the option is open as to whether to use a repeated measures analysis approach, or whether to analyse data at each timepoint independently. The repeated measures approach is inherently more complex, and requires some assumptions to be made about the variance/covariance structure within the data. On the other hand, in situations where the patterns of variation are similar across timepoints, a repeated measures analysis can give more power and will lead to a simpler interpretation of the data, since the least significant difference between two groups being compared will be the same at all timepoints. For an introduction to this topic in relation to dog telemetry studies, see Aylott et al. (2011).

Should Males and Females Be Analysed Separately or Together? This has been debated many times in the safety assessment context, with the primary driver towards combining sexes being a desire to reduce animal numbers and/or increase study power to detect differences, particularly for studies in larger animals where group sizes tend to be smaller. A commonly held view amongst statisticians in this area is that the decision has to be taken on a case by case basis, either analysing sexes separately, or generating estimators for each sex separately from a combined sexes analysis. For an introduction to this topic, see Wiklund et al. (2005).

This section has introduced a number of analysis issues particularly relevant for nonclinical safety studies and in the next section we will consider some industry trends. Chapter 10 covers many aspects in more detail.

9.5 Practical Considerations and Industry Trends

> It is always wise to look ahead, but difficult to look further than you can see. (Winston Churchill).

9.5.1 What Drives Statistical Changes in Nonclinical Safety?

In addition to having a good grasp of current practice, it is advisable for the nonclinical safety statistician to have one eye on the road ahead to be aware of the direction in which the pharmaceutical industry is travelling. We discuss here some of the many changes within the industry which can affect statistical practice. The first two changes relate to ways of working: firstly, an increased tendency for work to be conducted through collaboration with multiple partners and secondly, the growing capability to link data from multiple sources and to explore and learn from patterns found. Statistical practice will also be affected by the desire for continuous improvement, both in the statistical methods we use and also through the replacement, reduction and refinement (the "3Rs") of animal procedures to reduce the use of animals in research. Possible impacts are outlined below, and we conclude by considering the future of the nonclinical safety statistician.

9.5.2 Ways of Working: Collaboration with External Partners

One of the trends amongst large multi-national companies within the pharmaceutical industry is a move away from retaining the greater part of the drug discovery and development process internally, towards working increasingly with external partners, who may specialise in specific areas. A key driver for this may be to increase flexibility and the speed of access to breaking science, and to reduce fixed costs. An example of this from within nonclinical safety is when Contract Research Organisations (CROs) conduct some or all of the more standard GLP nonclinical safety studies on behalf of pharmaceutical developers. There are some specific considerations for statisticians working within this context.

In one example, where expert statistical resource sat only with the study initiator, the institution running the study proposed to conduct a statistical analysis suitable for a parallel groups design, when the requested study was a crossover design i.e. required comparisons to be made within animals. This can happen when, for example, the partner uses an automated analysis solution which, in order to be general enough to fit most situations, may not be the most powerful or may not be appropriate for a specific study design. The statistician should be prepared to help the study initiator to understand and evaluate the risks and the consequences of the proposed analysis, and to be part of a dialogue to agree a way forward. The critical thing here is to ensure any concerns are raised and addressed as early as possible, and that if a GLP study is involved, the quality assurance aspects are fully taken into account. Solutions will vary according to situation, and may involve the statistical analysis being conducted by the study initiator.

If the institutions working together both have professional statistical expertise, there is usually opportunity for dialogue between both parties to resolve any questions relating to study design or analysis. The statistician at the institution requesting the study will likely have a role alongside their scientific colleagues in assuring the quality of the study design and planned analyses as well as in monitoring the performance over time of frequently used study types (see Sect. 9.3.2). They should be aware that if they request, for example, extra visualisations or analyses from a provider, this can incur additional costs. The statistician from the study provider may need to work closely with and even challenge the study initiator to understand why they want the statistical analyses in a particular way, and to work towards a solution which meets the needs of both parties. Sometimes parts of a study run at a CRO, including the statistical analysis, may be subcontracted to another provider. It is usually most effective if the statistical experts from each party, wherever they physically sit, can discuss relevant matters directly with one another. It is always important to agree in detail the nature and format of data and results to be exchanged, along with data security arrangements.

9.5.3 Ways of Working: Looking for Patterns Across Datasets

Advances in information technology mean that it is now easier than it has ever been to search through large data sets, and to combine data from multiple sources. This provides scientists with new opportunities to look across results from multiple studies and to use the combined evidence to make predictions or generate hypotheses for testing. The data may be in the public domain, for example, advanced literature-searching techniques can be used to search for published data relating to particular compounds or study types. Alternatively, organisations may combine in-house data from a variety of models and assays to try and learn from patterns in past data. The goal may be to better understand something about future risks for new compounds or about translation of effects. This area of work can often be seen within the nonclinical setting as an area primarily for engaging informaticians, mathematical modellers and those skilled in simulation, rather than statistical experts.

The desire to fully utilise and learn from all data generated is of course to be applauded, and there is much to be gained from systems which draw together data from various repositories to allow side by side visualisation. However, caution is needed when making inferences or predictions based on this approach. Statisticians can help by providing the appropriate level of statistical thinking, to address aspects which can otherwise be easily overlooked. Two examples follow.

Sources of Potential Bias in the compiled data set(s) need to be identified and acknowledged; these may render the data inadequate for the purpose for which it has been collected. For example, if we want to explore how well findings in a specific type of preclinical toxicity study translate into the clinical setting, then however advanced our literature or database search methods may be, we will almost certainly lack data on those compounds found to have the most adverse effects in preclinical studies, where this has led to a decision not to proceed into the clinic.

Appropriate Estimates of Accuracy and Precision need to be provided alongside any predictions made. In any attempt to make a prediction about an outcome (e.g. the overall likelihood that a drug substance will pose an unacceptable risk for a specific clinical safety concern) based on combining a series of inputs (e.g. the results from a series of in vitro and in vivo tests used to better understand the hazard) some effort should be made to describe the uncertainty associated with the prediction. The technique used to do this can vary, with methods such as cross-validation and simulation being potentially useful. The important thing to avoid is the presentation of point estimate predictions as if they are 'truth'. Any scientist using predictions to make decisions, for example which potential drug substance to progress out of a set, needs to know how far apart two predictions for different drug-substances should be to give sufficient confidence that one is truly better than the other on the dimension being predicted.

9.5.4 Continuous Improvement: Statistical Methods Used

9.5.4.1 Dose Response Evaluation of Data

Many different types of toxicological studies evaluate the impact of a range of doses, generally referred to as low, intermediate and high doses. Traditionally, typical statistical analyses have tested either for a dose-related monotonic trend or made a series of pair-wise comparisons between each substance-dosed group and a concurrent vehicle control group. No attempt has been made to infer what would happen in between those dose levels which were actually tested in the experiment.

Interest has naturally arisen, however, in making greater use of dose-response evaluations through curve fitting, for example in evaluating genetic toxicity data (Gollapudi et al. 2013) Advantages include the possibility of testing an increased range of dose levels using adaptive designs, with smaller group sizes at each dose level, so as to better understand the overall dose response pattern. Disadvantages include the fact that higher doses may need to be tested to define the top of the dose-response curve, and this may not be ethical. The nonclinical statistician is advised to be aware of developments in this area in terms of regulatory views, available software, and discussions in safety scientist forums, in addition to developing a good understanding of issues in design and analysis linked to curve-fitting approaches. Statisticians with no prior experience of fitting curves to biological data, could look for example at (Motulsky and Christopoulos 2004) for a general introduction to the subject area.

9.5.4.2 Move to Use Effect Sizes and Confidence Intervals Rather Than p-Values

As has been observed already in this chapter, toxicology studies are noteworthy for their ability to generate large volumes of data on multiple endpoints, evaluated in a wide variety of ways. Evaluation ranges from qualitative assessment using expert biological knowledge and judgment, through to analyses which generate tables with row upon row of p-values. Often, p-values are displayed relative to cutoffs e.g. $p < 0.05$, but the actual size of effect, the direction of the effect, the nature of the test (one or two sided) and the error on the estimate i.e. the confidence interval around it, are not always displayed. Whilst the apparent simplicity of the p-value is appealing, nonclinical statisticians have long promoted the advantages of presenting effect sizes and confidence intervals, which provide the decision-making scientist with a much clearer picture of the study outcomes. Effect sizes are also helpful when aggregating data from multiple studies in a meta-analysis. An example is given by van der Spoel et al. (2011) of aggregating results of pig, dog and sheep studies investigating whether cardiac stem cell therapy is safe. If statisticians build on success to date, then the move to display effect sizes and confidence intervals routinely wherever this is appropriate will move from being valued in isolated pockets to being widely acknowledged as industry good practice.

9.5.4.3 Changes in How Historical Control Data Are Used

For safety studies, many organisations have a large amount of historical data indicating how control groups have behaved over weeks, months and even years. These are often referred to, in order to put the results of a current study into context, for example, by seeing how data from the current study compare with the distribution of data from similar historical control animals for the same parameter. Some comments have already been made about use and statistical monitoring of historical control data (see Sect. 9.3.2). There remains an opportunity, however, to reduce concurrent control group sizes in situations where rigorous standardisation of animal strain and study protocol means that historical control results, which are stable particularly in terms of variation between animals over time, could be integrated into and add power to the analysis of a new study; but this would require a change both to how we think about individual studies, and to the statistical analysis methods used. Bayesian statistics seems an obvious solution here, to incorporate prior beliefs based on years of observation and experience in a laboratory.

9.5.5 Continuous Improvement: Replacement, Reduction, Refinement (3Rs) in Animal Studies

Whilst there is a widely held view that many safety studies have been "standard" for a long time, continuous improvements aimed at reducing and refining animal studies which cannot yet be replaced mean that, in practice, the details of studies are constantly changing. An example here is the introduction of microsampling, that is, the ability to take and analyse blood samples of a much smaller volume than was possible historically. Microsampling has beneficial implications for animal welfare and for science. It reduces or removes the need for additional animals to be included in a study simply to measure exposure to a substance, since with microsampling, multiple samples of low volume can be taken from main study animals without exceeding blood sampling limits. It also allows direct matching of exposure data to clinical and pathological observations for individual animals. However, it is important to evaluate the impact of such a method change on the measurement of other study parameters, including other haematological or behavioural parameters. The statistician has a key role in the design and evaluation of appropriate validation studies, particularly where the main studies may be reasonably complex e.g. reproductive toxicology studies (Powles-Glover et al. 2014a, b).

Another industry trend aimed at reducing animal use is the inclusion of more and more endpoints into studies originally designed for a different purpose. Examples here are the increasing inclusion of safety pharmacology endpoints into general toxicity studies, to flag marked effects and allow for detailed follow up. Whilst there is ongoing scientific debate about the impact on the animals of the extra measurements, there are also specific issues of statistical relevance here. One example is that the general toxicity study may include more or fewer animals than the comparable safety pharmacology stand-alone study. It is important to understand the impact of this on the size of effects which the study has power to detect. Likewise, general toxicity studies tend to use parallel group designs, whilst a typical cardiovascular study in the dog using telemetry, for example, would make comparisons within an animal. Again, what is needed here is a clear presentation on the limits of what can be detected.

But in any attempt to replace, reduce, and refine the use of animals in safety studies, the statistician should not just be concerned with the properties of individual studies, but strive to keep in mind and engage with the scientific debates regarding the whole program of in vivo work. Two examples of topics hotly debated amongst statisticians and scientists are the use of inbred or outbred strains of animals, and the trade-off between making individual studies sensitive to signals through standardisation of the experiments (e.g. in rat models using the same supplier, age, sex, range of body weights etc.), versus ensuring some heterogeneity to increase the extent to which results will generalise and translate to other relevant situations. These issues may be particularly relevant for animal behavioural tests (Richter et al. 2010).

9.5.6 *The Future of the Nonclinical Safety Statistician*

This chapter has provided an overview of nonclinical drug safety testing, which aids the selection of safe compounds for clinical studies by identifying safety liabilities and their relationship to exposure, and where appropriate, reversibility. We have highlighted some of the ways in which statisticians can add value by engaging with their expert toxicology colleagues, not just to define appropriate statistical analyses for studies, but more broadly, engaging with the science and with design issues pertinent to whole programmes of work. Since much work conducted in this area is quantitative or semi quantitative, regulated, and somewhat standardised, the new statistician could be forgiven for thinking that they would have little scope for future development in such a setting. In this chapter we have set out to demonstrate that this is not the case, and to provide some tips on how to get engaged with this essential and continuously changing field of scientific learning. As long as technology continues to develop, replacements and refinements for animal studies continue to be sought, and clinical learning continues to inform the need for supporting nonclinical studies, there will be satisfying, interesting and valuable work for the nonclinical safety statistician to do.

References

Altman DG, Bland JM (1994a) Diagnostic tests 1: sensitivity and specificity. BMJ 308:1552

Altman DG, Bland JM (1994b) Diagnostic tests 2: predictive values. BMJ 309:102

Altman DG, Bland JM (1994c) Diagnostic tests 3: receiver operating characteristic plots. BMJ 309:188

Arrowsmith J, Miller P (2013) Trial watch: phase II and phase III attrition rates 2011–2012. Nat Rev Drug Discov 12(8):569

Aylott M, Bate S, Collins S, Jarvis P, Saul J (2011) Review of the statistical analysis of the dog telemetry study. Pharm Stat 10(3):236–249

Cook D, Brown D, Alexander R, March R, Morgan P, Satterthwaite G, Pangalos MN (2014) Lessons learned from the fate of AstraZeneca's drug pipeline: a five-dimensional framework. Nat Rev Drug Discov 13(6):419–431

Ettlin RA, Kuroda J, Plassmann S, Prentice DE (2010) Successful drug development despite adverse preclinical findings part 1: processes to address issues and most important findings. J Toxicol Pathol 23(4):189–211

Gollapudi BB, Johnson GE, Hernandez LG, Pottenger LH, Dearfield KL, Jeffrey AM, Julien E, Kim JH, Lovell DP, Macgregor JT, Moore MM, van Benthem J, White PA, Zeiger E, Thybaud V (2013) Quantitative approaches for assessing dose-response relationships in genetic toxicology studies. Environ Mol Mutagen 54(1):8–18

Hayes J, Doherty AT, Adkins DJ, Oldman K, O'Donovan MR (2009) The rat bone marrow micronucleus test–study design and statistical power. Mutagenesis 24(5):419–424

ICH (1994a) International conference on harmonisation safety guideline S3A – note for guidance on toxicokinetics: the assessment of systemic exposure in toxicity studies. http://www.ich.org/products/guidelines/safety/article/safety-guidelines.html. Accessed 20 Aug 2014

ICH (1994b) International conference on harmonisation safety guideline S3B – pharmacokinetics: guidance for repeated dose tissue distribution studies. http://www.ich.org/products/guidelines/safety/article/safety-guidelines.html. Accessed 20 Aug 2014

ICH (1995) International conference on harmonisation safety guideline S1A – need for carcinogenicity studies of pharmaceuticals. http://www.ich.org/products/guidelines/safety/article/safety-guidelines.html. Accessed 20 Aug 2014

ICH (1997) International conference on harmonisation safety guideline S1B – testing for carcinogenecity of pharmaceuticals. http://www.ich.org/products/guidelines/safety/article/safety-guidelines.html. Accessed 20 Aug 2014

ICH (1998) International conference on harmonisation safety guideline S4 – duration of chronic toxicity testing in animals (Rodent and Non Rodent Toxicity Testing). http://www.ich.org/products/guidelines/safety/article/safety-guidelines.html. Accessed 20 Aug 2014

ICH (2000a) International conference on harmonisation safety guideline S5(R2) – detection of toxicity to reproduction for medicinal products & toxicity to male fertility. http://www.ich.org/products/guidelines/safety/article/safety-guidelines.html. Accessed 20 Aug 2014

ICH (2000b) International conference on harmonisation safety guideline S7A – safety pharmacology studies for human pharmaceuticals. http://www.ich.org/products/guidelines/safety/article/safety-guidelines.html. Accessed 20 Aug 2014

ICH (2005a) International conference on harmonisation safety guideline S7B – the non-clinical evaluation of the potential for delayed ventricular repolarization (QT Interval Prolongation) by human pharmaceuticals. http://www.ich.org/products/guidelines/safety/article/safety-guidelines.html. Accessed 20 Aug 2014

ICH (2005b) International conference on harmonisation safety guideline S8 – immunotoxicity studies for human pharmaceuticals. http://www.ich.org/products/guidelines/safety/article/safety-guidelines.html. Accessed 20 Aug 2014

ICH (2008) International conference on harmonisation safety guideline S1C(R2) – dose selection for carcinogenicity studies of pharmaceuticals. http://www.ich.org/products/guidelines/safety/article/safety-guidelines.html. Accessed 20 Aug 2014

ICH (2009) International conference on harmonisation multidisciplinary guideline M3(R2) – nonclinical safety studies for the conduct of human clinical trials and marketing authorisation for pharmaceuticals. http://www.ich.org/products/guidelines/multidisciplinary/article/multidisciplinary-guidelines.html. Accessed 20 Aug 2014

ICH (2011a) International conference on harmonisation safety guideline S2(R1) – guidance on genotoxicity testing and data interpretation for pharmaceuticals intended for human use. http://www.ich.org/products/guidelines/safety/article/safety-guidelines.html. Accessed 20 Aug 2014

ICH (2011b) International conference on harmonisation safety guideline S6(R1) – preclinical safety evaluation of biotechnology-derived pharmaceuticals. http://www.ich.org/products/guidelines/safety/article/safety-guidelines.html. Accessed 20 Aug 2014

Jarvis P, Saul J, Aylott M, Bate S, Geys H, Sherington J (2011) An assessment of the statistical methods used to analyse toxicology studies. Pharm Stat 10(6):477–484

Motulsky H, Christopoulos A (2004) Fitting models to biological data using linear and nonlinear regression. http://www.graphpad.com/manuals/prism4/regressionbook.pdf. Accessed 20 Aug 2014

Pandher K, Leach MW, Burns-Naas LA (2012) Appropriate use of recovery groups in nonclinical toxicity studies: value in a science-driven case-by-case approach. Vet Pathol 49(2):357–361

Peers IS, South MC, Ceuppens PR, Bright JD, Pilling E (2014) Can you trust your animal study data? Nat Rev Drug Discov 13(7):560

Powles-Glover N, Kirk S, Jardine L, Clubb S, Stewart J (2014a) Assessment of haematological and clinical pathology effects of blood microsampling in suckling and weaned juvenile rats. Regul Toxicol Pharmacol 69(3):425–433

Powles-Glover N, Kirk S, Wilkinson C, Robinson S, Stewart J (2014b) Assessment of toxicological effects of blood microsampling in the vehicle dosed adult rat. Regul Toxicol Pharmacol 68(3):325–331

Pugsley MK, Authier S, Curtis MJ (2008) Principles of safety pharmacology. Br J Pharmacol 154(7):1382–1399

Richter SH, Garner JP, Auer C, Kunert J, Würbel H (2010) Systematic variation improves reproducibility of animal experiments. Nat Methods 7(3):167–168

Shirley E (1977) A non-parametric equivalent of Williams' test for contrasting increasing dose levels of a treatment. Biometrics 33(2):386–389

Sparrow SS, Robinson S, Bolam S, Bruce C, Danks A, Everett D, Fulcher S, Hill RE, Palmer H, Scott EW, Chapman KL (2011) Opportunities to minimise animal use in pharmaceutical regulatory general toxicology: a cross-company review. Regul Toxicol Pharmacol 61(2): 222–229

TripAdvisor (2009) Review. http://www.travel-library.com/hotels/north_america/usa/california/los_angeles/sls_hotel_at_beverly_hills_los_angeles.html. Accessed 14 July 2014

van der Spoel TI, Jansen of Lorkeers SJ, Agostoni P, van Belle E, Gyongyosi M, Sluijter JP, Cramer MJ, Doevendans PA, Chamuleau SA (2011) Human relevance of pre-clinical studies in stem cell therapy: systematic review and meta-analysis of large animal models of ischaemic heart disease. Cardiovasc Res 91(4):649–658

Wheeler DJ (2000) Understanding variation: the key to managing chaos. SPC, Knoxville, Tennessee

Wiklund SJ, Svens K, Palm M, Holland T (2005) Benefits of combining the sexes when evaluating data from toxicological studies. J Appl Toxicol 25(2):135–142

Williams DA (1971) A test for differences between treatment means when several dose levels are compared with a zero dose control. Biometrics 27(1):103–117

Chapter 10
General Toxicology, Safety Pharmacology, Reproductive Toxicology, and Juvenile Toxicology Studies

Steven A. Bailey, Dingzhou Li and David M. Potter

Abstract This chapter provides a survey of key nonclinical safety assays. For each study type, we discuss the typical study designs employed, including a summary of the type of endpoints collected. We then provide an overview of common statistical approaches in each setting. There are some general themes that are common across the study types (e.g., trend testing). At the same time, the different study types may have features that require special consideration (e.g., cross-over designs for safety pharmacology studies, intra-litter correlation in reproductive toxicology studies). While some of the design aspects of these studies are to some extent "fixed" by precedent across the industry, we do address sample size and power considerations, as this information can be valuable to understanding how statistical results can contribute to the overall interpretation of these studies. Finally, for any discussion of statistical approaches, there are likely to be multiple reasonable approaches. We've attempted to cover some of the more common approaches in detail, but we recognize that our treatment is not exhaustive. Where possible, we have provided references for further reading.

Keywords Toxicology • Juvenile toxicology • Reproductive toxicology • Safety pharmacology • Preclinical safety • Trend testing

10.1 Introduction

As was discussed in Chap. 9, a range of safety studies must be conducted prior to clinical trials. General toxicology, genetic toxicology and safety pharmacology studies, for example, must be conducted prior to Phase I clinical trials. The need for and timing of other safety studies, such as reproductive toxicology and carcino-

S.A. Bailey
Drug Safety R&D Statistics, Pfizer Inc, Andover, MA, USA

D. Li • D.M. Potter (✉)
Drug Safety R&D Statistics, Pfizer Inc, Groton, CT, USA
e-mail: david.m.potter@pfizer.com

L. Zhang (ed.), *Nonclinical Statistics for Pharmaceutical and Biotechnology Industries*, Statistics for Biology and Health, DOI 10.1007/978-3-319-23558-5_10

genicity studies, will depend on the type of drug (e.g., small molecule, biologic) and the intended patient population. These safety studies are termed "regulatory" studies because of requirements and guidance from regulatory agencies on the scope, duration, and timing of these studies. The three main regulatory bodies, the US Food and Drug Administration (FDA), the European Medicines Agency (EMA), and the Japanese Pharmaceutical and Medical Devices Agency (PMDA), differ in some instances in their recommendations regarding the safety studies needed to support various stages of clinical development. There have, however, been efforts to standardize and harmonize recommendations across the three regions (International Conference on Harmonization 2010). This chapter addresses general toxicology, safety pharmacology, reproductive toxicology, and juvenile toxicology studies.

10.2 General Toxicology

10.2.1 Overview

The overall objectives of general toxicology studies include identifying target organ toxicity (i.e., which organs are potentially affected), characterizing the dose–response or exposure-response relationship, assessing the potential reversibility of test article effects, and identifying possible endpoints to use to monitor adverse events in clinical trials. A wide range of endpoints are collected during these studies. Quantitative endpoints include body weight, food consumption, organ weights, and clinical pathology measurements. Clinical pathology can include hematology, serum chemistry, and urinalysis endpoints. In addition, more qualitative endpoints include clinical signs, gross pathology, and histopathology. Toxicokinetics (measurement of exposure to the test article) are included in these designs to monitor drug exposure levels, to aid with interpretation of findings in the study, and aid in selecting starting clinical doses.

An expectation of these studies is that they establish a dose–response relationship, from no effect to adverse effect. To do this, most studies include a control and at least three doses of the drug being evaluated, with doses chosen to exceed anticipated clinical doses, and the high dose chosen to induce toxicity in order to help identify target organ toxicity. The highest dose selected is typically the "maximum tolerated dose" (MTD) meaning that animals show evidence of toxicity (e.g., decreased body weight, changes in clinical signs) but do not experience mortality or morbidity. Dose selection and its impact on the effective and efficient use of animals is a critical consideration in these studies. If doses are chosen too high (potentially causing mortality), or too low to cause any toxicity, studies may need to be repeated.

The studies are typically conducted in a rodent species (e.g., rat) and a non-rodent species (e.g., dog, non-human primate (NHP)). For rodents, there are typically 10 animals per sex per treatment group. For non-rodents, 3–4 animals per sex per group is common. In many cases, exploratory (i.e., non-regulatory) studies are conducted in advance of the regulatory studies in order to inform the design of the regulatory studies. These studies are typically smaller (e.g., 5 rodents per sex per group, 1 non-rodent per sex per group).

The duration of a general toxicology study is chosen based on the clinical studies it supports. The first set of general toxicology studies are in support of Phase I clinical trials in humans. Initial "dose-range finding" (DRF) studies are non-regulatory studies that seek to identify the maximum tolerated dose (MTD) to inform dose selection in subsequent regulatory studies. Their duration is usually 2 weeks or less, but may be much longer depending on the half-life of the test article. Regulatory studies, typically up to 1 month in duration, are then conducted. To support Phase II studies, longer (typically up to 6 months) regulatory studies are required, again, in both rodents and non-rodents. To support Phase III studies and post-approval use in patients, regulatory studies up to 9 months in duration are conducted, usually in non-rodents. In addition, carcinogenicity assessments for lifetime exposure are conducted (typically 2 years in rats and 6 months in transgenic mice). Some recent research on the design and analysis of carcinogenicity studies is presented in Chap. 12.

For the most part, these studies are "multi-dose" or "repeat-dose" studies, meaning that the drug is administered regularly (e.g., daily) during the course of the study. "Acute" studies, by contrast, are characterized by either a one-time administration of the drug or possibly repeated doses in a short time-frame (e.g., 1 day).

The results of these studies, in conjunction with other nonclinical safety assessments, contribute to the identification of No Observed Adverse Effect Level (NOAEL). Definitions of the NOAEL vary somewhat, but in general it is taken to be the highest experimental dose without an adverse effect. Note that confidence in the determination of the NOAEL will depend on the study design. For an excellent review and discussion of the NOAEL and its limitations, as well as alternative approaches, see Dorato and Engelhardt (2005). Included in their article is a discussion of the Benchmark Dose (BMD) Method (BMD), introduced by Crump (1984). The idea is to fit a dose–response model to the study data and select through calibration (i.e., inverse prediction) the dose level that corresponds to a prespecified adverse response (e.g., a certain percentage increase over the control group response). Then, the lower bound on a confidence interval for the dose level is used as the identified dose. For a review of the BMD approach, see Filipsson et al. (2003).

10.2.2 Statistical Analysis Methods

10.2.2.1 Comparing Dose Groups to Control

Since these are parallel group designs, typically with a control and three dose groups, the analysis options are relatively straightforward. Analysis of variance

(ANOVA) methods can be used, with pairwise comparisons of each dose group back to control. Because the primary comparisons of interest are typically relative to a single control group, Dunnett's Test (Dunnett 1955) is often used. Trend testing methods are also used, taking advantage of the natural ordering of the dose groups in most studies. This approach allows for an overall assessment of a dose–response relationship, and with sequential testing variations (Tukey et al. 1985), also provides a way of estimating a "no statistical-significance of trend dose [NOSTASOT]", as Tukey described it. Assuming a control group and three dose groups (low, intermediate, and high), a sequential trend test could be conducted as shown in the flowchart in Fig. 10.1.

The advantage of trend testing methods is that for monotonic dose–response patterns, the methods are more sensitive than pairwise comparisons alone. Note that the NOSTASOT for a particular endpoint is being declared based on a *lack* of statistical significance, which could result due to lack of a true effect, or due to a lack of power to detect an effect.

There are several options for implementing trend testing. One common approach is to estimate linear contrasts in the context of an ANOVA. Consider the data shown in Table 10.1 and Fig. 10.2 for the liver enzyme alanine aminotransferase (ALT) from a hypothetical 1-month general toxicology study. Elevations in ALT often reflect changes in liver structure or function.

The data suggest an elevation in ALT levels with increasing dose. Note also that variation appears to increase with the magnitude of ALT. This is common for many clinical pathology parameters, and a log transformation is often appropriate. For this example, Levene's Test didn't suggest strong evidence of unequal variance ($p = 0.085$), and hence we analyze the data on the original scale. The overall F-test from an ANOVA indicates a significant difference among the groups ($F = 23$, $p < 0.001$, $df = 3.16$). The toxicologist is specifically interested in which dose groups differ from control, and so pairwise comparisons are conducted. Table 10.2 shows the results of pairwise comparisons using Dunnett's Test and a sequential trend test.

Table 10.3 shows the linear contrasts used for the trend test. Contrast 1 tests for an overall linear trend among all four dose groups. Contrast 2 tests for a linear trend among the control, low, and intermediate dose groups only, and it is only conducted if Contrast 1 is statistically significant. Similarly, Contrast 3 tests for a difference between the control and low dose group and is only conducted if Contrast 2 is statistically significant. For these data, both the high and intermediate doses would be declared significantly different from control, at the 5 % level. Note that using this sequential approach, testing at subsequent doses only occurs if the initial contrast is statistically significant. Hence, the overall (i.e., family-wise) error rate of the procedure is less than or equal to α.

In contrast, Dunnett's Test indicates a difference between the high dose group and control only, at the 5 % level. In general, in settings where monotonic dose–response patterns are expected, then trend testing methods will be more powerful than Dunnett's Test. Simulations can be used to assess the extent of the advantage. For example, assume that the true mean levels of ALT are (21,24,26,29) in the

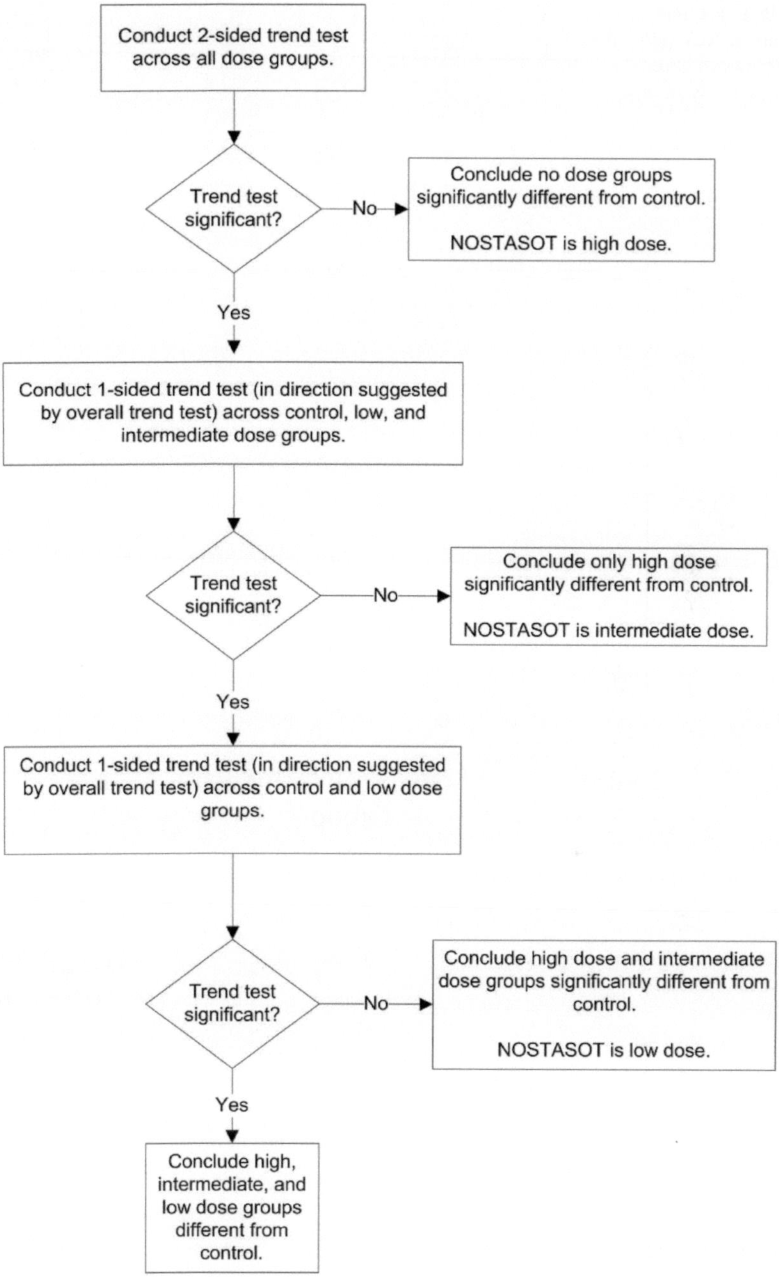

Fig. 10.1 Flow chart for sequential trend testing

Table 10.1 Example hypothetical ALT (U/L) data from a 1-month general toxicology study in rats

Control	Low dose	Intermediate dose	High dose
21.3	17.3	33.6	40.3
16.9	20.2	22.8	48.5
24.4	23.1	28.3	34.5
19.2	27.6	25.2	38.6
21.3	19.2	26.0	51.0

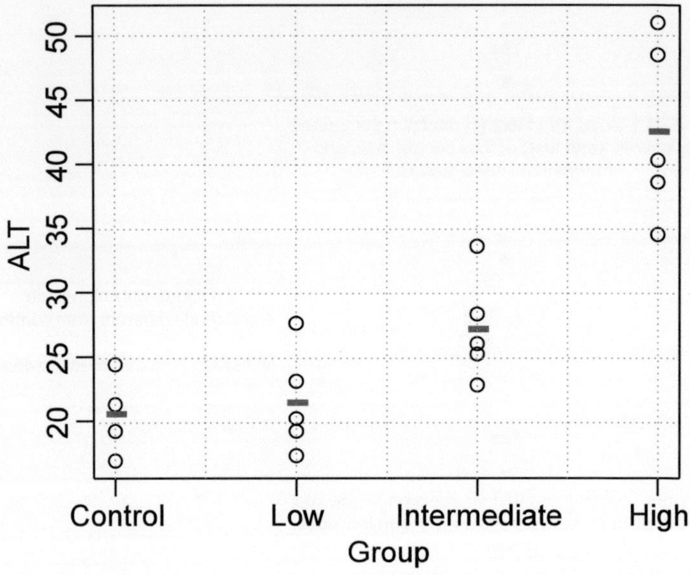

Fig. 10.2 Scatterplot of example ALT data

Table 10.2 Summary of pairwise comparisons for ALT

				P-value	
Treatment	N	Mean	SD	Dunnett	Trend
Control	5	20.6	2.8	–	–
Low	5	21.5	4.0	0.98	0.39
Intermediate	5	27.2	4.1	0.10	0.02
High	5	42.6	6.9	<0.01	<0.01

Table 10.3 Linear contrasts for trend testing with four treatment groups

	Contrast 1	Contrast 2	Contrast 3
Control	−3	−1	−1
Low	−1	0	1
Intermediate	1	1	0
High	3	0	0

Table 10.4 Power to detect differences between high dose group and control using sequential trend test and Dunnett's test

N per group	Power – Trend test (%)	Power – Dunnett's test (%)
3	32.0	17.4
4	42.7	25.9
5	52.9	33.8
6	61.2	42.3
7	70.0	51.6
8	75.7	58.5
9	80.4	63.9
10	84.4	69.2
15	96.1	89.5
20	98.9	96.2

Simulated group means = (21.24,26,29). Standard deviation = 6. 10,000 runs

control, low dose, intermediate dose, and high dose groups respectively, and that our estimate (from historical control data) of the standard deviation is 6. For a range of sample sizes, Table 10.4 shows the proportion of times the high dose group was significantly different from the control group at the 5 % level.

In addition to the linear contrasts approach, there are other methods for testing for trends. For example, some methods assume only a monotonic response. The null and alternative hypotheses in this setting are:

$$H_0 : \mu_1 = \mu_2 = \mu_3 = \mu_4$$

$$H_A : \mu_1 \leq \mu_2 \leq \mu_3 \leq \mu_4 \, (\text{or}\, \mu_1 \geq \mu_2 \geq \mu_3 \geq \mu_4)$$

with at least one strict inequality. One method due to Williams (1971, 1972) is basically a series of pairwise t-tests of each dose group versus control, based on *amalgamated* means. Because of this similarity to the traditional t test, it was termed the $\bar{t}$ test. The amalgamation procedure enforces a non-decreasing (or non-increasing) ordering of the sample means, using the *pooled adjacent violators (PAV) algorithm* Consider an example where the dose group means are (2,4,6,5) as shown in Fig. 10.3. The PAV algorithm moves left to right, checking for non-monotonicity among adjacent pairs. In this case, the intermediate and high dose groups violate monotonicity, so their means are averaged. The final group means are thus (2,4,5.5,5.5). In a second more extreme example, assume that the dose group means are (2,7,7,1), as shown in Fig. 10.4. In this case, the intermediate and high dose group means violate monotonicity, and so their means are pooled, resulting in group mean equal to (2,7,4,4). Because the low dose group mean now violates monotonicity when compared to the pooled intermediate and high dose group means, those three group means are pooled, resulting in (2,5,5,5) for the final amalgamated group means. In addition to the pairwise t-tests based on amalgamated

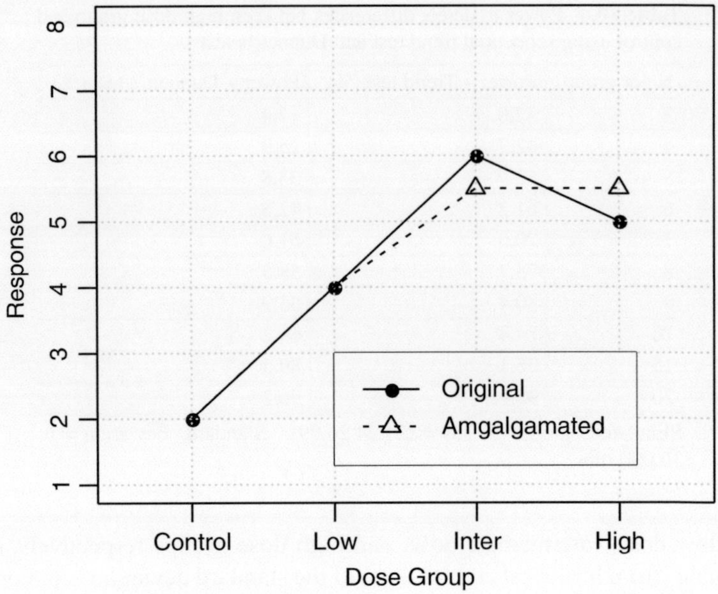

Fig. 10.3 Example of amalgamation procedure

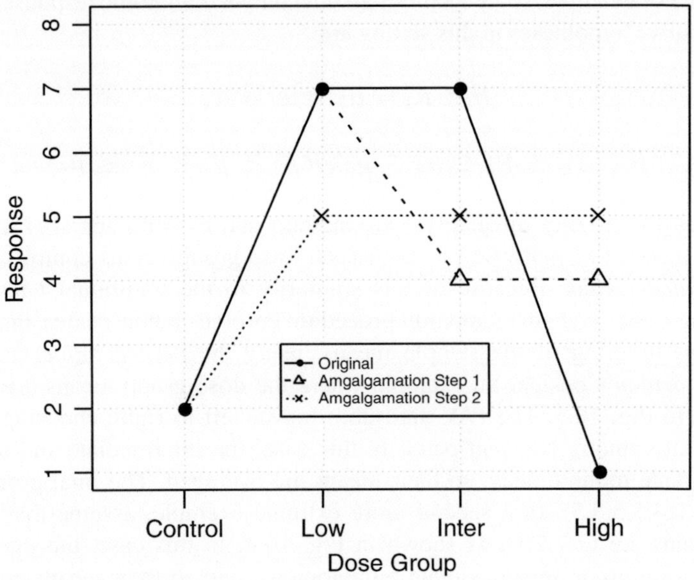

Fig. 10.4 Example of multiple-step amalgamation procedure

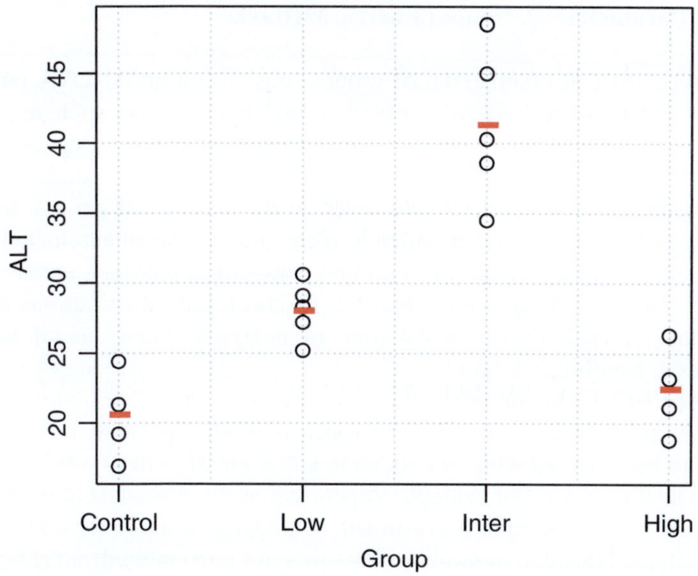

Fig. 10.5 Example of non-monotonic relationship with dose group

means due to Williams, there is also a trend test based on amalgamation means. This test, called the $\overline{E}^2$ test (Barlow et al. 1972).is essentially an overall ANOVA F-test based on amalgamated means. For more details on the $\bar{t}$ and $\overline{E}^2$ tests, including critical values and calculation of p-values, see the original papers, as well as Bailey (1998).

Occasionally in general toxicology studies, the observed pattern in group means with increasing dose is not monotonic. See, for example, data shown in Fig. 10.5. In these cases, sequential trend testing methods lead to one of two conclusions. In one case, the initial trend test will not be significant, so the sequential testing stops with no dose groups being declared different from control. In a second case, the initial trend test across all dose groups will be significant (due to the influence of the low and intermediate dose groups), leading to the high dose group being declared significant. In the case shown in Fig. 10.3, the initial trend test (using linear contrasts) is significant ($p = 0.016$), leading to the conclusion that the high dose group differs from control. In either of the two cases, interpretation can be challenging. One way of addressing this issue is to include a check for monotonicity upfront. One approach to testing for non-monotonicity is based on comparing the group means with the amalgamated means, using an F-test. See Healey (1999) for more details. If there is evidence of non-monotonicity, then the results from pairwise comparisons (say, using Dunnett's Test) could be provided instead. Alternatively, the approach taken by Bretz and Hothorn (2003) could be considered, in which multiple contrast tests (MCT's) are used to identify a potential trend only up to a given dose level.

10.2.2.2 Parametric vs. Nonparametric Methods

In a typical statistical comparison of groups (e.g., using an ANOVA model), the researcher checks (often visually) distributional assumptions such as normality of the residuals and equal variability across treatment groups. Depending on the assessment, it might make sense to transform the dependent variable (e.g., log) or use a nonparametric approach. However, in the analysis of general toxicology studies, the statistical methods are often implemented as part of automated systems; the same model will be run for multiple endpoints, possibly across multiple timepoints. Hence, it may not be feasible to check data distributions and other assumptions for each model at the time of analyses. There are at least three approaches to handling this issue.

One approach is to automate the assessment of the distributional or other assumptions prior to analysis. This approach is often represented as a "decision tree" (not to be confused with classification and regression trees used in predictive modeling) or flow chart. For example, depending on an initial test of normality, the data may or may not be rank transformed, or undergo some other transformation, prior to analysis. Note that in some systems we have encountered, data are analyzed nonparametrically if an initial test suggests a departure from the assumption of constant variability across treatment groups. Many nonparametric approaches (e.g., Wilcoxon, Kruskal–Wallis) still rely on the homogeneous variance assumption, however. A second approach is to evaluate historical control data to evaluate the distribution of each endpoint. Those variables that deviate appreciably from normality could be routinely analyzed using a log or rank transformation; these choices for each endpoint would be prespecified in the automated system. A third approach is to use a rank transformation for all parameters. This would guard against cases where an extreme value may mask a potential effect, and in general won't result in an appreciable loss of power. For example, using the setup in the previous section, with mean ALT equal to (21, 24, 26, 29), we can compare the power of a rank-based approach to the original, using the sequential trend test based on linear contrasts. The results are shown in Table 10.5 and suggest some loss of power with the rank-based approach, but typically only a few percentage points in this scenario.

In addition to applying contrast-based trend tests to the ranks of the data, there other nonparametric implementations of trend tests, including those due to Shirley (1977) and Jonckheere (1954).

10.2.2.3 Sex Effects

General toxicology studies are typically conducted in both sexes. Traditionally, statistical analyses have been conducted separately for each sex. There may be a gain in sensitivity by conducting a combined analysis including a model term for sex (i.e., as a block) in the ANOVA. The challenge arises if there is a statistical interaction between sex and dose for one or more parameters in a given study. It is not uncommon to observe sex-related differences in exposure due to differential

Table 10.5 Power comparison of parametric and rank-based sequential trend tests

	Based on original data values			Based on ranks		
N per group	Low dose (%)	Intermediate dose (%)	High dose (%)	Low dose (%)	Intermediate dose (%)	High dose (%)
3	15.6	24.4	32.0	15.9	24.9	31.7
4	17.7	30.0	42.7	17.9	30.0	41.3
5	19.3	35.4	52.9	18.7	34.4	50.4
6	21.6	39.6	61.2	21.2	39.2	58.9
7	23.0	45.4	70.0	22.9	44.7	67.6
8	26.4	49.6	75.7	26.0	48.7	73.0
9	27.5	53.6	80.4	26.7	53.1	78.0
10	29.4	57.0	84.4	28.5	56.4	82.7
15	38.2	73.0	96.1	37.4	71.6	95.2
20	47.6	83.5	98.9	46.5	82.3	98.6

True group means = [21,24,26,29]. Sequential linear contrast approach

metabolism or hormonal influence. In this case, the toxicologist will want to evaluate the impact of treatment separately for each gender. Again, since these analyses are often automated, the conventional approach has been to assume the potential for an interaction, and analyze by gender. This is an area for continued development.

10.2.2.4 Time Effects

Some endpoints, such as organ weights, can be collected only once in a general toxicology study. Others, such as body weights and food intake, are captured more frequently (e.g., weekly). Clinical chemistry, hematology, and urinalysis parameters are typically collected at the end of study, but may also be collected multiple times (e.g., monthly in a 3-month study). In some cases, and especially for large animal studies (e.g., NHP, dog), a baseline measurement (i.e., prior to any dosing) may be taken for each animal. In these studies, it is typical for the toxicologist and/or clinical pathologist to focus on changes from baseline values, rather than comparing control and test article-administered animals at each time point, especially when the sample sizes are small (e.g. 3/sex/group).

Incorporating a baseline adjustment (i.e., as a covariate) into an ANOVA model may, in some cases, improve the power of these analyses. However, it is important to evaluate the extent of correlation between baseline and follow-up measurements for each endpoint. A recent internal Pfizer study of clinical pathology control data from 20 GLP general toxicology studies in NHP's showed a substantial range in correlation (approximately 0.15–0.95) across about 45 endpoints. Some endpoints (e.g., ALT) had within-animal correlations above 0.90. Others (e.g., glucose) were in the range of 0.35. Overall, more than 1/3 of the endpoints had within-animal correlations below 0.5; for these endpoints, including a baseline covariate may actually reduce the sensitivity of statistical tests.

Table 10.6 Reference ranges for select clinical chemistry parameters, based on Wistar rats

Analyte	Range	Units
Glucose	91–218	mg/dL
Potassium	3.3–5.0	mmol/L
Cholesterol	30–71	mg/dL
Alanine Aminotransferase (ALT)	15–66	U/L

10.2.2.5 Reference Ranges

In addition to comparisons between dose groups and concurrent controls in a general toxicology study, for many endpoints there are well-established reference ranges, based on historical control data, against which to compare individual data values. A reference range is defined as an interval in which some percentage (e.g. 95 %) of an endpoint's values would fall, assuming a healthy population of subjects. These intervals can serve as the basis for determining whether individual drug-treated animals are unusual in their response. There are both parametric and nonparametric approaches to constructing these intervals. For the former case, the data (possibly transformed) are assumed to be normally distributed, and quantiles (e.g., 2.5 % and 97.5 %) are derived based on normal theory. In the nonparametric case, the sample quantiles are computed directly from the data. Some example reference ranges for Wistar Han IGS rats based on recent historical control data at Pfizer are shown in Table 10.6. These reference ranges were calculated using the EP Evaluator software (EP Evaluator 2005), which uses a nonparametric approach (Clinical and Laboratory Standards Institute 2000). When constructing reference ranges, it's important to note that there are species, strain, and age differences (e.g., a Wistar Han IGS rat is not the same as a Sprague–Dawley rat), reference ranges drift over time, and may be specific to a facility or testing platform.

An important step in computing reference ranges is to ensure that the samples used are relatively homogeneous with respect to key attributes like age, sex, and species, to the extent that these factors affect the normal range of values.

With a well-constructed historical control database, meaningful investigations into other possible sources of variation (e.g. seasonal) can be performed. See Sect. 9.3.2 in Chap. 9 for further information on plotting historical control data.

10.2.3 Sample Size and Power Considerations

Assessing the statistical power of general toxicology studies poses several challenges. In a simple two-sample comparison (one treatment group and one control group) for a given variable, a typical sample size calculation relies on an estimate of variability and a required difference to detect between group means. Estimating biological variability is relatively straightforward, given an adequate set of historical control data. Elucidating a single agreed-upon difference to detect for a given endpoint is often more difficult, as it may depend on a particular toxicologist's experience as well as the particular compound being studied and the disease area. In addition, a change in a single endpoint is rarely interpreted on its own; instead, the change is interpreted in the context of changes in other endpoints, both quantitative and qualitative. In this sense, the statistical results are univariate in nature, but the interpretation by the toxicologist is multivariate. Even having agreed on a difference to detect each endpoint, the question remains as to how to assess the suitability of the sample size relative to all of the collected endpoints (food consumption, body weights, organ weights, and clinical pathology data).

In general, the sample sizes used in general toxicology studies appear to be driven primarily by regulatory guidance and historical precedent. For example, an excellent review article by Sparrow et al. (2011) states:

> In regulatory general toxicology studies the animal numbers used are not driven by statistical input. There are several reasons for this, such as the potential hazards of a substance being unknown in advance of the studies being conducted. Therefore, there is no specific change that the study can be statistically powered to detect. In addition, the frequency of the potential hazard is unknown in the initial toxicology studies and may turn out to be a frequently occurring or a low incidence change.
>
> Due to the multifactorial nature of toxic changes, assessment of toxicity in all species is made by examination of the data generated for each individual animal by integration and correlation of in-life and post mortem findings. Experience has demonstrated that the numbers used and illustrated in this manuscript are sufficient to identify the most potential hazards, a dose/exposure response for the hazards and to generate data sets that are sufficient to provide study sponsors and regulators with information that allows decisions to be made about clinical trials and marketing.

Even with fixed sample sizes, the machinery of power calculations can still be used. That is, given a sample size and an estimate of variability, we can compute the minimum detectable difference (MDD) for each endpoint, assuming 80 % power. This can lead to fruitful discussions about the types of changes that can be meaningfully detected with statistical analyses.

10.3 Safety Pharmacology Studies

10.3.1 Overview

The main objective of safety pharmacology studies is to understand potential undesirable pharmacodynamic effects of a test article or an intervention on physiological functions in relation to exposure in the therapeutic range and above. A more comprehensive definition of safety pharmacology studies is given in ICH S7A guideline (International Conference on Harmonization 2001). We focus on the three major types of in vivo safety pharmacology studies: pulmonary-respiratory, cardiovascular (CV), neurofunctional (NF) experiments. Each type focuses on an important aspect of the possible acute adverse effects (typically within 48 h after dosing) caused by the test article or the intervention.

10.3.2 Pulmonary-Respiratory Studies

The focus of pulmonary-respiratory studies is to de-risk targets and test articles that may possess respiratory issues. The typical study design is a parallel group. Following a period of acclimation, unrestrained animals are placed in a whole body plethysmograph chamber for approximately 6 h. Originating from the pressure change in the chamber, the respiratory signal is routed through an amplifier to a data acquisition system, and the respiratory parameters are logged by the computer. The main parameters in the pulmonary-respiratory study are the tidal volume (i.e., the normal volume of air displaced between normal inhalation and exhalation), respiratory rate, and minute volume (i.e., the volume of air inhaled or exhaled from the lung per minute). The raw data (usually a data point every 5 s) are then averaged into some sequential time interval bins, including one at the baseline (i.e., prior to dosing), for subsequent statistical analysis.

10.3.3 Cardiovascular Studies

The primary goal of the CV studies is to determine if there is a CV risk associated with a test article or an intervention, measured by blood pressure, heart rate, and electrocardiogram readings. Typically, these endpoints are collected from "telemeterized" animals that are free to move about during the course of the study. A clear advantage of this technology is that it has minimal interference with the animal's normal function. Multi-channel signals are transmitted wirelessly from surgically-placed electrodes to the receiver in the monitoring room, giving a comprehensive recording of the cardiovascular changes in real time. At an adequate sampling rate, such a data acquisition system provides a more accurate picture of the dynamics in the physiology and may capture minute responses that would not be possible using non-ambulatory (i.e., recumbent) approaches, which require the animal to be restrained.

10.3.4 Neurofunctional Studies

Effects of the test article or the intervention on the central nervous system (CNS) are typically evaluated using assessments of motor activity, coordination, sensory/motor reflex responses, behavioral changes, and body temperature. Two key study types are the functional observation battery (FOB) and locomotor activity (LMA). FOB is the mainstay observational assay designed to identify points of CNS concern for follow-up. It consists of a battery of endpoints (the number varying from company to company, usually 20 to 30) covering different aspects of CNS issues, such as activity and excitability, and autonomic, neuromuscular, and sensory/motor responses. Most of the endpoints are binary (Yes/No), with the rest having more than two levels (e.g., Normal, Mild, and Severe). Upon completion of the FOB and body temperature assessment, the animal will be immediately placed into its assigned locomotor activity chamber and LMA testing will begin. LMA will include assessments of both horizontal (XY ambulation) and vertical movements (rearing).

10.3.5 Statistical Analyses

10.3.5.1 Overview

Although these studies measure different physiological aspects of the animal, the statistical analysis methods are similar across study types, depending on the nature of the endpoint of interest. For continuous endpoints, the main tool is ANOVA or ANCOVA. If a baseline or pre-dose measurement exists, ANCOVA with the baseline as a covariate is recommended as in general, it will be a more powerful approach (Senn 2007).

The discrete endpoints are typically analyzed using pairwise Fisher's exact tests or the Chi-square test. Since directionality is presumed for adverse events (i.e., Abnormal is always worse than Normal), comparisons between dose groups are made using one-tailed tests. Additionally, the Cochran–Armitage test or the Jonckheere–Terpstra test may be performed when the intent is to characterize a dose response relationship.

10.3.5.2 Super-Intervals

In safety pharmacology studies, particularly in CV studies, the raw data coming out of the data acquisition systems usually have high resolutions (at the sampling rate of 200 Hz there is a data point every 0.005 s). Even if the software uses a procedure to construct "moving-average" summaries of the individual recordings into longer intervals such as 15-min time bins, the process still generates a huge amount of data over a period of 24–48 h. Therefore, further coarse-graining is needed to reduce the data quantity without significant loss of fidelity in representing the characteristic physiology. To that end, the so-called super-interval binning method (Sivarajah et al. 2010) has been proposed to take into account the particular pharmacokinetics profile of the test article in a given experiment, and has been shown to have reasonably good performance in pilot studies using many known positive controls. With this method, each animal typically ends up with four to six observations per day per treatment. An example of super-intervals is shown in Fig. 10.6.

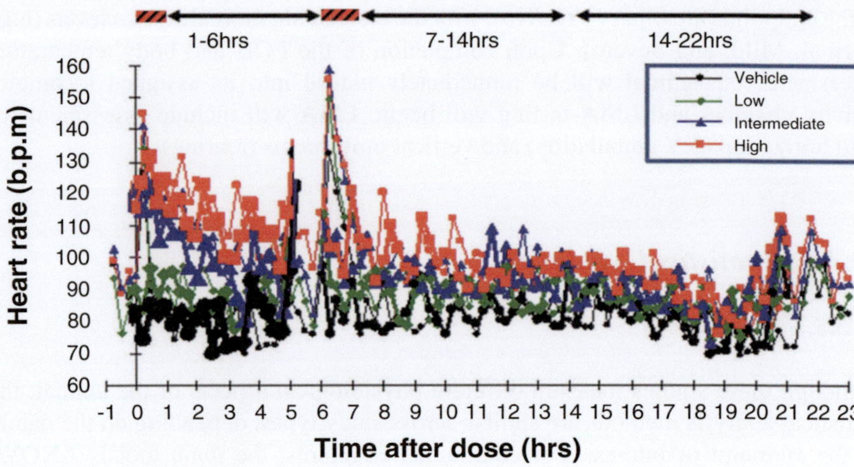

Fig. 10.6 Example of super-intervals in a cardiovascular study

10.3.5.3 Repeated Measures vs. the Cross-Sectional approach

The statistician has several options with the super-intervals. One approach is a repeated-measure ANOVA/ANCOVA, which takes into account the within-subject correlation across time intervals. Alternatively, ANOVA/ANCOVA models can be fit at each time interval (this approach is often called a "time-by-time" or "cross-sectional" approach). The rationale of either approach will be discussed in subsequent sections for each study type. An advantage of the time-by-time approach is that identifying and/or modeling the precise form of the correlation is not a concern. On the other hand, the approach may suffer from the loss of statistical power relative to the repeated measures approach (detailed discussion on this will be provided in Sect. 10.3.6).

The repeated-measure approach is fairly straightforward except when it comes to the choice of covariance structure. When the super-intervals have equal lengths and are evenly spaced, it is natural to assume an AR(1) structure between observations on the same animal. In contrast, when the super-intervals vary significantly in length and position, the AR(1) structure may not fit as well. One approach is to try several candidate structures and use criteria such as AIC or BIC to select the best model. However, this procedure may not be robust against data changes, meaning that missing data, replacement animals, or even the experimental day could change the overall structure. Furthermore, simulation studies have shown that using criteria such as AIC, one could still select an incorrect covariance structure. Furthermore, with small sample sizes in some safety pharmacology studies, we may not be able to afford to use an overly complex structure. With these considerations, it is common to start with the simplest structure: compound symmetry. Only when the estimate of the intra-class correlation is negative would we consider alternatives to compound symmetry. In fact, when negative intra-class correlation happens, it is usually due

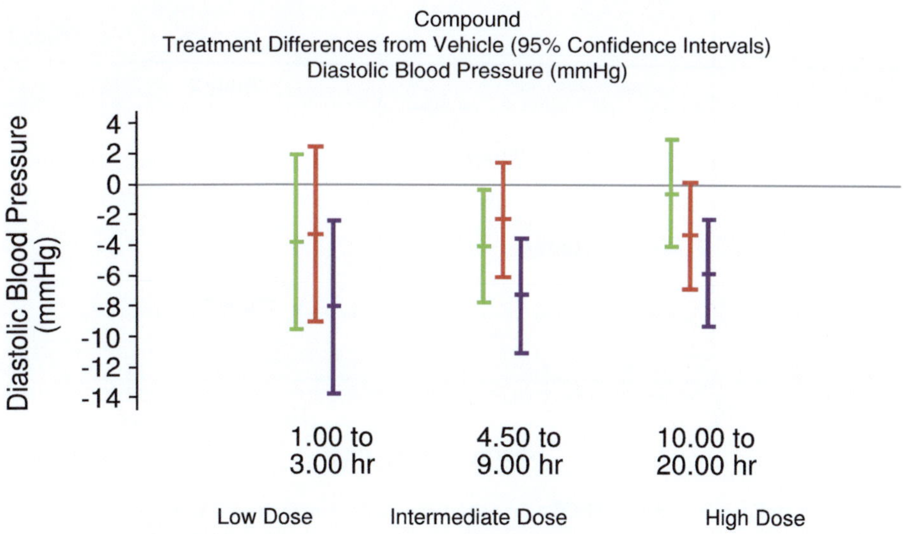

Fig. 10.7 Example confidence intervals from CV study

to one or two animals that behaved drastically differently from the rest. In this case, we may have to abandon the repeated-measure paradigm altogether and switch to the cross-sectional method.

10.3.5.4 The Use of Confidence Intervals

Although significance testing and p-values have been extensively used in assessing test article effects, confidence intervals are also recommended. Confidence intervals and significance testing are linked as they have the same components (e.g. difference in means and standard error of that difference); however, a confidence interval provides more information than a single p-value. An example of the 95 % confidence intervals for treatment effects in a hypothetical CV study is shown in Fig. 10.7. The statistical significance and magnitude of the treatment effect at each time interval as well as their evolution over time can be clearly identified from the figure.

Confidence intervals also serve as a useful bridge between statistical significance and biological relevance. In Fig. 10.8, each confidence interval corresponds to a hypothetical CV study. Consider a common CV endpoint, the QT interval. A drug-induced prolongation of the QT intervals can lead to serious adverse effects. Typically, a heart-rate adjusted version, the "corrected QT (QTc)" is used. Suppose a QTc change of 8 ms or higher is considered biologically important. Then our conclusions based on the confidence intervals are:

- Study 1 was both statistically significant and biologically relevant (indicating a QTc prolongation);

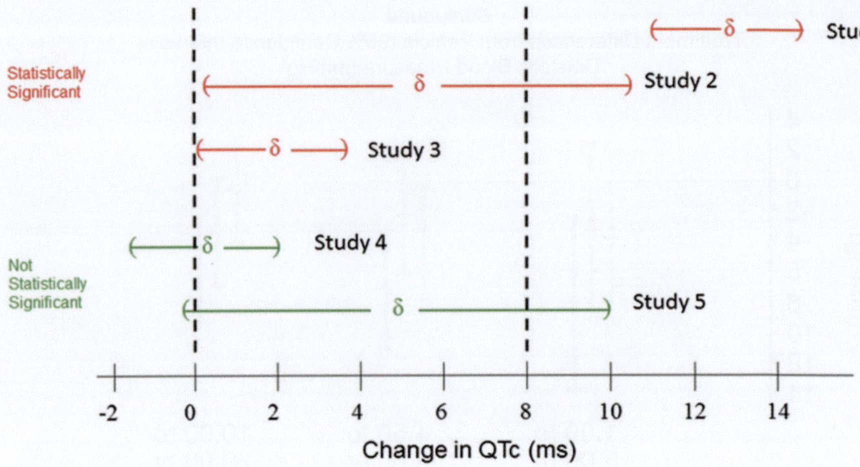

Fig. 10.8 Confidence intervals for QTc measurements from hypothetical CV studies

- Study 2 was statistically significant but its biological relevance was unknown;
- Study 3 was statistically significant but not biologically relevant;
- Study 4 was neither statistically significant nor biologically relevant;
- Study 5 was not statistically significant and its biological relevance was unknown.

Therefore, Studies 2 and 5 did not provide definitive information about the biological relevance of the QTc effect, and hence either further reduction in the measurement variability or a larger sample size is needed in those cases.

10.3.5.5 Trend and Monotonicity Testing

Many safety pharmacology endpoints are likely to demonstrate a monotonic dose response relationship. In these cases, greater statistical power can be achieved using trend testing methods. For that reason, a decision-tree type of analysis is sometimes adopted. Specifically, a monotonicity test (see some of the methods discussed in Sect. 10.2.2.1) is carried out upfront. If monotonicity cannot be rejected, then the significance of the treatment effect of each dose will be determined from a sequential trend test. On the other hand, if monotonicity is rejected, then the significance of the treatment effect will be based on pairwise comparisons (i.e., t-test) between each dose and the vehicle. This decision-tree method is able to detect treatment effects that would otherwise not be caught by either method (i.e., trend test or pairwise t-test).

Table 10.7 Latin squares design for crossover study

Animals	Treatment 1	Treatment 2	Treatment 3	Treatment 4
1	Vehicle	Low	Intermediate	High
2	Low	High	Vehicle	Intermediate
3	Intermediate	Vehicle	High	Low
4	High	Intermediate	Low	Vehicle

10.3.5.6 The Crossover Design

The crossover design is common in large animal safety pharmacology studies. In such design, each animal receives all the treatments in a randomized sequence, with a washout period between treatments. Such choice of design is mainly driven by the relatively large between-animal variability and smaller sample sizes. Nevertheless, the suitability of the crossover design is also determined by the pharmacokinetics of the test article. If the washout period is not sufficient for the test article to be eliminated from the animal, or in other words, if significant carryover effects are present between treatments, the crossover design is not recommended. For that reason, such a design is not usually adopted for biologics, which tend to have longer half-lives than small molecules.

There are three factors in the crossover design for a typical CV study: treatment (vehicle vs. various dose levels of a test article), period (the day or week when the animal is treated), and animal. Therefore, a Latin square is employed to balance all the factors (Table 10.7). Since a typical study has four doses (vehicle, low, intermediate, and high dose), the 4×4 square has become the most commonly used design. It then follows that the treatments take place on four different days and the number of animals is a multiple of four.

An improved design is the Williams square in which the first-order carryover effects are also balanced (The Latin square in Table 10.7 is also a Williams square). In other words, in such a design every treatment immediately follows the other treatments once and only once. Note that there are 24 different 4×4 Williams squares, the choice of which needs to be clearly conveyed and agreed upon with the study director. A study that has an odd number of doses requires two different Williams squares to balance the first-order carryover effects.

10.3.6 Power of Safety Pharmacology Studies

It is crucial to understand the power of a particular assay or animal model before applying it to routine studies. An underpowered design would lead to futility, whereas an overpowered one wastes resources, including animals. Moreover, the discriminant power is an important factor when developing a new assay or model that may replace an existing one; comparing the sensitivity of both assays is an important criterion in the decision. We will first discuss the power analysis for continuous variables as it is relatively more straightforward, the basic statistical tool for which is the non-central t- or F-distribution assuming the underlying distribution is normal. Next we will turn to categorical variables, whose power analysis is more complicated.

Besides the assay itself, the statistical design and analysis are also important components of the power analysis and sample size calculation. Since the major safety pharmacology study designs are either crossover or parallel groups, it follows that the comparison between within- and between-animal variability is critical to the experimental design.

In practice, a cohort of recent studies is selected to estimate the biological variability (between and/or within) from each study. To do this, an ANOVA-based model is fit to either the entire set of data or only the control arm of the data. Then, a representative metric (e.g., the median) of all the variability estimates can be used for the power analysis or sample size calculation. Note that the variability estimates from the control data may demonstrate a more identifiable distribution, but their magnitudes could be less than the ones in the treated data, and hence underestimate the overall variability. A remedy for this would be to inflate the estimate by some amount, e.g., using the 75th percentile instead of the median from all the studies.

Whenever baseline data are present, we would recommend doing the power analysis both with and without baseline adjustment and then plotting the power curves of both cases on the same graph. This visual comparison will serve as a useful tool to help educate the scientists on the importance of baseline adjustment and thus the benefits of using ANCOVA. An example of the power plot is given in Fig. 10.9 comparing the power of ANOVA to that of ANCOVA.

As mentioned in Sect. 10.3.5.3, the repeated-measures analysis can be more powerful than the cross-sectional approach. Given that the experimental unit (e.g., animal) is not changed between the two approaches, this claim may seem counter-intuitive. The key factor is the correlation among the observations from the same experimental unit. The benefit of the repeated-measure analysis increases when the correlation becomes smaller. In other words, when the observations from the same animal become more like uncorrelated samples, the repeated-measures analysis, which includes all the observations from the same animal, effectually enlarges the sample size and hence yields more power than the cross-sectional analysis which only takes one sample at a time. More detailed discussion of the statistical power of repeated-measures and split-plot designs can be found in Rochon (1991) and Bradley and Russell (1998).

When discussing power calculations, it may be necessary to remind the scientist that statistical power (or the minimum detectable difference, for that matter) derived in the above manner is a prospective estimate rather than a descriptive metric of the set of studies selected for the power analysis. Therefore, since we do not know the true treatment effect of these studies, one should not use the power or sample size estimate to "verify" that they apply to those studies. In other words, the power is a conditional probability (i.e., the probability that, with a significance test, a study is found to have a treatment effect size greater than a certain value out of all the studies

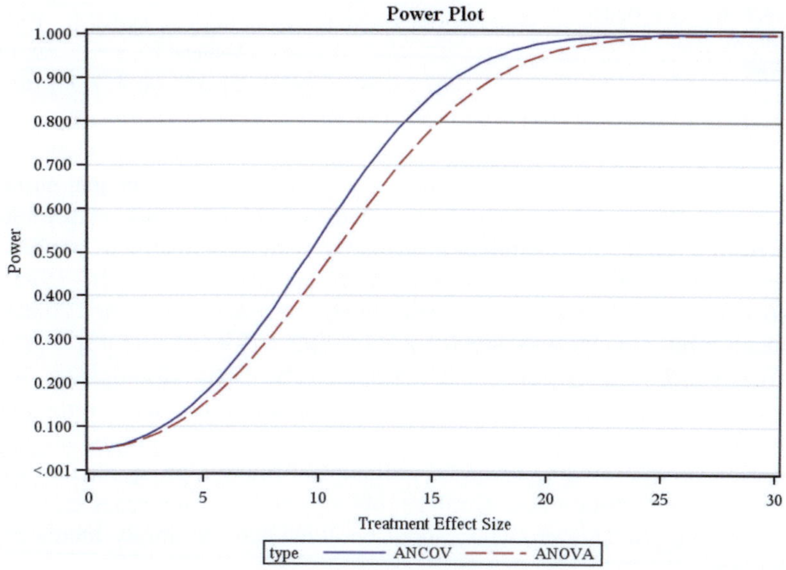

Fig. 10.9 Example of power comparison between ANOVA and ANCOVA models

whose true effect sizes are greater than that value) as opposed to an unconditional one (i.e., of all the studies conducted, regardless of their true treatment effect sizes). Based on our experience, misunderstandings sometimes occurred if such difference had not been clarified beforehand.

In contrast, the power analysis of categorical data from Safety Pharmacology studies can be more complex because: (1) the prior prevalence of some adverse events may not be known with sufficient precision, especially for rare events; and (2) it can be challenging to define a treatment effect "size" for categorical data. Power analysis of categorical data in many cases is based on the non-central distribution of the test statistic. A detailed discussion and examples are provided in Lachin (2011).

10.3.7 Other Issues in Safety Pharmacology Study Analysis

10.3.7.1 Missing Data

When a subset of the data is missing, the ordinary mean of the group differences may no longer be an unbiased estimate of the treatment effect. To address this, the use of fitted means (also known as least-squares means) is recommended for statistical reporting. The concept of fitted means can be challenging to communicate, and is often met with some resistance because the fitted means can differ from the ordinary means. However, the importance of using the fitted means can be

Table 10.8 Example study data from two-period parallel group study

	Period 1	Period 2
Vehicle	2, 5	9, 12, 15
Drug	6, 6, 7, 7	16

clearly explained using an example study with two factors and unequal number of observations at each combination of factors. The hypothetical endpoint in Table 10.8 comes from a single-dose parallel-group study with two treatment periods. The apparent means of the vehicle and drug groups are $(2+5+9+12+15)/5=8.6$ and $(6+6+7+7+16)/5=8.4$, respectively, whereas the fitted means of the vehicle and drug groups are $0.5\times[(2+5)/2+(9+12+15)/3]=7.75$ and $0.5\times[(6+6+7+7)/4+16]=11.25$, respectively. This discrepancy is also known as the Simpson's paradox. The fitted mean takes into account the unequal number of animals in each combination of the factors, and appropriately weights each combination, providing unbiased estimates of the true group means. Missing data in a crossover design pose similar problems, and in some cases can be more difficult to address. This problem cannot be alleviated by simply increasing the sample size.

10.3.7.2 When the Sample Size Is Too Small

For some investigative or exploratory Safety Pharmacology studies, a study design with very small sample sizes may be chosen. This may lead to significantly underpowered studies. Moreover, even at a seemingly moderate sample size, the power of some study designs can be much lower than other designs. For example, a common GLP Safety Pharmacology cross-over design includes four animals, four dose groups, and four periods. Suppose that the underlying within-animal standard deviation of QTc is 4 ms. Then such a design can provide a statistical power of 66 % to detect QTc changes of 8 ms and higher. In contrast, with a 2×2 crossover design including four animals, two dose groups and two periods, the power to detect the same magnitude of change is reduced to 46 %.

Baseline adjustment can be also questionable when the sample size is too small. With an extremely small n, there is like to be complete confounding between animal (as a block effect) and baseline. It is also possible that the post-dose data are positively correlated with baseline when all the animals are examined, whereas such correlation becomes negative within each individual animal.

10.3.7.3 Use of Toxicokinetics (TK) Data

It is recommended that the statistician have a good working knowledge of the terminology and basic concepts in TK. Occasionally, TK parameters need to be included in a statistical model to account for variation in exposure levels; they might also be indirectly compared to the statistical results to better understand the correlation between the exposure and treatment effect. Furthermore, TK/PK parameters play an important role in cross-species and preclinical-to-clinical translation studies, as drug effects across different species are more appropriately compared at similar exposures.

10.4 Reproductive Toxicology Studies

10.4.1 Overview

The study of reproductive and developmental toxicology in pharmaceutical development owes its beginnings to the marketing of Thalidomide in the late 1950s (McBride 1961; Weaver and Brunden 1998a). Thalidomide was marketed for respiratory infections, insomnia, coughs, colds, and headaches. Thalidomide was also taken routinely by pregnant mothers to control morning sickness, however at that point in time drug testing during pregnancy was not routinely performed. While Thalidomide was initially considered safe, experts estimate that by the time it was removed from the market in November 1961, Thalidomide use resulted in the birth of more than 10,000 children with serious birth defects, many of whom subsequently died prior to their first birthday. As a result, the hurdles required for drug approval were modified, and the study of reproductive toxicology was added as a requirement for drug approval in many countries throughout the world.

Reproductive toxicology focuses on the study of toxicities associated with the reproductive process. This includes effects on male or female sexual function and mating behavior, female fertility, maternal care, and pup growth and development.

Developmental toxicology, also known as teratology, focuses on the study of toxicities associated with the development of the offspring. This includes effects on fetal development such as birth weight and developmental birth defects. Some consider developmental toxicology to be a subset of reproductive toxicology. The term "Teratology" refers to developmental toxicology, but is sometimes erroneously used to refer to reproductive toxicology as well.

Studies are designed to evaluate both reproductive and developmental toxicity simultaneously when possible. Studies are generally performed in either rats, mice, and/or rabbits. Periodically, studies are performed in non-human primates, but due to cost and sample size issues, these studies are rare and only performed when studies in the usual species will not suffice (e.g., no biological action).

Three standard study types are performed. These study types are outlined in the International Conference on Harmonization (ICH) guidelines, which were originally published in 1992 and most recently modified in 2005 (International Conference on Harmonization 2005). The three study types are defined as follows:

1. Fertility, reproductive performance, and early embryonic development

   This study focuses on the period from prior to mating through implantation of the embryo in the uterus. Both males and females are dosed prior to mating, although sometimes a pair of studies where a single sex is dosed in each is performed. Animals are allowed to mate, and females are sacrificed during gestation at some point after the mid-point of gestation. This study is conducted in a single species, most likely rats.

2. Embryo-Fetal Developmental Toxicity

   This study focuses on the ability of the mothers to successfully carry the pregnancy, on potential birth defects in the offspring, and on toxicity during the period of organ formation and development of the offspring. Females are dosed starting after the implantation of the embryo, and are sacrificed just prior to the end of gestation. This study is conducted in two species, most likely rats and rabbits.

3. Perinatal and postnatal toxicity

   This study focuses on the late stages of pregnancy and on the early stages of the pup's life. Females are dosed starting after implantation and continues through lactation until the pups are weaned. This study is conducted in a single species, most likely rats.

10.4.2 Study Design and Endpoints Collected

Studies typically have a control group and three or four dosed groups. The high dose level is chosen with the intent of inducing a toxic effect at the high dose level on the dosed animal (mother or father). The lowest dose is chosen to have no observable adverse effects, and the middle doses are typically chosen equally spaced between the low and high dosage levels on a log-scale. If a toxic effect can be induced on one of the non-reproductive parameters collected on the parent (e.g. body weight), without having any negative effect on the offspring, then the drug is 'safe' from a reproductive toxicity standpoint. The logic is that the offspring are protected from reproductive toxicities because the parents would never be dosed at a level that high. Therefore, the presence of a treatment effect on a parental, non-reproductive parameter is not a detrimental finding for the study.

The endpoints collected in each of the study types listed above are outlined in Table 10.9.

Note that Table 10.9 is an example of the standard set of parameters collected in each of the study types. Different testing facilities, and even separate studies within a testing facility may modify this list depending on the study design, the anticipated action of the test compound, and other information that may be available during protocol preparation.

10.4.3 Statistical Analysis

In general, the statistical issues discussed previously in the general toxicology section (Sect. 10.2) such as normality of the data, equality of variances and the use of trend testing for dose–response are applicable for reproductive toxicology studies as well.

Table 10.9 Summary of typical endpoints in reproductive toxicity studies

Endpoint	Fertility, reproductive performance, and early embryonic development study	Embryo-fetal developmental toxicity study	Perinatal/postnatal toxicity study	Type of data
Paternal body weight	√			Continuous
Paternal food intake	√			Continuous
Maternal body weight	√	√	√	Continuous
Maternal food intake	√	√	√	Continuous
Estrous Cycles	√			Dichotomous and/or continuous
Mating Performance	√			Continuous (time to mate) and dichotomous (mating, fertility indices)
Hysterotomy findings	√	√		Counts and/ or proportions
Fetal weight		√		Continuous
Fetal Examination Findings (fetal malformations, variations, and retardations)		√		Proportions, and/ or dichotomous
Gestation length			√	Continuous
Pup findings at birth (live, dead)			√	Counts and/ or proportions
Pup survival			√	Proportions
Pups with milk in stomach			√	Proportions
Pup weight			√	Continuous

In addition, the issue of defining the appropriate experimental unit is critical for reproductive studies. The experimental unit is defined as the unit to which the treatment is applied (Chen 1998, 2000; Palmer 1974; Selwyn 1988; Staples and Haseman 1974; Weaver and Brunden 1998b). For reproductive studies, this is either the mother or the father. However, many of the parameters collected in these studies are collected on the offspring. It is well established that litter-mates are more similar than pups from different litters. Failure to account for the intra-litter correlation by treating the offspring data as independent experimental units will inflate the Type I error rate and result in an invalid statistical analysis. For some data types, the solution may be simple (e.g. a nested ANOVA design). However, for many of the data types collected in these studies (e.g. count data), the appropriate statistical approach may not be obvious.

From a statistical standpoint, another topic of interest is the varied data types that require analysis. In a general toxicity study, most parameters are continuous in nature. A reproductive toxicity study, on the other hand, is likely to contain continuous (e.g. body weights), count (e.g. number corpora lutea per litter), proportions (e.g. resorptions per litter), and dichotomous (e.g. mating index) data. Some standard statistical approaches generally utilized for each of these data types are outlined below.

10.4.3.1 Statistical Analysis of Continuous Endpoints

Most maternal and paternal endpoints (e.g., body weight, food intake, time to mate, gestation length, estrous cycle length) are continuous in nature. These parameters can be analyzed using the methods outlined previously in the general toxicology discussion.

Some parameters measured on the offspring (fetal weight, pup weight) are continuous in nature. However, because these parameters are not collected on independent animals, the analysis must account for this fact. This can be accomplished by the use of a nested ANOVA, using the litter as a factor in the model. Alternatively, sometimes a litter average is generated, and the standard general toxicology methods are applied to the litter averages.

In addition, for pup weight and fetal weight, the weight of the offspring is dependent on the size of the litter (Chen and Gaylor 1992; McCarthy 1967). In general, larger litters have smaller offspring. This can be addressed through the use of a covariate for the litter size in the model.

10.4.3.2 Statistical Analysis of Counts

Many of the parameters collected are counts of the number of occurrences per litter. Some examples of these are the number of corpora lutea per litter, number of implants, number of resorptions, number dead pups, and the number of live per litter.

It is important to note that for some of these parameters, there is a high percentage of zeroes among the litters in a study.

Count data can be analyzed using the generalized estimating equation approach (Liang and Zeger 1986; Zeger and Liang 1986). This approach is outlined in detail in Chen (1998). However, in some cases where there is a high proportion of 0 observations, the iterative fitting process required for this model may not converge to a solution. Because of this, count data are often analyzed using nonparametric rank-based methods such as a Kruskal–Wallis test or the Jonckheere trend test.

10.4.3.3 Statistical Analysis of Proportion Data

Many of the parameters collected are proportions of responders per litter. Whenever possible, the proportion per litter is a better endpoint for analysis than counts, since counts can also be influenced by changes in the litter size due to competing drug effects (Bailey 2008; Chen 1998). Some examples of these are proportion resorbed per litter, proportion live per litter, and proportion with malformations per litter (either individual malformations or some form of a combined category). In addition, similar to the count parameters, some parameters have a high percentage of zeroes among the litters in a study.

A common approach to the analysis of proportion data is to use a beta-binomial model (Williams 1975). Alternatively, this data can be analyzed using a generalized estimating equation approach (Chen 1998). Alternatively the data can be transformed using an arc-sine transformation and analyzed using standard ANOVA methods. Finally, a nonparametric rank-based approach such as a Kruskal–Wallis test or the Jonckheere trend test are often utilized for their robustness, simplicity and understandability.

10.4.3.4 Statistical Analysis of Dichotomous Endpoints

Parameters with dichotomous response data are generally maternal parameters, such as estrous cycles (cycling normally), and mating and fertility indices. Also, sometimes the presence of fetal examination findings in a litter (any fetus with a finding) is analyzed as dichotomous data.

Dichotomous data are expressed as a $2 \times C$ contingency table, where C is the number of groups in the study. The data are summarized as the percent responding per dosage group, and standard contingency table methods, such as the Chi-square test, Cochran–Armitage test, or the Fisher Exact Test are used for the analysis. Ciminera proposed a randomization test for dose–response trend that is an extension of the Fisher Exact test for histopathology data (Ciminera 1985) that would be applicable here as well.

10.4.3.5 Multivariate Approaches

Over the years there have been a number of publications proposing simultaneous analysis of multiple endpoints (Catalano et al. 1993; Catalano and Ryan 1992; Chen et al. 1991; Ryan 1992). One of the justifications for this type of approach is to reduce the number of analyses being performed on the study, thereby reducing the overall study-wide Type I error. This type of approach can also increase the power of detecting effects if the outcomes under consideration are due to a common biological mechanism.

These approaches have seen limited implementation in the pharmaceutical industry. The study scientists are reticent to identify parameters that they feel can be grouped for analysis without also considering the analysis results from each of the component analyses. Therefore, this would lead to more analyses being performed on a study, rather than less. Also, if a significant result is found, the study scientists want to know what components of the combination contributed to the significant result. While these approaches are nice in theory, they are too complex for many study scientists to understand and embrace.

10.5 Juvenile Toxicology Studies

Prior to 1994, drug testing was generally performed on adult animals in toxicity studies, and on adult subjects in clinical trials. If a drug was approved, it was then often prescribed to children, generally at lower doses, under the assumption that children would respond similarly to the compound as adults. However, due to differences in pharmacology, and differences in development (e.g. the human brain at birth is neurologically equivalent to a rat brain at 2 weeks of age), this assumption is often not true.

In 1994, in order to address this gap, the FDA encouraged manufacturers to survey existing data of marketed compounds to determine if there was sufficient evidence to support pediatric use information in the drug's label. When this did not markedly increase the number of products with adequate pediatric labeling, the FDA created a regulation, approved in December 1998, requiring that pediatric safety and effectiveness be addressed either through studies or through a waiver in NDAs and BLAs. At that point in time, Juvenile toxicology studies to support pediatric clinical trials started to become more common. The regulations were invalidated in 2002 by the federal courts, but then resurrected by Congress as the Pediatric Research Equity Act (PREA) in December 2003. The PREA was then renewed by Congress in 2007. Conduct of the nonclinical aspects of a pediatric program are governed by FDA guidelines issued in 2006 (U.S. Food and Drug Administration 2006).

Juvenile toxicology studies are similar in scope to regulatory toxicity studies, with the difference in the age of the animals at the initiation of dosing. The age and duration of dosing is dependent on the anticipated age of the target human population, and should correlate with the corresponding stage of development in

Table 10.10 Litter-mate intraclass correlation for selected parameters

Parameter	Intraclass correlation (%)[a]
Pup weights during weaning (days 0–21)	69–72
Clinical Pathology at weaning	19–55

[a]Intraclass correlation is the proportion of the variance that can be attributed to which litter the pup is from

the test animals (Zoetis and Walls 2003). Therefore it is possible that dosing may start as early as 1 day of age. This causes problems because there is the temptation, to minimize animal usage for ethical reasons, to treat the animals as independent units where in reality they are not. Examples of litter-mate intraclass correlation for various parameters, based on historical data, are listed below in Table 10.10. Due to this intraclass correlation, the total number of animals required to maintain sufficient power in the study is greater than a study where adult animals are dosed (Bailey 2006). For example, for a parameter that has a 50 % litter-mate intraclass correlation, a study with 40 pups in 10 litters (4 pups/litter) has an 'effective' sample size of 16. This means that these 40 pups give the same amount of information as 16 pups if the pups came from 16 independent litters.

In addition to the usual parameters collected in a regulatory toxicology study, neurobehavioral parameters are sometimes collected if there is reason to believe the test compound might affect brain function or development. The same issues listed above apply for these parameters as well. An effective analysis of these parameters can be difficult, because of the small sample sizes and the large variability inherent in these parameters.

Parameters collected in these studies are continuous in nature. Therefore, the methods outlined in the Reproductive Toxicology section for continuous parameter are also applicable for the analysis of Juvenile Toxicology studies.

References

Bailey SA (1998) Subchronic toxicity studies. In: Chow S-C, Liu J-P (eds) Design and analysis of animal studies in pharmaceutical development. Marcel Dekker, New York, pp 135–195

Bailey S (2006) Design and analysis issues in juvenile animal toxicity studies for pharmaceutical development: a statistician's perspective. Poster presented at the teratology society meetings, Tucson, AZ, 26 June 2006. Abstract: Birth Defects Research 76, p 383

Bailey S (2008) Relationships between litter size and resorptions, dead, and live fetuses: implications for statistical analysis and data interpretation. Poster presented at the teratology society meetings, Monterey, CA, 30 June 2008. Abstract: Birth Defects Research 82, p 352

Barlow RE, Bartholomew DJ, Bremner JM, Brunk HD (1972) Statistical inference under order restrictions. Wiley, New York

Bradley DR, Russell RL (1998) Some cautions regarding statistical power in split-plot designs. Behav Res Methods Insturm Comput 30(3):462–477

Bretz F, Hothorn LA (2003) Statistical analysis of monotone or non-monotone dose–response data from in vitro toxicological assays. Altern Lab Anim 31:81–96

Catalano PJ, Ryan LM (1992) Bivariate latent variable models for clustered discrete and continuous outcomes. J Am Stat Assoc 87:651–658

Catalano PJ, Scharfstein DO, Ryan LM, Kimmel CA, Kimmel GL (1993) Statistical model for fetal death, fetal weight, and malformation in developmental toxicity studies. Teratology 47: 281–290

Chen J (1998) Analysis of reproductive and developmental studies. In: Chow S, Liu J (eds) Design and analysis of animal studies in pharmaceutical development. Marcel Dekker, New York, pp 309–355

Chen J (2000) Reproductive studies. In: Chow S (ed) Encyclopedia of biopharmaceutical statistics. Marcel Dekker, New York, pp 445–453

Chen J, Gaylor D (1992) Correlations of developmental end points observed after 2,4,5-trichlorophenoxyacetic acid exposure in mice. Teratology 45:241–246

Chen JJ, Kodell RL, Howe RB, Gaylor DW (1991) Analysis of trinomial responses from reproductive and developmental toxicity experiments. Biometrics 47:1049–1058

Ciminera J (1985) Some issues in the design, evaluation, and interpretation of tumorigenicity studies in animals. In: Proceedings of the symposium on long-term animal carcinogenicity studies: a statistical perspective. American Statistical Association, Washington, DC, pp 26–35

Clinical and Laboratory Standards Institute (2000) How to define and determine reference intervals in the clinical laboratory: approved guideline, vol 2. CLSI, CLSI document C28-A2, Wayne

Crump K (1984) A new method for determining allowable daily intakes. Fundam Appl Toxicol 4:854–871

Dorato MA, Engelhardt JA (2005) The no-observed-adverse-effect-level in drug safety evaluations: use, issues, and definition(s). Regul Toxcol Pharmacol 42:265–274

Dunnett CW (1955) A multiple comparison procedure for comparing several treatments with a control. J Am Stat Assoc 50:1096–1121

EP Evaluator (2005) Release 7 (EE7). Computer software for evaluating clinical laboratory methods. David G. Rhoads Associates, Kennett Square

Filipsson AF, Sand S, Nilsson J, Victorin K (2003) The benchmark dose method—review of available models, and recommendations for application in health risk assessment. Crit Rev Toxicol 33(5):505–542

Healey GF (1999) The F1 approximate parametric test for monotonicity. Internal Technical Information Document ST9725, Department of Statistics, Huntingdon Life Sciences, Huntingdon

International Conference on Harmonization (2001) S7A – Safety pharmacology studies for human pharmaceuticals. http://www.fda.gov/downloads/drugs/guidancecomplianceregulatoryinformation/guidances/ucm074959.pdf. Accessed 1 Sept 2014

International Conference on Harmonization (2005) S5(R2) – detection of toxicity to reproduction for medicinal products & toxicity to male fertility. http://www.ich.org/fileadmin/Public_Web_Site/ICH_Products/Guidelines/Safety/S5_R2/Step4/S5_R2__Guideline.pdf. Accessed 1 Oct 2014

International Conference on Harmonization (2010) M3(R2) – nonclinical safety studies for the conduct of human clinical trials and marketing authorization for pharmaceuticals. http://www.fda.gov/Drugs/GuidanceComplianceRegulatoryInformation/Guidances/default.htm. Accessed 30 June 2014

Jonckheere AR (1954) A distribution-free k-sample test against ordered alternatives. Biometrika 41:133–145

Lachin J (2011) Power and sample size evaluation for the Cochran–Mantel–Haenszel mean score (Wilcoxon rank sum) test and the Cochran–Armitage test for trend. Stat Med 30:3057–3066

Liang KY, Zeger SL (1986) Longitudinal data analysis using generalized linear models. Biometrika 73:13–22

McBride WG (1961) Thalidomide and congenital abnormalities. Lancet 2:1138

McCarthy JC (1967) Effects of litter size and maternal weight on foetal and placental weight in mice. J Reprod Fertil 14:507–510

Palmer AK (1974) Statistical sampling and choice of sampling units. Teratology 10:301–302

Rochon J (1991) Sample size calculation for two-group repeated-measures experiments. Biometrics 47:1383–1398
Ryan L (1992) Quantitative risk assessment for developmental toxicity. Biometrics 48:163–174
Selwyn MR (1988) Preclinical safety development. In: Peace KE (ed) Biopharmaceutical statistics for drug development. Marcel Dekker, New York, pp 231–271
Shirley E (1977) A nonparametric equivalent of Williams' test for contrasting existing dose levels of a treatment. Biometrics 33:386–389
Senn S (2007) Statistical issues in drug development. Wiley, Chichester, pp 99–100
Sivarajah A1, Collins S, Sutton MR, Regan N, West H, Holbrook M, Edmunds N (2010) Cardiovascular safety assessments in the conscious telemetered dog: utilisation of super-intervals to enhance statistical power. J Pharmacol Toxicol Methods 62(1):12–19
Sparrow S, Robinson S, Bolan S, Bruce C, Danks A, Everett D, Fulcher S, Hill RE, Palmer H, Scott EW, Chapman KL (2011) Opportunities to minimize animal use in pharmaceutical regulatory general toxicology: a cross-company review. Regul Toxicol Pharmacol 61:222–229
Staples RE, Haseman JK (1974) Selection of appropriate experimental units in teratology. Teratology 9:259–260
Tukey JW, Ciminera JL, Heyse JF (1985) Testing the statistical certainty of a response to increasing doses of a drug. Biometrics 41(1):295–301
U.S. Food and Drug Administration (2006) Guidance for industry – nonclinical safety evaluation of pediatric drug products. U.S. Food and Drug Administration, Rockville
Weaver J, Brunden M (1998a) Design of developmental and reproductive toxicity studies. In: Chow S, Liu J (eds) Design and analysis of animal studies in pharmaceutical development. Marcel Dekker, New York, pp 291–308
Weaver J, Brunden M (1998b) The design of long term carcinogenicity studies. In: Chow S, Liu J (eds) Design and analysis of animal studies in pharmaceutical development. Marcel Dekker, New York, pp 227–258
Williams DA (1971) A test for differences between treatment means when several dose levels are compared with a zero dose control. Biometrics 27:103–117
Williams DA (1972) The comparison of several dose levels with a zero dose control. Biometrics 28:519–531
Williams DA (1975) The analysis of binary responses from toxicological experiments involving reproduction and teratogenicity. Biometrics 31:949–952
Zeger SL, Liang KY (1986) Longitudinal data for discrete and continuous outcomes. Biometrics 42:121–130
Zoetis T, Walls I (eds) (2003) Principles and practices for direct dosing of pre-weaning mammals in toxicity testing and research. A report of the ILSI risk science institute expert working group. ILSI, Washington DC

Chapter 11
Clinical Assays for Biological Macromolecules

Jianchun Zhang and Robert J. Kubiak

Abstract Assessments of pharmacokinetics (PK), pharmacodynamics (PD) and immunogenicity are indispensable parts of the development process for therapeutic biologics. It is, therefore, essential to develop suitable assays for these assessments. However, development of assays for biological macromolecules poses unique challenges. In this chapter, we review the scientific background of clinical assay development and validation, as well as the common statistical methods used during the life-cycle of assay development.

Keywords Anti-drug antibodies (ADA) • Assay validation • Calibration curve • Cut point • Four-parameter logistic model • Generalized least squares • Immunogenicity • Incurred sample reanalysis • Ligand-binding assay

11.1 Introduction

Evaluations of pharmacokinetics (PK) and pharmacodynamics (PD) and immunogenicity (IM) are key components of the biological drug development process. PK describes the time course of a drug in the body starting with liberation of the drug from its formulation, absorption and entering the blood stream, distribution throughout tissues of the body, metabolism, and finally elimination. This process is often abbreviated as LADME or, as it is the case with biotherapeutics which are administered directly into the blood stream, as ADME. The key piece of data needed for evaluation of PK is concentration of the drug in body fluids such as blood or urine. This can be accomplished by a variety of analytical methods, which for the sake of simplicity will here be referred to as PK measurements. PD is the study of physiological effects that the drug has on the body. Consequently, the term PD measurements refers to analytical methods required to quantify biomarkers related to the effects of the drug. With the advance of biotherapeutics such as peptides, recombinant proteins, and monoclonal antibodies to name just a few, the potential

J. Zhang (✉) • R.J. Kubiak
MedImmune, Gaithersburg, MD, USA
e-mail: zhangji@medimmune.com; kubiakr@medimmune.com

L. Zhang (ed.), *Nonclinical Statistics for Pharmaceutical and Biotechnology Industries*, Statistics for Biology and Health, DOI 10.1007/978-3-319-23558-5_11

of these molecules to elicit immune response in the host is a serious concern. Except for vaccines, an immune response against biotherapeutics is an undesired side-effect that may impact both PK and PD. Presence of anti-drug antibodies (ADAs) may directly alter the ability of the drug to interact with its target (so called neutralizing antibodies or NAbs) or alter the rate of elimination of the drug thus changing its PK profile. Apart from their impact on drug efficacy, the presence of ADAs may be associated with adverse effects, which is of great concern from the safety point of view. IM assays aim to detect, quantify and characterize ADAs formed during the course of drug treatment. PK, PD and IM form a dynamic system with data from PK, PD and IM being inter-dependent and complementing each other. The trio should be interpreted in the context of clinical outcomes including efficacy and safety.

There are several regulatory guidance documents (FDA 2009, 2014a, b; EMA 2007, 2011, 2012), white papers (Booth et al. 2015; DeSilva et al. 2003; Gupta et al. 2007, 2011; Lee et al. 2006; Mire-Sluis et al. 2004; Shankar et al. 2008). There are numerous publications relating to clinical assays required throughout the life-cycle of macromolecules including development, validation and in-study monitoring. Some of this literature provides a comprehensive scientific review of the many aspects of particular assays but with limited statistics exposure. Boulanger et al. (2010) give a comprehensive coverage of statistical methods used for method validation in general. Implementation of many of these methods is described in Boulanger et al. (2007). However, these manuscripts still lack some technical details of statistical methods which may be insufficient for researchers that need a better understanding of statistics relating to bioanalytical questions, and statisticians who want to up skill their knowledge and understanding. In this chapter, we intend to provide a balanced treatment of both the scientific background and statistical details relating to clinical assays (for both non-clinical and clinical studies). Due to the limited space, it is, however, impossible to consider every type and aspect of bioanalytical assays. The topics covered here are highly selective and reflect the authors' preference based on their experience. Interested readers are referred to the literature cited in the text for further exploration.

11.1.1 Background

11.1.1.1 Pharmacokinetic and Pharmacodynamic Measurements

PK analysis describes the fate of a drug introduced into the body and requires measurement of drug concentrations in biological matrices (e.g. serum, plasma, urine). During development of medicinal products, the PK parameters derived from these measurements are needed to establish safe and efficacious dosing regimens for patients. Understandably, the bioanalytical methods used to measure drug concentrations must be properly validated to ensure that the method is suitable for its intended purpose and can generate results that can stand up to scientific and regulatory scrutiny.

The physiological effects of a drug on the body are described as clinical endpoints. A clinical endpoint is defined as a characteristic(s) that reflects how the patient feels, functions or survives (Lee et al. 2006). PD biomarkers are often used as surrogates for clinical endpoints and are defined as a characteristic that is objectively measured and evaluated as an indicator of normal biologic processes, pathogenic processes, or pharmacological response to a therapeutic intervention (Lee et al. 2006). A PD biomarker can be patient's blood pressure, glucose levels, the level of the drug target activation or inhibition and potentially many other characteristics related to drug action on the body (see Chap. 14: Biomarker Evaluation, for further discussion).

11.1.1.2 Immunogenicity

Biopharmaceuticals have an inherent propensity to elicit an immune response in the treated patients. The consequences of the immune response can range from completely benign to potentially life threatening (Rosenberg and Worobec 2004, 2005a, b). Consequently, detection and monitoring of anti-drug antibodies (ADAs) during the life-cycle of drug development including post-marketing phases is of interest to stakeholders including sponsors, regulators and patients. Depending on how ADAs interact with the circulating drug in the body, an ADA can be further classified as non-neutralizing or neutralizing. Non-neutralizing ADAs interact with a drug in a way that does not prevent it from binding to its target. However, the non-neutralizing ADAs may impact efficacy of the drug by increasing or decreasing drug clearance. In the latter case, this may lead to enhanced exposure and potentially toxic effects. Neutralizing ADAs interfere with a drug binding to its target and prevent the drug from performing its intended function. Development of high levels of neutralizing ADAs may render the patient completely unresponsive to the biotherapeutic. ADAs of the IgE subclass are a minor component of the total ADA response and as such, this type of ADA is unlikely to have any effect on drug efficacy. However, the IgE antibodies may be directly associated with patients safety by being responsible for adverse events ranging from mild allergic reaction at the injection site to anaphylactic shock. Consequently, in addition to ADA detection, characterizing of NAb and ADA isotypes is of interest to the regulatory agencies.

There are a myriad of factors that may potentially affect drug immunogenicity. These factors are both patient-related and product-related. Because of this complexity, a risk-based approach (Kloks et al. 2015; Koren et al. 2008; FDA 2014a, b; EMA 2007, 2012; USP chapter <1106> 2014; Rosenberg and Worobec 2004, 2005a, b) is recommended to assess immunogenicity and its consequences.

11.1.2 Immunogenicity: Analytical Challenges

Bioanalytical assays aimed at detection, measurement and characterization of ADA pose multiple challenges. Antibodies are a major constituent of human serum and are present in concentrations ranging from 10 to 16 mg/mL; while clinical impact of ADA can manifest itself at concentrations of 250 ng/mL and lower (FDA 2009). Although all antibodies are fairly similar to each other in their general physical-chemical properties and structure, the human immune system can generate approximately 10^9 different antibody clones capable of recognizing a vast variety of antigens. The polyclonal nature of the immune response means that individual ADA clones can engage the drug in different ways by binding to different epitopes on the drug molecule. Consequently, the ADA response does not consist of a single entity but is highly heterogeneous and likely to vary from patient to patient and undergoes changes throughout the course of treatment. The majority of biotherapeutics are monoclonal antibodies which poses an additional challenge for detection of ADA since both drug and anti-drug antibodies are almost indistinguishable from the majority of endogenous immunoglobulins already in circulation. Presence of the biotherapeutic drug in samples obtained from subjects is likely to interfere with ADA assessment which further complicates the interpretation of immunogenicity data.

11.1.2.1 Tiered Testing

Due to the complex nature of immune response, immunogenicity assays are far less straightforward than assays for other analytes. The large number of samples from clinical trials that need to be tested for presence of ADAs requires a strategy that allows for rapid screening of all samples and then focusing the available resources on the samples with the highest probability of being positive in order to quantify and characterize the ADA. A tiered approach is typically deployed for this purpose, see Fig. 11.1. In the first tier all samples are subjected to a rapid screening assay to remove the samples that are negative. Samples with response at or above a certain critical level (cut point) are considered as potentially positive and retained for the next round of testing. The second tier of testing aims to distinguish specific-binding ADAs from non-specific signals and eliminate the majority of the false positives generated in the first tier. The third tier of testing consists of characterization of immune response and may include semi-quantification of ADA (titer), detection of neutralizing antibodies, isotyping and other assays based on specific needs of the drug development program.

11.1.2.1.1 Screening Assay

In the screening tier of immunogenicity testing, the samples responses are compared against the screening cut point. Samples with the response below the screening cut point are declared negative and excluded from further testing. Samples with responses at or above the screening cut point are defined as potentially positive and directed for additional testing in the confirmatory tier. Since samples below the screening cut point are never re-tested again, it is important to avoid false

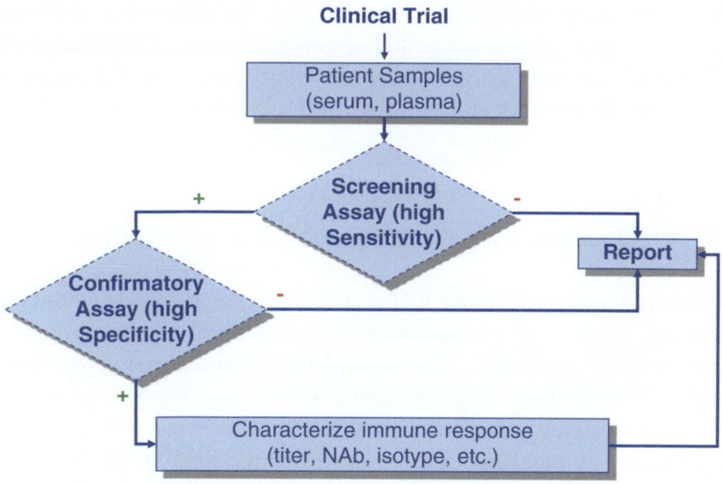

Fig. 11.1 Typical multi-tiered immunogenicity testing approach employed in clinical trials

negative classifications. Therefore, selection of an appropriate screening cut point involves a tradeoff between sensitivity and specificity. High specificity (a low false positive rate) is likely to result in low sensitivity (a high number of false negatives). Since immunogenicity may have impact on patient safety, risk management dictates that false positives are preferred to false negatives (Koren et al. 2008). Regulatory agencies and white papers recommend setting the screening cut point to generate 5 % of false positive classifications with the aim of decreasing the number of false negatives.

While there is a relationship between false positive and false negative rates, the latter cannot be clearly defined or measured. This is due to the lack of a "gold standard" in immunogenicity assays. As explained above, ADA is not a single and well defined situation but rather a nebulous and ever changing entity that consists of multiple antibody clones binding to multiple epitopes of the therapeutic macromolecule. For obvious reasons, one cannot immunize human subjects with the drug for the sole purpose of obtaining ADAs that could serve as a positive control in an assay. Antibodies against the drug of interest are raised in animals and are used as surrogate positive controls with the caveat that they are a partial and imperfect representation of an ADA developed in patients. Animal-derived polyclonal (recognizing multiple epitopes) or monoclonal (recognizing a single epitope) anti-drug antibodies are used as positive controls to establish such assay performance parameters as intra- and inter-assay precision, sensitivity and drug tolerance during pre-study assay validation, and for in-study monitoring of assay performance. The problem of false negatives is especially difficult to address since it is unclear what ADA levels may lead to clinically meaningful outcomes. Regulatory agencies recommend that clinical immunogenicity assays be able to detect 250–500 ng/mL ADA. However, it is conceivable that some antibodies can have a large impact at significantly lower concentrations (e.g. IgE isotype); while other may have no observable consequences at higher concentrations. Therefore, a risk-based strategy is particularly important to immunogenicity assessment (Kloks et al. 2015; Koren et al. 2008; Rosenberg and Worobec 2004, 2005a, b). Section 11.3 describes in detail current statistical methods used for determination of the screening cut point.

11.1.2.1.2 Confirmatory Assay

The screening tier tries to eliminate all true negatives and results in a pool of samples that are greatly "enriched" in true positives in addition to some (approximately 5 %) false positive responses. The second tier of testing should eliminate the vast majority of these false positives and provide unambiguous positive/negative classification. The assay commonly used for this purpose involves competitive inhibition of signal by unlabeled drug to show that the signal is specific to the drug. Any signal generated by ADA should be inhibited by addition of the drug, whilst a signal resulting from non-specific binding should not be inhibited or inhibited to a lesser extent. It is unclear what level of signal inhibition corresponds to a cutoff between specific and non-specific binding. Neyer et al. (2006) proposed the use of a paired t-test to compare the sample signal in presence and absence of the drug. Arguably, with a sufficiently large number of observations (repeated measurements), very small differences caused by the addition of the drug may become statistically significant, regardless of biological variability and clinical relevance. Another approach applied by some laboratories is to spike samples with low, medium and high amounts of the positive ADA and to treat them with a known amount of drug. A confirmatory cut point is then derived by statistical analysis of the signal differences between spiked and non-spiked samples. This approach is strongly dependent on the nature of the positive control ADA, with a possibility that ADAs derived from animals are not representative of the immune response in patients. This complicates the interpretation of immunogenicity data (Smith et al. 2011). The most common bioanalytical approach is to obtain inhibition data by spiking excess of drug into 50–100 individual drug-naïve samples and statistically determining a cutoff response corresponding to 1 % or 0.1 % false positive rate; see Sect. 11.3 for the description of related statistical methods.

11.1.2.1.3 Neutralizing Antibody Assay

Neutralizing antibodies (NAbs) interfere with the binding of a drug to its target and prevent the drug from eliciting the desired physiological effect. Presence of NAbs can prove to be especially dangerous if they neutralize activity of non-redundant endogenous protein in humans. The neutralizing effect of NAbs can be best detected using cell lines which respond to the drug in question. The direct action of a drug on cells may result in cell proliferation or cell death. In the absence of NAbs, the drug should have its full impact on cells, whilst in the presence of NAbs, the effect of a drug should diminish. Drugs may have indirect effects on cells when they enhance or prevent binding of soluble ligands to cell surfaces. When NAbs are absent, a drug can fully engage with the ligand and prevent it from exercising its physiological function. On the other hand, presence of NAbs leads to apparent enhancement of the ligand activity. Cell-based assays can prove extremely challenging due to lack of appropriate cell lines and difficulties associated with their growth and maintenance. Cell-based assays often suffer from low sensitivity, narrow dynamic range, and poor tolerance to circulating drug and serum proteins. Alternatively, ligand-binding assays can be used to analyze NAbs and typically are not subject to the disadvantages of cell-based assays. When a drug target is present on cell surface, cell-based assays are arguably best suited for detection of NAbs and evaluation of in vivo interactions between ADA, drug and target. At the same time, a legitimate argument can be made that ligand-binding assays with their higher sensitivity and robustness are ultimately more informative and better suited to the analysis of clinical samples.

11.1.3 Assay Platforms

An analytical platform utilizes certain unique physical-chemical properties of the analyte to detect and quantify it in a matrix of interest. A thorough understanding of these properties is indispensable in order to properly interpret bioanalytical data. Detailed discussion of the plethora of analytical techniques available to biopharmaceutical researcher is beyond the scope of this chapter and here we will focus only on ligand-binding assays, which are widely used for PK and PD measurements as well as for ADA detection.

11.1.3.1 Ligand-Binding Assays

Ligand-binding assays (LBAs) depend on interactions between a receptor and its ligand. In broader terms, the receptor-ligand pair can include combinations of various proteins (peptides, antibodies or their fragments, enzymes, receptors, etc.), nucleic acids (RNA, DNA), and relatively small molecules (e.g. steroids). These interactions are characterized by high affinity and specificity. Consequently, ligand-receptor complexes are formed in the presence of other similar species. These properties make ligand-binding assays ideal for detection and quantification of biological molecules in complex matrices. The large number of possible interactions allow for a variety of possible assay formats. The vast majority of ligand-binding assays require some sort of solid support to capture the analyte of interest from the sample and remove the excess of matrix. This solid support can be in the form of appropriately activated plastic surfaces (plates, tubes, biosensor chips etc.) or in form of beads or nanoparticles which allow application of fluidics. A general description of assay formats is shown in the Fig. 11.2.

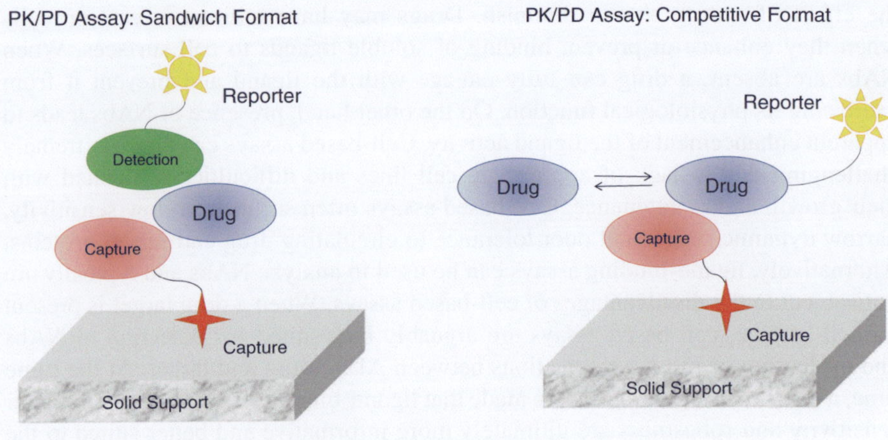

Fig. 11.2 General schematic description of ligand-binding assay platforms

11.1.3.1.1 Sandwich Formats

In sandwich formats, the drug is "sandwiched" between the capture and detection species. The capture species (e.g. anti-drug antibody, drug receptor) is immobilized on the solid support surface, which is incubated for a certain time with the sample to bind to the drug. The solid support is then washed to remove excess matrix components and subsequently incubated with the detection species (e.g. anti-drug antibody, drug receptor etc.). The detector is typically conjugated with a reporter capable of generating a quantifiable signal. In the earliest ligand-binding assays, the detection species carried a radioactive label such as ^{3}H, ^{35}S or ^{125}I. However, radioactive reagents are now used less often due to problems with their disposal and personnel safety issues. One of the most common LBA formats is an enzyme-linked immunosorbent assay (ELISA), where the detector is conjugated with an enzyme which turns over a substrate to generate signal (e.g. optical density or light emission). The amount of the bound reporter and the intensity of generated signal are proportional to the concentration of the drug in the sample. Ligand-binding assays often utilize the extremely high affinity of streptavidin and avidin for biotin. Strept(avidin)-biotin interactions can be used to bind the capture species (labeled with biotin) to the streptavidin-coated surface. Similarly, the detection species can be labeled with biotin to capture streptavidin conjugated with the reporter.

11.1.3.1.2 Competitive Formats

In competitive formats, the drug present in the sample competes for the capture species with the drug labeled with reporter. The nature of interactions is the same as that in sandwich formats but the signal intensity is inversely proportional to the drug concentration with maximum signal generated by blank matrix. In competitive assays only one species is required to bind to the drug. Consequently, these assay formats are not as specific as the sandwich assays where a drug is recognized by two different binding partners.

11.1.3.2 Immunogenicity Assay Formats

The heterogeneity of immune responses poses enormous problems for the design of assay formats. Ideally, a screening assay should be capable of detecting all immunoglobulin classes (IgG, IgM, IgA, IgD and IgE). An assay should also be able to demonstrate sufficient sensitivity and have a low level of false positive classifications. To detect ADA, the drug used in an assay must be as close as possible to its native form, since addition of molecules that facilitate capture or detection of ADA (labeling with radioactive isotopes, biotinylation, ruthenylation or passive absorption on solid support) may block or distort the existing epitopes and conceivably introduce others resulting in non-specific signals. The necessity of using a drug as a reagent to detect presence of ADA means that immunogenicity assays are susceptible to interference by the circulating drug. Preferably, immunogenicity samples should be collected when the drug is washed out from circulation but this is not always feasible. A solid understanding of the advantages and limitations of an assay format can aid the design of immunogenicity assessment programs as well as interpretation of resulting data. The most common formats for immunogenicity assays are briefly described below (Fig. 11.3).

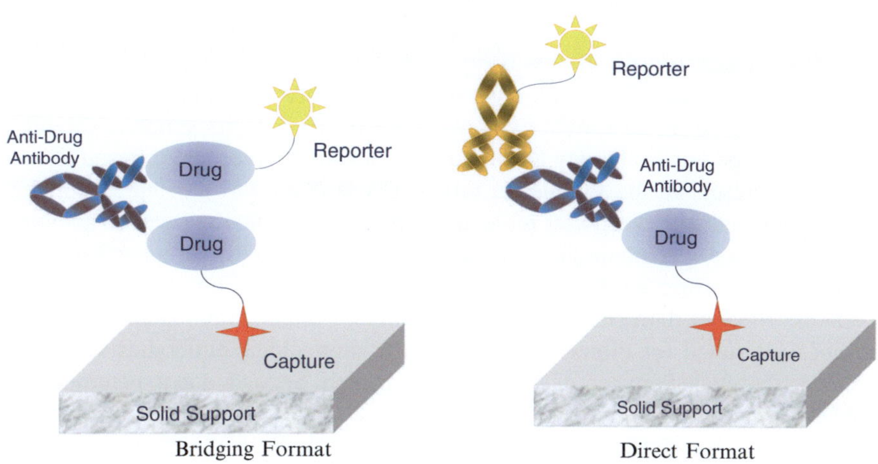

Fig. 11.3 Different ADA assay formats

11.1.3.2.1 Bridging Formats

Bridging assays are arguably the most common immunogenicity assay format. This format depends on the ability of ADA to bind to more than one molecule of the drug at the same time. Two different forms of the drug are needed for the assay: one that allows capture onto a solid surface and one that can serve as a reporter for generation of a signal. Upon binding of ADA with the two different forms of drug, the resulting antibody-drug complex can be captured and detected. Capture of the drug is accomplished by either passive absorption on solid support, or by labeling the drug with biotin and capturing it on streptavidin or avidin-coated surfaces. The most commonly used platforms for this type of assay are ELISA and electrochemiluminescent immunoassay (ECLIA), where the signal is measured either as optical density (ELISA) or as light emitted by a ruthenium chelate reporter (ECLIA).

Popularity of the bridging assays is due to their high sensitivity and capability to detect all immunoglobulin classes and most isotypes. These assays can be used for detection of ADA in different species which allows use of the same format for non-clinical and clinical studies. The major disadvantage of this format is that it does not detect antibodies directly; rather it detects multivalent species that are capable of binding to more than one molecule of the drug at the same time. For this reason, bridging formats can be considered as excellent screening assays with an inherent potential for generation of false positive responses.

11.1.3.2.2 Direct Formats

The unique characteristic of this assay format is its direct or specific detection of ADA. In contrast to the bridging assays that utilize the multivalent nature of ADA binding, the direct format explicitly detects ADA. In its simplest variant, a drug is captured (e.g. passively absorbed) onto a solid surface and allowed to bind with ADA present in the sample. In turn, ADA is detected by using an anti-IgG antibody labeled with a reporter molecule. Alternatively, the drug can be anchored to a surface via a specific monoclonal antibody or by biotin (for streptavidin-coated surfaces). As before, ADA is detected by an anti-IgG antibody conjugated with a suitable reporter. Direct formats suffer from two major draw-backs. First, they are not well suited for biotherapeutics that are monoclonal antibodies since they cannot differentiate between ADA and the drug itself. Second, due to heterogeneity of ADAs, it is difficult to detect all possible subclasses (e.g. IgG, IgM, IgE).

11.1.4 Assay Development and Validation

During assay development, appropriate assay platforms suitable for requirements of the program under development need to be determined along with various assay conditions such as: reagents, minimal required dilution (MRD), incubation times, etc. Design of experiments (DOE) is particularly useful to find the optimal combination of various assay conditions. At the completion of assay development, Standard Operating Procedure (SOP) is established with detailed description of the analytical procedure. A validation protocol specifies the assay parameters to be evaluated and sets a priori acceptance criteria that need to be met in order to demonstrate that the analytical method is fit for its intended purpose. The exact assay parameters that need to be validated together with their acceptance criteria depend on the type of assay (e.g. quantitative or semi-quantitative). Analytical challenges include, but are not limited to, analyte stability during sample handling, potential interfering factors, and cross-reactive species. At completion of the validation step, the SOP is finalized for in-study use. A detailed description of the validation experiments with results is presented in a validation report. For compliance purposes, the level of details included in the validation report should be sufficient to independently reconstruct the study at a later date.

For quantitative assays such as a PK assay, the main validation parameters are sensitivity, specificity/selectivity, linearity, dynamic range, accuracy, precision, limit of detection (LOD), limit of quantification (LOQ), stability and robustness. Some of these parameters are defined below. Readers are referred to the regulatory documents and industry white papers for more details (EMA 2011; FDA 2014b; Booth et al. 2015; DeSilva et al. 2003; Gupta et al. 2011; Kelley and DeSilva 2007; Lee et al. 2006; Miller et al. 2001; Shankar et al. 2008; USP chapter <1106> 2014).

Accuracy Describes the closeness of the analyte concentration as determined by the analytical method to the nominal concentration. Accuracy can be determined within an analytical run (intra-assay) and across analytical runs (inter-assay).

Precision Describes the closeness of the repeated individual measurements of the analyte (intra- and inter-assay) and is typically expressed as coefficient of variation.

Sensitivity In bioanalytical sciences, sensitivity typically refers to the lowest analyte concentration that can be detected (limit of detection LOD) or quantified (lower limit of quantification LOQ) by a given analytical method.

Dynamic Range Typically refers to the range of the concentrations between the lower and upper limits of quantification.

Selectivity/Specificity Selectivity describes the ability of the analytical method to accurately detect the analyte of interest in the presence of other components of a biological matrix. Specificity can be considered as a subset of selectivity as it describes the method's ability to detect only the analyte of interest in the presence of other similar analytes. In ligand-binding assays, the term specificity is often used interchangeably with cross-reactivity.

Stability Describes analyte stability when exposed to different situations that may be encountered during sample handling and analytical procedures. Stability evaluation typically includes testing of multiple freeze-thaw cycles, prolonged storage at room temperature, and stability during refrigeration and freezing.

Due to the broad definition of biomarkers, PD measurements can face more challenges than PK measurements. When biomarkers are well defined (e.g. steroids) and a suitable reference standard is available, results from the PD assay are definitive and can be validated in the same way as PK assays. When the reference standard is not available in purified form, or is not representative of the endogenous forms of the biomarker, a PD assay generates continuous responses which are semi-quantitative and expressed as the intensity of the generated signal. Purely qualitative assays generate categorical data which lack proportionality to the concentration of the analyte in the sample (Lee et al. 2006).

Immunogenicity assays are a good example of semi-quantitative assays. Due to the lack of a reference standard, the results from ADA assays cannot be presented as concentrations and are typically expressed as titer or minimum dilutions that render the sample negative (below the assay cut point). Consequently, the assay cut points, and especially the screening cut point is a critical assay parameter. Although many assay parameters have the same meaning as those for quantitative assays, some are different. For example, the assay sensitivity is defined as the lowest concentration at which the control antibody preparation consistently produces a positive result, which is closely related to the cut point. Accuracy is typically defined as the portion of true classifications (positive and negative) from all classifications of quality control samples.

The rest of this chapter is organized as follows. Section 11.2 introduces the statistical aspects of PK assay development and validation with the focus on calibration curve, accuracy and precision as well as incurred sample reanalysis. Section 11.3 describes the various statistical methods for determination of a cut point, including recent developments.

11.2 PK/PD Assay Development and Validation

In this section, we focus on PK/PD assay development and validation. These aspects are guided scientifically by regulatory guidance such as EMA (2011) and FDA (2014b) as well as industry white papers by DeSilva et al. (2003) and Booth et al. (2015). Lee et al. (2006) covers PD/biomarker assays. Booth et al. (2015), DeSilva et al. (2003), Kelley and Desilva (2007), DeSilva and Bowsher (2010) cover many aspects of LBA assays with technical details including assay reagent selection, reference material, stability, specificity, selectivity, calibration curve, accuracy, precision and many others. These are recommended sources for practitioners.

For quantitative PK or PD assays, a typical analytical run consists of calibration standards, matrix blank (processed matrix sample without the analyte), a set of appropriate quality controls, and finally, the study samples to be analyzed. Calibration standards are prepared by spiking a known amount of the analyte into an appropriate matrix to generate a concentration-response relationship from which the concentrations of the study samples can be interpolated. The blank matrix sample shows the response of the processed matrix which ideally should be below that of the lowest calibration standard. In cases where an endogenous analyte is present in the sample, its response may be subtracted from the responses of the calibration standards. The quality control samples consist of an analyte spiked into the assay matrix at the low, medium and high levels of the quantitative range of the assay. Additional quality controls may be prepared at the LLOQ and ULOQ (Lower and Upper Limits of Quantification, respectively). The concentrations of the quality control samples are interpolated from the calibration curve and compared with their respective nominal concentrations. Preferably, the quality control samples should be treated in the same way as the study samples and mimic them in every respect (storage, processing and treatment in the bioanalytical assay). In order for an analytical run to be accepted and results from the unknown samples reported, the calibration standards and quality controls must meet a set of pre-defined acceptance criteria. During analytical method validation, quality control samples are used to determine accuracy and precision.

11.2.1 Calibration Curve

A calibration curve is the description of the relationship between signal from the analyte of interest and its concentration, and is used to measure unknown concentrations of study samples given their signal readings. The curve is prepared by utilizing reference material where known concentrations of analyte are spiked into an assay matrix. The concentration of individual calibration points are established during method development, confirmed during the validation stage, and used for in-study sample testing. The calibration curves are needed for all PK and PD (biomarker) measurements, when well characterized reference standards are available. The concentration-response relationship in ligand binding assays (LBA) is well known to have a nonlinear mean response and non-constant variance. DeSilva et al. (2003), DeSilva and Bowsher (2010) and Boulanger et al. (2010) have described how to establish calibration curves in general. They recommend that during method development, the number of standard points and replicates should be large enough to allow the selection of the most appropriate calibration model; see Table 3 of DeSilva et al. (2003) for the design of calibration experiments for the different stages of method validation. A more statistically rigorous method is to apply design of experiment (DOE) method.

Several publications such as DeSilva et al. (2003) and DeSilva and Bowsher (2010) provide details relating to finding and fitting appropriate calibration curves without too many statistical details. The basic calibration curve is a four-parameter logistic model which has the following form:

$$y_j = d + \frac{a - d}{1 + (x_j/c)^b} + \varepsilon_j$$

where y_j is the j-th observed response at concentration x_j and b is the slope parameter, a is the response at zero concentration and d is the response at an infinite concentration with $b > 0$, c is the inflection point around which the mean calibration curve is symmetric; it is the concentration corresponding to the mean response halfway between a and d. If $d > a$, then the concentration-response relationship is increasing; when $a > d$, the relationship is decreasing. Sometimes, the parameters a and d have alternative interpretations. In this case, a is the maximal response and d is a minimal response; when $b>0$, the concentration-response relationship is decreasing and vise-versa when $b < 0$. The four-parameter logistic model is believed to be valid for competitive assays while a more general five-parameter logistic model better characterizes the concentration-response relationship of non-competitive assays. The five-parameter logistic model has the following form:

$$y_j = d + \frac{a - d}{\left(1 + (x_j/c)^b\right)^g} + \varepsilon_j$$

This model allows for an asymmetric concentration-response curve by adding an additional parameter g, which effectively allows the mean response function to approach the minimal or maximal response at a different rate (Findlay and Dillard 2007). In both of these models, ε_j is the intra-assay error term.

For the above calibration models, the variance of the error term is often not constant but changes across the concentration range. This phenomenon is known as heteroscedasticity. Taking into account this variance heteroscedasticity generally can improve the overall quality of model fitting as well as the accuracy and precision of the resulting calibration, as emphasized by Findlay and Dillard (2007). For a regression model with constant variance, a least-squares method is often adopted. To account for the variance heteroscedasticity, a weighted least-squares method can be used for parameter estimation. The idea is to place less weight on responses with higher variability. Mathematically, consider the following general regression problem:

$$y_j = f(x_j, \beta) + \sigma w_j^{-1/2} \varepsilon_j$$

The weighted least-squares method is to find the estimation of β by minimizing:

$$\sum w_j \left(y_j - f(x_j, \beta)\right)^2.$$

Instead of using the standard deviation at each concentration as weights, which is discouraged especially when the number of replicates is small, various weighting function forms are proposed. O'Connell et al. (1993), for example, considers a specific type of weighting function which works well empirically for immunoassay calibration curves. Specifically, they consider the following functional form:

$$y_j = f(x_j, \beta) + \sigma g[f(x_j, \beta)]\varepsilon_j$$

with $g[f(x_j, \beta)] = (f(x_j, \beta))^\theta$. The variance at concentration x_j is a power function of the mean response. Essentially, the task of model fitting during the development stage is to find the best appropriate calibration curve with an appropriate weighting function. Computationally, however, the model fitting procedure is much more complex than that of linear models. In the remainder of this section, we will discuss the statistical details of model building for non-linear calibration models.

11.2.1.1 Calibration from a Single Run

The parameter estimation of a nonlinear calibration curve with non-constant variance is not trivial. Unlike the linear regression model, there is no closed-form solution and iteration is often needed. Here, we introduce the approach proposed in O'Connell et al. (1993) and Davidian and Giltinan (1995). A more systematic exposition of the non-linear regression model in general can be found in Seber and Wild (1989) and Bates and Watts (1988).

A Generalized Least Squares (GLS) with variance function estimation is preferred since it does not assume distribution of the error term and is relatively robust to distribution misspecification. When the error term is assumed to be normally distributed, the GLS estimation is the same as the maximum likelihood estimation. Since there is no closed-form solution for the parameter estimations, numerical iteration is warranted. The GLS estimation starts with an initial estimator of β, usually the un-weighted least squares estimator. The variance parameters are estimated by utilizing the relationship between the residuals from the least squares regression and the concentration, using the likelihood method. The resultant weights are then used to obtain an updated weighted least-squares estimate of β. This procedure repeats until a convergence criterion is met. Mathematical details can be found in Appendix.

11.2.1.1.1 Calibration and Precision Profile

Once parameter estimates are obtained, a residual plot and lack-of-fit test may be applied to check the adequacy of the fitted model. To select from the different candidate models, the bias (accuracy) and precision of the calibrated standard points are compared. For a model under consideration, the fitted model can be used to estimate the unknown concentration of a sample with, say, M replications. Denote the mean response of these M replications by y_0, then the unknown concentration is estimated as

$$\hat{x} = h\left(y_0, \hat{\beta}\right),$$

where $h = f^{-1}$ is the inverse function. Since both y_0 and $\hat{\beta}$ are random (and independent), the approximate variance of $\hat{x}$ can be obtained by the Delta method (a first-order Taylor expansion):

$$\mathrm{Var}(\hat{x}) = \frac{h_y^2(y_0,\beta)\,\sigma^2 g^2(y_0,\beta)}{M} + h_\beta^T(y_0,\beta)\,V h_\beta(y_0,\beta) \qquad (11.1)$$

where h_β and h_y are the derivatives with respect to β and y, respectively. V is the covariance matrix for the estimator of β. The estimation of this approximated variance can be obtained by replacing unknown quantities in the above formula with their corresponding estimates.

The precision profile (Ekins and Edwards 1983) for an immunoassay is the relationship between the concentration and its precision of the calibrated concentration (in terms of standard error or coefficient of variation) across the full range of concentrations of interest. For a single batch, the estimated standard error or %CV is plotted against the concentration and is known as the precision profile. It is noted that this precision profile refers to the intra-assay precision and does not take into account the inter-assay variability. The intra-assay precision profile, as shown in Fig. 11.4, gives the working range of concentrations with acceptable %CV. It is particularly useful during the assay development stage, when the assay needs to be optimized against various assay conditions. It may also be used to select the appropriate calibration model among the many candidate models.

11.2.1.2 Calibration from Multiple Runs

Typically, multiple assay runs are performed for an analytical method. Instead of estimating the calibration curve for each run, it is preferable to pool data across all assay runs. This benefits the variance function estimation. However, this does not

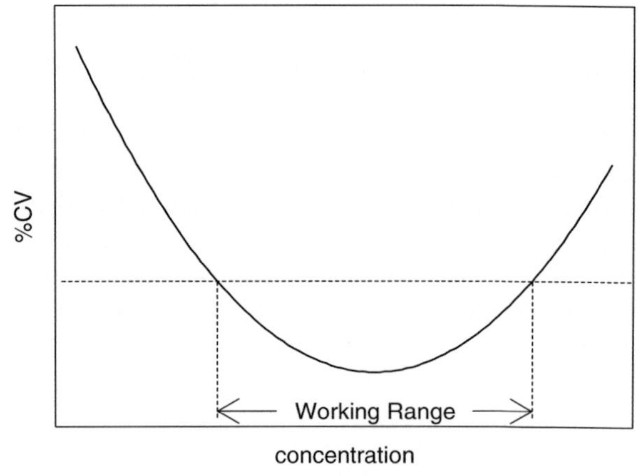

Fig. 11.4 An example of precision profile for assessing intra-assay variability

mean calculating the average at each concentration; instead, individual data from all the runs are fit by a single model. The calibration curve model can be rewritten as

$$y_{ij} = f\left(x_{ij}, \beta_i\right) + \sigma g\left[f\left(x_{ij}, \beta_i\right)\right] \varepsilon_{ij}$$

where i indicates the i-th run. This model is nothing but the collection of calibration models from each run. By writing the model in this way, and assuming the variance function parameters are constant across runs, it is expected that more accurate estimation of the variance function parameters will be achieved. Giltinan and Davidian (1994) and Davidian and Giltinan (1995) argue that the variance function parameter estimation from a single run can vary substantially from run to run even though it is theoretically assumed to be the same. Consequently, the precision profile may be less accurate. There is a simple way, however, to pool data from all the runs to obtain a better estimate of the variance function parameters. This only requires a slight change to the GLS algorithm for a single run. See Appendix for the parameter estimation with data from multiple runs.

Davidian and Giltinan (1995) and Zeng and Davidian (1997) consider the nonlinear mixed-effect model by assuming that β_i follows a multivariate normal distribution with mean β and covariance matrix D. They use empirical Bayes method to estimate run-specific parameters. Simulation studies show that there is a notable gain in efficiency of parameter estimation. The nonlinear mixed-effect model is widely used in many areas; however, application to assay calibration seems to be rare in practice.

11.2.2 Accuracy, Precision, Limit of Quantification and Acceptance Criteria

For each assay run, the relative error (%RE) is defined as

$$\%RE = 100 * \frac{\text{calibrated concentration} - \text{true concentration}}{\text{true concentration}}$$

and the precision is just the square root of the estimated variance of the calibrated concentration, while the coefficient of variation (%CV) is the ratio of precision to concentration. As recommended by DeSilva et al. (2003) and DeSilva and Bowsher (2010), the appropriateness of a candidate model in the development stage is judged by comparing the %RE and %CV. Specifically, they recommend that the absolute %RE for each intra-run standard points should be less than 20 % for most of the points. Across all the runs, the absolute average %RE be less than 10 % and the %CV be less than 15 % for all the standard points, the latter %CV is best understood as the inter-assay %CV.

DeSilva and Bowsher (2010) and Kelley and Desilva (2007) point out that the precision profile can be used to assess the limits of quantification. However, they caution that it is necessary to obtain precision profile of data from multiple runs, since the variability in LBA arises mainly from inter-assay variability.Calibrated estimation of standard points from all the runs may be used to construct an inter-assay precision profile or total error profile (Lee et al. 2006). Sadler (2008) review existing methods to estimate the inter-assay precision profile for data from several runs. One of these methods is polynomial regression of the following form:

$$y = (\alpha + \beta x)^{\gamma}$$

Alternatively, the inter-assay variance or total error may be estimated from an ANOVA model.

Once the method has been optimized (including the calibration model) during assay development, it should be confirmed during the validation stage. The calibration model should be confirmed in a minimum of six independent validation runs. The standard concentrations should not be changed once the assay validation has started. The general recommendation is that, for a standard curve to be acceptable, the %RE should be within 20 % for at least 75 % of the standard points. The cumulative mean %RE and %CV from all runs should be within 20 % for each standard concentration. Validation samples should be prepared at least at five different concentrations preferably evenly spanning the dynamic range from the anticipated LLOQ to ULOQ. The LLOQ and ULOQ are determined by the lowest and highest validation samples for which %CV and %RE are both less than 20 % (25 % at LLOQ) and total error (sum of %CV and absolute %RE) be less than 30 %. The actual LLOQ and ULOQ concentrations should be determined from validation samples by back-calculation from the calibration curve.

Common recommendations for assay validation acceptance criteria are that the inter-assay %CV and absolute mean %RE both be within 20 % (25 % at LLOQ). The total error should be less than 30 % (40 % at LLOQ). From a statistical point of view, this acceptance criterion can yield an uncontrolled risk of rejecting a suitable method (producer's risk) and accepting an unsuitable method (consumer's risk). Miller et al. (2001) discuss several statistical acceptance criteria such as equivalence test and prediction interval. Boulanger et al. (2010) and Boulanger et al. (2007) discuss using β-expectation tolerance interval as acceptance criteria, Hoffman and Kringle (2007) instead consider β-content tolerance intervals.

Once validation is successfully completed, the analytical method is deemed fit for its intended purpose. In-study runs are monitored by QC samples. For LBA methods, the so-called 4-6-30 rule is often used as a criterion to accept or reject a run. The rule means that 4 out 6 QC samples should be within 30 % of their nominal values. FDA guidance (FDA 2014b), however, recommends a 4-6-20 rule. In addition, at least 50 % of QCs at each concentration should be within 20 % of their nominal values. Boulanger et al. (2007) discuss the effect of different rules on the impact of acceptance probability.

11.2.3 Incurred Sample Reanalysis

Recently, the incurred sample reanalysis (ISR) has been added as an integral part of bioanalytical method life cycle management as required by some regulatory agencies to assure that the validated assay can generate reproducible results for PK studies. In ISR, a selected number of study samples representing all time points and dose groups (with the exception of placebo or samples originally testing below the LLOQ) are reanalyzed and compared with the original result. Failure of ISR may indicate presence of inherent problems with the analytical method and its unsuitability for supporting pharmacokinetic measurements. Currently ISR is restricted only to definitive quantitative assays with proper calibration standards.

Since the American Association of Pharmaceutical Scientists (AAPS)/FDA bioanalytical workshop in 2006, 2008 and 2013, this topic has been widely discussed in the literature. Focus of the discussion has shifted from whether ISR is needed to how ISR should be properly conducted. Several publications have been devoted to discussion of ISR. Rocci et al. (2007) provide detailed recommendations from sample selection to statistical methods for conducting ISR analysis. In brief, a decision should be made regarding which study needs to be selected for ISR. Then individual samples (not a single pooled sample or validation samples) should be selected from PK studies covering a wide range of doses. Understandably, both accuracy and precision play an important role in reproducibility. Traditional methods such as the paired t-test and Pearson's correlation are not recommended because these methods fail to take both accuracy and precision into account simultaneously. The Bland-Altman plot (Bland and Altman 1986) is a well-known convenient tool to visualize the limits of agreement between the original results and results from reanalysis of incurred samples. To assess the reproducibility of incurred samples, Rocci et al. (2007) recommends assessing the accuracy and precision separately. Specifically, the mean and standard deviation of the pair-wise differences (or relative difference) for individual samples are calculated. A 95 % confidence interval for the mean of differences is constructed and the confidence bounds are plotted on the Bland-Altman graph for accuracy assessment. If the value of 0 is beyond the confidence interval, significant bias can be declared. For precision assessment, it is typically required that two thirds of the samples should have pair-wise differences that are within the limits of agreement. The limit of agreement is calculated as an average of the pair-wise differences plus or minus one times the standard deviation. While this method is convenient, Lytle et al. (2009) suggest that the limits of agreement should be replaced by a tolerance interval to guarantee that the limit of agreement criterion is met with high confidence. Bland and Altman (2007) further considered limits of agreement for repeated measurements, which is typically the case for PK studies.

11.3 Immunogenicity Assay and Cut Point

11.3.1 Standard Approach

A key parameter during assay validation is the determination of the cut point. According to FDA guidance (FDA 2009), USP chapter <1106> (2014) and Shankar et al. (2008), the cut points should be statistically derived from data generated during pre-study validation.

11.3.1.1 Screening Cut Point

For the screening assay cut point experiment, typically at least 50 subjects are needed for clinical studies and 25–30 subjects are needed for nonclinical studies. Samples should be from drug-naïve donors from the targeted disease population whenever possible. The experimental design for the establishment of cut point should be well balanced to account for potential confounding factors in order to accurately estimate different sources of variability such as inter-sample, inter-run, intra-run, inter-analyst, etc. (Zhong and Zhou 2014). Shankar et al. (2008) as well as Devanarayan and Tovey (2011) give examples of a balanced design. In this section, it is assumed that the cut point design is appropriately conducted and the data are readily available for statistical analysis. In facilitating the understanding for both bioanalytical scientists and statisticians, we will first describe the "standard" approach by Shankar et al. (2008) for cut point determination. Some recent developments in the literature then follow.

The "standard" approach has been described in details in the appendix of Shankar et al. (2008) and further explained in Devanarayan and Tovey (2011). A cut point is typically established during the validation study and applied to in-study samples. Depending on the nature of cut point data, there are three types of screening cut point, namely: (a) fixed cut point; (b) floating cut point; (c) dynamic cut point. The fixed cut point is determined during the validation stage and applied directly to in-study samples until there is a justification for change. The floating cut point is a value obtained by applying a normalization factor which is determined from the validation data, to the background value during the in-study stage. In other words, the normalization factor can be considered fixed from validation stage to in-study stage and it may be called cut point factor. The dynamic cut point is a cut point which may change between instruments or between analysts, etc. It is often tedious or impossible to implement and further investigation is warranted. In some cases, an analyst specific or instrument specific fixed or floating cut point may be used instead.

According to Shankar et al. (2008), the process of determining the screening cut point consists of several steps: (1) investigate distribution and exclude outliers; (2) compare assay run means and variances to determine the type of cut point; (3) calculate the screening cut point. A schematic diagram is presented in Fig. 1 of Shankar et al. (2008).

11.3.1.1.1 Investigate Distribution and Exclude Outliers

The first step is to choose an appropriate data transformation such as logarithmic transformation. Once the transformation is determined, all subsequent analysis of data, such as outlier evaluation, comparisons of means and variances across assay runs, cut point calculations should be performed on the transformed scale. The purpose of selection of an appropriate transformation is based on the intuition that the data would be more symmetric and closer to a normal distribution on the transformed scale. Devanarayan and Tovey (2011) recommend that the selection of a transformation be based on averaged data across all the runs. The skewness parameter may be estimated from the data. The Shapiro–Walk normality test can be performed to test for normality assumption. If the skewness parameter and normality test indicate that the data under logarithmic transformation (or other transformation) is more symmetric and more normal, which is typically the case, the logarithmic transformation should be used on all the data values.

Once an appropriate transformation is found, outliers should be removed before determining the cut point. Intuitively, an outlier from the average data indicate that the donor or sample has consistent higher or lower than normal values and is thus a biological outlier. Biological outliers are outliers due to heterogeneity of the donor/sample population. For example, samples with pre-existing anti-drug antibodies will manifest extremely high signals and would be presumably flagged by the outlier removal procedures. For the same donor, the value from one run may be unusually higher or lower than the values from the other runs. Such a value is suspected to be an analytical outlier. The removal of outliers is not a trivial task. Devanarayan and Tovey (2011) recommend that Tukey's Box-plot method be used for outlier removal. Specifically, they recommend that the outlier boxplot criteria to be applied to the data from each assay run as well as to the average data values across the assay runs. Outliers from the averaged data are biological outliers while outliers identified from each assay run are analytical outliers. Zhang et al. (2013) argue that the analytical outliers should be removed before the biological outliers. See Sect. 11.3.2 for details.

11.3.1.1.2 Compare Assay Run Means and Variances to Determine Which Type of Cut Point Is Appropriate

The white paper by Shankar et al. (2008) recommends conducting an ANOVA and testing the equality of assay run means as well as the homogeneity of variances across assay runs. Depending on the test result, a fixed or floating cut point may be used. If the assay run means and variances are not statistically different between runs, the fixed cut point may be used; the floating cut point can be also used in this case. If the assay run means are statistically different but the variance are not statistically different among runs, the floating cut point can be used. When assay variances are statistically different, the dynamic cut point may be used.

Table 11.1 ANOVA table for testing runs mean equality

Source of variation	Sum of squares (SS)	Degree of freedom	Mean SS (MS)	Expected MS
Inter-donor	$SST = J\sum\left(\bar{y}_{.j} - \bar{y}_{..}\right)^2$	J-1	MST=SST/(J-1)	$\sigma^2 + I\sum\beta_j^2/(J-1)$
Intra-run	$SSE = \sum\sum\left(y_{ij} - \bar{y}_{.j}\right)^2$	IJ-J	MSE=SSE/(IJ-J)	σ^2
Total	$SSTO = \sum\sum\left(y_{ij} - \bar{y}_{..}\right)^2$	IJ-1		

Consider the ANOVA model:

$$y_{ij} = \mu + \beta_j + \varepsilon_{ij}$$

where y_{ij} denote the signal readout of i-th donor at j-th run. Assume there are I donors in total with each being tested J times (runs), resulting in IJ signal readouts in total. To test for mean equality of β_j s (being 0), the ANOVA F-test can be used. The ANOVA table is shown in Table 11.1. The F-test is the ratio of MST and MSE. As indicated by the expected mean squares, the F ratio will be close to 1 when the β_j s are equal (to 0). The F-test statistic follows an F distribution with J-1 and (IJ-J) degrees of freedom.

Levene's test (Levene 1960) can be applied to test homogeneity of variances across runs. Levene's test is often used before a comparison of means. Let $z_{ij} = \left|y_{ij} - y_{.j}\right|$ be the absolute deviations of y_{ij} with respect to their corresponding run mean $y_{.j}$. Levene's test determines whether the expected values of the absolute deviations for the J runs are equal. Levene's test is simply the F test statistic used to test equality of the means by replacing y_{ij} with z_{ij}. The resulting test statistic approximately follows an F distribution with the same degrees of freedom as those for testing mean equality.

11.3.1.1.3 Calculate the Screening Cut Point

Depending on the distribution of the data, a parametric or nonparametric method may be used to determine the cut point. As an alternative to the parametric method, a robust parametric method can be applied by down-weighting the extreme values instead of treating them as outliers.

If the data appear to be normally distributed after appropriate transformation and outlier removal, the parametric method can be used. A simple procedure would be to estimate the standard deviation from each run and then pool them together. The method of averaging the results across all the runs for each donor is discouraged as it does not take all the intra-run variability into consideration and consequently will underestimate the true cut point, because typically only one run is to be conducted for each of the in-study donors. Formally, the mean and standard deviation (SD) can be estimated from the ANOVA model mentioned above. The cut point is then calculated as mean + 1.645SD. For robust method, the median and 1.483 times

MAD, where MAD is the median absolute deviation, play the roles of mean and standard deviation. A more rigorous way is the mixed-effect ANOVA model:

$$y_{ij} = \mu + \alpha_i + \beta_j + \varepsilon_{ij} \tag{11.2}$$

where α_i s are random and normally distributed with mean 0 and variance σ_α^2, which represents the donor-to-donor variability. β_j s are either fixed-effects or random effects and normally distributed with mean 0 and variance σ_β^2, which represents the run-to-run variability. The error term has mean 0 and variance σ^2. The cut point is then calculated as mean + 1.645SD. However, SD should incorporate both the inter-donor variability (biological variability) and intra-assay variability (analytical variability), and the SD is calculated as $\sqrt{\sigma_\alpha^2 + \sigma^2}$. For fixed cut point, the β_j's can be dropped from the above model. For floating cut point, the resulting cut point factor is determined by subtracting the cut point from the average of negative control across runs. The floating cut point for an in-study assay is then the run mean of the negative control plus the cut point factor. If the log-transformation is applied, the cut point factor is obtained by dividing the floating cut point by the average of negative control and the in-study cut point is the product of cut point factor and average of negative control of the in-study assay run. If the normality assumption is violated, the nonparametric approach may be used. The 95th nonparametric quantile is then calculated accordingly.

One underlying assumption of the floating cut point is that, for the cut point factor to be constant across the runs, the run-specific negative control should be able to correct for the run-to-run variability. In other words, the signal corrected for the negative control background should behave similarly to the uncorrected signal in the fixed cut point case. While this assumption may be hard to verify, a less stringent version of this is to assume that β_js and their corresponding negative control values demonstrate a similar trend across runs. This assumption may be verified by evaluating the significance of the correlation and linear relationship of the negative control readings from each of the assay runs versus the average of the individual donor samples from the corresponding runs.

11.3.1.2 Confirmatory Cut Point

Confirmatory cut point characterizes the second tier of immunogenicity testing and corresponds to the minimum signal inhibition with unlabeled drug that signifies specific binding events. In the "standard approach", the signal inhibition data are generated together with the screening cut point data by analyzing the same drug-naïve samples in presence and absence of excess unlabeled drug. The signal change upon addition of drug is typically expressed as %inhibition:

$$\%Inhibition = 100\%\left[1 - \left(\frac{Signal\ in\ Presence\ of\ Drug}{Signal\ in\ Absence\ of\ Drug}\right)\right]$$

The %inhibition values may occasionally be negative and therefore not amenable to logarithmic transformation. A ratio of signal in absence of drug to signal in presence of drug may be preferred for use in statistical analyses.

$$IR = \frac{Signal\ in\ Absence\ of\ Drug}{Signal\ in\ Presence\ of\ Drug}$$

The *IR* values are always positive and increase in linear fashion with absolute signal change. Therefore, *IR* can be subjected to log-transformation and is more suitable for analysis of intra- and inter-assay precision.

The confirmatory cut point should also be statistically derived. Calculation of confirmatory cut point is similar to that of fixed screening cut point. Appropriate transformation and outlier removal is recommended. If transformation is not possible due to negative values of %inhibition, *IR* can be used instead. The confirmatory false positive rate is set to be 1 % or 0.1 %, as recommended by Shankar et al. (2008).

11.3.2 Recent Developments

While the standard approach recommended by the white paper is detailed and easy to follow for practitioners, there are still some aspects that are currently being debated as shown in the recent literature.

11.3.2.1 Data normalization

Zhang et al. (2013) argue that relying on an ANOVA F-test as described in Sect. 11.3.1.1 in order to choose between using fixed or floating cut point should be discouraged because insignificant results may be due to large biological variability and significant results may be due to large sample size. As a result, they suggest always using a floating cut point. In addition, they recommend that the raw individual data should be normalized by dividing by the average of the negative controls on that plate. Compared to the data without normalization, the run-to-run variability after data normalization is reduced dramatically without diminishing the donor-to-donor variability. If the normalized data variability is homogeneous across runs, the cut point factor can be considered as a "fixed" cut point and may be calculated the same way as the usual fixed cut point. A mixed-effect model $y_{ij} = \mu + \alpha_i + \varepsilon_{ij}$ can be applied with the help of a statistician. If the data are normally distributed, the resulting cut point is calculated as mean + 1.645SD, where SD is the square root of total variability.

11.3.2.2 Data Structure

Sometimes, scientists may calculate a cut point directly from all data values without taking the data structure of repeated measurement into account. Alternatively, the cut point from each run is calculated and the final cut point is obtained by taking an average of these cut points across multiple runs. Statistician may question the soundness of this practice. Hutson (2003) compares these two approaches in the context of nonparametric estimation under a one-way ANOVA setting and concludes that the two approaches produce similar results and are asymptotically equivalent. However, the approach of first taking the average across runs to obtain a single data point for each donor is a biased estimator and should be avoided in practice. It is conjectured that the same conclusion is applicable to the parametric method.

11.3.2.3 Outliers

In the white paper (Shankar et al. 2008), it is mentioned that the Boxplot method can be used for outlier detection. However, it is not clear how to remove biological outliers and analytical outliers. Devanarayan and Tovey (2011) describe the procedure as follows: (1) for each sample calculate the mean value from all runs; (2) apply the Boxplot method to the sample means and remove the samples whose sample means are outside the Boxplot fences; (3) after removing all samples identified from (2), apply the Boxplot method for each run and remove values outside the fences. Samples identified in step (2) are claimed to be biological outliers and values identified in step (3) are analytical outliers. This approach is easy to implement. However, Zhang et al. (2013) show that the approach cannot identify outliers correctly. They consider the following hypothetical scenario. Suppose 48 donors are analyzed in four runs and there is no run-to-run variability (upper left panel in Fig. 11.5). Now assume that due to analytical variation, the signal value of donor 1 (lower left panel in Fig. 11.5) in run 2 is the same as that of donor 2 in runs 1, 3 and 4; and the signal value of donor 2 in run 2 is the same as that of donor 1 in runs 1, 3 and 4. As a result, the values of donor 1 and 2 in run 2 are considerably different from value of the same donor from other runs and could be considered as analytical outliers. Now follow the above procedure and apply the Boxplot method to sample means (upper right panel in Fig. 11.5). Two donors are identified as biological outliers and all values of these two donors from the four runs are removed. Next, apply the Boxplot method to each run of the remaining data (lower right panel in Fig. 11.5), it is clear that the two analytical outliers in run 2 are well within the fences and not flagged. The reason is that analytical outlier should be detected based on analytical variability. But in this procedure, when applying the Boxplot method to each run, analytical outlier detection is based on donor-to-donor variability. Zhang et al. (2013) argue that analytical outliers should be detected before biological outliers.

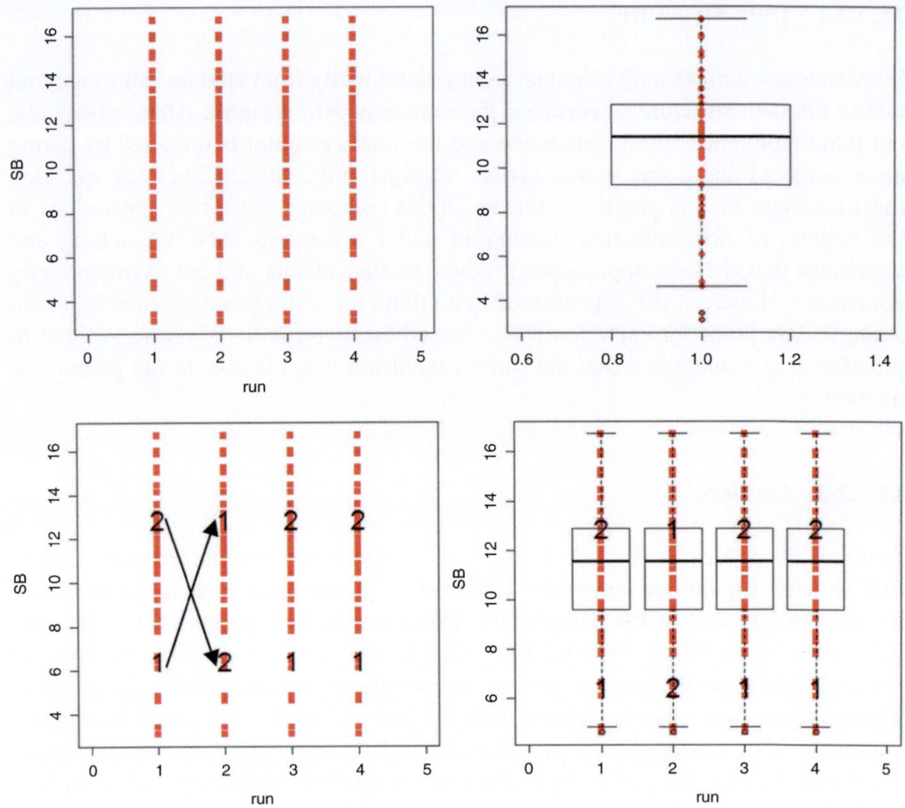

Fig. 11.5 Outlier removal. *Source*: Zhang et al. (2013)

11.3.2.4 Non-normal Distributions

Sometimes, data is not normally distributed even after transformation. The last resort would be to use the nonparametric method. Although the nonparametric method is convenient because it is distribution-free, it requires a moderately large sample size to achieve estimation efficiency compared to the parametric method. Motivated by this, Schlain et al. (2010) proposed a three-parameter gamma distribution to fit cut point data. The three-parameter gamma density function is given by the following equation:

$$f(y) = \frac{(y-\gamma)^{\alpha-1} \exp(-(y-\gamma)/\beta)}{\beta^{\alpha}\Gamma(\alpha)}, \ \alpha > 0, \ \beta > 0, \ y > \gamma$$

where α is the shape parameter, β is the scale parameter and γ is the threshold parameter. The distribution is unimodal and right-skewed as is the case with log-normal distributions. When the shape parameter α goes to infinity, the gamma distribution degenerates to a normal distribution. Maximum likelihood estimation (MLE) may be used to obtain parameter estimations through iteration since a close-form solution is not available. However, it is noted that when the shape parameter is close to 1, the MLE is unstable and thus not recommended. Given that the method of moment estimation is less accurate than MLE, Cohen and Whitten (1986) recommend using modified moment estimation (MME), especially when the shape parameter is small. In MME, the central third moment is replaced with the expectation of cumulative distribution function of the minimal order statistics, namely, $E[F(Y_{(1)})] = 1/(n+1)$. For extensive discussion of gamma distribution in general and its parameter estimation, see Johnson et al. (1994).

Simulation studies conducted by Schlain et al. (2010) show that if the data follow the gamma distribution, the method based on normal distribution can over-estimate the false positive rate by as much as 3 %. On the other hand, if the data follow log-normal distribution, normal-based results could under-estimate the false positive rate by as large as 2.3 % on average. The nonparametric method has much more variability than that of gamma distribution when the sample size is small. With a moderate sample size, the loss of efficiency of nonparametric method depends on the specific underlying true distribution. Therefore, impact on the false positive rate should be evaluated case-by-case.

11.3.2.5 Prediction Interval vs. Tolerance Interval

The quantile (or percentile) as a cut point makes intuitive sense in targeting, say 5 % false positive rate. However, in the normal distribution case, the estimator $\text{mean} + 1.645\sqrt{\text{total variability}}$ is neither an unbiased estimator of the true quantile nor does it give a 5 % false positive rate on average. Hoffman and Berger (2011) compare different cut point estimators including nonparametric quantile, parametric quantile, and upper prediction bound assuming the underlying model is a normal random effect model. A prediction bound or prediction interval is an estimate of a bound/interval in which future observations will fall with a certain probability. Their simulation studies show that the upper prediction bound best maintains the targeted false positive rate on average while the variation around the targeted false positive rate is minimal among the methods mentioned above. The parametric quantile fails to maintain the targeted false positive rate when the sample size is not large enough. For these reasons, Hoffman and Berger (2011), Zhong and Zhou (2014) recommend using prediction bound for cut point determination.

While the prediction bound maintains the targeted false positive rate in the average sense, there is a 50 % chance that it will give rise to a false positive rate being less than or more than the targeted false positive rate. Such deviation can sometimes be immense. Hoffman and Berger (2011) argue that a "confidence-level" cut point approach may be warranted to ensure that the chance of being less than the targeted false positive rate is small. This leads to the concept of tolerance bound. The tolerance bound is such that with some confidence level, a specified proportion of population samples will be within the bound. The one-sided tolerance bound is nothing but a one-sided confidence bound for the unknown quantile parameter. Consider independently and identically distributed (IID) data $X_1, X_2, \ldots X_n$ with unknown distribution. Then, the (1-β) level 1-α content one-sided confidence bound $L(X_1, X_2, \ldots X_n)$ is such that

$$\Pr\left[\Pr\left(X > L\left(X_1, X_2, \ldots X_n\right)\right) \geq 1 - \alpha\right] \geq 1 - \beta$$

The above formula can be re-written as $\Pr\left[q\left(\alpha\right) \geq L\left(X_1, X_2, \ldots X_n\right)\right] \geq 1 - \beta$, or $\Pr\left[FPR \geq \alpha\right] \geq 1 - \beta$. The former indicates that the lower tolerance bound is de facto the lower confidence bound for q(α), the (1-α)-th quantile. While the latter guarantees the false positive rate has $1 - \beta$ confidence of being greater than the targeted false positive rate α. In practice, the data are no longer IID (independently and identically distributed), but correlated. The tolerance bound for random effect models such as model (11.2) has been well studied in the literature. Krishnamoorthy and Mathew (2009) reviewed different approaches to constructing the tolerance bound for various models such as one-way and two-way random effect models. Shen et al. (2015) investigate the performance of tolerance bound in terms of maintaining targeted false positive rate.

11.3.2.6 Confirmatory cut point

Wakshull and Coleman (2011) propose that the confirmatory assay should be orthogonal to the screening assay, i.e., provide information that is independent of the screening results. Such orthogonality should manifest itself in lack of correlation between the screening and confirmatory results. If the screening and confirmatory results are highly correlated, the confirmatory assay becomes redundant since it provides limited new information beyond that obtained in the screening assay. Kubiak et al. (2013) demonstrate that a high degree of correlation between the two assays is expected due to the very experimental design where signal in absence of drug is measured essentially twice; first in the screening assay and then again in the confirmatory assay. Consequently, the confirmatory tier consisting of competitive inhibition cannot be considered as the definitive assay for positive/negative classification of immunogenicity samples but at best provides supplementary information to the screening assay.

A.1 Appendix

A.1.1 Generalized Least Squares Estimation for Nonlinear Regression Model with No-Constant Variance

The GLS algorithm is an iterative procedure. At the beginning, let $k = 0$ and set the least square estimator $\hat{\beta}_{LS}$ as $\hat{\beta}^{(0)}$. In general, for the k-th iteration, perform as follows:

1. Given $\hat{\beta}^{(k)}$, estimate θ and σ^2 by the method of pseudo-likelihood (PL). The pseudo-likelihood function is:

$$PL\left(\hat{\beta}^{(k)}, \theta, \sigma^2\right) = -N\log(\sigma) - \sum \log g\left(x_i, \hat{\beta}^{(k)}, \theta\right) - \frac{1}{2\sigma^2}\sum_i \frac{\left(y_{ij} - f\left(x_i, \hat{\beta}^{(k)}\right)\right)^2}{g^2\left(x_i, \hat{\beta}^{(k)}, \theta\right)}$$

   The variance parameters are estimated by minimizing the pseudo-likelihood using numerical method. The weights are thus: $\hat{w}_i = g^{-2}\left(x_i, \hat{\beta}^{(k)}, \hat{\theta}\right)$.
2. Use the estimated weights from step 1 to obtain an updated estimator of β by minimizing

$$\sum_i \hat{w}_i \left[y_{ij} - f\left(x_i, \hat{\beta}^{(k)}\right)\right]^2$$

   Denote the resultant estimator as $\hat{\beta}^{(k+1)}$.
   The procedure stops after certain steps when the parameter estimates converge.
   When there are multiple runs, estimate β same as in a single run.
The pseudo-likelihood, however, becomes:

$$PL\left(\theta, \sigma^2\right) = \sum PL\left(\hat{\beta}_i^{(k)}, \theta, \sigma^2\right).$$

The other part of the algorithm remains the same as in the single run.

References

Bates DM, Watts DG (1988) Nonlinear regression analysis and its applications. John Wiles & Sons, New York

Bland JM, Altman DG (1986) Statistical methods for assessing agreement between two methods of clinical measurement. Lancet 1:307–310

Bland JM, Altman DG (2007) Agreement between methods of measurement with multiple observations per individual. J Biopharm Stat 17(4):571–582

Booth B, Arnold ME, DeSilva B et al (2015) Workshop report: Crystal City V – quantitative bioanalytical method validation and implementation: the 2013 revised FDA guidance. AAPS J 17:277–288

Boulanger R, Devanaryan V, Dewe W, Smith W (2007) Statistical considerations in analytical method validation. In: Dmitrienko A, Huang-Stein C, D'agostino R (eds) Pharmaceutical statistics in SAS: a practical guide. SAS Institute, North Carolina

Boulanger R, Devanaryan V, Dewe W, Smith W (2010) Statistical considerations in the validation of ligand-binding assays. In: Khan MN, Findlay JW (eds). Ligand-binding assays: development, validation and implementation in the drug development arena. Wiley, New Jersey

Cohen AC, Whitten BJ (1986) Modified moment estimation for the three-parameter gamma distribution. J Qual Technol 18:53–62

Davidian M, Giltinan DM (1995) Nonlinear models for repeated measurement data. Chapman & London

DeSilva B, Bowsher R (2010) Validation of ligand-binding assays to support pharmacokinetic assessments of biotherapeutics. In: Khan M, Findlay J (eds) Ligand-binding assays, development, validation, and implementation in the drug development arena. Wiley, New Jersey

DeSilva B, Smith W, Weiner R et al (2003) Recommendations for the bioanalytical method validation of ligand-binding assays to support pharmacokinetic assessments of macromolecules. Pharm Res 20:1885–1900

Devanarayan V, Tovey M (2011) Cut point and performance characteristics for anti-drug antibody assays. In: Tovey MG (ed) Detection and quantification of antibodies to biopharmaceuticals: practical and applied considerations. Wiley, New York, pp 289–308

Dudley RA, Edwards P, Ekins RP, Finney DJ, McKenzie IG, Raab GM, Rodbard D, Rodgers RP (1985) Guidelines for immunoassay data processing. Clin Chem 31(8):1264–1271

Ekins RP, Edwards PR (1983) The precision profile: its use in assay design, assessment and quality control. In: Hunter WM, Corrie JE (eds) Immunoassays for clinical chemistry. Churchill Livingstone, New York

European Medicines Agency (EMA) (2007) Guideline on immunogenicity assessment of biotechnology-derived therapeutic proteins. EMEA/CHMP/BMWP/14327/2006

European Medicines Agency (EMA) (2011) Guideline on bioanalytical method validation. EMEA/CHMP/EWP/192217/2009

European Medicines Agency (EMA) (2012) Guideline on immunogenicity assessment of monoclonal antibodies intended for in vivo clinical use. EMA/CHMP/BMWP/86289/2010

FDA (2009) Draft guidance for industry: assay development for immunogenicity testing of therapeutic proteins. Food and Drug Administration, Silver Spring, Maryland

FDA (2014a) Guidance for industry: immunogenicity assessment for therapeutic protein products. FDA, US Department of Health and Human Services

FDA (2014b) Guidance for industry: analytical procedures and methods validation for drugs and biologics. Food and Drug Administration, Silver Spring, Maryland

Findlay JWA, Dillard RF (2007) Appropriate calibration curve fitting in ligand binding assays. AAPS J 9:E260–E267

Giltinan DM, Davidian M (1994) Assays for recombinant proteins: a problem in nonlinear calibration. Stat Med 13:1165–1179

Gupta S, Indelicato SR, Jethwa V et al (2007) Recommendations for the design, optimization and qualification of cell-based assays used for the detection of neutralizing antibody responses elicited to biological therapeutics. J Immunol Methods 321:1–18

Gupta S, Devanarayan V, Finco-Kent D et al (2011) Recommendations for the validation of cell-based assays used for the detection of neutralizing antibody immune responses elicited against biological therapeutics. J Pharm Biomed Anal 55:878–888

Hoffman D, Berger M (2011) Statistical considerations for calculation of immunogenicity screening assay cut points. J Immunol Methods 373:200–208

Hoffman D, Kringle R (2007) A total error approach for the validation of quantitative analytical methods. Pharm Res 24(6):1157–1164

Hutson AD (2003) Nonparametric estimation of normal ranges given one-way ANOVA random effects assumptions. Stat Probab Lett 64:415–424

Johnson NL, Kotz S, Balakrishnan N (1994) Continuous univariate distributions, 2nd edn. Wiley, New York

Kelley M, Desilva B (2007) Key elements of bioanalytical method validation for macromolecules. AAPS J 9:E156–E163

Kloks C, Berger C, Cortez P et al (2015) A fit-fir-purpose strategy for the risk-based immunogenicity testing of biotherapeutics: a European industry perspective. J Immunol Methods 417:1–9

Koren E, Smith HW, Shores E, Shankar G, Finco-Kent D, Rup B, Barrett YC, Devanarayan V, Gorovits B, Gupta S, Parish T, Quarmby V, Moxness M, Swanson SJ, Taniguchi G, Zuckerman LA, Stebbins CC, Mire-Sluis A (2008) Recommendations on-risk-based strategies for detection and characterization of antibodies against biotechnology products. J Immunol Methods 333:1–9

Krishnamoorthy K, Mathew T (2009) Statistical tolerance regions. Wiley, Hoboken

Kubiak R et al (2013) Correlation of Screening and confirmatory results in tiered immunogenicity testing by solution-phase bridging assays. J Pharm Biomed Anal 74:235–245

Lee JW, Devanarayan V, Barrett YC et al (2006) Fit-for-purpose method development and validation of successful biomarker measurement. Pharm Res 23:312–328

Levene H (1960) Robust tests for equality of variances. In: Olkin I (ed) Contributions to probability and statistics. Standford University Press, California

Lytle FE, Julian RK, Tabert AM (2009) Incurred sample reanalysis: enhancing the Bland-Altman approach with tolerance intervals. Bioanalysis 1(4):705–714

Miller KJ, Bowsher RR, Celniker A et al (2001) Workshop on bioanalytical methods validation for macromolecules: summary report. Pharm Res 18:1373–1383

Mire-Sluis AR, Barrett YC, Devanarayan V et al (2004) Recommendations for the design and optimization of immunoassays used in the detection of host antibodies against biotechnology products. J Immunol Methods 289:1–16

Neyer L et al (2006) Confirming human antibody responses to a therapeutic monoclonal antibody using a statistical approach. J Immunol Methods 315:80–87

O'Connell et al (1993) Calibration and assay development using the four-parameter logistic model. Chemom Intell Lab Syst 20:97–114

Rocci ML, Devanarayan V, Haughey DB, Jardieu P (2007) Confirmatory reanalysis of incurred bioanalytical samples. The AAPS journal 9(3):336–343

Rosenberg AS, Worobec A (2004) A risk-based approach to immunogenicity concerns of therapeutic protein products. Part 1: Considering consequences of the immune response to a protein. BioPharm Int 17:22–26

Rosenberg AS, Worobec A (2005a) A risk-based approach to immunogenicity concerns of therapeutic protein products. Part 2: Considering host-specific and product-specific factors impacting immunogenicity. BioPharm Int 17:34–42

Rosenberg AS, Worobec A (2005b) A risk-based approach to immunogenicity concerns of therapeutic protein products. Part 3: effects of manufacturing changes on immunogenicity studies. BioPharm Int 17:32–36

Sadler WA (2008) Imprecision profiling. Clin Biochem 29:S33–S36

Schlain B, Amaravadi L, Donley J et al (2010) A novel gamma-fitting statistical method for anti-drug antibody assays to establish assay cut points for data with non-normal distribution. J Immunol Methods 352:161–168

Seber GAF, Wild CJ (1989) Nonlinear regression. Wiley, New York

Shankar G, Devanarayan V, Amaravadi L et al (2008) Recommendations for the validation of immunoassays used for detection of host antibodies against biotechnology products. J Pharm Biomed Anal 48:1267–1281

Shen M, Dong X, Tsong Y (2015) Statistical evaluation of several methods for cut point determination of immunogenicity screening assay. J Biopharm Stat 25(2):269–279

Smith HW, Moxness M, Marsden R (2011) Summary of confirmation cut point discussions. AAPS J 13:227–229

USP chapter <1106> (2014) Immunogenicity assays-design and validation of immunoassays to detect anti-drug antibodies. United States Pharmacopeia, Version 37

Wakshull E, Coleman D (2011) Confirmatory immunogenicity assays. In: Tovey M (ed) Detection and quantification of antibodies to biopharmaceuticals: practical and applied considerations. Wiley, New York, pp 103–117

Zeng Q, Davidian M (1997) Calibration inference based on multiple runs of an immunoassay. Biometrics 53:1304–1317

Zhang L, Zhang J, Kubiak R, Yang H (2013) Statistical methods and tool for cut point analysis in immunogenicity assays. J Immunol Methods 389:79–87

Zhong Z, Zhou L (2014) Good practices for statistical estimation of cut point. In: Immunogenicity assay development, validation and implementation. Future Medicine, pp 51–60

Chapter 12
Recent Research Projects by the FDA's Pharmacology and Toxicology Statistics Team

Karl K. Lin, Matthew T. Jackson, Min Min, Mohammad Atiar Rahman, and Steven F. Thomson

Abstract In addition to regular review work, the Pharmacology and Toxicology Statistics Team in CDER/FDA is actively engaged in a number of research projects. In this chapter we summarize some of our recent investigations and findings.

We have conducted a simulation study (discussed in Sect. 12.2) to evaluate the increase in Type 2 error attributable to the adoption by some non-statistical scientists within the agency of more stringent decision criteria than those we have recommended for the determination of statistically significant carcinogenicity findings in long term rodent bioassays. In many cases, the probability of a Type 2 error is inflated by a factor of 1.5 or more.

A second simulation study (Sect. 12.3) has found that both the Type 1 and Type 2 error rates are highly sensitive to experimental design. In particular, designs using a dual vehicle control group are more powerful than designs using the same number of animals but a single vehicle control group, but this increase in power comes at the expense of a greatly inflated Type 1 error rate.

Since the column totals of the tables of permutations of animals to treatment groups cannot be presumed to be fixed, the exact methods used in the Cochran-Armitage test are not applicable to the poly-k test for trend. Section 12.4 presents simple examples showing all possible permutations of animals, and procedures for computing the probabilities of the individual permutations to obtain the exact p-values. Section 12.5 builds on this by proposing an exact ratio poly-k test method using samples of possible permutations of animals. The proposed ratio poly-k

---

The article reflects the views of the authors and should not be construed to represent FDA's views or policies.

K.K. Lin (✉) • M. Min • M.A. Rahman • S.F. Thomson
US Food and Drug Administration, Center for Drug Evaluation and Research, Office of Translational Sciences, Office of Biostatistics, Division of Biometrics 6, Silver Spring, MD, USA
e-mail: karl.lin@fda.hhs.gov

M.T. Jackson
Formerly of FDA/CDER/OTS/OB/DB6, Silver Spring, MD, USA

L. Zhang (ed.), *Nonclinical Statistics for Pharmaceutical and Biotechnology Industries*, Statistics for Biology and Health, DOI 10.1007/978-3-319-23558-5_12

test does not assume fixed column sums and uses the procedure in Bieler and Williams (Biometrics 49(3):793–801, 1993) to obtain the null variance estimate of the adjusted quantal tumor response estimate. Results of simulations show that the modified exact poly-3 method has similar sizes and levels of power compared to the method proposed in Mancuso et al. (Biometrics 58:403–412, 2002) that also uses samples of permutations but uses the binomial null variance estimate of the adjusted response rates and is based on the assumption of fixed column sums.

Bayesians attempt to model not only the statistical data generating process as in the frequentist statistics, but also to model knowledge about the parameters governing that process. Section 12.6 includes a short review of possible reasons for adopting a Bayesian approach, and examples of survival and carcinogenicity analyses.

Keywords Bayesian methods in nonclinical biostatistics • Carcinogenicity studies • Consumer's risk • Exact poly-3 trend tests • Experimental designs • Finite dimensional logistic model • Finite dimensional proportional • Hazards model • Multiplicity adjustment • Nonparametric Bayesian analysis • Permutational distribution • Producer's risk

12.1 Introduction

It is required by law that the sponsor of a new drug that is intended for chronic use by patients for certain indications conduct carcinogenicity studies in animals to assess the carcinogenic potential of the drug. These studies are reviewed independently within CDER/FDA. These independent reviews are conducted by interdisciplinary groups. The statistical component of such a review includes an assessment of the design and conduct of the study, and a complete reanalysis of all statistical data. This work is performed by members of the Pharmacology and Toxicology (Pharm/Tox) Statistics Team. However, the decision of whether the drug should be considered a potential carcinogen is based on more than just statistical evidence, and as such the statistical review comprises just one part of the FDA internal decision process.

In 2001, the FDA released a draft document entitled "Guidance for Industry; Statistical aspects of the design, analysis, and interpretation of chronic rodent carcinogenicity studies of pharmaceuticals" (US Food and Drug Administration—Center for Drug Evaluation and Research 2001). This guidance was issued in the Federal Register (Tuesday, May 8 2001, Vol. 66 No. 89) and was reviewed by the public over a 90 day comment period in 2001. Sixteen comments were received from drug companies, professional organizations of the pharmaceutical industry, and individual experts from the U.S., Europe, and Japan; these are available in FDA Docket No. 01D0194. Great efforts are being made within the FDA to finalize this document.

The draft guidance describes the general process and methods used by the Pharm/Tox team in their reviews. These methods are also used widely by drug companies in the U.S. and abroad. In addition, the team also draws on results from other published research.

In addition to writing reviews, the statistical Pharm/Tox team contributes to the development of FDA policy and conducts statistical research. Indeed, research published by members of the team (Lin 1998; Lin and Ali 2006; Lin and Rahman 1998; Lin et al. 2010; Rahman and Lin 2008, 2009, 2010; Rahman and Tiwari 2012) is often incorporated into the team's statistical reviews. More generally, in their efforts to keep abreast of new advancements in this area and to improve the quality of their reviews, members of the team have conducted regulatory research studies in collaborations with experts within and outside the agency.

The purposes of this chapter are twofold. One is to provide updates of results of some research studies that have been presented or published previously (Sects. 12.4 and 12.6). The other is to share with the professional community the results of some research studies that have not been presented or published (Sects. 12.2, 12.3 and 12.5).

In this chapter, results of five recent research projects by members of the Pharm/Tox Statistics Team are presented. Sections 12.2 and 12.3 describe two simulation studies investigating the effect on Type 1 and Type 2 error of varying the decision rules (Sect. 12.2) and experimental design (Sect. 12.3). Sections 12.4 and 12.5 discuss the development of exact methods for the poly-k test for a dose response relationship. Finally, Sect. 12.6 is a general discussion of the use of Bayesian methods in reviews of carcinogenicity studies.

12.2 An Evaluation of the Alternative Statistical Decision Rules for the Interpretation of Study Results

12.2.1 Introduction

It is specifically recommended in the draft guidance document (US Food and Drug Administration—Center for Drug Evaluation and Research 2001) that when evaluating the carcinogenic potential of a drug:

1. Trend tests, which have been extensively studied in the literature (Lin 1995, 1997, 1998, 2000a,b; Lin and Ali 1994, 2006; Lin and Rahman 1998; Rahman and Lin 2008, 2009), should be the primary tests used.
2. Pairwise tests may be used in lieu of trend tests, but only in those rare cases where they are deemed more appropriate than the trend tests.

The reason for preferring the trend test to the pairwise test is that, under most circumstances, the trend test will be more powerful than the pairwise test (for any given significance level). Note that the above guidance document recommends that only one test, either the trend test or the pairwise test, is to be used to conclude a statistically significant carcinogenic effect.

In the context of carcinogenicity studies (and safety studies in general), the Type 1 error rate is primarily a measure of the producer's risk; if a drug is withheld from market due to an incorrect finding of a carcinogenic effect (or even if its usage is merely curtailed), then it is the producer of the drug who faces the greatest loss. Conversely, the Type 2 error rate is primarily a measure of the consumer's risk, as it is the consumer who stands to suffer in the event that a truly carcinogenic drug is brought to market without the carcinogenic effect being reported. Proceeding from this philosophical stance (and the similar position of Center for Drug Evaluation and Research 2005), the draft guidance (US Food and Drug Administration—Center for Drug Evaluation and Research 2001) recommends a goal of maximizing power while keeping the overall (study-wise) false positive rate at approximately 10 %. In order to achieve this goal, a collection of significance thresholds are recommended. These thresholds, presented in Table 12.1, are grounded in Lin and Rahman (1998) and Rahman and Lin (2008) (for the trend test) and Haseman (1983, 1984) (for the pairwise test).

However, this goal has not been universally accepted. There is a desire on the part of some non-statistical scientists within the agency to restrict positive findings to those where there is statistical evidence of *both* a positive dose response relationship *and* an increased incidence in the high dose group compared to the control group. In other words, a joint test is desired. This is not an intrinsically unreasonable position. Nonetheless, every test needs significance thresholds, and since the only significance thresholds included in US Food and Drug Administration—Center for Drug Evaluation and Research (2001) are for single tests, it is natural (but incorrect!) for non-statistical scientists to construct a joint test using these thresholds. We will refer to this decision rule as the *joint test* rule. See Table 12.2.

We are very concerned about the ramifications of the use of this rule. While the trend and pairwise test are clearly not independent, their association is far from perfect. Accordingly, the requirement that both tests yield individually statistically significant results necessarily results in a more conservative test than either the trend test or the pairwise test alone (at the same significance thresholds). The purpose of this section is to present the results of our simulation study showing a serious consequence of the adoption of this rule: a huge inflation of the false negative rate (i.e., the consumer's risk) for the final interpretation of the carcinogenicity potential of a new drug.

12.2.2 Design of Simulation Study

The objective of this study is to conduct a simulation study to evaluate the inflation of the false negative rate resulting from the joint test (compared with the trend test alone).

We modeled survival and tumor data using Weibull distributions (see Eqs. (12.1) and (12.2)). The values of the parameters A, B, C, and D, were taken from the landmark National Toxicology Program (NTP) study by Dinse (1985) (see

Table 12.1 Recommended significance levels for the trend test or the pairwise comparisons (US Food and Drug Administration—Center for Drug Evaluation and Research 2001: Lines 1093–1094 on page 30)

	Tests for positive trend	Control-high pairwise comparisons (one-tailed)
Standard 2-year studies with 2 species and 2 sexes	Common and rare tumors are tested at 0.005 and 0.025 significance levels, respectively	Common and rare tumors are tested at 0.01 and 0.05 significance levels, respectively
Alternative ICH studies (one 2-year study in one species and one short- or medium- term study, two sexes)	Common and rare tumors are tested at 0.01 and 0.05 significance levels respectively	Under development and not yet available

Note: Following Haseman (1983), a tumor is classified as a rare tumor if it has a background rate of less than 1 %, and is classified as a common tumor otherwise

Table 12.2 The joint test rule (not recommended!)

	Trend test	Pairwise test
Rare tumor	0.025	0.050
Common tumor	0.005	0.010

Note: The joint test is positive for a particular tumor endpoint only if both the pairwise test between the high dose and control groups and the dose response (trend) test are significant at the levels indicated above

Tables 12.3 and 12.4). Values of these parameters were chosen to vary four different factors, ultimately resulting in 36 different sets of simulation conditions.

The factors used in the NTP study were defined as follows:

1. Low or high tumor background rate: The prevalence rate at 2 years in the control group is 5 % (low) or 20 % (high).
2. Tumors appear early or late: The prevalence rate of the control group at 1.5 years is 50 % (appearing early) or 10 % (appearing late) of the prevalence rate at 2 years.
3. No dose effect, a small dose effect, or a large dose effect on tumor prevalence: The prevalence of the high dose group at 2 years minus the prevalence of the control group at 2 years is 0 % (no effect), or 10 % (small effect), or 20 % (large effect).
4. No dose effect, a small dose effect, or a large dose effect on mortality: The expected proportion of animals alive in the high dose group at 2 years is 70 % (no effect), 40 % (small effect), or 10 % (large effect). The expected proportion of animals alive in the control group at 2 years is taken as 70 %.

However, there are important differences between the NTP design described above and the design used in our simulation study. Whereas the NTP study simulated three treatment groups with doses $x = 0$, $x = 1$, and $x = 2$ (called the *control*, *low*, and *high* dose groups), our study used four treatment groups (with doses $x = 0$,

Table 12.3 Data generation parameters for the Weibull models for time to tumor onset (Dinse 1985)

	Model description				Weibull parameters			
Simulation conditions	Background tumor rate		Tumor appearance[a]	Dose effect[b,c]	*A*	*B*	*C*	*D*
1, 13, 25	Low	0.05	Early	None	17	2	6.78×10^{-6}	0
2, 14, 26	Low	0.05	Early	Small	17	2	6.78×10^{-6}	7.36×10^{-6}
3, 15, 27	Low	0.05	Early	Large	17	2	6.78×10^{-6}	1.561×10^{-5}
4, 16, 28	Low	0.05	Late	None	56	3	4.65×10^{-7}	0
5, 17, 29	Low	0.05	Late	Small	56	3	4.65×10^{-7}	5.025×10^{-7}
6, 18, 30	Low	0.05	Late	Large	56	3	4.65×10^{-7}	1.0675×10^{-6}
7, 19, 31	High	0.20	Early	None	21	2	3.24×10^{-5}	0
8, 20, 32	High	0.20	Early	Small	21	2	3.24×10^{-5}	9.7×10^{-6}
9, 21, 33	High	0.20	Early	Large	21	2	3.24×10^{-5}	2.09×10^{-5}
10, 22, 34	High	0.20	Late	None	57	3	2.15×10^{-6}	0
11, 23, 35	High	0.20	Late	Small	57	3	2.15×10^{-6}	6.45×10^{-7}
12, 24, 36	High	0.20	Late	Large	57	3	2.15×10^{-6}	1.383×10^{-6}

Notes on factors used in the simulation by Dinse (1985):
[a]Tumors appear early or late: The prevalence rate of the control group at 1.5 years is 50 % (appearing early) or 10 % (appearing late) of the prevalence rate at 2 years
[b]No effect, a small effect, or a large effect on tumor prevalence: The prevalence of the high dose group ($x = 2$) at 2 years minus the prevalence of the control group at 2 years is 0 % (none effect), or 10 % (small effect), or 20 % (large effect)
[c]It is also be noted that for our study, the percentage differences corresponding to those in note b are 0, 15, and 28 % (for the high dose group with $x = 3$)

Table 12.4 Data generation parameters for the Weibull models for time to death (Dinse 1985)

		Weibull parameters			
Simulation conditions	Drug effect on death[a,b]	*A*	*B*	*C*	*D*
1–12	None	0	4	3.05×10^{-9}	0
13–24	Small	0	4	3.05×10^{-9}	2.390×10^{-9}
25–36	Large	0	4	3.05×10^{-9}	8.325×10^{-9}

Notes on factors used in the simulation by Dinse (1985):
[a]No effect, a small effect, or a large effect on mortality: The expected proportion of animals alive in the high dose group ($x = 2$) at 2 years is 70 % (none), 40 % (smalleffect), or 10 % (large effect). The expected proportion of animals alive in the control group at 2 years is taken as 70 %
[b]It is also be noted that for our study, the survival probabilities corresponding to those in note d are 70, 30, and 4 % (for the high dose group with $x = 3$)

$x = 1$, $x = 2$, and $x = 3$, called the *control*, *low*, *mid*, and *high* dose groups respectively). Since the values of the parameters *A*, *B*, *C*, and *D* used were the same in the two studies (see Tables 12.3 and 12.4), the characterizations of the effect of the dose level on tumorigenesis and mortality, factors 3 and 4, apply to the dose level $x = 2$, i.e., to the mid dose level. To recast these descriptions in terms of the effect at the $x = 3$ (high dose) level, factors 3 and 4 become factors 3′ and 4′:

3′ No dose effect, a small dose effect, or a large dose effect on tumor prevalence: The prevalence of the high dose group at 2 years minus the prevalence of the control group at 2 years is 0 % (no effect), or approximately 15 % (small effect), or approximately 28 % (large effect).
4′ No dose effect, a small dose effect, or a large dose effect on mortality: The expected proportion of animals alive in the high dose group at 2 years is 70 % (no effect), 30 % (small effect), or 4 % (large effect). The expected proportion of animals alive in the control group at 2 years is taken as 70 %.

These differences can be expected to have the following effects on the Type 2 error rates for our study (relative to the NTP study):

- The higher tumorigenesis rates in the high dose groups should help to reduce the false negative rates (or to increase the levels of power) of statistical tests.
- On the other hand, higher levels of mortality will reduce the effective sample size and thus tend to increase the false negative rates (or to decrease the levels of power).[1]

In our study, tumor data were generated for 4 treatment groups with equally spaced increasing doses (i.e., $x = 0$, $x = 1$, $x = 2$, and $x = 3$). There were 50 animals per group. The study duration was 2 years (104 weeks), and all animals surviving after 104 weeks were terminally sacrificed. All tumors were assumed to be incidental.

The tumor detection time (T_0) (measured in weeks) and the time to natural death (T_1) of an animal receiving dose level x were modeled by four parameter Weibull distributions:

$$S(t,x) = P[T_i > t|X = x] = \begin{cases} e^{-(C+Dx)(t-A)^B} & \text{if} t > A \\ 1 & \text{if} t \leq A \end{cases} \tag{12.1}$$

where A is the location parameter, B is the shape parameter, C is the baseline scale parameter, and D is the dose effect parameter. Tables 12.3 and 12.4 list the sets of values for these parameters used in Dinse (1985).

The prevalence function for incidental tumors equals the cumulative function of time to tumor onset, i.e.,

$$P(t|x) = \Pr[T_0 \leq t|X = x] = 1 - S(t,x). \tag{12.2}$$

Each of the 36 simulation conditions described in Tables 12.3 and 12.4 was simulated 10,000 times. For each simulation, 200 animals were generated; each animal was assigned to a dose group (50 animals per group) and had a tumor

[1] Although this tendency is not absolute. See the discussion in footnote 8 on page 313.

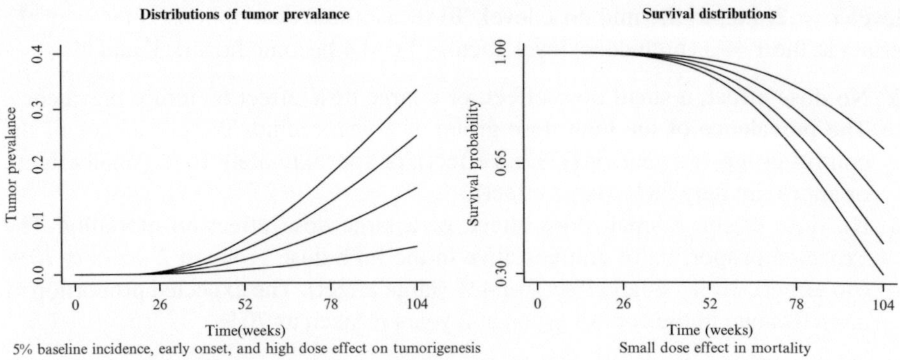

Fig. 12.1 Sample tumor prevalence and mortality curves

onset time (T_0) and death time (T_1) simulated using Eq. (12.1). The actual time of death (T) for each animal was defined as the minimum of T_1 and 104 weeks, i.e., $T = \min\{T_1, 104\}$. The animal developed the tumor (i.e., became a tumor bearing animal (TBA)) only if the time to tumor onset did not exceed the time to death. The actual tumor detection time was assumed to be the time of death T. Animals in the same dose group were equally likely to develop the tumor in their life times. It was assumed that tumors were developed independently of each other. The first panel of Fig. 12.1 graphically represents the Weibull models used to generate the tumor prevalence data when the background tumor rate is low, the dose effect on tumor prevalence is large, and the tumor appears early (the model used in simulation conditions 3, 15, and 27). The second panel graphically represents the Weibull models used to generate the survival data when the dose effect on mortality is small (simulation conditions 13–24). The age-adjusted Peto method to test for a dose response relationship (Peto et al. 1980) and the age adjusted Fisher exact test for pairwise differences in tumor incidence, (in each case using the NTP partition of time intervals[2]), were applied to calculate p-values.

Three rules for determining if a test of the drug effect on development of a given tumor type was statistically significant were applied to the simulated data. They were:

1. Requiring a statistically significant result in the trend test alone. This is the rule recommended in US Food and Drug Administration—Center for Drug Evaluation and Research (2001).
2. Requiring statistically significant results both in the trend test and in any of the three pairwise comparison tests (control versus low, control versus medium, control versus high).

[2]The NTP partition divides the 104 week study into the following subintervals: 0–52 weeks, 53–78 weeks, 79–92 weeks, 93–104 weeks, and terminal sacrifice.

3. Requiring statistically significant results both in the trend test and in the control versus high group pairwise comparison test. This is the joint test rule.

In each case, it was assumed that the tests were being conducted as part of a standard two-species study. The rules for rare tumor types were used when the incidence rate in the control group were below 1%; otherwise the rules for common tumors types were used.

After simulating and analyzing tumor data 10,000 times for each of the 36 sets of simulation conditions, the Type 1 and Type 2 error rates were estimated.

12.2.3 *Results of the Simulation Study*

Since we are simultaneously considering both models where the null hypothesis is true (so that there is no genuine dose effect on tumor incidence) and models where it is false (where there is a genuine dose effect), we need terminology that can apply equally well to both of these cases. For any given set of simulation conditions, the *retention rate* is the probability of retaining the null hypothesis. If the null hypothesis is true, then this rate is the probability of a *true negative*, and is 1 − thefalsepositiverate (Type I error). If the null hypothesis is false, then the retention rate is the probability of a *false negative* or Type 2 error. In this case, it is 1 − power. Correspondingly, the *rejection rate* is 1 − theretentionrate, and is the probability that the null hypothesis is rejected. It is either the false positive rate (if the null hypothesis is true) or the level of power (if the alternative hypothesis is true). The results (retention rates and percent changes of retention rates) of the simulation study are presented in Table 12.5.

Results of the evaluation of Type 1 error patterns in the study conducted and reported in Dinse (1985) show that the Peto test without continuity correction and with the partition of time intervals of the study duration proposed by NTP (see Footnote 2) yields attained false positive rates close to the nominal levels (0.05 and 0.01) used in the test. That means that the test is a good one that is neither conservative nor anti-conservative.

The evaluation of Type 1 error patterns found by this simulation study is done by using the rates at which the null hypothesis was rejected under those simulation conditions for which there was no dose effect on tumor rate (simulation conditions 1, 4, 7, 10, 13, 16, 19, 22, 25, 28, 31, and 34 in Table 12.5). The tumor types were classified as rare or common based on the incidence rate of the concurrent control. The results of this simulation study show a very interesting pattern in levels of attained Type 1 error. The attained levels of Type 1 error under various simulation conditions divided into two groups. The division was by the factor of background rate, either 20 or 5 %. The attained Type1 levels of the first group were around 0.005. The attained Type 1 error rates for the second group were around 0.015. The observed results and pattern of the attained Type 1 errors make sense. For the simulated conditions with 20 % background rate, probably almost all of the 10,000

Table 12.5 Estimated retention rates under three decision rules

Simulation condition	Simulation condition properties				Retention probabilities			% Change in retention rate	
	Dose effect on mortality	Tumor appearance time	Dose effect on tumor prevalence	Tumor background rate	Trend test only	Trend test and H/C	Trend test and any pairwise	Trend test and H/C	Trend test and any pairwise
1	No	Early	No	0.05	0.984	0.9934	0.9919	0.9553	0.8028
2	No	Early	Small	0.05	0.6283	0.7084	0.6957	12.75	10.73
3	No	Early	Large	0.05	0.1313	0.178	0.1595	35.57	21.48
4	No	Late	No	0.05	0.9827	0.9927	0.9915	1.018	0.8955
5	No	Late	Small	0.05	0.6314	0.7208	0.7076	14.16	12.07
6	No	Late	Large	0.05	0.1408	0.2018	0.1811	43.32	28.62
7	No	Early	No	0.2	0.9953	0.9979	0.9974	0.2612	0.211
8	No	Early	Small	0.2	0.8377	0.8805	0.8715	5.109	4.035
9	No	Early	Large	0.2	0.3424	0.427	0.398	24.71	16.24
10	No	Late	No	0.2	0.9952	0.9972	0.9972	0.201	0.201
11	No	Late	Small	0.2	0.8399	0.8869	0.8772	5.596	4.441
12	No	Late	Large	0.2	0.3754	0.4864	0.4565	29.57	21.6
13	Small	Early	No	0.05	0.9855	0.9985	0.9978	1.319	1.248
14	Small	Early	Small	0.05	0.6967	0.8465	0.8324	21.5	19.48
15	Small	Early	Large	0.05	0.2152	0.4112	0.3574	91.08	66.08
16	Small	Late	No	0.05	0.9819	0.9991	0.9977	1.752	1.609
17	Small	Late	Small	0.05	0.722	0.9161	0.8903	26.88	23.31
18	Small	Late	Large	0.05	0.2682	0.6794	0.6021	153.3	124.5
19	Small	Early	No	0.2	0.9948	0.9996	0.9995	0.4825	0.4725
20	Small	Early	Small	0.2	0.8753	0.9694	0.9606	10.75	9.745

21	Small	Early	Large	0.2	0.4649	0.7564	0.711	62.7	52.94
22	Small	Late	No	0.2	0.9961	0.9999	0.9996	0.3815	0.3514
23	Small	Late	Small	0.2	0.8935	0.9939	0.9885	11.24	10.63
24	Small	Late	Large	0.2	0.538	0.9455	0.9095	75.74	69.05
25	Large	Early	No	0.05	0.9856	0.9994	0.9989	1.4	1.349
26	Large	Early	Small	0.05	0.8381	0.9587	0.948	14.39	13.11
27	Large	Early	Large	0.05	0.5358	0.8133	0.7796	51.79	45.5
28	Large	Late	No	0.05	0.9828	1	1	1.75	1.75
29	Large	Late	Small	0.05	0.8675	0.996	0.9886	14.81	13.96
30	Large	Late	Large	0.05	0.6447	0.9807	0.9428	52.12	46.24
31	Large	Early	No	0.2	0.994	1	1	0.6036	0.6036
32	Large	Early	Small	0.2	0.9414	0.9994	0.9985	6.161	6.065
33	Large	Early	Large	0.2	0.7445	0.9823	0.97	31.94	30.29
34	Large	Late	No	0.2	0.9956	1	1	0.4419	0.4419
35	Large	Late	Small	0.2	0.9585	1	0.9999	4.33	4.319
36	Large	Late	Large	0.2	0.835	0.9998	0.9989	19.74	19.63

Note: The estimated retention rates of the simulation conditions where the null hypothesis is true, (conditions 1, 4, 7, 10, 13, 16, 19, 22, 25, 28, 31, and 34), are the probabilities of not committing a Type 1 error. For the remaining simulation conditions, the rates are the Type 2 error rates generated datasets (each dataset containing tumor and survival data of four treatment groups of 50 animals each group) will have a tumor rate of equal to or greater than 1 % (the definition of a common tumor) in the control group. The attained levels of Type 1 error rates under various simulated conditions in this group are close to the nominal significance levels for common tumors.[3]

The attained Type 1 error rates for the other group were between the nominal levels of significance of 0.005 (for the trend test for common tumors) and 0.025 (for the trend test for rare tumors) and not around 0.005. The reason for this phenomenon is that, though the background rate in the simulated conditions for this group was 5 % that is considered as a rate for a common tumor, some of the 10,000 generated datasets had tumor rates less than 1 % in the control group. For this subset of the 10,000 datasets, the nominal level of 0.025 was used in the trend test. See Sect. 12.3.4.3 for a more detailed discussion of this factor.

As mentioned previously, the main objective of our study is the evaluation of the Type 2 error rate under various conditions. As was expected, the Type 2 error (or false negative) rates resulting from the joint test decision rule are higher than those from the procedure recommended in the guidance document of using trend test alone. This is due to the fact that in statistical theory the false positive rate (measuring the producer's risk in the regulatory review of toxicology studies) and the false negative rate (measuring the consumer's risk) run in the opposite direction; use of the joint test decision rule will cut down the former rate only at the expense of inflating the latter rate.

The estimated false negative rates resulting from the extensive simulation study under the three decision rules listed in Sect. 12.2.2 are shown in Table 12.5. The last two columns of the table show the percentage changes in the retention rates of decision rules (2) and (3) respectively, compared to those of (1). For those simulation conditions where the null hypothesis is true (1, 4, 7, 10, 13, 16, 19, 22, 25, 28, 31, and 34), these values measure the percentage change in the probability of not committing a Type 1 error. For the remaining simulation conditions, these values measure the inflation of the Type 2 error rate attributable to the adoption of the more stringent rules.

The magnitude of the inflation of false negative rate resulting from the joint test decision rule of requiring statistically significant results in both the trend test and the C-H (High versus Control) pairwise comparison test depends on all the four factors,

[3] It becomes more complicated in the evaluation of conservativeness or anti-conservativeness of the joint test (simultaneous combination of the trend test and the pairwise comparison test) under the agency practice. This is so because the two tests are not independent since the pairwise comparison tests used a subset (a half) of the data used in the trend test. Theoretically, if the trend test and the pairwise comparison test are actually independent and are tested at 0.005 and 0.01 levels of significance, respectively for the effect of a common tumor type, then the nominal level of significance of the joint tests should be $0.005 \times 0.010 = 0.00005$. Some of the levels of attained Type I error of the joint tests are larger than 0.00005 due to the dependence of the two tests that were applied simultaneously. To evaluate this nominal rate directly would require estimation of the association between the two tests. Results of the simulation study, as expected, show that the attained levels of type I error (1-retention probability under the simulation conditions in which there is no drug effect on tumor prevalence) are smaller than those of the trend test alone.

namely, drug effect on mortality, background tumor rate, time of tumor appearance, and drug effect on tumor incidence considered in the simulation that are also listed in the notes at the bottom of Tables 12.3 and 12.4.

12.2.4 Discussion

Results of the simulation study show that the factor of the effect of the dose on tumor prevalence rates has the largest impact on the inflation of the false negative rate when both the trend test and the C-H pairwise comparison tests are required to be statistically significant simultaneously in order to conclude that the effect is statistically significant. The inflations are most serious in the situations in which the dose has a large effect on tumor prevalence. The inflation can be as high as 153.3 %. The actual Type 2 error rate can be more than double.

The above finding is the most alarming result among those from our simulation study. When the dose of a new test drug has large effects on tumor prevalence (up to 28 % difference in incidence rates between the high dose and the control groups), it is a clear indication that the drug is carcinogenic. Exactly in these most important situations the joint test decision rule causes the most serious inflations of the false negative error rate (or the most serious reductions in statistical power to detect the true carcinogenic effect). The net result of this alarming finding is that using the levels of significance recommended for the trend test alone and the pairwise test alone in the joint test decision rule can multiply the probability of failure to detect a true carcinogenic effect by a factor up to two or even more, compared with the procedure based on the result of the trend test alone.

It is true that the results in Table 12.5 show that, for the situations in which the dose has a small effect (up to 15 % difference in incidence rates between the high dose and the control groups) on tumor prevalence, the increases of false negative rates caused by the joint test decision rule are not much more than those from using the trend test alone (increases can be up to 27 %). However, this observation does not imply that the joint test decision rule is justified. The reason is that standard carcinogenicity studies use small group sizes as a surrogate for a large population with low tumor incidence. There is very little power (i.e., the false negative rates are close to 100 %) for a statistical test using a small level of significance such as 0.005 to detect any true carcinogenicity effect because of low tumor incidence rates. In those situations there will be little room for the further increase in the false negative rate no matter how many additional tests are put on top of the original trend test.

It might be argued that the large inflations in false negative rates in the use of the joint test over the use of the trend test alone could be due to the large dose effect on death (only 4 % animals alive at 2 years introduced by the additional treatment group with $x = 3$. The argument may sound valid since as mentioned previously, the decrease of percentage of animals alive at 2 years increases the false negative rates. We are aware of the small number of alive animals at 2 years in the simulation condition of large dose effect on death caused by using the built-in Weibull model described in Dinse (1985) and by including the additional group with $x = 3$ in the our study.

However, as mentioned previously, our main interest in this study is to evaluate the percentages of inflation in false negative rates attributable to the use of the joint test compared with the trend test alone. The false negative rates in the joint test and in the trend test alone are certainly also of our interest. They are not the main interest. So the issue of excessive mortality under the simulation condition of a large dose effect on death should not be a major issue in this study since it has similar impacts on false negative rates in both the joint test and the trend test alone. Furthermore, it is seen from Table 12.5 that the largest inflations (63–153 %) in the false negative rate happened under the conditions in which the dose effect on death is small (30 % alive animals at 2 years) rather than under the condition in which the effect is large (4 % alive animals at 2 years).

The extremely large false negative rates in the above simulated situations caused by the nature (low cancer rates and small group sample sizes) of a carcinogenicity experiment, reinforce the important arguments that it is necessary to allow an overall (for a compound across studies in two species and two sexes) false positive rate of about 10 % to raise the power (or to reduce the false negative rate) of an individual statistical test. This important finding of the simulation study clearly supports our big concern about failing to detect carcinogenic effects in the use of the joint test decision rule in determining the statistical significance of the carcinogenicity of a new drug. Again, the producer's risk using a trend alone is known at the level of significance used (0.5 % for a common tumor and 2.5 % for a rare tumor in a two-species study) and is small in relation to the consumer's risk that can be 100 or 200 times the level of the known producer's risk. The levels of significance recommended in the guidance for industry document were developed with the consideration of those situations in which the carcinogenicity experiment has great limitations. Trying to cut down only the producer's risk (false positive rate in toxicology studies) beyond that which safeguards against the huge consumer's risk (false negative rates in toxicology studies) is not consistent with the FDA mission as a regulatory agency that has the duty to protect the health and well-being of the American general public.

As mentioned previously, the decision rules (levels of significance) recommended in US Food and Drug Administration—Center for Drug Evaluation and Research (2001) are for trend tests alone and for pairwise comparisons alone, and not for the joint test. To meet the desire of some non-statistical scientists within the agency to require statistically significant results for both the trend test and the C-H pairwise comparison simultaneously to conclude that the effect on the development of a given tumor/organ combination as statistically significant, and still to consider the special nature of standard carcinogenicity studies (i.e., using small group sizes as a surrogate of a large population with low a tumor incidence endpoint), we have conducted additional studies and proposed new sets of significance levels for a joint test along with some updates of the previously recommended ones. These are presented in Table 12.6. We have found that the use of these new levels keeps the overall false positive rate (for the joint test) to approximately 10 % again for a compound across studies in two species and two sexes.

Table 12.6 Recommended decision rules (levels of significance) for controlling the overall false positive rates for various statistical tests performed and submission types

Submission type		Tumor type	Decision rule: Trend test alone	Decision rule: Pairwise test alone	Decision rule: Joint test: Trend test	Decision rule: Joint test: Pairwise test
Standard 2 year study with two sexes and two species		Common	0.005	0.01	0.005	0.05
		Rare	0.025	0.05	0.025	0.10
Alternative ICH Studies (One 2-year study in one species and one short- or medium-term alternative study, two sexes)	Two-year study	Common	0.005	0.01	0.005	0.05
		Rare	0.025	0.05	0.025	0.10
	Short- or medium-term alternative study	Common	0.05	0.05	0.05	0.05
		Rare	0.05	0.05	0.05	0.05
Standard 2 year studies with two sexes and one species		Common	0.01	0.025	0.01	0.05
		Rare	0.05	0.10	0.05	0.10

12.3 The Relationship Between Experimental Design and Error Rates

In this section, we describe the results of a second simulation study. The aim of the simulation study discussed in Sect. 12.2 was to compare decision rules, evaluating the impact on error rates of the adoption of the joint test rule (which is more conservative than the trend test rule recommended in US Food and Drug Administration—Center for Drug Evaluation and Research (2001)—see Tables 12.1 and 12.2). By contrast, this second study, which was conducted independently, compares the effects of the use of different experimental designs on the error rates, all under the same decision rule. The decision rule used in this study is the joint test rule (Table 12.2) since, despite the absence of any theoretical justification for its use, this rule is currently used by non-statistical scientists within the agency as the basis for labeling and other regulatory decisions.[4]

We first consider the nature of the various hypotheses under consideration, and the associated error rates (Sect. 12.3.1). This provides us with the terminology to express our motivation (Sect. 12.3.2). We then describe in detail the four designs that have been compared (Sect. 12.3.3), and the simulation models used to test these designs (Sect. 12.3.4). The results of the simulation are discussed in Sects. 12.3.5 (power) and 12.3.6 (Type 1 error). We conclude with a brief discussion (Sect. 12.3.7).

12.3.1 Endpoints, Hypotheses, and Error Rates

The task of analyzing data from long term rodent bioassays is complicated by a severe multiplicity problem. But it is not quite the case that we are merely faced with a multitude of equally important tumor endpoints. Rather, we are faced with a hierarchy of hypotheses.

- At the lowest level, we have individual tumor types, and some selected tumor combinations, associated with null hypotheses of the form:

  > Administration of the test article is not associated with an increase in the incidence rate of *malignant astrocytomas of the brain* in *female rats*.

  We call such hypotheses the *local null hypotheses*.
- The next level of the hierarchy is the *experiment* level. A standard study includes four experiments: on male mice, on female mice, on male rats, and on female rats. Each of these experiments is analyzed independently, leading to four *global null hypotheses* of the form:

[4] See, for instance, the discussion of osteosarcomas and osteomas in female mice in Center for Drug Evaluation and Research (2103).

> There is no organ–tumor pair, or reasonable combination of organ-tumor pairs, for which administration of the test article is positively associated with tumorigenesis in *male mice*.

Note that some studies consist of just two experiments in a single species (or, very rarely, in two species and a single sex).

- The highest level of the hierarchy of hypotheses is the *study* level. There is a single study-wise null hypothesis:

> For none of the experiments conducted is the corresponding global null hypothesis false.

For any given local null hypothesis, the probability of rejecting that null hypothesis is called either the *local false positive rate* (LFPR) or the *local power*, depending on whether the null hypothesis is in fact true. If all the local null hypotheses in a given experiment are true, then the *global false positive rate* (GFPR) for that experiment is the probability of rejecting the global null hypothesis, and can be estimated from the various estimates for the LFPRs for the endpoints under consideration.[5] The goal of the multiplicity adjustments in US Food and Drug Administration—Center for Drug Evaluation and Research (2001) is to maintain the study-wise false positive rate at about 10 %. Since most studies consist of four independent experiments, we consider our target level for false positives to be a GFPR of approximately 2.5 %.[6]

The calculation of a GFPR from the LFPR depends on the relationship between the local and global null hypotheses. We capture this relationship with the notion of a tumor *spectrum*: If $\mathcal{T}$ is the parameter space for tumor types, then a spectrum is a function $S : \mathcal{T} \rightarrow \mathbb{N}$; $S(t)$ is the number of independent tumor types being tested with parameter value t. In our case, $\mathcal{T}$ is one dimensional: under the global null hypothesis we assume that each tumor can be characterized by its background prevalence rate.[7]

In our simulations, we generate estimates for the power and LFPR for three different classes of tumor:

1. *Rare* tumors have a background prevalence rate (i.e., the lifetime incidence rate among those animals who do not die from causes unrelated to the particular tumor type before the end of the study, typically at 104 weeks) of 0.5 %.

---

[5]This calculation assumes that all endpoints are independent. This assumption is not strictly true, especially when considering combinations of endpoints. However, it is reasonable to assume that the endpoints are close enough to being independent that the resulting estimate of the GFPR is accurate enough for our purposes.

[6]In fact, if four independent experiments are to have a combined study-wise false positive rate of 10 %, then it suffices for them individually to have GFPRs of $1 - (1 - 0.1)^{1/4} = 0.026$. However, since it is not practical to calibrate the GFPR so precisely, there is no practical distinction between target GFPRs of 0.025 and 0.026.

[7]More sophisticated models might treat $\mathcal{T}$ as higher dimensional. For example, in the simulation study in Sect. 12.2, the parameter space $\mathcal{T}$ is two dimensional, with the two dimensions representing the background prevalence rate and the tumor onset time. (Although Eq. (12.1) has *three* independent parameters (not counting the dose response parameter D), the parameters A and B are not varied independently—see Table 12.3.)

2. *Common* tumors have a background prevalence rate of 2 %.
3. *Very common* tumors have a background prevalence rate of 10 %.

A tumor spectrum for us therefore consists of a triple $\langle n_1, n_2, n_3 \rangle$, indicating that the global null hypothesis is the conjunction of $n_1 + n_2 + n_3$ local null hypotheses, and asserts the absence of a treatment effect on tumorigenicity for n_1 rare, n_2 common, and n_3 very common independent tumor endpoints.

Given such a spectrum, and under any given set of conditions, the GFPR is easy to calculate from the LFPR estimates for the three tumor types under those conditions:

$$\mathrm{GFPR} = 1 - \prod_{i=1}^{3} (1 - F_i)^{n_i} \tag{12.3}$$

where F_i is the estimated LFPR for the i-th class of tumors. Since our desired false positive rates are phrased in terms of the study-wise false positive rate (which we want to keep to a level of approximately 10 %), we are more concerned with the GFPR than the LFPR.

Global power is slightly harder to calculate, since it is a function of a specific global alternate hypothesis. It is unclear what a realistic global alternative hypothesis might look like, except that a global alternative hypothesis is likely to be the conjunction of a very small number of local alternative hypotheses with a large number of local null hypotheses. Accordingly, we focus our attention on the local power.

In summary then, the two quantities that we most wish to estimate are the local power and the GFPR.

12.3.2 Motivation

For any given experimental design, there is a clear and well understood trade-off between the Type 1 rate (the false positive rate) and the Type 2 error rate (1 minus the power): by adjusting the rejection region for a test, usually by manipulating the significance thresholds, the test can be made more or less conservative. A more conservative test has a lower false positive rate, but only at the expense of a higher Type 2 error rate (i.e., lower power), while a more liberal test lowers the Type 2 error rate at the cost of raising the Type 1 error rate. Finding an appropriate balance of Type 1 and Type 2 errors is an important part of the statistical design for an experiment, and requires a consideration of the relative costs of the two types of error. It is generally acknowledged (see Center for Drug Evaluation and Research 2005) that for safety studies this balance should prioritize Type 2 error control.

However, this trade-off applies only to a fixed experimental design; by adjusting the design, it may be possible to simultaneously improve both Type 1 *and* Type 2 error rates.[8] Beyond this general principle, there is a particular reason to suspect that adjusting the design might affect error rates for carcinogenicity studies. It has been shown (Lin and Rahman 1998; Rahman and Lin 2008) that, using the trend test alone and the significance thresholds in Table 12.1, the study-wise false positive rate for rodent carcinogenicity studies is approximately 10 %. However, under this decision rule, the nominal false positive rate for a single rare tumor type is 2.5 %. Given that each study includes dozens of rare endpoints, the tests must be strongly over-conservative for rare tumor types[9]; the decision rules in US Food and Drug Administration—Center for Drug Evaluation and Research (2001) rely heavily on this over-conservativeness in order to keep the GFPR to an acceptable level. But this sort of over-conservativeness is exactly the sort of phenomenon that one would expect to be quite sensitive to changes in study design.

12.3.3 Designs Compared

To get a sense of the designs currently in use, we conducted a brief investigation of 32 recent submissions, and drew the following general conclusions:

- While most designs use a single vehicle control group, a substantial proportion do use two duplicate vehicle control groups.
- A large majority of designs use three treated groups.
- The total number of animals used can vary considerably, but is typically between 250 and 300.
- The "traditional" design of four equal groups of 50 is still in use, but is not common; most designs use larger samples of animals.

Bearing these observations in mind, we compare four designs, outlined in Table 12.7. Three of these designs (D1–2 and D4) utilize the same number of animals (260), so that any effects due to differences in the disposition of the animals will not be obscured by differences due to overall sample size.

[8]It is a familiar result for asymptotic tests that simply increasing the sample size improves power while maintaining the Type 1 error rate at the nominal level. (This principle also applies to rare event data except that exact tests are frequently over-conservative, meaning that increases in the sample size can actually increase the Type 1 error rate even while keeping the rate below the nominal level, and that power can sometimes decrease as the sample size increases—see Chernick and Liu 2002). However, the inclusion of large numbers of extra animals is an inelegant (and expensive) way to shift the ROC curve; we are interested in modifications to the experimental design that leave the overall number of animals unchanged.

[9]The observation that the trend tests is strongly over-conservative for rare tumors is not at odds with the finding in Dinse (1985) that the trend test is not over-conservative for tumors with a background prevalence rate of 5 or 20 %. As the expected number of tumors increases, one expects exact tests to converge to the asymptotic tests, and the LFPRs to converge to the nominal value of α.

Table 12.7 Experimental designs considered

Design number	Number of animals per group				
	Control	Low	Mid	High	Total
D1	65	65	65	65	260
D2	104	52	52	52	260
D3	50	50	50	50	200
D4	60	50	100	50	260

The first two designs, D1 and D2, are representative of designs currently in use. Design D1 uses four equal groups of 65 animals whereas design D2 uses a larger control group (104 animals) and three equal dose groups (52 animals each). This is equivalent to a design with five equal groups, comprising two identical vehicle control groups (which, since they are identical, may be safely combined) and three treated groups.

The third design tested (D3) is the "traditional" 200 animal design. Although D3 uses fewer animals than the other designs (but is otherwise similar to D1), it has been included to enable comparison with the many simulation studies and investigations which use this design, such as that described in Sect. 12.2, and in Dinse (1985), Portier and Hoel (1983), Lin and Rahman (1998), and Rahman and Lin (2008)

In light of the investigation (Jackson 2015) of the possible benefits of unbalanced designs (where the animals are not allocated equally to the various dose groups), we have also included an unbalanced design for comparison. This design (D4) follows the suggestions of Portier and Hoel (1983):

> … we feel that a design with 50 to 60 of the experimental animals at control ($d_0 = 0$), 40 to 60 of the animals at the MTD ($d_3 = 1$) and the remaining animals allocated as one-third to a group given a dose of 10–30 % MTD ($d_1 = 0.25$ seems best) and two-thirds to a group given a dose of 50 % MTD ($d_2 = 0.5$). No less than 150 experimental animals should be used, and more than 300 animals is generally wasteful. An acceptable number of animals would be 200.

Accordingly, 60 animals have been allocated to the control group and 50 to the high dose group, with the remaining 150 animals allocated 2:1 to the mid and low dose groups.

12.3.4 *Statistical Methodology*

12.3.4.1 Simulation Schema

We have conducted two separate simulation studies. The first study was designed to compare the (local) power of the four designs to detect genuine increases in tumorigenicity for the three tumor types (rare, common, and very common) described in Sect. 12.3.1. In each case, about fifty different effect sizes (measured as the odds ratio for tumor incidence between a high dose and control animals at 104 weeks) were tested 1000 times. While 1000 simulations are not adequate to accurately estimate the power for a particular effect size (we can expect a margin of error in the estimate of approximately 3 %), we may still form an accurate impression of the general shape of the power curves.

The second simulation study was aimed at evaluating false positive (Type 1 error) rates. The immediate focus was on the LFPR, the rate at which individual organ-tumor endpoints for which there is no genuine effect are falsely found to be targets of a carcinogenic effect. Because local false positives are very rare, and because imprecision in the estimate of the local false positive rate is amplified when computing the global false positive rate, each simulation scenario has been repeated at least 250,000 times. The resulting estimates are amalgamated to compute the GFPR by appealing to independence and applying Eq. (12.3) to three different tumor spectra.

For both of the simulation studies, data were simulated using a competing risks model. The two competing hazards were tumorigenesis and death due to a non-tumor cause.

- Since these simulations are intended to evaluate power and GFPRs under fairly optimal circumstances, only one toxicity model has been considered: the hazard function for non-tumor death has the form $h_M(t) = \lambda t(\mu x + 1)$, where x is the dose and t is the time. The parameters λ and μ are chosen so that the probabilities of a control animal and a high dose animal experiencing non-tumor death before the scheduled termination date are 0.4 and 0.7 respectively.
- Tumor onset time is modeled according to the poly-3 assumptions. This means that for any given animal, the probability of tumorigenesis before time t has the form $P[T \leq t] = \lambda t^3$ where the parameter λ is a measure of the animal's tumor risk, and so depends on the dose x, the background prevalence rate (i.e., the tumor incidence rate when $x = 0$), and the dose effect on tumorigenesis to be simulated. (In the case of the LFPR simulations, it is assumed that there is no dose effect on tumorigenesis, and λ therefore depends on the background prevalence rate alone).

Although these simulations were devised independently of those in Sect. 12.2, the resulting models are in practice quite similar. Tumor onset times modeled by this approach are very similar to those of the "early onset" models, although non-tumor mortality times tend to be earlier than those simulated in Sect. 12.2. The effect of this difference is likely to be a small reduction in power (and LFPRs) in the present model compared with those used in Sect. 12.2.

12.3.4.2 Decision Rule

As noted above, we are initially concerned with estimating local power and LFPRs. Accordingly, under each scenario, we simulate data for a single 24 month experiment (*male mice*, for example), and a single tumor endpoint (*cortical cell carcinoma of the adrenal gland*, for example). Each set of simulated data includes a death time for each animal and information about whether the animal developed a tumor. From these data, two poly-3 tests (see Sect. 12.4 and Bailer and Portier 1988; Bieler and Williams 1993; US Food and Drug Administration—Center for Drug Evaluation and Research 2001) are conducted: a trend test across all groups, and a pairwise test between the control and high dose groups. As we are using the joint test rule (discussed at length in Sect. 12.2.1), the null hypothesis of no tumorigenic effect is rejected only when *both* the trend and pairwise tests yield individually significant results, at the levels indicated in Table 12.2.

12.3.4.3 Misclassification

The use of the observed incidence rate in the control group to classify a tumor as rare or common is potentially problematic. There is clearly a substantial likelihood that common tumors (with a background prevalence rate of 2 %) will be misclassified as rare, and judged against the "wrong" significance thresholds. Given the difference in the significance thresholds between those used for rare and for common tumors, it is to be expected that this misclassification effect could have an appreciable liberalizing effect on the decision rules used. Furthermore, this liberalizing effect will be amplified by the fact that misclassification is positively associated with low p-values.[10] This effect was noted, discussed, and even quantified (albeit for different decision rules and simulation scenarios than those used here) in Westfall and Soper (1998).

The probability of misclassification is dependent on both the background prevalence rate of the tumor and the number of animals in the control group. Since D2 has more than 100 animals in the control group, an experiment using this design will treat a particular tumor endpoint as rare if there is no more than one tumor bearing control animal; the other designs will consider an endpoint to be rare only if no tumor bearing control animals are found at all. The effects of this difference are seen in Fig. 12.2.

Under more traditional circumstances, given an exact test procedure and a fixed set of significance standards, one would expect the false positive rate to increase asymptotically to the nominal level as the expected event count increased. However, given the misclassification effect in this context, we expect something different; for tumors with a 2 % background prevalence rate (and for those with a 5 % background rate in the simulation study in Sect. 12.2), the LFPR can be anticipated to converge to a value below the nominal significance level for rare tumors, but *above* the nominal significance level for common tumors. For tumors with a 10 % background prevalence rate, by contrast, we can expect the LFPR to be somewhat closer to the nominal value for common tumors.[11]

[10]For this reason, the converse effect is of much less concern; misclassification of rare tumors as common is positively associated with large p-values so the cases where misclassification occurs are unlikely to be significant at even the rare tumor thresholds.

[11]The actual nominal value of the joint test is hard to evaluate. See footnote 3 on page 306.

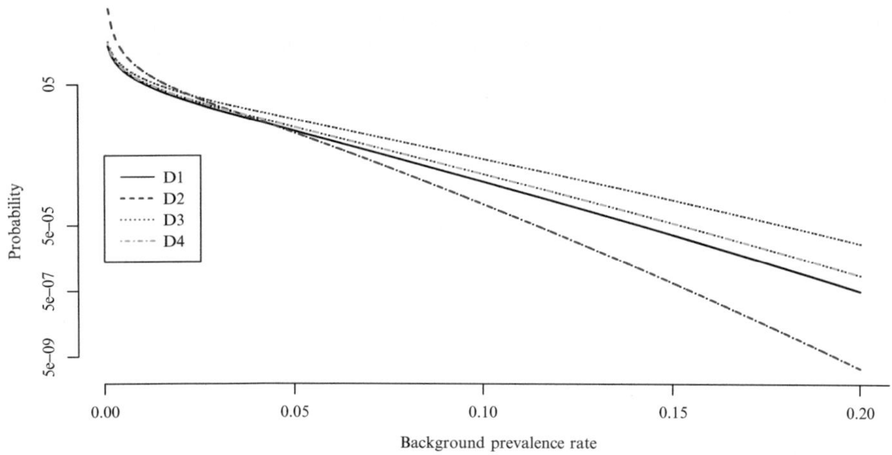

Fig. 12.2 Probability of classifying a tumor type as rare

It is uncertain whether the differential effect of misclassification on the four designs should be viewed as intrinsic to the designs, or an additional, unequal source of noise. However, given the paucity of relevant historical control data (Rahman and Lin 2009), and that a two tiered decision rule is in use, there seems to be little alternative to this method for now.[12] Accordingly, this is the most commonly used method for classifying tumors as rare or common, and we have elected to treat it as an intrinsic feature of the statistical design.

Nonetheless, it should also be remembered that statistical analysis is only one stage in the FDA review process, and that pharmacology and toxicology reviewers are free to exercise their professional prerogative and overturn the empirical determination. This is especially likely for the rarest and commonest tumors. More generally though, it is apparent that this misclassification effect must be taken into account when designing, conducting, and interpreting any simulations to evaluate carcinogenicity studies.

12.3.5 Power

The results of the power simulations are shown in Fig. 12.3.

Designs D1 and D2 are clearly more powerful than D3 and D4. For very common tumors, there is little difference between the two, but for both rare and common tumors, design D2 appears to be appreciably more powerful than D1. For rare

[12]Although it is to be hoped that in the longer term the use of the SEND data standard (Clinical Data Interchange Standards Consortium (CDISC) 2011) will enable the more efficient construction of large historical control databases.

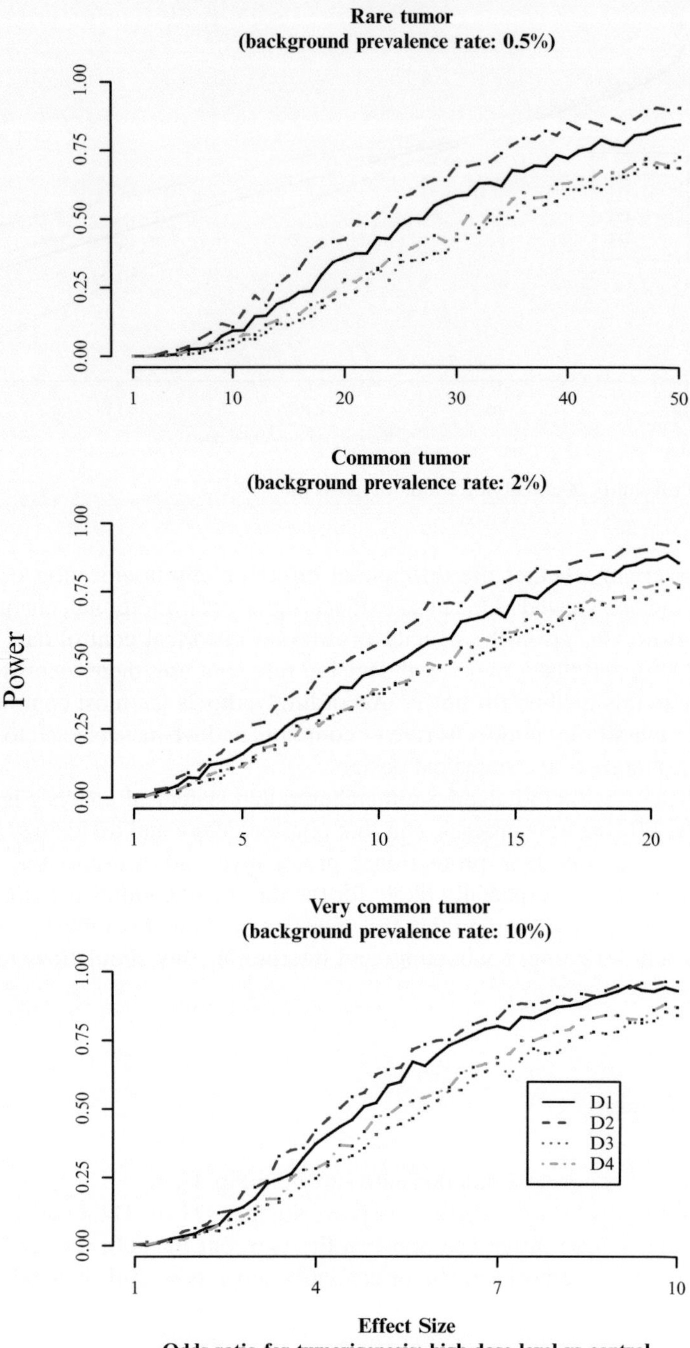

Fig. 12.3 Estimated power

tumors and an effect size of 30 (corresponding to a risk difference (RD) between the control group and the high dose group of 12.6 %), D1 and D2 have approximately 60 and 70 % power respectively. For common tumors and an effect size of 10 (RD=14.9%), D1 and D2 have approximately 45 and 55 % power respectively. More generally, Fig. 12.3 suggests that design D2 delivers about 10 % more power than D1 across a fairly wide range of scenarios of interest.

Since it uses the fewest animals, it is not surprising that D3 is the least powerful of the four designs. Direct comparison of D3 and D1 (which are similar except for the fact that D1 uses 30 % more animals in each group) shows the benefit in power that an increased sample size can bring.

That said, it is striking that the design D4, with 260 animals, is barely more powerful than D3. As we have seen from our comparison of D1 with D3, adding animals in such a way that the groups remain equal in size does increase the power of the design (absent any sawtooth effects). Furthermore, adding animals unequally to the groups can improve the power even more (see Jackson 2015). However, in the case of design D4, the extra animals (compared with D3) were not added with the goal of improving power. Indeed, the notion of optimality which D4 is intended to satisfy is quite different from our narrow goal of maximizing power while keeping the GFPR to approximately 2.5 % (Portier and Hoel 1983):

> For our purposes, an optimal experimental design is a design that minimizes the mean-squared error of the maximum likelihood estimate of the virtually safe dose from the Armitage-Doll multistage model and maintains a high power for the detection of increased carcinogenic response.

In addition, the intended maintenance of "a high power for the detection of increased carcinogenic response" was predicated on a decision rule using the trend test alone, with a significance threshold of 0.05—a much more liberal testing regime even than that recommended in US Food and Drug Administration—Center for Drug Evaluation and Research (2001), let alone than the more conservative joint test rule used here.

It is worth noting that the unbalanced approach of Design D4 is almost the anthesis of that proposed in Jackson (2015); in the latter, power is maximized by concentrating animals in the control and high dose groups, whereas in D4 they are concentrated in the intermediate groups.

12.3.6 False Positive Rate

12.3.6.1 The Local False Positive Rate

For each design, at least 250,000 simulations were conducted to estimate the rate at which the local null hypothesis is rejected when the tumor hazard is unchanged across dose groups. The resulting estimates of the LFPR (with 95 % confidence intervals) for each of the four designs and three tumor types (rare, common, and very common) are shown in Table 12.8.

Table 12.8 Local false positive rates (%) with 95 % confidence intervals

Design	Background prevalence rate					
	0.5 % (rare)		2 % (common)		10 % (very common)	
D1	0.0024	(0.0009,0.0052)	0.2264	(0.2078, 0.2450)	0.3000	(0.2786, 0.3214)
D2	0.0424	(0.0343,0.0505)	0.5928	(0.5627, 0.6229)	0.3792	(0.3551, 0.4033)
D3	0.0012	(0.0002,0.0035)	0.1008	(0.0884, 0.1132)	0.4528	(0.4265, 0.4791)
D4	0.0011	(0.0000,0.0024)	0.1048	(0.0926, 0.1169)	0.2070	(0.1899, 0.2241)

Table 12.9 Probability of misclassification

Design	Lifetime tumor incidence rate		
	0.5 % (rare)	2 % (common)	10 % (very common)
D1	0.278	0.269	0.001
D2	0.096	0.382	<0.001
D3	0.222	0.364	0.005
D4	0.260	0.298	0.002

Generally speaking, we can expect two aspects of design to affect the LFPRs. As the sample size increases, the expected number of tumors also increases, which in turn means that exact tests will behave more like asymptotic tests. In particular, we can expect the LFPRs to converge to the nominal α-level as the sample size increases. Since the tests are exact, this means that the LFPRs will tend to increase (with sample size), and since we know that the tests are strongly over-conservative for design D3 (see Sect. 12.3.2) we know that there is considerable room for growth from the levels associated with this design.

The second effect will tend to act in the opposite direction. The determination of significance thresholds depends on the number of tumor bearing animals (TBAs) in the control group: for a design with fewer than one hundred control animals, the tumor type will be considered rare just in the case that no control animals develop the tumor. (see Sect. 12.3.4.3). Thus, increasing the number of control animals (while keeping this number below 100) will increase the likelihood of a tumor type being classified as common. Since the significance thresholds for common tumors are lower than for rare tumors, this effect will make designs with more animals *more* conservative. (This reasoning does not apply to D2 which has more than 100 control animals. Since for this design, at last two tumor bearing control animals must be found in order for a tumor type to be considered common, D2 is more likely than the other designs to classify tumors as rare.) See Table 12.9 to see tumor misclassification rates for the different designs.[13]

[13]For computational reasons, these calculations use the lifetime tumor incidence rate rather than the background prevalence rate used elsewhere in this chapter.

12.3.6.1.1 Comparison of D1 and D2

The most striking feature of Table 12.8 is the difference between D1 and D2. As discussed in Sect. 12.3.3, both of these designs are in regular use, and are treated similarly for analysis purposes. However, D2 is prone to far higher false positive rates than D1, raising doubts about whether it is reasonable to treat the two designs interchangeably.

For rare tumor types, the LFPR rate for D1 is so low as to be almost negligible (approximately 1 in 40,000). Even if there are 200 such endpoints, their combined false positive rate (12.3) will be under 0.5 %. As a result, genuinely rare tumor endpoints do not contribute much to the GFPR under this design. The LFPR for D2 is about 17 higher than that for D1. While still small in absolute terms, and still strongly over-conservative, this rate is high enough that it would only take a relatively small number of rare endpoints to generate an unacceptably high combined false positive rate. In other words, despite being over-conservative, the LFPR for D2 is not over-conservative enough to meet our goals for the GFPR. See Sect. 12.3.6.2 for a more detailed discussion.

Compounding this problem, the LFPR for common tumors under D2 is exceedingly high; almost 0.6 %. This level is actually above the nominal rate for the trend test for common tumors (0.5 %), and so must be at least partially attributable to the misclassification effect noted above. The LFPR for common tumors under D1 is much lower (about 40 % of the rate under D2), but still far higher than that for rare tumors.

The LFPRs for very common tumors are much closer to each other than the rates for less common tumors (although D1 still has a higher LFPR than D1). This confirms the idea that D2's very high LFPR for common tumors is largely due to misclassification, as this misclassification effect would be expected to be less pronounced for the very common tumors (see Table 12.9).

12.3.6.1.2 Comparison of D1 and D3

For both rare and common tumors, D1 has a considerably higher LFPR than D3; about twice as high in both cases. This difference is to be expected, although it is still unsettling that an increase of just 30 % in the sample size (from D3) can result in a doubling of the LFPRs. As the background rate of tumors increases, the difference in over-conservativeness of the designs becomes less influential than D3's greater tendency toward misclassification as rare, so that D3 actually exhibits a higher LFPR than D1 for very common tumors (see Fig. 12.2).

12.3.6.1.3 Comparison of D3 and D4

The designs D3 and D4 have comparable LFPRs for rare and common tumors, although D4 has a noticeably lower LFPR for very common tumors than D3.

This is somewhat surprising since D4 was not proposed with the specific goal of constraining the Type 1 error rate. Overall, we can say that of the four designs considered, D4 has the lowest false positive rates.

12.3.6.2 The Global False Positive Rate

As discussed in Sect. 12.3.1, a tumor spectrum must be selected before the GFPR may be estimated. However, the selection of a realistic tumor spectrum is difficult. The endpoints under consideration are the *potential* tumor types which a pathologist may report if they are found. This list is not fixed from study to study. For example, one pathologist might group tumors found in the cervix or uterus together, whereas another might report these separately. Similarly, one pathologist might group hibernomas (lipomas of brown adipose tissue) with classic (white adipose tissue) lipomas under the general heading *lipoma—adipose tissue*, whereas another might not. Nonetheless, these variations are relatively minor, and the requirement that pathologists conduct a "complete necropsy," the widely accepted suggestions of Bergman et al. (2003), and the forthcoming adoption of the SEND data standard (Clinical Data Interchange Standards Consortium (CDISC) 2011) all ensure a reasonable level of uniformity.

However, the data from individual studies only reference tumor types which were actually found in those studies, so that many rare tumor endpoints are not mentioned at all in study reports. The data presented in the compilations by Charles River Laboratories (Giknis and Clifford 2004, 2005) are helpful in this regard. On the basis of the data in these tabulations, we have chosen three tumor spectra (shown in Table 12.10) which seem to form a plausible range for a typical tumor spectrum.[14]

For two of the three spectra, the GFPR for D1 is close to the target level of 2.5 %, and even for S3, it is "only" twice the desired level. However, the GFPRs for D2 consistently exceed the target GFPR by a very wide margin. The GFPR for D3 is close to the target, confirming the results of Sect. 12.2. The conservativeness of D4 carries over to the calculation of the GFPR which is actually below the target level for the thinner spectra S1 and S2, and close to 2.5 % for S3.

Table 12.10 Estimated global false positive rates

Spectrum	Endpoints				Global FPR (%)			
number	0.5 %	2 %	10 %	Total	D1	D2	D3	D4
S1	66	6	3	75	2.39	7.23	2.03	1.31
S2	100	6	4	110	2.76	8.91	2.51	1.56
S3	140	15	5	160	5.13	15.43	3.87	2.73

[14] Insofar as we know what is typical. These spectra should only be taken to represent a range of plausible scenarios, and not assumed to be in any way definitive.

Some care must be taken when interpreting these results. The LFPR for a tumor type with a true background prevalence rate of, say, 0.1 % is likely to be somewhat lower than that for a tumor with a background prevalence rate of 0.5 % (our rare tumor exemplar). This is unlikely to affect the estimates of the GFPRs for D1, D3, and D4, since the LFPR for rare tumors is so low for these designs that the GFPR is largely insensitive to the number of rare endpoints in the spectrum. This is not true of D2, though. A spectrum of 100 independent tumor types, half with a background prevalence rate of 0.1 % and half with a rate of 0.5 % might yield an appreciably lower GFPR rate under D2 than a spectrum of 100 independent tumor types each with a rate of 0.5 %. However, this effect can only explain a small part of the difference in GFPR between D1 and D2. For example, even for S2, with just 10 non-rare endpoints, the 6 common and 4 very common endpoints between them contribute over 40 % of the GFPR: if *all* the rare endpoints were disregarded for D2, the GFPR for S2 would still be 3.9 %; above the 2.5 % target and higher than the GFPR of any of the other designs.

12.3.7 Discussion

12.3.7.1 Comparison of Currently Used Designs

There are clearly substantial differences between designs D1 and D2. Figure 12.3 shows that D2 has appreciably more power than D1 over a range of meaningful scenarios—in many cases close to 10 % more—although the difference is slight for very common tumors. But as demonstrated in Tables 12.8 and 12.10, D2 also suffers from a considerably worse false positive rate than D1. Taken together, these two observations lead to the conclusion that D2 is *more liberal* than D1.

It is reasonable to consider phasing out the use of D2 altogether. The duplicate control design was originally introduced to test for the effects of extra-binomial within-study variability between the two concurrent control groups (Baldrick and Reeve 2007; Haseman et al. 1986; US Food and Drug Administration—Center for Drug Evaluation and Research 2001); any significant differences between the control groups were an indication of a failure of experimental conduct, such as when two groups of cages are subject to different environmental conditions. However, such comparisons between control groups are clearly underpowered to detect such environmental effects, especially given how hard it is to detect even moderately strong *treatment* effects.

If D2 is to be retained as an acceptable design, the fact that it is more liberal than D1 needs to be taken into account. At the very least, it seems inappropriate to continue to use the same significance thresholds for designs D1 and D2.

Designs D1 and D3 are very similar, differing only in the total number of animals used; both designs distribute animals equally among the single vehicle and three dose groups. Differences between these designs' statistical properties are therefore entirely attributable to differences in sample size. D1 is appreciably more powerful than D3 across a wide range of scenarios, but also suffers from slightly higher false positive rates. In the case of D3, the GFPR tends to be well below the target rate of 2.5 %, confirming the findings of Sect. 12.2 that there is room to use more liberal significance thresholds (see Table 12.6), and thereby increase power. However, this does not appear to be the case for the more widely used D1 design.

In general, the differences between the three designs' statistical properties are not negligible, meaning that our ability to draw conclusions about one the behavior of design from the study of the other is limited. This is especially problematic since a great deal of our understanding of rodent carcinogenicity studies has its foundations in studies of design D3, but this design is fading from popularity.

12.3.7.2 The Design D4

The unbalanced design D4 was not optimized to maximize power using the sort of decision rule currently in place, and so it is not surprising that it is substantially less powerful than the other 260 animal designs. However, it is striking that the distribution of animals does yield a very low false positive rate (comparable to D3, although better for very common tumors). Compared with the traditional 200 animal design, then, the addition of 60 animals to D4 yields a real but modest benefit in both power and Type 1 error. It is reasonable to think that these extra animals instead provide considerable added information about the virtually safe dose (that being the intent behind this design), but testing this notion is outside the scope of this study.

12.3.7.3 The Balance Between Type 1 and Type 2 Error

The goal of achieving satisfactory power whilst keeping the GFPR to approximately 2.5 % is difficult to achieve, even aside from the ambiguity over what exactly constitutes "satisfactory power".

Inspection of Fig. 12.3 shows that for a tumor with a background prevalence rate of 0.5 %, design D1 delivers at least 50 % power when the effect size is above 27 (RiskDifference(RD) = 11.4%), and 75 % when the effect size is above 42 (RD = 16.9%). For a tumor with a background prevalence rate of 2 %, the power is above 50 % when the effect size is above 11 (RD = 16.3%) and above 75 % when the effect size is above about 16 (RD = 22.6%).

But even if we conclude that these levels of power are adequate, we are faced with the fact that the GFPR for D1 is generally somewhat above the target rate of 2.5 %. Lowering the significance thresholds to further limit the GFPR could only be done at the expense of the power, and so seems unwise, given that these studies are essentially safety studies.

In Jackson (2015), an alternative designs which delivers considerably more power than even D2 (the most powerful design considered here) while simultaneously lowering the GFPR rate to a level similar to or below that associated with D1 is investigated.

12.4 The Exact Poly-*k* Test

12.4.1 Introduction

The poly-*k* method is a mortality adjusted trend test for tumor incidence. This method was originally suggested in Bailer and Portier (1988), and was improved in Bieler and Williams (1993).

As some tumors may have long latency periods, animals with shorter life spans may face a disproportionately reduced risk of tumor onset. The poly-*k* method suggests correcting this problem by adjusting the number n_i of animals at risk in the *i*th dose group to compensate for early deaths. Operationally, the *j*th animal in the *i*th dose group gets a score $w_{ij} \leq 1$; this score is 1 if the animal lives for the full study period (T), or develops the tumor type being tested before dying. Conversely, if this animal dies at the time $t_{ij} < T$ before the end of the study without developing the tumor being tested, it gets a score of

$$w_{ij} = \left(\frac{t_{ij}}{T}\right)^k < 1.$$

The adjusted group size for Group *i* is then defined as

$$n_i^* = \sum_i w_{ij}.$$

As an interpretation, an animal with score $w_{ij} = 1$ can be considered as a whole animal, while an animal with score $w_{ij} < 1$ can be considered as a partial animal. The adjusted group size n_i^* is equal to n_i (the original group size) if all the animals in the group either survive until the end of the study or develops at least one tumor of the type being tested; otherwise the adjusted group size is less than n_i, except for some marginal cases due to rounding. These adjusted group sizes are then used to perform trend and pairwise (between treated groups and the control group) tests of tumor incidence rates using the Cochran-Armitage test procedure (Armitage 1955).

One critical point to consider when using the poly-*k* test is the choice of the appropriate value of *k*, which depends on the tumor incidence pattern with the increased dose. For long term 104 week standard rat and mouse studies, a value of k = 3 is suggested in the literature.[15] In this case we refer to the procedure as the *poly-3 test*. It should be noted that the assumption for Cochran-Armitage test is that the marginal total n_i is fixed. However, in this case n_i^* is a random variable. As a result the calculation of the variance of the test statistic needs to be modified. An estimate of this variance, using the delta method and the weighted least squares technique is suggested in Bieler and Williams (1993).

[15] Portier et al. (1986) recommends $k = 3$, although other values have been investigated (Gebregziabher and Hoel 2009; Moon et al. 2003). However, as noted in Gebregziabher and Hoel (2009), it appears that the tests are largely insensitive to the choice of *k*.

It may be noted that unlike the methods suggested in Peto et al. (1980), the poly-k analysis is independent of tumor context of observation information (i.e. if the tumor was observed on incidental or fatal context), which is a major advantage of this method over the Peto method.

12.4.2 The Exact Poly-k Method

The outcome of the experiment, for a specific tumor endpoint, can typically be summarized by a results table such as Table 12.11. Replacing the number of animals in each cell by the corresponding adjusted group sizes, we get a new results table, Table 12.12.

A simple-minded exact poly-k test can now be conducted performing the Cochran-Armitage test using the data in Table 12.12. However, in order to use the exact Cochran-Armitage test, the row and column totals for all permuted configurations of the observed table must be fixed. Since calculation of the n_i^* terms (the column totals) depends on the survival pattern of the animals, these terms cannot be assumed to be fixed, and the naïve use of the Cochran-Armitage test is not correct. For an appropriate exact test the adjusted column totals must be recalculated for every permutation of all animals.

We illustrate this method by considering a simple example:

12.4.2.1 Illustrative Example

Consider an experiment with two dose groups and five animals per group, continued up to 104 weeks. Suppose that the observed data for a specific tumor type are as

Table 12.11 Results table for single endpoint without survival adjustments

Group number	0	1	⋯	i	⋯	r
Dose level	$d_0 = 0$	d_1	⋯	d_i	⋯	d_r
Original group size	n_0	n_1	⋯	n_i	⋯	n_r
TBAs	x_0	x_1	⋯	x_i	⋯	x_r
Non tumor bearing animals (NTBAs)	$(n_0 - x_0)$	$(n_1 - x_1)$	⋯	$(n_i - x_i)$	⋯	$(n_r - x_r)$

Table 12.12 Results table for single endpoint with survival adjustments

Group number	0	1	⋯	i	⋯	r
Dose level	$d_0 = 0$	d_1	⋯	d_i	⋯	d_r
Adjusted group size	n_0^*	n_1^*	⋯	n_i^*	⋯	n_r^*
TBAs	x_0	x_1	⋯	x_i	⋯	x_r
Adjusted NTBAs	$(n_0^* - x_0)$	$(n_1^* - x_1)$	⋯	$(n_i^* - x_i)$	⋯	$(n_r^* - x_r)$

Table 12.13 Raw output for example

Animal number	A1	A2	A3	A4	A5	A6	A7	A8	A9	A10
Dose level	0	0	0	0	0	1	1	1	1	1
Time of death (weeks)	80	104	104	96	104	74	98	104	50	104
Tumor code	0	0	0	0	0	1	0	0	0	1
Poly-3 weight	0.46	1.00	1.00	0.79	1.00	1.00	0.84	1.00	0.11	1.00

Table 12.14 Example of results table without survival adjustments

Group number	0	1	
Dose level	0	1	Total
Original group size	5	5	10
TBAs	0	2	2
Non tumor bearing animals	5	3	8

Table 12.15 Example of results table with survival adjustments

Group number	0	1	
Dose level	0	1	Total
Adjusted group size (rounded)	4	4	8
TBAs	0	2	2
Adjusted non tumor bearing animals	4	2	6

shown in Table 12.13. Calculating the adjusted group sizes, and rounding so we may use discrete tests, we have

$$n_0^* = \text{Round}\,(0.46 + 1.00 + 1.00 + 0.79 + 1.00) = \text{Round}(4.25) = 4$$
$$n_1^* = \text{Round}\,(1.00 + 0.84 + 1.00 + 0.11 + 1.00) = \text{Round}(3.95) = 4.$$

Tables 12.14 and 12.15 summarize this data for analysis in the styles of Tables 12.11 and 12.12. The table p values for the asymptotic Cochran-Armitage tests are $p = 0.06681$ for the unadjusted data and $p = 0.06332$ for the adjusted data.

12.4.2.2 Exact Method

An exact methodology for the poly-k test is possible. In the following we will describe this method.

The approach is combinatorial; we consider permutations of the animals among the dose group. Each permutation generates a summary table of the form of Table 12.15 (although many permutations can generate similar tables). For each table, we calculate the test statistic T defined by:

$$T = \sum_{i=0}^{r} x_i d_i \tag{12.4}$$

Table 12.16 Output after a typical permutation

Animal number	A1	A3	A7	A9	A10	A2	A4	A5	A6	A8
Dose level	0	0	0	0	0	1	1	1	1	1
Time of death (weeks)	80	104	98	50	104	104	96	104	74	104
Tumor code	0	0	0	0	1	0	0	0	1	0
Poly-3 weight	0.46	1.00	0.84	0.11	1.00	1.00	0.79	1.00	0.36	1.00

Table 12.17 Summary table for permuted data

Group number	0	1	
Dose level	0	1	Total
Adjusted group size (rounded)	3	4	7
TBAs	1	1	2
Adjusted non tumor bearing animals	2	3	5

The table p-value of a given outcome with test statistic $T = t$ can be calculated using the hypergeometric distribution.

We define an equivalence relation on permutations, saying that two permutations p_1 and p_2 are equivalent ($p_1 \sim p_2$) if they both allocate the same number of TBAs to each dose group. It is clear that the test statistic T respects this equivalence relation, but it is possible to have $T_{p_1} = T_{p_2}$ without $p_1 \sim p_2$ (Table 12.16).

For this permutation, we have

$$n_0^* = \text{Round}\,(0.46 + 1.00 + 0.84 + 0.11 + 1.00) = \text{Round}(3.40) = 3$$

$$n_1^* = \text{Round}\,(1.00 + 0.7865 + 1.00 + 0.3602 + 1.00) = \text{Round}(4.14) = 4.$$

The resulting summary table is displayed in Table 12.17:

We now calculate the probability that a randomly selected permutation induces a particular table of permuted values (such as Table 12.17). The random variables x_0 and x_1 can be assumed to be drawn from a random hypergeometric distribution $g(n_0^*, n_1^*)$. The probability of a given hypergeometric distribution's realization depends on the random parameters n_0^* and n_1^*, and so can be denoted $\Pr\left[g(n_0^*, n_1^*)\right]$. The probability of observing a particular table is therefore:

$$\Pr\left[x_0, x_1, n_0^*, n_1^*, g(n_0^*, n_1^*)\right] = \Pr\left[g(n_0^*, n_1^*)\right] \cdot \Pr\left[x_0, x_1, n_0^*, n_1^* \middle| g(n_0^*, n_1^*)\right] \quad (12.5)$$

where

$$\Pr\left[x_0, x_1, n_0^*, n_1^* \middle| g(n_0^*, n_1^*)\right] = \frac{\binom{n_0^*}{x_0}\binom{n_1^*}{x_1}}{\binom{n^*}{x}} \Big/ C\,(x_0, x_1) \quad (12.6)$$

with $C(x_0, x_1)$ as the total number of ways in which x TBAs and $n - x$ NTBAs can be arranged into groups of size n_0 and n_1 such that exactly x_0 of the TBAs are in Group 0. This quantity is given by:

$$C(x_0, x_1) = \binom{x}{x_0}\binom{n-x}{n_0-x_0}\binom{x-x_0}{x_1}\binom{(n-x)-(n_0-x_0)}{n_1-x_1}. \tag{12.7}$$

Note that $C(x_0, x_1)$ is equal to the total number of ways in which n number of animals can be arranged so that x_0 of the tumor bearing animals are in Group 0, and the remaining x_1 are in Group 1. For $r + 1$ treatment groups the general formula for $C(x_0, x_1, \ldots, x_r)$ is

$$C(\mathbf{x}) = \prod_{i=0}^{r-1}\binom{x - \sum_{k<i} x_k}{x_i}\binom{(n-x) - \sum_{k<i}(n_k - x_k)}{n_i - x_i}. \tag{12.8}$$

with $x = x_0 + x_1 + \ldots + x_r$, $x_{-1} = -x_0$, and $n_{-1} = -n_0$.

For example, for Table 12.14, with $x_{-1} = 0$, $x_0 = 0$, $x_1 = 2$, $n_{-1} = -5$, $n_0 = 5$, and $n_1 = 5$, we have $x = 0 + 2 = 2$, and $n = 5 + 5 = 10$, and

$$C(0,2) = \binom{2}{0}\binom{10-2}{5-0}\binom{(2-0)}{2}\binom{(10-2)-(5-0)}{5-2} = 56.$$

The complete list of these 56 arrangements of 10 animals is given in Table 12.18. For each permutation p in this list, the test statistic T_p is equal to $0 \times 0 + 2 \times 12$. Furthermore, since there are just two groups, this is an exhaustive list of all permutations satisfying $T_p = 2$. The probability that $T = 2$ is therefore calculated by adding the probabilities for each of these 56 permutations, using Eq. (12.5).

As a simple choice, the $\Pr[g(n_0^*, n_1^*)]$ can be taken to be equal for each $g(n_0^*, n_1^*)$, and can be determined from the identity

$$\begin{aligned}&\sum \Pr\left[x_0, x_1, n_0^*, n_1^*, g(n_0^*, n_1^*)\right]\\ &\quad = \sum \Pr\left[g(n_0^*, n_1^*)\right] \Pr\left[x_0, x_1, n_0^*, n_1^* | g(n_0^*, n_1^*)\right] = 1\end{aligned} \tag{12.9}$$

so that

$$\Pr\left[g(n_0^*, n_1^*)\right] = \frac{1}{\sum \Pr\left[x_0, x_1, n_0^*, n_1^* | g(n_0^*, n_1^*)\right]}. \tag{12.10}$$

In this respect, the factor $\Pr\left[g(n_0^*, n_1^*)\right]$ can also be considered as a normalizing factor. Therefore, operationally we first calculate $\sum \Pr\left[x_0, x_1, n_0^*, n_1^* | g(n_0^*, n_1^*)\right]$ and then normalize using $\Pr\left[g(n_0^*, n_1^*)\right]$.

It may be noted here that for the use of the hyper geometric distribution, although we round the n_j^*s to their nearest values, it is possible to use the ceiling or floor functions instead. This is a matter of individual discretion. One's choice will have

Table 12.18 All possible arrangements of 10 animals at risk with 0 TBAs in Group 0 and 2 TBAs in Group 1

Permutation	Animal numbers										n_0^*	n_1^*	n_0^*	n_1^*
number	Group 0					Group 1					Exact		Rounded	
1	1	2	3	4	5	6	7	8	9	10	4.24	3.95	4	4
2	1	2	3	4	9	5	6	7	8	10	3.35	4.84	3	5
3	1	2	3	4	8	5	6	7	9	10	4.24	3.95	4	4
4	1	2	3	8	9	4	5	6	7	10	3.57	4.62	4	5
5	1	2	3	4	7	5	6	8	9	10	4.08	4.11	4	4
6	1	2	3	7	8	4	5	6	9	10	4.29	3.9	4	4
7	1	2	3	7	9	4	5	6	8	10	3.4	4.79	3	5
8	1	2	7	8	9	3	4	5	6	10	3.4	4.79	3	5
9	1	2	3	5	7	4	6	8	9	10	4.29	3.9	4	4
10	1	2	3	5	8	4	6	7	9	10	4.46	3.73	4	4
11	1	2	3	5	9	4	6	7	8	10	3.57	4.62	4	5
12	1	2	5	8	9	3	4	6	7	10	3.57	4.62	4	5
13	1	2	5	7	8	3	4	6	9	10	4.29	3.9	4	4
14	1	2	5	7	9	3	4	6	8	10	3.4	4.79	3	5
15	1	5	7	8	9	2	3	4	6	10	3.4	4.79	3	5
16	1	2	4	5	9	3	6	7	8	10	3.35	4.84	3	5
17	1	2	4	5	8	3	6	7	9	10	4.24	3.95	4	4
18	1	2	4	5	7	3	6	8	9	10	4.08	4.11	4	4
19	1	2	4	7	9	3	5	6	8	10	3.19	5	3	5
20	1	2	4	7	8	3	5	6	9	10	4.08	4.11	4	4
21	1	2	4	8	9	3	5	6	7	10	3.35	4.84	3	5
22	1	4	7	8	9	2	3	5	6	10	3.19	5	3	5
23	1	4	5	8	9	2	3	6	7	10	3.35	4.84	3	5
24	1	4	5	7	8	2	3	6	9	10	4.08	4.11	4	4
25	1	4	5	7	9	2	3	6	8	10	3.19	5	3	5
26	4	5	7	8	9	1	2	3	6	10	3.73	4.46	4	4
27	1	3	4	8	9	2	5	6	7	10	3.35	4.84	3	5
28	1	3	4	7	8	2	5	6	9	10	4.08	4.11	4	4
29	1	3	4	7	9	2	5	6	8	10	3.19	5	3	5
30	1	3	4	5	7	2	6	8	9	10	4.08	4.11	4	4
31	1	3	4	5	8	2	6	7	9	10	4.24	3.95	4	4
32	1	3	4	5	9	2	6	7	8	10	3.35	4.84	3	5
33	1	3	5	7	9	2	4	6	8	10	3.4	4.79	3	5
34	1	3	5	7	8	2	4	6	9	10	4.29	3.9	4	4
35	1	3	5	8	9	2	4	6	7	10	3.57	4.62	4	5
36	1	3	7	8	9	2	4	5	6	10	3.4	4.79	3	5
37	3	5	7	8	9	1	2	4	6	10	3.95	4.24	4	4
38	3	4	7	8	9	1	2	5	6	10	3.73	4.46	4	4
39	3	4	5	8	9	1	2	6	7	10	3.9	4.29	4	4
40	3	4	5	7	8	1	2	6	9	10	4.62	3.57	5	4

(continued)

Table 12.18 (continued)

Permutation	Animal numbers										n_0^*	n_1^*	n_0^*	n_1^*
number	Group 0					Group 1					Exact		Rounded	
41	3	4	5	7	9	1	2	6	8	10	3.73	4.46	4	4
42	2	3	7	8	9	1	4	5	6	10	3.95	4.24	4	4
43	2	3	5	8	9	1	4	6	7	10	4.11	4.08	4	4
44	2	3	5	7	8	1	4	6	9	10	4.84	3.35	5	3
45	2	3	5	7	9	1	4	6	8	10	3.95	4.24	4	4
46	2	3	4	5	9	1	6	7	8	10	3.9	4.29	4	4
47	2	3	4	5	8	1	6	7	9	10	4.79	3.4	5	3
48	2	3	4	5	7	1	6	8	9	10	4.62	3.57	5	4
49	2	3	4	7	9	1	5	6	8	10	3.73	4.46	4	4
50	2	3	4	7	8	1	5	6	9	10	4.62	3.57	5	4
51	2	3	4	8	9	1	5	6	7	10	3.9	4.29	4	4
52	2	4	5	7	9	1	3	6	8	10	3.73	4.46	4	4
53	2	4	5	7	8	1	3	6	9	10	4.62	3.57	5	4
54	2	4	5	8	9	1	3	6	7	10	3.9	4.29	4	4
55	2	4	7	8	9	1	3	5	6	10	3.73	4.46	4	4
56	2	5	7	8	9	1	3	4	6	10	3.95	4.24	4	4

some effects on the calculation of the p-value if the sample size is very small (as would have been the case with our example, with five animals per group). However for moderately large sample size this choice will have minimal effect. It should be noted that, as discussed in Sect. 12.3.3, the regular carcinogenicity studies have 50–70 animals per group.

12.4.3 Second Example

We further illustrate the use and properties of our method with another example:

12.4.3.1 Framework of Second Example

12.4.3.1.1 Design

Two dose groups with dose levels 0 and 1. Twelve animals, with animal numbers $1, \ldots, 12$ are randomly divided into two treatment groups, G0 and G1, of six animals each. Terminal sacrifice is at Week 104. The example data are presented in Table 12.19. The observed value of the test statistic T is $1 \times 0 + 4 \times 1 = 4$.

As described in Sect. 12.4.2.1, we can calculate the distribution of T; the distribution is shown in Table 12.20 Since the observed value of T is 4, the p-value

Table 12.19 Observed data from second example

Animal number	1	2	3	4	5	6	7	8	9	10	11	12
Dose group	G1	G0	G1	G0	G1	G1	G1	G1	G0	G0	G0	G0
Tumor code	0	0	1	0	1	0	1	1	0	0	1	0
Survival time (Week)	78	104	55	104	104	104	65	85	104	50	45	104

Table 12.20 Distribution of test statistic for second example

	PDF	
T	Poly-3 adjusted test	Unadjusted Cochran-Armitage test
0	0.00973	0.00758
1	0.11932	0.11364
2	0.37095	0.37879
3	0.37095	0.37879
4	0.11932	0.11364
5	0.00973	0.00758

of the test is $\Pr[T \geq 4]$. Using the exact poly-3 test, we get a p-value of 0.12905. For the unadjusted Cochran-Armitage test, the p-value is $0.11364+0.00758 = 0.12122$. For the asymptotic one-tailed test (using StatXact), it is 0.05124.

It should be noted that this method is based on an extensive computational procedure requiring the evaluation of all possible permutations of the animals to $r + 1$ dose groups. This computational complexity is a big challenge for the application of the proposed method in the data analysis of real studies. However, some modifications of commercially available software for the calculations of the probabilities of the hypergeometric distribution may facilitate these calculations.

12.5 Modified Exact Poly-3 Method

Since the exact poly-3 method described in Sects. 12.4.2.1 and 12.4.3 has severe computational limitations when we have group sizes of 50 or larger, the alternative and more practical way is to use the permutation sample to estimate p-values. A survival-adjusted exact randomization trend test procedure (Mancuso et al. 2002) has been proposed to use the permutation sample to estimate the p-values. The test is carried out by using PROC STRATIFY with fixed row and column sums assumptions. In order to reduce biases caused by the assumptions of fixed column and row sums using PROC STRATIFY and binomial null variance estimate from Mancuso et al. (2002), we are proposing a modified exact poly-3 trend test that can be regarded as an exact version of the poly-3 test (Bieler and Williams 1993).

12.5.1 Motivating Problem

As discussed earlier (see Sect. 12.4.2, and especially Table 12.13), animal survival time is not a fixed quantity. The adjusted quantal response estimates, $p_i^* = x_i/n_i^*$, are actually ratios of linear statistics. Hence, the numerators and denominators of these estimates are both subject to random variation.

12.5.2 Permutational Distribution for the Modified Poly-3 Test

12.5.2.1 The Modified Poly-3 Trend Test

The quantal response tests that focus on crude lifetime tumor incidence rates and make no adjustment for differences in survival experiences across dose groups are often biased, since they implicitly assume that all animals are at equal risk of developing a tumor over the course of study. As mentioned in Sect. 12.4.1, in order to address this issue, Bailer and Portier (1988) introduced a modification to the Cochran-Armitage test for trend that adjusts for differences in treatment lethality while requiring no assumptions regarding tumor lethality or changes to study design. This poly-3 trend test incorporates a weighting scheme that allows fractional information into the analysis for animals not at full risk for tumor development. This weighting scheme essentially modifies the denominators of the crude quantal response estimates of lifetime tumor incidence to more closely approximate the total number of animal years at risk in each experimental group.

Using this weighting scheme and the notation used previously, we define $p_i^* = x_i/n_i^*$ as the *adjusted quantal response estimate of lifetime tumor incidence in group i*; and

$$p^* = \frac{\sum_i x_i}{\sum_i n_i^*} \tag{12.11}$$

as the *experiment-wide adjusted quantal response estimate of lifetime tumor incidence*; and $q_i^* = 1 - p_i^*$ and $q^* = 1 - p^*$. As previously stated, binomial null variance estimates of p_i^* do not apply since they are based on the assumption that the number of animals at risk is fixed. However, by using a Taylor expansion, a pooled null variance estimate for p_i^* can be found:

$$\text{var}_0\left(p_i^*\right) \approx \left(\frac{n_i}{n_i^*}\right)^2 \cdot \frac{\sum_i \sum_j (r_{ij} - \bar{r}_i)^2}{n - (g+1)}. \tag{12.12}$$

where $n = \sum_i n_i$, $r_{ij} = x_i - p_i^* w_{ij}$, and $g + 1$ is the number of experimental groups in the study.

A computational formula for the modified Cochran-Armitage test statistic, which will be referred to as the *ratio test* proposed in Bieler and Williams (1993) and denoted by Z_r is given as follows:

$$Z_r = \frac{\sum_i a_i p_i^* d_i - \left(\sum_i a_i d_i\right)\left(\sum_i a_i p_i^*\right) / \sum_i a_i}{\sqrt{C\left(\sum_i a_i d_i^2 - \left(\sum_i a_i d_i\right)^2 / \sum_i a_i\right)}} \tag{12.13}$$

where

$$C = \sum_i \sum_j \frac{\left(r_{ij} - \bar{r}_{ij}\right)^2}{n - (g+1)} \quad a_i = \frac{n_i}{\left(n_i^*\right)^2}.$$

12.5.3 Permutational Distribution for the Modified Poly-3 Test

Exact methods are preferable for sparse data. Permutation tests consider all possible assignments of animals to dose groups as equally likely, while fixing the rest of the information obtained in the experiment. Under the null hypothesis of no treatment effect, this results in an exact conditional distribution of the test statistic when intercurrent mortality patterns are equal across groups, and it is asymptotically correct when the mortality patterns are unequal (Fairweather et al. 1998; Heimann and Neuhaus 1998). Given a data set, consider all, say M, possible allocations of animals to groups while keeping the observed data for each animal fixed. Corresponding to these M arrangements, we may obtain M values of the test statistic. The permutational distribution of the modified poly-3 test statistic results from assigning equal probability to each of these M values. Letting Z_r^* be the observed value, the p-value is the proportion of the M values that are at least as extreme as Z_r^*. By exhaustive enumeration, the computation for the p-value using the permutational distribution for the test is straightforward and efficient if the number of animals in the study is small. For data involving large numbers of subjects, the p-value associated with the permutational distribution of the test statistic may be approximated by a sample of the set of all permutations.

12.5.4 Simulations and Results

We conducted a Monte Carlo simulation study to evaluate the following tests: the exact version of the modified poly-3 test (Bieler and Williams 1993), the exact version of the poly-3 test (Bailer and Portier 1988) and PROC STRATIFY in two simulation designs (Dinse 1985; Portier et al. 1986). For each configuration, 10,000 simulated data sets were generated and tested by various methods at the nominal significance level $\alpha = 0.05$. Additionally, these methods were tested against the significance thresholds described in Table 12.1 for a standard study (where, as in Sects. 12.2 and 12.3, rarity was determined by the incidence rate in the control group).

In conducting the exact versions of the modified or the not-modified poly-3 test using the permutational distribution, p-values were estimated from samples of 5000 permutations.

12.5.4.1 Monte Carlo Simulation Design 1

A typical bioassay design with four groups of 50 animals each and an experimental duration of 104 weeks is used in the study. The design is simulated to have a single terminal sacrifice at the end of the experiment, as in the customary long-term rodent bioassay. The dose levels used are (0,1,2,3) across groups. The three independent variables T_0 (time to tumor onset), T_2 (time from tumor onset until death from the tumor), and T_1 (time until death from a competing risk) are used to model animal tumorigenicity data. These variables are generated from the modified Weibull distributions used by Portier et al. (1986) and others in the literature (Ahn and Kodell 1995; Chang et al. 2000; Kodell and Ahn 1997; Kodell et al. 1994). The survival function for T_0 is

$$S(t) = \exp\left[-\delta_1(1/104)^{\delta_2}\right].$$

with $\delta_2 \in \{1.5, 3, 6\}$ and δ_1 chosen so that the probability of tumor onset by the end of the study attains the desired rate. Since the study is concerned with rare events, tumor rates between 0.01 and 0.15 are used.

The survival function for T_1 is

$$Q(t) = \exp\left[-\phi\left(\gamma_1 t + \gamma_2 t^{\gamma_3}\right)\right] \quad (12.14)$$

with $\gamma_1 = 10^{-4}$, $\gamma_2 = 10^{-16}$ and $\gamma_3 = 7.425531$, and the value ϕ chosen such that the competing risks survival rate with respect to all causes of death except for the tumor of interest at 104 weeks is either 0.5 for all groups or (0.5, 0.4, 0.3, 0.2) across groups. The control survival rate chosen represents the one recently observed in the NTP studies for male Fischer 344 rats (Haseman et al. 1998), although it is somewhat below average for B6C3F$_1$ mice and F344/N female rats in the NTP feeding studies.

For simplicity, the survival function for T_2 has the same form as $Q(t)$ with the same values of γ_1, γ_2 and γ_3.

12.5.4.2 Results of First Monte Carlo Simulation

The results of the first set of simulations are presented in Table 12.21 (common tumors) and Table 12.22 (rare tumors). In the following tables, Proposed refers to our proposed method and Mancuso refers to the method described in Mancuso et al. (2002). For trend test, the multiplicity adjustment decision rule (referred as Adjusted α in the following tables) is common and rare tumors are tested at 0:005 and 0:025 significance levels, respectively.

Table 12.21 Null hypothesis rejection rates for Monte Carlo simulation of modified poly-3 trend test and the poly-3 test described in Bailer and Portier (1988) and Mancuso et al. (2002)

	Tumor incidence				Decision rule			
					Adjusted α		$\alpha = 0.05$	
	Control	Low	Mid	High	Proposed method	Mancuso method	Proposed method	Mancuso method
Size	0.01	0.01	0.01	0.01	0.0314	0.0334	0.0631	0.0657
	0.02	0.02	0.02	0.02	0.0100	0.0100	0.0650	0.0637
	0.03	0.03	0.03	0.03	0.0083	0.0086	0.0644	0.0643
	0.04	0.04	0.04	0.04	0.0083	0.0080	0.0712	0.0704
	0.05	0.05	0.05	0.05	0.0080	0.0080	0.0741	0.0731
	0.075	0.075	0.075	0.075	0.0111	0.0106	0.0846	0.0838
	0.10	0.10	0.10	0.10	0.0123	0.0121	0.0875	0.0861
Power	0.01	0.02	0.06	0.12	0.86	0.94	0.89	0.96
	0.02	0.02	0.14	0.14	0.84	0.84	0.89	0.88
	0.01	0.03	0.14	0.12	0.69	0.70	0.78	0.78
	0.01	0.01	0.01	0.14	0.90	0.99	0.93	0.99
	0.01	0.12	0.10	0.14	0.44	0.45	0.59	0.59
	0.01	0.14	0.14	0.12	0.19	0.19	0.32	0.31

Table 12.22 Summary sizes of tests of rare tumors with spontaneous incidence rates 0–1 % in simulation design I

	Tumor incidence				Decision rule			
					Adjusted α		$\alpha = 0.05$	
	Control	Low	Mid	High	Proposed method	Mancuso method	Proposed method	Mancuso method
Size	0.001	0.001	0.001	0.001	0.0209	0.0335	0.0315	0.0442
	0.002	0.002	0.002	0.002	0.0307	0.0481	0.0473	0.0649
	0.003	0.003	0.003	0.003	0.0348	0.0541	0.0549	0.0716
	0.004	0.004	0.004	0.004	0.0360	0.0512	0.0569	0.0690
	0.005	0.005	0.005	0.005	0.0386	0.0500	0.0592	0.0690
	0.006	0.006	0.006	0.006	0.0372	0.0471	0.0598	0.0685
	0.0075	0.0075	0.0075	00.075	0.0356	0.0405	0.0601	0.0655
	0.009	0.009	0.009	0.009	0.0346	0.0376	0.0648	0.0676

These results in Table 12.21 indicate that there were inflations of Type 1 error under both adjusted α and 0.05 significance levels in all cases. Both the proposed Bieler and Williams (1993) and Mancuso's methods in Mancuso et al. (2002) have very similar power. The results in the Table 12.22 show that in the four treatment group experimental design with rare tumors with background incidence rates between 0 and 1 %, Type 1 errors were inflated in both the proposed and Mancuso's methods similar to the inflated Type 1 errors of the common tumors. But the proposed method has slightly more control over Type 1 error compared to Mancuso's method.

12.5.4.3 Second Set of Monte Carlo Simulations

We also conducted a second set of simulations to investigate the modified poly-3 trend test. For this set, thirty-six simulation models were obtained by varying the levels of the four factors described in Dinse (1985) and presented in Sect. 12.2.

In general we observed that:

1. By using multiplicity adjustment, sizes are all less than 0.041. An inflation of size still occurs in 8 out of 12 cases: seven in the range 0.0071–0.03 and one at 0.041 based on adjusted levels of significance.
2. When a small dose effect on tumor rate, a high background tumor rate and later tumor appearance were simulated, power appeared to be lower.
3. Some of the high background tumor rate cases have very low levels of power. In these cases, using just 5000 replicates to estimate the p-values was inadequate.

12.5.5 *Test Results Using Some NDA Datasets*

Three NDA submissions were randomly chosen from the set of submissions reviewed by the authors to compare the proposed method with PROC STRATIFY. The permutation sample sizes were all around 6000. There were about tenfold differences in p-values between the proposed and Mancuso's methods. We did further investigations on the significant cases to see how different permutation sample sizes would have an impact on p-value calculations for the trend test. Permutation samples of size of 1,000,000 were used. However, the p-values of samples of size of 1,000,000 were not much different from the p-values using the permutation samples size around 6000. The only changes were in the 3rd, 4th or 5th decimal places of the p-values. Since there are very large numbers of permutations (usually in the magnitude of 10^{100} permutations with for 50 animals per groups in a 4 group experimental design), computational limitations are a very real concern for all of these methods.

12.6 A Short Review of the General Bayesian Approaches to Possible Survival and Carcinogenicity Analyses

12.6.1 Points We Want to Make Before We Get Going

- Bayesian methods base conclusions solely on the posterior distribution of the parameters.
- It may be that an improper prior (i.e. a distribution that is not a proper density) is useful, provided it results in a valid posterior.
- Conclusions about a single parameter are based on the posterior distribution for that parameter found by integrating out all other parameters, including any nuisance parameters. This is a very general procedure, but may seem inflexible in comparison to the variety of frequentist methods.
- An approach to Bayesian hypothesis testing is consider the null hypothesis as a Bernoulli variable, and to use observed data to construct a posterior for the probability that the hypothesis is in fact true.

A central Bayesian result having far-reaching implications for both Bayesian and frequentist statistical analysis is the so-called *likelihood principle*:

> Unless results are based only on the data that were actually observed and not those that could have occurred, results can be improved.

12.6.2 Bayesianism and Nonclinical Biostatistics

Historically, except for those cases where researchers have been able to exploit conjugate priors (families of distributions for which if a prior is in the family, then any posterior will also be in the family) Bayesianism has been limited by the computational complexity of calculating posterior distributions, especially after repeated updating. However, as computational power has increased, more areas of statistics, including nonclinical biostatistics, have seen Bayesian methods become viable.

It is generally the case that as more data is collected, the influence of the initial prior diminishes, and the posterior becomes primarily a reflection of the observed empirical data. When working with large datasets, this is reassuring, and addresses the common criticism that Bayesian methods are founded on an unjustifiable choice of a prior. However, as has already been noted repeatedly in this chapter, one of the greatest challenges of reviewing rodent carcinogenicity data is the rarity of the events. So in contrast to the reassuring asymptotic case, Bayesian analyses of such data are more, rather than less, sensitive to the choice of a prior. For these smaller sample sizes, we need to use so-called *noninformative*, *vague*, or *objective* priors that do not dominate the data. The so-called *reference prior* method described in Bernardo (1979) and Bernardo and Smith (1994) is an automatic procedure for generating such priors.

In the analysis of most rodent carcinogenicity studies there are two primary goals:

1. To analyze the effect of the compound under study on survival.
2. To analyze the effect of the compound on the development of neoplasms.

In the typical frequentist testing of carcinogenicity hypotheses or survival the usual null hypothesis is that some set of parameters (typically slope parameters, such as D in the Weibull parameterization described in Sect. 12.2.2) are equal to zero. Testing this hypothesis in the manner described above (Sect. 12.6.1), we are interested in the posterior distribution of the random event that D is identically equal to 0.

The following proposed Bayesian analyses are intended to be illustrative only, and not prescriptive.

12.6.3 Notational Conventions for Examples

In the examples below, we adopt the following notational conventions:

There are I tumor types (as discussed in Sect. 12.3.6.2), J animals, and K dose groups. Animal j is a member of group $\kappa(j)$, and the total number of animals in group k is denoted n_k. Without loss of generality, let $k = 0$ denote the control group. The animals in group k are treated with dose d_k (so $d_0 = 0$). The maximum time in the study is denoted T, and the time at which animal j leaves the study (either through natural death or sacrifice) is denoted t_j.

12.6.4 Survival Analysis Example: Finite Dimensional Proportional Hazards Model

The probability of an animal surviving past time t is given by the survival function $S(t) = \Pr(T > t)$. Let $f(t)$ denote the density of T. The instantaneous hazard function is $h(t) = f(t)/S(t)$, and the cumulative hazard H is defined by:

$$H(t) = \int_0^t h(u)\mathrm{d}u.$$

The following identities follow immediately:

$$f(t) = h(t)S(t) \quad \ln(S(t)) = -H(t) \quad S(t) = \mathrm{e}^{-H(t)} \quad f(t) = \mathrm{e}^{-H(t)}.$$

The standard Cox regression form of the proportional hazards model for such survival models specifies the hazard function:

$$h(t|\mathbf{x}) = h_0(t)\mathrm{e}^{\mathbf{x}^\top\boldsymbol{\beta}}.$$

Then treatment effects can be investigated by assessing the differential effects of treatment in the $e^{\mathbf{x}^T\boldsymbol{\beta}}$ term. Among other possible specifications, this can reflect a trend over dose or individual dose effects.

Statistical inference on survival is based on proposing a probability model for $S(t)$ or one of its derivations. The probability model is defined so that hypotheses to be investigated are specified as parameters in the model. A frequentist analysis takes parameters as fixed and assesses the likelihood of the observed data. A Bayesian analysis starts by noting that parameters are not known, and assumes that a prior distribution is a natural measure of this lack of exact knowledge. Then the Bayesian analysis assesses the impact of the actual observed data on this prior.

Frequentist analysis of the Cox model uses asymptotics to analyze the linear predictor (and by extension the hazard ratio), but disregards the baseline hazard h_0.[16] By contrast, a Bayesian analysis requires priors on all parameters, including the baseline hazard. In this example, we consider a finite dimensional space of possible baseline hazard functions, namely the piecewise step functions; i.e., hazard functions of the form

$$h_0(t) = \sum_{m=0}^{M} \lambda_m \mathbf{I}_{(a_m, a_{m+1}]}(t). \tag{12.15}$$

Without loss of generality, we may assume $0 = a_0 < a_1 \ldots < a_M < a_{M+1} = T$.

In the formulation above, the baseline hazard is confounded with the specification of treatment effects, i.e., a multiplicative constant can be moved to either the baseline hazard or the term with covariates. The dose effect at level k is represented by the scalar β_k, interpreted as the log of the hazard ratio relative to the control group. Note that $\beta_0 = 0$. We thus have $K - 1$ unknown scalars, together with the unknown baseline hazard function $h_0(t)$. The model could be simplified further by assuming

$$\beta_k = \eta d_k + \mu \tag{12.16}$$

for $k > 0$ i.e., by assuming a simple linear trend in dose.

Given that $a_m < t \leq a_{m+1}$, the integrated cumulative hazard for an animal in group k may be written as:

$$H_0(t) = e^{\beta_k} \int_0^t h_0(u)\, du = e^{\beta_k} \left(\left(\sum_{n=0}^{m-1} \lambda_n (a_{n+1} - a_n) \right) + \lambda_m (t - a_m) \right)$$

and the likelihood for subject j can be written

$$L_j(\boldsymbol{\beta}) \propto \begin{cases} e^{-H_0(t_j)} & \text{if the } j^{\text{th}} \text{ subject is censored at time } t_j \\ \lambda_{\kappa(j)} e^{\beta_{\kappa(j)} - H_0(t_j)} & \text{if the } j^{\text{th}} \text{ subject fails at time } t_j. \end{cases} \tag{12.17}$$

[16] For this reason it is often called a *semiparametric* model.

Note that in the case that we use the model described in Eq. (12.16), the parameters in the likelihood function shown in Eq. (12.17) will be η and μ, rather than the parameter vector $\boldsymbol{\beta}$.

Because this looks like a sample of exponential inter-arrival times, we would expect the simple fail/not fail distributions to correspond to Poisson random variables. For subject j censored or failed at time t_j define γ_{jm} by

$$\gamma_{jm} = \begin{cases} \lambda_m \left(a_{m+1} - a_m\right) & \text{for } t_j > a_{m+1} \\ \lambda_m \left(t_j - a_m\right) & \text{for } a_m < t_j \leq a_{m+1} \\ 0 & \text{otherwise.} \end{cases}$$

Thus

$$S(t) = \mathrm{e}^{-H(t)} = \prod_{\{m \mid a_m \leq t_j\}}^{M} \exp\left(-\mathrm{e}^{\beta_{\kappa(j)}} \gamma_{jm}\right).$$

Furthermore, $(t_j - a_m)$ is constant with respect to the parameters, and hence can be incorporated in the likelihood for subjects who fail by multiplying λ_j by this difference. Thus for subject j, the likelihood can also be written as:

$$L_j(\boldsymbol{\beta}) \propto \begin{cases} \prod_{m=1}^{M} \exp\left(-\mathrm{e}^{\beta_{\kappa(j)}} \gamma_{jm}\right) & \text{if the } j^{\text{th}} \text{ subject is censored at time } t_j \\ \gamma_{jm} \mathrm{e}^{\beta_{\kappa(j)}} \prod_{m=1}^{M} \exp\left(-\mathrm{e}^{\beta_{\kappa(j)}} \gamma_{jm}\right) & \text{if the } j^{\text{th}} \text{ subject fails at time } t_j. \end{cases}$$

Although it looks messy, this is the likelihood of T independent Poisson random variables with mean $\mathrm{e}^{\beta_{\kappa(j)}} \gamma_{jm}$ where all responses are zero. This is only a computational convenience but allows easy estimation of the appropriate parameters using standard software (e.g., Lunn et al. 2000—see Sect. 12.6.8). Thus we need to specify an appropriate prior for the baseline hazard. Note that the baseline hazard is essentially the hazard of the control group. A gamma prior would be skewed to the right and would seem to be an appropriate choice. The standard 2 year study could be broken down into twelve 2 month periods. Sacrifice or accidental death could be treated as a reduction in the risk set, but not as a mortality event. In most circumstances we would might prefer a specification of an increasing hazard (again easily specified in WinBUGS or OpenBUGS (Lunn et al. 2000).

12.6.5 Carcinogenicity Example: Finite Dimensional Logistic Model

A logistic model is easy to implement in OpenBUGS or WinBUGS (Lunn et al. 2000). For this analysis we define mixed two-stage/three-stage hierarchical models for tests of trend and pairwise comparisons.

For testing trend, we define θ_{ij} to be the probability that tumor i is found in subject j, and we build the following model:

$$\text{logit}(\theta_{ij}) = \alpha_i + \beta_i d_{\kappa(j)} + \gamma_i \ln(t_j) + \delta_j \tag{12.18}$$

where δ_j is the individual random subject effect.[17]

We assign model priors:

$$\begin{aligned}
\alpha_i &\sim \text{N}(\mu_\alpha, \sigma_\alpha^2) \\
\beta_i &\sim \pi_i \mathbf{I}_{[0]} + (1 - \pi_i)\text{N}(\mu_\beta, \sigma_\beta^2) \\
\pi_i &\sim \text{Beta}(\phi, \psi) \\
\gamma_i &\sim \text{N}(\mu_\gamma, \sigma_\gamma^2)
\end{aligned}$$

for $i = 1, \ldots, I$ and a random subject effect

$$\delta_j \sim \text{N}(\mu_\delta, \sigma_\delta^2)$$

for $j = 1, \ldots, J$. For computational convenience, we typically define $\mu_\alpha = \mu_\beta = \mu_\gamma = \mu_\delta = 0$ and $\sigma_\delta^2 = \sigma_\alpha^2 + \sigma_\beta^2 = \sigma_g^2 = \sigma_s^2 = 100$, $\sigma_\alpha^2, \sigma_\beta^2, \sigma_g^2 \sim$ InverseGamma$(1, 3)$.

The model for the pairwise comparison between group k and the control group is similar:

$$\text{logit}(\theta_{ij}) = \alpha_i + \beta_{ik} \mathbf{I}_{\{\kappa(j)=k\}} + \gamma_i \ln(t_j) + \delta_j. \tag{12.19}$$

Our priors have the form:

$$\begin{aligned}
\pi_{ik} &\sim \text{Beta}(\phi, \psi) \\
\beta_{ik} &\sim \pi_{ik} \mathbf{I}_{[0]} + (1 - \pi_{ik})\text{N}(\mu_{\beta_k}, \sigma_{\beta_k}^2)
\end{aligned}$$

for $j = 1, \ldots, J$ and $i = 1, \ldots, n_t$. Note that with this parameterization, for $k = 2, \ldots, K$, the β_{ik} terms represent the deviation of treatment effect from the controls. These should represent reasonably well dispersed priors on parameters.

[17] It is interesting to note that this model implies that the odds of tumorigenesis are proportional to $t_j^{\gamma_i}$, which (when the probability of tumorigenesis is low) is essentially equivalent to the poly-k assumption discussed elsewhere in this chapter.

12.6.6 Survival Analysis Example: Nonparametric Bayesian Analysis

Some applications involve increasing numbers of parameters or even infinite dimensional problems. Perhaps the knowledge about the parameter could follow a probability distribution *not* indexed by small set of parameters. For example, instead of something like a simple normal distribution indexed with a mean, μ and variance, σ^2, the family could be say one of the continuous location family distributions or possibly even the inclusive continuous probability distributions. In a simple misnomer such problems have come to be called "Bayesian Nonparametrics." The challenge is not the fact that there are no parameters, but rather that there are far too many. Since it seems to be quite difficult to specify priors with content in infinite dimensional space it seems more appropriate to work with objective priors that cover much of the parameter space.

One of many possible standard models for the survival function is to model the logarithm of the survival with a normal distribution, i.e. to specify that T_i follows a lognormal distribution. However, the typical Bayesian nonparametric model takes such a specification and uses it as a baseline function to be perturbed to "robustify" the model using a so-called Dependent Dirichlet Process (DDP) as the prior on this space of probability distributions. This function represents the prior using a so-called Dependent Dirichlet Process (DDP) as the prior on this space of probability distributions, which uses a mixture of normal distributions weighted by a Dirichlet process on the normal parameters. The prior is defined as a Dirichlet process where the baseline distribution models the linear parameters, where has the linear mean parameters has a normal distribution as prior and the variance parameters with a Gamma distribution. The prior of the precision parameter of the Dirichlet process is specified as a gamma distribution. The priors for the other hyperparameters in this function are conjugate distributions. Following the notation of Jara et al. (2014), we can write:

$$
\begin{aligned}
\ln(T_i) &= t_i|\mathbf{f}_{X_i} \sim \mathbf{f}_{X_i} \\
\mathbf{f}_{X_i} &= \int \mathrm{N}(X_i\boldsymbol{\beta}, \sigma^2)G(\mathrm{d}\boldsymbol{\beta}\mathrm{d}\sigma^2) \\
G|\alpha, G_0 &\sim DP(\alpha G_0)
\end{aligned}
$$

Typically distributions of the hyperparameters above can be specified as follows:

$$
\begin{aligned}
G_0 &= \mathrm{N}(\beta|\mu_b, s_b)\Gamma\left(\sigma^2|\frac{\tau_1}{2}, \frac{\tau_2}{2}\right) \\
\alpha|a_0, b_0 &\sim \mathrm{Gamma}(a_0, b_0) \\
\mu_b|m_0, s_0 &\sim \mathrm{N}(m_0, S_0) \\
s_b|\nu, \Psi &\sim \mathrm{InvWishart}(\nu, \Psi) \\
\tau_2|\tau_{s_1}, \tau_{s_2} &\sim \mathrm{Gamma}(\tau_{s_1}, \tau_{s_2})
\end{aligned}
$$

See, for instance De Iorio et al. (2009). The parameterization used to compare doses can be captured by a dummy coding, as in the finite dimensional example (see Sect. 12.6.5).

12.6.7 Carcinogenicity Example: Nonparametric Logistic Model

A similar model to the one in Example 12.6.5 takes the baseline distribution as a logistic distribution. The nonparametric Bayesian approach treats an actual probability distribution as one of the parameters. This distribution is then sampled from an infinite dimensional space of possible distributions, which is both mathematically challenging and where, unlike most finite dimensional parameters, it is difficult to specify appropriate prior distributions. Thus one attempts to specify robust priors on the slope and treatment differences that have a small impact on the result. The baseline model follows a simple logit model for tests of trend and pairwise comparisons. For testing trend, we define p_{ijk} as the probability of tumor type i being found in subject j in treatment group k. That is, with $i = 1$ to n_t tumors and $j = 1$ to n_s animals, and dose d_k, leaving the experiment at time $t - j$ and subject effect δ_j:

$$\text{logit}(p_{ijjk}) = \alpha_i + \beta_i d_k + \gamma_i t_j + \delta_j \tag{12.20}$$

with assigned model priors:

$$\begin{aligned} \alpha_i &\sim \text{N}(\mu_{\alpha_i}, \sigma_\alpha^2) \\ \beta &\sim \text{N}(\mu_B, \sigma_\beta^2) \\ \gamma_i &\sim \text{N}(\mu_\beta, \sigma_g^2) \end{aligned}$$

But now, instead of directly specifying that the animal random effect δ_j, we specify the distribution as a Dirichlet process (DP) on the space of distributions.

$$\begin{aligned} \delta_i | G &\sim G \\ G|\alpha, G_0 &\sim \text{DP}(\alpha G_0) \\ G_0 &\sim \text{N}(\mu, \Sigma) \end{aligned}$$

Note that care seems to be needed to ensure that parameters are identified. Again, this is a simple application of the function in the `DPpackage` (Jara et al. 2014) in R (R Core Team 2012).

12.6.8 Software

Prior to the development of various Markov Chain Monte Carlo methods, actual software for doing a Bayesian analysis was largely limited to approximate solutions or even insisting on so-called conjugate priors. While many statisticians could see the philosophical advantages of Bayesian methodology, the lack of good methods of computing the posterior limited the application of these methods. All that has changed with the development of so called Markov Chain Monte Carlo methods, in their most simple form similar to so-called importance sampling.

For most problems in "classical" Bayesian analyses the user is faced with a plethora of choices in packages or programs. WinBUGS and its more recent descendant OpenBUGS (Lunn et al. 2000) are probably the oldest and most used general programs ("BUGS" stands for *Bayesian analysis Using Gibbs Sampling*). The various versions of the manuals have the warning in quite noticeable type "WARNING: MCMC can be dangerous." The point is that MCMC methods work well when they move around through the appropriate parameter space reaching near all feasible points. When they are stuck in a region of the parameter space and can not leave the MCMC methods can fail. WinBUGS includes several diagnostics for this type of behavior.

Several SAS procedures have options for a Bayesian analysis, usually with default, but quite reasonable priors. For more general use, PROC MCMC, provides a detailed analysis including extensive diagnostics on the MCMC Markov Chains, but, as with all very general procedures, requires careful coding of priors and likelihoods. It includes extensive, possibly nearly exhaustive, diagnostics for the MCMC behavior.

R users (R Core Team 2012) have a number of packages for Bayesian Analysis available to them. `LaplacesDemon` is a very general package. `MCMCpack` includes a relatively long list of functions for MCMC analysis. `BayesSurv` has a number of R functions for survival models. Last but certainly not least, in this short and by no means exhaustive list, `DPpackage` (Jara et al. 2014) is a very general collection of functions for Nonparametric Bayesian Analysis and is undoubtedly currently the easiest way to implement such models.

Acknowledgements The authors would like to acknowledge the support of Yi Tsong, the director of Division of Biometrics 6, in FDA/CDER/OTS/OB, while researching the work contained in this chapter.

References

Ahn H, Kodell R (1995) Estimation and testing of tumor incidence rates in experiments lacking cause-of-death data. Biom J 37:745–765

Armitage P (1955) Tests for linear trends in proportions and frequencies. Biometrics 11(3): 375–386

Bailer AJ, Portier CJ (1988) Effects of treatment-induced mortality and tumor-induced mortality on tests for carcinogenicity in small samples. Biometrics 44(2):417–431

Baldrick P, Reeve L (2007) Carcinogenicity evaluation: comparison of tumor data from dual control groups in the cd-1 mouse. Toxicol Pathol 35(4):562–575

Bergman CL, Adler RR, Morton DG, Regan KS, Yano BL (2003) Recommended tissue list for histopathologic examination in repeat-dose toxicity and carcinogenicity studies: a proposal of the society of toxicologic pathology (stp). Toxicol Pathol 31(2):252–253

Bernardo JM (1979) Reference posterior distributions for Bayesian inference. J R Stat Soc Ser B Methodol 41(2):113–147

Bernardo JM, Smith AFM (1994) Bayesian statistics. Wiley, Chichester

Bieler GS, Williams RL (1993) Ratio estimates, the delta method, and quantal response tests for increased carcinogenicity. Biometrics 49(3):793–801

Center for Drug Evaluation and Research (2005) Reviewer guidance: conducting a clinical safety review of a new product application and preparing a report on the review. United States Food and Drug Administration

Center for Drug Evaluation and Research (2103) Pharmacology review—NDA 205437 (otzela). Technical report, US Food and Drug Administration. http://www.accessdata.fda.gov/drugsatfda_docs/nda/2014/205437Orig1s000PharmR.pdf

Chang J, Ahn H, Chen J (2000) On sequential closed testing procedures for a comparison of dose groups with a control. Commun Stat Theory Methods 29:941–956

Chernick MR, Liu CY (2002) The saw-toothed behavior of power versus sample size and software solutions. Am Stat 56(2):149–155

Clinical Data Interchange Standards Consortium (CDISC) (2011) Standard for exchange of nonclinical data implementation guide: nonclinical studies version 3.0

De Iorio M, Johnson WO, Müller P, Rosner GL (2009) Bayesian nonparametric nonproportional hazards survival modeling. Biometrics 65(3):762–771. doi:10.1111/j.1541-0420.2008.01166.x. http://dx.doi.org/10.1111/j.1541-0420.2008.01166.x

Dinse GE (1985) Testing for a trend in tumor prevalence rates: I. nonlethal tumors. Biometrics 41(3):751

Fairweather WR, Bhattacharyya A, Ceuppens PR, Heimann G, Hothorn LA, Kodell RL, Lin KK, Mager H, Middleton BJ, Slob W, Soper KA, Stallard N, Venture J, Wright J (1998) Biostatistical methodology in carcinogenicity studies. Drug Inf J 32:401–421

Gebregziabher M, Hoel D (2009) Applications of the poly-*k* statistical test to life-time cancer bioassay studies. Hum Ecol Risk Assess 15(5):858–875

Giknis MLA, Clifford CB (2004) Compilation of spontaneous neoplastic lesions and survival in Crl:CD® rats from control groups. Charles River Laboratories, Worcester

Giknis MLA, Clifford CB (2005) Spontaneous neoplastic lesions in the CrlCD-1(ICR) mouse in control groups from 18 month and 2 year studies. Charles River Laboratories, Worcester

Haseman J (1983) A reexamination of false-positive rates carcinogenesis studies. Fundam Appl Toxicol 3(4):334–343

Haseman J (1984) Statistical issues in the design, analysis and interpretation of animal carcinogenicity studies. Environ Health Perspect 58:385–392

Haseman J, Winbush J, O'Donnel M (1986) Use of dual control groups to estimate false positive rates in laboratory animal carcinogenicity studies. Fundam Appl Toxicol 7:573–584

Haseman JK, Hailey JR, Morris RW (1998) Spontaneous neoplasm incidences in fischer 344 rats and b6c3f1 mice in two-year carcinogenicity studies: a national toxicology program update. Toxicol Pathol 26(3):428–441

Heimann G, Neuhaus G (1998) Permutational distribution of the log-rank statistic under random censorship with applications to carcinogenicity assays. Biometrics 54:168–184

Jackson MT (2015) Improving the power of long term rodent bioassays by adjusting the experimental design. Regul Toxicol Pharmacol 72(2):231–342. http://dx.doi.org/10.1016/j.yrtph.2015.04.011

Jara A, Hanson T, Quintana F, Mueller P, Rosner G (2014) Package dppackage. http://www.mat.puc.cl/~ajara

Kodell R, Ahn H (1997) An age-adjusted trend test for the tumor incidence rate for multiple-sacrifice experiments. Biometrics 53:1467–1474

Kodell R, Chen J, Moore G (1994) Comparing distributions of time to onset of disease in animal tumorigenicity experiments. Commun Stat Theory Methods 23:959–980

Lin KK (1995) A regulatory perspective on statistical methods for analyzing new drug carcinogenicity study data. Bio/Pharam Q 1(2):19–20

Lin KK (1997) Control of overall false positive rates in animal carcinogenicity studies of pharmaceuticals. Presentation, 1997 FDA Forum on Regulatory Science, Bethesda MD

Lin KK (1998) CDER/FDA formats for submission of animal carcinogenicity study data. Drug Inf J 32:43–52

Lin KK (2000a) Carcinogenicity studies of pharmaceuticals. In: Chow SC (ed) Encylopedia of biopharmaceutical statistics, 3rd edn. Encylopedia of biopharmaceutical statistics. CRC Press, Boca Raton, pp 88–103

Lin KK (2000b) Progress report on the guidance for industry for statistical aspects of the design, analysis, and interpretation of chronic rodent carcinogenicity studies of pharmaceuticals. J Biopharm Stat 10(4):481–501

Lin KK, Ali MW (1994) Statistical review and evaluation of animal carcinogenicity studies of pharmaceuticals. In: Buncher CR, Tsay JY (eds) Statistics in the pharmaceutical industry, 2nd edn. Marcel Dekker, New York

Lin KK, Ali MW (2006) Statistical review and evaluation of animal carcinogenicity studies of pharmaceuticals. In: Buncher CR, Tsay JY (eds) Statistics in the pharmaceutical industry, 3rd edn. Chapman & Hall, Boca Raton, pp 17–54

Lin KK, Rahman MA (1998) Overall false positive rates in tests for linear trend in tumor incidence in animal carcinogenicity studies of new drugs. J Biopharm Stat 8(1):1–15

Lin KK, Thomson SF, Rahman MA (2010) The design and statistical analysis of toxicology studies. In: Jagadeesh G, Murthy S, Gupta Y, Prakash A (eds) Biomedical research: from ideation to publications, 1st edn. Wolters Kluwer, New Delhi

Lunn D, Thomas A, Best N, Spiegelhalter D (2000) Winbugs—a bayesian modelling framework: concepts, structure, and extensibility. Stat Comput 10:325–337

Mancuso J, Ahn H, Chen J, Mancuso J (2002) Age-adjusted exact trend tests in the event of rare occurrences. Biometrics 58:403–412

Moon H, Ahn H, Kodell RL, Lee JJ (2003) Estimation of *k* for the poly-*k* test with application to animal carcinogenicity studies. Stat Med 22(16):2619–2636

Peto R, Pike MC, Day NE, Gray RG, Lee PN, Parish S, Peto J, Richards S, Wahrendorf J (1980) Guidelines for simple, sensitive significance tests for carcinogenic effects in long-term animal experiments. IARC Monogr Eval Carcinog Risk Chem Hum Suppl NIL (2 Suppl):311–426

Portier C, Hoel D (1983) Optimal design of the chronic animal bioassay. J Toxicol Environ Health 12(1):1–19

Portier C, Hedges J, Hoel D (1986) Age-specific models of mortality and tumor onset for historical control animals in the national toxicology programs carcinogenicity experiments. Cancer Res 46:4372–4378

R Core Team (2012) R: A language and environment for statistical computing. http://www.R-project.org/

Rahman MA, Lin KK (2008) A comparison of false positive rates of Peto and poly-3 methods for long-term carcinogenicity data analysis using multiple comparison adjustment method suggested by Lin and Rahman. J Biopharm Stat 18(5):949–958

Rahman MA, Lin KK (2009) Design and analysis of chronic carcinogenicity studies of pharmaceuticals in rodents. In: Peace KE (ed) Design and analysis of clinical trials with time-to-event endpoints. Chapman & Hall/CRC Biostatistics series. Taylor & Francis, Boca Raton

Rahman MA, Lin KK (2010) Statistics in pharmacology. In: Jagadeesh G, Murthy S, Gupta Y, Prakash A (eds) Biomedical research: from ideation to publications, 1st edn. Wolters Kluwer, New Delhi

Rahman MA, Tiwari RC (2012) Pairwise comparisons in the analysis of carcinogenicity data. Health 4:910–918

US Food and Drug Administration—Center for Drug Evaluation and Research (2001) Guidance for industry: statistical aspects of the design, analysis, and interpretation of chronic rodent carcinogenicity studies of pharmaceuticals. US Department of Health and Human Services, unfinalized – draft only

Westfall P, Soper K (1998) Weighted multiplicity adjustments for animal carcinogenicity tests. J Biopharm Stat 8(1):23–44

Chapter 13
Design and Evaluation of Drug Combination Studies

Lanju Zhang and Hyuna Yang

Abstract Drug combination studies are generally conducted to look for synergistic effects. In this chapter, we discuss typical study design and analysis methods for drug combination studies, with a focus on in vitro experiments. Two reference models Loewe additivity and Bliss independence are used for synergy evaluation. A new index based on Bliss independence is introduced, comparable to the interaction index based on Loewe Additivity. An example data set is used to demonstrate the implementation of these analysis methods. In the final discussion, we point out some future research areas.

Keywords Bliss independence • Dose response curve • Drug combination • Interaction index • Loewe Additivity • Synergy

13.1 Introduction

Drug combination studies are conducted in many fields. In the pharmaceutical industry, as drug research and development costs have been skyrocketing and the attrition rate has been climbing up, companies are looking to combine several compounds as a new treatment for diseases. This is especially the case in oncology and infectious disease therapeutic areas. In agriculture, scientists may seek to combine several synergistic chemicals for a more effective pesticide. In environmental sciences, environmentalists may also study whether a toxicity synergy can result from combination of chemical pollutants discharged together. Scientists are looking for therapeutic synergy in the first two scenarios and toxic synergy in the third scenario.

This chapter will focus on drug combination studies in the pharmaceutical industry. Just as mutual funds provide more investment options by combining different individual stocks, combination therapies offer scientists a discovery tool that can lead to many more treatment options. However, drug combination also

L. Zhang (✉) • H. Yang
AbbVie Inc., North Chicago, IL, USA
e-mail: Lanju.zhang@abbvie.com; hyuna.yang@abbvie.com

L. Zhang (ed.), *Nonclinical Statistics for Pharmaceutical and Biotechnology Industries*, Statistics for Biology and Health, DOI 10.1007/978-3-319-23558-5_13

poses challenges to the drug discovery and development process. For example, in addition to studying each of the combined compounds, one needs to test a factorial combination of many doses of both compounds. The first question is how to design such experiments. Another challenge is how to evaluate the combined effect. As can be seen shortly, defining synergy is based on a reference model, which has several options but no one has been convincingly established as the best. Then the same study result can lead to different conclusions using different reference models.

Drug combination studies are conducted in different phases of drug discovery and development. Early in drug discovery, after a disease target is identified, scientists screen compounds that may have a target related. There are two ways of screening. One is unguided screening, in which some compounds sampled from a library are combined and studied. In this case, scientists only have a limited understanding about the mechanism of action of these compounds to narrow down candidate compounds in the library. The other is guided screening, where the mechanism of action of a candidate is well characterized and shows some effect on the target. Then based on their knowledge of biology or disease, scientists hypothesize compounds whose mechanism of action is understood and projected to be synergistic with the compound of interest. Then combination studies of these compound pairs are designed and conducted. At this stage, drug combination studies are typically in vitro, so multiple doses, usually no less than 4, of each compound can be used in the combinations, resulting in a factorial matrix of 4×4 or even higher dimension. In this case, a dose response curve of each monotherapy can be well modeled and synergy can be evaluated based on either Loewe additivity (Loewe 1928) or Bliss independence (Bliss 1939), which will be introduced in a later section.

If a compound pair is found to be synergistic in vitro, it moves to in vivo testing. In this stage, only two or at most three dose levels of each can usually be studied, limited by the number of animals appropriate for such studies. Studies typically use a simple factorial design. However, it is hard to evaluate synergy using methods as those for in vitro studies. One option is the min test (Hung 2000), which tests the hypothesis that the combination is superior to both single agents. This test is also used to show the efficacy of a drug combination in clinical trials. It may be helpful to note that a *drug* usually means a compound that has been approved by a regulatory agency for marketing. Before approval, it is called a *compound* or *drug candidate*. However, we use compound and drug interchangeably in this chapter.

In this chapter, we will discuss experimental designs for drug combination studies and different analysis methods for study data, with a focus on in vitro studies and a passing discussion of in vivo and clinical studies. We will propose an analysis method for combinations which we recommend as good practice to interpret study results.

Before we move to the next section, we will note a brief history of drug combination studies. In a broad sense, combination therapies can be dated back thousands of years ago when the Chinese started herbal medicines which typically included a combination of several herbs. However, herbal medicines were different from modern combination therapies because the dose response of each active herb in the combination was not characterized and the dose of each herb in the combination was based on empirical knowledge and often varied from one prescriber to another. Systematic evaluation of drug synergy started in the early twentieth century when Loewe defined a reference model Loewe additivity (Loewe 1928) and Bliss defined an independence reference model (Bliss 1939). Later a wealth of literature has been developed by pharmacologists, statisticians, chemists, and biologists. Excellent reviews include Berenbaum (1989) and Greco et al. (1995).

13.2 Study Design

Studies at different stages use different designs. For in vitro studies, several designs have been proposed. One is the fixed ratio (ray) design, in which one can select a dose D_1 of drug 1 and a dose D_2 of drug 2 (e.g., their EC_{50}s), and a fraction c, to obtain a combination dose $D = c\, D_1 + (1\text{-}c)\, D_2$. Then a series of concentrations or dilutions of this combination forms a ray as in Fig. 13.1. The combination fraction c can take different values, resulting in other rays. For a detailed discussion of this design, see Tallarida (2000) and Straetemans et al. (2005). For its optimal version, see Donev and Tobias (2011).

This design is typically used in industry for in-vitro studies. A third design (Tan et al. 2009) is an optimal one based on a uniform measure that meets some power requirement for testing departure from Loewe additivity.

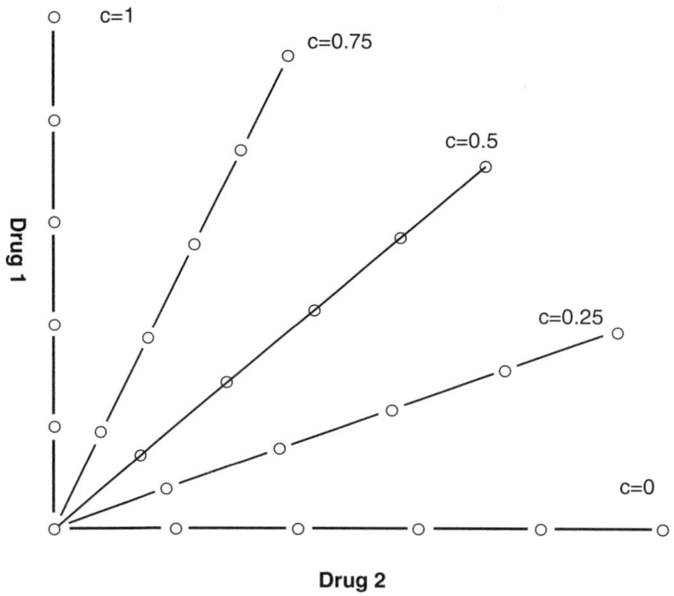

Fig. 13.1 A ray or fixed ratio design for two drugs in combination studies

For in vivo studies, the number of doses is limited to typically two per compound to avoid using an excessive number of animals. Therefore, usually a factorial design is used.

13.3 Analysis Methods

13.3.1 In Vitro Studies

In Vitro studies typically use a 96-well plate with a number of doses of each monotherapy and their combinations. The experiment can be replicated on, for example, three plates. A factorial/checkerboard design is often used in the industry. Here we illustrate the study with an example from Harbron (2010) and the data is reproduced in Table 13.1. Nine doses were tested for each monotherapy with three-fold spacing, of which the six lowest doses were tested in combination based on a factorial design. The test was replicated on three plates, resulting in three growth inhibition values. We use the three-parameter logistic model to model the monotherapy dose response curves.

$$y = f(d) = \frac{E\ d^m}{d^m + (EC_{50})^m} \tag{13.1}$$

where y is the drug response at dose d, E the maximum drug response (tumor inhibition), m the slope parameter, and EC_{50} the dose that produces 50 % of the maximum drug response. Note that another parameter can be added to represent the minimum drug response, which we assume to be 0. The model was fit to monotherapy data of drug 1 and drug 2 in Table 13.1 and displayed in Fig. 13.2.

For the data in Table 13.1, the question of interest is, "Are these two drugs synergistic?"

This turns out to be a question that has been extensively explored for almost a century and will continue to be researched. The difficulty can be illustrated with the following example. Suppose two people A and B need to plant and water a given number of trees in a park. A and B can finish planting and watering in 10 h and 6 h respectively by oneself. Now suppose they work together. If each still does his/her own planting and watering, then together you can expect they will finish in $1/(1/10 + 1/6) = 3.75$ h. However, if they finish in a shorter time, say, 3.5 h, then we will say there is synergy from their teamwork. On the other hand, if they finish in a longer time, say, 5 h, then there is antagonism from their teamwork. In this case the expected or reference teamwork time is 3.75 h. Now consider another case where person A is better at planting and person B is better at watering. Let us assume it takes person A to plant all trees in 3 h and person B to water all trees in 3 h. Then they can finish planting and watering in 3 hours. So in this case, 3 h is the reference

Table 13.1 An in vitro drug combination data (reproduced from the data table in Harbron (2010))

	Drug 2										
		0	0.037	0.11	0.33	1	3	9	27	81	243
Drug 1	0		3.1	1	1	8.5	13.3	23.7	53.1	78.9	93.5
			1.5	1	8.8	1	14.7	30.2	59	82.9	98.8
			1	1	5.9	4.5	18.1	42.5	62	81.5	86.2
	0.037	1	1	1	1	8.4	21.8	38.5			
		5.8	2	1	2.9	10	4.7	34.9			
		1	1	1	4.2	7.6	9.5	35.2			
	0.11	1	1	2.6	1	5.4	22.2	32.8			
		1	1	1	2.5	9.8	22.5	34.8			
		1	1	9.2	2	8.9	15.6	30.4			
	0.33	1	1	1	4.7	8.5	22.5	37.9			
		4.2	6.2	4.9	6.3	12.3	19.8	41.7			
		13.3	6.1	9.5	5.6	7.2	15.9	34.3			
	1	1.9	16	3.4	21.2	22.9	34	52.9			
		4.2	6	6.6	19.6	23.4	37.7	46.4			
		5.7	15.8	15.5	14.7	26.4	42.1	53.9			
	3	20.6	41.1	49.4	43	50.5	55.8	66.8			
		31.7	42.1	50.4	48.3	40	56.6	59.2			
		23.9	43.1	51.3	46.1	52.5	61.8	64.2			
	9	56.2	69.2	66.8	76.8	84.7	75.6	77.5			
		58.5	82.1	83.5	83.4	79.3	68.6	77.6			
		66.6	71.1	72.8	83.1	84	85.5	79.8			
	27	89.4									
		84.9									
		85.8									
	81	92.9									
		97.6									
		90.9									
	243	99									
		93.7									
		99									

teamwork time. If together they finish in less than (or more than) 3 h, then their teamwork is synergistic (or antagonistic). For example, if they finish in 3.5 h, then there is antagonism from their combination work. Therefore, evaluation of synergy depends on a reference value which in turn is dependent on the way these two people work.

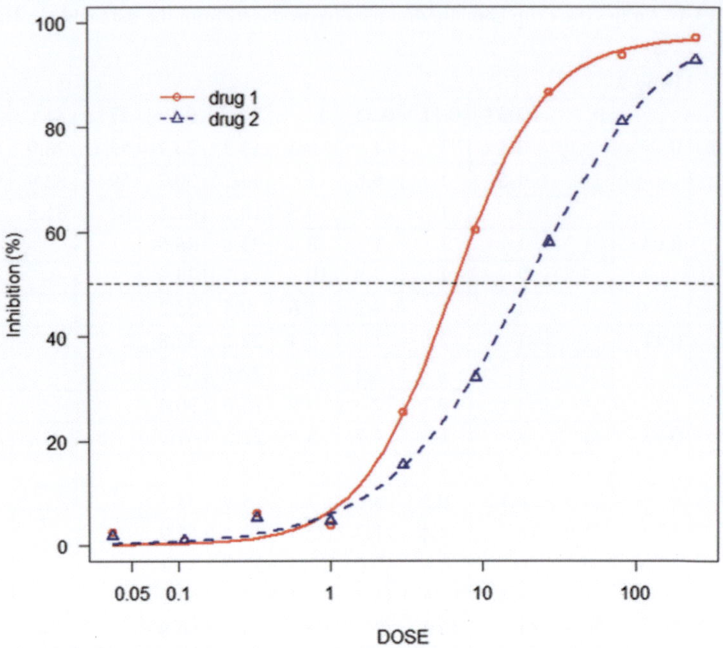

Fig. 13.2 Three parameter logistic model fitting for monotherapy data of drug 1 and drug 2

13.3.1.1 Reference Models

We will see that evaluation of the synergy of a drug combination also needs a reference model. There are different choices of reference model that may lead to different conclusions for the same drug combination data. Some suggest that choice of reference model be determined by the compounds' mechanism of action (Greco et al. 1995). Even if this is the case, sometimes we know the mechanism of action of both compounds, but other times we do not know. In the above example, if we do not know about these two people's working efficiency, but only know they together finish in 3.5 h, then we cannot tell whether or not there is synergy. In the following we discuss reference models for drug combinations, their relationship and how/when to apply them.

13.3.1.1.1 Loewe Additivity

This reference model dates back to the early twentieth century (Loewe 1928). Suppose drug 1 and drug 2 have monotherapy dose response curves, respectively,$y_i = f_i\left(d_i\right),\ i = 1,\ 2$, as defined in (13.1), where d_i is the dose of drug i. If a combination of the two drugs with doses d_1 and d_2 produces a response y, and,

$$\frac{d_1}{D_1(y)} + \frac{d_2}{D_2(y)} = \tau \tag{13.2}$$

where $D_i(y) = f_i^{-1}(y), \quad i = 1, \; 2$ is the dose for monotherapy drug i to generate the same response y, then we say the two drugs are synergistic (additive, antagonistic) when $\tau < 1(=1, >1)$ at combination d_1 and d_2. Several comments are in order for this definition. First, τ is also called interaction index (Berenbaum 1989). Second, the effect of drug combination is evaluated locally. In other words, τ depends on d_1 and d_2. It is very likely that two drugs are synergistic at one combination, but additive or antagonistic in other combinations. Third, when τ =1, we can re-write the equation as,

$$d_1 + \frac{D_1(y)}{D_2(y)} d_2 = D_1(y)$$

where $\rho(y) = D_1(y)/D_2(y)$ is called relative potency between drug 1 and drug 2 at response y. Then amount of drug 2, d_2, can be considered as equivalent to $(D_1(y) - d_1)$ in terms of drug 1, with consideration of their relative potency, to produce a response y. If $\rho(y)$ does not depend on y, we say these two drugs have a constant relative potency. So this definition of additivity applies both when relative potency is constant and when it is not constant. In the latter case, remember to use the relative potency at response y from the combination, which is often misunderstood. For example, Perterson and Novick (2007, pp. 131–132) argue that this definition produces contradicting results when relative potency is not constant, but they did not use the relative potency at the response y produced by the combination d_1 and d_2. Instead, they used relative potency at $f_1(d_1)$ in one derivation and $f_2(d_2)$ in the other derivation. A detailed explanation of Loewe additivity with varying relative potency can be found in Plummer and Short (1990). Finally, the definition can be presented graphically with an isobologram as in Fig. 13.3. The response used in the figure is $y = 50$ %. If a combination of the two drugs at doses (d_1, d_2) also produces a response $y = 50$ %, then the two drugs are synergistic (additive, antagonistic) at these two doses if the point (d_1, d_2) falls below (on, above) the line in Fig. 13.3.

There are two types of synergy evaluation methods based on Loewe additivity. One is to estimate the interaction index and find its confidence interval (Greco et al. 1995; Lee and Kong 2009; Harbron 2010). Instead of estimating the index at each combination, Zhao et al. (2012) constructed an interaction index surface with a confidence band for the drug combination region ($a < d_1 < b$, $c < d_2 < d$), which provides synergy information for tested or untested combinations in this region in a continuum. The other type is a response surface approach and a parameter(s) in the response model measures the departure from additivity (Plummer and Short 1990; Greco et al. 1995; Kong and Lee 2006). Although these two types are in essence equivalent, the former gives the interaction index more explicitly. In this chapter we will take the first method since it is consistent with use of isobolograms, with which most scientists are familiar.

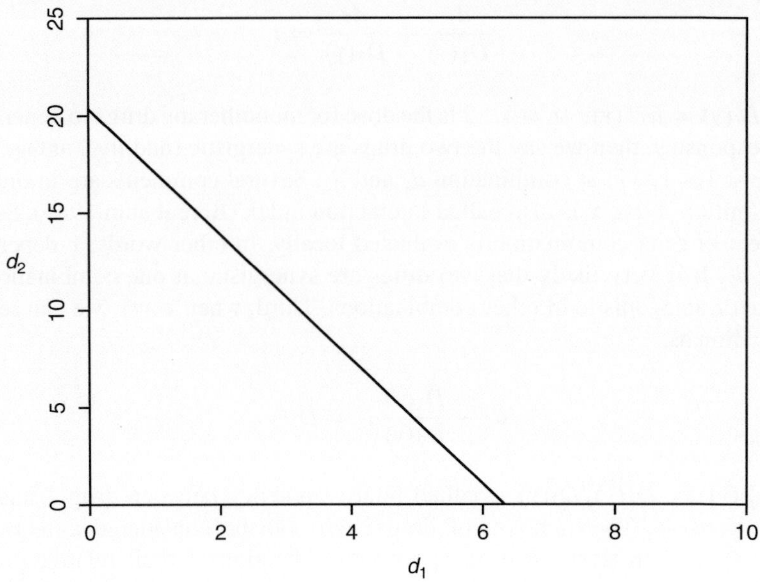

Fig. 13.3 Isobologram at 50 % inhibition. The two drugs are synergistic (antagonistic) at any combination of (d_1, d_2) below (above) the line and producing 50 % inhibition

13.3.1.2 Bliss Independence

This reference model was proposed by Bliss (1939). Assuming two drugs act on the same target independently, the Bliss independence model can be presented as follows,

$$p_{12} - (p_1 + p_2 - p_1 * p_2) \begin{cases} > 0, & \text{Synergistic} \\ = 0, & \text{Independence} \\ < 0, & \text{Antagonistic} \end{cases} \tag{13.3}$$

where p_{12}, p_1, and p_2 are observed inhibition of combining drug 1 and drug 2 at doses (d_1, d_2), monotherapy drug 1 at dose d_1, and monotherapy drug 2 at dose d_2, respectively. For example, in Table 13.1, drug 1 had average inhibition $p_1 = 60.4\%$ at $d_1 = 9$, drug 2 had average inhibition $p_2 = 15.4$ % at dose $d_2 = 3$, and a combination of drug 1 and drug 2 at (9, 3) resulted in an average inhibition of $p_{12} = 76.6$ %, which was larger than $p_1 + p_2 - p_1 * p_2 = 66.5$ %, the two drug are synergistic at combination (9, 3).

Note that

$$p_{12} - (p_1 + p_2 - p_1 * p_2) = 0 \tag{13.4}$$

is the probability statement for a union of two independent events (Casella and Berger 2002). It can be interpreted as: the expected chance, under the assumption of Bliss independence, of a disease target being affected by a combination of drug 1 at dose d_1 and drug 2 at dose d_2 is equal to the sum of chances of the disease target being affected by monotherapies drug 1 and drug 2 at the same doses, respectively, minus the chance of the disease target being affected by both drugs simultaneously. In fact, these p's are the same as "y" in Eq. (13.1). We use p here for its connection to probabilities.

At the time of writing, there is no index defined based on Bliss independence in the literature. In the following we introduce a new index. Equation (13.4) can be rewritten as,

$$1 - p_{12} = (1 - p_1)(1 - p_2)$$

Similar to the interaction index based on Loewe additivity, we define the independence index as,

$$\frac{1 - p_{12}}{(1 - p_1)(1 - p_2)} = \gamma$$

and

$$\gamma \begin{cases} < 1, & \text{Synergistic} \\ = 1, & \text{Independence} \\ > 1, & \text{Antagonistic} \end{cases}$$

To implement the independence index, we suggest using log-transformation for a better model fitting.

$$\log(1 - p_{12}) = \eta\left(\log(1 - p_1) + \log(1 - p_2)\right)$$

or $\frac{\log(1-p_{12})}{\log(1-p_1)+\log(1-p_2)} = \eta$, and now

$$\eta \begin{cases} > 1, & \text{Synergistic} \\ = 1, & \text{Independence} \\ < 1, & \text{Antagonistic} \end{cases}$$

since $0 < 1 - p_{12} < 1$, and $\log(1 - p_{12}) < 0$. With introduction of this new index, analysis methods based on Loewe additivity and Bliss independence can be conducted in a similar fashion and both results can be presented as an index.

13.3.1.3 Analysis of an Example

Here we use the data in Table 13.1 to demonstrate analysis methods based on the two reference models discussed in the last two subsections. For the Loewe additivity reference model, we recommend Harbron's unified approach (Harbron 2010). It can model an overall interaction index for the whole experimental dose space of the two drugs, interaction indices for each dose level of drug 1 or drug 2, or interaction indices at each combination of two drugs. Since these models are hierarchical, one can do model selection by testing extra residual reduction as in typical regression methods. The model does not only give a point estimate of the interaction indices, but also provides a p-value or interval estimate. A core R program for the model fitting is available in the appendix of that paper.

First we fitted model (13.2) with dose response model in (13.1). The result was $\widehat{\tau} = 0.605$ with a p-value <0.0001. Therefore, overall the two drugs were statistically significantly synergistic.

Next we fitted a model with an interaction index at each combination. The model is given in Eq. (13.5).

$$\frac{d_1}{D_1(y)} + \frac{d_2}{D_2(y)} = \tau\,(d_1, d_2) \tag{13.5}$$

The results are shown in Fig. 13.4, where indices in color were statistically significant (at significance level 0.05). This figure revealed that the two drugs were significantly synergistic at high doses of drug 1.

Other index pattern can be modeled, for example, for each dose level. An interaction index surface can also be constructed, as in Zhao et al. (2012).

Next we analyze the same data based on the Bliss independence reference model. We first use an approach that is simple and often used by scientists. Since at each dose combination there were three replicates, we can obtain three values of the difference on the left hand side of (13.3). A t-test can be used to test whether the difference is significantly different from zero with the three values. The mean difference at each combination is depicted in Fig. 13.5, with numbers in bold, italic, and color font statistically significantly larger than zero (at significance level 0.05). These numbers represent synergistic dose combinations. If a mean difference is statistically significantly smaller than zero, then corresponding dose combination is antagonistic. There is no such combination based on Table 13.1. Similar to Fig. 13.4, Fig. 13.5 also revealed that the two drugs were synergistic at higher dose levels of drug 1. However, there were more significantly synergistic dose combinations based on Loewe Additivity than those based on Bliss independence.

In this analysis, a t-test was conducted at each drug combination. Since there were only three replicates, the power is obviously limited. One may use another significance level instead of 0.05. A better approach is to analyze all the data in one model. One way to do this was achieved in Zhao et al. (2014).

Finally we analyze this data set with the independence index. This can also use Harbron's framework and determine an overall independence index and an

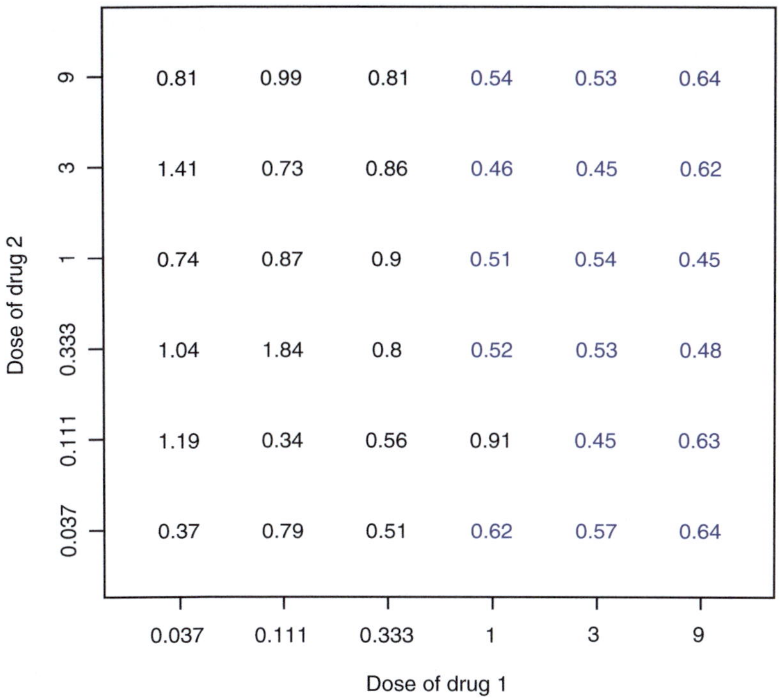

Fig. 13.4 Estimates of interaction index at each combination. Statistically significant synergistic effects are shown in *blue*. (Note these estimates differ slightly from the figure in Harbron's (2010) paper)

independence index at each dose combination. The Overall independece index was estimated to be $\widehat{\eta} = 1.52$ with a p-value <0.0001, indicating two drugs were statistically significantly synergistic. Independence indices at all dose combinations are shown at Fig. 13.6, revealing almost the same synergy results as the interaction index method.

13.3.1.4 Relationship Between Loewe Additivity and Bliss Independence

Obviously for a given data set, using different reference models can give different synergy result. Which one should a scientist take? If the mechanism of action (MoA) is ascertained, then a choice can be made: if the two drugs have the same MoA, then the Loewe additivity model should be used; on the other hand, if the two drugs have independent MoAs, then the Bliss independence model should be used. However, in most situations, the MoA of two drugs is usually not ascertained. So it is interesting to know the relationship between these two reference models, irrespective of MoA of the two drugs.

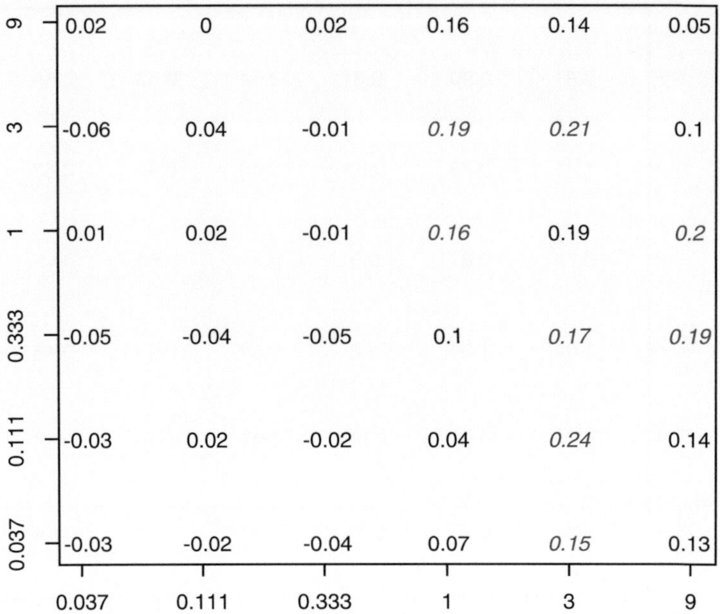

Fig. 13.5 Observed minus expected mean difference at each combination. Statistically significant synergistic effects are shown in *blue*

Berenbaum (1989) showed that the two reference models were equivalent if the dose response of two drugs follows an exponential model. Drescher and Boedeker (1995) extended this comparison to other models, including probit, logit and Weibull models. However, for these models, Loewe additivity and Bliss independence may or may not give the same conclusion (synergistic or not).

We suggest one should analyze the same data set with both reference models. Then depending on the context of the analysis, one can choose a conservative result. The context could be to look for a therapeutic effect. For example, in most drug discovery settings, scientists want to find drug combinations that show synergy. In this case, one should choose dose combinations only if both reference models show a synergistic result. The other context could be to look for a toxic effect. For example, in interaction test for toxic pollutants, one wants to avoid dose combinations showing synergy. In this situation, one should conclude synergy as long as one reference model result demonstrates so.

13.3.2 Min Test for In Vivo or Clinical Studies

For in vivo studies or clinical trials, a factorial design is usually used. Since there are only a few dose levels per drug, it is hard to characterize a dose response

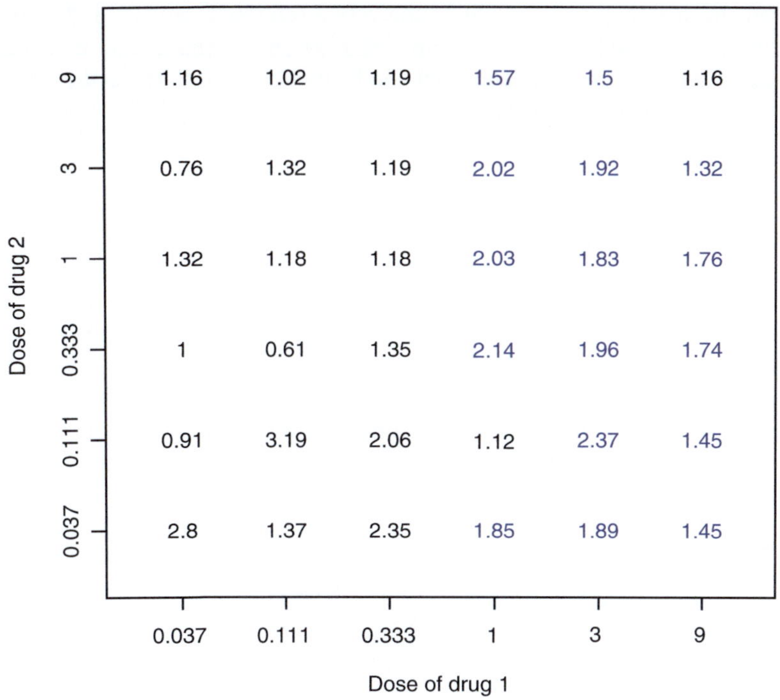

Fig. 13.6 Estimates of independence index η at each combination. Statistically significant synergistic effects are shown in *blue*

curve for monotherapies. Therefore, a min testis often used for data analysis which tests whether the effect of a dose combination, say (d_1, d_2), is significantly larger than the effect of drug 1 at d_1 AND drug 2 at d_2. Therefore analysis for such studies does not require a reference model. However, multiplicity adjustment may be needed to control the overall type I error rate since this test is conducted at multiple combinations. For details on this test and analysis methods for such studies, refer to Hung (2000) and references therein.

13.4 Discussions and Conclusions

The challenge of how to evaluate synergy of drug combinations has a long history. Synergy evaluation requires a reference model that is usually based on the monotherapy drug dose response relationship and describes the expected effect of two drug combinations when there is no interaction between them. Loewe additivity and Bliss independence, albeit criticized frequently from one perspective or another, have been the two central competing reference models, as Greco et al. (1995)

reviewed. In this chapter, we defined these two reference models and introduced some analysis methods based on them. We also introduced a new independence index based on Bliss independence that can be modeled similarly to the interaction index based on Loewe additivity.

There is still no "standard" method to evaluate synergy. In fact, Greco (2010) recently asked some experts in this field to evaluate the same two data sets, and without surprise conclusions differed. Readers are strongly encouraged to go through the articles to appreciate analysis methods other than those discussed in this chapter. A method of particular interest is nonlinear blending, proposed by Perterson and Novick (2007). This method takes a mixture experimental design by fixing the total amount of two drugs and uses the idea of min test with a response surface model to evaluate synergy. A combination is synergistic if its response is higher than both monotherapy responses. So this method does not define a reference model such as Loewe additivity and Bliss independence. A problem with this approach is that when the potency of the two drugs is dramatically different, it is difficult to study a fixed amount of both drugs. For example, in their example (Fig. 9 on Page 141) combining FLG and AZT, the former has an EC_{50} of about 7.2 and the latter has an EC_{50} of 0.04. It may be difficult to even increase the dose of AZT to the level of 7.2. A remedy may be to use doses scaled by EC_{50}.

There are significantly fewer articles in the literature using Bliss independence reference models. One reason may be that its simplicity does not warrant any publication. Another reason is that there is not an index for this reference model. However, this method is more often used by scientists, especially in Oncology, for the same reason of simplicity. More statistical rigor should be brought into analysis methods using this model, instead of making decisions based on the difference in Eq. (13.3). Zhao et al. (2014) is such an attempt. In this chapter, we introduced the independence index that provides a similar framework for synergy evaluation as the interaction index.

Another research area is to further study the relationship between these two reference models. Berenbaum (1989) and Drescher and Boedeker (1995) have identified some situations in which the relationship between these two reference models can be established. The relationship depends on the dose response models used. More work is needed to study the relationship between the interaction index and the independence index and to further understand how to interpret such a relationship in the context of synergy evaluation.

Ideally, there should be some "standard" practice for synergy evaluation. We mentioned some principles along this line, such as choosing a reference model based on mechanism of action (MoA) if the MoA is clear, or analyzing data with both reference models and making decisions based on therapeutic or toxic context.

References

Berenbaum MC (1989) What is synergy? Pharmacol Rev 41:93–141

Bliss CI (1939) The toxicity of poisons applied jointly. Ann Appl Biol 26:585–615

Casella G, Berger RL (2002) Statistical inference (second edition) Duxbury/Thomson Learning, Pacific Grove, CA

Donev AN, Tobias RD (2011) Optimal serial dilutions designs for drug discovery experiments. J Biopharm Stat 21:484–497

Drescher K, Boedeker W (1995) Assessment of the combined effects of substances: the relationship between concentration addition and independent action. Biometrics 51:716–730

Greco WR (2010) Concentration-effect modeling of single agents and combinations of agents. Frontier Biosci, preface

Greco WR, Bravo G, Parsons JC (1995) The search for synergy: a critical review from a response surface perspective. Pharmacol Rev 47:331–385

Harbron C (2010) A flexible unified approach to the analysis of pre-clinical combination studies. Stat Med 29:1746–1756

Hung HMJ (2000) Evaluation of a combination drug with multiple doses in unbalanced factorial design clinical trials. Stat Med 19:2079–2087

Kong M, Lee JJ (2006) A generalized response surface model with varying relative potency for assessing drug interaction. Biometrics 62:986–995

Lee JJ, Kong M (2009) Confidence intervals of interaction index for assessing multiple drug interaction. Stat Biopharm Res 1:4–17

Loewe S (1928) Die quantitation probleme der pharmakologie. Ergeb Physiol Biol Chem Exp Pharmkol 27:47–187

Perterson JJ, Novick SJ (2007) Nonlinear blending: a useful general concept for the assessment of combination drug synergy. J Receptor Sig Transduct 27:125–146

Plummer JL, Short TG (1990) Statistical modeling of the effects of drug combinations. J Pharmacol Methods 23:297–309

Straetemans R, O'Brien T, Wouters L, Van Dun J, Janicot M, Bijnens L, Burzykowski T, Aerts M (2005) Design and analysis of drug combination experiments. Biom J 47:299–308

Tallarida RJ (2000) Drug synergy and dose-effect data analysis. Chapman and Hall/CRC, Boca Raton, FL

Tan MT, Fang H, Tian G (2009) Dose and sample size determination for multi-drug combination studies. Stat Biopharma Res 1:301–316

Zhao W, Zhang L, Zeng L, Yang H (2012) A two-stage response surface approach to modeling drug interaction. Stat Biopham Res 4:375–383

Zhao W, Sachsenmeier K, Zhang L, Sult E, Hollingsworth RE, Yang H (2014) A new Bliss independence model to analyze drug combination data. J Biomol Screen 19:817–821

Chapter 14
Biomarkers

Chris Harbron

Abstract Biomarkers are playing an increasingly important role throughout many aspects of the pharmaceutical discovery and development pipeline. They have many differing roles and applications and statistics plays a critical role in their discovery, validation or qualification and how they are applied and utilised. In this chapter we shall discuss what biomarkers are, the types of data that they generate and the impact on the subsequent statistical analysis, paying particular attention to the avoidance of false positives in biomarker discovery and confirming the technical performance of assays measuring biomarkers.

Keywords Biomarkers • Multivariate analysis • Variability • Cross-validation • Concordance

14.1 What Is a Biomarker?

At its simplest the word "biomarker" can be decomposed into a measure or marker of a biological process. Various different definitions have been proposed to expand upon this, for example one of the most frequently quoted is the National Institutes of Health's (Atkinson et al. 2001, p. 91) definition of a biomarker as "a characteristic that is objectively measured and evaluated as an indicator of normal biological processes, pathogenic processes or pharmacological responses to a therapeutic intervention".

The word biomarker covers a huge range of different measurements, substrates being measured, technologies to perform the measurement and ways in which they can be applied and utilised. One key feature of biomarkers is that they are measured to make a decision or to provide additional information to assist and inform in the making of a decision. This decision may be at many different levels for many different stakeholders, for example:

- a physician determining the treatment with the most appropriate risk-benefit profile for an individual patient based upon a personalised healthcare biomarker

C. Harbron (✉)
Roche Products Ltd, Welwyn Garden City, UK
e-mail: chris.harbron@roche.com

L. Zhang (ed.), *Nonclinical Statistics for Pharmaceutical and Biotechnology Industries*, Statistics for Biology and Health, DOI 10.1007/978-3-319-23558-5_14

- a regulatory body deciding if a development compound can be progressed into man based upon a pre-clinical safety biomarker
- or a pharmaceutical company deciding whether to invest in moving a compound into a later stage of clinical testing by looking at the response to treatment of a Proof Of Mechanism marker

14.1.1 Types of Biomarkers: Uses and Applications

Biomarkers can be categorised in many different ways. Note that these are not mutually exclusive: the same biomarker may fulfil many roles in different scenarios or sometimes within the same scenario.

Various different classifications of biomarkers have been proposed and discussed, for example by Frank and Hargreaves (2003) and Jenkins et al. (2011). Here we will not focus on these definitions, but just provide an overview of some of the most common applications:

Longitudinal or Pharmacodynamic Biomarkers measure dynamic changes in response to a treatment intervention, these can be used to show evidence of the activity of a compound.

The following types of biomarkers provide evidence, in increasing levels, of a compound's interaction with the disease and its potential for clinical efficacy: *Proof Of Mechanism Biomarkers* demonstrate modulation of the drug target; *Proof Of Principal Biomarkers* demonstrate modulation of the cellular phenotype and a *Proof Of Concept Biomarker* demonstrates some therapeutic benefit to a patient (for example an imaging biomarker showing an improvement in the patients' condition) whilst not being a regulatory clinical endpoint. Surrogate endpoints are the next increment in these levels of evidence supporting a compounds' efficacy. For an endpoint to have demonstrated surrogacy it has to have reached strict criteria (Prentice 1989) requiring a large body of evidence.

Dynamic biomarkers can also apply to the safety of a compound at any stage of development. At all stages of pre-clinical and clinical development, the number of individual animals or patients treated will be small compared to the size of the population which will ultimately be treated meaning there is a sizeable risk that a rare adverse event, or evidence that an adverse event is treatment related, will not be observed even in large phase 3 programmes. Safety biomarkers can either alert to potential risks or assist in confirming or otherwise the relationship of the action of the drug to any observed or postulated adverse events.

Another main application, underpinning the growth area of interest of what is variously termed personalised or stratified or precision medicine, are biomarkers explaining and being able to predict the variation in clinical outcome between individuals. These fall into two classes: *Prognostic Biomarkers* which predict the prognosis or likely outcome of disease irrespective of treatment and *Predictive*

Biomarkers which predict the outcome of disease following a specific treatment. These correspond to biomarker main effects and treatment-by-biomarker interactions within a statistical model.

Frequently biomarkers are both prognostic and predictive, as if there are multiple forms of a disease which can be distinguished by a marker, with only one of the forms being amenable to treatment by a certain compound, the underlying prognosis of these differing forms of the disease is also liable to be different.

Predictive biomarkers are frequently based upon measurements taken prior to treatment to determine the most appropriate therapy. However, an alternative "trial-and-error" strategy which may be appropriate in non-acute scenarios with low-risk therapies is to use early response, either in a recognised endpoint, a surrogate endpoint or a pharmacodynamic endpoint, as a predictor of long term response.

14.1.2 Types of Biomarker Data: Binary/Categorical/Continuous

Reported biomarkers will generally either be continuous (e.g. the expression of a gene or concentration of an analyte), binary or categorical (e.g. the presence or absence of a specific mutation) or ordered categorical (e.g. 0, 1+, 2+, 3+ immunohistochemical staining). Statistically, these will require to be treated differently in models, for example adopting either a regression or an ANOVA approach when the biomarker is a response variable. However frequently these different classes of data aren't as distinct as they may initially be presented. For example in contrast with inherited host genetics where mutations will generally either be expressed or not, a mutation test in a heterogeneous tumour will return a percentage of cells exhibiting the mutation—a continuous measure. Similarly whilst IHC markers are typically categorised into 3 or 4 levels they are measuring a continuous underlying level of protein expression, which is sometimes captured by an H-score. There is frequently a tendency to categorise or dichotomise biomarker data and indeed before a biomarker can be utilised as a patient selection tool it will have to be dichotomised into biomarker negative and biomarker positive populations, but this will always lead to a loss of information and power. As a general principal the underlying continuous measure should be sought out and the decision of where to place a cut-off to dichotomise this measure should be delayed for as long as possible, both to maintain power in the meantime and to allow for the final choice of threshold to be based upon as much information as possible.

14.1.3 Types of Biomarker Data: Univariate/Multivariate

In some situations there may be a single biomarker, typically with strong scientific rationale, to be measured and analysed, whilst in other situations there may be multiple biomarkers, either measured independently or generated simultaneously from a multiplex technology, extending up to many thousands of markers with gene expression microarrays or even millions of genetic markers. When the data is multivariate this may be analysed adopting either a univariate or a multivariate approach or often most fruitfully a combination of the two, this is discussed further in the biomarker discovery section below.

It should also be borne in mind that many univariate biomarkers are measured alongside other variables such as positive and negative controls and the presented result is the outcome of a normalisation procedure. This means that even these apparently univariate variables are the result of a combination of multiple variables and the details of how the normalisation is performed will impact the final result. This is discussed in more detail within the pre-processing section below.

14.1.4 Pre-processing of Biomarker Data

One hidden feature of much biomarker data is that the presented values are frequently not directly measured, but are the result of a number of pre-processing steps. As a general rule the more complex the technology, the greater the degree of data pre-processing that will need to be applied before commencing analysis. These are frequently complex procedures and can develop in an evolutionary manner as issues are discovered and addressed. As an example the Affymetrix gene expression microarray platform has generated in excess of 30 different methods. Clearly this creates the potential for confusion as the application of different pre-processing algorithms will generate different results with the potential to generate different conclusions.

The methods for data pre-processing are bespoke to each technology, but typically will include steps such as background correction, positive control normalisation and housekeeping normalisation. These steps will reduce the effects of both unavoidable technical variability, for example small variations in the quantity of analyte being studied, as well as nuisance biological variability for example related to the quantity or the amount of degradation that has occurred within the biological sample. The objective of a pre-processing algorithm is to generate a measure reflecting the true underlying biological variability that we are interested in rather than any artefact induced throughout the sampling, storage and processing stages. Some simple diagnostic plots can help to assess the effectiveness of the pre-processing algorithm, one of the most powerful is plotting the final normalised biomarker expression against a measure of the sample quality, for example a housekeeping gene. This should be uncorrelated as we would expect that the true biological variability is unrelated to any processing artefacts. This may also highlight outlying poor quality samples which should be treated with caution in any downstream analysis.

14.2 Biomarker Discovery

14.2.1 Data and Biological Support

For a biomarker to have maximal credibility it should be supported both by experimental data and by a wider scientific body of understanding. Differing levels of confidence in the prior knowledge, may lead to different strategies in the identification of biomarkers. With a strong initial focussed hypothesis, there may only be a need to measure the known biomarker as a specific hypothesis test. Alternatively in the absence of much or any strong prior knowledge, a hypothesis free approach looking at a large number of potential biomarkers may be more appropriate. With highly dimensional data such as from microarrays the wealth of data could make finding a few relevant biomarkers a needle in a haystack activity, where any true biomarker would have to be highly significant in order to stand out amongst the false positives. Between these two extremes are alternative strategies, for example a candidate gene approach analysing just a selected subset of markers with enhanced biological rationale or a mixed analysis looking at a single or limited set of candidate markers as a primary analysis, but also a wider more comprehensive set of markers as a secondary or exploratory analysis. This could potentially be a beneficial area for a Bayesian approach, using prior knowledge to place prior likelihoods for each gene.

14.2.2 Multivariate Techniques and Algorithms

There are a large number of variously called multivariate, data mining or machine learning techniques or algorithms. These algorithms generate mathematical models which combine the results from a number of variables to generate a prediction of another variable, typically an outcome. They have the properties that they can cope with situations with more variables than observations, although this power must be used with proper caution to avoid overfitting. That is fitting an overly complex model to the data, modelling noise, which then does not apply more generally to other data sets, and may result in false claims of accuracy. These can be applied in two different ways: to generate predictive models which based upon the measured values predict an outcome or property of a new sample, or variable selection—by observing which variables contribute most to the model and taking these forward for further investigation without being concerned about how the algorithm was combining the individual variables.

A huge number of different data mining algorithms have been developed, each with their own sets of properties. A non-exhaustive selection of these are listed below:

14.2.2.1 Regression Based Methods

Standard regression based methods become unstable and subsequently fall over when the number of variables approaches or exceeds the number of observations and can also become unstable when there are high levels of correlation between the predictor variables.

The *Elastic Net* (Zou and Hastie 2005), including the *LASSO* and *ridge regression* as special cases, minimises the sum of squares of the residuals as in regression with an additional term penalising the size of the coefficients in the fitted model. The balance between the fit to the training set and the size of the coefficients is determined by a tuning parameter, λ, which ranges from an overfitted model perfectly fitting all data points to a null model with all coefficients set to zero when the term for the size of the model coefficients dominates the goodness of fit component. The level of this tuning parameter is set by cross-validation to achieve an appropriate balance between goodness of fit to the training set whilst avoiding overfitting. An elastic net model typically sets the regression coefficients of most variables equal to zero, generating a parsimonious model and acting as an efficient variable selector.

Partial Least Squares (PLS) identifies the linear combination of X-variables which has the greatest covariance with the response variable(s), lying part way between Principal Components Analysis (PCA) which identifies the linear combination explaining the greatest proportion of total variation of the variables independent of outcome and regression selecting the linear combination with the greatest correlation with outcome. In contrast to elastic nets, most variables within a PLS model will have a non-zero coefficient.

Tree Based Methods form the basis of several other data mining algorithms. A standard decision tree, has the advantage of generating a simple readily visibly communicable model, but is a purely heuristic approach with no guarantee of optimality and the final model is frequently one of many different but similarly performing models where there is no reason to believe any one will be more correct or give improved predictions on subsequent data than any other.

PartDSA (Molinaro et al. 2010) attempts to get closer to an optimal tree by using a forward, backward and swapping selection algorithm to generate trees rather than the normal iterative growth.

One of the issues with decision trees is that there are typically a large number of potential trees which are roughly as good as each other within the training data, and none of which are definitively correct. The one that gets selected may be optimal for this particular dataset, but would not necessarily translate to giving optimal predictions in future datasets. To address this methods have developed which generate a large collection or ensemble of trees which are averaged over to give the final prediction, giving a much more robust prediction.

Random Forests (Breiman 2001) introduce two sources of variability into generating each individual tree. The datasets are perturbed by generating a bootstrap sample of the same size of the original data for each individual tree, and only a random sample of variables are considered for each individual split. These also provide computational advantages, reducing the computational requirements by only examining a subset of variables at each step and the bootstrapping of the data allows out-of-bag estimation of error rates by considering those observations which did not contribute to each individual tree without the much more computationally intensive requirement of cross-validation.

Gradient Boosting Machines (Friedman 2002) also generate an ensemble of trees, but by a different iterative mechanism. Each tree is grown optimally, but then only a shrunken version of the tree predictions is used, by dividing the prediction by a shrinkage parameter. Subsequent trees are fitted to the residuals from the sum of all previous shrunken trees. Eventually this process would converge to a model perfectly fitting the training dataset, so cross-validation is applied to determine the optimal number of trees used within the final model.

These methods can be sensitive to some of the tuning parameters within the algorithm, in particular the parameter which restricts the minimum size of the leaves at the end of each individual tree and in random forests the *mtry* parameter determining the number of variables to consider for each split. The GBM shrinkage parameter has to be balanced between being as small as possible to generate the best models against the practical requirements for computational time, especially if employing a cross-validation strategy.

Support Vector Machines (Cortes and Vapnik 1995) takes a different approach by identifying the widest multidimensional hyperplane that gives the maximum separation between members of different classes.

Nearest Neighbours looks at the closest observations in the training set to a new observation and predicts a class for the new observation based upon a voting scheme. This is most effective in lower dimensional scenarios and in very low dimensional scenarios can behave similarly to random forests.

Genetic Function Approximators mimic an evolutionary process with mathematical equations. These have a tendency to generate complex functions which may not be robust when applied to subsequent data sets.

Whenever using a multivariate analysis, performing a univariate gene-by-gene analysis, i.e. analysing each variable in turn as if it was the only biomarker collected, alongside is frequently highly informative. Although counterexamples can artificially be constructed, in practice it is unlikely that there will be strong signals within a multivariate analysis without some of the signal being visible within the univariate analyses. Similarly strong signals in a set of univariate analyses will naturally translate to a multivariate algorithm. The *False Discovery Rate* (Storey 2002) provides a valuable quantification of the results of a set of univariate analyses by estimating the proportion of the variables giving an unadjusted significant result which are false positives and with a larger number of variables provides a more pragmatic summary of results than adopting a stricter family-wise error rate control.

Initial data visualisation plays a key role prior to any multivariate modelling. Principal Components Analysis (PCA) is a valuable tool, identifying the key overall sources of variability within the data as well as multivariate outliers. Observations can be coloured both by nuisance parameters such as data source, processing batch or sample quality to identify if there are likely to be issues with the data, or by the parameters being modelled to give an early indication of the likely degree of success of any modelling activities.

Clustering is also sometimes used as an exploratory tool. However, these can be sensitive to the choice of clustering algorithm and distance metrics employed. Underlying this is the fact that the majority of data sets don't fall into neat clusters, but form a continuous distribution and so any attempt to forcibly separate them into distinct clusters is artificial.

The FDA set up the MAQC-II project (Shi et al. 2010) as a collaborative effort to try and establish best practice in the generation of predictive models, particularly to data generated from microarrays although it is reasonable to assume the conclusions are more widely applicable. In MAQC-II 13 microarray data sets had predictive models fitted by 36 separate analysis teams where they concluded that some data sets contained more information that could be modelled than others: "Some endpoints are highly predictive ... provided that sound modeling procedures are used. Other endpoints are inherently difficult to predict regardless of the model development protocol" (p. 834) and that the skill and approach of the analysts had more impact than the choice of one modelling technique above another and there is no universal best modelling technique which is uniformly better than all others : "There are clear differences in proficiency between data analysis teams correlated with the level of experience of the team.", "Many models with similar performance can be developed from a given data set.", "Simple data analysis methods often perform as well as more complicated approaches", "Applying good modelling practices appeared to be more important than the actual choice of a particular algorithm" (p. 834).

14.2.3 Cross-Validation

The concept of validation or replication is to demonstrate a wider applicability of the finding from analysis of any individual dataset by demonstrating that a finding is also seen within another independent data set. The more complex the finding is, particularly if it is involving multivariate data invoking a complex algorithm or based upon a complex technology requiring many processing steps, the more critical this is.

Whilst validation in an independent dataset will and should remain the gold standard approach, other strategies can be employed to provide an early view of the performance of a biomarker within an independent data set which may be valuable in the frequent situation where there isn't an abundance of appropriate datasets. The simplest approach is to split the data set into training and test sets, where the model is fitted to the training set and tested on the test set. This will understate the difference to a truly independent data set as two parts of the same dataset will be more similar than two separate datasets, but provides a useful initial indication. However splitting the dataset in this manner will lower the power of detecting an optimal biomarker in the training set as it is a smaller subset of the complete data, and also requires retaining a large enough test set to have the power to validate the findings from the training set. There are many possible ways which a data set could be split into training and test sets, and each of these will potentially generate different results, both in the discovery and the validation phases. To maintain credibility for any findings it is therefore critical to specify up front before starting any analysis exactly how the data will be split.

Cross-validation provides a method for efficiently performing independent testing of a model within the same dataset used to generate the model, which can be viewed as repeatedly splitting the data into training and test sets. A proportion ($1/k$) of the data is selected as a test set and the remainder of the data analysed as a training set to generate a model which is tested on the left out test set. This process is repeated k times until all observations have been excluded from the training set and predicted as members of the test set, and the results aggregated across all k test sets. k can vary from 2 up to the number of observations n (leave-one-out cross validation). Typically values in the range of 5–10-fold cross validation is considered optimal. However, as with the test-training set approach, different divisions of the data into k-folds will give different results. This suggests an approach where the cross-validation is repeated many times each time with different splits in order to reduce this variability.

Cross-validation is used in two different ways. One as discussed above is to estimate the error associated with a model derived from a particular dataset. The other is to optimise a modelling parameter, for example the number of trees in a Gradient Boosting Model or the lambda parameter balancing model fit against coefficient size in an elastic net, where the model is chosen to minimise the cross-validation error. The error from this optimisation will be optimistically biased as the particular parameter or model chosen is that which is optimal for the training dataset. To also obtain an unbiased error estimate, a two-stage cross-validation process must be deployed where the cross-validation to select the optimal parameter is considered part of the modelling process and an additional level of cross-validation is wrapped around the whole process including the first level of cross-validation.

A subtle but important distinction between cross-validation and the training-test set approaches is that the later generates a specific outcome model, and it is this specific model which is examined in the test set. In contrast cross-validation examines the process for generating the outcome and how well this process typically performs and then extrapolates this to the performance of applying this process to generate a model from the complete dataset. The specific model which is generated by the full-dataset analysis isn't directly tested.

The training-test set and cross-validation approaches are unlikely to capture the variability that will be seen moving to a different data set or from a study to a real-world situation with many more sources of variability. Although a cross-validated estimate of error will be more representative than the model fit error which will be optimistically biased, this error is likely to increase when tested in a completely independent dataset.

14.2.4 Publicly Available Datasets: Invaluable Assets with Hidden Dangers

There is a very welcome movement of making experimental data publicly available, either as supplementary data linked to publications or in public repositories such as the Gene Expression Omnibus (GEO: http://www.ncbi.nlm.nih.gov/geo). Indeed some journals are now making this a pre-requisite for publication. This wide sharing and availability of data can only be good for advancing science and this will facilitate both biomarker discovery and testing of biomarker discoveries across a range of independent data sources. However accessing external data sources without fully understanding how the data were collected and any hidden features within the data should be treated with caution. Baggerly and Coombes (2008) provide an example where they demonstrated that an apparent multivariate prediction reported in the literature which had subsequently been deployed into clinical trials could be explained by batch processing effects within the data set.

There is also a similar danger from combining data from several different sources into a single large dataset. Frequently the largest differences in both biomarkers and clinical phenotypes will be between-dataset variation. Failure to be aware of and to subsequently account for this confounding may lead to false results which may exist between datasets but are not visible within datasets and do not extrapolate more widely.

14.3 Technical Performance of Biomarkers

Whatever their intended application, understanding the operating characteristics of a biomarker, its intrinsic level of variability and any biases that it might be subject to is critical. It can be shown that the benefits of a personalised medicine approach enriching by a biomarker rapidly diminish with increasing levels of variability in the biomarker. Similarly for biomarkers being used as analysis endpoints, increasing levels of variability reduces the power of detecting responses or differences between treatment groups.

The ultimate aim is that a biomarker will give a result reflecting the true current biology of the individual, independent of the particular sample taken and the processing route taken. This is a necessary but not sufficient prerequisite for clinical utility, i.e. that the biomarker measures something useful that has been demonstrated to aid clinical decision making.

14.3.1 Sources of Variability

Biomarkers are subject to many sources of variability, broadly these can be split into two categories: technical variability and biological variability. Once identified there is, theoretically at least, generally the potential to reduce technical variability by tightening up processes, whilst biological variability is frequently something we have to live with and where the most frequent mitigating action is to address with a greater number or density of samples.

Whilst individual sources of variability may be examined separately, the total variability that will be observed in practice is the sum of all the contributing variabilities. This means that if there are a number of moderately sized sources of variation, whilst none of these on their own may be a show stopper, in combination their total variability may limit the information available from the biomarker assay.

14.3.2 Technical Variability

Factors potentially contributing to technical variability cover all steps that occur in the entire process between the patient and the final biomarker result. These include sampling processes, sample thickness, storage and transport conditions, and inter-machine, reagent batch, reader and operator variability.

In the not unusual situation of limited numbers or quantities of samples, approaches such as fractional factorial experimental design may provide more efficient ways to investigate the variation associated with a range of potential sources of variability.

14.3.3 Biological Variability

Biological variability can appear in several different ways each potentially requiring different mitigation strategies.

14.3.3.1 Short Term Temporal Variability

Many biological processes are known to vary on a daily cycle, reflecting waking and sleeping patterns as well as food intake. For this reason, wherever possible, samples should be taken from subjects at equivalent times of day throughout the study.

14.3.3.2 Long Term Temporal Variability

Longer term biological fluctuations can also occur, for example responses to temperature changes throughout the course of a year. In oncology tumours will evolve over time either naturally or developing resistance to treatment so that archival tissue samples, especially those seen in patients who have undergone multiple lines of therapy since the sample was taken, may not be representative of the current disease within a patient. The concordance of archival and fresh samples should be checked and if there is a high discordance consideration should be made of the validity of using archival samples for treatment decisions.

14.3.3.3 Spatial Variability

In oncology different samples taken from the same individual may display heterogeneity. This variability may occur within a tumour, some tissue types e.g. gastric frequently display a large degree of heterogeneity, or may occur between the primary tumour and metastases.

14.3.4 Impact of Variability on Design

With all of these types of variability, statistics has a key role in identifying and describing the levels and types of variability present, and the impact of this variability on the design of studies utilising the biomarker and how the biomarker will be deployed in practice.

Where the agreement between historical and current samples is low this may require obtaining fresh samples from a patient instead of relying on archival samples. Similarly heterogeneity between types of tumour tissue may suggest the tissue type that should be sampled.

Where there is considerable spatial tumour heterogeneity this may require using several samples, for example when using small biopsy samples, the HER2 test for gastric cancer recommends examining 7–8 evaluable specimens from different regions of the tumour, calling a patient HER2 positive and eligible to receive Herceptin™ if any one of these samples shows a positive result.

14.3.5 Acceptance Criteria

Before embarking on a study of variability it is useful to establish acceptance criteria, that is levels of performance which the biomarker assay or diagnostic should achieve. Ideally these should be linked to the clinical impact of any incorrect outcome in either direction, however in advance of knowing the clinical data this can be hard to establish in advance as the relationship of the biomarker to clinical response is likely not to have been established at this stage. An alternative is to observe what levels of variability have been observed previously with this or similar technologies in comparable scenarios.

14.3.6 Comparison to Gold Standard

Where an assay for a biomarker is well-established and can be considered a gold standard, then any new assay for this biomarker should be compared to the current assay and be expected to demonstrate high levels of concordance.

In the absence of a gold standard, for example a new biomarker, then this comparison is not possible, although comparing with other markers to understand the relationships between different markers can be illuminating and help to build confidence if the findings correspond with biological understanding.

The development of an improved biomarker assay, for example with greater sensitivity than previous assays provides an interesting challenge. In this situation some differences to the previous assay are to be expected and hoped for as they represent the improvement from using the new assay. In this scenario it is just changes in the wrong direction, that is a loss of sensitivity which should be identified for concern.

14.3.7 Evaluating Biomarker Performance

Within personalised medicine, biomarkers are frequently dichotomised to a binary measure corresponding to treat or don't treat with a particular drug. This may be a natural binary split e.g. a single nucleotide polymorphism mutation is present or not, may be a split based on a clear biomodality of a continuous measure for example (oestrogen receptor) ER positive or negative in breast cancer or may be an optimal split of a continuous variable derived from an observed relationship to clinical response for example the categories in the Genomic Health Mammaprint tool. Understanding the variability is critical in these situations, it is clearly desirable that if a patient would be prescribed a certain treatment in one centre on Monday, will they be prescribed the same treatment in a different centre on Tuesday.

Table 14.1 Summary statistics for a binary classification test

		Gold standard		
		Positive	Negative	
New biomarker	Positive	TP	FP	PPV = TP/(TP + FP)
	Negative	FN	TN	NPV = TN/(TN + FN)
		Sensitivity = TP/(TP + FN)	Specificity = TN/(TN + FP)	

TP true positives, *FP* false positives, *FN* false negatives, *TN* true negatives, *PPV* positive predictive value, *NPV* negative predictive value

In these cases the relationship to a gold standard is frequently summarised by sensitivity and specificity, respectively the proportion of true positives and true negatives which are correctly identified by the new test. Positive and negative predictive values are complementary concepts, respectively the proportion of positive and negative results from the new test which are correct predictions. Confusingly clinical validity is also often summarised by the same terms where the gold standard test is replaced by clinical outcome, and a positive or negative test correspond to responding and non-responding patients respectively. Table 14.1 below shows these concepts.

The Receiver Operating Characteristic (ROC) curve provides a summary method for examining the performance of a biomarker in terms of its sensitivity and specificity over a range of potential cutoffs that could be used to dichotomise the population into biomarker positives and negatives. These can then be summarised by an area under the ROC curve (AUC), which ends up being a scaled version of the Mann–Whitney statistic comparing the biomarker between the two groups. Recently the theoretical basis of the AUC has been criticised with Hand (2009) proposing the H-measure as an alternative with improved properties.

14.3.8 Evaluating Concordance

A typical concordance study will assess a number of samples a number of times, depending upon the source of variability being studied. A simple overall agreement can be calculated by counting the proportion of times the ratings from the same sample agrees. However, this is highly dependent upon the overall prevalence, for example if predicting a rare disease the majority of samples will be negative so a new diagnostics could attain a high overall agreement by scoring every single sample as negative, which is not very informative. The Kappa statistic provides a chance corrected measure of agreement. The original Cohen's kappa was calculated for two raters and allowed for the two raters to have different marginal distributions. Fleiss's Kappa assumes a common marginal distribution for all raters, allowing extension to any number of assessors. Krippendorf's Alpha provides a method for calculating Fleiss's Kappa for incomplete data with missing values.

One issue can be that as concordance studies will always be of limited size, only a limited range of the total variability that may be observed in practical use will be observed, both in terms of the range of samples being observed and the range of conditions, e.g. pathologists being observed. In fact because of the attention given to aspects such as training in these studies, there is a risk that these studies will understate the true level of variability that will be observed when the test is more widely exposed to the greater range of samples and conditions that will be encountered in real scenarios.

14.4 Summary

Biomarkers are playing an increasingly important role throughout many aspects of the pharmaceutical discovery and development pipeline, being applied in many different ways. Statistics have multiple inputs which are critical to the discovery, validation or qualification and application of biomarkers and how they are applied and utilised.

Fundamentally biomarkers generate data which can be analysed using the same statistical best practice and professional judgement as any other data or endpoint. But as with any analysis understanding the context, how the results will be interpreted and applied, how the data was generated and any features that will influence the data, is critical.

References

Atkinson AJ, Colburn WA, DeGruttola VG, DeMets DL, Downing GJ, Hoth DF, Oates JA, Peck CC, Schooley RT, Spilker BA, Woodcock J, Zeger SL (2001) Biomarkers and surrogate endpoints: preferred definitions and conceptual framework. Clin Pharmacol Ther 69:89–95

Baggerly K, Coombes K (2008) Run batch effects potentially compromise the usefulness of genomic signatures for ovarian cancer. J Clin Oncol 26:1186–1187

Breiman L (2001) Random forests. Mach Learn 45:5–32

Cortes C, Vapnik V (1995) Support-vector networks. Mach Learn 20:273–297

Frank R, Hargreaves R (2003) Clinical biomarkers in drug discovery and development. Nat Rev Drug Discov 2:566–580

Friedman JH (2002) Stochastic gradient boosting. Comput Stat Data Anal 38:367–378

Hand DJ (2009) Measuring classier performance: a coherent alternative to the area under the ROC curve. Mach Learn 123:77–103

Jenkins M, Flynn A, Smart T, Harbron C, Sabin T, Ratnayake J, Delmar P, Herath A, Jarvis P, Matcham J (2011) A statistician's perspective on biomarkers in drug development. Pharm Stat 6:494–507

Molinaro AM, Lostritto K, van der Laan M (2010) partDSA: deletion/substitution/addition algorithm for partitioning the covariate space in prediction. Bioinformatics 26:1357–1363

Prentice R (1989) Surrogate endpoints in clinical trials: definition and operational criteria. Stat Med 8:431–440

Shi L et al (2010) The MicroArray Quality Control (MAQC)-II study of common practices for the development and validation of microarray-based predictive models. Nat Biotechnol 28:827–838
Storey J (2002) A direct approach to false discovery rates. J R Stat Soc Ser B 64(3):479–498
Zou H, Hastie T (2005) Regularization and variable selection via the elastic net. J R Stat Soc Ser B 67:301–320

Part IV
Statistical Methods for CMC

Chapter 15
Overview of Drug Development and Statistical Tools for Manufacturing and Testing

John Peterson and Stan Altan

Abstract An essential part of the application for marketing approval of a new drug or therapeutic biological product submitted to regulatory authorities is the Chemistry Manufacturing and Controls (CMC) section. It presents the sponsor company's documentation of sufficient scientific and engineering knowledge to manufacture the product with consistent quality that provides defined clinical efficacy with an acceptable safety profile. The CMC section includes three main parts: (1) Chemical Development (synthesis of a new molecular entity (NME), purification of a new biologic entity (NBE); (2) Pharmaceutical Development (comprised of formulation and process development); (3) Analytical Development (analytical methods for physical, chemical, biological characterization). Specifications are established during development and constitute an important part of the CMC section in defining the product's quality requirements. This chapter provides an overview of the drug development process, and some statistical tools useful in support of CMC studies. This chapter aims to set the stage for the subsequent 11 chapters in the CMC section of this book which delve in greater detail into important CMC related statistical topics.

Keywords Chemistry • Manufacturing and Controls • Quality by design • Bioassay • Probability of conformance • Bayesian method • Drug development

15.1 Introduction to Drug Development, Manufacturing and Analytical Testing

The Chemistry Manufacturing and Controls (CMC) aspects of drug development is an essential component of the overall process necessary for sponsor companies

---

J. Peterson (✉)
Statistical Sciences Group, GlaxoSmithKline Pharmaceuticals, Upper Providence, PA 19426, USA
e-mail: john.peterson@gsk.com

S. Altan
Nonclinical Statistics and Computing, Janssen R&D LLC, Raritan, NJ 08869, USA

L. Zhang (ed.), *Nonclinical Statistics for Pharmaceutical and Biotechnology Industries*, Statistics for Biology and Health, DOI 10.1007/978-3-319-23558-5_15

to obtain approval to market a drug in the United States and elsewhere in the world. In the United States, such approval requires the filing of a New Drug Application (NDA) or a Biologics License Applications (BLA) to the Federal Drug Administration (FDA). We will not delve into the details of the drug approval process and associated interactions with regulatory agencies but suffice it to say that the CMC section is a critical part of the NDA submission. It includes three main areas: (1) Chemical Development (synthesis of a new molecular entity (NME) or new chemical entity (NCE), purification of a new biologic entity (NBE), also referred to as an Active Pharmaceutical Ingredient (API) or biological API or Drug Substance),[1] (2) Pharmaceutical Development (comprised of formulation and process development), and (3) Analytical Development (analytical methods for physical, chemical/biological characterization).

Early CMC studies follow on the heels of the decision to develop a NME based on an expectation that the NME has an important therapeutic effect and is safe in humans. These early studies occur prior to the filing of the Investigational New Drug Exemption (IND) application with the FDA, which is necessary to study the NME in humans. At this point, initial toxicology and animal pharmacology studies are initiated to characterize the safety profile of the compound and its physiologic effects in animal models. At the same time, the early formulation investigations are taking place, to develop a mixture of the NME plus additional materials, known as excipients, which will provide the scientific understanding to find a unique formula and dosage form (e.g. solids such as a tablet or capsule, injectables such as a prefilled syringe, nasal spray and inhalation solution) to maintain its chemical stability and bioavailability suitable for human dosing in the early clinical trials. Such a mixture is referred to as the drug product, distinct from the Drug Substance which is the raw NME.

These early CMC studies, following the filing of the IND, lead to more extensive studies to refine the formulation, and to characterize its chemical and physical properties in increasingly greater detail. In addition, engineering and process studies are carried out to determine an optimal choice of manufacturing conditions as well as the development of state of the art chemical and biological analytical tools to measure the potency, purity and other characteristics important to assuring the identity of the drug product. Equally important, parallel clinical studies are being carried out in humans referred to by their stage of development: Phase 1 (single and multiple dose tolerance studies), Phase 2 (initial efficacy and pharmacokinetic studies) and Phase 3 (active controlled efficacy and long term safety studies) clinical trials. The drug product that patients are dosed with during these three

---

[1]A division of NMEs into large and small molecules is commonly found in the CMC literature. Small molecules are the classical drug mainly chemically synthesized. They are easily manufactured as tablets or capsules. Biological NMEs such as monoclonal antibodies are proteins that are either identical to or closely match endogenous proteins that play a key role in a disease process. They are manufactured as an injection or an infusion typically. Although small molecule drugs still make up 90 % of the drugs on the market, large molecule biological products are becoming increasingly more important.

phases of clinical studies are manufactured at small scale sizes following Good Manufacturing Practices (GMP), a set of government mandated rules governing the design, monitoring and control of pharmaceutical products to assure quality and enforced through the FDA. The analytical testing of the drug products are required to follow Good Laboratory Practices (GLP) rules to assure identity and support safety and quality. The latter is required under Code of Federal Regulations Title 21 (FDA 2015c). Both of these requirements point to the heavy regulation under which pharmaceutical products are manufactured and tested. It is impossible to be a CMC statistician and not appreciate the role of government regulation in the manufacture and testing of drug products, arguably the most heavily regulated industry in the world. There are corresponding regulatory rules which apply to the conduct of the clinical studies as well. Regulatory aspects of drug manufacture are discussed in greater detail in Chaps. 2 and 19.

The question of quality of a drug product falls squarely within the purview of CMC studies. What exactly does quality mean in this context? What is the role of Statistics in defining quality? What statistical tools are appropriate to answer these questions? In varying degrees of granularity, each of the chapters in this section of the book discuss statistical tools intended to address some important topic in the CMC arena: assay development, quality control, experimental design especially in the context of Quality by Design (QbD), process validation, statistical process and quality control, acceptance testing, stability modeling, dissolution, content uniformity, chemometrics and comparability studies. What they share in common is that in the background, there is a specification driving the need for the specific statistical tools discussed. It is beyond the scope of this chapter to discuss specifications and how they are derived, there is greater detail on this in Chaps. 18, 19 and 20, but one can say that conformance to specifications can be understood as the fundamental requirement for assuring the identity and quality of the drug product. The conformance is both a regulatory as well as a commercial requirement.

The CMC scientific studies and the statistical tools employed to design and elucidate the results of the CMC studies are intended to help define quality of the drug product through control of important drug product characteristics known as critical quality attributes (CQA). The study of these CQAs through appropriate statistical modeling leads to a greater scientific understanding of the product and process, as envisioned by the Quality by Design (QbD) paradigm set forth in the International Congress of Harmonization (ICH) guidances: Q8(R2) Pharmaceutical Development (2009), Q9 Quality Risk Management (2006) and Q10 Pharmaceutical Quality System (2009).

Modern statistical modeling approaches exploiting a Bayesian perspective have established a clear basis to answer the fundamental question of risk control as a probability statement in relation to process manufacturing factors (critical process parameters) and formulation changes linked to CQA outcomes (Peterson 2008). A set of CQA boundaries would be defined referred to as a *design space* within which a manufacturer could move without having to file a regulatory supplement seeking formal permission to modify the process. These concepts are completely consistent with the modernization of the pharmaceutical industry called for by the cGMPs for the twenty-first century initiative laid out in 2004 and later regulatory guidances (FDA 2015a, b). More on this topic is discussed in Chap. 18.

The follow-up to QbD and Design Space is process validation as elaborated on in great detail in the 2011 FDA guidance (FDA 2011). Two components of a formal process validation protocol are referred to as Stage 2-Process Performance Qualification (PPQ) and Stage 3-Continuous Process Verification (CPV). The notion of a life cycle approach to the manufacture of drug products is the intent of the 2011 guidance and proposes a more extensive role for statistical justification of moving from one stage of process validation to the next. An extensive discussion on the technical steps involved in Process Validation is given in Chap. 19.

Notions of quality in relation to pharmaceutical products are being increasingly related to "fit for use" concepts first elaborated on by Juran (1988). This calls for clear linkages of quality attributes to customer requirements, or ultimately in the context of pharmaceutical products, patient safety and efficacy. Much discussion has been held on establishing these linkages, especially in a QbD context, but so far, the necessary steps to provide such a clear connection have remained elusive. "Clinically relevant specifications" is a common topic of industry –regulatory agency conferences (Marroum 2012; Sharp 2012; Lostritto 2014) but exactly how we establish these in relation to CMC product attributes is still a work in progress.

One essential question relating to quality of a drug product concerns the amount of API being delivered to the patient. One might ask how close do individual dosage units deliver API to the stated amount. This is referred to as a content uniformity question. It is discussed in detail in Chap. 24. Analytical methods are developed and validated in connection with this question. Method development and validation are discussed in Chaps. 16 and 17. For small molecules, HPLC methods for potency assessment are typical, and for large molecules, a cell based bioassay would be developed. In both cases, an experimental design such as a Gauge R&R (also referred to as a "Gage" R&R) study could be used to study the total variability present in the measurement system. This is an important effort since the uncertainty associated with any analytical determination is a critical consideration (see ASTM E2655-14 2014) and has consequences on inferential statements related to drug product quality. In the case of a bioassay, average potency of a batch of biological product would be estimated using standard modeling methods relying on dose response curves and comparing the test sample to a standard curve. The use of a Gauge R&R design or a suitable block design (Altan and Shoung 2008) including analyst, run and plates and possibly other fixed and random effects provides a coherent approach to calculating estimates of the uncertainty associated with a reportable value based on 1 or more analytical determinations. An understanding of the uncertainty associated with such reportable values is an important consideration in both proposing and assessing drug product specifications.

The new Office of Pharmaceutical Quality (OPQ) established in 2014 has been charged with the responsibility of establishing quality metrics by which companies can evaluate their degree of "quality culture". The thinking is that notions of quality have to imbue a company from top down and every employee engaged in the manufacture of a drug product has to regard their job as in some way related to the furtherance of quality. The notion of quality metrics as a regulatory and commercial enterprise is still in its infancy, with considerable ongoing discussions being held at various joint industry FDA conferences (International Society of Professional Engineers 2015). It's clear that this major initiative by the FDA will evolve over the coming years and entail considerable accommodation by industry to pursue heightened standards in manufacturing quality. Readers should consult the current regulatory and scientific literature to stay abreast of its developments.

In the following sections of this chapter, we turn to a selection of statistical tools by way of introduction to their application to CMC studies, and pave the way for subsequent chapters to elaborate in greater detail on the topics opened up here.

15.2 Overview of Traditional CMC Statistical Methods

The traditional collection of statistical methods used to support drug product development and manufacturing are essentially the same as what might be called "industrial statistics". These involve statistical methods for process optimization, process validation, process monitoring, acceptance sampling, and manufacturing risk assessment.

While most of the statistical methods we touch on in this section have been around for decades, introduction of the concept of ICH Q8 "design space" and more modern manufacturing practices has created a need for statistical methods with better multivariate predictive capability. The concept of ICH Q8 "design space" has generated a need for improved assessment of the probability of conformance to specifications, particularly for processes with multiple responses. The introduction of modern pharmaceutical manufacturing and process monitoring technologies has produced a greater need to be able to handle large data sets comprised of many correlated measurements. An overview of the statistical methods that can address these two needs are given in Sects. 15.2.2 and 15.2.3, respectively. Subsequent Sects. 15.2.4–15.2.9 touch on important topics related to traditional categories of industrial statistical methods that have found good use in pharmaceutical development, manufacturing and testing. These sections serve as an introduction to subsequent chapters of the CMC section of this book which expand in much greater detail the topics touched on here.

15.2.1 Factorial and Fractional Factorial Designs

Often a pharmaceutical process may have many controllable factors that could *potentially* contribute to manipulation that will lead to process improvement. However, it is often the case that some factors will have a small or negligible effect over the range of experimental conditions to be considered. It is sometimes the case that only a small subset of factors acting together will produce substantial process improvement, but we may not know which these are among the many potential factors. As such, it is prudent to begin an experimental campaign with an efficient factor screening design to determine, in part, the key factors for use in process optimization techniques, such as response surface methodology. These factor screening experiments need to be efficient so that there are sufficient resources remaining for process optimization, once the critical factors are identified. In fact, some industrial statisticians recommend that no more than 25 % of available experimental resources should be used for the first set of experiments (Montgomery 2009, p. 556).

Initially, experimenters will want to make a list of potential factors that could affect a process quality attribute (or attributes), and then perform an experiment (or experiments) to screen out those factors that have little or no effect upon the process. If the number of potential factors is not too large (2 to 4 say) an experimenter may consider doing a full factorial design, provided he or she has the required resources. For a screening design, it is often best to use only 2 (or possibly 3) levels for each factor. A popular factorial screening design is the 2^k design. Here, each factor level is assigned only two levels. For k factors, the design has 2^k combinations of runs that must be executed. A full factorial design in k factors can model all linear and interaction terms up to the one k-way interaction. Often though, only linear and pairwise interactions are modeled, unless there is prior concern about the presence of a higher-order interaction.

The factorial design analysis typically involves doing a preliminary fit of the full model and then checking model assumptions. This will involve a residual analysis to check for possible outliers and normality of the residuals. Statistically insignificant factors should be considered for removal from the model, but care must be taken, particularly if the residuals show large variability in the data. Low power to detect practically significant factors could then be a concern. Typical regression model selection criteria (such as AIC or BIC), can also be used to discard unimportant factors. Of course, practical scientific background knowledge also plays a part in such decisions.

Often, the experimenter will also perform one or more "center point" runs. These are runs executed at the center point of all factors that are quantitative. If all the factors are quantitative, there would be only one center point whose coordinates are at the center of each factor (halfway between the factor's lower and upper limits). Factor ranges are an important consideration in sufficiently characterizing the linear or quadratic surface. A design with three quantitative factors and one qualitative factor (e.g. catalyst type) might have center points, one at each level of the qualitative factors. The purpose of the center point runs are to provide information about possible lack-of-fit from the type of model that is fit for the factorial design. Such designs typically support models with only first order and pairwise interactions terms such as

$$y = \beta_0 + \sum_{i=1}^{k} \beta_i x_i + \sum_{i<j} \beta_{ij} x_i x_j + e. \tag{15.1}$$

If the investigator is concerned about some kind of departure from the model in (15.1), the center points can provide a test for such a departure, although it provides little information as to the exact type of departure (del Castillo 2007, pp. 411–412).

As the number of factors to be studied increases, full factorial designs quickly become too costly. If we are willing to assume that higher order interactions are negligible, then one should consider fractional factorial designs. Negligible higher order interactions are associated with a certain degree of smoothness on the underlying response surface. If we believe that the underlying process is not very volatile as we change the factors across the experimental region, then the assumption of negligible higher order interactions may be reasonable. Fractional factorial designs are often certain fractions of a 2^k factorial design. They have the design form $2^{k\text{-}p}$, where k is the number of factors and p is associated with the particular fraction of the 2^k design. For example, a $2^{5\text{-}1}$ design denotes a half-replicate of a 2^5 design, while $2^{6\text{-}2}$ denotes a one-fourth replicate of a 2^6 design.

For fractional factorial designs, some interaction terms are confounded (or aliased) with possibly first order or other interaction terms. This means that certain interaction terms cannot be estimated apart from other terms. If (negligible) higher order interactions terms are confounded with first order or lower order interaction terms, then we may still be able to make practical inferences from a fractional factorial design analysis. Fractional factorial designs can be categorized into how well they can "resolve" factor effects. These categories are called design resolution categories. For a resolution III design, no main effects (first order terms) are confounded with any other main effect, but main effects are aliased with two-factor interactions and two-factor interactions are aliased with each other. For a resolution IV design, no main effect is aliased with any other main effect or with any two-factor interaction, but two-factor interactions are aliased with each other. For a resolution V design, no main effect or two-factor interaction is aliased with any other main effect or two-factor interaction, but two-factor interactions are aliased with three-factor interactions. Resolution III and IV designs are particularly good for factor screening as they efficiently provide useful information about main effects and some information about two-factor interactions (Montgomery 2009, chapter 13).

Recently, a special class of screening design has gathered some popularity. It is called a "definitive screening design". It requires only twice the number of runs as there are factors in the experiment. The analysis of these designs is straightforward if only main effects or main and pure-quadratic effects are active. The analysis becomes more challenging if both two-factor interactions and pure-quadratic effects are active because these may be correlated (partially aliased) (Jones and Nachtsheim 2011). If a factor's effect is strongly curvilinear, a fractional factorial design may miss this effect and screen out that factor. If there are more than six factors, but three or fewer are active, then the definitive screening design is capable of estimating a full quadratic response surface model in those few active factors. In this case, a follow-up experiment may not be needed for a response surface analysis.

15.2.2 Response Surfaces: Experimental Design and Optimization

Factorial and fractional factorial designs are useful for factor screening so that experimenters can identify the key factors that affect their process. It may be the case the results of a factorial or a fractional factorial design produce sufficient process improvement that the pharmaceutical scientist or chemical engineer will decide to terminate any further process improvement experiments. However, it is important to understand that a response surface follow up experiment may not only produce further process improvement, but it will also produce further process understanding by way of the response surface. If a process requires modification of factor settings (e.g. due to an uncontrollable change in another factor) it may not be clear how to make such a change without a response surface, particularly if the only experimental information available is from a fractional factorial design.

The results of the factor screening experiments may provide some indication that the process optimum may be within the region of the screening experiment. This could be the case, for example, if the response from center-point runs are better than all of the responses at the factorial design points. However, it is probably more likely that the results of the screening design point towards a process optimum that is towards the edge, or outside of, the factor screening experimental region. If this is the case, and if the results of the screening experiment provide no credible evidence of response surface curvature within the original region, then the experimenter should consider the method of "steepest ascent/descent" for moving further towards the process optimum. The classical method of steepest ascent/descent uses a linear surface approximation to move further towards the process optimum. Often, the original screening design may provide sufficient experimental points to support a linear surface. Clearly, statistical and practically significant factor interactions and/or a significant test for curvature indicates that a second-order response surface model needs to be built as the linear surface is not an adequate approximation. The method of steepest ascent/descent is a form of a "ridge analysis" (in this case for a linear surface), to be reviewed briefly below. See Myers et al. (2009, chapter 5) for further details on steepest ascent.

Special experimental designs exist for developing second-order response surface models. If it appears from the screening design that the process optimum may well be within the original experimental region, then a central composite design may be able to be developed by building upon some of the original screening design points (Myers et al. 2009, pp. 192–193). The Box–Behnken design (Box and Behnken 1960) is also an efficient design for building second-order response surfaces.

Once a second-order response surface model is developed, analytical and graphical techniques exist for exploring the surface to determine the nature of the "optimum". The basic form for the second-order response surface is

$$y = \beta_0 + \sum_{i=1}^{k} \beta_i x_i + \sum_{j=1}^{k} \sum_{i=1}^{k} \beta_{ij} x_i x_j + e, \tag{15.2}$$

where y is the response variable, $x_1, \ldots, x_k$ are the factors. It is often useful to express the equation in (15.2) in the vector–matrix form as

$$y = \beta_0 + \boldsymbol{x}'\boldsymbol{\beta} + \boldsymbol{x}'\mathbf{B}\boldsymbol{x} + e, \tag{15.3}$$

where $\boldsymbol{x} = (x_1, \ldots, x_k)'$, $\boldsymbol{\beta} = (\beta_1, \ldots, \beta_k)'$, and $\mathbf{B}$ is a symmetric matrix whose ith diagonal element is β_{ii} and whose (i, j)th-off diagonal element is $\frac{1}{2}\beta_{ij}$. The form in (15.3) is useful in that the matrix $\mathbf{B}$ is helpful in determining shape characteristics of the quadratic surface. If $\mathbf{B}$ is positive definite (i.e. all eigenvalues are positive), then the surface is convex (upward), while if $\mathbf{B}$ is negative definite (i.e. all eigenvalues are negative), then the surface is concave (downward). If $\mathbf{B}$ is positive or negative definite, then the surface associated with (15.3) has a stationary point that is a global minimum or maximum, respectively. If, however, some of the eigenvalues of $\mathbf{B}$ change sign, then the surface in (15.3) is that of a "saddle surface", which has a stationary point, but no global optimum (maximum or minimum). The stationary point, $\boldsymbol{x}_0$, is that point at which the gradient vector of the response surface is stationary (i.e. all elements of the gradient vector are zero). It follows then that

$$\boldsymbol{x}_0 = -\frac{1}{2}\mathbf{B}^{-1}\boldsymbol{\beta}.$$

Further insight into the nature of a quadratic response surface can be assessed by doing a "canonical analysis" (Myers et al. 2009, chapter 6). A canonical analysis invokes a coordinate transformation that replaces the equation in (15.2) with

$$\widehat{y} = \widehat{y}_s + \sum_{i=1}^{k} \lambda_i w_i^2, \tag{15.4}$$

where $\hat{y}_s$ is the predicted response at the estimated stationary point, $\widehat{\boldsymbol{x}}_0 = -\frac{1}{2}\widehat{\mathbf{B}}^{-1}\widehat{\boldsymbol{\beta}}$, and $\lambda_1, \ldots, \lambda_k$ are the eigenvalues of $\widehat{\mathbf{B}}$. The variables, $w_1, \ldots, w_k$, are known as the canonical variables, where $\mathbf{w} = (w_1, \ldots, w_k)' = \mathbf{P}'(\boldsymbol{x} - \widehat{\boldsymbol{x}}_0)$ and $\mathbf{P}$ is such that $\mathbf{P}'\widehat{\mathbf{B}}\mathbf{P} = \boldsymbol{\Lambda}$ and $\boldsymbol{\Lambda} = diag(\lambda_1, \ldots, \lambda_k)$. Here, $\mathbf{P}$ is a $k \times k$ matrix of normalized eigenvectors associated with $\widehat{\mathbf{B}}$.

One can see from (15.4) that if $|\lambda_i|$ is small, then moving in the direction vector given by $(0,..,0, \lambda_i, 0, \ldots, 0)'$ (in the $\boldsymbol{w}$-space) will result in little change from the stationary point. Such movements can be useful if, for example, process cost can be reduced for conditions somewhat away from the stationary point, but on or close to the line intersecting $\widehat{\boldsymbol{x}}_0$ along the $(0,.., 0, \lambda_i, 0, \ldots, 0)'$ direction in the $\boldsymbol{w}$-space.

However, the response surface shape may indicate that a global optimum is not within the regions covered by experimentation thus far. Or, it may be that operation at the estimated stationary point, $\widehat{\boldsymbol{x}}_0$, is considered too costly. In such cases, it may be useful to conduct a "ridge analysis". This is a situation where we want to do constrained optimization, staying basically within the experimental region. For a classical ridge analysis (Hoerl 1964), we optimize the quadratic response surface on spheres (of varying radii) centered at the origin of the response surface design region. In fact, one can consider ridge analysis to be the second-order response surface analogue of the steepest ascent/descent method for linear response surfaces. Draper (1963) provides an algorithmic procedure for producing a ridge analysis for estimated quadratic response surfaces. Peterson (1993) generalizes the ridge analysis approach to account for the model parameter uncertainty and to also make it apply to a wider class of models which are only linear in the model parameters, e.g. many mixture experiment models (Cornell 2002).

15.2.3 Ruggedness and Robustness

Once an analytical assay, or other type of process, has been optimized, one may want to assess the sensitivity of the assay or process to minor departures from the set process conditions. Here, we call this analysis a "ruggedness" assessment. Some of these conditions may involve factors for which the process is known to be sensitive, while other conditions may involve factors that (through testing or process knowledge) are known to be less influential. In a ruggedness assessment, the experimenter has to keep in mind how much control is possible for each factor. If one has tight control on all of the sensitive factors, then the process may indeed be rugged against deviations that commonly occur in manufacturing or in practical use (e.g. as with a validation assay). Ruggedness evaluations are often done on important assays that are used quite frequently.

Much of the classical literature on ruggedness evaluations for assays involve performing screening designs with many factors over carefully chosen (typically small) ranges. See, for example, Vander Heyden et al. (2001). The purpose of such an experiment is to see if any factor effects are statistically (and practically) significant. However, such experiments are typically not designed to have pre-specified power to detect certain effects with high probability. Furthermore, a typical factorial analysis of variance does not capture the probability that future responses from a process over this ruggedness experimental region will be within specifications. To address this issue, Peterson and Yahyah (2009) apply a Bayesian predictive approach to quantify the maximum and minimum probabilities that a future response will meet specifications within the ruggedness experimental region.

If the maximum probability is too small, the process is not considered rugged. However, if the minimum probability is sufficiently large, then the process is considered rugged.

This problem of process sensitivity to certain factors can sometimes be addressed by moving the process set conditions to a point that, while sub-optimal, produces a process that is more robust to minor deviations from the set point. This can sometimes be achieved by exploiting factor interactions, or from an analysis of a response surface. In some situations, however, the process is sensitive to factors which are noisy. This may be particularly true in manufacturing where process factors cannot be controlled as accurately as on the laboratory scale.

When a process will be ultimately subject to noisy factors (often called noise variables), a robustness analysis can be employed known as "robust parameter design" (Myers et al. 2009, chapter 10). Robust parameter design has found productive applications in industries involving automobile, processed food, detergent, and computer chip manufacturing. However, it has to date, not been widely employed in the pharmaceutical industry, although applications to the pharmaceutical industry are starting to appear in the manufacturing literature (Cho and Shin 2012; Shin et al. 2014).

The basic idea behind robust parameter design is that the process at manufacturing scale has both noise factors and controllable factors. For example, noise factors might be temperature (deviation from set point) and moisture, while controllable factors might be processing time and a set point associated with temperature. If any of the noise variables interact with the one or more of the controllable factors, then it may be possible to reduce the transmission of variation produced by the noise variables. This may result in a reduction in variation about a process target, which in turn will increase the probability of meeting specification for that process. There is quite a large literature on statistical methods associated with robust parameter design involving a univariate process response. (See Myers et al. 2009, chapter 10 for references.) However, there are fewer articles on robust parameter design method for multiple response processes (e.g. Miró-Quesada et al. 2004). Miró-Quesada et al. (2004) introduce a Bayesian predictive approach that is widely applicable to both single and multiple response robust parameter design problems.

15.2.4 Process Capability

During or directly following process optimization, the experimenter should also consider some aspect of process capability analysis. Such an analysis involves assessing the distribution of the process response over its specification interval (or region) and the probability of meeting specifications. Assessment of process capability involves a joint assessment of both the process mean and variance, as well as the variation of the process responses about the quality target, which in turn is related to the probability of meeting specifications. Further process capability should also be assessed during pilot plant and manufacturing as part of the process monitoring activities. This is because the distribution of process responses may involve previously unforeseen temporal effects associated with sequential trends, the day of the week, etc. Process capability is clearly important because a process may be optimized, but it may not be "capable", i.e. it may have an unacceptable probability of meeting specifications.

Process capability indices have become popular as a way to succinctly quantify process capability. The C_p index has the form

$$\frac{USL - LSL}{6\sigma},$$

where USL = "upper specification limit", LSL = "lower specification limit", and σ equals the process standard deviation. The C_p index is estimated by substituting the estimated standard deviation, $\widehat{\sigma}$, for σ. As one can clearly see, the larger the C_p index, the better is the process capability. However, the C_p index does not take into account where the process mean is located relative the specification limits. A more sensitive process capability index is denoted as C_{pk} which has the form

$$\min\left(\frac{USL - \mu}{3\sigma}, \frac{\mu - LSL}{3\sigma}\right).$$

The C_{pk} index is estimated by substituting the estimates for μ and σ for their respective population parameters appearing in the C_{pk} definition. The magnitude of C_p relative to C_{pk} is a measure of how off center a process is relative to its target.

Statistical inference relative to process capability indices require critical care. Such index estimates may have rather wide sampling variability. In addition, the process capability index can be misleading if the process is not in control (i.e. stable and predictable). It is hoped that before or during the process optimization phase that a process can be brought into control. However, validation of such control may need to be confirmed during the actual running of the process over time using statistical process monitoring techniques (Montgomery 2009, p. 364). In addition, process capability indices have in the past received criticism for trying to represent a multifaceted idea (process capability) as one single number (Nelson 1992; Kotz and Johnson 2002). While this criticism has some merit (and is applicable to any other single statistic), it is only valid if such indices are reported to the exclusion of other aspects of the distribution of process responses.

15.2.5 Measurement System Capability

Measurement capability is an important aspect of any quality system. If measurements are poor (e.g. very noisy and/or biased) process improvement may be slow and difficult. As such, it is important to be able to assess the capability of a measurement system, and improve it if necessary. This section will review concepts and methods for measurement system capability assessment and improvement.

Two important concepts in measurement capability analysis are "repeatability" and "reproducibility". Repeatability is the variation associated with repeated measures on the same unit under *identical* conditions, while reproducibility is the variation associated when units are measures under *different* natural process conditions, such as different operators, time periods, etc. A measurement system with good capability is able to easily distinguish between good and bad units.

A simple model for measurement systems analysis (MSA) is

$$Y = T + e,$$

where Y is the observed measurement on the system, T is the true value of the system response (for a single unit, e.g. a batch or tablet), and e is the error difference. (It is assumed here that Y, T and e are stochastically independent.) In MSA the total variance of this system is typically represented by

$$\sigma^2_{\text{Total}} = \sigma^2_{\text{Process}} + \sigma^2_{\text{Gauge}},$$

where $\sigma^2_{\text{Total}} = \text{var}(Y)$, $\sigma^2_{\text{Process}} = \text{var}(T)$, and $\sigma^2_{\text{Gauge}} = \text{var}(e)$. Clearly, accurate measuring devices are associated with a small "gauge" variance. It is also possible to (additively) decompose the gauge variance into two natural variance components, $\sigma^2_{\text{Repeatability}}$ and $\sigma^2_{\text{Reproducibility}}$. Here, "reproducibility" is the variability due to different process conditions (e.g. different operators, time periods, or environments), while "repeatability" is the variation due to the gauge (i.e. measuring device) itself. The experiment used to measure the components of σ^2_{Gauge} is typically called a "gauge R&R" study.

The estimation of variance components associated with a gauge R&R study are applied not only to pharmaceutical manufacturing process, but also to important assays. If a process or assay shows unacceptable variation, a careful variance components analysis may help to uncover the key source (or sources) of variation responsible for poor process or assay performance.

15.2.6 Statistical Process Control

In pharmaceutical manufacturing or in routine assay utilization, such processes will tend to drift over time and will eventually fall out of a state of control. All CMC processes have some natural (noise) variation to which they are subject. If this source of underlying variation is natural and historically known, this is typically called "common cause" variation. Such variation is "stationary" in the sense that it is random and does not drift or change abruptly over time. A process that is in control will be subject only to common cause variation. However, other types of variation will eventually creep in and affect a process. These sources of variation are known as "special cause" variation. Typical sources of special cause variation are: machine error, operator error, or contaminated raw materials. Special cause variation is often large when compared to common cause variation. Special cause variation can take on many forms. For example, it may appear as a one-time large deviation or as a gradual trend that eventually pushes a process out of control. Statistical process control (SPC) is a methodology for timely detection of special cause variation and more generally for obtaining a better understanding of process variation, particularly with regard to temporal effects. See Chap. 20 for additional discussion on this and related topics.

As a practical matter, it is important to remember that SPC only provides detection of special cause variation. Operator or engineering action will be needed to eliminate these special causes of variation, so that the process can be brought back into a state of control. Identification of an assignable cause for the special cause variation may require further statistical analysis or experimentation (e.g. a variance components analysis).

An SPC chart is used to monitor a process so that efficient detection of special cause variation can be obtained. An SPC chart consists of a "center line" that represents the mean of the quality statistic being measured when the process is in a state of control. The quality statistic is typically not just one single measured process response but rather the average of a group of the same quality responses chosen within a close time frame. It is expected that the common cause variation will induce random variation of this statistic about the center line. The SPC chart also has upper and lower control limits. These two control limits are chosen (with respect to the amount of common cause variation) so that nearly all of the SPC statistic values over time will fall between them. If the statistic values being measured over time vary randomly about the center line and stay within the control limits, then the process is in state of control and no action is necessary. In fact, any action to try to improve a process that is subject only to common cause variation may only increase the variation of that process. If however the SPC statistic values start to fall outside of the control limits or behave in a systematic or nonrandom manner about the center line, then this suggests that the process is out of control.

SPC methodology involves a variety of statistical tools for developing a control chart to meet the needs of the process and the manufacturer. Specification of the control limits is one of the most important decisions to be made in creating a control chart. Such specification should be made relative to the distribution of the quality statistic being measured, e.g. the sample mean. If the quality statistic being measured is a sample mean, then it is common practice to place the control limits at an estimated "3 sigma" distance from the center line. Here, sigma refers to the standard deviation of the distribution of the sample mean, not the population of individual quality responses. Three-sigma control limits have historically performed well in practice for many industries.

Another critical choice in control chart development involves collection of data in "rational subgroups". The strategy for the selection of rational subgroups should be such that data will be sampled in subgroups, so that if special causes are present, the chance for differences between subgroups to appear will be maximized, while the chance for differences within a subgroup will be minimized. The strategy for one's rational subgroup definition is very important and may depend upon one or more aspects of the process. How the rational subgroups are defined may affect the detection properties of the SPC chart. For example sampling several measurements at a specific discrete time points throughout the day will help the SPC chart to detect monotone shifts in the process. However, randomly sampling all process output across a sampling interval will result in a different rational group strategy, which may be better at detecting process shifts that go out-of-control and then back in between prespecified time points.

In addition to rational subgroups, it is important to pay attention to various patterns on the control chart. A control chart may indicate an out-of-control situation when one or more points lie outside of the control limits or when a nonrandom pattern of points appears. Such a pattern may or may not be monotone within a given run of points. The problem of pattern recognition associated with an out-of-control process requires both the use of statistical tools and knowledge about the process. A general statistical tool to analyze possible patterns of non-randomness is the runs test (Kenett and Zacks 2014, chapter 9).

As one might expect, having multiple rules for detecting out-of-control trends or patterns can lead to an increase in false alarms, particularly if a process is assumed to be out-of-control if at least one, out of many such rules, provides such indication. It may be possible to adjust the false positive rate for the simultaneous use of such rules, but this may be difficult as many such rules are not statistically independent. For example, one rule may be "one or more points outside of the control limits" while another rule may be "six points in a row with a monotone trend". This is a situation where computer simulation can help to provide some insights regarding the probabilities of false alarms when using multiple rules for detecting out-of-control trends or patterns.

Typical control chart development is divided into two phases, Phase I and Phase II. The purpose of Phase I is to develop the center line and control limits for the chart, as well as to ascertain if the process is in control when operating in the sequential setting as intended. Phase I may also be a time when the process requires further tweaking to be brought into a state of control. The classical Shewhart control charts are generally very effective for Phase I because they are easy to develop and interpret, and are effective for detecting large changes or prolonged shifts in the process. In phase II, the process is now assumed to be reasonably stable so that phase II is primarily a phase of process monitoring. Here, we expect more subtle changes in the process over time, and so more refined SPC charts may be employed such as cumulative sum and Exponentially Weighted Moving Average (EMWA) control charts (Montgomery 2009, chapter 9).

The notion of "average run length" (ARL) is a good measure for evaluation a process in Phase II. The ARL associated with an SPC chart or an SPC method is the expected number of points that must be plotted before detecting an out-of-control situation. For the classical Shewhart control chart, the $\text{ARL} = 1/p$, where p is the probability that any point exceeds the control limits. This probability, p, can often be increased by taking a larger sample at each observation point. Another useful, and related measure, is the "average time to signal" (ATS). If samples are taken that are t hours apart, the ATS = ARLt. We can always reduce the ATS by increasing the process sampling frequency. In addition, the ARL and ATS can be improved by judicious use of more refined control chart methodology (in some cases through cumulative sum or EWMA control charts). For some control charts (e.g. Shewhart) the run time distribution is skewed so that the mean of the distribution may not be a good measure. As such, some analysts prefer to report percentiles of the run length distribution instead.

Many processes involve multiple quality measurements. As such, there are many situations where one needs to monitor multiple quality characteristics simultaneously. However, statistical process monitoring to detect out-of-control processes or trends away from target control requires special care, and possibly multiple SPC charts of different types. The naive use of a standard SPC chart for each of several measured quality characteristics can lead to false alarms as well as missing out-of-control process responses. False alarms can happen more often than expected with a single SPC chart because the probability of *at least* one false alarm out of several can be noticeably greater than the false alarm rate on an individual SPC chart. In addition, multivariate outliers can be missed with the use of only individual control charts. Because of this, special process monitoring methods have been developed for multivariate process monitoring and control.

One of the first methods to address multivariate process monitoring is the Hotelling T^2 control chart. This chart involves plotting a version of the Hotelling T^2 statistic against an upper control limit chi-square or F critical value. The Hotelling T^2 statistic can be modified to address either the individual or grouped data situation. However, an out-of-control signal by the Hotelling T^2 statistic does not provide any indication of what particular quality response or responses are responsible. In addition to SPC charts for individual quality responses, one can plot the statistics, $d_j = T^2 - T^2_{-j}$, where T^2_{-j} is the value of the T^2 statistic but with the jth quality response omitted.

Several other univariate control chart procedures have been generalized to the multivariate setting. As stated above, EWMA charts were developed to better detect small changes in the process mean (for a single quality response type). Analogously, multivariate EWMA chart methods have been developed to detect small shifts in a mean vector. In addition, some procedures have been developed to monitor the multivariate process variation by using statistics which are functions of the sample variance-covariance matrix. See Montgomery (2009, pp 516–517) for details.

When the number of quality responses starts to become large (more than 10 say), standard multivariate control chart methods start to become less efficient in that the ARL also increases. This is because any shifts in one or two response types become diluted in the large space of all of the quality responses. In such cases, it may be helpful to try to reduce the dimensionality if the problem by projecting the high-dimensional data into a lower dimensional subspace. One approach is to use principal components. For example, if the first two principal components account for a large proportion of the variation, one can plot the principal component scores labeled by their order within the process as given by the recorded data vectors. This is called a trajectory plot. Another approach is to collect the first few important principal components and then apply the multivariate EWMA chart approach to them (Scranton et al. 1996).

15.2.7 Acceptance Sampling

When lots of raw materials arrive at a manufacturing plant, it is typical to inspect such lots for defects or some measure related to raw material quality. In addition, pharmaceutical companies often inspect newly manufactured lots of product before making a decision about whether or not to release the product lot or patch for further processing or public consumption.

However, it is useful to note that the primary purpose of acceptance sampling is to sentence lots as acceptable or not; it is not to create a formal estimate of lot quality. In fact, most acceptance sampling procedures are not designed for estimation of lot quality Montgomery (2009). Acceptance sampling should not be a substitute for process monitoring and control. Nonetheless, the use of acceptance sampling plans over time produces a history of information which reflects on the quality of the process producing the lots or batches. In addition, this may provide motivation for process improvement work if too many lots or batches are rejected. See Chap. 20 for additional discussions on this and related ideas.

Acceptance sampling can be divided into two categories related to item description, attribute plans and variable plans. Attributes are quality characteristics that have discrete "accept" or "reject" levels (e.g. "defective" or "acceptable"). Variables sampling plans involve a quality characteristic that is measured on a continuous scale. Acceptance sampling plans can also be categorized according to their sequential nature. For example, a "single sampling" plan takes one sample of n units from a lot and a decision is made based on that one sample to sentence the lot as acceptable or not. A "double sampling" plan works in two stages. A first sample is taken and a decision is made to (i) accept the lot, (ii) reject the lot, or (iii) take a second sample from the lot. If the second sample is taken, the information from both the first and second sample is used to accept or reject the lot. It is also possible to have multi-stage sampling plans that are generalizations of a two-stage sampling plan, whereby more than two samples from the lot are required to make a decision. It should be noted that acceptance sampling plans (aside from 100 %) inspection usually always involve some sort of random sampling.

A useful characterization of the performance of an acceptance sampling plan is the operating-characteristic (OC) curve. This curve is a plot of the "probability of accepting the lot" vs. the "lot fraction defective". Clearly, this should be a monotone decreasing curve. The location and shape of this curve displays the discriminatory power of the sampling plan. To better understand how this works, consider the attributes sampling situation where the lot size, N, is very large so that for a random sample of size n ($n << N$) the number of defectives has approximately a binomial

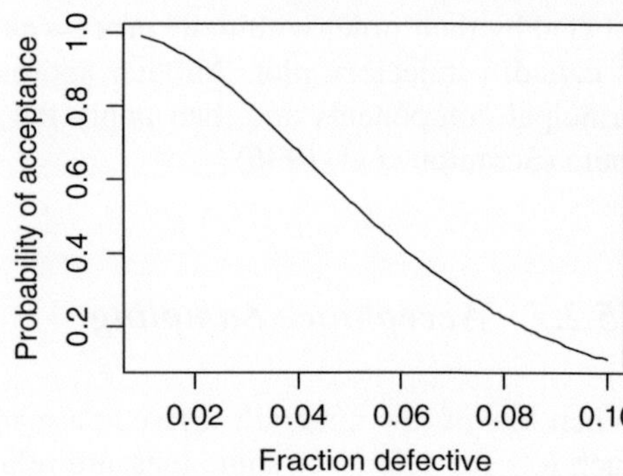

Fig. 15.1 Operating characteristic (OC) curve for a sample size of $n = 50$ and $d^* = 2$ acceptance number for defectives

distribution. We assume here that π is the fraction of defective items in the lot. If D is the number of defective items found out of a sample of size n, the D has a binomial distribution with probability function,

$$\Pr(D = d; n, \pi) = \binom{n}{d} \pi^d (1 - \pi)^{n-d}.$$

Suppose that we accept the lot if $D \leq d^*$. We call d^* the acceptance number. Then the probability of accepting the lot is

$$p(\pi) = \Pr\left(D \leq d^*; n, \pi\right) = \sum_{d=0}^{d^*} \binom{n}{d} \pi^d (1 - \pi)^{n-d}.$$

For specific values of n and d^*, one can plot $p(\pi)$ vs. π to obtain the OC curve. For example if $n = 50$ and $d^* = 2$, then the OC curve is shown in Fig. 15.1 below.

If the sample size, n, increases (keeping d^* proportional to n), then the slope (in the neighborhood of the curve inflection) will become steeper. This indicates more discriminatory power for the sampling plan.

Often the statistician or quality engineer will focus on certain points on the OC curve. For example, they may be interested knowing what level of lot quality (fraction defective) would be associated with a high probability of acceptance (e.g. 0.95). Or, they (and consumers) may be interested in the level of lot quality associated with a low probability of acceptance. A sampling plan is often established with regard to an acceptable quality level (AQL). The AQL is the lowest level of quality for the manufacturer's process that would be consider acceptable as a process average. The sampling plan will typically be designed so that the OC curve shows a high probability of acceptance at the AQL. It is important to note that the AQL is not usually intended to be a specification on the product Montgomery (2009). It is simply a standard against which to sentence lots. See Montgomery (2009) for an overview of OC curves and AQL's associated with double or multiple sampling plans.

With regard to variable sampling plans, there are advantages and disadvantages of which one should be aware. The main advantage of the variables sampling plan is that essentially the same OC curve can be obtained, but with a smaller sample size, than for the corresponding attributes sampling plan. This is because there is more information in variables sampling than in attribute sampling. However, for a variables sampling plan, the probability distribution of the quality characteristic being used must be known, at least up to a good approximation. If the true distribution of the quality characteristic for the variables sampling plan deviates enough from the assumed distribution, then serious departures of the computed OC from the real underlying one may arise, leading to rather biased decisions about lot sentencing. A typical assumption is that the quality characteristic for a variables sampling plan has a normal distribution. But, this assumption should be checked carefully. See Montgomery (2009) for an overview of variables sampling plans.

15.2.8 Failure Mode and Reliability Assessment

Process and product reliability assessment involve two basic activities: (i) identification of failure modes and (ii) quantification of reliabilities associated with these failure modes. Identification of all key failure modes is important. "Failure modes" means the ways, or modes, in which something might fail. Their identification is important. Suppose all but one key failure model is identified, and much effort has gone into quantifying the reliabilities of the (identified) potential modes of failure. Nonetheless, a process can still be likely to produce a seriously flawed product because one of the key failure models was not identified.

A popular approach to identification of failure modes and their subsequent reliability quantification is called "failure mode and effects analysis" (FMEA). Teamwork involving key representatives related to different aspects of a manufacturing process is important for executing an FMEA (Breyfogle 2003). As part of the FMEA process, a team works to identify potential failure modes for design functions or process requirements. The teams then assign a severity measure to each identified failure mode. They also assign a measure of the frequency of occurrence to the failure mode, along with a measure of the likelihood of detection. After this analysis, the team typically calculates a "risk priority number" (RPN), which is the product of three numbers: the severity measure, the frequency measure, and the likelihood of detection measure. Typically, the RPN values are used to prioritize process improvement efforts, with teams tackling the potential failure models with the highest RPN's. However, RPN's have been criticized as sometimes misleading because they do not strictly follow the rules of arithmetic (Wheeler 2011).

The problem with RPN's is that the severity measure, the frequency measure, and the likelihood of detection measure are based upon ordinal measures, i.e. they are based upon ordered rankings. For example, the severity score may have levels such as: "no discernible effect", "very minor", "minor", "very low, "low", "moderate", "severe', "very severe". The frequency-of- failure score may have levels such as:

"unlikely", "relatively few failures", "occasional failures", "frequent failures", and "persistent failures". Likewise, the likelihood-of-detection score may have levels such as: "almost certain", "very high", "high", "moderately high", "moderate", "low", "very low", "remote", "cannot detect failure". However, multiplication of these ordinal values is improper because the measures are not on an "interval ratio scale". This does not mean that the failure modes identification process is flawed; it only means that the RPN's can be misleading. If ordinal scales must be used, it is better to prioritize the situations first by severity, then by occurrence within each level of severity, and finally by detectability within each combination of severity and occurrence (Wheeler 2011). More formally, it is better to model the reliability of a process by using the identified failure modes and the laws of probability. This can be done using a Bayesian network (Garcìa and Gilabert 2011).

The quantification of reliability probabilities for a process can be a complex and/or fragile undertaking. By a "reliability probability" we simply mean the probability of desired event (e.g. meeting one or more process specifications). For a complex, multi-stage process the quantification of the overall reliability probability can involve sophisticated modeling of conditional and unconditional distributions. See for example Barlow and Proschan (1975). A more subtle point is that even if sophisticated probability analyses are completed, the actual process reliability and the estimated one may in some situations be rather different. If a reliability analysis is based upon carefully designed experiments, providing adequate information, some probability-based reliability calculations may be fairly accurate. However, for practical reasons, probability based reliability calculations may be conditional on the properties of raw materials from a specific supplier or upon the functioning of a specific piece of equipment. If either or both of these change, the reliability probability may change abruptly. Hence, reliability probabilities are at best only good for the near future, and require re-validation from time to time.

15.2.9 Experimental Design and Modeling Considerations in Bioassay Potency Testing

As mentioned previously, experimental design considerations and modeling strategies are a critical part of potency testing carried out through bioassays, frequently using in vitro or cell based bioassays using microtiter plates or in some cases, in vivo or live animal bioassays (Finney 1978). The balanced incomplete block design (BIBD) given in Table 15.1 is an example of a statistical design which could be used to estimate variance components attributable to analyst, run and plate for the given bioassay. One standard microtiter plate consists of wells located across 8 rows and 12 columns, 96 wells total, with each well providing a measurement in relation to a known or unknown concentration of a biologic material. The layout of a 96 well plate with rows and columns marked as A–H and 1–12 respectively is given in Fig. 15.2.

Table 15.1 BIBD design for a Gauge R&R study of a biologic

Analyst	Run	Plate	Sample concentration (%)					
			25	50	67	100	150	200
1	1	1	X	X	X			
		2	X				X	X
	2	3			X	X	X	
		4		X	X			X
	3	5	X	X		X		
		6		X			X	X
	4	7			X	X		X
		8	X			X		X
	5	9		X		X	X	
		10	X		X		X	
2	1	1				X	X	X
		2		X	X	X		
	2	3	X	X				X
		4	X			X	X	
	3	5			X		X	X
		6	X		X	X		
	4	7	X	X			X	
		8		X	X		X	
	5	9	X		X			X
		10		X		X		X

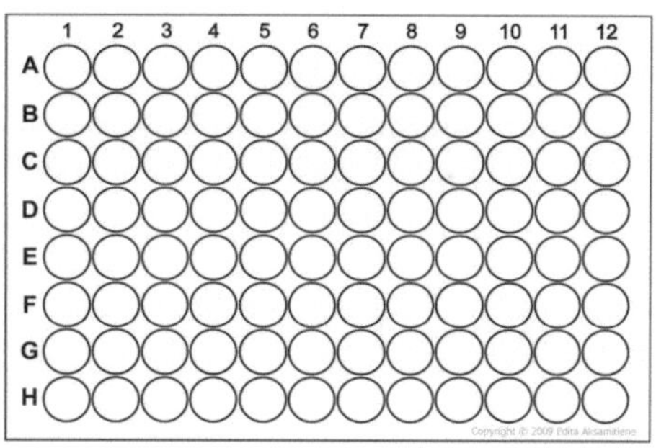

Fig. 15.2 96-Well plate layout

The given BIBD also permitted the investigation of dilutional linearity. Dilutional linearity relates to the extent of recovery across a range of known concentrations. For example, known samples of 25 %, 50 %, and so on could be plated out and tested to see if the measured concentrations are reasonably close to their known concentrations following the scheme laid out in Table 15.1. This is an example of two analysts, each analyst performing five runs and each run consisting of three plates, each plate accommodating three samples (known concentrations) plus a standard. Each plate would consist of a set of dilutions comparing the test samples to a standard curve and calculating a relative potency. The design could be extended to include additional analysts to improve the estimation of analyst as a random component and additional factors such as Laboratory by repeating the design across multiple laboratories.

Typically, samples (concentrations) would be plated out according to a serial dilution scheme across columns 1–12, with pairs of contiguous rows starting with A–B corresponding to replicates of a given sample. In this case, the BIBD design calls for three samples to be present on each plate plus a standard curve, where each plate can be thought of as a block in relation to samples. Samples are plated out as if they were at 100 % concentration. The measured concentrations across all plates according to the given BIBD are related to linearity, accuracy (closeness of % recovery to target) and precision. The allocation of samples to pairs of rows across the 20 plates is an important design question. Ideally, one would seek to balance row locations (pairs of rows) on the plate with sample concentration, to orthogonalize location within plates effects with plate and sample concentration effects. One can imagine a latin square design across plates, arranged in such a way as to provide at least proportionality across plates for row and sample effects. It is not known whether a general construction method exists for the interweaving of a latin square with a BIBD to provide such balance as we have in this design. In such a situation, one can construct an allocation that achieves near-orthogonality, by allocating each sample concentration to each pair of row locations (A–B, C–D, E–F, G–H) as close to an equal number of times as possible across plates. In this design, each sample is present in half of the plates. Therefore, the best one can do is six permutations of 2,2,3,3 corresponding to number of plates with a given sample in locations A–B, C–D, E–F, G–H. An example of such a design is given in Table 15.2 where 1,2,3,4,5,6 correspond to sample concentrations 25,50,67,100,150,200 % of target respectively, and S stands for Standard.

The overall statistical modeling and analysis would be carried out in two steps. First, potency estimates will be generated based on the constrained four-parameter logistic model, followed by a linear mixed model incorporating within and between plate variance terms to produce the final combined estimates. Other important aspects related to potency estimation in the context of bioassays is given by Davidian and Giltinan 1993, Giltinan 1998, Lansky 2002, O'Connell et al 1993 and Rodbard et al 1994.

15.2.9.1 Statistical Model for Potency Estimation

For potency estimation purposes, assume a heteroscedastic model $y_i = f(x_i, \beta) + \varepsilon_i$ where y_i denotes the independent responses (count, OD, etc) at concentration x_i, $f(x_i, \beta)$ is typically the four parameter logistic given by:

$$f(x_i, \beta) = \beta_2 + \frac{\beta_1 - \beta_2}{1 + \left(\frac{x_i}{\beta_3}\right)^{\beta_4}} = \beta_2 + \frac{\beta_1 - \beta_2}{1 + \exp\left(\beta_4 \left[\log(x_i) - \log(\beta_3)\right]\right)} \tag{15.5}$$

Table 15.2 Allocation of treatments to plate locations

Analyst	Run	Plate	Plate row A–B	C–D	E–F	G–H
1	1	1	S	1	2	3
		2	5	S	1	6
	2	3	3	4	S	5
		4	2	3	6	S
	3	5	S	2	4	1
		6	6	S	5	2
	4	7	3	6	S	4
		8	4	1	6	S
	5	9	S	5	2	4
		10	1	S	3	5
2	1	1	4	5	S	6
		2	2	3	4	S
	2	3	S	6	1	2
		4	5	S	1	4
	3	5	6	5	S	3
		6	4	1	3	S
	4	7	S	2	5	1
		8	3	S	2	5
	5	9	1	6	S	3
		10	2	4	6	S

where $\beta = (\beta_1, \beta_2, \beta_3, \beta_4)$ and $\beta_1 =$ asymptote as the concentration $x \rightarrow 0$ (for $\beta_4 > 0$), $\beta_2 =$ asymptote as $x \rightarrow \infty$, $\beta_3 =$ concentration corresponding to response halfway between the asymptotes, $\beta_4 =$ slope parameter. One could argue that these parameter values are fixed constants related to physical properties of the molecule but subject to variation due to known and unknown factors. The mean response is given by $E(y_i) = \mu_i = f(x_i, \beta)$. Then $\text{Var}(y_i) = \sigma^2 g^2\{f(x_i, \beta), \gamma\}$, where $g^2\{f(x_i, \beta), \gamma\}$ is referred to as the variance function with parameter γ and expresses the heteroscedasticity, σ^2 is a scale parameter. The least squares method used to estimate the parameters given in (15.5) requires weighting as a consequence of the heteroscedastic model. The weights are provided by the variance function and therefore its correct specification is critical not only to the estimation procedure itself, but also to the calculation of standard errors for the parameter estimates. For the purposes of this discussion, the power of the mean function will be used relating variance to mean as $g^2\{f(x_i, \beta), \gamma\} = \mu_i^{\gamma}$.

Generalized least squares (GLS) method is commonly used to estimate the parameters. The method is as follows: (i) estimate β by a preliminary unweighted fit using ordinary least squares, (ii) use the residuals from the preliminary fit to estimate γ and σ, (iii) based on the estimates from (ii), form new weights, and re-estimate β using weighted least squares chosen to minimize the objective function given by

$$O_{GLS} = \sum \left[\frac{(y_i - \widehat{\mu}_i)^2}{\sigma^2 \widehat{\mu}_i^{\gamma}} + \log\left(\sigma^2 \widehat{\mu}_i^{\gamma}\right) \right] \tag{15.6}$$

where $\widehat{\mu}_i = f(x_i, \widehat{\beta})$ and $\widehat{\beta}_i$ is the estimator of β_i from the previous step and return to step (ii) iterating until convergence. The use of (15.6) to estimate γ and σ leads to pseudo-likelihood estimates of the parameters (Giltinan and Ruppert 1989).

15.2.9.2 The Constrained Four Parameter Logistic Model

The term 'constrained' implies that p ($p > 1$) curves are being estimated, and that 1 or more parameters may be common across the curves. Suppose $p = 2$, such as when a test preparation is being compared to a standard. Extend Eq. (15.5) to accommodate both curves using a piecewise regression notation. Let s,t index standard and test respectively and assume $\beta_{s1} = \beta_{t1}$, $\beta_{s2} = \beta_{t2}$, $\beta_{s4} = \beta_{t4}$ letting only β_{s3} and β_{t3} vary as a consequence of the conditions of similarity (Finney 1978). Further, let $\beta^*_s = \log\beta_{s3}$ and $\beta^*_t = \log\beta_{t3}$ and since the log relative potency $\rho_t = \beta^*_s - \beta^*_t$, a relative potency parameter can be incorporated yielding the model:

$$y_i = \beta_2 + \frac{\beta_1 - \beta_2}{1 + \exp\beta_4 \left\{I_s \log x_i + I_t \left(\log x_i + \rho_t\right) - \beta^*{}_s\right\}} + e_i \tag{15.7}$$

where the concentrations $x_i's$, and responses $y_i's$ are arranged according to indicator variables I_s and I_t denoting standard and test, respectively. Estimates of the parameters given in (15.7) and their estimated covariance matrix are produced as part of the GLS algorithm. The model is easily extended to the case of more than a single test vs. a standard.

Statistical testing of similarity is a requirement as a consequence of validity considerations. In the case of the common parallel line linear model with fixed parameters, the classical approach (Finney 1978) relies on the principle of extra sums of squares. Extensions of this approach to the non-linear mixed model case have not been reported in the literature although Lansky (1999) mentions a test for parallelism as the essential test in relation to a split-block design. More modern approaches rely on an equivalence test to establish similarity. The equivalence testing is applied to the two asymptote parameters and the shape parameter. A good discussion on equivalence testing to establish similarity is found in the USP 38/NF 33 (2015) and Lansky (2014).

Mixed effects modeling can be applied to combine estimates across plates to produce an average potency across tests, variance components and address linearity. An appropriate mixed effects model would lead to potency estimates adjusted for between and within plate sources of variation (Altan et al. 2002). The variance component estimates of run, plate and residual variance coming out of the mixed model would be combined to report intermediate precision. Repeatability would be the residual error. These estimates can be combined in various ways to report an assay format uncertainty term corresponding to a reportable value. These ideas are discussed in greater detail in USP 38/NF 33 <1033> 2015.

15.3 Bayesian Applications: Probability of Conformance and ICH Q8 Design Space

Process optimization and reliability assessment are key issues for CMC statistics. Because Bayesian methods directly lead to predictive distributions, they are in turn useful for reliability assessment and process optimization by way of reliability assessment. This is true for both single and multiple response processes.

15.3.1 Probability of Conformance and Process Capability

An important issue in quality-by-design is being able to calculate the probability that a process will meet its specifications. This is important, because a process may have been "optimized" by means of its associated mean response surface, but this does not mean that the process is likely to meet its specifications. In fact, a traditional approach of using "overlapping mean responses" for optimizing processes with multiple quality attributes can result in misleading inferences for assessing the probability of conformance and ICH Q8 design space (Peterson and Lief 2010).

A natural and easy-to-understand approach to process capability, particularly for multiple response processes, is to simply compute the probability that the process will meet all of its specifications simultaneously. This can be written from a Bayesian perspective as

$$\Pr\left(\boldsymbol{Y} \in S \mid data\right), \tag{15.8}$$

where the probability in (15.8) is based upon the posterior predictive distribution (unconditional on the model parameters). This notion for process capability was first put forth by Bernardo and Irony (1996). This concept was extended to regression models by Peterson (2004), thereby creating the function

$$p\left(\boldsymbol{x}\right) = \Pr\left(\boldsymbol{Y} \in S \mid \boldsymbol{x}, data\right), \tag{15.9}$$

where $\boldsymbol{x}$ is a vector of process factors and $\boldsymbol{Y}$ is related to $\boldsymbol{x}$ by way of a stochastic response surface model. By optimizing the $p(\boldsymbol{x})$ function, one can optimize the process in a way that optimizes a measure of process capability. As it turns out, this Bayesian predictive approach to process optimization adapts itself quite easily to solve many process optimization problems that have challenging complexities.

For multivariate problems, one would typically need to fit different model forms to each response-type, so as to not over fit. This is because some response types may require simple models, while others may require more complex model forms. If (parametrically) linear models are being used, different model forms will result in what are called "seemingly unrelated regression" (SUR) models (Zellner 1962; Srivastava and Giles 1987). Fortunately, SUR models are easy to analyze using

Gibbs sampling (Percy 1992; Peterson et al. 2009b). The Bayesm R package (Rossi 2012) has a function, rsurGibbs, which can be used to sample from the posterior for a SUR model.

The process optimization approach in (15.9) can be easily modified to solve the multiple-response robust parameter design problem (Miró-Quesada et al. 2004). For the robust parameter design problem, some of the factors are noisy. The idea is to adjust the controllable factors so as to reduce the variation transmitted by the noisy factors. The Bayesian predictive approach is able to do this in a very natural way by averaging over the noise distribution to obtain

$$p(\boldsymbol{x}_c) = E_{\boldsymbol{x}_n}\{\Pr(\boldsymbol{Y} \in S \,|\, \boldsymbol{x}_c,\ \boldsymbol{x}_n, data)\} = \Pr(\boldsymbol{Y} \in S \,|\, \boldsymbol{x}_c, data) \qquad (15.10)$$

where $\boldsymbol{x}_n$ is a vector of noise variables and $\boldsymbol{x}_c$ is a vector of the controllable factors. Optimization of $p(\boldsymbol{x}_c)$ in (15.10) adjusts the controllable factor levels so that noise variable error transmission is dampened in just the right way as to optimize the probability of conformance to specifications.

The posterior predictive optimization approach can also be generalized to (univariate or multivariate) mixed-effect models. For example, one can optimize the probability of conformance for a (univariate or multivariate) split-plot experiment. The most technically difficult part here is sampling from the posterior of the model parameters. Once that has been accomplished, optimization of the $p(\boldsymbol{x})$ is fairly straightforward.

WinBUGS (Lunn et al. 2000) or OpenBUGS (Thomas et al. 2006) software can be used to sample from the posterior for SUR models or mixed-effect SUR models that also contain random effects. In addition, the R package, MCMCglmm (Hadfield 2010) can also be used to sample from the posterior for such models. It is even possible to use Bayesian model averaging (Press 2003, chapter 13) to account for uncertainty of model form in the $p(\boldsymbol{x})$ function. See articles by Rajagopal and del Castillo (2005), Rajagopal et al. (2005), and Ng (2010) for details.

The textbook, *Process Optimization – A Statistical Approach*, (del Castillo 2007) is the only one that gives an introduction to process optimization using the Bayesian predictive approach. Del Castillo (2007) lists seven advantages of using the Bayesian predictive approach to process optimization. These are:

1. It considers the uncertainty of the model parameters. (Many response surface techniques do not do this.)
2. It considers the correlation among the responses. (Classical desirability function approaches (e.g. Harrington 1965; Derringer and Suich 1980) do not do this. However, Chiao and Hamada (2001) and Peterson (2008) show that the joint probability of conformance to specifications can be strongly dependent upon the correlation structure of the regression residuals.)
3. Informative prior information can be used, as well as non-informative priors.
4. A well-calibrated (factor) region ("sweet spot") of acceptable probability of conformance to specifications can be constructed in a straightforward manner. (The classical "overlapping means" approach to constructing a sweet spot can result in a region with quite low probabilities of meeting specification (Peterson and Lief 2010).

5. It can be used for more general optimizations such as $p(\boldsymbol{x}) = \Pr(D(\boldsymbol{y}) \geq D*|\boldsymbol{x}, data)$, where $D(\boldsymbol{y})$ is a desirability function (Derringer and Suich 1980; Harrington 1965). Note, however, that such desirability function approaches as typically used in classical response surface optimization do not provide a measure of how likely it is that the desirability will exceed a pre-specified amount, e.g. D^*.
6. It allows one to perform a pre-posterior analysis if the optimal probability of conformance is too low. Here, by "pre-posterior analysis" we mean that one can simulate additional data from a fitted model, and then see how much this additional information changes the posterior probability of conformance to specifications. If the change goes from say, 0.83 to 0.96 and 0.96 is acceptable, then this suggests that our process is adequate, and that we may only need additional data for confirmation purposes. If, however, after the addition of substantially more "data", the change goes from 0.83 to 0.88 and 0.88 is not acceptable, then this indicates that we may need to improve the process itself.
7. It is easy to add noise variables, thereby providing a solution to the multivariate robust parameter design problem.

It is also useful to note that Bayesian inference has been successfully applied to a variety of CMC problems. For example, the Bayesian predictive approach has been applied to complex sampling algorithms that are sometimes used in the pharmaceutical industry. For example, it may be desired to compute the probability of a certain drug passing a multi-stage USP test, whereby several tablets are sampled and tested, with the possibility that more may need to be sampled and tested. LeBlond and Mockus (2014) provide an example of quantifying the probability of passing a compendial standard for content uniformity. The Bayesian approach of computing relevant probabilities has also been applied to some assay validation problems (Novick et al. 2011, 2012). Recently, the Bayesian approach has been applied to dissolution curve comparisons. See LeBlond et al. (2015) for a review and Novick et al. (2015) for an innovative application.

15.3.2 Probability of Conformance and ICH Q8 Design Space

The ICH Q8 Guidance on Pharmaceutical Development has proposed the notion of a "design space". This is defined as "the multidimensional combination and interaction of input variables (e.g., material attributes) and process parameters that have been demonstrated to provide assurance of quality". One can think of a design space as a collection of manufacturing recipes each of which should be likely to meet quality specifications. However, for actually constructing a design space, it is helpful to have a mathematical definition, that is the design space equation. From a Bayesian predictive perspective, one can define the ICH Q8 design space as

$$\{\boldsymbol{x} : \Pr(\boldsymbol{Y} \in S|\boldsymbol{x}, data) \geq R\},$$

where, as before, $\boldsymbol{x}$ is a vector of process factors and $\boldsymbol{Y}$ is a vector of quality response-types (also referred to as quality attributes) (Peterson 2008). Here, R is a pre-specified reliability level. See also Peterson (2009).

It is curious to note that Appendix 2 of the ICH Q8 Guidance gives an example of a design space that appears to be of the form of overlapping mean response surfaces. However, as shown by Peterson and Lief (2010) such a design space may have very low probability of simultaneously meeting all specifications! It may be that this statistical issue was a blind spot for the developers of the ICH Q8 Guidance. Fortunately, further publications supporting a Bayesian approach to ICH Q8 design space have appeared in the literature. See for example Stockdale and Cheng (2009), Peterson et al. (2009a, b), Castagnoli et al. (2010), Peterson and Kenett (2011), LeBrun et al. (2012), Maeda et al. (2012), Crump et al. (2013), Mockus et al. (2014), Gong et al. (2015), Chavez et al. (2015), Chatzizaharia and Hatziavramidis (2015), and LeBrun et al. (2015).

ICH Q8 design space concepts have been applied not only to pharmaceutical manufacturing, but also to pharmaceutical assay development. Thus one could have a collection of assay conditions that have been demonstrated to be likely to have quality attributes that meet the assay specifications. A Bayesian approach to design space for assay robustness is described in Peterson and Yahyah (2009). Subsequent articles have appeared in the literature: Mbinze et al. (2012), LeBrun et al. (2013), Hubert et al. (2014), Dispas et al. (2014a, b), and Rozet et al. (2012, 2015).

References

Altan S, Shoung J (2008) Block designs in method transfer experiments. J Biopharm Stat 18: 996–1004

Altan S, Manola A, Davidian M, Raghavarao D (2002) The constrained four parameter logistic model. Dev Biol Standard 107:71–76

ASTM E2655-14 (2014) Standard guide for reporting uncertainty of test results and use of the term measurement uncertainty in ASTM test methods. ASTM International, West Conshohocken. www.astm.org

Barlow RE, Proschan F (1975) Statistical theory of reliability and life testing – probability models. Holt, Rinehart, and Winston, Inc., New York

Bernardo JM, Irony TZ (1996) A general multivariate Bayesian process capability index. Statistician 45:487–502

Box GEP, Behnken DW (1960) Some new three-level designs for the study of quantitative variables. Technometrics 2:455–475

Breyfogle FW (2003) Implementing six sigma, 2nd edn. Wiley, Hoboken

Castagnoli C, Yahyah M, Cimarosti Z, Peterson JJ (2010) Application of quality by design prinicpals for the definition of a robust crystallization process for casopitant mesylate. Org Process Res Dev 14(6):1407–1419

Chatzizaharia KA, Hatziavramidis TD (2015) Dissolution efficiency and design space for an oral pharmaceutical product in tablet form. Ind Eng Chem Res. doi:10.1021/ie5050567

Chavez P-F, Lebrun P, Sacré P-Y, De Bleye C, Netchacovitch L, Cuypers S, Mantanus J, Motte H, Schubert M, Evrard B, Hubert P, Ziemons E (2015) Optimization of a pharmaceutical tablet formulation based on a design space approach and using vibrational spectroscopy as PAT tool. Int J Pharm 486:13–20

Chiao C, Hamada M (2001) Analyzing experiments with correlated multiple responses. J Qual Technol 33:451–465

Cho BR, Shin S (2012) Quality improvement and robust design methods to a pharmaceutical research and development. Math Prob Eng. http://dx.doi.org/10.1155/2012/193246

Cornell JA (2002) Experiments with mixtures – designs, models, and the analysis of mixture data, 3rd edn. Wiley, New York

Crump BR, Goss C, Lovelace T, Lewis R, Peterson J (2013) Influence of reaction parameters on the first principles reaction rate modeling of a nitro reduction. Org Process Res Dev 17(10): 1277–1286

Davidian M, Giltinan D (1993) Some simple methods for estimating intra-individual variability in nonlinear mixed effects models. Biometrics 49:59–73

Del Castillo E (2007) Process optimization—a statistical approach. Springer, New York

Derringer G, Suich R (1980) Simultaneous optimization of several response variables. J Qual Technol 12:214–219

Dispas A, Lebrun P, Andri B, Rozet E, Hubert P (2014a) Robust method optimization strategy—a useful tool for method transfer: the case of SFC. J Pharm Biomed Anal 88:519–524

Dispas A, Lebrun P, Ziemons E, Marini R, Rozet E, Hubert PH (2014) Evaluation of the quantitative performances of supercritical fluidchromatography: from method development to validation. J Chromatogr A 1353:78–88

Draper NR (1963) Ridge analysis of response surfaces. Technometrics 5:469–479

Food and Drug Administration (2011) Center for drugs evaluation research. Guidance for industry process validation: general principles and practices. http://www.fda.gov/downloads/Drugs/Guidances/UCM070336.pdf

FDA (2015a) Pharmaceutical quality/manufacturing standards (CGMP) http://www.fda.gov/Drugs/GuidanceComplianceRegulatoryInformation/Guidances/ucm064971.htm

FDA (2015b) Pharmaceutical cGMPS for the 21st century — a risk-based approach: second progress report and implementation plan. http://www.fda.gov/Drugs/DevelopmentApprovalProcess/Manufacturing/QuestionsandAnswersonCurrentGoodManufacturingPracticescGMPforDrugs/UCM071836

FDA (2015c) CFR – code of federal regulations title 21. http://www.accessdata.fda.gov/scripts/cdrh/cfdocs/cfcfr/cfrsearch.cfm

Finney DJ (1978) Statistical method in biological assay, 3rd edn. Charles Griffin, New York

Garcìa A, Gilabert E (2011) Mapping FMEA into Bayesian networks. Int J Performability Eng 7(6):525–537

Giltinan D (1998) Statistical issues in assay development and use. Adv Exp Med Biol 445:173–190

Giltinan DM, Ruppert D (1989) Fitting heteroscedastic regression models to individual pharmacokinetic data using standard statistical software. J Pharmacokinet Biopharm 17:601–614

Gong X, Chen H, Pan J, Qu H (2015) Optimization of *Panax notoginseng* extraction process using a design space approach. Sep Purif Technol 141:197–206

Hadfield JD (2010) MCMC methods for multi-response generalized linear mixed models: the MCMCglmm R package. J Stat Softw 33(2):1–22

Harrington EC (1965) The desirability function. Ind Qual Control 21:494–498

Hoerl A (1964) Ridge analysis. Chem Eng Symp Ser 60:67–77

Hubert C, LeBrun P, Houari S, Ziemons E, Rozet E, Hubert P (2014) Improvement of a stability-indicating method by quality-by-design versus quality-by-testing: a case of a learning process. J Pharm Biomed Anal 88:401–409

International Society of Professional Engineers (2015) Quality metrics summit: driving quality through data and knowledge. Baltimore. http://www.ispe.org/2015-quality-metrics-summit

Jones B, Nachtsheim CJ (2011) A class of three-level designs for definitive screening in the presence of second-order effects. J Qual Technol 43(1):1–15

Juran JM (1988) Juran on planning for quality. The Free Press, New York

Kenett RS, Zacks S (2014) Modern industrial statistics: with applications in R, MINITAB and JMP. Wiley, Chichester

Kotz S, Johnson NL (2002) Process capability indices – a review, 1992–2000. J Qual Technol 34(1):2–19

Lansky D (1999) Validation of bioassays for quality control. Brown F, Mire-Sluis AR (eds): biological characterization and assay of cytokines and growth factors. Dev Biol Stand Basel, Karger 97:157–168

Lansky D (2002) Strip plot designs, mixed models, and comparisons between linear and non-linear models in microtitre plate bioassays. Dev Biol Stand 107:11–23

Lansky D (2014) Near-universal similarity bounds for bioassays. In: Conference on statistical and data management approaches for biotechnology drug development, August 27–28, 2014, Rockville, MD

LeBlond D, Mockus L (2014) The posterior probability of passing a compendial standard, part 1: uniformity of dosage units. Stat Biopharm Res 6(3):270–286

LeBlond D, Altan S, Novick S, Peterson J, Shen Y, Yang H (2015) In-vitro dissolution curve comparisons: a critique of current practice. Dissolut Technol, To appear http://www.qualitydigest.com/inside/quality-insider-article/problems-risk-priority-numbers.html

Lebrun P, Krier F, Mantanus J, Grohganz H, Yang M, Rozet E, Boulanger B, Evrard B, Rantanen J, Hubert P (2012) Design space approach in the optimization of the spray-drying process. Eur J Pharm Biopharm 80:226–234

Lebrun P, Boulanger B, Debrus B, Lambert P, Hubert P (2013) A Bayesian design space for analytical methods based on multivariate models and predictions. J Biopharm Stat 23(6): 1330–1351

Lebrun P, Giacoletti K, Scherder T, Rozet E, Boulanger B (2015) A quality by design approach for longitudinal quality attributes. J Biopharm Stat 25:247–259

Lostritto R (2014) Clinically relevant specifications (CRS): a regulatory perspective. IFPAC Conference, Baltimore

Lunn DJ, Thomas A, Best N, Spiegelhalter D (2000) WinBUGS – a Bayesian modelling framework: concepts, structure, and extensibility. Stat Comput 10:325–337

Maeda J, Suzuki T, Takayama K (2012) Novel method for constructing a large-scale design space in lubrication process by using Bayesian estimation based on the reliability of a scale-up rule. Chem Pharm Bull 60(9):1155–1163

Marroum PJ (2012) Clinically relevant dissolution methods and specifications. http://www.americanpharmaceuticalreview.com/Featured-Articles/38389-Clinically-Relevant-Dissolution-Methods-and-Specifications/

Mbinze JK, Lebrun P, Debrus B, Dispas A, Kalenda N, Mavar Tayey Mbay J, Schofield T, Boulanger B, Rozet E, Huberta P, Marini RD (2012) Application of an innovative design space optimization strategy to the development of liquid chromatographic methods to combat potentially counterfeit nonsteroidal anti-inflammatory drugs. J Chromatogr A 1263:113–124

Miró-Quesada G, del Castillo E, Peterson JJ (2004) A Bayesian approach for multiple response surface optimization in the presence of noise variables. J Appl Stat 31:251–270

Mockus L, Peterson JJ, Lainez JM, Reklaitis GV (2014) Batch-to-batch variation: a key component for modeling chemical manufacturing processes. Org Process Res Dev. doi:10.1021/op500244f

Montgomery DC (2009) Introduction to statistical quality control, 6th edn. Wiley, Hoboken

Myers RH, Montgomery DC, Anderson-Cook CM (2009) Response surface methodology - product and process optimization using designed experiments, 3rd edn. Wiley, Hoboken

Nelson PR (1992) Editorial. J Qual Technol 24(4):175

Ng SH (2010) A Bayesian model averaging approach to multiple-response optimization. J Qual Technol 42(1):52–68

Novick SJ, Chiswell K, Peterson JJ (2011) A Bayesian approach to show assay equivalence with replicate responses over a specified concentration range. Stat Biopharm Res 4(2):102–117

Novick SJ, Yang H, Peterson JJ (2012) A Bayesian approach to parallelism in testing in bioassay. Stat Biopharm Res 4(4):357–374

Novick SJ, Shen Y, Yang H, Peterson JJ, LeBlond D, Altan S (2015) Dissolution curve comparisons through the F_2 parameter, a Bayesian extension of the f_2 statistic. J Biopharm Stat 25(2): 351–371

O'Connell MA, Belanger BA, Haaland PD (1993) Calibration and assay development using the four-parameter logistic model. Chemometr Intell Lab Syst 20:97–114

Percy DF (1992) Prediction for seemingly unrelated regressions. J R Stat Soc Ser B 54:243–252

Peterson JJ (1993) A general approach to ridge analysis with confidence intervals. Technometrics 35:204–214

Peterson JJ (2004) A posterior predictive approach to multiple response surface optimization. J Qual Technol 36:139–153

Peterson JJ (2008) A Bayesian approach to the ICH Q8 definition of design space. J Biopharm Stat 18:959–975

Peterson JJ (2009) What your ICH Q8 design space needs: a multivariate predictive distribution. Pharm Manuf 8(10):23–28

Peterson J, Kenett R (2011) Modeling opportunities for statisticians supporting quality by design efforts for pharmaceutical development & manufacturing. Biopharm Rep 8(2):6–16

Peterson JJ, Lief K (2010) The ICH Q8 definition of design space: a comparison of the overlapping means and the bayesian predictive approaches. Stat Biopharm Res 2:249–259

Peterson JJ, Yahyah M (2009) A Bayesian design space approach to robustness and system suitability for pharmaceutical assays and other processes. Stat Biopharm Res 1(4):441–449

Peterson JJ, Snee RD, McAllister PR, Schofield TL, Carella AJ (2009a) Statistics in the pharmaceutical development and manufacturing (with discussion). J Qual Technol 41:111–147

Peterson JJ, Miro-Quesada G, del Castillo E (2009b) A Bayesian reliability approach to multiple response optimization with seemingly unrelated regression models. J Qual Technol Quant Manag 6(4):353–369

Press SJ (2003) Subjective and objective Bayesian statistics, 2nd edn. Wiley, Hoboken

Q10 Pharmaceutical Quality System (2009) International conference on harmonizaiton, guidance for industry. http://www.fda.gov/downloads/Drugs/GuidanceComplianceRegulatoryInformation/Guidances/ucm073517.pdf

Q8(R2) Pharmaceutical Development (2009) International conference on harmonizaiton guidance for industry. http://www.ich.org/fileadmin/Public_Web_Site/ICH_Products/Guidelines/Quality/Q8_R1/Step4/Q8_R2_Guideline.pdf

Q9 Quality Risk Management (2006) International conference on harmonizaiton, guidance for industry. http://www.fda.gov/downloads/Drugs/.\kern\fontdimen3\font.\kern\fontdimen3\font.\kern\fontdimen3\font/Guidances/ucm073511.pdf

Rajagopal R, del Castillo E (2005) Model-robust process optimization using Bayesian model averaging. Technometrics 47(2):152–163

Rajagopal R, del Castillo E, Peterson JJ (2005) Model and distribution-robust process optimization with noise factors. J Qual Technol 37:210–222; Corrigendum 38, p 83

Rodbard D, Lenox RH, Wray HL, Ramaith D (1994) Statistical quality control and routine data processing for radioimmunoassay. Clin Chem 20/10:1255–1270

Rossi P (2012) Bayesm: Bayesian inference for marketing/micro-econometrics. R package version 2.2-5. http://CRAN.R-project.org/package=bayesm

Rozet E, Lebrun P, Debrus B, Philippe Hubert P (2012) New methodology for the development of chromatographic methods with bioanalytical application. Bioanalysis 4(7):755–758

Rozet E, Lebrun P, Michiels J-F, Sondag P, Scherder T, Boulanger B (2015) Analytical procedure validation and the quality by design paradigm. J Biopharm Stat 25:260–268

Scranton R, Runger GC, Keats JB, Montgomery DC (1996) Efficient shift detection using exponentially weighted moving average control charts and principal components. Qual Reliab Eng Int 12(3):165–172

Sharp SS (2012) Establishing clinically relevant drug product specifications: FDA perspective FDA/ONDQA/biopharmaceutics 2012 AAPS annual meeting and exposition, Chicago, IL. http://www.fda.gov/downloads/AboutFDA/CentersOffices/OfficeofMedicalProductsandTobacco/CDER/UCM341185.pdf

Shin S, Truong NKV, Goethals PL, Cho BR, Jeong SH (2014) Robust design modeling and optimization of a multi-response time series for a pharmaceutical process. Int J Adv Manuf Technol 74:1017–1031

Srivastava VK, Giles DEA (1987) Seemingly unrelated regression equations models. Marcel Dekker, New York

Stockdale G, Cheng A (2009) Finding design space and reliable operating region using a multivariate Bayesian approach with experimental design. Qual Technol Quant Manag 6(4):391–408

Thomas A, O'Hara B, Ligges U, Sturtz S (2006) Making BUGS open. R News 6(1):12–17

USP 38/NF 33 (2015) Design and development of biological assays <1032>, vol 1. United States Pharmacopeial Convention, Rockville, pp 769–796

USP 38/NF 33 <1033> (2015) Biological assay validation, vol 1. United States Pharmacopeial Convention, Rockville, pp 796–807

Vander Heyden, Y., Nijhuis, A., Smeyers-Verbeke, J., Vandeginste, B. G. M., and Massart, D. L. (2001). "Guidance for Robustness/Ruggedness Tests in Methods Validation," *Journal of Pharmaceutical and Biomedical Analysis*, 24, 723–753. 441, 444, 445, 446

Wheeler DJ (2011) Problems with risk priority numbers. Qual Dig. http://www.qualitydigest.com/inside/quality-insider-article/problems-risk-priority-numbers.html

Zellner A (1962) An efficient method of estimating seemingly unrelated regression equations and tests for aggregation bias. J Am Stat Assoc 57:348–368

Chapter 16
Assay Validation

Perceval Sondag, Pierre Lebrun, Eric Rozet, and Bruno Boulanger

Abstract Recently, the lifecycle management concept for analytical procedures was introduced. It is strongly related to the Quality by Design concept given in the ICH-Q8 guidance. This contrasts with ICH-Q2 recommendations that only focus on the validation step to evaluate the performance of an analytical procedure. ICH-Q2's well-known check-list approach fails to provide assurance of the quality of future results with respect to the intended use of the procedure.

In this chapter, we propose and evaluate several decision rules to align quality of results to the objective of an analytical procedure using a risk-based approach. The β-expectation tolerance interval on the reportable result is shown to be the best way to assess whether a procedure will deliver quality results while maintaining a reasonable compromise between producer and patient risks. The β-expectation tolerance interval had previously been recognized in several papers as an excellent expression of the uncertainty of measurements. In addition, the accuracy profile is shown to be a simple way to apply a decision rule over the entire dosing range envisaged for the assay or for each potency or concentration level, based on the β-expectation tolerance interval.

Keywords Assay validation • Quality by Design for analytical method development • Analytical target profile • Beta expectation tolerance interval

16.1 Introduction

Since the publication of the Quality by Design (QbD) concepts in ICH-Q8 in 2009 (ICH 2009) for the development and validation of processes, there is an increasingly convergent agreement that the same paradigm also applies to bioassays and analytical methods. Recently, several publications introduced the lifecycle management concept of analytical methods, an approach closely related to QbD (Borman et al. 2007; Schweitzer et al. 2010; Nethercote et al. 2012; Rozet et al. 2015).

P. Sondag (✉) • P. Lebrun • E. Rozet • B. Boulanger
Arlenda S.A., 93 Chaussèe Verte, 4470, Saint-Georges, Belgium
e-mail: perceval.sondag@arlenda.com

L. Zhang (ed.), *Nonclinical Statistics for Pharmaceutical and Biotechnology Industries*, Statistics for Biology and Health, DOI 10.1007/978-3-319-23558-5_16

Indeed, both QbD and lifecycle management processes start with the identification of the objectives and requirements, then knowledge building during method development, validation/qualification, and finally, development of a control strategy to permit a continued improvement. This contrasts with ICH-Q2 recommendations that only focus on the validation step. Proposing a lifecycle management paradigm is relevant but may pose some statistical challenges if one wants to properly integrate the objectives, development, validation, and the control strategy. This chapter will focus on the application of decision rules to realize this integration and ensure risk control throughout the life-cycle of an assay.

16.2 Define the Objective of an Assay

Defining the objective and expectations or requirements should be the starting point for all assays and analytical methods. As stated in ICH-Q2, "*the objective of validation of an analytical procedure is to demonstrate that it is suitable for its intended purpose*". Of course the purpose should be clearly identified and defined to ensure that fitness can be assessed, e.g. supporting the release of a product, a stability study, or the development phases. Practically speaking this is often "blue sky" thinking because the reality of the development of a new bio-pharmaceutical product is not as linear as one might hope. Indeed the product and its production process are developed in parallel with the assays supporting them and therefore the "purpose" to be served is de facto an evolving target. For instance, at the start of development of a new product, it is difficult to define the required precision to support optimizing the production process. This complex co-evolution of both processes and assays too often results in undefined assay objectives and requirements. This leaves it to the analyst alone to define quality criteria for the assays based on good analytical practices and standards of performance in his or her respective field of expertise. The consequence is that common quality criteria for assays are not linked to the purpose of the assay. Conversely, process developers (i.e. analytical laboratory customers) often have only vague and evolving notions about their needs and specifications and find it challenging to communicate upfront to the analytical laboratories the analytical requirements needed to serve their purpose. This lack of alignment and communication gap between the users of the results and the developers of the measurement systems often contributes to poor understanding and mismanagement of decision risk by both analytical and process development personnel.

16.2.1 Analytical Target Profile

The starting point of an analytical lifecycle compliant paradigm is the Analytical Target Profile (ATP), which defines the intended purpose of the assay. The ATP defines the analyte or analytes to be measured, in which matrix, over what concentration range(s) as well as the required method performance criteria and specifications. These criteria and specifications should be linked to the intended purpose of an assay. This means that the performance criteria should be primarily based on the properties of the reported analytical results and not on the technology used to obtain those results. As stated by the USP Validation Panel in their stimuli paper (USP 2013), "*Results generated using analytical procedures provide the basis for key decisions regarding compliance with regulatory, compendial, and manufacturing limits. The results are applied against Decision Rules that give a prescription for the acceptance or rejection of a product based on the measurement result, its uncertainty, and acceptance criteria, taking into account the acceptable level of the probability of making a wrong decision*". This makes clear that the central focus of the ATP is the reported result, its acceptance criteria, and its uncertainty or equivalently the associated probability a result may not meet the acceptance criteria.

The same stimuli proposes an ATP template such as the following: "*The procedure should be able to quantify* [*analyte*] *in presence of* [*X, Y, Z*] *over a range of A% to B% of the nominal concentration with an accuracy and uncertainty so that the reportable result falls within* $\pm$*C% of the true value with at least 90% probability determined with 95% confidence*". Beyond the classical criteria such as accuracy and precision of the reportable result, the concept of probability, is introduced to evaluate the trust in the result. Probability is also known as quality level in other regulations, such as International Organization for Standardization (ISO) standards (ISO 1999). These quantitative performance requirements included in the ATP then serve as validation acceptance limits that must be achieved by the analytical procedure during the validation phase. By analogy with process validation, the analytical validation activity may also be referred to as Stage 2 qualification (FDA 2011). As triggered by ICH Q9 (ICH 2005), several authors have gone further by including in the definition of the ATP, the maximum acceptable probability of making improper decisions using the results generated by analytical procedures (Borman et al. 2007; Schweitzer et al. 2010; Nethercote et al. 2012).

16.3 Critical Quality Attribute for Assays

The Critical Quality Attributes (CQAs) of an assay are the measurable quantities by which the quality of the developed assay can be judged. In other words, the CQAs provide quantitative metrics for a fit-for-purpose decision. Historically, most analysts have taken their assay CQAs from the list of validation criteria found in ICH-Q2. This list includes method selectivity, (e.g., chromatographic resolution or separation criteria), the run time of the analysis, the signal to noise ratio, the precision and the trueness of the assay, the linearity of the results, the lower limit of quantification and the analytical range of the analytical method. The challenge then is to define fit-for-purpose-relevant specifications or acceptance limits from such CQAs. In the absence of strong guidance from product and process developers, specifications or acceptance limits for each of the validation criteria were arbitrarily chosen. An example of such an arbitrary criterion, which is largely irrelevant to the quality of the reported result, is the minimum R^2 value, e.g. 0.98, required to accept a calibration curve. This example illustrates the difficulty of establishing meaningful fit-for-purpose CQAs and acceptance limits that provide for proper management of analytical decision risks. Such an arbitrary, check-box approach to CQA and acceptance limit identification fails to close the gap between the analytical laboratories and the users or customers of the results.

Here, a procedure is proposed, that identifies the CQAs that most directly govern analytical decision making over the life-cycle of an assay. We distinguish these CQAs from other performance criteria that need to be reported to ensure that common scientific and industry standards have been satisfied.

16.3.1 Objective of an Assay

The objective of an assay is to produce reportable results that accurately estimate unknown quantities μ that the laboratory has to determine. Note that accuracy of a single measurement should not be confused with the bias—or trueness—of an assay. As indicated in ICH-Q2 and FDA Guidance for Industry on Bioanalytical Method Validation (2001), "*The accuracy expresses the closeness of agreement between the value which is accepted either as a conventional true value or an accepted reference value and the value found*", i.e. $x_i - \mu, (x_i - \mu)/\mu,\ x_i$, x_i being *one* reportable result as an estimate of μ. Bias—or trueness—of an assay on the other hand is the difference between the average of many reportable results (i.e. an infinite number) and μ.

Therefore one expects an assay to give results x_i's for which the relative difference (φ) from the true value (μ) of the sample is likely to be sufficiently small, for example within predefined acceptance limits λ. Stated more formally, the true probability π that a reportable result's relative error lies within acceptance limits [-λ, +λ] should be greater than a specified quality (probability) level π_{min} (Hubert et al. 2004; Boulanger et al. 2007, 2010).

$$\pi = P(-\lambda < \varphi < \lambda) > \pi_{min}. \tag{16.1}$$

In the metrology literature, λ is often referred to as the target measurement uncertainty. When the true bias δ and the true precision σ of the assay are known, then the quality level of the method is straightforward to compute under the normal error assumption, i.e.

$$\pi = P(|\varphi| < \lambda) = P\left(\frac{-\lambda - \delta}{\sigma} < Z < \frac{\lambda - \delta}{\sigma}\right), \tag{16.2}$$

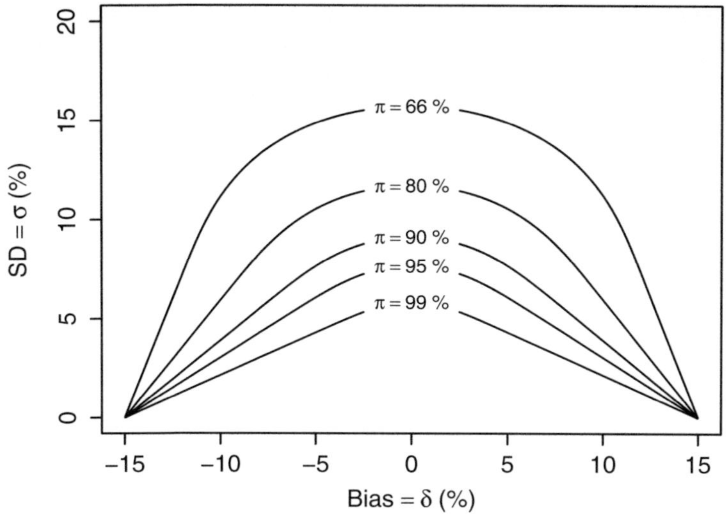

Fig. 16.1 Acceptance region of analytical methods as a function of the method bias and precision when $\lambda = 15$ %

where Z is a standard normal random variable. Note how Eq. (16.2) relates the primary CQA π with the traditional metrics δ and σ that are discussed in ICH Q2.

This leads to a definition of the "acceptance region", i.e. the set of performance parameters $\{\delta, \sigma\}$ such that the quality level π is greater than π_{min}. Figure 16.1 shows, below the curves, the acceptance region for various values of π_{min} (99, 95, 90, 80 and 66.7 %) when the λ acceptance limits are fixed at [−15 %, +15 %] as recommended for example with bioanalytical methods (FDA 2001). Logically, as can be seen from Fig. 16.1, the greater the true standard deviation σ of the assay or the greater the bias δ, the less likely a result's relative error will fall within the acceptance limits.

Therefore, for a given quality level π_{min}, all methods whose true performance criteria $\{\delta, \sigma\}$ are inside the corresponding acceptance region should be likely accepted after validation experiments, while methods with performance outside the same regions should in most case be rejected. The higher is the required quality level π_{min}, say 99 %, the smaller is the region representing potential assays able to fulfil this requirement.

16.3.2 Objective of Validation of an Assay

The objective of a validation or qualification phase is to collect, using proper experimental designs, information about the performance of the assay. Performance measure estimates of bias and precision permit a calculation of the probability that the assay will provide, in the future, a given result's relative error within the acceptance limits [-λ, +λ]. The expected probability $\widehat{\pi}$ of a future result's relative error that will fall within the acceptance limits, given the estimated performance of the assay is:

$$\hat{\pi} = E_{\hat{\delta},\hat{\sigma}} \left\{ P\left(|\varphi| < \lambda\right) \middle| \hat{\delta}, \hat{\sigma} \right\}. \tag{16.3}$$

There are several potential statistical approaches available for calculating the probability of future results from unknown samples falling within the specifications. This is further discussed in Sect. 16.4.

16.4 Decision Rules About Primary CQA

16.4.1 Suitable Decision Rule

Good decision-making during validation implies rejecting analytical methods whose true performance (δ, σ) are not included in the acceptance region (i.e., of Fig. 16.1) while accepting analytical methods whose performance (δ, σ) are in the acceptance region, for a given quality level π_{min}. However, as the true bias and the precision are unknown and estimated from data, a decision rule has to be defined as a function of performance parameter estimates. Most common decision rules for analytical method validation are presented in Sects. 16.4.1 and 16.4.2.

Figure 16.2b represents a sketch of a suitable decision rule for an acceptance level $\pi_{min} = 80$ % and a true bias $\delta = 6$ % as a function of true σ. The vertical line is at $\sigma = 10$ %, i.e. the top of acceptance region in Fig. 16.2a for the corresponding $\pi_{min} = 80$ % and bias $\delta = 6$ %. That is, a perfect decision rule should accept all methods with $\sigma \leq 10\%$ and reject all methods with $\sigma > 10\%$. However, every decision rule involves risk. For this reason, the acceptance line in Fig. 16.2b is S-shaped rather than appearing as an ideal step function. The red shaded area represents the laboratory risk because these are methods that are truly acceptable but rejected by the decision rule, while the green shaded area represents the consumer risk because these are the methods which are truly not acceptable, but are accepted. The purpose is therefore to propose a unique decision rule that will minimize both risks, for a given, reasonable, and effective sample size.

Historically the decision about the validity of an assay was based on the estimated bias and precision. The decision was driven by frequentist hypothesis testing concepts on parameters. An assay was declared as acceptable if the estimated bias and the variance were smaller than some arbitrary limits using appropriate hypothesis testing methods and corresponding type-I error levels. Such a decision rule cannot be recommended since the primary CQA of an assay is the accuracy of its future results and not its current estimated bias and precision performance. In addition, unknown complexities are introduced by defining acceptance limits on

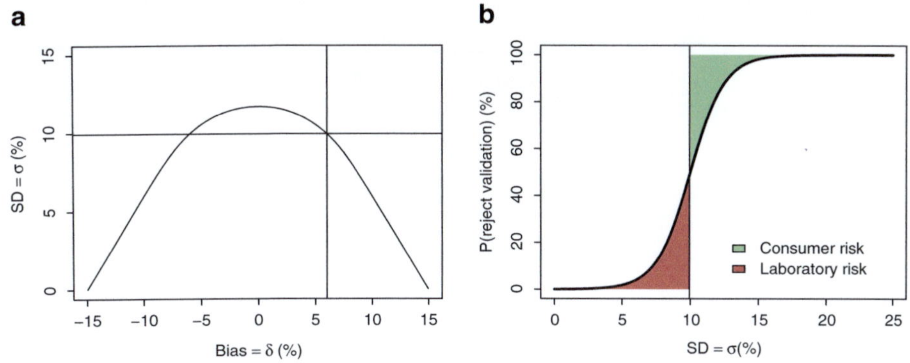

Fig. 16.2 (**a**, **b**): Example of suitable "decision rule" for an acceptance level $\pi_{min} = 80\ \%$ and a true bias $\delta = 6\ \%$ as a function of true σ

bias and precision, when the objective is all about the quality of the results. The user of the results will make a decision about a batch, such as whether to release a batch, or about the stability of a product, based on the individual reportable results. This decision will not be based on the estimated performance of the assay that generated the results. There are several established statistical solutions to relate the estimated performances to the quality of the results. They subtly differ depending on how the assay validation rule is formulated.

Note that following decision rules are based on the assumption that the true relative error of a measurement can be obtained. That is, that the true value is known. In practice, it might not always be the case and an additional uncertainty on the reference value must be taken into account. This is beyond the scope of this chapter.

16.4.2 Risk-Based Decision Rules

The first family of solutions utilizes a risk-based formulation: what is the risk of observing future reportable results that will be unacceptably far from the true value of the unknown sample? The direct and natural option is to use the predictive distribution of future results' relative error and to compute the predictive probability of falling within the acceptance limits. This can be obtained using Bayesian modelling and can be easily derived whatever the underlying distribution of the results and for a wide variety of experimental designs, including nested and unbalanced designs. Under the assumption of normally distributed results, non-informative priors and for a one-level nested design, the posterior predictive distribution is a *t*-distribution (Fig. 16.3). This distribution forms the basis of the β-expectation tolerance interval (Box and Tiao 1973; Hamada et al. 2004; Lebrun et al. 2013). Fortunately, *t*-distribution functions are incorporated into

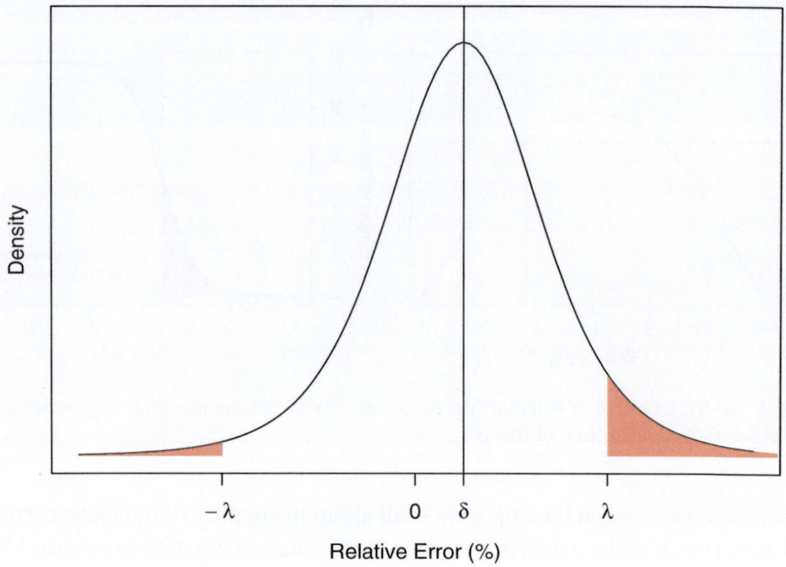

Fig. 16.3 Representation of a non-centered t-distribution. *Shaded red* area represents the probability that a result's relative error will fall outside the acceptance limits

widely available spreadsheet packages. This avoids the need for complex numerical methods or computer programing. When working with the predictive distribution of future results' relative error, the risk-based decision to declare an assay as valid is made when the predictive probability of falling inside the acceptance limits is greater or equal to the predefined quality level π_{min}.

16.4.3 Interval-Based Decision Rules

The other family of solutions is based on the tolerance intervals that should be totally included within the acceptance limits for declaring an assay as valid. Again two options are possible, the first one being based on the β-expectation tolerance interval (Guttmann 1970; Mee 1984; Hubert et al. 2007a, b), which is in fact a prediction interval. The second option is to use the β-content, γ-confidence tolerance interval (Mee 1984; Hoffman and Kringle 2005), often referred to as the beta-gamma tolerance interval. This interval is commonly used in practice but not recommended for decision making in validation of assays, as it will be shown.

16.5 Predictive Distribution and Tolerance Intervals

16.5.1 Bayesian Predictive Distribution

Let's first envisage the problem in a simple case, i.e. assuming all measurements are drawn from a unique normally distributed population having a single component of variance σ^2. In that case, assuming identical, independent, normally distributed errors:

$$\Phi_i \sim N\left(\mu + \delta,\ \sigma^2\right). \tag{16.4}$$

As already stated, the true performance metrics of the method (δ, σ) are unknown and must be estimated during the validation experiments.

The posterior distribution of the performance parameters of the method is, according to Bayes rule, proportional to the product of the likelihood [Eq. (16.1)] and the prior distribution on the parameters, i.e.:

$$P\left(\delta, \sigma^2 \middle| data\right) \propto P\left(data \middle| \delta, \sigma^2\right) * P\left(\delta, \sigma^2\right), \tag{16.5}$$

where data refers to a set of n independent errors, φ_i, obtained during validation.

To compare with the frequentist estimates, non-informative independent priors are assumed on the parameters. In that case, one can show that the joint posterior on the performance parameters is:

$$P\left(\delta, \sigma^2 \middle| data\right) \propto \sigma^{\frac{2}{n}-1} \exp\left\{-\frac{1}{2\sigma}\left[(n-1)\, s^2 + n\left(\overline{\phi} - \delta\right)^2\right]\right\} \tag{16.6}$$

where $\overline{\phi}$ is the average of all n data values, s is the usual estimate of σ, and $\sigma^{\frac{2}{n}-1} \exp\left\{-\frac{1}{2\sigma}\left[(n-1)\, s^2 + n\left(\overline{\phi} - \delta\right)^2\right]\right\}$ is proportional to the likelihood function of the parameters of a normal distribution.

Given the model and the posterior parameter distributions, the challenge in validation is to estimate the distribution of future measurement errors ϕ^*? This can be addressed by computing the posterior predictive distribution of the future values of ϕ^* conditionally on the available information, i.e. the data obtained during the validation experiments. Again, using Bayesian developments, one can show that the distribution of future measurement errors ϕ^* is:

$$P\left(\phi^* \middle| data\right) = \int\int P\left(\phi^* \middle| \delta, \sigma^2, data\right) * P\left(\delta, \sigma^2 \middle| data\right)\, d\delta\, d\sigma^2, \tag{16.7}$$

where $P\left(\phi^* \middle| \delta, \sigma^2, data\right)$ is the likelihood [Eq. (16.4)] for given values of the parameters and $P\left(\delta, \sigma^2 \middle| data\right)$ is the posterior distribution of the model parameters

[Eq. (16.6)]. In the specific case of identically distributed normal errors, the double integral can be calculated and the resulting density is:

$$P\left(\phi^{*}\middle|data\right) = t_{n-1}\left(\overline{\phi}, \left(1 + \frac{1}{n}\right)s^{2}\right). \tag{16.8}$$

Therefore, when a one-level nested design is used, then the predictive distribution of future measures ϕ^{*} is a t-distribution. In this case, the shortest two sided β% interval from those t-distributions, is equivalent to the β-expectation tolerance interval of Eq. (16.9) described in the next section. An important conclusion from the above derivation is that, in this case, an interval derived from the posterior predictive distribution [Eq. (16.8)] will be identical to the usual frequentist prediction interval for a future value. Therefore the usual frequentist interval also has a Bayesian interpretation—that the true value ϕ^{*} is contained within the interval with probability β, given estimates of performances of the assay and the uncertainty of those estimates. Then, if that interval is within our acceptance limits, then, each future error has *at least* a probability β to fall within the acceptance limits. Notice the emphasis on "at least". As will be shown below, the interval approach may be conservative relative to a risk based decision rule, such as described in Sect. 16.4.2 above, when sample sizes are large. More detail on Bayesian statistics and predictive distribution can be found in Gelman et al. (2014).

16.5.2 Tolerance Intervals

The β-expectation tolerance interval is defined as follows:

$$E_{\hat{\delta},\hat{\sigma}}\left\{P_{\varphi}\left(\hat{\delta} - k_{E}\hat{\sigma} < \varphi < \hat{\delta} + k_{E}\hat{\sigma}\middle|\hat{\delta},\hat{\sigma}\right)\right\} = \beta, \tag{16.9}$$

where $\boldsymbol{k_E} = \boldsymbol{t}_{f,\frac{1+\pi_{min}}{2}}, \sqrt{1+\frac{1}{n_E}}$, and $\boldsymbol{n_E} = \frac{\hat{\sigma}^2}{\hat{V}(\hat{\delta})}$ is the effective sample size. The $1/n_E$ correction takes into account the uncertainty of the estimate of the mean while the t-distribution takes into account the uncertainty on the estimate of variance σ^2. In the simple case, i.e. all measurements from the validation experiment are considered drawn from the same population, then $\boldsymbol{n_E} = \boldsymbol{n}$ and $\boldsymbol{f} = \boldsymbol{n} - 1$.

In a balanced nested design case, i.e., when measurements come from validation experiments with I runs and J replicates per run, then, an approximation for n_e and f is (Mee 1984):

$$n_E = \frac{I\,(MSA + (J-1)\,MSE)}{MSA}, \quad \boldsymbol{f} = \frac{(\boldsymbol{MSA} + (\boldsymbol{J}-1)\,\boldsymbol{MSE})^2}{\frac{1}{I-1}\boldsymbol{MSA}^2 + \frac{J-1}{I}\boldsymbol{MSE}^2} \tag{16.10}$$

and *MSA* and *MSE* are the mean squared errors corresponding to between and within runs, respectively.

The β-content, γ-confidence tolerance interval is defined as follows:

$$P_{\hat{\delta},\hat{\sigma}}\left\{P_{\varphi}\left(\hat{\delta}-k_c\hat{\sigma}<\varphi<\hat{\delta}+k_c\hat{\sigma}\middle|\hat{\delta},\hat{\sigma}\right)\geq\beta\right\}=\gamma, \tag{16.11}$$

where

$$k_c=z_{\frac{1+\pi_{min}}{2}}\sqrt{1+\frac{1}{\hat{\sigma}^2}\sqrt{\sum_{j=1}^{2}c_j^2H_j^2MS_j^2}}\sqrt{1+\frac{1}{n_e}}, \tag{16.12}$$

$c_1=\frac{1}{J}$ and $c_2=1-\frac{1}{J}$ and $\mathrm{H_j}=\frac{1}{F_{1-\gamma;c_j;\infty}}-1$, where $z_{\frac{1+\pi_{min}}{2}}$ is the $\left(\frac{1+\pi_{\min}}{2}\right)^{th}$ quantile of a standard normal distribution and $F_{1-\gamma;c_j,\,\infty}$ is the $(1-\gamma)^{th}$ quantile of a F distribution with degrees of freedom $c_j,\ \infty$.

As with the β-expectation tolerance interval, the $1/n_E$ correction takes into account the uncertainty of the estimate of the mean. The uncertainty on the estimate of the variance σ^2 is approximate depending on the specific design of the validation experiment (Hoffman and Kringle 2005).

For the simple case, i.e. all measurements from the validation experiment is considered drawn from the same population. Then the first square root in the k_c expression can be evaluated as follows:

$$\sqrt{1+\frac{1}{\hat{\sigma}^2}\sqrt{\sum_{j=1}^{2}c_j^2H_j^2MS_j^2}}=\sqrt{\frac{n-1}{\chi^2_{1-\gamma,n-1}}}. \tag{16.13}$$

For the balanced, one-way nested design case, i.e. when measurements come from validation experiments with I runs and J replicates within run, then the first root in the expression for k_c of Eq. (16.12) is approximated by:

$$\sqrt{1+\frac{1}{\hat{\sigma}^2}\sqrt{\sum_{j=1}^{2}c_j^2H_j^2MS_j^2}}=\sqrt{1+\frac{1}{\hat{\sigma}^2}\sqrt{H_1^2\left(\frac{1}{J}\right)^2MSA^2+H_2^2\left(1-\frac{1}{J}\right)^2MSE^2}}. \tag{16.14}$$

16.6 Performance of Decision Rules

To evaluate the relative performance of the proposed decision methods, let's assume that $\pi_{\min}$ is 80 % and acceptance limits are [−15 %, +15 %]. The region of true good methods where each measurement relative error has a probability of 80 % to fall within [−15 %, 15 %] is inside the black bell shaped curve of Fig. 16.2a.

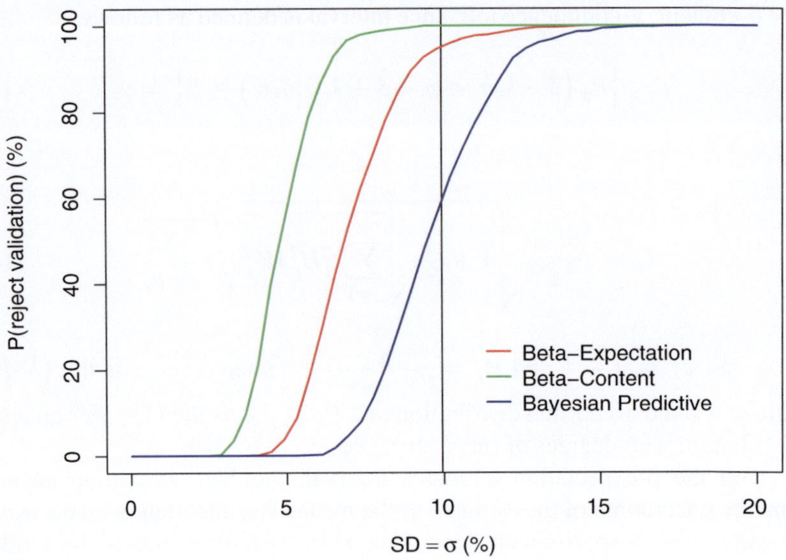

Fig. 16.4 Probability to reject after validation experiments with 5 runs and 5 replicates per run using the β-content, γ-confidence tolerance interval (*green line*), the β-expectation tolerance interval (*red line*) and the risk-based approach based on Bayesian predictive distribution (*blue line*) as a function of the %RSD assuming a true bias of 6 %. The *vertical line* at ∼10 % is the maximum true %RSD that will give 80 % of relative errors within [−15 %, 15 %]

Simulations are carried out based on a balanced one-way nested design with 5 runs ($I = 5$) and 5 replicates per run ($J = 5$). This is a design commonly used in the validation of assays. It is assumed that the true between-runs variance is equal to the true within-runs variances, i.e.:

$$R = \frac{\sigma_{Bewteen-Runs}}{\sigma_{Within-Runs}} = 1 \tag{16.15}$$

In addition, the simulations assume a true bias δ equal to 6 %. In later sections simulation results will be presented covering a range of bias and precision values.

Figure 16.4 shows the probability of rejecting the analytical method after validation of experiments using the three decision methods presented as a function of the true precision of the assay assuming a true bias 6 %. The black vertical line is the maximal true SD value to guarantee each error has a probability to fall within the acceptance limits greater or equal to $\pi_{\min}$. As can be seen, with the 80 %-content, 95 %-confidence tolerance interval, acceptable methods whose true precision are in the range 7–10 % have more than 95 % chance to be rejected while still acceptable. The 80 %-content, 95 %-confidence tolerance interval clearly eliminates the consumer risk at the full expense of the laboratory risk. The fact it minimizes the consumer risk has already been noted by Hoffman and Kringle (2005). However, with the 80 %-expectation tolerance interval, truly acceptable assays have more chance to be accepted and the risk of accepting inappropriate methods remain limited. Assays with a SD greater than 10 % have very limited chance to be accepted with practical $5 \times 5 = 25$ samples in validation experiments. The 80 %-expectation tolerance interval however reduces considerably the laboratory risk without major increase of the consumer risk. On the other hand, the use of the predictive distribution to compute the predictive probability of each future result's relative error falling within the acceptance limits indicates a consumer risk that is too large to be recommended.

This consumer risk using the β-expectation tolerance interval can be reduced by increasing the number of experiments performed in validation.

Another, but similar way to compare the various decision methods in the balanced one-way nested design case is by assessing the impact of sample size. For instance, the probability of accepting an assay can be computed and compared after either $5 \times 5 = 25$ or $100 \times 100 = 10{,}000$ measurements (at a single dose) as a function of the true proportion of relative errors within the acceptance limits, assuming again a true bias of 6 %. This simulation is shown for 5×5 format in Fig. 16.5.

It appears that with the 80 %-content 95 %-confidence tolerance interval there is no chance to accept a method that is expected to provide 80 % of relative errors within acceptance limits. Only methods with 90 % or more of relative errors within the limits may have a small chance to be accepted using the β-content, γ-confidence tolerance Interval. On the other hand, when using the 80 %-expectation tolerance interval as decision rule, good methods have a reasonable chance to be accepted and there are very limited chances to accept a method that has less than 70 % of future relative error within the acceptance limits. The consumer risk is therefore limited. In addition, consumer risk is further limited by the inclusion of QC test samples during routine analyses.

Another way to compare the three decision rules: (1) β-expectation tolerance interval; (2) β-Content, γ-confidence tolerance interval, and (3) predictive probability, is to represent the region of true analytical method performances that have at least 10 % chance to be accepted. These regions for each respective decision rule are shown in Fig. 16.6. As already noted, assays whose true performance is inside the black contour is acceptable and a decision rule whose performances are close to, but still inside that limit, is suitable. This is clearly the case when using the β-expectation tolerance interval (red line) that is the closest to the black contour line. Using the β-content, γ-confidence tolerance interval, however, results in a wide range of truly acceptable method performances that are not accepted. This conservative performance of the β-content, γ-confidence tolerance interval will certainly not be suitable for laboratories, given the high producer risk it induces. On the other hand, the rule based on predictive probability to be greater than the minimal quality level π_{min}, results in excessive risk of truly unacceptable assays being accepted.

These simulations results clearly favor the use of the β-expectation tolerance interval instead of the β-content, γ-confidence tolerance interval in the face of

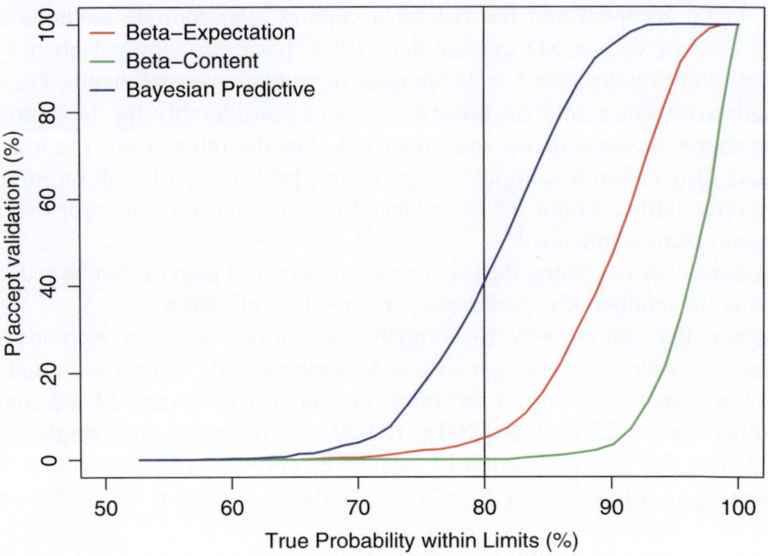

Fig. 16.5 Probability to accept validation after experiments with 5 runs and 5 replicates per run using the β-content, γ-confidence tolerance interval (*green line*), the β-expectation tolerance interval (*red line*) and the risk-based approach based on Bayesian predictive distribution (*blue line*) as a function of the true probability, assuming a true bias of 6 %. The *vertical line* at 80 % is the minimal quality level π_{min} that a measurement error falls within [15 %, +15 %]

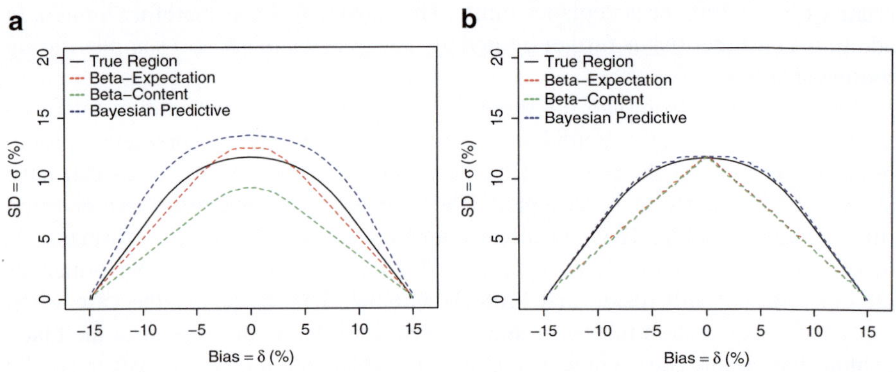

Fig. 16.6 Acceptance regions as in Fig. 16.2a. The *black line* delimitates the region of true good method performances giving 80 % of relative error within [−15 %, 15 %] acceptance limits. The *red dotted line* delimitates the region of method performances that will be rejected in 10 % of the cases using the β-expectation tolerance interval, the *green line* is the region for the β-content, γ-confidence tolerance interval and the *blue dotted line* is the region for the risk-based approach based on Bayesian predictive distribution. The *left panel* is for a 5 × 5 = 25 results and the *right panel* is for a 100 × 100 = 10,000 results

uncertainty when the number of experiments is limited as is commonly the case in analytical laboratories. Given the limited sample sizes that are practical in laboratories, the best pragmatic trade-off that allows controlling the laboratory cost without putting at risk the customer is to use the β-expectation tolerance interval.

When the sample size is large, as in the right panel of Fig. 16.6, then, as expected, the decision rule based on the predictive probability converges to the true region of acceptable methods. This decision rule has the best asymptotic properties. When sample size increases, both tolerance interval decision rules converge because the confidence becomes very high. But with the interval-based decision rules, both tolerance intervals do not converge to the true region but rather to a more conservative triangular region embedded within the true acceptance region. The triangular shape of the acceptance region is a consequence of the two one-sided interval testing approach. For finite sample size, the β-expectation tolerance interval appears to be the recommended and easiest solution to balance consumer and producer risks given that neither tolerance interval possesses superior asymptotic properties.

An additional reason for using the β-expectation tolerance interval is the fact the width of the interval is in fact equal to the Uncertainty of measurement as recommended by the ISO standard 21478 (ISO 2010; Feinberg et al. 2004).

16.7 Metric for Optimization of an Assay

The above rules are to be applied over the whole dosing range envisaged for the assay or for each potency or concentration level. This is known as the Accuracy Profile as published by Hubert et al. in 2004 for the first time. The Accuracy Profile is created by computing the β-expectation tolerance intervals by level and connecting the upper and lower limits. It provides an overall view of the future capability of an assay over the intended range and therefore becomes a key metric to optimize during assay development. As an example, Fig. 16.7 shows side by side two different calibration curves—four-parameter logistic model or quadratic polynomial model- for an ELISA assay with the corresponding Accuracy Profiles, respectively.

Instead of evaluating the quality of the fit using classical statistics such as maximum likelihood related criteria or the common root mean square error of prediction, a calibration model is selected if it improves the accuracy of the reportable results (Mantanus et al. 2010). Indeed the usual quality of fit statistics minimizes the error vertically (Y) while the objective of the assay is to minimize the error in the concentration or potency scale. It is also important to notice that the Accuracy Profile combines into one graphic and one metric the trueness (bias), the precision, the uncertainty about future results and the limit of quantification (LOQ). Indeed as it can be seen in Fig. 16.7b, the LOQ is reached when the Accuracy Profile crosses the Acceptance Limits.

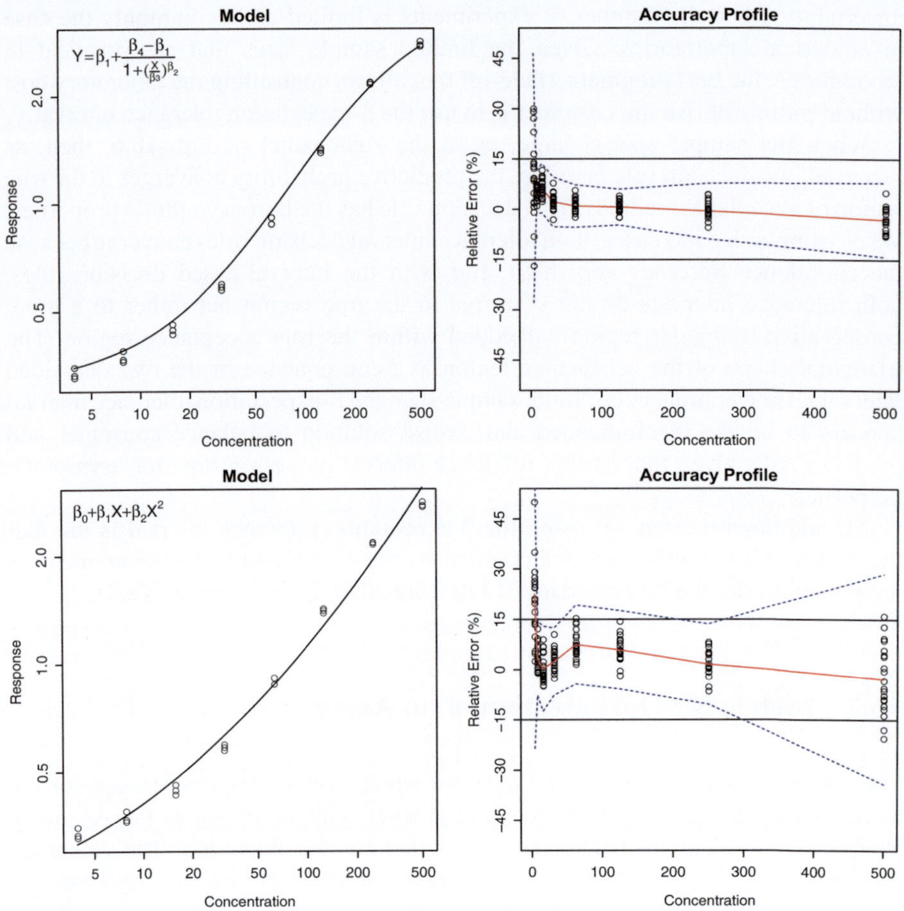

Fig. 16.7 (**a**) (*above*) and (**b**) (*bottom*): Calibration curves for an ELISA assay and the resulting accuracy profile. (**a**): the four-parameter logistic model. (**b**): quadratic polynomial. The *red line* connects the means or the trueness. The *blue dotted lines* connect the upper and lower β-expectation tolerance intervals and the *black dotted lines* are the acceptance limits

During the development of an assay, it is therefore recommended that operating conditions are identified, that will make the Accuracy Profile as narrow as possible over the largest range of values, concentrations or potencies, using an easy to read graphic, as illustrated in Fig. 16.7.

16.8 Conclusions

The key objective of an assay is to be capable, in the future, to report results with relative error falling within some predefined acceptance limits. The β-expectation tolerance interval has been shown to be the optimal decision rule under uncertainty when the sample size is limited as is the case in analytical work. This approach was introduced by Hubert et al. in 2004. Subsequently it has been widely cited and used in many publications with success. When applied to a range of concentration of potency levels, the resulting Accuracy Profile provides in a snapshot prediction of the quality of future results. This also becomes a reliable metric to optimize an assay and can be used when choosing a calibration model.

Acknowledgements The authors would like to thank David Leblond and Stan Altan for their useful comments that significantly increased the quality of this chapter.

References

Borman P, Nethercote P, Chatfield M, Thompson D, Truman K (2007) The application of quality by design to analytical methods. Pharm Technol. http://www.pharmtech.com/application-quality-design-analytical-methods

Boulanger B, Dewé W, Gilbert A, Govaerts B, Maumy M (2007) Risk management for analytical methods based on the total error concept: conciliating the objectives of the pre-study and in-study validation phases. Chemom Intell Lab Syst 86:198–207

Boulanger B, Devanaryan V, Dewe W (2010) Statistical considerations in the validation of ligand binding assays. In: Development and validation of ligand-binding assays. Wiley, New York

Box GE, Tiao GC (1973) Bayesian inference in statistical analysis. Wiley, Boston

Feinberg M, Boulanger B, Dewe W, Hubert P (2004) New advances in method validation and measurement uncertainty aimed at improving the quality of chemical data. Anal Bioanal Chem 380(3):502–514

Gelman A, Carlin J, Stern H, Dunson D, Venturi A, Rubin D (2014) Bayesian data analysis, section 3.2, 3rd edn. CRC, Boca Raton. ISBN 978-1-4389-4095-5

Guttmann I (1970) Statistical tolerance regions: classical and Bayesian. Griffin

Hamada M, Johnson V, Moore LM, Wendelberger J (2004) Bayesian prediction intervals and their relationship to tolerance intervals. Technometrics 46(4):452–459

Hoffman D, Kringle R (2005) Two-sided tolerance intervals for balanced and unbalanced random effects models. J Biopharm Stat 15(2):283–293

Hubert P, Nguyen-Huu J-J, Boulanger B, Chapuzet E, Chiap P, Cohen N, Compagnon P-A, Dewé W, Feinberg M, Lallier M et al (2004) Harmonization of strategies for the validation of quantitative analytical procedures: a SFSTP proposal – part I. J Pharm Biomed Anal 36(3):579–586

Hubert P, Nguyen-Huu J-J, Boulanger B, Chapuzet E, Chiap P, Cohen N, Compagnon P-A, Dewé W, Feinberg M, Lallier M et al (2007a) Harmonization of strategies for the validation of quantitative analytical procedures: a SFSTP proposal – part II. J Pharm Biomed Anal 45(1):70–81

Hubert P, Nguyen-Huu J-J, Boulanger B, Chapuzet E, Cohen N, Compagnon P-A, Dewé W, Feinberg M, Laurentie M, Mercier N et al (2007b) Harmonization of strategies for the validation of quantitative analytical procedures: a SFSTP proposal – part III. J Pharm Biomed Anal 45(1):82–96

International Conference on Harmonization (ICH) of Technical Requirements for Registration of Pharmaceuticals for Human Use: Topic Q8 (R2). Pharmaceutical Development (2009)

International Conference on Harmonization (ICH) of Technical Requirements for Registration of Pharmaceuticals for Human Use: Topic Q9. Quality Risk Management (2005)

International Organization for Standardization, ISO 2589-1: Sampling procedures for inspection by attributes – part 1: Sampling schemes indexed by acceptance quality limit (AQL) for lot-by-lot inspection (1999)

International Organization for Standardization, ISO 21748: Guidance for the use of repeatability, reproducibility and trueness estimates in measurement uncertainty estimation (2010)

Lebrun P, Boulanger B, Debrus B, Lambert P, Hubert P (2013) A Bayesian design space for analytical methods based on multivariate models and predictions. J Biopharm Stat 23(6):1330–1351

Mantanus J, Ziémons E, Lebrun P, Rozet E, Klinkenberg R, Streel B, Evrard B, Hubert P (2010) Active content determination of non-coated pharmaceutical pellets by near infrared spectroscopy: method development, validation and reliability evaluation. Talanta 80:1750–1757

Mee RW (1984) β-expectation and β-content tolerance limits for balanced one-way ANOVA random model. Technometrics 26(3):251–254

Nethercote P, Ermer J (2012) Quality by Design for analytical methods: implications for method validation and transfer. Pharm Technol 36(10):74–79

Rozet E, Lebrun P, Michiels J-F, Sondag P, Scherder T, Boulanger B (2015) Analytical procedure validation and the quality by design paradigm. J Biopharm Stat 25(2):260–268

Schweitzer M, Pohl M, Hanna-Brown M, Nethercote P, Borman P, Hansen G, Smith K, Larew J (2010) Implications and opportunities of applying QbD principles to analytical measurements. Pharm Technol 34(2):52–59

U.S. Food and Drug Administration. Department of Health and Human Services: Guidance for Industry, Bioanalytical Method Validation (2001)

U.S. Food and Drug Administration. Department of Health and Human Services: Process validation: General Principles and Practices (2011)

USP Validation and Verification Expert Panel (2013) Lifecycle management of analytical procedures: method development, procedure performance qualification, and procedure performance validation, IVT

Chapter 17
Lifecycle Approach to Bioassay

Timothy Schofield

Abstract Biological assay or bioassay is an analytical method used to measure the biological activity or potency of a biopharmaceutical or a vaccine. CMC statisticians contribute to the design, development, analysis and validation of bioassays. Skills in linear and nonlinear mixed effects modeling, design of experiments (DOE) and equivalence testing are essential to that support. This chapter describes statistical methods used to support bioassay and provides references for further explorations.

Keywords Bioassay • Potency • Similarity • Equivalence test • Bioassay validation • Bioassay characterization

17.1 Introduction

Biological assay or bioassay is an analytical method used to measure the biological activity or potency of a biopharmaceutical or a vaccine. Bioassays are distinguished from physical/chemical assays in using animals (in vivo bioassay) or cell culture (in vitro and ex vivo bioassay) to elicit a biological response. These methods are uniquely challenging due to the variability inherent in biological systems. While not classical bioassay, immunochemical methods such as enzyme linked immunoassay (ELISA) used to measure binding activity of a biotherapeutic or vaccine, product related impurities such as host cell proteins, and immunological response in clinical subjects are designed, analyzed and validated using similar statistical approaches. Statisticians contribute to the development and design of bioassays as well as the validation and maintenance of these methods. This chapter will describe the statistical designs and analyses used to process bioassay response data, to develop and validate bioassays, and to maintain a bioassay throughout its lifecycle.

T. Schofield (✉)
MedImmune, LLC, One MedImmune Way, Gaithersburg, MD 20878, USA
e-mail: schofieldt@medimmune.com

L. Zhang (ed.), *Nonclinical Statistics for Pharmaceutical and Biotechnology Industries*, Statistics for Biology and Health, DOI 10.1007/978-3-319-23558-5_17

17.2 Bioassay Designs and Models

17.2.1 Design Considerations

The underlying design of a bioassay is a concentration series of a test material. Replicates are typically measured at each concentration in the series. These include replicate animals in an in vivo bioassay or replicate wells of a micro-titer plate in an in vitro bioassay. A key statistical consideration in the allocation of experimental units to concentrations in a series is the independence of responses among the units. The correlation structure among the units lends itself to different bioassay variabilities. In general, independence among the experimental units results in more efficient management of bioassay variability. Thus a strategy of making independent concentration series is better than replicating concentrations from a single series. This along with randomization (see randomization below) helps reduce the variability of potency measurement.

Considerations in the construction of a series (number of concentrations and concentration interval) include the nature of the response (that is, the type of data), the intended measure of potency (for example, the half maximal effective concentration EC_{50}, or relative potency), the projected range of potencies and the mathematical model which will be utilized to translate the responses to a measure of potency. Bioassay response can be either continuous or quantal. Continuous response is associated with the readout from an instrument such as luminescence which is generated as a result of an intra-cellular reaction to a drug or vaccine. Quantal response is dichotomous (positive or negative) resulting in the number of animals or wells of a plate that are positive to a concentration of a drug or vaccine. Some laboratories convert continuous to quantal response by establishing a positive threshold, then re-expressing response as the number that exceed the threshold. This re-expression should be avoided due to the loss of information and resulting increase in variability from converting continuous to dichotomous data.

Many bioassays are performed in blocks. Blocking factors include plates in an in vitro bioassay or cages in an in vivo bioassay. Locations on a plate are also blocks. Responses may vary block to block or within a block. This variation can be ameliorated statistically through randomization. Randomization converts the block effect, or systematic variability, to random variability which can be managed through replication. Block effects are studied during bioassay development in uniformity experiments, where a fixed concentration of a material is administered to experimental units (wells of a plate or individual animals) and patterns in the responses are explored. A typical block factor is the edges of a plate where extremes of conditions such as heat and humidity can impact response. This is avoided by dispensing a placebo in the wells along the edges of each plate. Block effects may also be observed across the rows or columns of a plate when the reactions are time sensitive and operations are performed in sequence along one or the other of these dimensions (for example, serial pipetting across rows of a plate). It should be noted that the lack of pattern during uniformity studies does not preclude block effects in future performance of the bioassay. Randomization should be considered whenever possible. Other approaches for addressing block effects include strategic placement of a reference standard or assay control.

While complete randomization across experimental units is ideal, this is seldom implemented when manual operations such as serial pipetting is performed in the bioassay. Simple randomization schemes using split-plot or strip-plot designs help reduce the systematic variability (serial dilution error, see Lansky 2002) in bioassay measurement. Illustrations of randomized designs are given in the United States Pharmacopeia (USP) General Chapter <1032> *Design and Development of Biological Assays*.

Replication is performed to reduce the random variability in the reportable value of the bioassay. The reportable value is defined as the potency value of a test material which will be compared to the acceptance criterion (that is, specification if in place). Replication may be performed within an assay (intra-assay or intra-run) and/or between assays (inter-assay or inter-run). The definition of an assay or a run can be complex and relates to the blocks within the bioassay. As discussed above (see independence) replication between assays is usually more effective in reducing the variability of the reportable value.

A form of replication within an assay is multiple independent concentration series of a material. This is contrasted to the practice of performing a single concentration series, then aliquoting each concentration to multiple experimental units. This is commonly called pseudo-replication and is less effective at reducing variability than independent series. Factors which comprise an assay or a run might also be replicated. Those factors include technicians, instruments and plates. Further discussion of inter-assay replication is given in the Sect. 17.3.2.

17.2.2 Bioassay Models

Nonlinear and linear models are used to fit bioassay responses. The choice of type of model and model equation are guided by the theoretical kinetics of bioassay response as well as practical considerations.

17.2.2.1 Nonlinear Models

Mathematical models are used in bioassay to transform responses across a series of concentrations of a test material into a measure of biological activity or potency. Nonlinear or linear models are used to fit the responses of the test material alone or together with a reference material. The choice of bioassay model should be made in conjunction with the design considerations above.

Responses across a concentration series are governed by a kinetics principle which typically yields a sigmoid (S-shaped) curve (Fig. 17.1).

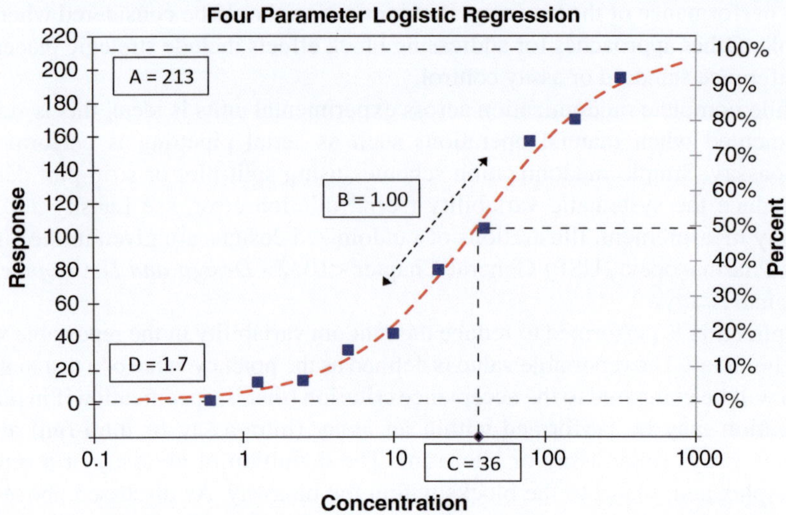

Fig. 17.1 Four parameter logistic regression with asymptotes A and D, 50th percentile C and slope B

A model derived from Michaelis-Menten kinetics is the four-parameter logistic regression equation:

$$Y = D + (A - D) \,/\, \left(1 + \left(\frac{X}{C}\right)^{B}\right). \tag{17.1}$$

In this version of the model the parameters A and D represent the upper and lower asymptotes, respectively, C is the concentration corresponding to half the range in the asymptotes, $(A - D)/2$, and B is a slope parameter. The parameter C is sometimes denoted EC_{50} for the half maximal effective concentration or IC_{50} for the half maximal inhibitory concentration, and is used as a measure of the absolute potency of a test material in some laboratories. A five parameter logistic model incorporates a parameter for asymmetry of the concentration response curve, and is discussed in Gottschalk (2005).

USP General Chapter <1032> on *Design and Development of Biological Assays* recommends the use of a reference standard to report the relative potency of a test material to the standard. In this design the reference standard is assayed alongside the test material to reduce the variability due to influences of the biological system on assay responses. The design of such a bioassay is called a parallel curve design (Fig. 17.2).

In the model associated with this design the test material is assumed to behave as a dilution (shifted right, requiring higher concentrations to achieve the same response profile as the standard) or as a concentration (shifted left, requiring lower concentrations) of the reference standard. It is further assumed that the shift is

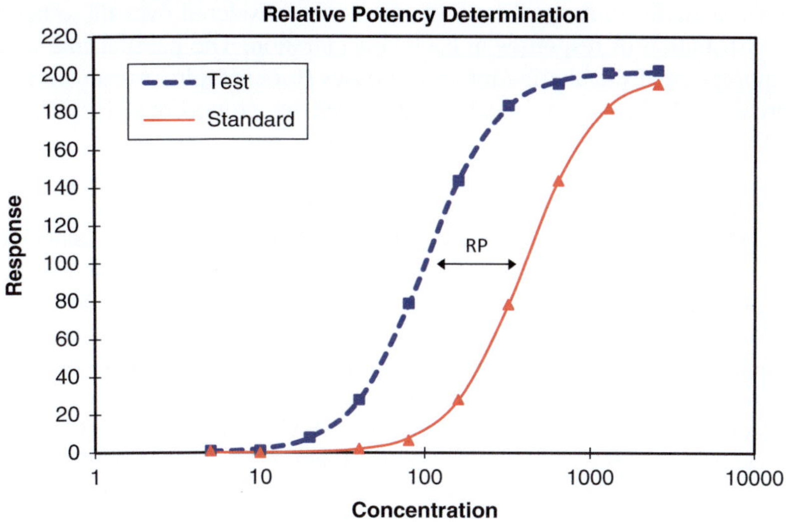

Fig. 17.2 Parallel curve design yielding a relative potency (RP) of test to standard

constant across the range of responses in the assay. This condition of a constant shift is called similarity or parallelism between the test and reference curves. A test of similarity is performed prior to the calculation of relative potency (see Sect. 17.2.4).

As discussed in Sect. 17.2.1, the number of concentrations and concentration increment should account for the projected range of potencies and the mathematical model which will be utilized to translate the responses to the measure of potency. When the four parameter logistic model is used, sufficient numbers of concentrations should be planned to assess similarity (two points across the range of potencies on the asymptotes) and to obtain a robust estimate of relative potency (four points across the range of potencies in the approximately linear region of the curve). The combined requirements usually result in selection of 10–12 concentrations to span the range to be tested. Some laboratories adjust the concentration series to satisfy the assumptions when the test material is expected to vary significantly from the standard (for example, forced degradation samples).

A special case of nonlinear modeling is applied to quantal responses (number of positives). Analyses can be performed after a linearizing transformation of the responses (probit or logit transformation). Similar considerations are used for the design of quantal response bioassay as parallel curve bioassay. The models used to assess similarity and to fit the data will be discussed in Sect. 17.2.3.

17.2.2.2 Linear Models

Among the linear models used to fit bioassay data are parallel line and slope ratio models. Both models use a reference standard to calculate the relative potency of the test material to the standard. One may be considered over the other based on the distribution of responses at each concentration. The parallel line model is more appropriate if the distribution of responses (for example, instrument readout) is approximately log-normal, and characterized by proportional increase in the variability (standard deviation) of the responses with increase in the level of response. The slope ratio model might be used if the distribution is approximately normal and the variability is constant across levels.

A design consideration associated with these two models is the concentration scaling. For reasons of the regression characteristics of either model (that is, symmetric weighting across concentrations), geometric scaling should be used to create the series for the parallel-line model while arithmetic scaling should be used in conjunction with the slope ratio model. The conditions required to support a slope ratio model are infrequently met in current bioassays. Further details on the analysis of slope ratio bioassays are given in USP General Chapter <1032> *Design and Development of Biological Assays* and <1034> *Analysis of Biological Assays.*

Linear models are derived from the nonlinear model in one of two ways. One is based on selection of concentrations in the "linear" region of the full concentration response curve (Fig. 17.3).

The linear region is typically selected to contain concentrations yielding 10–90 % or 20–80 % of maximal response, and should include at least three or four concentrations to satisfy the requirements for the assessment of curvature and parallelism as well as the determination of relative potency. This is managed through the selection of the concentration scaling in correspondence to the steepness of the linear portion of the concentration response curve, as well as consideration of the expected potencies to be tested. As with parallel curve design the laboratory may vary the concentration series of a test material which is expected to significantly vary from the reference standard (for example, forced degradation samples).

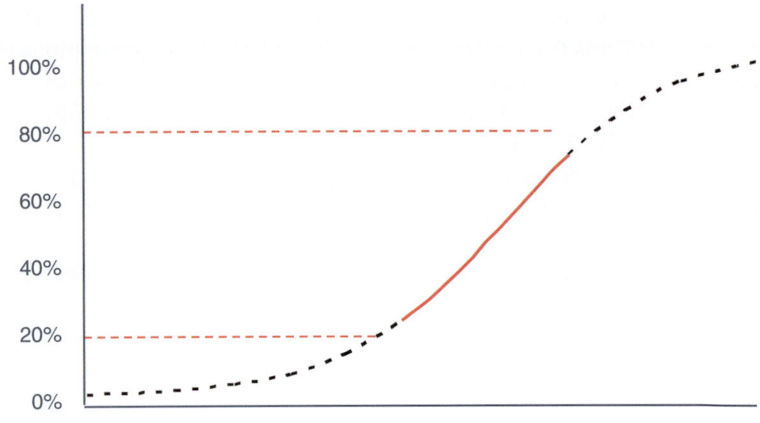

Fig. 17.3 Approximately linear region of concentration response curve (20–80 %)

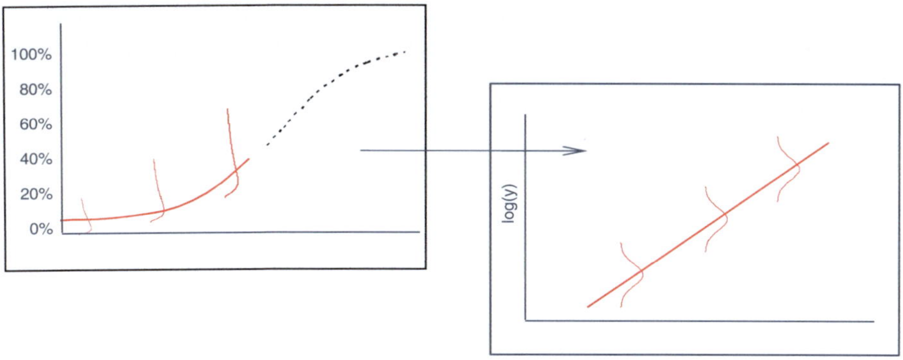

Fig. 17.4 Log transformation of the lower region of curve linearizes the curve and normalizes the residuals

A second linear model derives from the lower region of the full concentration response curve (Fig. 17.4).

This lower region can be linearized using a log-log fit to the data. This approach has optimal properties as will be discussed in Sect. 17.2.2.3.

17.2.2.3 Selection of a Model

The selection of a model depends upon a number of considerations including technical constraints, efficiency, variability of the biological system and regulatory requirements. Technical constraints include such things as the numbers of wells across a row or a column of a plate or availability of data processing software. Efficiency is a consideration for bioassays used for high throughput screening or to support large process development studies. The variability of the biological system should be considered in conjunction with the ability to effectively manage the concentration series across the range of samples which will be tested in the bioassay. Regulatory authorities may differ in their preference towards one model over another.

Technical, efficiency and regulatory considerations notwithstanding, nonlinear models may be preferred over their linear counterparts. This is because the nonlinear models relate better to the kinetics principles underlying bioassay concentration response than their linear approximations. Between the linear approximations selection of the lower portion of the concentration response profile has practical as well as statistical advantages. One of these advantages stems from the distributions of most bioassay responses. Responses such as luminescence are typically log-normally distributed. Thus a log-log fit to the curve both linearizes the curve and transforms response into an appropriate scale for analysis. Another advantage stems from the variability of the biological system and the range of potencies. When the biological system or potency range varies the linear region shifts towards one or another of the asymptotic regions of the curve. These shifts induce parallelism failures which can result in biased estimates of potency for a test material (see Sect. 17.2.5).

The absolute potency of the reference standard may be used as a system suitability test rather than (or in addition) to calculate relative potency. This may be appropriate when the biological system is tightly controlled, and determination of relative potency may increase variability due to compounding the variability of the standard together with the variability of the test material. Use of the reference standard for relative potency determination or as a system suitability test can be assessed utilizing correlation analysis on paired measurements [$C_i' = \log(C_i)$ where i indexes the test (T) and standard (S)] in the bioassay. The paired measurements might be made as duplicate series of the reference standard as part of the determination of an acceptance criterion for similarity (see Sect. 17.2.4). Using a formula for the log of the relative potency (see Sect. 17.2.6), $M = C_S' - C_T'$, the variance of M is given by:

$$Var(M) = Var\left(C_S'\right) + Var\left(C_T'\right) - 2 \cdot Cov\left(C_S', C_T'\right). \quad (17.2)$$

The variability of relative potency determination may exceed that of C_T when:

$$Var\left(C_T'\right) > 2 \cdot Cov\left(C_S', C_T'\right). \quad (17.3)$$

In this case the laboratory might prefer to report C_T rather than relative potency. Care should be taken, however, to consider the lifecycle of the bioassay. Absolute potency is susceptible to differences in conditions and reagents between laboratories. Relative potency should be used if the method will be transferred to other laboratories.

17.2.3 Model Fitting

Nonlinear or linear regression models are used to fit bioassay responses, assess similarity and determine potency. Standard statistical approaches to fitting and assessing these models will be discussed. These include consideration of the scale of responses, assessment of outliers and goodness-of-fit of the statistical model.

17.2.3.1 Scale of Responses

As in typical regression the assumptions of the method should be evaluated prior to fitting a model. Key assumptions are that the residuals from the regression are approximately normally distributed and the variability of responses is homogeneous across levels. Many bioassays yield non-normal responses owing to the bound of 0

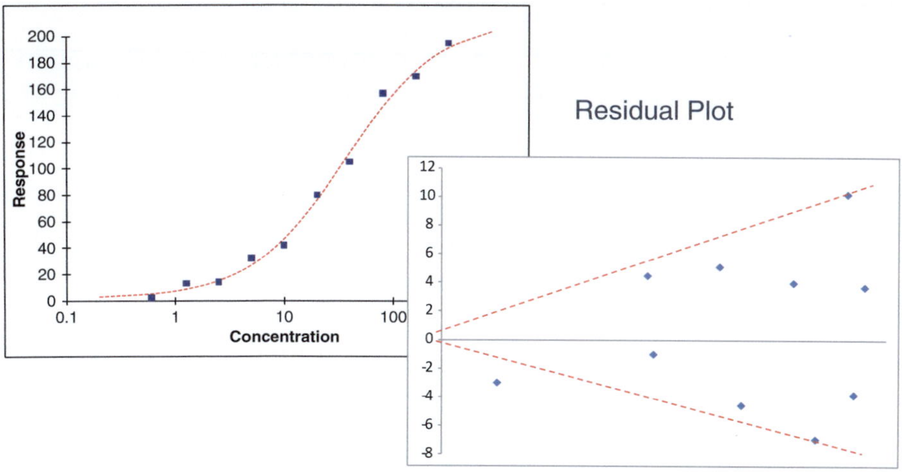

Fig. 17.5 Residual plot showing increasing variability with increasing concentration

on response measurement and the inherent nature of the biological response. In many bioassays the scale of the responses is approximately log-normal. Inattention to the scale of responses can lead to instability of procedures for assessing similarity and outliers as well as increases in variability of potency measurement.

As the conditions of non-normality of responses and heterogeneity of variability are related, these can be assessed through simple statistical tools such as residual plots or through replication studies. The residual plot (Fig. 17.5) from a curve or series of curves exhibiting log-normal (heterogeneous variability) behavior is characterized by increasing spread in residuals with increasing level of response (or concentration for increasing concentration response bioassays).

A study with independent replicates at each concentration may also reveal heterogeneous variability (Fig. 17.6).

Several approaches can be utilized to mitigate the impact of heterogeneity of variability on potency determination. Two of these are to perform weighted regression or to transform the responses. A weighted regression is performed by minimizing the weighted sum of squares for error.

$$Minimize\ SSE = \sum_i w_i \left(y_i - \hat{y}_i\right)^2 \qquad (17.4)$$

where

$$\widehat{Y}_i = \widehat{D} + \left(\widehat{A} - \widehat{D}\right) \Big/ \left(1 + \left(\frac{X_i}{\widehat{C}}\right)^{\widehat{B}}\right)$$

$$w_i = 1/\operatorname{var}\left(\widehat{y}_i\right).$$

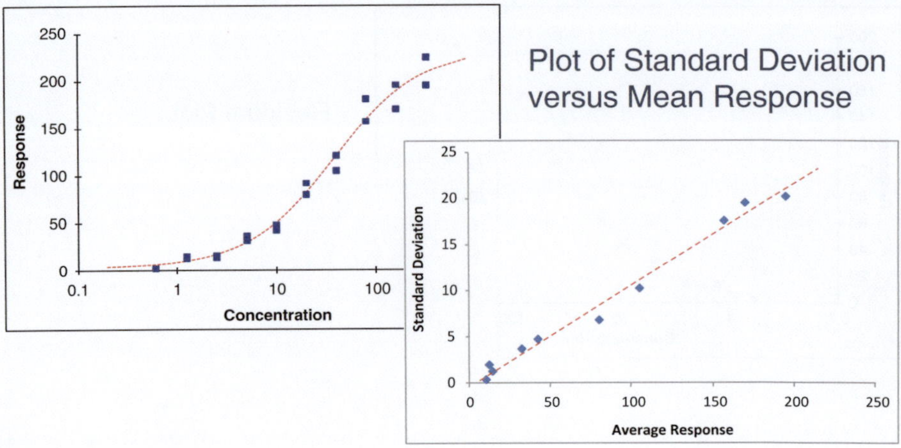

Fig. 17.6 Plot of standard deviation versus average response

Alternatively a transformation can be performed using a Box-Cox transformation of the responses (the residuals).

$$y_i^{(\lambda)} = \begin{cases} \left(y_i^{\lambda} - 1\right)/\lambda & \mathit{if}\ \lambda \neq 0 \\ \ln\left(y_i\right) & \mathit{if}\ \lambda = 0 \end{cases} \tag{17.5}$$

The log-likelihood of the power function can be used to read off the power λ and 95 % confidence interval. If the confidence interval includes 0 this is evidence towards log transformation.

Transformation may be preferred to weighted regression because transformation usually generates responses which meet conditions both of normality and homogeneity of variability. Transformation also reduces the apparent excess variability in responses in the region of the upper asymptote of the fitted model. That variability may cause excess false positive assessments of similarity and can mask important properties in the data such as a "hook" or prozone effect. In principle the transformed data are unlikely to follow Michaelis-Menten kinetics; however, this departure from the theoretical kinetics is typically outweighed by the improved statistical properties of the transformed data. Probit and logit analysis of quantal response bioassay data are performed using recursive weighted regression.

17.2.3.2 Outlier Analysis

Outlier analysis is sometimes performed on replicates to detect individual responses which may be due to mechanical error (missed well, pipetting error, etc.) or some other special cause variation. Approaches include replication based and model based outlier methods.

Replication based outlier methods look at groupings of replicates to determine one or more potential outliers. It is important to assess scale of response prior to outlier analysis. In particular log-normally distributed responses may appear to contain outliers due to the skewed nature in the distribution of responses. Most commonly employed methods such as Grubb's test or Dixon's test described in USP Chapter <111> Design and Analysis of Biological Assays, assume normality of the underlying distribution. Other methods such as range charting and residual analysis use an estimate of the response variability to assess when a range among replicates or a residual is beyond usual performance in the bioassay. These methods should be restricted to replicates which are independent and not linked to replicates in other groupings. Pseudo-replicates derived from aliquots of a common sample concentration may be assessed as a group. True replicates derived from independent dilutions series should not be assessed as a group. This is because the special cause variation may derive from the series rather than the individual replicate. Finally replication based outlier methods are typically insensitive due to the small number of replicates.

Model based outlier methods utilize the residuals from the fits of an appropriate bioassay model to the concentration response data (Fig. 17.7). As these depend upon an appropriate model, goodness-of-fit of the model should be performed prior to outlier analysis (see Sect. 17.2.3.3).

Simple methods such as dividing the residual by an estimate of the root mean square error after removal of the suspect outlier are available in some software packages, while graphical methods such as residual plots or probability plots may serve this purpose. While methods for calculating standardized residuals (that is, residuals divided by their standard error) are documented for linear models, these are more difficult to calculate for nonlinear models. Additional considerations for the assessment of outliers include:

- For parallel curve or parallel line bioassay the residuals should be obtained from the unconstrained (individual parameter) fits to the curves.

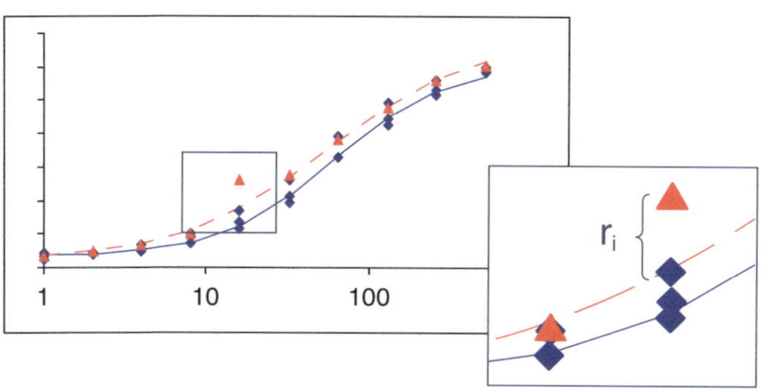

Fig. 17.7 Example of a model residual

- Care should be taken to assess an outlier against the appropriate estimate of variability (for example, pseudo replicates should be assessed using a different estimate of variability than replicates derived from different series).

The actions taken when one or more outliers are detected vary. Some laboratories will exclude the individual outliers while others may exclude a dilution or dilutions associated with the outliers. The lab should assess the impact of exclusion of the outlier on similarity testing and relative potency determination. If removal of the outlier shifts the decision regarding similarity or compliance with the potency specification, care should be taken to err on the side of the conservative decision. In spite of the availability of outlier tools a lab may choose to forego the practice of outlier analysis after performing an assessment of the impact of outliers on the determination of similarity and potency in a bioassay.

17.2.3.3 Goodness-of-Fit

Goodness-of-fit (GOF) of the bioassay model constitutes a test or set of tests to establish that the model used to fit the responses provides an acceptable description of the data. For linear models statistical measures such as R-squared are used to assess GOF. It can be shown, however, that curvilinear data can yield a satisfactory R-squared (see Anscombe 1973; Chatterjee and Firat 2007). The European Pharmacopeia (EP) Chapter 5.3 describes a test of curvature to establish that a linear model provides an acceptable fit to the data. That test suffers, however, from addressing the wrong hypothesis. The conclusion from the test is that there is insufficient evidence to conclude curvature, which is not the same as concluding that there is acceptable linearity. An amended version of the EP test might be to establish a metric of curvature (for example, the quadratic coefficient) and a margin of acceptability for the metric. This is analogous to the approach which is described for similarity or parallelism (see Sect. 17.2.4).

Additional heuristic approaches can be applied to both linear and nonlinear models. A plot of the residuals will reveal patterns which may be expected to occur when the model generates a poor fit to the data (for example, quadratic curvature for the linear model or runs of positive and negative residuals for a nonlinear model). Finally the lab may address lack of fit during development by assuring the fit with a higher order model such as a quadratic or nonlinear model.

17.2.4 Test of Similarity

A test of similarity (also called test of parallelism in the case of a linear model) is used in the processing of bioassay data to warrant that the kinetics of a test material is similar to that of a standard (except for a shift along the concentration scale). This is a requirement in parallel line and parallel curve bioassay (Fig. 17.8), but

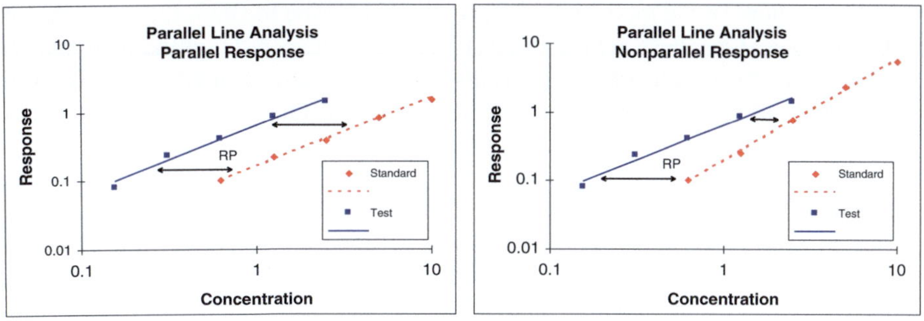

Fig. 17.8 Parallel and nonparallel concentration responses between test and standard materials

should also be assessed against the control when an absolute measure of potency (for example, EC_{50}) is reported. The test is used both biologically to establish that the kinetics of the test material has not changed due to an alteration in its primary or secondary structure, and/or operationally to warrant the measurement of relative potency.

A test of similarity was traditionally performed as part of the analysis of variance (see EP 5.3) using an F-test statistic that is constructed by comparison of the error sums of squares of the unconstrained model (individual fits to the test and standard data) and constrained model (parameters constrained to be equal for the test and the standard). The curves are determined to be equivalent when the F-test statistic is less than an appropriate percentile of the F-distribution. As previously, the conclusion when the test requirement is satisfied is that there is insufficient evidence to conclude that the curves are dissimilar. This is not the same as concluding that the curves are similar.

An alternative method described in USP <1032> is to test similarity using an equivalence approach. This begins with choosing a metric of similarity. In parallel line bioassay this might be the difference or the ratio of slopes between the linear fits to the test and standard data. The ratio may be preferred as it is unitless and can be formulated as a percent difference relative to the standard slope. A metric of similarity for a slope ratio bioassay is the difference in intercepts from the linear fits.

A metric (or metrics) of similarity for parallel curve bioassay is complicated by the fact that there are multiple parameters which describe the shape of the curve (for example, A, B and D in four parameter logistic regression). Methods for reducing these to one or two metrics follow:

- A chi-square test is described in Gottschalk and Dunn (2005). The chi-square test statistic is the numerator of the F-test statistic described in EP 5.3, and is proposed because of the reduced sensitivity to the precision of the unconstrained fit. As described in the article the chi-square test statistic is compared to the appropriate percentile of the chi-square distribution. An alternative is to use the chi-square statistic as a metric for equivalence testing.

- Two functions of the model parameters which describe meaningful properties of the biological response are the slope (B) and the range (A-D) of the fitted curve. The difference (or ratio) between the fits to the test and standard data may be used as metrics of similarity between the curves. Because these are likely to be highly correlated, one of these or a multivariate combination should be considered.
- In bioassays with increasing response with increasing concentration the lower asymptote (D) estimates the background response of the biological system. Since this is a property of the system and not a property of the test and standard materials the metrics may be reduced to the maximum (A) and the slope (B). Similar consideration should be given to the correlation between A and B as above.

Once a metric (or metrics) has been selected an equivalence margin and rule are established to test similarity. USP <1032> *Design and Development of Biological Assays* describes four approaches.

1. The distribution of the metric can be estimated from a sampling of bioassay runs containing a pair of samples. The pair may be duplicates of the reference standard (sometimes referred to as ref-ref pairs) or the reference standard and a control. Statistical process control (SPC) limits can be developed based on the data or on simulations made from properties which affect the similarity metric(s). Since the designation of test and reference is arbitrary, the maximum difference of the percentile from the expected value (zero if the difference, one if the ratio) should be used as the margin. The bounds for the ratio may be set geometrically to account for the behavior of a ratio statistic. A test sample will be judged similar if the calculated value for the metric falls within the equivalence margin.
2. A modification of the first approach is based on a margin calculated from a confidence interval on the similarity metric. This is appropriate when there is sufficient data to construct an interval. As before the margin is the bound on the interval which differs most from the expected value. A test sample will be judged similar if the confidence interval on the metric falls within the equivalence margin.
3. Where samples exist which represent "known" similarity failures, these form test-ref pairs which generate a distribution of the metric(s) that can be used together with the distribution formed from ref-ref pairs to discriminate the two. Methods such as receiver operating characteristic (ROC) curves can be used to determine a threshold. A test sample will be judged similar if the metric is less than or equal to the threshold.
4. A threshold can be determined if a correlation can be established between the similarity metric and an important change in the product being tested (for example, an important conformational change in the molecule or in vivo change in safety or efficacy). The estimated correlation relationship or simulations can be used to determine a threshold which forecasts a meaningful shift in the product characteristic. The uncertainty in the prediction relationship can be used to help ensure conformance to satisfactory product integrity. A test sample will be judged similar if the metric or the confidence bound on the metric is less than or equal to the threshold.

The first two approaches are limited in that there is no control of the false pass rate (that is, the rate of missing meaningful non-similarity) and the margins are based strictly on bioassay variability. The second approach helps to control the false failure rate if the rule is amended to conclude that the test material is non-similar if the interval falls completely outside the margin, and the decision is undetermined if the interval includes the margin. Further discussion of these approaches is contained in USP <1032>.

17.2.5 Caution on the Application of Suitability Testing

Suitability approaches such as similarity testing should be used cautiously when the action taken is to terminate data processing and/or to perform a retest. Some forms of the rules will erroneously conclude that the test sample is dissimilar from the standard when the data are insufficient to make the assessment (for example, the concentration response for the test does not reach one or the other of the asymptotes). This is not a similarity failure of the sample but rather a design failure or mis-specification of the bioassay range. Failures to generate data in one or the other of the asymptotic regions occur in hyper- and hypo-potent samples. Performing a retest will result in generating a potency (reportable value) which is biased towards the middle of the potency range. The range of the bioassay should extend to individual assays which are assured to provide data to assess similarity. Alternatively the series of assays which generate a reportable value can be re-performed using a concentration range of the test sample that has been adjusted to provide acceptable data to assess similarity and calculate potency.

17.2.6 Potency Determination

The assessment of similarity and calculation of potency should be undertaken in consideration of the design of the bioassay, including the format utilized to obtain the reportable value for the test sample. This is particularly important when assays are conducted in blocks or when there are blocks nested within an assay. For the most part different analyses will generate identical point estimates for the model parameters. However, the format structure will impact the confidence interval on the metric of similarity when this is used, and on the estimated variability associated with potency determination.

The potency or relative potency of a test material is calculated as a function of the model parameters after an appropriate assessment of goodness-of-fit and similarity has been concluded. The relative potency is calculated as given in Table 17.1.

Table 17.1 Determination of relative potency for various bioassay models

Model	Parameterization	Relative potency (RP)[a]
Parallel line	$y_S = a_S + b \cdot \ln(x)$ $y_T = a_T + b \cdot \ln(x)$	$RP = \exp\left(\frac{a_T - a_S}{b}\right)$
Parallel curve	$y_S = D + (A - D)/\left(1 + \left(\frac{x}{C_S}\right)\right)^B$ $y_T = D + (A - D)/\left(1 + \left(\frac{x}{C_T}\right)\right)^B$	$RP = \frac{C_S}{C_T}$
Slope ratio	$y_S = a + b_S \cdot \ln(x)$ $y_T = a + b_T \cdot \ln(x)$	$RP = \frac{b_T}{b_S}$
Parallel quantal response[b]	$p_S = \dfrac{1}{1 + (x/EC_{50}(S))^{-b}}$ $p_T = \dfrac{1}{1 + (x/EC_{50}(T))^{-b}}$	$RP = \frac{EC_{50}(S)}{EC_{50}(T)}$

[a]RP associated with increasing concentration response; parameters associated with T and S are transposed for decreasing response. Absolute potency of the test material is given by C_T and $EC_{50}(T)$ in the parallel curve and parallel quantal response models respectively
[b]Parameterization equations are represented using the logit transformation; a probit (normal equivalent deviate) transformation may also be used

Continuous responses are represented as y while concentration is symbolized x in the equations. Quantal response is represented as p, the proportion of experimental units which are positive at each concentration.

The distribution of potency (hereafter referring either to absolute potency or relative potency) measurements is impacted by the data analysis. For many products the manufacturing distribution is dominated by the variability of the bioassay. The data analysis for some of the models is performed in the log(x) scale, and/or the resulting potency is a ratio of the model parameters. In these cases a log transformation of potency may be required to achieve an approximately normal distribution. Some consequences of this are:

- The reportable value which may be calculated as a combination of several independent assays should be calculated using log transformation to obtain the geometric mean potency of the test material.
- The product specification should be geometrically scaled (for example, 80–125%) to account for the skewness induced by retransforming potency back from the log to the linear scale.
- Limits on the bioassay control should be likewise geometrically scaled.
- Bioassay validation calculations including capability analysis, sample size determination and data analysis should be performed using log transformation.
- Analyses performed on bioassay data should be performed using log transformation. It is noteworthy that for stability analysis the log transformation is useful both to achieve normality and to linearize the stability kinetics (that is, first-order kinetics).

Potency should be reported with an appropriate measure of uncertainty of the estimate. This is ideally a confidence interval or standard error on the potency. Model based and sample based methods for determining a confidence interval on relative potency estimates are described in USP General Chapter <1034> *Analysis of Biological Assays*. Model based confidence intervals require that the interval consider all factors that influence the model's estimate of precision. Sample based estimates of potency can be combined and also reported with a confidence interval. The sample based estimates should share as few bioassay factors as possible. Thus estimates made against the same concentration series of the reference standard or estimates derived off the same plate of the bioassay are not independent, leading to an inaccurate assessment of the variability of the combined potency. This may be difficult to accomplish when factors which impact potency determination change slowly (for example, a key reagent such as culture media may be used for several months before replacement). The effects of many of these factors are mitigated with the use of a reference standard to report relative potency. Robust design (for example, parallel curve rather than parallel line) also helps protect against factors which impact the biological system.

Methods for combining independent relative potency estimates are described in USP General Chapter <1034> *Analysis of Biological Assays*. These range from determining a standard confidence interval for the log relative potency then transforming back by the anti-log, to weighted averages of log relative potency. Similar methods can be applied to combining independent absolute potency estimates.

17.3 Bioassay Development and Validation

As a general principle a bioassay should be "fit for use" to make decisions throughout the lifecycle of the product. In the early stages of the product lifecycle the bioassay is used to screen molecules and stabilizer formulations, support process development and release and monitor stability of clinical materials. During commercial manufacture the bioassay continues to be used to release intermediates and final product, and to manage materials throughout their shelf-life. This lifecycle approach to bioassay development and validation is similar to the principles expressed in the quality by design (QbD) paradigm for product development and validation, and is clearly articulated in the USP Pharmacopeial Forum stimulus article titled *Lifecycle Management of Analytical Procedures: Method Development, Procedure Performance Qualification, and Procedure Performance Verification* (Martin et al. 2014).

The bioassay may be considered a manufacturing process which produces a product, measurements which are utilized by decision makers (the customer) during development and commercial control. Similar to the quality target product profile (qTPP defined in ICH Q8(R2) (2009)) which defines the quality and business requirements for the product, the analytical target profile (ATP) defines the requirements for bioassay measurements. Concepts such as quality risk management, knowledge management and analytical control strategy apply similarly.

Thus the strategies and the tools associated with a QbD approach to product development can be utilized throughout the bioassay lifecycle. Two key milestones in the bioassay lifecycle are robustness studies which are performed during development to define the bioassay "design space," and bioassay validation (procedure performance qualification) which is carried out prior to product licensure to demonstrate that the bioassay is fit for use to support commercial manufacture and control.

17.3.1 Bioassay Robustness

Robustness studies are performed during development and contribute to the overall quality and understanding of the bioassay. These studies are typically performed prior to validation as a basis for documenting the parameter (factor) settings which are specified in the bioassay standard operating procedure (SOP). Univariate or multivariate design of experiments (DOE) is used to explore relationships between the factor settings and the method performance attributes. Univariate experiments are usually reserved for factors which are ancillary to the method SOP such as sample age, culture media age, etc. Multivariate DOE is performed on the intra-assay factors which might impact performance of the bioassay.

Bioassay precision is the primary attribute addressed during development and validation. Performance should be assessed across the variety of samples which may be tested in the bioassay. This includes process intermediates which contain substances which may interfere with response in the bioassay (selectivity), and degraded samples which may perform dis-similarly to the standard due to changes in the higher order structure of the molecule. Sensitivity to interfering substances and changes in structure may manifest itself either as a change in potency or lack of similarity of the sample to the reference standard.

A multivariate DOE is undertaken in a series of steps: (1) map the bioassay process using tools such as an Ishikawa diagram then perform a risk assessment on the method factors; (2) screen (using Plackett-Burman or other highly fractionated designs) factors which have been identified in the risk assessment as potentially impacting bioassay performance; and (3) follow up with a response surface design to model a mathematical relationship between the "significant factors" from the screening study and the performance attributes of the method.

Standard approaches for analyzing the data from the screening study can be used, such as graphical analysis using Pareto or normal probability plots alone or together with calculation of a margin of error (see Vandeer Heyden 2001). These suffer, however, from ignoring the magnitude of the effect and its impact on method performance. The individual effects may instead be transformed into variance components using the relationship $s_i^2 = E_i^2/2$, where E_i is equal to the effect due to factor i. The cumulative impact of the factors with and without the larger components can be explored to determine whether these have an important impact on the overall variability. The analysis should be performed against the target variability defined in the ATP.

Those factors which have been identified in the screening study will be carried into a response surface design such as a central composite or Box-Bencken design.

The mathematical model can be used to establish the region in the study factors which yield acceptable precision. This can be described simply as the region (or inscribed ranges) which intersect the acceptable precision or through modeling and simulation the region with acceptable predictive probability of meeting the method requirements (Gelman et al. 2014).

The approaches discussed thus far utilize the robustness study to define the region in the assay parameters which yield acceptable precision. An alternative approach is to perform the robustness study to confirm acceptable performance within specified ranges of the parameters. In this regard the focus is not on which, if any, parameters require control and the control ranges, but instead the acceptance of specified ranges. In this approach the need to resolve main effects and interactions is relaxed and the design is instead used as a strategic basis for organizing runs across multiple factors. The design provides a strategic sampling of measurements which are analyzed to show that the bioassay variability (simple standard deviation or variance components) meets the method performance requirement.

17.3.2 Bioassay Validation

Bioassay validation should be carried out after method development and before testing either key late phase development materials or commercial product. The validation approach described in USP General Chapter <1033> *Biological Assay Validation* is a precision study performed using "ruggedness factors" which represent key qualitative variables (for example, analyst, reagent lot, instrument) which change over the lifecycle of the bioassay. The process map and risk assessment performed in support of the robustness study can be used to identify key factors for study. The validation may also include robustness factors (that is, factors which can be controlled in the bioassay such as incubation temperature) when there is a potential interaction between a robustness factor and a ruggedness factor such as a heat sensitive reagent. Careful selection of the factors which are included in the validation help ensure that the estimates of relative accuracy and intermediate precision are representative of long term performance of the bioassay. Co-validation between multiple testing laboratories may be performed when the bioassay is run at several sites or when the bioassay is likely to be transferred to a regulatory laboratory. This inter-laboratory component of a validation is called reproducibility.

While not explicitly discussed in the USP chapter any method which utilizes a dilution or concentration series of a test sample or reference standard may follow the approach discussed here. This includes calibration methods where a single dilution of a test sample is interpolated off of a reference standard curve as well as methods which use a single concentration of the test sample and reference to report the relative response of the test material. In these cases an additional goal of the validation is to establish the similarity of concentration response of test samples to the standard. A common case of these types of methods are clinical assays which measure the antibody titer in a blood sample. This is commonly addressed by using a dilution of the test sample where the matrix components no longer impacts the measurement in the bioassay.

17.3.2.1 Validation Acceptance Criteria

Validation acceptance criteria should be established to demonstrate that the bioassay is fit for use. This is usually addressed through the acceptance criterion on potency in the product specification. USP <1033> illustrates the use of a process capability index as a basis for criteria on relative bias (RB) and release assay variability (σ_{RA}).

$$Cpm = \frac{USL - LSL}{6\sqrt{\sigma^2_{\Pr oduct} + RB^2 + \sigma^2_{RA}}}, \tag{17.6}$$

where USL and LSL are the upper and lower release specification limits respectively and $\sigma^2_{\Pr oduct}$ is the target product variance. The relationship between release assay variability and the intermediate precision of the bioassay (σ_{IP}) is given by $\sigma^2_{RA} = \sigma^2_{IP}/k$ where k is the number of independent assays performed to obtain the reportable value. A single combination or a region of combinations (Hoffman and Kringle 2007) of relative bias and intermediate precision can be established which satisfy the condition Cpm > c according to the business practices of the company. Values such as c = 0.67 or c = 1.00 might be considered to restrict the false failure rate (probability of an out of specification result) to 5 % (2-sigma) and 0.3 % (3-sigma) respectively. An acceptance criterion for percent relative bias is specified as $\pm AC\,\%$ and for intermediate precision as $\leq \%GCV$ (see Sect. 17.3.2.3 for the definition of %GCV) which together meet the condition on Cpm. The rate of falsely passing an out-of-specification lot should be addressed in the product release limits. Approaches for mitigating the risk of a poor decision in other applications of the bioassay will be discussed in Sect. 17.3.3.

Alternative approaches for bioassay validation have been discussed in the literature (Hubert et al. 2007). These include the use of a β-content tolerance interval which combines relative bias and release assay variability to establish the likely range (with probability $1 - \alpha$) of a proportion (β) of future measurements. These approaches aim at future measurements rather than validation parameters as the basis for establishing fitness for use of the bioassay. The validation acceptance criterion will be that the β-content tolerance interval falls within LSL to USL, or that the predictive probability that future measurements fall within the limits is greater than or equal to $1 - \alpha$.

17.3.2.2 Validation Study Design

The design of the bioassay validation should include the number and types of samples which will be included in the study (for example, product and process intermediates, stability samples) as well as the number of independent assays and conditions which will be tested in each assay run. USP <1033> recommends using five-levels of a test sample to cover a range of potencies in the bioassay. That range should cover the acceptance criterion range for the product (drug substance and drug product) specification, but might extend beyond this to warrant testing of forced degradation samples and to cover the range of individual assays (see Sect. 17.2.5). Fewer than five-levels may be used when the laboratory is secure in its knowledge of the performance of the bioassay at the extremes of the range. Failure to meet the validation acceptance criteria at these extremes may limit the utility of the bioassay to test samples which vary from target.

The number of assays performed in the validation may be governed by the nature of the design as well as the experimental risks associated with the study. Factorial designs are used to incorporate multiple ruggedness factors into the study (for example, analyst, reagent lot), which may be combined with nested factors to replicate significant operational components (for example, plates). The two types of experimental risk which might be controlled through the study size are the risk that a parameter meets its acceptance criterion for an "invalid assay" (α), and the risk that the parameter will fail its criterion when the bioassay is "valid" (β). A sample size for relative bias at a single concentration is:

$$n \geq \frac{\left(t_{\alpha,df} + t_{\beta/2,df}\right)^2 \sigma_{IP}^2}{(\theta - \delta)^2}, \tag{17.7}$$

where $t_{\alpha,df}$ and $t_{\beta/2,df}$ are distributional points from a Student's t-distribution, σ_{IP} is the acceptance criterion on intermediate precision, θ is the acceptance criterion for relative accuracy, and δ is a correction on θ which accounts for a non-zero estimate of relative bias in the study. Note that this calculation requires a recursive solution because the degrees of freedom (*df*) are a function of *n*.

The sample size can be amended to account for pooling the data across levels to assess relative accuracy. Mathematically this is equivalent to averaging the relative responses (observed response divided by expected response) across the k-levels. When all levels have been tested together in each assay run, the variability of the average is given by:

$$\sigma_{Pooled}^2 = \sigma_{Between}^2 + \frac{\sigma_{Within}^2}{k}, \tag{17.8}$$

where $\sigma_{Between}$ and σ_{Within} represent the between-assay and within-assay components of variability respectively. Preliminary estimates of $\sigma_{Between}$ and σ_{Within}, or a mixture which sums (in the squares) to σ_{IP}^2 (square of the acceptance criterion on intermediate precision) can be used in the calculation of sample size.

The validation study design described thus far treats an individual assay (run) as the experimental unit (that is, individual assays will be subject to the combinations of factor conditions). Some laboratories use the format of the reportable value (that is, average across i-assay runs) as the experimental unit in the study. In this case, as with pooling across levels, the sample size should be amended to account for the predicted variability of the reportable value.

The added efficiency of treating the individual assay run as the validation experimental unit allows the laboratory to redistribute resources towards including additional ruggedness and robustness factors and/or levels in the design. This also makes available information which can be used towards developing bioassay formats for other uses of the bioassay such as process improvements and stability assessments. Care should be taken, however, when the series which is performed to obtain a reportable value has unique operational properties which affect variability. This may be the case when the series is performed in short sequence, leading to operational artifacts such as fatigue. Those operations which potentially impact method variability should be preserved in order to obtain representative estimates of bioassay variability.

17.3.2.3 Validation Data Analysis

The validation data analysis should be conducted according to the design of the study. Data (potency measurements) should be collected to more significant digits than is specified for an individual assay run or the reportable value in order to minimize the risk of rounding error on study conclusions (note that this should be coupled with other calculations such as that of the reportable value or stability evaluations). Data should be rounded in the calculation of intermediate precision when calculations such as reportable value are made from rounded values in the database. In this way the estimates of precision preserve the impact of rounding error on variability. As described previously, all analyses should be performed on the log of potency measurements. Calculations can be done on data at each level of the test sample or combined levels. Analysis using a mixed model with restricted maximum likelihood estimation (REML) is recommended.

Variance components can be estimated on individual study factors or in the composite as between-assay variability. These will give approximately the same solution when no single factor or factors contribute to the majority of the variability. A factor or factors which stand out should be earmarked for corrective action. That corrective action may be to manage the factor (for example, enhanced training if analyst is a significant factor, or qualification if a key reagent stands out in the analysis) or through strategic replication in the design of the reportable value (see Sect. 17.3.3).

Variance component estimates can be utilized to report the intermediate precision of the bioassay,

$$\widehat{\sigma}_{IP}^2 = \sum_i \widehat{\sigma}_i^2, \tag{17.9}$$

where $\widehat{\sigma}_i^2$ is the variance component estimate for the *ith* factor (or interaction).

While repeatability or within-assay variability is typically reported as the variance component associated with *error*, this has little if any use in practice. Nevertheless, some laboratories report this to meet expectations of worldwide regulatory authorities.

USP <1033> recommends reporting precision estimates as the percent geometric coefficient of variation (*%GCV*). The *%GCV* is analogous to %CV and was introduced by Kirkwood 1979, as a measure of variability for geometrically scaled measurements. If M represents the average of natural log transformed potency and s_{ln} the standard deviation of the transformed data, then Kirkwood defines the geometric standard deviation as $GSD = e^{s_{\ln}}$, and a single standard deviation interval on potency is:

$$e^{M \pm s_{\ln}} = (GM \div GSD, GM \cdot GSD)\,, \tag{17.10}$$

where GM is the geometric mean potency (the anti-log of M). Other intervals (for example, prediction intervals, confidence intervals) are defined in like manner, with division and multiplication by the appropriate function of s_{ln}. The *%GCV* is:

$$\%GCV = 100 \cdot (GSD - 1) = 100 \cdot (e^{s_{\ln}} - 1)\,\%. \tag{17.11}$$

This and other measures of variability associated with the log-normal distribution are discussed in Tan (2005).

In its composite formulation the intermediate precision at each level in the bioassay can be expressed as a function of estimated within-assay variation ($\widehat{\sigma}_W^2$) and between-assay variation ($\widehat{\sigma}_B^2$):

$$\widehat{\sigma}_{IP}^2 = \widehat{\sigma}_B^2 + \widehat{\sigma}_W^2. \tag{17.12}$$

This may be reported as $\%GCV = 100 \cdot \left(e^{\widehat{\sigma}_{IP}} - 1\right)\%$. The bioassay is considered to be valid at each level if the point estimate on intermediate precision meets the acceptance criterion $\leq IP\%$. Alternatively, the laboratory may require that the one-sided confidence bound on intermediate precision meets the acceptance criterion to provide statistical evidence that the requirement is met. Methods for estimating intermediate precision across levels will be discussed later in this section.

The equations for calculating $\widehat{\sigma}_{IP}^2$ permit development of a confidence bound on intermediate precision *%GCV*. In the simple case of *n-assays* with *r-replicates* per assay let:

$$\bar{y} = \sum_{i=1}^{n}\sum_{j=1}^{r} y_{ij}/nr, \ \ \bar{y}_i = \sum_{j=1}^{r} y_{ij}/r,$$

$$S_1^2 = r\sum_{i=1}^{n}(\bar{y}_i - \bar{y})^2/(n-1) \tag{17.13}$$

$$S_2^2 = \sum_{i=1}^{n}\sum_{j=1}^{r}(y_{ij} - \bar{y}_i)^2/n(r-1).$$

Then,

$$\widehat{\sigma}_{IP}^2 = \left(\frac{1}{r}\right)S_1^2 + \left(1 - \frac{1}{r}\right)S_2^2. \tag{17.14}$$

A modified large-sample upper confidence bound recommended in Nijhuis and Van den Heuvel (2007) for σ_{ip}^2 is:

$$UB = \widehat{\sigma}_{IP}^2 + \sqrt{\left[H_1 S_1^2\left(\tfrac{1}{r}\right)\right]^2 + \left[H_2 S_2^2\left(1 - \tfrac{1}{r}\right)\right]^2}$$

$$H_1 = \frac{n-1}{\chi_{\alpha,n-1}^2} - 1 \tag{17.15}$$

$$H_2 = \frac{n(r-1)}{\chi_{\alpha,n(r-1)}^2} - 1,$$

where $\chi_{\alpha,n-1}^2$ and $\chi_{\alpha,n(r-1)}^2$ are distributional points from the chi-square distribution. The upper bound on intermediate precision $\%GCV$ becomes $100\left(\sqrt{UB} - 1\right)\%$.

While the validation study wasn't powered to control the upper bound on intermediate precision, this provides information which may be used to develop a provisional format until more information becomes available (see Sect. 17.3.3).

Relative bias or percent relative bias ($\%RB$) is calculated at each level as follows:

$$\%RB_i = 100\left(\frac{\bar{y}_i - \mu_i}{\mu_i}\right) = 100\left(\frac{\bar{y}_i}{\mu_i} - 1\right)\%, \tag{17.16}$$

where μ_i is the expected value of the ith level in the validation. Using the construction equations above a $100(1 - 2\alpha)\%$ confidence interval (corresponding to a two one-sided test, or TOST, performed at a level of significance α) can be calculated on μ_i and substituted into the equation for $\%RB_i$:

$$\bar{y}_i \pm t_{1-\alpha,n-1}\sqrt{\frac{S_1^2}{nr}}. \tag{17.17}$$

The bioassay meets the validation acceptance criterion at the ith level when the *100(1 − 2α)%* confidence interval falls wholly within the acceptance criterion ±AC%.

In the validation study test samples at different levels are typically tested together in the same assay (i.e., potencies are reported relative to a common reference curve). In this case the analysis of relative accuracy at individual levels does not account for the correlation in results across levels. An overall analysis of the bioassay validation data can be performed when there is evidence that the relative bias is uniform across levels. While a statistical test of the interaction between assay and level can be performed to assess uniformity across levels, this is flawed in not being able to conclude uniformity (i.e., equivalent relative bias across levels). Similar approaches might be applied as in the assessment of curvature (see Sect. 17.2.3.3) wherein the data can be examined for expected departures from uniformity across levels. For parallel line analysis using the linear portion of the concentration response curve the failure mode for non-uniformity is drift in responses towards the asymptotes yielding underestimates of potency at the extreme levels. Examination of the data for this pattern and sensitivity analysis to address potential impacts to quality decisions might be used justify use of the data across the range of responses.

Alternatively confidence bounds on intermediate precision and relative accuracy determined across levels can be used as added protection against the assumption of uniformity. The degree of non-uniformity of relative bias across concentrations will be manifest in an increase in the apparent variability of the bioassay, and thus wider confidence bounds on the validation parameters. A mixed effects model associating level (L, a fixed effect) and assay (a, a random effect) is given by:

$$y_{ijk} = \mu + L_i + a_j + \varepsilon_{ijk}, \tag{17.18}$$

where $a_j \sim N\left(0, \sigma_B^2\right)$ and $\varepsilon_{ijk} \sim N\left(0, \sigma_{aL}^2 + \sigma_W^2\right)$. Thus the variability associated with the interaction of assay and level is pooled into the variability associated with within-assay replicates resulting in an increase in the size of the confidence interval on relative bias. Likewise the increased variability will be reflected in the estimation of intermediate precision:

$$\widehat{\sigma}_{IP}^2 = \widehat{\sigma}_B^2 + \widehat{\sigma}_{aL}^2 + \widehat{\sigma}_W^2. \tag{17.20}$$

17.3.3 Bioassay Characterization

When the bioassay validation has been conducted as a precision study the estimated variance components can be utilized to forecast the variability of different formats of the reportable value design. In the simplest case this is the number of independent assays (i) and number of replicates within each assay (j).

$$\widehat{\sigma}_{ij}^2 = \frac{\widehat{\sigma}_B^2}{i} + \frac{\widehat{\sigma}_W^2}{ij}, \tag{17.21}$$

where $\widehat{\sigma}_B^2$ and $\widehat{\sigma}_W^2$ are the between-assay and within-assay variance component estimates. The laboratory can use this formula to determine the most efficient format which meets the requirement for acceptable precision of the reportable value. Thus, for example, formats using a single replicate in 3 independent assays (3 × 1) and 2 replicates in 2 independent assay (2 × 2) may both meet the precision requirement, but the 2 × 2 format is more efficient in taking less time with less impact on reference standard supply.

An upper confidence bound on the format variability can be derived similarly as for intermediate precision above (see Burdick et al. 2005). Using the constructs from before:

$$\begin{gathered}\widehat{\sigma}_{ij}^2 = \tfrac{S_1^2 - S_2^2}{ir} + \tfrac{S_2^2}{ij} = \left(\tfrac{1}{ir}\right) S_1^2 + \left(\tfrac{1}{ij} - \tfrac{1}{ir}\right) S_2^2, \\ UB = \widehat{\sigma}_{IP}^2 + \sqrt{\left[H_1 S_1^2 \left(\tfrac{1}{ir}\right)\right]^2 + \left[H_2 S_2^2 \left(\tfrac{1}{ij} - \tfrac{1}{ir}\right)\right]^2},\end{gathered} \tag{17.22}$$

where H_1 and H_2 are calculated as above. This solution is appropriate if r (the number of replicates performed in the precision study) is greater than or equal to j (the number of replicates used to generate the reportable value). Solutions for $r < j$ are given in Burdick et al. (2005).

Other uses of the bioassay such as to support process development and stability assessment frequently involve comparing intra-assay results or inter-assay results. The critical difference (CD) associated with comparing intra-assay results, such as those obtained from a process experiment tested together in the same assays is that associated with relative potency determination:

$$CD_{intra-comparison} \cong 2\widehat{\sigma}_{ij}, \tag{17.23}$$

where the value 2 is an approximation to an appropriate distributional value associated with degrees of freedom related to the estimation of $\widehat{\sigma}_{ij}$ and adjustments appropriate to the nature of the inference (e.g., prediction range versus tolerance range, adjustment for multiplicity, etc.). The critical difference associated with comparing inter-assay results, such as those obtained for separate time points from a stability study is:

$$CD_{inter-comparison} \cong 2\sqrt{2}\widehat{\sigma}_{ij}. \tag{17.24}$$

The bioassay format can be modified to meet the goals (acceptance criteria) of these and other more complex statistical designs.

17.3.4 Continued Performance Verification

The bioassay validation provides an informed snapshot of the long term performance of the method. The estimate of intermediate precision can be verified through strategic assessment of appropriate controls in the procedure. A system of ongoing statistical process control provides a basis for revealing unexpected factors which impact bioassay performance as well as periodic reassessment of the intermediate precision of the bioassay.

A lifecycle approach to bioassay validation is not limited to basic SPC. Lifetime events such as method transfer, reagent and reference standard qualification, and method bridging should be carried out in a manner which ensures continued reliability of the method. An equivalence approach to study these events helps manage the risks to product quality as well as supply. Similar to validation acceptance criteria equivalence criteria can be determined using a process capability approach, while the design and analysis of study data may follow the approaches set forth in Sect. 17.3.2.

Acknowledgements I would like to thank Rick Burdick for his guidance in formulating confidence bounds on intermediate precision and format variability as well as his comments throughout the chapter. I would also like to thank Stan Altan for his careful review of the chapter and for his suggestions which added further clarity to some of the content.

References

Anscombe F (1973) Graphs in statistical analysis. Am Stat 27(1):17–21

Burdick RK, Borror CM, Montgomery DC (2005) Design and Analysis of Gauge R&R Studies. SIAM 143–148

Chatterjee S, Firat A (2007) Generating data with identical statistics but dissimilar graphics: a follow up to the Anscombe dataset. Am Stat 61(3):248–254

European Pharmacopeia 8th Edition, Chapter 5.3, Statistical analysis of results of biological assays and tests

Gelman A et al (2014) Bayesian data analysis. Chapman and Hall, Boca Raton

Gottschalk PG, Dunn JR (2005) The five-parameter logistic: a characterization and comparison with the four-parameter logistic. Anal Biochem 343(1):54–65

Gottschalk PG, Dunn JR (2005) Measuring parallelism, linearity, and relative potency in bioassay and immunoassay data. J Biopharm Stat 15:437–463

Hoffman D, Kringle R (2007) A total error approach for the validation of quantitative analytical methods. Pharm Res 24(6):1157–64

Hubert P et al (2007) Harmonization of strategies for the validation of quantitative analytical procedures A SFSTP proposal–part III. J Pharm Biomed Anal 45:82–96

ICH Q8(R2) Pharmaceutical development (2009)

Kirkwood T (1979) Geometric means and measures of dispersion, letter to the editor of biometrics 35: 908–909

Lansky D (2002) Strip-plot designs, mixed models, and comparisons between linear and non-linear models for microtitre plate bioassays. Dev Biol 107:11–23

Martin GP et al (2014) Stimuli to the revision process, lifecycle management of analytical procedures: method development, procedure performance qualification, and procedure performance verification Pharmacopeial Forum 39(5)

Nijius M, Van den Heuval ER (2007) Closed-form confidence intervals on measures of precision for an interlaboratory study. J Biopharm Stat 17(1):123–142

Tan C (2005) RSD and other variability measures of the lognormal distribution. Pharmaceopeial Forum 31(2):653–5

USP Chapter <111> (2015) Design and analysis of biological assays, USP 38 – NF30

USP General Chapter <1032> (2015) Design and development of biological assays, USP 38 – NF30

USP General Chapter <1033> (2015) Biological assay validation, USP 38 – NF30

USP General Chapter <1034> (2015) Analysis of biological assays, USP 38 – NF30

Chapter 18
Quality by Design: Building Quality into Products and Processes

Ronald D. Snee

Abstract The US Food and Drug Administration introduced the pharmaceutical and biotech industries to Quality by Design (QbD) in 2004. While new to these industries QbD had been used and found effective in many other industries for more than 40 years. This chapter discusses the "What", "Why" and "How" of QbD. The focus is on the building blocks of QbD and the statistical concepts, methods and tools that enable the effective implementation of the approach. Numerous case studies from pharma and biotech are used to illustrate the approaches.

Keywords Process design • Process development • Process robustness • Design space

18.1 It's About Quality Stupid

During the 1992 presidential election James Carville, campaign strategist for the Clinton election team recognized that "It's the Economy Stupid" suggesting that the key issue on the minds of Americans was the U.S. economy. Clinton revised the thrust of his campaign and, as they say the rest is history.

Similarly today in the twenty-first century, the key issue on the minds of US Food and Drug Administration (FDA) and Pharmaceutical and Biotech companies is quality. Driven by greater global competition and the growing impact of information technology, the pharmaceutical industry faces a need to improve its performance. Speed to market, product quality, regulatory compliance, cost reduction, waste, and cycle time are among the concerns that must be addressed in a systematic, focused, and sustainable manner. Quality by design (QbD), an approach introduced by the US Food and Drug Administration, provides an effective tool for addressing these concerns.

This chapter discusses the key steps for implementing QbD and the associated statistical concepts, methods and tools that can be used to implement the different

R.D. Snee, Ph.D. (✉)
Snee Associates, LLC, Newark, DE, USA
e-mail: Ron@SneeAssociates.com

L. Zhang (ed.), *Nonclinical Statistics for Pharmaceutical and Biotechnology Industries*, Statistics for Biology and Health, DOI 10.1007/978-3-319-23558-5_18

steps. A focus is placed on how to develop the process understanding that leads to a useful design space, process-control methods, and characterization of process risk. Product and process life-cycle model validation is also addressed. Integrating these concepts provides a holistic approach for effectively designing and improving products and processes.

18.2 Quality by Design: Its Origins and Building Blocks

Janet Woodcock, director of FDA's Center for Drug Evaluation and Research, defined the desired state of pharmaceutical manufacturing as "a maximally efficient, agile, flexible pharmaceutical manufacturing sector that reliably produces high-quality drug products without extensive regulatory oversight". QbD has been suggested as the route to achieving Woodcock's vision. ICH Q8 (R1) Step 2 defines QbD as a

> Systematic approach to development that begins with predefined objectives, emphasizes product and process understanding and process control, and is based on sound science and quality risk management (ICH 2005).

QbD is not new. The story begins in the 1920s with Sir Ronald A. Fisher. Fisher founded the field of statistical design of experiments (DOE) while working on agricultural and biological research studies in Rothamstead Experiment station in England. Fisher was frustrated that existing data collected without any structured plan or design did not yield any useful results. He published the first book on the subject (Fisher 1935). DOE is, of course, a critical building block of QbD.

Fast forward some 25 years and we find that in the late 1940s and early 1950s that DOE as a discipline beginning to gain acceptance by industry. The need in this case was how to effectively experiment with industrial processes. The big advance, paradigm shift in modern terms, came with the publication of the paper, "On the Experimental Attainment of Optimum Conditions" by Box and Wilson (1951). This paper addressed the problem (the need) of optimization of processes, which led to the idea of an "operating window", which in pharma parlance is called the "design space". Box and Wilson were working at Imperial Chemical Industries researching how to optimize chemical production processes. Their approach later became known as response surface methodology (RSM).

So how did the design space idea reach pharma? In the late 1960s and early 1970s, Joe Schwartz, a scientist at Merck saw the value of using RSM and process optimization in the development of formulation processes. He continued his research at the University of the Sciences in Philadelphia and educated several graduates on the subject who went to work in the pharma industry (Schwartz et al. 1973).

But the RSM approach didn't catch on in pharma like it did in the chemical and other process industries. Apparently the process improvement need wasn't yet identified in pharma. In the early part of this century, however, the FDA saw the need for pharma to improve its processes and developed the idea of QbD as the overarching system to aid the process.

One of the central leaders of the effort at the FDA was Ajaz Hussain, who was aware of Schwartz's work and deepened his knowledge of the subject by communicating with Schwartz to learn about the approach from the master himself (Hussain 2009). So we see that QbD, like many useful approaches is not completely new, building on the ideas of others. Indeed we stand on the shoulders of giants; in this case, Fisher, Box, Schwartz and Hussain. Clearly QbD, with its design space built on a sound foundation.

From an operational perspective, QbD is a systematic and scientific approach to product and process design and development that uses the following:

- Multivariate data acquisition and modeling to identify and understand the critical sources of variability
- Process-control techniques to ensure product quality and accurate and reliable prediction of patient safety and product efficacy
- Product and process design space established for raw-material properties, process parameters, machine parameters, environmental factors, and other conditions to enable risk management
- Control space for formulation and process factors that affect product performance.

QbD is useful for improving existing products, developing and improving analytical methods, and developing new products. The crux is implementing QbD in a cost-effective manner. The following issues are critical in that assessment:

- Recognition that the end result of successful implementation of QbD are the design space, process-control methodology and estimates of risk levels
- A strategy for identifying the critical process parameters that define the design space
- The creation of robust products and processes that sustain the performance of the product and process over time
- The use of change-management techniques to enable the cultural change required for success and long-term sustainability.

A lack of understanding of QbD in its entirety is a large stumbling block to its use. Stephen Covey, chairman of the Covey Leadership Center, points out that a successful strategy for any endeavor is to "begin with the end in mind" (Covey 1989). Following Covey's advice, the first step of QbD is to understand the critical outputs of QbD and then identify the critical building blocks of QbD, namely, improving process understanding and control to reduce risk. The outputs of design space, process-control procedures, and the risk level (both quantitative and qualitative risk assessment) are consistent with this approach (ICH 2005).

Before results can be realized, however, the building blocks of QbD (Snee 2009a, b) need to be assembled (see Fig. 18.1 and Table 18.1). The definition of QbD stated above calls for "predefined objectives" which is referred to as the "Quality Target Product Profile" (QTPP). Operationally the QTPP is defined as "a prospective

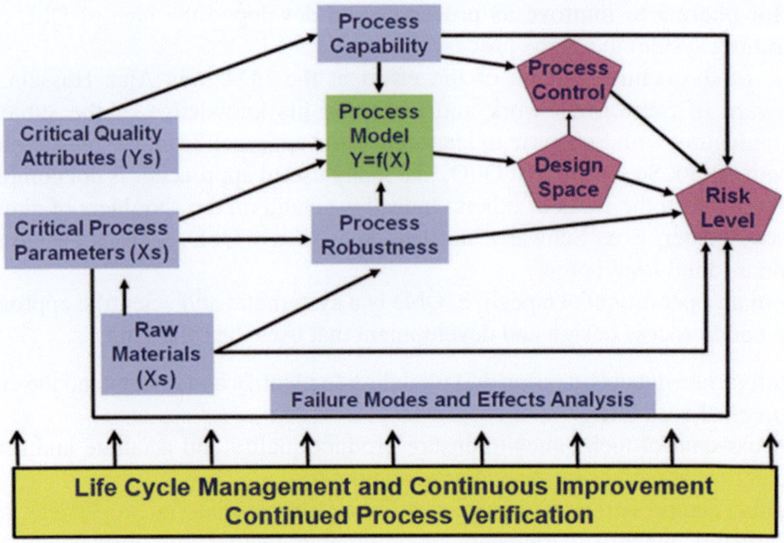

Fig. 18.1 Building blocks of quality by design

summary of the quality characteristics of a drug product that will be ideally achieved to assure the desired quality, taking into account safety and efficacy of the drug product" (ICH 2005).

The QbD building blocks that enable the QTPP to be realized are outlined below:

- Identify critical quality attributes (CQAs)
- Characterize raw-material variation
- Identify critical process parameters (CPPs)
- Characterize design space
- Ensure process capability, control, and robustness
- Identify analytical method capability, control, and robustness
- Create process-model monitoring and maintenance
- Offer risk analysis and management
- Implement Life cycle management: Continuous improvement and continued process verification.

Attention must be paid to the product formulation, manufacturing process, and analytical methods. Measurement is a process that needs to be designed, improved, and controlled just as any other process. The QbD building blocks provide a picture of the critical elements of the roadmap. It is critical to success to recognize how the building blocks are linked and sequenced over time. Figure 18.1 provides a roadmap for implementing QbD telling us that QbD builds as the product and process is developed; hence QbD is a sequential approach. The building blocks are created and assembled using the principles of Statistical Engineering (Hoerl and Snee 2010) which provides a framework for approaching large initiatives such as QbD.

Table 18.1 Descriptions of the building blocks of quality by design

Building block	Description
Critical quality attributes (CQAs)	The critical process output measurements linked to patient needs.
Critical process parameters (CPPs)	The process inputs (active pharmaceutical ingredient and excipients), control, and environmental factors that have major effects on the CQAs.
Raw-materials factors	Include the stability and capability of raw-material manufacturing processes that affect process robustness, process capability, and process stability.
Process model	A quantitative picture of the process based on fundamental and statistical relationships that predict the CQA results.
Design space	The combinations of input variables and process parameters that provide quality assurance.
Process and measurement capability	Tracks process performance relative to CQA specifications and provides measurement repeatability and reproducibility regarding CQAs.
Process and measurement robustness	The ability of the process and measurement system to perform when faced with uncontrolled variation in process, input, and environmental variables.
Process and measurement control	Control procedures, including statistical process control, that keep the process and the measurement system on target and within the desired variation.
Failure-modes-and-effects analysis (FMEA) of the CPPs	Examines raw-material variables, identifies how the process can fail, and, reveals after appropriate controls and fixes are in place, the areas of the process that remain at greatest risk of failing.
Risk level	A function of the design space, FMEA results, and process and measurement capability, control, and robustness.

18.3 Process Understanding: Critical to Process Development, Operation and Improvement

Process understanding is fundamental to the QbD approach. Indeed process understanding is an integral part of the definition of QbD. Regulatory flexibility comes from showing that a given process is well understood. According to FDA (2004), a process is generally considered to be well understood when the following conditions are met:

1. All critical sources of variability are identified and explained
2. Variability is managed by the process
3. Product-quality attributes can be accurately and reliably predicted within the design space established for the materials used, process parameters, manufacturing, environmental and other conditions.

Process understanding is needed not only for product and process development, but also for successful technology transfer from development to manufacturing and from site-to-site, which includes transfer to contract manufacturing organizations (Alaedini et al. 2007; Snee 2006; Snee et al. 2009a). It is very difficult, if not impossible, to *successfully and effectively* create, operate, improve, or transfer a process that is not understood.

The importance of process understanding is illustrated by the following case. A new solid-dose, 24-hour controlled-release product for pain management had been approved but not yet validated because it had encountered wide variations in its dissolution rate. The manufacturer did not know whether the dissolution problems were related to the active pharmaceutical ingredient (API), the excipients, or to variables in the manufacturing process—or to some combination of these factors.

Frustrated with the lack of process understanding, the manufacturer narrowed the range of possible causes of the unacceptable dissolution rate to nine potential variables—four properties of the raw material and five process variables. The team used a designed experiment (DOE) to screen out irrelevant variables and to find the best operating values for the critical variables (Snee et al. 2009b).

The analysis showed that one process variable exerted the greatest influence on dissolution and that other process and raw material variables and their interactions also played a key role. The importance of the process variable with the largest effect had been unknown prior to this experiment even after more than 8 years of development work. This enhanced process understanding enabled the company to define the design space and the product was successfully validated and launched.

This example illustrates the criticality of process understanding. The FDA noted the importance of process understanding when they released "Guidance for Industry: PAT – A Framework for Pharmaceutical Development, Manufacturing and Quality Assurance", (FDA 2004). The FDA was responding to the realities of the pharmaceutical and biotech industries; namely that Pharma/Biotech needs to improve operations and speed up product development. Compliance continues to be an issue and risks must be identified, quantified and reduced. The root causes of many compliance issues relate to processes that are neither well understood nor well controlled.

Prediction of process and product performance requires some form of a model, $Y = f(X)$. In this conceptual model, Y is the process outputs such as Critical Quality Attributes (CQAs) of the product and X denotes the various process and environmental variables that have an effect of the process outputs, often referred to as Critical Process Parameters (CPPs). Models may be empirical, developed from data, or mechanistic, based on first principles.

In developing process understanding it is helpful to create a process schematic such as the one for a compression process shown in Fig. 18.2. Here we see the process outputs (Ys), Process inputs (Xs), Process control variables (Xs) and environmental variables (Xs). The goal is to produce a process model of the form $Y = f(Xs)$ that will accurately predict process performance as measured by the Ys (CQAs).

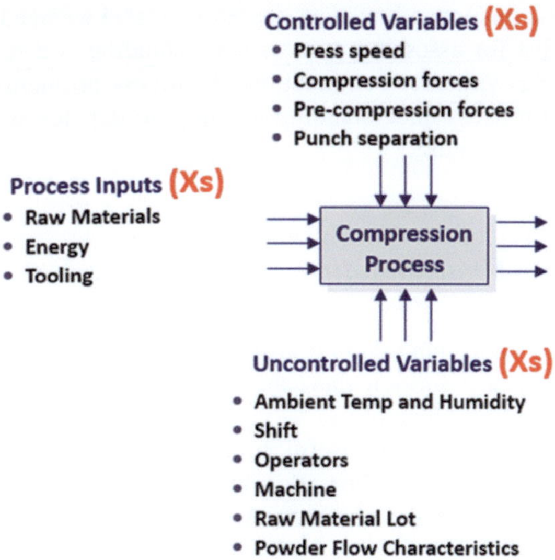

Fig. 18.2 Process schematic showing process inputs, control variables environmental variables and outputs. Developing the model Y = f(X) enables prediction of future process performance

McCurdy et al. (2010) provide an example of such models developed for a roller compaction process. Among the models reported was a model in which tablet potency relative standard deviation (RSD) was increased by increasing mill screen size (SS) and decreased with increasing roller force (RF) and gap width (GW). They reported a quantitative model for the relationship:

$$\text{Log (Tablet Potency RSD)} = -0.15 - 0.08\ (\text{RF}) - 0.06\ (\text{GW}) + 0.06\ (\text{SS})$$

Process understanding summarized and codified in the form of the process model, conceptually represented as Y = f(X), can contain any number of variables (Xs). These models typically include linear, interaction and curvature terms as well as other types of mathematical functions.

At a strategic level, a way to assess process understanding is to observe how the process is operating. When process understanding is adequate the following will be observed:

- Stable processes (in statistical control) are capable of producing product that meets specifications
- Little firefighting and heroic efforts required to keep the process on target
- Processes are running at the designed speed with little waste
- Processes are operating with the expected efficiency and cost structure
- Employee job satisfaction and customer satisfaction is high
- Process performance is predictable

To assess the state of process understanding at an operational level we need a list of desired characteristics. A list for assessing process understanding is discussed in the following sections along with the identification of process problems that frequently result from lack of process understanding and how to develop process understanding including what tools to use.

18.3.1 Assessing Process Understanding

The FDA definition of process understanding is useful at a high level but a more descriptive definition is needed; a definition that can be used to determine if a process is understood at an operational level.

Table 18.2 lists the characteristics that are useful in determining when process understanding exists for a given process. First it is important that the critical variables (Xs) that drive the process are known. Such variables are typically called critical process parameters (CPP). It is helpful to broaden this definition to include both input and environmental variables as well as process variables; sometimes referred as the "knobs" on the process.

It is important to know the critical environmental variables (uncontrolled noise variables), such as ambient conditions and raw material lot variation, can have a major effect on the process output (Ys). Designing the process to be insensitive to these uncontrolled variations results in a "robust" process.

Measurement systems are in place and the amount of measurement repeatability and reproducibility is known for both output (Y) and input (X) parameters. The measurement systems need to be robust to minor and inevitable variations in how the procedures are used to implement the methods on a routine basis. This critical aspect or process understanding is often overlooked in the development process. Gage Repeatability and Reproducibility studies and method robustness investigations are essential to proper understanding of the measurement systems.

Table 18.2 Characteristics of process understanding

• Critical process parameters (Xs) that drive the process are known and used to construct the process design space and process control approach.
• Critical environmental and uncontrolled (noise) variables that affect the critical quality attributes (Ys) are known and used to design the process to be insensitive to these uncontrolled variations (robustness)
• Robust measurement systems are in place and the measurement repeatability and reproducibility is known for all critical quality attributes (Ys) and critical process parameters (Xs)
• Process capability is known
• Process failure modes are known and removed or mitigated
• Process control procedures and plans are in place

Process capability studies involving the estimation of process capability and process performance indices (Cp, Cpk, Pp and Ppk) are useful in establishing process capability. Sample size is a critical issue here. From a statistical perspective 30 samples is the minimum for assessing process capability; much more useful indices are developed from samples on 60–90 observations. In Chap. 20 the authors recommend sample sizes of 100–200 for reasonable size confidence intervals.

In assessing the various sources of risk in the process, it is essential that the potential process failure modes be known. This is greatly aided by performing a failure modes and effects analysis at the beginning of the development process and as part of the validation of the product formulation and process selected for commercialization.

Process control procedures and plans should be in place. This will help assure that the process remains on target at the desired process settings. This control procedure should also include a periodic verification of the process model, $Y = f(X)$, used to develop the design space. This is also recommended by the FDA's Process Validation Guidance (FDA 2011).

18.3.2 Process Problems Are Typically due to Lack of Process Understanding

Although it is "a blinding flash of the obvious", it often overlooked that when you have a process problem it is due to a lack of process understanding. When a process problem occurs you often hear "Who did it; who do we blame"? Or "How do we get it fixed as soon as possible?" Juran emphasized that 85 % of the problems are due to the process and the remaining 15 % are due to the people who operate the process (Juran and DeFeo 2010).

While a sense of urgency in fixing process problems is appropriate some better questions to ask are "How did the process fail?" and "What do we know about this process; do we have adequate understanding of how this process works"?

Table 18.3 summarizes some examples of process problems and how new process understanding lead to significant improvements; sometimes in unexpected areas. Note that these examples cover a wide range of manufacturing and non-manufacturing issues including capacity shortfalls, defective batches, process interruptions, batch release time and report error rates. All were significant problems in terms of both financial and process performance. The increased process understanding resulted in significant improvements.

18.3.3 How Do We Develop Process Understanding?

Consistent with the FDA (2004) definition of process understanding noted previously in this chapter, we see in Fig. 18.3 that a critical first step in developing process

Table 18.3 Descriptions of tools used for developing process understanding

Tool	Description
Process map	A schematic of a process showing process steps and process inputs and outputs.
Cause-and-effect matrix	A prioritization matrix that enables one to select those process input variables that have the greatest effect on the process output variables.
Measurement systems analysis	A study of the measurement system, typically using Gage R&R* studies, to quantify the measurement of repeatability and reproducibility.
Capability study	An analysis of process variation versus process specifications to assess the ability of the process to meet specifications.
Failure-mode-and-effects analysis	An analytical approach for identifying process problems by prioritizing failure modes and their causes.
Multivariate study	A study that samples the process as it operates and, by statistical and graphical analysis, that identifies the important controlled and uncontrolled (i.e., noise) variables.
Design of experiments	A method of experimentation that identifies, with minimum testing, how key process input variables affect the output of the process.
Control plan	A document that summarizes the results of a Six-Sigma project and aids the operator in controlling the process.

[a]R&R refers to repeatability and reproducibility

Fig. 18.3 Developing and using process understanding

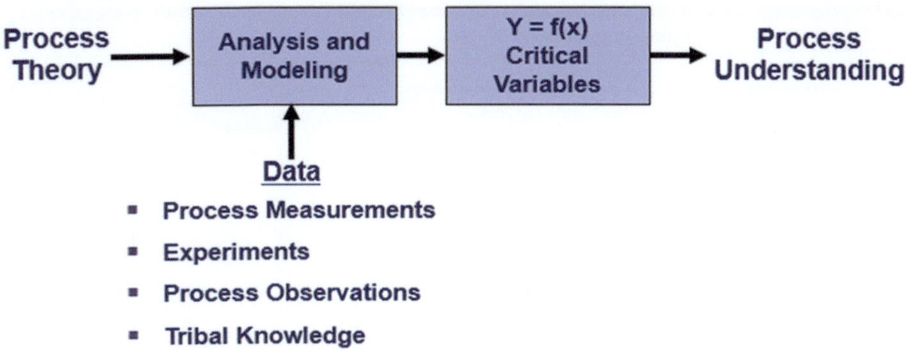

Fig. 18.4 Routes to process understanding

understanding is to recognize that process understanding is related to process variation. As you analyze process variation and identify root causes of the variation, you increase your understanding of the process. Process risk is an increasing function of process variation and a decreasing function of process understanding. Increasing process understanding reduces process risk and increases compliance.

In Fig. 18.4 we see that analyzing the process by combining process theory and data (measurements and observations, experiments and tribal knowledge in the form of what the organization knows about the process). Science and engineering theory when interpreted in the light of data enhances process understanding and results in more science and engineering being used in understanding, improving and operating the process.

The integration of theory and data produces a process model, Y = f(X), and identifies the critical variables that have a major effect on process performance. Fortunately there are typically only 3–6 critical variables. This finding is based on the Pareto principle (80 % of the variation is due to 20 % of the causes) and experience of analyzing numerous processes in a variety of environments by many different investigators (Juran and DeFeo 2010).

18.3.4 What Tools Do We Use to Develop Process Understanding?

Process analysis is strongly data based, creating the need for data-based tools for the collection and analysis of data and knowledge-based tools that help us collect information on process knowledge (Fig. 18.5 and Table 18.4). We are fortunate that all the tools needed to develop process understanding described above are provided by QbD and Process Analytical Technology (FDA 2004) and Lean Six Sigma methodologies (Snee and Hoerl 2003; Snee 2007).

It all starts with a team which includes a variety of skills including formulation science, process engineering, data management and statistics. In my experience

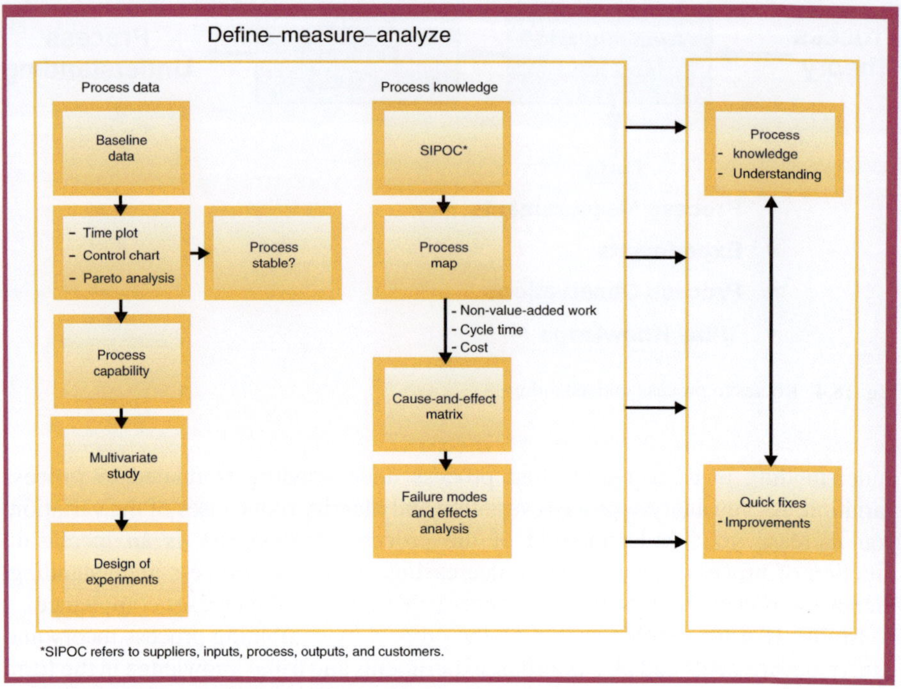

Fig. 18.5 Tools for developing process understanding

Improvement teams often have limited formulation science and data management skills. Process knowledge tools include the process flow chart, value stream map, cause and effect matrix and failure modes and effects analysis (FMEA).

The data-based tools include design of experiments, regression analysis, analysis of variance, measurement system analysis and statistical process control. The DMAIC (Define, Measure, Analyze, Improve and Control) process improvement framework and its tools are particularly useful for solving process problems. A natural by-product of using DMAIC is the development of process knowledge and understanding, which flow from the linking and sequencing of the DMAIC tools. Development of process understanding is built into the method (Fig. 18.5).

18.4 Design Space

Although all the building blocks of QbD are important, the creation and use of the design space is arguably the most important aspect of QbD. The design space is the

> Multidimensional combination and interaction of input variables (e.g., material attributes) and process parameters that have been demonstrated to ensure quality.

Table 18.4 Process understanding leads to improved process performance: some examples

Problem	New process understanding	Result of improvements based on new process understanding
Batch release takes too long	Batch record review system flow improved. Source of review bottleneck identified.	Batch release time reduced 35–55 % resulting in inventory savings of $5MM and $200 k/year cost reduction
Low capacity not able to meet market demand	Yield greatly affected by media lot variation. New raw material specifications needed.	Yield Increased 25 %
Batch defect rate too high	Better mixing operation needed including: methods and rate of ingredient addition, revised location of mixing impeller, tighter specs for mixing speeds and times and greater consistency is blender set-up.	Defect rate significantly reduced saving $750 k/year
Process interruptions too frequent	Root cause was Inadequate supporting systems including, lack of spare parts, missing batch record forms and lack of standard operating procedures.	Process interruptions reduced 67 % saving $1.7MM/year
Report error rate too high	Report developer not checking spelling, fact accuracy and grammar.	Error rate reduced 70 %

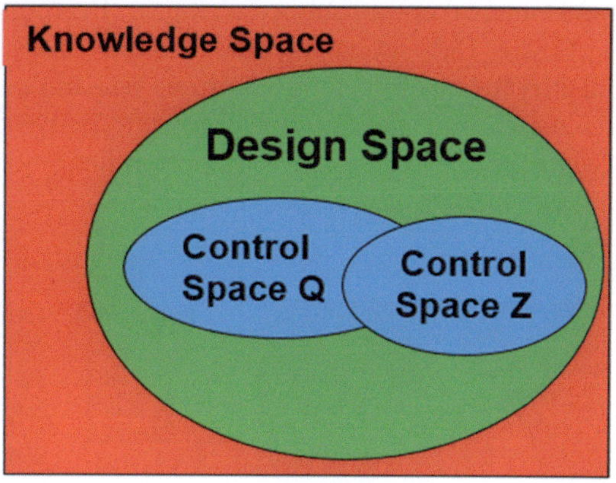

Fig. 18.6 Predictor variable spaces—knowledge, design and control

The relation between the knowledge, design and control spaces are shown schematically in Fig. 18.6. A process can have more than one control space. The Control Space is the region or point in the design space at which the process is operated. This space is also sometimes referred to as the Normal Operating Region (NOR). A process can have more than one control space within the design space.

A key question is how to create the design space, particularly when products are often locked into a design space before the process is well understood. The following two-phase approach is recommended:

- Create the design space during the development phase by focusing on minimizing risk and paying close attention to collecting the data that are most critically needed to speed up development and to understand the risk levels involved
- After the process has been moved into manufacturing, collect data during process operation to refine the process model, design space, and control space as additional data become available over time.

Continued Process Verification (CPV) from Stage 3 of the FDA Process Validation Guidance (FDA 2011) is very effective in implementing the second phase of this approach. CPV and process monitoring is an important building block of QbD and will be discussed in greater detail later in this chapter.

The following examples illustrate the concepts behind the design space. Fundamental to the construction of the design space is having a quantitative model, Y = f(X), for the product or process being studied. Figure 18.7 shows a contour plot for dissolution (spec > 80 %) and friability (spec < 2 %) as a function of two process parameters. We find the combination of process parameters that will satisfy both the dissolution and friability specifications simultaneously by overlaying the contour plots as shown in Fig. 18.8. This approach, referred to as the Overlapping Means Approach (OMA) by Peterson and Lief (2010) will be discussed later in further detail. The location of the desired control space can also be found using mathematical optimization techniques (Derringer and Suich 1980).

Figure 18.9 shows another example of overlaid contour plots being used to identify the combinations of particle size and excipient concentration that will meet the dissolution specifications. This plot makes it easy to see how the design space (white area) relates to the variation in the process variables. The design space has more flexibility with respect to excipient concentration than particle size.

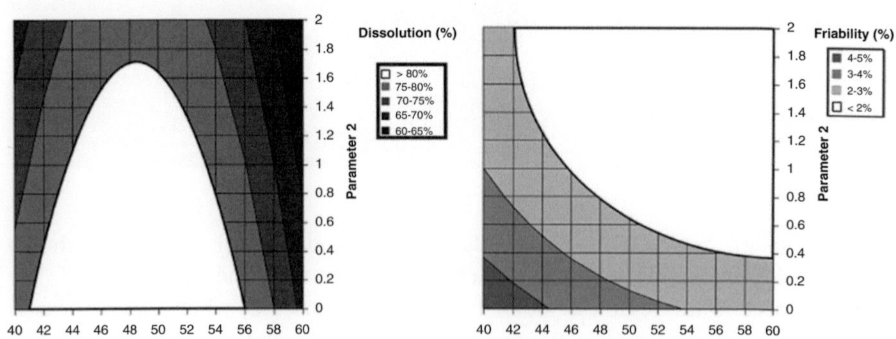

Fig. 18.7 Contour plots of dissolution and friability as a function of process parameters 1 and 2

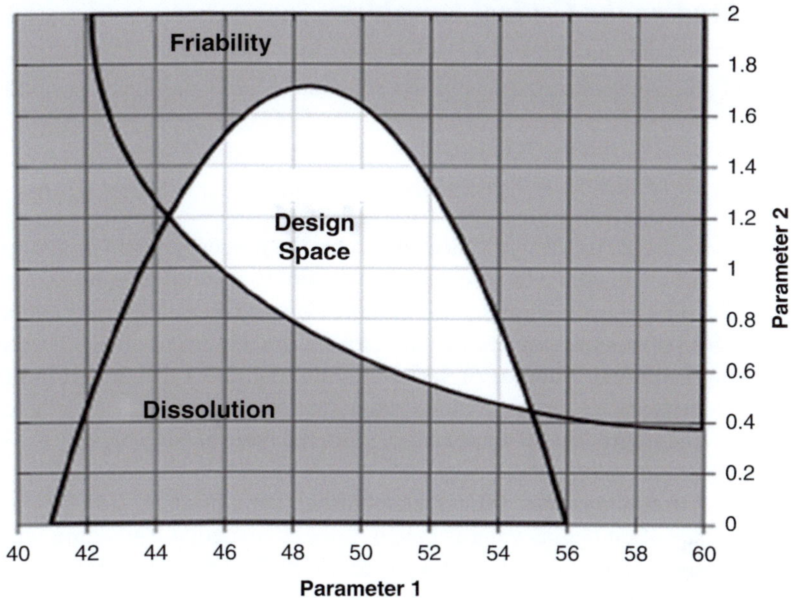

Fig. 18.8 Design space comprised of the overlap region of ranges for friability and dissolution

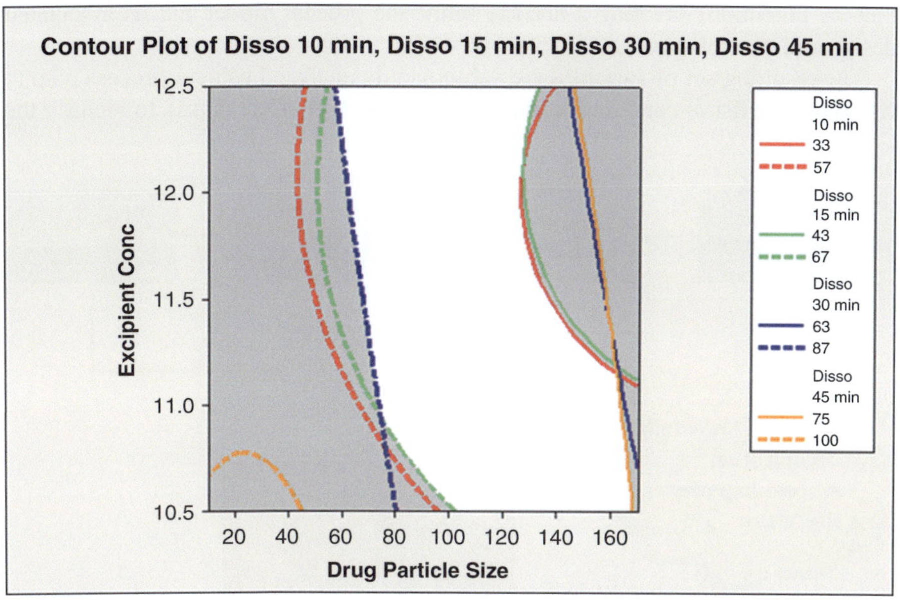

Fig. 18.9 Design space region where product will be in specifications

18.4.1 Finding the Critical Variables

Product and process understanding are fundamental to QbD and the development of the model

$$Y = f(x_1, x_2, \ldots, x_p)$$

which is used to create the design space and create the process control methodology. The question is how to create the process model quickly without missing any important variables. Getting the right set of variables (i.e., critical process parameters, input variables such as raw-material characteristics and environmental variables) in the beginning is critical. Sources of variability and risk can be obtained in several ways. Interactions between raw-material characteristics and process variables are ever-present and more easily understood with the use of statistically designed experiments.

Figure 18.10 contains the critical elements of the approach. Identifying the critical variables often begins what is called "tribal knowledge," meaning what the organization knows about the product and process under study. This information is combined with the knowledge gained in development and scale-up, a mechanistic understanding of the chemistry involved, literature searches, and historical experience. The search for critical variables is a continuing endeavor throughout the life of the product and process. Conditions change, and new knowledge is developed, thereby potentially creating a need to refine the process model and its associated design and control spaces.

The resulting set of variables are subsequently analyzed using a process map to round out the list of candidate variables, the cause-and-effect matrix to identify the

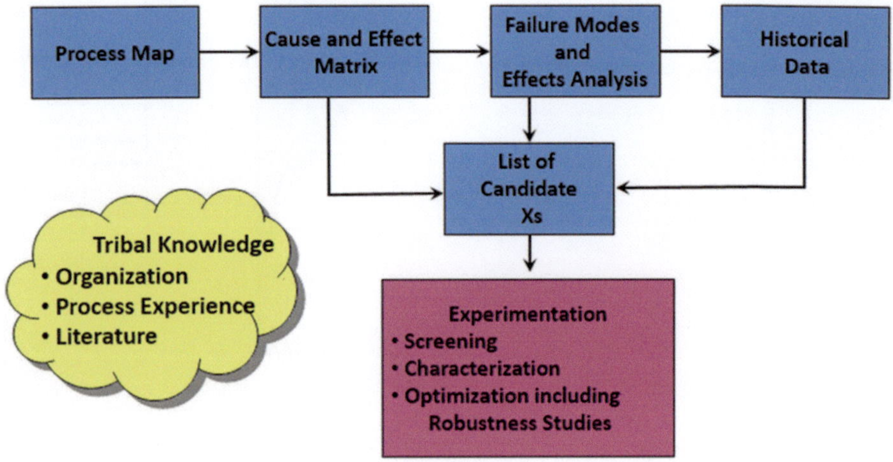

Fig. 18.10 Developing the list of candidate variables (Xs)

Characteristic	Screening	Characterization	Optimization
No. of Factors	More than 6	3-6	2-5
Desired Information	Critical Factors	Understand how System Works	Prediction Equation, Optimization, **Design Space**
Model Form	Linear or Main Effects	Linear and Interaction Effects	Linear, Interaction and Curvilinear Effects
Experiment Design	Plackett-Burman Fractional-Factorials	Full and Fractional Factorials	Response Surface

Fig. 18.11 Comparison of experimental environments

high-priority variables, and the FMEA to identify how the process can fail. This work will identify those variables that require measurement system analysis and those variables that require further experimentation (Hulbert et al. 2008).

Identifying potential variables typically results in a long list of candidate variables, so a strategy for prioritizing the list is needed. In the author's experience and that of others, the DOE-based strategy-of-experimentation approach (see Fig. 18.11), developed at DuPont Company in Wilmington, DE, is a very effective approach (Pfeiffer 1988). Developing an understanding of the experimental environment and matching the strategy to the environment is fundamental to this approach. A three-phase strategy (i.e., screening, characterization, and optimization) and two-phase strategy (i.e., screening followed by optimization and characterization followed by optimization) are the most effective. In almost all cases, an optimization experiment is run to develop the model for the system that will be used to define the design space and the control space.

The confirmation (i.e., validity check), through the experimentation model used to construct the design space and control space is fundamental to this approach. Confirmation experiments are conducted during the development phase. The model is confirmed periodically as the process operates over time. This ongoing confirmation is essential to ensure that the process has not changed and that the design and control spaces are still valid. The ongoing confirmation of the model happens during the second phase of the development process, as previously described.

The screening—characterization—optimization (SCO) strategy is illustrated by the work of Yan and Le-he (2007) who describe a fermentation optimization study that uses the screening followed by optimization strategy. In this investigation 12 process variables were optimized. The first experiment used a 16-run

Plackett-Burman screening design (1946) to study the effects of the 12 variables. The four variables with the largest effects were studied subsequently in a 16-run optimization experiment. The optimized conditions produced an enzyme activity that was 54 % higher than the operations produced at the beginning of the experimentation work.

The SCO Strategy in fact embodies seven strategies developed using single and multiple combinations of the screening, characterization and optimization phases. The end result of each of these sequences is a completed project. There is no guarantee of success in a given instance, only that SCO strategy will "raise your batting average" (Snee 2009c). It enables the user to get the right data in the right amount at the right time.

The SCO Strategy works because of its underlying theory that has been tested and enhanced through use as summarized below:

- Pareto principle applies; the majority of the variation is due to a few causes (Xs). The general wisdom is that most systems are driven by 3–6 variables
- Plan for more than one experiment as process knowledge builds over time through sequential experimentation
- Experimental environment defines the appropriate design. Understand the environment and the appropriate design will be easier to create
- Multiple variables (inputs Xs and outputs Ys) must be studied simultaneously to develop deep process knowledge
- Most response functions can be adequately approximated within the region of interest by 1st and 2nd order polynomial models (Taylor Series expansion). Complex effects such as extreme curvature, cubic functions and interactions involving more than two variables rarely exist.
- Effects of process instability can be reduced by using randomization and blocking

The **SCO Strategy** helps minimize the amount of data collected by recognizing the phases of experimentation. Utilizing the different phases of experimentation results in the total amount of experimentation being done in "bites". These bites allow subject matter expertise and judgment to be utilized more frequently and certainly at the end of each phase.

18.4.2 Formulation Studies

Attempting to optimize formulations involving multiple components by varying one component at a time is, at best a low yield strategy. A formulation scientist's experience is typical. He reported that "an 11 component formulation was studied. A formulation that worked was found. Unfortunately it was a very painful experience; it took a long time filled with uncertainty, anxiety and lots of stress. What was worse is we were never sure that we were even close to the best formulation".

There must be a better way and fortunately there is (Cornell 2002; Snee and Piepel 2013). The DOE and Strategy of Experimentation approaches described above can also be used effectively in the development of formulations where the formula ingredients are expressed on a percent basis and add to 100 % (or 1.0 when the ingredients are expressed in proportions).

The concepts apply including:

- Statistical designs (called mixture designs) are used to collect the data
  - Screening experiments are used to identify the ingredient that are having the largest effects on the formulation performance
  - Optimization designs are used to identify the design space
- Graphical analyses are performed to study the effects of the components
- Models are fit to the data to describe the response surface and construct contour plots
- Design spaces are constructed using contour plots and mathematical optimization
- Confirmation experiments are conducted to verify the performance of the selected formulations
- There is no limit on the number of components that can be investigated and almost any combination of component (wide and narrow) ranges can be studied.

As when experimenting with process variables the approach of creating formulations using designed experiments (mixture designs in the case of formulations) enables the user to obtain the right data in right amount in the right time.

18.4.3 Bayesian Design Space Predictive Approach

The design space concept is a major step forward beyond the "One-Factor-at-a-Time" approach to the development and improvement of products and processes. Development of a design space requires a deep understanding of the product and process involved. It is this understanding that the FDA looks for in regulatory QbD filings. Construction of the design space also requires the creation of quantitative models based on data collected from appropriately designed experiments.

Peterson and Lief (2010) emphasized a major limitation to the design space construction process as is commonly practiced known as the Overlapping Means Approach (OMA). The approach has been widely used since at least the 1960s. OMA, which was described above, is based on the use of regression models to construct contour plots (overlapping when more than one response is involved) and mathematical optimization to find the combination of predictor variables that will produce product within specifications.

Predictions of OMA are based on the predicted means responses. As a result, at the edge of the design space, approximately 50 % (or more) of the time, the product is predicted to be outside specifications, assuming the process variation is normally distributed. Moving the process target closer to the center of the design space certainly reduces this probability. Peterson and Lief (2010) illustrate a subtle but important risk issue with the OMA. For a design space based upon the

OMA involving multiple response types, it is possible for such a design space to contain points near the boundary of intersecting mean response surface contours that correspond to probabilities much less than 0.5 of meeting all specifications.

To avoid this limitation, Peterson (2008) and Peterson and Lief (2010) recommend a Bayesian based approach which calculates the probability that product specifications will be met at different combinations of the predictor variables. The resulting approach is quantitative, risk based and utilizes prior product and process information. These are desirable characteristics mentioned quite often in regulatory guidance. The Bayesian predictive approach uses design of experiment to collect the needed data just as with the OMA. The difference is in the method of design space construction, not in the experimentation involved.

Peterson and Lief (2010) discuss an example which illustrates the Bayesian approach and shows how it compares to the OMA. The experiment involved three responses (spec limits): Y1 = disintegration time (<15 min), Y2 = friability (<0.8 %) and Y3 = hardness (8–14 kp). The six predictor variables involved were: X1 = Amount Water Added, X2 = Water Addition Rate, X3 = Wet Massing Time, X4 = Main Compression Force, X5 = Main Compression/Precompression Ratio and X6 = Speed.

Factor X6 was found to have no effect on any response and Factor X5 effect was small and set at the center point for design space calculations. The resulting design space is shown in Fig. 18.12 in which the OMA design space is shown in yellow and the predictive design space with minimum reliability value = 0.8 shown in black. As expected, the predictive space is smaller than the OMA design space. The worst probability of meeting all specifications for this OMA design space was found to be 0.15.

One of the practical challenges of the Bayesian approach is that the computational capabilities have not as yet been implemented in widely available software packages such as JMP, Design Expert and Minitab. This situation will likely improve over time. Peterson and Lief (2010) discuss several alternative computational approaches that are available in other software packages and provide references to additional information and uses of the Bayesian approach.

In the meantime there are two things that should be considered when using the OMA. First the selected control space and associated process target(s) should be verified with confirmation experiments. This is a critical step in good experimental practice. The confirmation experiments should be designed to estimate the probability of exceeding specifications. This is a characteristic of both the OMA and the Bayesian approach.

Second, Stage 3 of the FDA Process Validation Guidance (FDA 2011) calls for "Continued Process Verification" which is discussed in the next section. Stage 3 recommends a lifecycle approach to process monitoring. Such an approach will quickly detect when product is out of specification and should be designed to provide an early warning regarding when such a situation might occur.

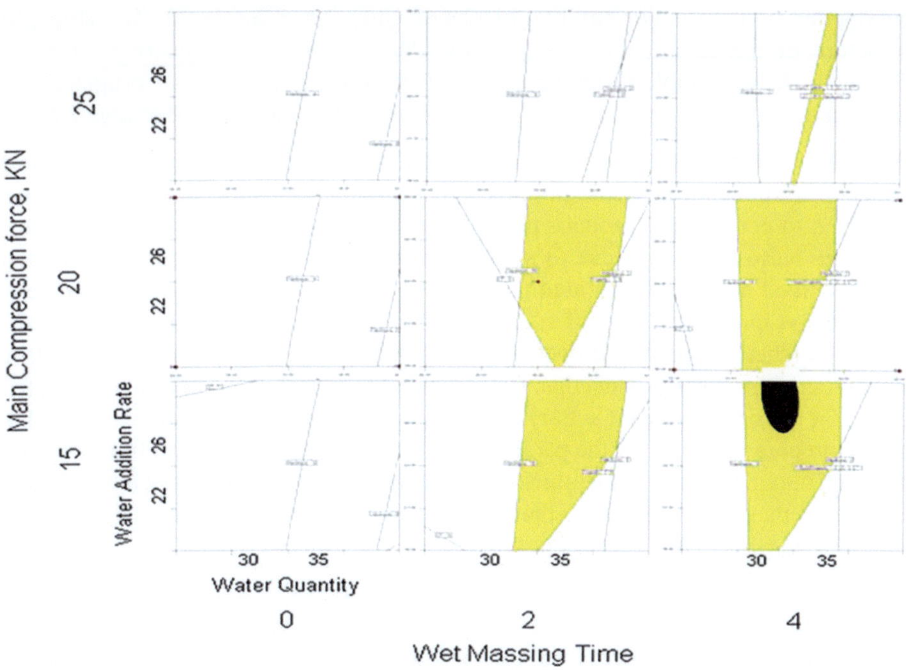

Fig. 18.12 Design space—*yellow* region based on OMA; *black* region based on Bayesian predictive approach

18.5 Continued Process Verification

Routine, ongoing assessment of process performance and product quality is crucial to ensuring that high quality pharmaceuticals reach the patient in a timely fashion. This need is addressed in Stage 3 of the FDA Process Validation Guidance (FDA 2011) which calls for Continued Process Verification (CPV). A systems approach to CPV, including key challenges, the role of quality by design (QbD), and how to operate the system effectively is discussed in this section. A detailed discussion of the FDA Process Validation Guidance is the subject of Chap. 19.

In traditional solid-dosage pharmaceutical manufacturing, process data are routinely analyzed at two points in time to assess process stability and capability. During the production of each batch, process operators and quality-control departments collect data to ensure stability and capability and take appropriate remedial actions when needed. On a less frequent basis (i.e., monthly or quarterly), batch-to-batch variation is analyzed based on product parameters to assess the long-term stability and capability of the process.

Process Monitoring, Control and Capability The industry and regulatory focus on QbD places an even greater emphasis on the quality of pharmaceutical products and the performance of the pharmaceutical-manufacturing processes. Process and product control are major building blocks of QbD (Snee 2009a, b). The strategic structure of the International Society for Pharmaceutical Engineering's (ISPE) Product Lifecycle Implementation Plan (PQLI) lists the "process performance and product quality monitoring system" as one of its critical elements (Berridge 2009).

Process Stability and Capability are central to assessing the performance of any process. Manufacturing processes that are stable and capable over time can be expected to consistently produce product that is within specifications and thereby cause no harm to patients due to nonconforming product. Stability and capability are described as follows (Montgomery 2013a, b): A stable manufacturing process is a process that is in a state of statistical control as each batch is being produced and as batches are produced over time. A process in a state of statistical control consistently produces product that varies within the process control limits; typically set at the process average (X-Bar) plus and minus three standard deviations (SD) of the process variation for the parameter of interest. Separate control limits are set for each parameter (e.g., tablet thickness and hardness). Any sample value that falls outside of these limits indicates that the process may not be in a state of statistical control.

A capable process is one that consistently produces tablets that are within specifications for all tablet parameters (Montgomery 2013a, b). A process capability analysis compares the process variation to the lower and upper specification limits for the product. A broadly used measure of process capability is the Ppk index, or process performance index, which is discussed in greater detail later in this chapter.

A production process can include any one of the four combinations of stability and capability (Hoerl and Snee 2012): stable and capable (desired state), stable and incapable, unstable and capable, and unstable and incapable (worst possible situation).

Process stability and capability are typically evaluated at least twice:

1) During the production of each batch to ensure that the process is in control and to identify when process adjustments are needed. Some key questions that need to be addressed during this analysis include:

    - Is the batch production process stable during the production of the batch with no trends, shifts, or cycles present?
    - Is the process capable of meeting specifications (i.e., are the process-capability indices acceptable)?
    - Is the within-batch sampling variation small, indicating a stable production process?

2) Monthly or quarterly to ensure batch-to-batch control throughout a given year and between years. Some important questions that should be addressed during this analysis include:

    - Is the batch-to-batch variation stable from year to year and within years with no shifts, trends, or cycles present?
    - Is the batch-to-batch variation small?

These two analyses also help to assess the robustness of the process.

Control Limit Versus Specification Limit Control limits are calculated from process data and applied to the process. Control limits are used to assess the stability of the process and to determine the need for process adjustments when out of control samples are detected. On the other hand, specification limits apply to the product. Specification limits are used to assess the capability of the process to produce a product that has the desired properties and characteristics.

18.5.1 A Systems Approach for Continued Process Verification

It is generally agreed in industry that a systems-based approach enables operations to be more efficient and sustainable. Schematics of such systems are shown in Fig. 18.13 for monitoring individual batches and in Fig. 18.14 for monitoring batch-to-batch variation (Snee and Hoerl 2003, 2005; Snee and Gardner 2008). The systems underlying Figs. 18.13 and 18.14 have the following characteristics:

- Data are periodically collected from the process. Pharmaceutical manufacturing processes are often monitored using 30–60 min samples.
- These data are used to monitor processes for stability and capability using control charts, process capability indices, analysis of variance, time plots, boxplots, and histograms.
- The analysis identifies when process adjustments are needed to get the process back on target.

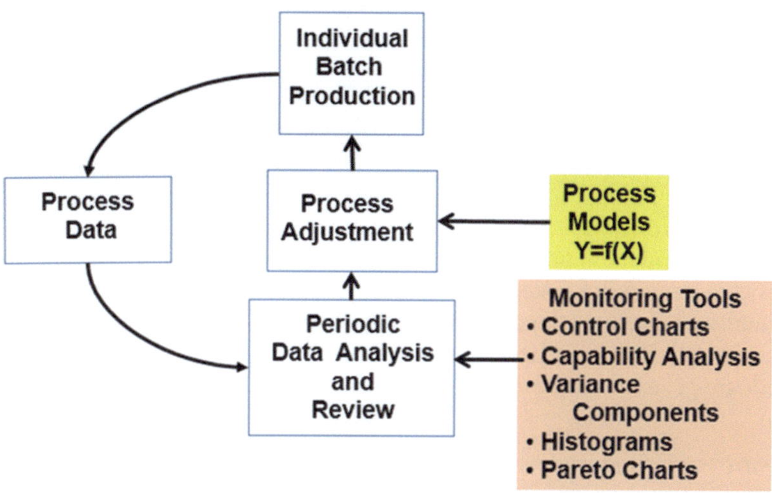

Fig. 18.13 Framework example for monitoring process stability and capability for Individual batches

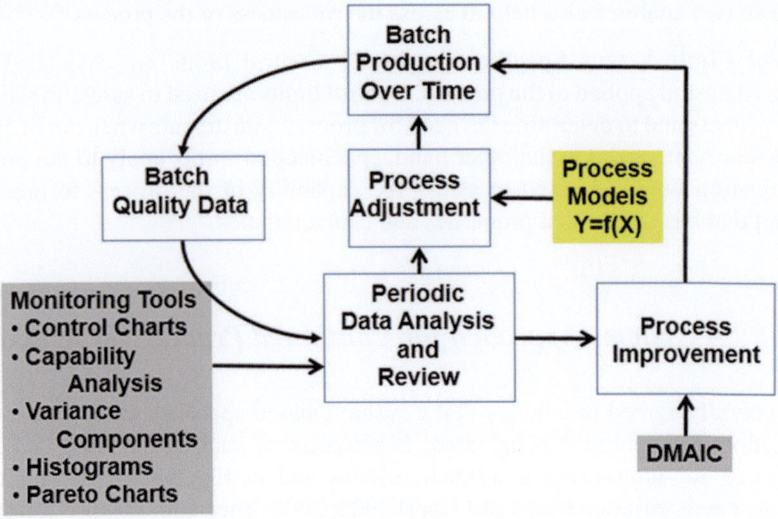

Fig. 18.14 Framework example for monitoring batch-to-batch variation over time

- Records are kept on the types of problems identified. As significant problems are identified or problems begin to appear on a regular basis, the resulting issues and documentation are incorporated into process-improvement activities to develop permanent solutions.

Process improvement can be effectively completed using the DMAIC (define, measure, analyze, improve, control) problem-solving and process-improvement framework (Snee and Hoerl 2003, 2005). The use of the tools in this framework is discussed in the next section.

18.5.2 Assessment Tools

As a general principle, it is rare that a manufacturing process that is stable and capable will produce a product that is out of specification. The primary purpose of a process monitoring system is to address the question: Is this process capable of consistently producing product that is within specifications over time? The statistical analyses conducted to answer this question are briefly described below. These methods are generally accepted and well documented in the literature (Montgomery 2013a).

Control-Chart Analysis A control-chart analysis is used to assess the stability of a process over time. The Shewhart chart has been widely used to assess process stability since the 1930s. Other types of control charts are also useful for monitoring processes (Montgomery 2013a).

A stable process is a predictable process; a process whose product will vary within a stated set of limits. A stable process is sometimes referred to as being in "a state of statistical control" (Montgomery 2013a). A stable process has no sources of special-cause variation—that is, effects of variables are outside the process but have an effect on the performance of the process (e.g., process operators, ambient temperature and humidity, raw material lot).

The most commonly used indicator of special-cause variation is a process that has product measurements outside of the control limits which are typically set at X-Bar plus and minus three SD of the process variation for the parameter of interest.

For example a process may be producing tablets with an average hardness of 4.0 kp and a standard deviation of 0.3 kp. The control limits are thus $4.0 \pm 3(0.3)$ for a range of 3.1–4.9. Any tablet sample outside of that range is an indication that the process average may have changed and a process adjustment may be needed. Separate control limits are set for each parameter.

Figure 18.15 shows a control chart for a total weight of 10 tablets manufactured by two tablet presses. The graphic demonstrates that both presses are stable and in control and producing tablets with the same average weight and variation.

Figure 18.16 shows a control chart for assay values for batches produced over a 3-year period. The process is stable through the middle of Year 2 and begins to decrease in Year 3. When a process adjustment is made, the batch assay values return close to the values observed in Year 1. This example is interesting because a process shift is shown, but none of the batch assay values are close to the assay specifications of 90–110 %.

Out-of-specification (OOS) and out-of-control (OOC) values require attention and sometimes an "investigation" is needed. These values are not always caused by

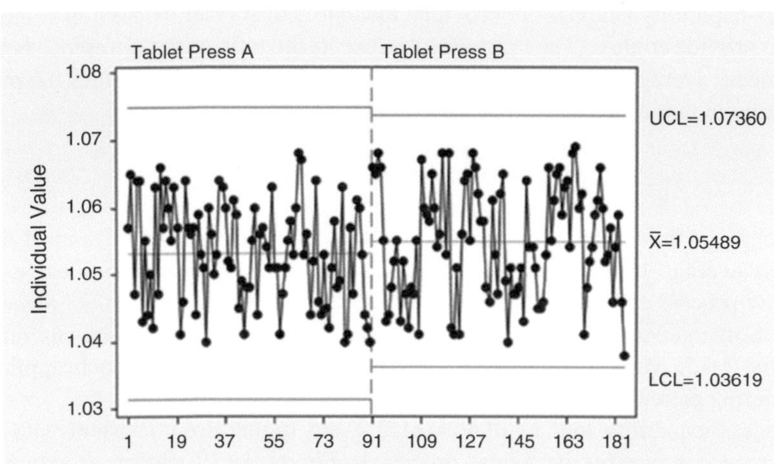

Fig. 18.15 Control chart showing batch tablet weight produced using two presses (*A* and *B*). UCL is upper control limit and LCL is lower control limit

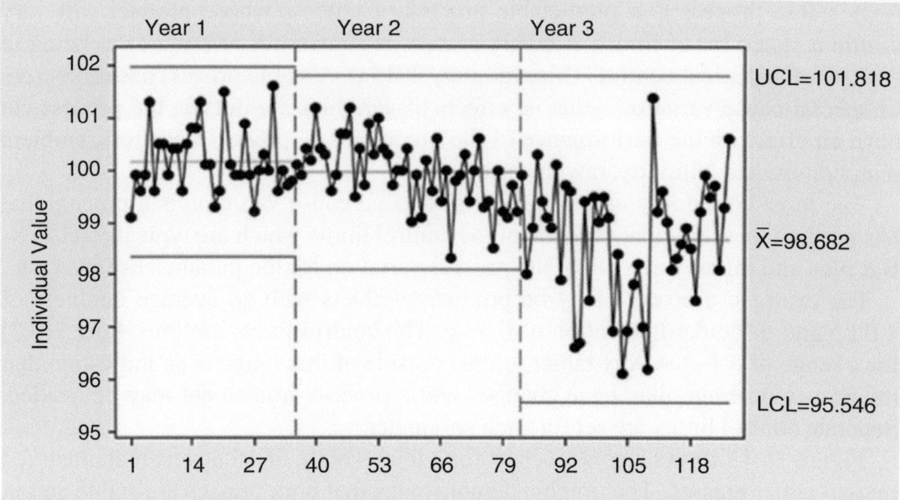

Fig. 18.16 Control chart showing assay values of batches produced over 3 years. UCL is upper control limit and LCL is lower control limit

manufacturing problems, and may be caused by sampling errors, testing errors, or human administration errors such as recording or data keying. The causes of OOS and OOC measurements should be carefully considered when interpreting the OOC and OOS values and deciding on appropriate action.

Process-Capability Analysis A process-capability analysis is conducted to determine the ability of the process to meet product specifications. The details of process-capability analysis are provided in Chap. 20. A brief discussion is included here to provide context. The Ppk value represents the ratio of the difference between the process average and the nearest specification divided by three times the process standard deviation (SD).

$$\mathbf{Ppk = Min\,(A,\ B)\ /\ 3SD}$$

Where **A** is the upper specification minus the process average and **B** is the process average minus the lower specification. Two main statistics are used to measure process capability: percent of the measurements OOS and the process Ppk value. Some general guidelines for interpretation of the Ppk value are summarized in Table 18.5. More specific interpretation may be created for each application considering patient and process risk levels.

Process capability indices of Ppk of 2.0 and higher are consistent with high-performance processes or robust processes. Figure 18.17 shows an example of process capability for tablet weight. In this case, the Ppk value is 1.91 (an excellent category) which is based on the distribution of tablet weights being a considerable distance from the lower and upper specifications for tablet weight.

Table 18.5 Summary of process performance index (Ppk) interpretation

Rating	Capability index
Excellent	More than 1.50
Good	1.33–1.50
Acceptable	1.00–1.33
Poor	Less than 1.00

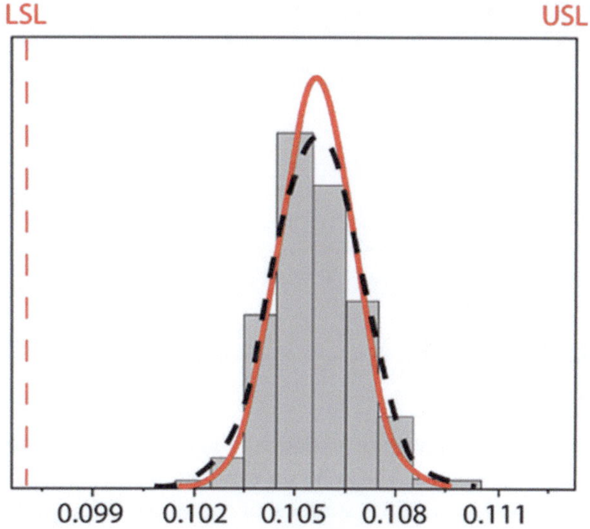

Fig. 18.17 Process capability for a tablet weight of Ppk = 1.91. LSL is lower specification limit and USL is upper specification limit

When a process is robust, small process upsets will not create OOS product. Accordingly, small OOC signals do not result in OSS product. A process is said to be "robust" if its performance is not significantly uninfluenced by variations in process inputs (e.g., raw material lot), process variables (e.g., press force and speed), and environmental variables (e.g., ambient temperature and humidity).

Analyzing Process Variation Another way to assess process stability is to study the variation in process performance that is caused by potential special-cause variation (e.g., tablet presses, raw-material lots, and process operating teams). Analysis of variance (ANOVA) enables one to identify variables that can increase variation in tablet parameters and that may produce OOS product.

The boxplot in Fig. 18.18 shows the distribution of tablet hardness values for a batch of tablets produced by two different tablet presses (X and Y). Press X has a wider hardness distribution than Press Y, yet none of the hardness values are outside the hardness specification of 1–6 kp. With this data in mind, the process operators can determine whether to make process adjustments.

After statistical significance of a comparison (e.g., average of Tablet Press X versus average of Tablet Press Y) is established, the practical significance of the difference in average values must be considered. This assessment is frequently carried out by expressing the observed difference in average values as a percentage of the overall process average. Subject-matter expertise is used to evaluate the practical importance of the observed percent difference.

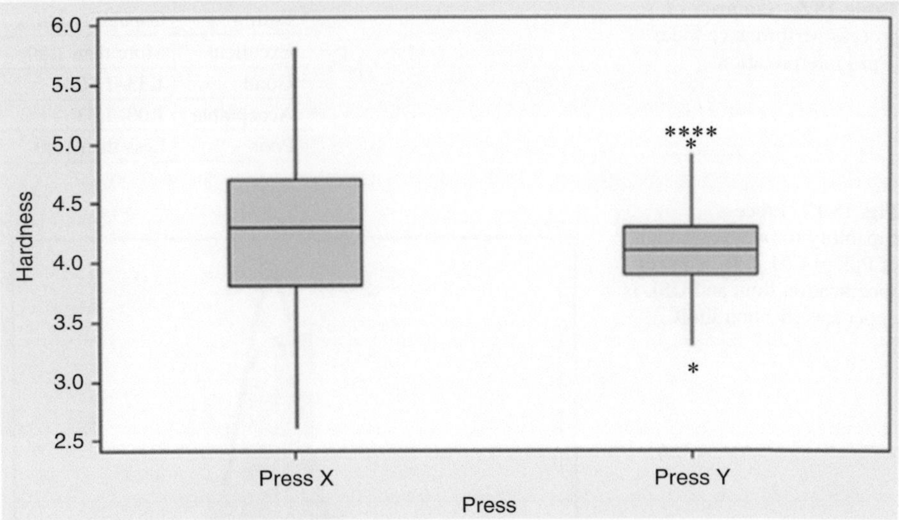

Fig. 18.18 Boxplots showing tablet hardness for two tablet presses (X and Y)

Nested analysis of variance is another form of ANOVA used to assess process stability. Nested ANOVA can estimate the portion of the total variation in the data attributed to various sources of variation (Montgomery 2013a, b). Typically, the larger the percent of the total variation attributed to a source of variation, the more important is the source of variation. Low amounts (<30 %) of long-term variation as determined by a nested ANOVA indicate a stable process (Snee and Hoerl 2012).

18.5.3 Creating Process Monitoring Systems

There are important considerations to take into account when designing, implementing, and operating process monitoring systems. The first of these is process understanding—that is, a deep knowledge of the variables that drive the process, which enables the accurate prediction of process performance. Effective application of QbD will result in better process understanding.

As mentioned previously, it is crucial to understand the sources and magnitudes of measurement variation, in particular the repeatability and reducibility of the measurement of the process parameters. Gage R&R studies are an effective method for measuring the repeatability and reproducibility of the measurement methods used (Montgomery 2013a). Ruggedness studies are effective for determining method robustness for measuring typical variations that occur during the routine use of the method (Schweitzer et al. 2010).

A systematic method is necessary to keep track of special-cause variation and to determine whether a systemic problem exists. This information can then be used to improve the process.

Two problems often observed related to monitoring systems include data not being analyzed routinely, and not taking action when significant sources of variation are identified. Regular management review and accountability can address both problems.

When a systematic approach is used, including regular review and action, the result is effective process monitoring. More importantly, high-quality pharmaceuticals are provided to the patient.

18.6 Process and Product Robustness

Process and product robustness are critical to the performance of the process over time. Design space should incorporate the results of robustness experiments. In general terms, *robust* means that the performance of the entity is not affected by the uncontrolled variation it encounters. Accordingly, a product is robust if its performance is not affected by uncontrolled variations in raw materials, manufacture, distribution, use, and disposal (PQRI 2006). Examples of robust products are user-friendly computers and software, pharmaceuticals that have no side effects regardless of how or when they are administered and medical instruments for home use. In all three of these instances it is desirable to have the product robust to the inexperience of the user.

A process is robust if its performance is not affected by uncontrolled variation in process inputs, process variables, and environmental variables. Robust processes are those processes that perform well when faced by large variations in input variables such as raw-material characteristics and variation due to environmental variables such as differences in ambient conditions, operating teams and equipment. It is also important that the process be insensitive to variation in the levels of the critical process parameters (CPP) such as equipment speed, material flow rates and process temperature.

Robustness studies can be carried out at various times during the development and operation of the process. Some example are summarized in Table 18.6.

PQRI (2006) describes a robustness study associated with the development of a tablet press. The effects of two variables are evaluated; compression pressure and press speed. The goal is to maximize the average tablet dissolution and minimize the variation in tablet dissolution at a fixed combination of compression pressure and press speed. The results for the 10-run central composite design (one point was replicated) are shown in Table 18.7. The average and standard deviation is based on 20 tablets collected at each point in the design. In Fig. 18.19 we see that desired settings are found to be in the area of Compression pressure of 250 and a press Speed of approximately 150–180. These settings are based on a requirement of a relative standard deviation (RSD = 100(standard deviation/average)) less than 4 %.

Table 18.6 Opportunities for using process robustness studies

Stage of development and operation	Robustness study
Process development	Assessment of the robustness of process to variations in control variables (Xs)
Manufacturing process operation	• Assessing robustness of process to minor variations in control variables (Xs) in the region of the process control point (target) • Can be done for a new process or an existing process
Environmental variable robustness	• Assessing robustness of the process to variables outside the process such as: ambient conditions, process inputs and use of product by customer: business or end user
Test method robustness	• Evaluate effects of small variations in how the method is used.

Table 18.7 Tablet press robustness study to identify compression pressure and press speed that will maximize tablet dissolution and minimize tablet dissolution variation

Run order	Compression pressure	Press speed	Dissolution (AVG)	Dissolution (StdDev)	Dissolution RSD (%)
1	350	160	83.12	2.14	2.57
2	150	160	81.54	2.40	2.94
3	250	280	96.05	3.73	3.88
4	150	260	80.38	6.18	7.68
5	390	210	69.32	6.08	8.77
6	250	140	94.81	1.14	1.20
7	250	210	96.27	3.59	3.73
8	250	210	94.27	6.37	6.76
9	110	210	70.76	4.03	5.70
10	350	260	83.71	7.10	8.48

This is somewhat unique situation in that average dissolution is affected only by compression pressure while dissolution variation (measured by standard deviation and relative standard deviation) was affected only by Press Speed.

Humphries et al. (1979) describe a process development study, the goal of which is to create an automated procedure for measuring ammonia in body fluids. The method sensitivity was not only maximized but it was also found that the manufacturing process was robust to the control variables. A 15-run face-centered-cube design was used to study the three critical process parameters: pH, buffer molarity and enzyme concentration. The response surface analysis showed that buffer molarity had no effect over the range studied in the experiment.

The sensitivity of the method was maximized within the experimental region and the area around the maximum was found to be very flat. This flatness showed that the sizeable variations in pH and enzyme concentration had little effect on the sensitivity of the method indication that variations in pH and enzyme concentration could be tolerated during the manufacturing of the reagents associated with the method.

Kelly et al. (1997) describe a robustness study of an existing fermentation purification process. The effects of small variations in five manufacturing process

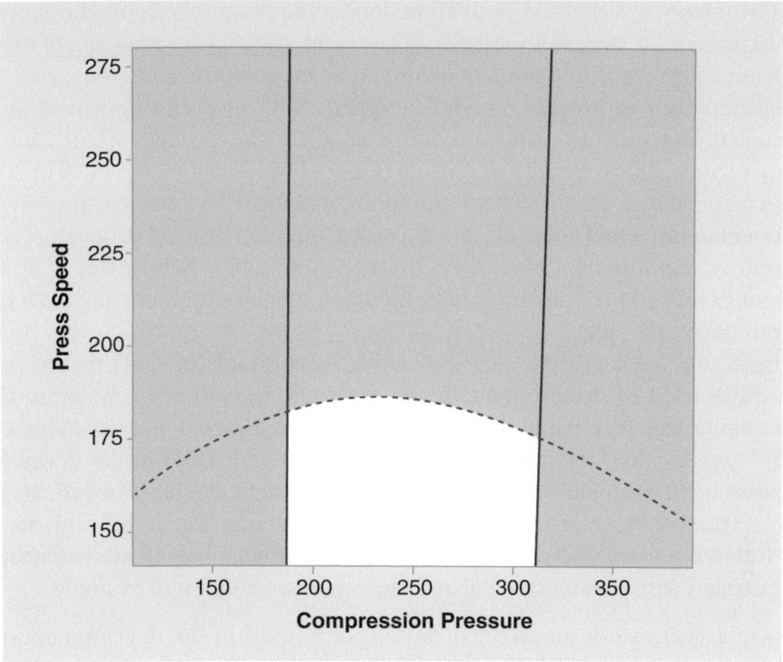

Fig. 18.19 Tablet press robustness study—contour plot shows that desired settings are compression pressure of 250 and press speed of 150–180

variables on the product recovery (%) and purity (%) were studied using a 16-run half-fraction of a five- factor factorial design. It was concluded that, over the ranges of the variables studied, the process was robust with respect to purity. Recovery was a different story with tighter control needed for three variables in order for the process to have the required performance.

An example of test method robustness is discussed in the following section. Robustness studies involving environmental variables are discussed by Box, Hunter and Hunter (2005).

18.7 QbD in Test Method Development and Validation

A characteristic of good science is good data. Quality data are arguably more important today than ever before. Data are used to develop products and processes, control our manufacturing processes (Snee 2010) and improve products and processes when needed. Quality data also reduces the risk of poor process performance and defective pharmaceuticals reaching patients.

Unfortunately test method evaluation appears to be an overlooked opportunity. In my experience a large number of measurement systems are inadequate resulting in poor decisions regarding product quality, process performance, costs, etc.

Measurement is a process that is developed, controlled and improved just like a manufacturing process. Indeed, quality data are the product of measurement processes (Snee 2005). Quality by Design (QbD), introduced by the FDA in 2005, is focused on the development, control and improvement of processes. Data are central to QbD and in turn QbD concepts, methods and tools can be used to develop, control and improve measurement processes (Borman et al. 2007; Schweitzer et al. 2010) As a result QbD and test methods have a complementary relationship; each can be used to improve the other.

This section discusses the concepts, methods and tools of QbD that have been successfully used to design, control and improve measurement systems. Earlier it was mentioned that using QbD in product and process development begins with the Quality Product Target Profile. Similarly using QbD in the development of measurement systems begins with the development of the "Analytical Target Profile". The specific approaches used are summarized in Table 18.8 and discussed in the following paragraphs. The concepts and methods involved are be introduced and illustrated with pharmaceutical and biotech case studies and examples.

Design of Experiments an effective QbD tool is used in the development of test methods to create the operability region for the method by first running a screening design to test the effects of various candidate test method variables. There are typically a large number of variables tested in the screening design to reduce the risk of missing any important variables. The variables found to have the largest effects (both positive and negative) are studied in a subsequent optimization experiment, the output of which is operating window for the method which serves the same function as the Design Space for a product or process. As a result we refer to this as the "Test Method Design Space".

Table 18.8 Quality by design methods for creating and improving test methods

Methods and tools	Analysis purpose
QbD approach	Speed up method development and reduce risk
Design of experiments using screening and optimization experiments	Method development including creation of test design space
Gage repeatability and reproducibility studies	Improve measurement quality
Method robustness studies	Create methods robust to small variations in how the method is used
Blind control sample	Continued verification of method repeatability and reproducibility over time
Process variation studies	Assess process variation to determine the relative contributions of the manufacturing process, sampling procedure and test method to the observed process variation

In a recent test method development project, 11 variables were studied in 24 runs using a Plackett-Burman screening design. The four variables with the largest effects were evaluated further in a 28-run optimization experiment producing the design space for the method. The next step in the development was to assess the effects of raw material variations.

Test Method Repeatability and Reproducibility is an important assessment once the method has been developed initially. Thus is done with using a Gage Repeatability and Reproducibility study referred to as a Gage R&R Study.

In the study 5–10 samples are evaluated by 2–4 analysts using 2–4 repeat tests sometimes involving 2–4 test instruments. Output from such a study produces quantitative measures of repeatability, reproducibility and measurement resolution. These statistics are then used to evaluate the value of the method to be used for product release and process improvement. The variance estimates obtained are also often used to design sampling plans to monitor the performance of the process going forward.

Test Method Ruggedness Sometimes we find that as a test method is used the observed variation in the test results becomes too large. What do I do now you ask? One possibility is to evaluate the measurement process/procedure for ruggedness (ASTM 1989; Wernimont 1985). Measurement method is "rugged," if it is *immune* to modest (and inevitable) departures from the conditions specified in the method (Youden 1961).

Ruggedness (sometimes called robustness) tests study the effects of small variations in the how the method is used. There are other sources of variation in a measurement method in addition to instruments and analysts which are typically the subject of Gage R&R studies. Such variables include raw material sources and method variables such as time and temperature. Ruggedness can be evaluated using two-level fractional-factorial designs including Plackett-Burman designs (Box et al. 2005; Montgomery 2013b).

A test method is said to be rugged if none of the variables studied have a significant effect. When significant effects are found a common fix is to rewrite the SOP to restrict the variation in the variables to a range over which the variable will not have a great effect on the performance of the test method.

Lewis et al. (1999) conducted a robustness study of a dissolution test method to determine the sensitivity of the dissolution measurement procedure to small changes in its execution. A large batch of tablets was used to assure uniformity of dissolution. Eight factors involved in the test procedure (five method variables, filter position, instrument and analyst) were varied over a small range using a 12-run Plackett-Burman design.

The statistical analysis of the data showed that none of the variables had a significant effect on the dissolution measurement. It was concluded that the test method was robust. The observed method dissolution time had a standard deviation is 1.9 min and a relative standard deviation of 10.7 %.

This example illustrates the flexibility of the two-level designs (Plackett-Burman and fractional-factorial) in conducting test method robustness studies. The eight variable studies were a mixture of both qualitative variables (test method variables) and qualitative variables (filter position, instrument and analyst). Studying a combination of qualitative and quantitative variables is a common occurrence in test method robustness studies.

Process Variation Studies Sometimes when the process variation is perceived to be too high it is not uncommon to think that the measurement is the root cause. Sometimes this is the case but often it is not. In such situations there are typically three source of variation that may contribute to the problem: the manufacturing process and the sampling process as well as the test method (Snee 1983).

In two instances that I'm aware of the sampling method was the issue. In one case the variation was too high because the sampling procedure was not followed. When the correct method was used the sampling variance dropped by 30 %. In another case each batch was sampled 3 times. When the process variance study was run sampling contributed only 6 % of the total variance. The Standard Operating Procedures were changed immediately to reduce the samples to two per batch; thereby cutting sampling and testing costs by one-third. A study was also initiated to see if one sample per batch would be sufficient.

Test Method Continued Verification The FDA Process Validation Guidance calls for Continued Process Verification which includes the test methods. An effective way to assess the long-term stability of a test method is to periodically submit "blind control" samples (also referred to as reference samples) from a common source for analysis along with routine production samples in a way that the analyst cannot determine the difference between the production samples and the control samples. Nunnally and McConnell (2007) conclude ". . . there is no better way to understand the true variability of the analytical method".

The control samples are typically tested 2–3 times (depending on the test method) at a given point in time. The sample averages are plotted on a control chart to evaluate the stability (reproducibility) of the method. The standard deviations of the repeat tests done on the samples are plotted on a control chart to assess the stability of the repeatability of the test method.

Another useful analysis to perform is an analysis of variance of the control sample data and compute the percent long-term variation which measures the stability of the test method over time. Long term variation variance components < 30 % are generally considered good with larger valves suggesting the method may be having reproducibility issues (Snee and Hoerl 2012).

It is concluded that using QbD concepts, methods and tools improves test method performance and reduces the risk of poor manufacturing process performance and defective pharmaceuticals reaching patients. Risk is reduced as the accuracy, repeatability and reproducibility increases. Reduced variation is a critical characteristic of good data quality as reduced variation results in reduced risk.

Screening experiments followed by optimization studies is an effective way to design effective test methods. A measurement process can be controlled using control samples and control chart and analysis of variance techniques. Measurement quality can be improved using Gage Repeatability and Reproducibility studies. Robust measurement systems can be created using statistical design of experiments. Product variation studies that separate sampling and process variation from test method variation is an effective way to determine the root cause of process variation problems.

18.8 Getting Started

Using QbD creates a paradigm shift from the usual approach to development (Snee et al. 2008b). QbD thus represents a cultural change that must be addressed with change-management techniques such as the eight-stage change model developed by Kotter (1996) and associated change-management tools (Kotter 1996; Holman et al. 2007). The importance of change management is not recognized by many organizations that deploy QbD and other improvement techniques. As a result, the promise of the associated change initiative is seldom realized even in minor terms.

The first step is to recognize that people, when first exposed to QbD will have legitimate concerns. Typically, three types of barriers are encountered: Technical—It won't work here; Financial—We can't afford it; and Psychological—It's too painful to change; the organization is not ready for this.

These legitimate concerns must be acknowledged and addressed. It is helpful to enable people to recognize the value of QbD which includes: Faster speed to market, reduced manufacturing costs, reduced regulatory burden and better allocation of resources.

People must also understand the trends in the world-wide business of Pharma and Biotech such as: competition is tougher than ever before, products are more complex, and the use of QbD is expanding throughout the industry. Organizations are recognizing the value of a focus on quality.

One of the best approaches to dealing with the concerns is to design the deployment of QbD in such a way that it produces important results quickly. Nothing succeeds like success. Kotter's change model tells us that we need to "generate short-term wins" (see Step 6 of Kotter's model). Short-term wins demonstrate two things: (1) QbD works in our environment and (2) QbD can produce significant results quickly.

The eight stages in Kotter's change model and some illustrative suggestions for implementation are shown in Table 18.9. Careful thought to introducing QbD to an organization can increase the speed and effectiveness of the deployment. First it is better to "start small and think big". Introduce a few QbD building blocks at the beginning and then add others as the organization seems ready to consider still other ideas.

Table 18.9 Kotter's eight stages of successful change

Stage	Activity	Illustrative example for XYZ pharma
1	Establish a sense of urgency	QbD is essential for XYZ pharma to get ahead and stay ahead
2	Create a guiding coalition	Champions for QbD are identified and working together
3	Develop a vision and strategy	What QbD will look like at XYZ pharma and what we will do to be successful
4	Communicate the change vision	Plan using variety of media
5	Empower employees for broad based action	Aggressive results oriented QbD skill development at all organizational levels
6	Generate short-term wins	Complete successful projects in 2–6 months
7	Consolidate gains and produce more change	Start next wave of projects and create annual plan
8	Anchor new approaches in the culture	Champions, skilled practitioners and infrastructure are developed

A good place to start is the use of the "design space" followed by the "continued process verification system". In every case identify some "demonstration projects" that will yield important results quickly. Provide training to build needed skills and create the supporting infrastructure including: Strategy, plans and goals; definitions of roles and responsibilities; and communication processes.

Of critical importance is a system of periodic management reviews of the QbD initiative (what's working and what isn't) and measurement of results. It is important to balance development speed and risk with getting the right data at the right time in the right amount.

18.9 New Directions in Statistical Research on QbD

The Bayesian approach to design space construction proposed by Peterson (2008) and discussed in Peterson and Lief (2010) is a major opportunity for additional statistical research. Design space construction is very important to the fast and effective development of pharmaceutical and biotech products and processes. Emphasis can be effectively placed on the development of roadmaps for using the Bayesian approach. This is particularly important for pharmaceutical scientists who are not familiar with the various statistical methods and software approached involved. The concepts, methods and tools of Statistical Engineering should be helpful in the development of such an approach (Hoerl and Snee 2010).

Another important topic for further research is the question of "scale up"—going from lab scale to commercial scale. Statistical model results typically represent lab scale. Projecting the results of statistical models to large scale commercial processes can be a risky process due to the inherent risks associated with the extrapolation of empirical models. Scale up is often based in large part on engineering fundamentals and process understanding. Emphasis can be effectively placed on the development of roadmaps to successfully complete scale up including data required and issues and pitfalls such work should be prepared to address.

18.10 Conclusion

As shown in other industries and more recently in the pharmaceutical industry, QbD is an effective method for developing new products and processes. QbD enables effective technology transfer and the optimization and improvement of existing processes. QbD works because it fosters process understanding that is fundamental to the creation of the design and control spaces and to sustain performance. The design space is critical to the success of QbD because it produces the following: performance on target and within specification at minimal cost with fewer defective batches and deviations; greater flexibility in process operation; and the ability to optimize manufacturing operations without facing additional regulatory filings or scrutiny.

This process understanding enables the reduction of process variation and the creation of process-control systems that are based on sound science and quality risk-management systems.

QbD works beyond development and manufacturing to include functions such as technology transfer, change control, deviation reduction, and analytical methods development and improvement. All work is a process, and QbD is an effective method for improving processes. The use of QbD will no doubt broaden in the future, and its application in the life sciences is almost without bounds.

References

Alaedini P, Snee RD, Hagen BW (2007) Technology transfer by design—using Lean Six Sigma to improve the process. Contract Pharm 9(4):4–9

ASTM (1989) Standard Guide for Conducting Ruggedness Tests, E1169. American Society for Testing and Materials, Philadelphia, PA

Berridge JC (2009) PQLI—what is it? Pharm Eng 29:36–39

Borman P et al (2007) Application of quality by design to analytical methods. Pharm Technol 31:142–152

Box GEP, Wilson KB (1951) On the experimental attainment of optimum conditions. J R Stat Soc Series B 13:1–4

Box GEP, Hunter JS, Hunter WG (2005) Statistics for experimenters, 2nd edn. Wiley, New York, pp 345–353

Cornell JA (2002) Experiments with mixtures: designs, models and analysis of mixture data, 3rd edn. Wiley, New York

Covey SR (1989) The 7 habits of highly effective people. Simon and Schuster, New York

Derringer G, Suich R (1980) Simultaneous optimization of several response variables. J Qual Technol 12(214–219):251–253
FDA (2004) PAT—a framework for pharmaceutical development, manufacturing and quality assurance, Rockville, MD
FDA (2011) Guidance for Industry: Process Validation: General Principles and Practices, US Food and Drug Administration, Rockville, MD
Fisher RA (1935) The design of experiments. Oliver and Boyd, London
Hoerl RW, Snee RD (2010) Statistical thinking and methods in quality improvement: a look to the future. Qual Eng 22(3):119–139
Hoerl RW, Snee RD (2012) Statistical thinking: improving business performance, 2nd edn. Wiley, Hoboken
Holman P, Devane T, Cady S (2007) The change handbook—the definitive resource on today's best methods for engaging whole systems. Berrett-Koehler Publishers, San Francisco
Hulbert MH et al (2008) Risk management in pharmaceutical product development—white paper prepared by the PhRMA Drug Product Technology Group. J Pharm Innov 3(1):227–248
Humphries BA, Snee RD, Melnychuk M, Donegan EJ (1979) Automated enzymatic assay for plasma ammonia. Clin Chem 25:26–30
Hussain AS (2009) Professor Joseph B. Schwartz's Contributions to the FDA's PAT and QbD Initiatives, Presented at the Joseph B. Schwartz Memorial Conference, University of the Sciences, Philadelphia, PA, March 2009
ICH (2005) Q8(RI) ICH Harmonized Tripartite Guideline: Pharmaceutical Development, Geneva, Switzerland
Juran JM, DeFeo JA (2010) Quality control handbook, 6th edn. McGraw-Hill, New York
Kelly B, Jennings P, Wright R, Briasco C (1997) Demonstrating process robustness for chromatographic purification of a recombinant protein. BioPharm Int 10:36–47
Kotter JP (1996) Leading change. Harvard Business School Press, Boston
Lewis GA, Mathieu D, Phan-Tan-Luu R (1999) Pharmaceutical experimental design. Marcel–Dekker, New York
McCurdy V, am Ende MT, Busch FR, Mustakis J, Rose R, Berry MR (2010) Quality by design using integrated active pharmaceutical ingredient—drug product approach to development. Pharm Eng, 28–29
Montgomery DC (2013a) Introduction to statistical quality control, 7th edn. Wiley, New York
Montgomery DC (2013b) Design and analysis of experiments, 8th edn. Wiley, New York (Chapter 13)
Nunnally BK, McConnell JS (2007) Six sigma in the pharmaceutical industry: understanding, reducing, and controlling variation in pharmaceuticals and biologics. CRC, Boca Raton
Peterson JJ (2008) A Bayesian approach to the ICH Q8 definition of design space. J Biopharm Stat 18:959–975
Peterson JJ, Lief K (2010) The ICH Q8 definition of the overlapping means and Bayesian predictive approaches. Stat Biopharm Res 2(2):249–259
Pfeiffer CG (1988) Planning efficient and effective experiments. Mater Eng 35–39
Plackett RL, Burman JP (1946) The design of optimum multifactorial experiments. Biometrika 33(4):305–325
PQRI (2006) Product Quality Research Institute (PQRI) Robustness Workgroup, Process Robustness—A PQRI White Paper. Pharm Eng 26(6): 1–11
Schwartz JB, Flamholz JR, Press RH (1973) Computer optimization of pharmaceutical formulations I: General procedure: II Application in troubleshooting. J Pharm Sci 62:1165–1170
Schweitzer M, Pohl M, Hanna-Brown M, Nethercote P, Borman P, Hansen G, Smith K, Larew J (2010) Implications and opportunities of applying Qbd principles to analytical measurements. Pharm Tech 34:52–59
Snee RD (1983) Graphical analysis of process variation studies. J Quality Technology 15:76–88
Snee RD (2005) Are we making decisions in a fog? The measurement process must be continually measured, monitored and improved. Qual Prog 75–77

Snee RD (2006) Lean Six Sigma and outsourcing—don't outsource a process you don't understand. Contract Pharm 8(8):4–10
Snee RD (2007) Use DMAIC to make improvement part of how we work. Qual Prog 52–54
Snee RD (2009a) Quality by design: four years and three myths later. Pharm Process 26(1):14–16
Snee RD (2009b) Building a framework for quality by design. Pharm Technol 33(10), web exclusive, (October 2009)
Snee RD (2009c) "Raising Your Batting Average" remember the importance of strategy in experimentation. Qual Prog 64–68
Snee RD (2010) Crucial considerations in monitoring process performance and product quality. Pharm Technol 34:38–40
Snee RD, Gardner EC (2008) Putting it all together—continuous improvement is better than postponed perfection. Qual Prog 56–59
Snee RD, Hoerl RW (2003) Leading six sigma—a step by step guide. FT Press/Prentice Hall, New York
Snee RD, Hoerl RW (2005) Six sigma beyond the factory floor—deployment strategies for financial services, health care, and the rest of the real economy. Prentice Hall, New York
Snee RD, Hoerl RW (2012) Going on feel: monitor and improve process stability to make customers happy. Qual Prog 45:39–41
Snee RD, Piepel GF (2013) Assessing component effects in formulation systems. Qual Eng 25(1):46–53
Snee RD, Reilly WJ, Meyers CA (2009a) International technology transfer by design. Int Pharm Ind 1(1):4–10
Snee RD, Cini P, Kamm JJ, Meyers C (2009b) Quality by design—shortening the path to acceptance. Pharm Process 20–24
Wernimont G (1985) Evaluation of ruggedness of an analytical process. In: Spendley W (ed) Use of statistics to develop and evaluate analytical methods. AOAC, Arlington, pp 78–82
Yan L, Le-he M (2007) Optimization of fermentation conditions for P450 BM-3 monooxygenase production by hybrid design methodology. J Zhejian Univ Sci B 8(1):27–32
Youden J (1961) Systematic errors in physical constants. Phys Today 14(9):32–42

Ronald D. Snee, **Ph.D.** is founder and president of Snee Associates, a firm dedicated to the successful implementation of process and organizational improvement initiatives. He provides guidance to senior executives in their pursuit of improved business performance using Quality by Design, Lean Six Sigma, and other improvement approaches that produce bottom line results. He worked at the DuPont Company for 24 years prior to initiating his consulting career. While at DuPont he served in a number of positions including Statistical Consultant Manager and Manager of Clinical Statistics. He also serves as an Adjunct Professor in the Pharmaceutical programs at Temple University and at Rutgers University. Ron received his BA from Washington and Jefferson College and MS and PhD degrees from Rutgers University. He is an academician in the International Academy for Quality and a fellow of the American Society of Quality, the American Statistical Association, and the American Association for the Advancement of Science. He has been awarded ASQ's Shewhart and Grant Medals, and ASA's Deming Lecture Award and Dixon Statistical Consulting Excellence Award as well as more than 30 other awards and honors. He is a frequent speaker and has published five books and more than 255 papers in the fields of performance improvement, quality, management, and statistics. He can be reached at Ron@SneeAssociates.com.

Chapter 19
Process Validation in the Twenty-First Century

Helen Strickland and Stan Altan

Abstract Process validation is a continuous series of activities that are aligned to the product lifecycle; it is no longer a one-off activity. The process validation lifecycle consists of three major stages: (1) process design, (2) process qualification, and (3) continued process verification. Each stage corresponds to an increasing level of scientific understanding of the product and the manufacturing process. A systematic process validation plan integrates science, risk management, and statistics to collect and evaluate appropriate data throughout the product lifecycle. The process validation plan must fulfill the requirement to provide documented evidence of the manufacturing process' ability to consistently produce finished drug product that meets specifications in relation to identity, strength, quality and purity. The evidence is established through statistical tools that seek to identify, detect and control sources of variation that impact on the critical quality attributes of the finished drug product. Acceptance sampling, tolerance intervals, control charting and sample size justification through standard and Bayesian approaches are fundamental elements of a comprehensive lifecycle approach to process validation. This chapter provides an overview of regulatory and statistical aspects of process validation as set forth in the 2011 FDA guidance on process validation and other guidances with emphasis on Stage 2—Process Performance Qualification and Stage 3—Continuous Process Verification.

Keywords Statistical process management • cGMP • Acceptance sampling • Statistical process control • Statistical quality control • Quality by Design

---

H. Strickland (✉)
GSK, Zebulon, NC, USA
e-mail: helen.n.strickland@gsk.com

S. Altan
Janssen R&D LLC, Raritan, NJ, USA

L. Zhang (ed.), *Nonclinical Statistics for Pharmaceutical and Biotechnology Industries*, Statistics for Biology and Health, DOI 10.1007/978-3-319-23558-5_19

19.1 Introduction

The term "Process Validation" (PV) describes the set of Chemistry, Manufacturing and Controls (CMC) related scientific, engineering and statistical activities necessary for a sponsor company to demonstrate the ability to manufacture a drug product. It is the documented ability to manufacture drug product in conformance with quality requirements related to clinical outcomes and is comprised of a series of activities taking place over the lifecycle of the product and process. The Food and Drug Administration's (FDA) definition is "the collection and evaluation of data, from the process design stage through commercial production, which establishes scientific evidence that a process is capable of consistently delivering quality product". The European Medicines Agency (EMA) definition also states that "documented evidence that the process, operated within established parameters, can perform effectively and reproducibly to produce a medicinal product meeting its predetermined specifications and quality attributes" is required (EMA 2014).

The FDA 2011 Process Validation guidance, entitled "Process Validation: General Principles and Practices", established an overall lifecycle approach comprised of three stages. Each stage corresponds to an increasing level of scientific understanding of the product and the manufacturing process.

Stage 1, Process Design (PD), is carried out during the early development phase and consists largely of designed experiments (DoEs) relating critical process parameters (CPP) and formulation factors to important outcomes known as critical quality attributes (CQA). Its goal is to define the commercial manufacturing process and to develop the process control strategy. This is discussed in detail in Chap. 18. Stage 2, Process Qualification (PQ), contains two elements; (1). design of the facility and qualification of the equipment and utilities, and (2). process performance qualification (PPQ). Stage 2 PPQ, is focused on inherent variability. Stage 3, Continuous Process Verification (CPV), can be divided into two phases: (1). initial commercial CPV, and (2). routine commercial CPV. There is not an official reference to two CPV phases; however, the document recommends "continued monitoring and sampling of process parameters and quality attributes at the level established during the process qualification stage until sufficient data are available to generate significant variability estimates", which then "provide the basis for establishing levels and frequency of routine sampling and monitoring." The focus shifts in moving into CPV from inherent variability to total variability with an emphasis on batch to batch variability.

A meaningful and systematic process validation plan integrates science, risk management, and statistics to collect and evaluate appropriate data throughout the product lifecycle. The goals of the process validation plan are to fulfill the requirement to provide documented evidence of the ability to identify, detect and control sources of variation that impact on the manufacturing process' ability to consistently produce finished drug product in relation to identity, strength, quality and purity.

The remainder of this chapter provides an overview of regulatory and statistical aspects of process validation as set forth in these and other guidances with emphasis on Stages 2 and 3.

19.2 Regulatory and Statutory Requirements for Process Validation

Process validation is required by Good Manufacturing Practices (GMP) rules. These rules relate to the scientific and engineering controls required for drug manufacture. In the United States, GMPs for drugs (finished pharmaceuticals and components) is a legally enforceable requirement under statutory requirements of section 501(a)(2)(B) of the Federal Food, Drug, and Cosmetic Act (Title 21 U.S.Code 351(a)(2)(B)). The current Good Manufacturing Practices (cGMP) regulations in CFR Title 21 Section 211 for finished pharmaceuticals require the application of quality control Statistics.

The role of Statistics from a manufacturing perspective in process validation is simply the application of statistical process management tools involving appropriate collection and evaluation of the data relating to quality measures of the manufacture of pharmaceutical products. The 2011 FDA guidance contains the FDA's current thinking on how to achieve the GMP statutory and regulatory requirements regarding process validation. The ICH *Q8(R2) Pharmaceutical Development*, *Q9 Quality Risk Management*, and *Q10 Pharmaceutical Quality System* provide additional regulatory insight as to how this could be achieved.

The FDA regulations describing current good manufacturing practices for finished pharmaceuticals are provided in 21 CFR sections 210 and 211. The following is a list of some of the specific cGMP regulations that require "manufacturing processes be designed and controlled to assure that in-process materials and the finished product meet predetermined quality requirements and do so consistently and reliably" (FDA 2011):

- 21 CFR 211.160 (b) (3): Samples must represent the batch under analysis.
- 21 CFR 211.165 (c) and (d): The sampling plan must result in statistical confidence.
- 21 CFR 211.165 (c): The batch must meet its predetermined specifications.
- 21 CFR 211.165 (d): Acceptance criteria for the sampling and testing conducted by the quality control unit shall be adequate to assure that batches of drug products meet each appropriate specification and appropriate statistical quality control criteria as a condition for their approval and release. The statistical quality control criteria shall include appropriate acceptance levels and/or appropriate rejection levels.
- 21 CFR 211.166 (a) (1) Sample size and test intervals based on statistical criteria for each attribute examined to assure valid estimates of stability.

- 21 CFR 211.110 (b) In-process specifications must be consistent with drug product specifications. It shall be derived from previous acceptable process average and process variability estimates where possible, and determined by the application of suitable statistical procedures.
- 21 CFR 211.180 (e) Product quality data and manufacturing process must be periodically reviewed.

The EMA describes the principles and practices of product and process validation lifecycle in three documents. The first document ICH Q8(R2) Pharmaceutical Development provides guidance on the first stage of the product lifecycle and process design. The second document, "Guideline on process validation for finished products-information and data to be provided in regulatory submissions" (EMA 2014), provides guidance on the validation of the manufacturing process which is considered the second stage in the product lifecycle. The third document, Annex 15: Qualification and Validation (EudraLex 2015), provides guidance on the third stage of the product lifecycle. The EMA refers to the third stage as on-going process verification (OPV).

The stages described in the 2011 FDA guidance can be regarded as a subset of the ICH Q10 Product Lifecycle stages. The ICH Q10 guidance refers to a four stage product lifecycle as compared with the FDA Process Validation guidance which refers to a three stage PV lifecycle that "aligns process validation activities with a product lifecycle concept" (FDA 2011). The ICH Q10 product lifecycle stages are (1) Pharmaceutical Development, (2) Technology Transfer, (3) Commercial Manufacture and (4) Product Discontinuation compared with the previously mentioned FDA process validation lifecycle stages: (1) Process Design, (2) Process Qualification, (3) Continued Process Verification. The process validation lifecycle activities begin in the latter part of the pharmaceutical development phase and stop at the end of the commercial phase. Additional regulatory perspective on process validation is found in ICH Q11 Development and Manufacture of Drug Substance (ICH 2012). ICH Q12 Lifecycle Management is in writing now and will provide more detailed guidance on continued process verification.

19.2.1 Quality by Design Applied to Analytical Method Development

Quality by Design and risk management can also be applied to analytical method development using the same concepts applied to processes validation described in ICH Q8, Q9 and Q10. A clear discussion on these ideas can be found in a Stimuli paper published by the United States Pharmacopeia (USP 2014). The lifecycle approach begins with an ATP (Analytical Target Profile), the equivalent of the Target Product Profile (TPP) for processes. Critical quality attributes are expressed through a reportable value and acceptance criteria to meet a fit-for-purpose requirement. Accuracy and measurement uncertainty limits are understood as conferring validity to an analytical procedure. An understanding of the relative contributions of analytical measurement uncertainty, process variability and other sources of variability to the total variability observed in process validation studies is an important aspect of the overall demonstration of scientific knowledge regarding the product and manufacturing process.

There is further elaboration on this approach in the proposed USP General Chapter <1210> Statistical Tools for Method Validation, a statistical companion chapter to USP General Chapter <1225> (USP 2014). An FDA draft guidance "Analytical Procedures and Methods Validation for Drugs and Biologics" (FDA 2014) provides a regulatory perspective. It is expected that additional discussion and publications will be forthcoming as the lifecycle approach is applied to analytical method development.

19.3 The Process Validation Lifecycle Concept and the Quality Framework

The lifecycle approach to process validation requires the outcome and conclusions from each stage to directly influence the data collection and data evaluation activities in the subsequent stage. Representative data must be collected and evaluated to confirm that in-process and batch release testing plans provide a high level of statistical confidence that the process produces product that is suitable for patient use and is in a state of control. Statistically based demonstration tests including sample size and sample acceptance criteria must be established for passing through each of the process validation lifecycle stages.

Following QbD principles, the product requirements would be clearly stated in the Quality Target Product Profile (QTPP). As given in ICH Q8(R2), the requirements provided in the QTPP would correspond to "the quality characteristics" of the drug product and drug substance that would be needed "to ensure the desired quality, taking into account safety and efficacy of the drug product." A risk management strategy would be coupled with the data evaluation steps to detect undesirable variation and unplanned departures from expected outcomes of the process (ICH 2005). Important statistical tools that have relevance to risk management during process validation can be found in Chaps. 16–18, 20 and 21.

19.4 Process Validation Lifecycle Statistics

ICH Q8(R2), Q9 and Q10 guidances provide references to several statistical tools that are commonly used such as design of experiments, regression analysis, multivariate analysis, statistical process control, statistical quality control, control charts, process capability and process performance assessments. Chapter 18 Quality by Design covers the application of those statistical methods employed primarily during product development and process design. The remainder of this chapter covers the application of statistical methodologies useful for Process Performance Qualification (PPQ) and Continued Process Verification (CPV).

19.4.1 Statistical Process Control (SPC) and Statistical Quality Control (SQC)

The statistical methods applied during PPQ and CPV are Statistical Process Control (SPC) and Statistical Quality Control (SQC) techniques. SPC and SQC utilize similar statistical tools; however, distinct differences in the objectives exist between the two applications as discussed by the International Standards Organization (ISO, 11462–1:2001). SPC is the application of statistical techniques and/or statistical or stochastic control algorithms to analyze process inputs (independent variables) with the intent to: (1) increase knowledge about a process; (2) steer a process to behave in the desired way; (3) characterize variation of final-product parameters to guide improvements in process performance (ISO 11462–1:2001).

The basic tool for SPC is the control chart. Lagging indicator (product based) control charts rather than leading indicator (process based) control charts are most common within the pharmaceutical industry. Product based control is also referred to as "after the event statistical product control" in ISO/TR 18532:2009. The reason that lagging indicator control charts are used more frequently is that the technical linkage between process inputs and process parameters (i.e. input characteristics and CPPs) and product characteristics (i.e. CQAs) are generally not definitively identified through prior knowledge, scientific first principles or Design of Experiments (ISO 18532:2009). SQC is the application of statistical tools, primarily control charting and acceptance sampling methodologies, to analyze the quality characteristics of the process outputs (i.e. dependent variables) with the intent to demonstrate product quality stability and/or product quality acceptability.

19.4.2 Conformance to Specification

Product quality acceptability is most often referred to as "conformance to specification" in the pharmaceutical industry. Conformance to specification implies that the drug substance or drug product will meet the listed acceptance criteria when tested by an appropriately referenced analytical procedure from batch release to expiry (ICH Q6A 1999). For many product quality characteristics, the European Pharmacopoeia, United States Pharmacopoeia, and Japanese Pharmacopoeia define public quality standards (i.e., analytical procedure, sample size and acceptance criteria) that a drug product is expected to meet if tested during the registered shelf-life. Unless otherwise stated, the EMA and the Japanese Ministry of Health, Labour and Welfare (MHLW) accept the use of compendial test standards as batch release criteria. The USP states that "similarity to statistical procedures may seem to suggest an intent to make inference to some larger group of units, but in all cases, statements about whether the compendial standard is met apply only to the units tested" (USP/NF 2014).

US regulation 21 CFR 211.165 subchapters (c) and (d) indicate that batch release criteria must be a statistical sampling plan (sample size, sample location, sampling frequency, and acceptance criteria) that provides statistical confidence that the batch meets specifications, and "includes appropriate acceptance level and/or appropriate rejection levels". With some exceptions, the sampling plan included in the product specifications for most product quality characteristics was not selected to achieve a pre-specified confidence nor does it include the acceptable or rejectable quality levels. Companies would be advised to carry out appropriate risk analyses incorporating stability studies should they choose to use compendial tests for product release purposes.

The requirements for the application of suitable statistical procedures to demonstrate conformance to specification is not new, the way to achieve this requirement has evolved. Therefore, it has become the responsibility of the statistician to provide guidance on how to justify the sampling plan and the acceptance criteria at PPQ, CPV initial commercial release, and CPV routine commercial release. Sections 19.5 and 19.6 of this chapter as well as Chaps. 23 and 24 provide information on how to determine statistically based batch release criteria. The methods discussed in the aforementioned sections assume that the established product release specifications incorporate any necessary requirements as a result of time and storage related product performance changes. Statistical models discussed in Chap. 22 should be considered to characterize CQA product changes during shelf life for this purpose.

In addition to acceptance sampling, the International Standards Organization (ISO) recommends what is referred to as the "guarantee system" for demonstrating conformance to specifications (ISO/TR 18532:2009). Acceptance sampling by itself is the least efficient approach as a decision making procedure because it is completely determined by the information from the sample that was chosen. In contrast, the guarantee system for assuring conformance to specification relies on all of the knowledge that was established during process design linking CPPs to CQAs. It builds in the necessary assurance of continued suitability and capability of the process.

The ICH Quality Implementation Working Group (IWG) discuss four important data elements, similar to principles of the ISO "guarantee system", for assessing whether or not a batch may be released to market. These consist of: (1). reviewing regulatory compliance data, (2). system data related to environmental, facility, utilities, and equipment for the current batch, (3). product-related data based on the manufacturing process, and (4). product-related data from quality control (ICH 2011). ICH Quality IWG emphasize that the use of an enhanced control strategy would allow conformance to specification to be demonstrated through: (1). increased process monitoring of all system related data such that all deviations or atypical events are identified and corrected, (2). increased use of SPC for process inputs and CPPs, and (3). increased control charting of in-process and final drug quality characteristics (CQAs) with less reliance on SQC acceptance sampling data (ICH 2011). The statistician can contribute to achieving this by utilizing the methods and concepts discussed in Chaps. 18, 20 and 21.

19.4.3 Acceptance Sampling

Acceptance sampling plans are considered statistically based sampling plans or probability based sampling plans. ISO states that a statistically based sampling plan is "... based on experience with the product and the process, where producer and consumer risks can be statistically evaluated, provided that random sampling and a predefined set of rules for varying sample size and sampling frequency are used" (ISO 2859–10:2006). American Standards for Testing and Materials (ASTM) defines a probability based sampling plan as a "sampling plan which makes use of the theory of probability to combine a suitable procedure for selecting sample items with an appropriate procedure for summarizing the test results so that inferences may be drawn and risks calculated from the test results by the theory of probability" (ASTM E456–08). Details on how to construct acceptance sampling plans are discussed in Chap. 20.

Acceptance sampling plans for inspection of qualitative and quantitative process outputs and product quality characteristics are available through voluntary consensus organizations such as ISO, ASTM and ANSI/ASQ. The theory and concepts of acceptance sampling plans are well-established; however, they are not well understood or used consistently by the pharmaceutical industry for purposes for which they were developed. There is relatively little discussion in regulatory guidances on the application of acceptance sampling plans for batch release for example although recent discussions by FDA representatives have emphasized the need to justify sampling plans through Operating Characteristic Curve calculations (Iyer 2014). Acceptance sampling plans have been developed for both attributes (binary outcome) and variables (continuous measurement outcome). Note that a critical quality attribute (CQAs) can be either a variable or a discrete attribute.

Standard acceptance sampling plans are indexed on either the acceptance quality limit [1](AQL) or the limiting quality (LQ) value. AQL and LQ are indices of acceptance sampling inspection plans. AQL sampling plans are intended to be used for lot-by-lot inspection or continuing series of lots which are indexed on producer's risk quality (ASQ 2008). AQL sampling plans are best used when a process has been demonstrated to be in a state of control and the current inspection batch or lot has

[1]The initials AQL are also referred to as "Acceptable Quality Level" (see Chap. 20). The meaning of the initials AQL was changed in international standards to Acceptance Quality Limit to more accurately describe its function.

been deemed to be from a process in a state of control. This is the expectation at routine commercial release CPV stage. LQ sampling plans are intended to be used for isolated lots which are indexed on the consumer's risk quality with a consumer risk typically less than 10 % (ISO 2859-2:1985). LQ sampling plans should be used when the process has not been validated, has not been demonstrated to be in a state of control, or for a process that was in a state of control but has given a signal that it may no longer be in a state of control. This would be the case for PPQ and the initial commercial release CPV stage.

For any given AQL sampling plan, the producer's risk should correspond to the probability of acceptance at the value of the stated AQL for which the plan has been indexed. This point on the Operating Characteristic (OC) Curve is also known as the producer's risk point. A consumer's risk point can also be associated with an AQL indexed sampling plan if the OC Curve is provided. For any given LQ sampling plan, the consumer's risk should correspond to the probability of acceptance at the value of the stated LQ for which the plan has been indexed. This point on the OC Curve is also known as the consumer's risk point (ISO 3534-2:2006). A producer's risk point can also be associated with an LQ indexed sampling plan if the OC Curve is provided. Additional details on OC curves and their construction can be found in Chap. 20, Acceptance Sampling.

Acceptance sampling plans provided in the published standards are indexed on a relatively small set of AQLs and LQs (ISO/TR 8550-1:2007, ISO/TR 8550-2:2007). Customized acceptance sampling plans can be easily developed. Examples of different methods for developing customized acceptance sampling plans will be given in the PPQ and CPV data collection evaluation strategy sections. Unless otherwise stated, standard acceptance sampling by variables are applicable to characteristics whose probability distributions are normal, at least a good approximation to normal, or transformed to approximately normal.

19.4.4 Process Stability and Process Capability

Process stability as related to a "state of control" and process capability have statistical definitions. However, the ICH Guidelines provide the following non-statistical definitions:

State of Control: A condition in which the set of controls consistently provides assurance of continued process performance and product quality (ICH Q10 2008).

Capability of a Process: Ability of a process to realize a product that will fulfill the requirements of that product (ICH Q10 2008).

The ability to adequately assess statistical control and process capability/performance during the routine release stage of CPV is dependent upon the information that was obtained on the probability distributions of the process inputs, process parameters and process outputs (ISO/TR 18532:2009). As discussed in

Chap. 21, control charting is one of the primary tools for SPC. It is used to distinguish between inherent and special cause variation and to assess process stability. A stable process is one that is in a state of statistical control. Traditionally a process is said to be in a state of statistical control if the probability distribution of the characteristic of interest is constant over time (Woodall 2000). A more "generalized" definition of statistical control now extends to cases for which an underlying statistical model of the quality characteristic is stable over time (Woodall 2000) which could include autocorrelation and multiple variance components related to batch effects. The view espoused by Woodall is particularly important for the pharmaceutical industry where anticipated stable batch effects related to changes in starting materials and environmental conditions can form an independent variance component.

The term "process capability" is routinely thought of as just a capability index or a process performance metric. However, process capability indices (PCIs) and process performance metrics are derived from the output of process capability or process performance analyses. PCIs are used for determining the ability of a process to comply with specifications and they are used to estimate the percentage of results outside the specification limits. PCIs describe the tail behavior of the probability distribution of the characteristic of interest; therefore, the probability distribution (normal, non-normal, skewed, uniform, etc) selected to model the process characteristic is essential for computing the PCIs (ISO 22514–2: 2014).

The PCI is the ratio of the estimated process location and dispersion of the probability distribution selected to model the process to a reference specification interval (ISO 22514–1: 2009). Quite often, PCIs are computed without performing the necessary performance and capability analyses. One important contribution of a statistician is to ensure that the PCIs are not computed without conducting the appropriate process performance and process capability analyses. Chapter 20 provides instructions for conducting process performance and process capability analyses. In addition, voluntary consensus standards such as ISO 22514 Parts 1 to 8 and ASTM E2281-08a provide extensive detail for correctly performing the studies.

19.4.5 Statistical Intervals

Statistical intervals are useful throughout the process validation lifecycle. Most university Statistics courses focus on the use of statistical intervals related to location and variation parameters. Less attention is given to statistical intervals constructed to contain a proportion of a population or intervals for probability of an event.

A heightened interest in the application of the one-sided tolerance bounds has arisen in pharmaceutical applications for assessing quality and conformance to specification. The advantage of this interval is that it allows for the calculation of a well-defined quality statement in regard to consumer's risk and consumer's risk quality as well as the producer's risk and producer's risk quality. One-sided tolerance intervals have been used in acceptance sampling for the last 50 years. The statistician just starting to deal with acceptance sampling may not be aware of the wide-spread use of this interval. Several papers regarding the application of the one-sided tolerance bound for acceptance sampling and calculation of the tolerance factors have been published. Owen refers to the tolerance factors as constants for "one-sided sampling plan and tolerance limits" denoted by k (Owens 1964). Hahn and Meeker (1991) refer to it as the factor for calculating normally distributed one-sided 100 % (1-alpha) tolerance bounds. ISO refers to this tolerance factor as the "one-sided tolerance limit" factor for unknown variances denoted by k4 (ISO 16269–6: 2005). Novick et al. (2009) discuss the usage of the one-sided lower tolerance bound and the one-sided upper tolerance bound as a two-one sided tolerance interval. This interval is referred to as the two one-sided parametric tolerance interval test (PTI-TOST). As discussed in Chap. 23 it is used as a two-tier acceptance sampling plan where .the overall alpha was apportioned, according to the sample sizes in each tier, using the Pocock alpha spending function. The k-factors used in ISO 3951:2005 for the "Single sampling plans of Form k for normal inspection: s–method" correspond to a PTI-TOST for a single tier test. Krishnamoorthy and Mathew (2009) have provided a comprehensive discussion of the theoretical development of tolerance intervals and tolerance regions and computational algorithms with numerous practical uses and examples.

19.4.6 Miscellaneous Intervals

ASTM E2709 is a general application of ASTM E2810. It was referenced in the 2011 FDA Process Validation guidance Document as one of many possible statistical methods to be used for evaluating product quality throughout the process validation lifecycle. Bergum (1990) published a statistical methodology for constructing acceptance criteria that when applied to a random and representative sample, would allow at a pre-determined stated confidence level $100(1-\alpha)\%$, that there is at least a stated probability (P) that any future sample taken from the sampled population (i.e. batch) would pass the older UDU criteria given prior to USP 28-NF 23. For ease of implementation, Bergum wrote a software program in SAS known as "Content Uniformity and Dissolution Acceptance Limit" (CuDAL). CuDAL Version 2 is based on the UDU test prescribed in chapter <905> of the current USP 37-NF 32. In 2011, the CuDAL method was adopted as a voluntary consensus standard, ASTM E2810 "Standard Practice for Demonstrating Capability to Comply with the Test for Uniformity of Dosage Units".

19.4.7 General Sampling Considerations

19.4.7.1 Leading Indicators

Leading indicators (i.e. process based) that have been characterized and identified as part of process design should not require enhanced sampling; however, the data generated throughout the manufacturing of the product should be collected for the purpose of assessing the within batch profile of each CPP. This data would be reviewed at the end of PPQ, at the end of the initial commercial release and at appropriately determined frequency during routine commercial release CPV. The conclusions obtained from the statistical review of the routine commercial release should facilitate and minimize the effort required to perform annual product reviews. This data would be used to demonstrate process stability.

19.4.7.2 Lagging Indicators

The sampling, testing and monitoring of lagging indicators (quality attributes) for batches manufactured during PPQ, should be performed under the assumption that each batch is unique and has been manufactured in isolation. Therefore, the selection of sample size, sample location/stratum and sampling frequency is driven by the need to: (1). adequately characterize the probability distribution of each batch individually, (2). demonstrate within batch control and consistency for each batch, and (3). state with a high degree of confidence that any PPQ batch of undesirable quality would fail the PPQ acceptance criteria. The sampling scheme for each batch must allow for the identification and characterization of significant sources of within batch variation. For example, representative stratified random sampling should be used for a process with multiple process input streams or multiple process output streams.

The data collection and data evaluation plan for lagging indicators during initial commercial CPV, may require a continuation of enhanced sampling and testing so that "sufficient data is generated to enable precise and robust inter-batch and intra-batch variation estimates" (FDA 2011). The data obtained during PPQ and CPV will be used to determine the type of control chart to be used during the routine CPV stage for each lagging indicator.

19.5 Process Performance Qualification (Element 2 of Process Qualification Stage)

Process performance qualification (PPQ), as defined in the FDA's 2011 PV guidance, is the second element of Process Qualification (PQ). The goal of PPQ is to demonstrate that the commercial manufacturing process performs as expected upon implementation of the process control strategy. Completion of the first PQ element, facility design and qualification of utilities and equipment (PQ Element 2), is outside the scope of this chapter. However it can be stated that it precedes PPQ and should be documented and summarized in a report that clearly states the qualification criteria. Use of appropriate statistical methods can add value to the qualification of the utilities and equipment activities.

This section presents PPQ lifecycle objectives and data collection and evaluation activities. It assumes that the process design stage has been properly completed. This would include properly documented process control strategy steps. The focus is on the data collection and data evaluation options for PPQ, which are very much dependent not only on the specifics of the product but on the questions of interest. The primary question of interest has two parts: (1). how effective is the proposed process control strategy in allowing the process to operate as expected, and (2). does the output of the process conform to specifications? This two-part question corresponds to the assessment of process stability and product acceptability. It also relates to the data collection and data evaluation plans for CPPs linked to CQAs, process inputs linked to CQAs and intermediate process outputs linked to CQAs (final process outputs). PPQ will have a higher level of sampling, meaning additional testing, and greater scrutiny of process performance than would be typical of routine commercial production.

Process based inferences imply that SPC of leading indicators is utilized and appropriate technical linkages between process inputs (CPPs) and product characteristics (CQAs) have been identified through prior knowledge, scientific first principles, or DOE. Enhanced data collection and data evaluation for the CPPs during PPQ would be for retrospective monitoring purposes. Enhanced data collection and data evaluation for prospective control purposes during PPQ should not be performed as the intent is to show via monitoring that the control strategy used for the leading indicators works as intended.

If the technical linkages between CPPs and CQAs have not been identified as is the case for most legacy products, then enhanced data collection and evaluation of the CPPs should be performed. The intent would be to characterize the CPPs and establish control charts. This could be accomplished by following Little's recommendations for implementation of effective statistical process control (Little 2001).

If the control strategy incorporates product based lagging indicators, additional data collection and data evaluation activities are required during PPQ. The product based lagging indicators may be intermediate process outputs or final process outputs linked to final product quality characteristics. In either case these process outputs may be deemed CQAs, depending on the results of a quality risk assessment. The amount of enhanced sampling or monitoring will vary depending on what was conducted during PD. In-process outputs with control limits do not require additional control data collection and data evaluation for prospective control purposes during PPQ. However, data should be collected to perform additional monitoring with the intent to demonstrate that the lagging indicator control strategy works as intended. Only the in-process outputs that have limits that relate to finished product quality specifications, or finished product quality characteristics require an enhanced conformance to specification assessment. Intermediate process outputs that were not adequately studied during PD but could be used as potential product based lagging indicators would require additional characterization through control charts during PPQ.

Characterization of the lagging indicator involves determining the probability distribution of the process output and determining the sources of variability. The sampling plan should be developed using all relevant process knowledge available at the time of PPQ. The following example represents the data collection and data evaluation plan for a product quality characteristic (i.e. a lagging indicator) with the purpose of demonstrating conformance to specification. The distribution of the product quality characteristic was during PD and was demonstrated to be binomial.

This example illustrates how acceptance sampling inspection by attributes (qualitative) can be used during PPQ. The registered limit for appearance defects is 1.0 %. Based on internal company expectations, a consumer's risk quality of 0.65 % at a consumer's risk of 5 % was selected. To ensure with a high degree of confidence that the manufacturing process did not produce individual batches with appearance defects exceeding 0.65 %, an acceptance sampling plan indexed on the limiting quality for isolated lots was needed during PPQ. The manufacturer examined published plans from the various voluntary consensus standards such as the following:

- ISO 2859–1: Sampling procedures for inspection by attributes—Part 1: Sampling schemes indexed by acceptance quality limit (AQL) for lot-by-lot inspection,
- ISO 2859–2: Sampling procedures for inspection by attributes—Part 2: Sampling plans indexed by limiting quality (LQ) for isolated lot inspection
- ASTM E2334 Practice for Setting an Upper Confidence Bound For a Fraction or Number of Non-Conforming items, or a Rate of Occurrence for Non-conformities, Using Attribute Data, When There is a Zero Response in the Sampling.

Based on the acceptance sampling plans published in the voluntary consensus standards, the manufacturer determined the consumer's risk points (CRP) and the producer's risk points (PRP) of each potential acceptance sampling plan (see Table 19.1). Multiple options exist for determining CRPs, PRPs, and other points of interest for acceptance sampling plans. One option that is appropriate for single sampling plans is to compute the conservative exact 100(1-α)% upper confidence bound on the percent non-conforming (p) for a binomial population, where n is the sample size, x is the number of nonconforming units observed in the sample, and $F_{(1-\alpha;\, 2x+2,\, 2n-2x)}$ is the 100(1-α)th percentile of the F distribution with (2x + 2) and (2n − 2x) degrees of freedom. Then

$$p = \left(1 + \frac{n-x}{(x+1)\,F_{(1-\alpha;\, 2x+2,\, 2n-2x)}}\right)^{-1} * 100$$

is the equation yielding the upper confidence bound on p (Hahn and Meeker 1991) and is based on the Clopper–Pearson method.

Table 19.1 Consumer's and producer's risk quality points for potential PPQ acceptance sampling plans

Sampling plan Sample size Accept number	Consumer's risk quality Consumer's risk = 5 % Quality level corresponding to 5 % probability of acceptance, % defective	Producer's risk quality Producer's risk = 5 % Quality level corresponding to 95 % probability of acceptance, % defective
460,0	0.649	0.011
500,0	0.597	0.010
500,1	0.945	0.071
500,2	1.25	0.163
800,0	0.374	0.006
800,1	0.591	0.044
800,2	0.785	0.102
1250,0	0.239	0.004
1250,1	0.379	0.028
1250,2	0.503	0.065
1250,3	0.619	0.109

For other tablet products that are manufactured using a similar process, the estimated defect rate is 0.025 %. The manufacturer will implement for the first time leading indicator (process based) control charts and lagging indicator (product based) control charts to control the components of the process that affects the appearance defect rate.

The tablets were film-coated in a single drum operation; therefore it was deemed appropriate that the samples could be randomly selected at the end of the film coating unit operation. The PPQ acceptance criteria required 3 or less appearance defects in 1250 randomly selected finished product tablets as directed in a plan given in ISO 2859–1:1999 (see Table 19.1). It was determined that four consecutive batches would be used during PPQ. If the process control strategy performs as expected and the defect rate for the process remains at 0.025 %, then there is an approximately 99.96 % probability that any given manufactured batch would meet the (1250, 3) accept criteria. If the process control strategy performed as expected with the leading and lagging indicators associated with the generation of appearance provided significant evidence for process operating in a state of control then there is approximately 99.9 % probability that 4 consecutive batches would pass the (1250, 3) criterion. If the defect rate were still constant but increased to 0.10 % there is a 15 % chance that at least one of the batches would fail during PPQ. Keeping with the same assumptions, there is a 95 % probability that PPQ would fail if the appearance defect rate increased to 0.30 %.

The results of the four batches met the acceptance sampling plan criteria. For Batch 1, Batch 3 and Batch 4 no appearance defects were observed and only one appearance defect was observed in the 1250 inspected for Batch 2. The results from the monitoring of the leading process based indictors (CPPs) and the lagging product based (process output characteristics) demonstrated that the process control strategy performed as expected. Since all relevant data supported that the four batches were manufactured from a consistent and well controlled process, the appearance data was pooled to provide an estimate of the process defect rate and a 95 % one-sided upper confidence bound on the percent non-conforming of the process, 0.02 % and 0.095 %, respectively.

In practice, when multiple characteristics are being assessed from the same sample, for example the 1250 tablets mentioned previously, the acceptance rule would be adjusted to reflect the desired level of quality for each characteristic. For example, Table 19.1 shows 4 different quality levels for a sample size of 1250, and these can be referenced depending on the desired quality level for each characteristic of interest.

As a result of a successful PPQ, the manufacturer committed to enhanced data collection and evaluation on an additional 25 batches. The number of batches = 25 for enhanced data collection during stage 3A CPV was chosen to provide a high confidence of the batch to batch standard deviation estimate being within a reasonably close interval of the true value. As shown in Fig. 19.1 a sample size of 25 batches provides an (1-α)*100 % = 85 % confidence of yielding an estimate within 20 % of the true value subject to assumptions of normality and independence. Under the normality assumption, the confidence level given by 1-α is calculated by solving for α in the following probability statement for a desired interval of closeness to the true value (defined by r as a proportion of the true value) with degrees of freedom ν (n = number of batches = ν + 1) and s_n^2 is an unbiased estimator of σ^2:

$$\Pr\left\{1 - r < \sqrt{\frac{s_n^2}{\sigma^2}} < 1 + r\right\}$$

$$= 1 - \alpha \text{ where } 1 - r = \sqrt{\frac{\chi^2_{\nu,\,\alpha/2}}{\nu}} \text{ and } 1 + r = \sqrt{\frac{\chi^2_{\nu,\,1-\alpha/2}}{\nu}}$$

and $\chi^2_{\nu,\gamma}$ is the γth percentile value of the chi-square distribution with ν degrees of freedom.

Similar calculations can be made to reflect levels of risk and closeness to the true value tailored to the product and process as deemed appropriate.

The manufacturer continued enhanced monitoring of the leading and lagging indicators to confirm that the within batch behavior was similar to that observed during PPQ and to better characterize the between batch variation. The conformance to specification plan for the appearance defect, used a single acceptance sampling plan (800, 2) where there was greater than an 85 % probability of failure if the true defect rate exceeded 0.65 % and greater than a 95 % probability of acceptance if the defect rate was less than 0.10 %. If the true appearance defect was appropriately controlled with the leading and lagging indicators at less than 0.025 % then there

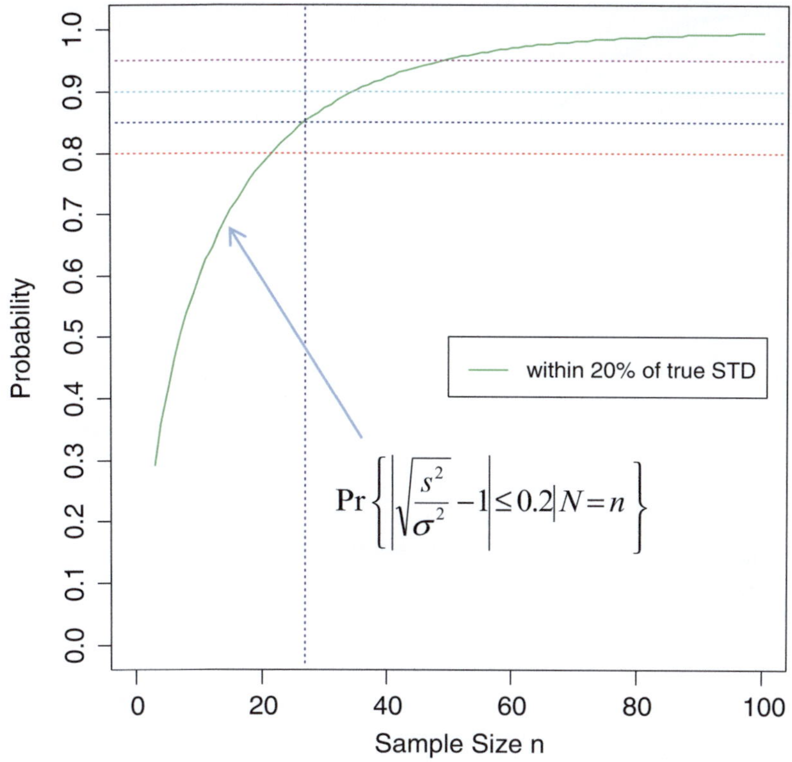

Fig. 19.1 Probability curve of [0.80 < Ratio of Sample SD to True SD < 1.20] vs sample size under assumption of normality

would be at least a 99 % probability of completing the CPV initial commercial release campaign of 25 batches without a conformance to specification failure.

If success is achieved during the initial CPV stage, the plan for routine commercial release would be to use a double sampling plan. For this double sampling plan, there would be greater than 95 % probability of acceptance for the first tier if the defect rate was less than 0.10 % with a consumer's risk of less than 10 % if the defect rate was greater than 0.65 %. This coincides with using a sampling plan for a continuing series of lots in the first tier and if failure occurs in the first tier, then switching to an isolated lot test with the consumer's risk quality of 0.65 % at consumer's risk of 10 %. An observed failure in the first tier during the CPV routine release stage given a successful initial commercial release CPV would be indicative of a batch defect rate significantly greater than 0.025 %.

The OC Curves for the PPQ, the CPV initial and CPV routine release acceptance sampling plans are provided in Fig. 19.2. After meeting the PPQ acceptance sampling criteria for all four batches, the acceptance sampling plan was changed to inspect 800 units and accept if no more than 2 nonconforming units were observed during the CPV initial commercial release stage. A further reduction in the

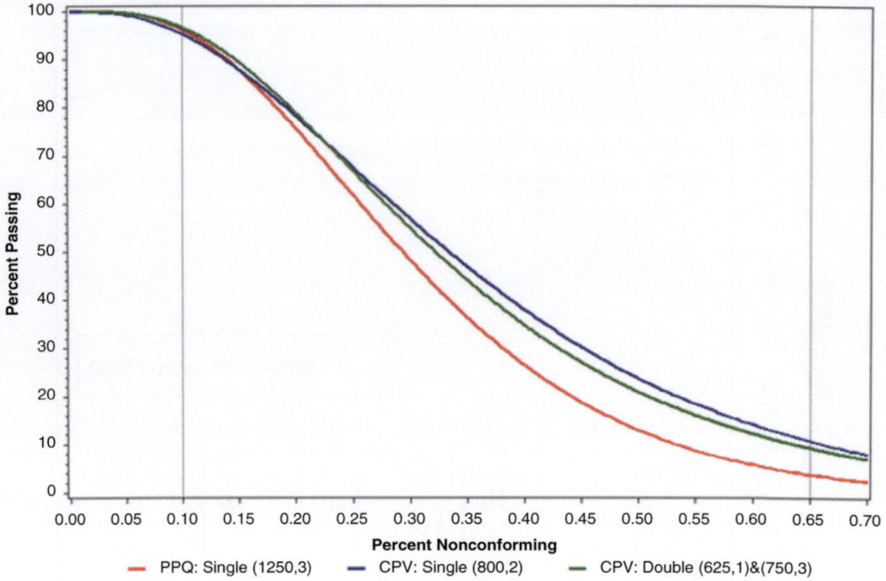

Fig. 19.2 OC curve for comparing PPQ and CPV routine release acceptance sampling plans

sample size would be allowed in the routine release stage of CPV, after successfully achieving the requirements for CPV initial commercial release.

19.5.1 Justification for Number of PPQ Batches

The FDA Process Validation guidance section on Stage 2: Process Performance Qualification states "The number of samples should be adequate to provide sufficient statistical confidence of quality both within a batch and between batches. The confidence level selected can be based on risk analysis as it relates to the particular attribute under examination." This makes explicit the need for a statistical justification of the number of batches chosen for PPQ. It also pertains to how many samples to take to characterize a given batch. These are essentially sample size estimation questions. The usual approach of specifying risks and confidence levels apply but must be refined to address the special circumstances of process validation. We understand risk in this context as the consequence of uncertainties associated with the lack of complete understanding of the product and the process. The remainder of this section will deal with the question of determining and justifying the number of PPQ batches. The question of number of samples to collect for a given batch was discussed in the previous section on acceptance sampling.

The International Society of Pharmaceutical Engineering (ISPE) issued a discussion paper describing three approaches to the question of number of PPQ batches

(Bryder et al. 2014). Each approach relates sample size to process and product knowledge and risk levels. These approaches do not represent a consensus position within the industry given how early the industry still is in embracing the principles enunciated in the 2011 guidance. They do form a reasonable starting point for further discussion and evolution. It is also important to emphasize that an assessment of between batch variability is generally not available until Stage 3 of the process validation lifecycle, so the question of number of batches to choose in PPQ is an especially thorny problem.

The ISPE Approach 1 is not based on a statistical argument but rather on judgment regarding the anticipated level of risk of process performance. It allows for 3, 5 or 10 lots corresponding to low, moderate or high levels of risk, respectively.

Approach 2 is based on a target process confidence and capability. It imposes a criterion that observed process capability Cpk ≥ 1. Lewis et al. (2014) showed through simulations that the proposed criterion would be difficult to achieve. For example, if the true Cpk $= 1$, then there is about a 50 % chance that the observed Cpk will exceed 1, regardless of the number of batches. It was also shown that to achieve an 80 % probability of meeting the criterion with batch sizes of 3,5 and 10, true underlying Cpk's of about 1.45, 1.35 and 1.25, respectively, would be required. How many pharmaceutical processes can reasonably be anticipated to operate at these levels of capability? Is 80 % adequate for probability of success from a company's control strategy perspective? If there are multiple CQAs, the proposed criterion is even more difficult to meet.

Approach 3 is based on expected coverage calculated through a property of order statistics. The expected probability that a future observation is between the observed minimum, $y_{(1)}$, and the observed maximum, $y_{(n)}$, is $\frac{n-1}{n+1}$ and n is the total sample size. If there are multiple CQAs, the probability of joint inclusion could be much lower.

None of the ISPE approaches rely on an explicit estimate of batch-to-batch variation or on prior knowledge of process performance expressed through a statistical statement. The latter criticism is particularly important in this context given that the PV guidance 2011 states "The decision to begin commercial distribution should be supported by data from commercial-scale batches. Data from laboratory and pilot studies can provide additional assurance that the commercial manufacturing process performs as expected." This clearly anticipates a Bayesian approach.

Lewis et al. (2014) proposed a probability of batch success approach to determining and justifying PPQ sample size. The advantages of the method are that it addresses the possibility of multiple CQAs in a natural way and allows a very flexible implementation in both statistical methodology (e.g. frequentist or Bayesian methodology) as well as risk-based approaches. The justification of the PPQ sample size is based on the probability, p, that a batch is within specifications. In this way it imposes no new process performance criteria.

The Bayesian approach with a conjugate prior begins by defining $x \sim$ Binomial (n, p), where x is the number of batches falling within specification and n is the number of batches manufactured. Let the prior distribution for p be a beta distribution with parameters α and β. In this formulation, α can be thought of as the number of prior successes, β as the number of prior failures, and $\alpha + \beta$ is the number of prior batches. The non-informative Jeffreys prior uses $\alpha = \beta = ½$.

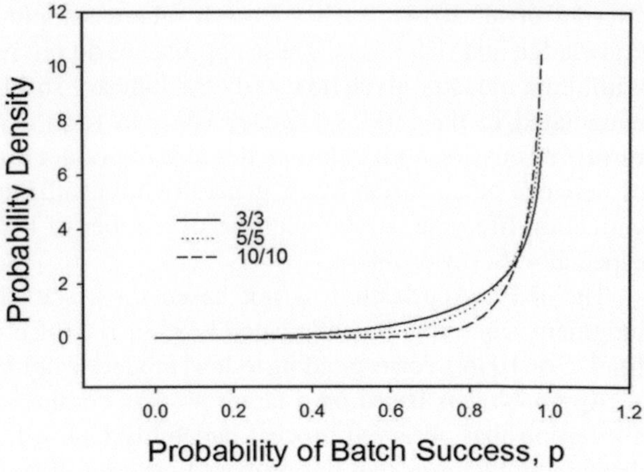

Fig. 19.3 Jeffreys prior and posteriors for 3/3, 5/5 and 10/10 Successes

The prior mean is $\alpha/(\alpha+\beta)$. Then it is known that the posterior distribution for p is also a beta distribution, with parameters $\alpha' = \alpha + x$ and $\beta' = \beta + n - x$. The posterior predictive distribution is a beta-binomial distribution given by $\Pr(Y = y|\alpha, \beta, n, x) = \binom{n}{y} \frac{B(\alpha' + y, \beta' + n - y)}{B(\alpha', \beta')}$ where α, β,x and n are as defined previously, and m is the number of PPQ batches to be manufactured. The posterior predictive probability that a single future batch is within specification is $(\alpha + x)/(\alpha + \beta + n)$.

Given the above Bayesian framework for probability of success of future batches, Fig. 19.3 shows the posterior distributions for 3 successful batches in 3 trials, 5 successful batches in 5 trials, and 10 successful batches in 10 trials.

Let's consider a hypothetical example. Assume that 10 clinical batches have been successfully manufactured, with no batch failures. Suppose that the goal is to defend the claim of a probability of batch manufacturing success of 0.95 or greater with approximately 80 % confidence. We can use the aforementioned posterior beta distribution to choose increasing values of the parameter n until we achieve the desired confidence level. In this case, 4 additional batches would be necessary with no batch failures, for a total of 14 successful batches in 14 trials. The predictive probability of a future single batch success would be $14.5/15 = 0.967$.

Lewis et al. (2014) also provided several "strawman" hypothetical examples of how this approach could be used in practice to justify the number of PPQ batches in relation to risk. These were:

1. hold lower confidence limit for p constant, vary confidence level with residual risk,
2. hold confidence level constant, vary lower confidence limit for p with residual risk,
3. hold parameters of prior distribution constant, vary posterior mean and 0.2 quantile with risk,
4. hold posterior mean and 0.2 quantile constant, vary prior distribution parameters with risk.

Table 19.2 Risk level and number of batches in relation to confidence level and lower bound on p when holding the lower confidence limit approximately constant

Risk level	Number batches	Confidence level	pL
Minimal	1	0.69	0.803
Low	3	0.80	0.809
Moderate	7	0.90	0.810
High	11	0.95	0.803

Table 19.3 Risk level and number of batches in relation to confidence level and lower bound on p when confidence level is held constant

Risk level	Number batches	Confidence level	pL
Minimal	1	0.80	0.585
Low	3	0.80	0.809
Moderate	7	0.80	0.908
High	14	0.80	0.952

Table 19.4 Risk level and number of batches in relation to confidence level holding the prior distribution constant

Risk level	Number of batches	Prior beta (α, β)	Posterior mean	Posterior 0.2 quantile
Minimal	1	(½, ½)	0.750	0.528
Low	4	(½, ½)	0.900	0.825
Moderate	8	(½, ½)	0.944	0.905
High	16	(½, ½)	0.971	0.951

In all the strawman examples, the calculations are with respect to n successful batches of n batches produced. The question of how to assimilate information related to batch failures is an important issue in this context. One should give careful thought to whether the failed batches belong to the population of batches arising from the manufacturing process of interest. If so, some accommodation may be required in the parameters of the prior distribution on p.

An example of strawman model 1 is given in Table 19.2. It gives the number of batches required when holding the lower confidence limit for p approximately constant, and varies the confidence level in relation to risk. The lower confidence limit for p was obtained using the Wilson score procedure.

An example of strawman model 2 is given in Table 19.3. It gives the number of batches required when holding the confidence level constant, and varying the lower confidence limit for p in relation to risk. The lower confidence limit for p was obtained using the Wilson score procedure.

An example of strawman model 3 is given in Table 19.4. It gives the number of batches required when holding the prior distribution constant, and varying the posterior mean and 0.2 quantile in relation to risk.

An example of strawman model 4 is given in Table 19.5. It gives the number of batches required when holding the posterior mean and 0.2 quantile constant, and varying the prior distribution in relation to risk.

Table 19.5 Risk level and number of batches in relation to confidence level holding the prior distribution constant

Risk level	Number of batches	Prior beta (α, β)	Posterior mean	Posterior 0.2 quantile
Minimal	1	(8½, ½)	0.950	0.915
Low	3	(6½, ½)	0.950	0.915
Moderate	6	(3½, ½)	0.950	0.915
High	9	(½, ½)	0.950	0.915

Yang (2013) further developed a Bayesian method related to probability of batch success addressing the question of sample size estimation for number of batches during PPQ. Yang's approach was similar in concept to the approach laid out by Lewis et al. (2014), Yang assumes a beta error distribution for process performance data from Stage 1 and combines with expected outcomes of Stage 2 PPQ to derive a posterior probability for future batches to meet specification. Yang also proposed a Bayesian definition for *quality assurance* as the posterior probability for test results of CQAs of a future batch meeting specification. Yang gives a comprehensive discussion and examples highlighting the advantages of the Bayesian approach over the traditional frequentist approach.

In this section we reviewed the ISPE's discussion paper giving three approaches for justifying PPQ sample size. Approach 1 may be reasonable, but is not statistically based. Approach 2 imposes a difficult-to-meet criterion on observed process capability. Approach 3 is based on a property of order statistics; it might not be practical when there are several CQAs. For these reasons, the method given by Lewis et al. (2014) that bases PPQ sample size on the probability that a batch is within specification embedded in a Bayesian framework has many advantages in practice. It imposes no new process performance criteria. It naturally incorporates any number of CQAs and permits a flexible implementation permitting a variety of process performance criteria to guide decision making. We hasten to add that this subject is an evolving science, and as stated earlier, no industry standard exists as of the writing of this section. Over the coming years, new ideas with additional approaches will be proposed and the novice statistician should consult current publications to gain a clearer insight into the state of knowledge regarding PPQ sample size estimation.

19.6 Continued Process Verification

Continued process verification (CPV) starts at the time commercial distribution of product begins. This occurs after successfully completing PPQ. The EMA refers to continued process verification as ongoing process verification (OPV). Ongoing process verification is defined as "documented evidence that the process remains in a state of control during commercial manufacture". The ICH Q8 guidance

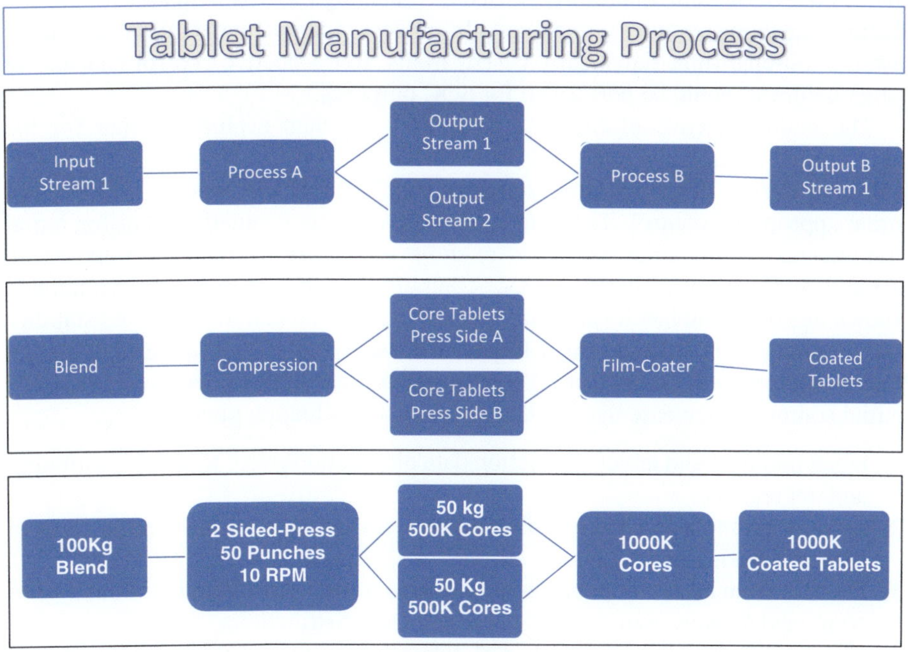

Fig. 19.4 Example tablet process for sampling considerations

also refers to continuous process verification, not to be confused with continued process verification, as "an alternative approach to process validation in which manufacturing process performance is continuously monitored and evaluated." Continued process verification requires manufacturers to maintain the process in a state of control over the life of the process. Keeping this perspective in mind, we build further upon what we learned during PPQ with a hypothetical example of how we might conduct a CPV exercise.

The hypothetical example involves a tablet manufacturing compression process to demonstrate data collection and data evaluation options for CPV (initial and routine commercial release). The compression process, as illustrated in Fig. 19.4 has one input stream (one granulation blend) and two output streams (double sided-press). The example includes a leading process based indicator (compression force as a CPP), a lagging product based indicator (hardness) and a final process output (dissolution). The execution of a well-designed PPQ protocol confirmed that the commercial manufacturing process performed as expected. Prior to the execution of the PPQ protocol, the facility, utilities and equipment were successfully qualified and all personnel were appropriately trained.

A leading indicator (CPP), compression force, was identified during process design as a prospective control for tablet hardness. It had been shown that if uncontrolled, it resulted in undesirable variation in tablet dissolution. The sampling and testing of the tablet hardness is used as a lagging process indicator to detect a disturbance to the compression process. In-process monitoring would be carried out using traditional (non-statistically based) limits. Retrospective control charting of tablet hardness would be performed for SQC purposes.

The main objective of the data collection and data evaluation plan for the CPV initial commercial release is to expand the body of knowledge that was collected during PPQ regarding the total variability of the process and demonstrate appropriate control. The data collection and data evaluation plan for initial commercial release as described is to more accurately estimate between batch variability. The between batch variance component combined with the accumulated knowledge of within batch variability are critical to demonstrating process stability and product acceptability in order to pave the way for routine commercial release. The following conclusions from the prior stages were used to determine the CPV initial commercial release data collection and data evaluation plan:

1) tablet hardness and tablet dissolution data distributions were tested for normality and not rejected,
2) average hardness and average dissolution were similar to values predicted from the PD model,
3) the data from each compression output stream were examined individually and combined to show comparability between their distributions,
4) during PPQ an adequate number of subgroups were sampled to assure representativeness of the batch and to permit an assessment of the state of statistical control,
5) a mixed effects model accounting for between and within batch variance components found within batch homogeneity across batches and a non-negligible between batch variance component,
6) acceptance criteria were applied to the dissolution results that demonstrated conformance to specification.

These conclusions supported a reduction in tablet hardness and tablet dissolution sampling and testing. The leading indicator (compression force) and the lagging indicator (tablet hardness) control charts from PPQ confirmed that the control strategy achieved within batch statistical stability. Product acceptability in relation to tablet dissolution behavior was demonstrated by applying acceptance criteria that provided assurance that the product conformed to specification.

In connection with the concept of conformance to specification, it is important to point out that many drug product specifications are based on compendial requirements or process capability. Recent discussions by regulators and others (Marroum 2012; Sharp 2012) have advocated the development of "clinically relevant" specifications. It is difficult to foresee moving forward how compendial requirements will be integrated with the notion of "clinically relevant specifications". Keeping in mind that this is an evolving science, specifications based on compendial requirements or process capability resemble acceptance sampling plans, in the sense that specific sample size and acceptance criteria are stipulated. These types of specifications overlook the role of operating characteristic curves in assessing consumer risk or producer risk. Therefore, sample size and acceptance criteria considerations may differ greatly across companies when enhanced sampling is conducted. This difficulty would be eliminated if product specifications were based on clinical relevance that were driven by a Quality Target Product Profile (QTPP) containing explicit quality requirements related to the population characteristics of the CQA.

There are various approaches that allow the acceptance criteria and sample size to be associated with statistical confidence and coverage statements as desired by the regulators and required by the cGMPs. One approach utilizes ASTM E2709 entitled "***Standard Practice for Demonstrating Capability to Comply with an Acceptance Procedure***". Other approaches utilize different types of confidence intervals and parametric tolerance intervals. Many of the variables acceptance sampling plans indexed on AQLs and LQs, provided in voluntary consensus standards are also based on parametric tolerance intervals.

The ASTM E2709 provides a general methodology for evaluating the probability of passing an acceptance procedure in terms of a pre-specified statistical confidence level and coverage. Chapter 23, Assessing Content Uniformity, provides a more in depth discussion of this methodology applied to the current ICH UDU test. It is discussed in specific detail applied to the ICH UDU test in ASTM E2810.

The parametric tolerance interval approach is another methodology for demonstrating product acceptability. It requires a pre-specified target interval associated with an appropriate choice of confidence and coverage. The manufacturer would justify the selection of the target interval, desired confidence and coverage based on risk considerations. There are two types of parametric tolerance intervals that are typically used to demonstrate conformance to specification:

A) Parametric Two-sided Tolerance Interval Plan (PTS-TI) interval for appropriate sample size using k value for stated confidence and coverage desired, and
B) Parametric Two One-sided Tolerance Interval PTI-TOST interval for appropriate sample size using k-value for stated confidence and coverage.

The PTS-TI controls for the proportion in the center of the distribution and the PTI-TOST controls for the proportion in each of the tails of the distribution. For both cases, a normal distribution is assumed. The PTI-TOST approach is discussed in Chap. 23 as the PTI test, where the confidence level is 95 % and the proportion in the tails is 6.25 % with a target interval of [80,120] %Label Claim (LC) for an inhalation product. A target interval of [85, 115]% LC has been used in some instances for oral solid dosage products.

Continuing the hypothetical CPV exercise example, Tables 19.6 and 19.7 provide example data collection and data evaluation plans for tablet hardness and tablet dissolution for PPQ and CPV (initial and routine), respectively. Chosen acceptance criteria corresponded to 95 % confidence that not more than 0.3 % of tablets will be less than D% dissolved in 15 min. These are shown in Table 19.8. Consumer risk protection is afforded in the sense that if the manufactured batch contained 0.3 % or more of the tablets whose in vitro dissolution was less than D% at 15 min, there is a 95 % or more probability of failing the acceptance criteria.

Table 19.6 Data collection and data evaluation example for tablet hardness

	PPQ	CPV Initial commercial	CPV routine commercial
Number of stratum or locations	30	15	15
Number of tablets	20 per stratum by output stream	20 per stratum by output stream	10 per stratum by output stream
Number of batches	4	25 Initial commercial	Routine commercial
In process control criteria applied to each subgroup during compression. Pre-defined action would be taken if criteria not met.	$\overline{X}_{20} \in [H_L, H_U]$	$\overline{X}_{20} \in [H_L, H_U]$	$\overline{X}_{20} \in [H_L, H_U]$
	$X_i \in [H_{L*}, H_{U*}]$	$X_i \in [H_{L*}, H_{U*}]$	$X_i \in [H_{L*}, H_{U*}]$
Characterize	Within batch variation by batch and output stream.	Between batch variation and within batch	NA
Monitor	Retrospective use of control charts where 1) subgroup is stratum by output stream and batch.	Retrospective control charts where 1) subgroup is stratum by output stream and batch. 2) subgroup is combined output stream for each stratum 3) subgroup is batch	Retrospective control charts where 1) subgroup is batch

This tablet weight in-process control limits illustrates the use of some non-statistically based control strategies

ASTM E2709 can also be used to calculate acceptance criteria for a prescribed lower probability bound and confidence level. The criteria established by the ASTM E2709 procedure would lead to a prescribed probability of passing the compendial test with the given confidence level. Although the probability of passing the compendial test is assured, the interpretation of confidence and coverage for this approach is not the same as for the PTI test.

It was shown with the hypothetical example that there is a natural progression from CPV initial commercial release to routine commercial release that results from process knowledge and process understanding accumulated from PPQ and CPV initial commercial release. At this point, the mixed effects model previously fit at the end of PPQ would be refit to more precisely estimate the between and within batch variance components. Those estimates would form the basis for further risk calculations justifying the reduced sampling protocol for routine release. As seen in Tables 19.6 and 19.7, routine commercial release represents a reduced sampling protocol with identical statistical tools as CPV 3A. It was shown that process validation requires the application of statistical monitoring and risk control tools throughout the life of the product in an environment of perpetual learning.

Table 19.7 Data collection and data evaluation example for tablet dissolution

Validation lifecycle stage	PPQ	CPV initial commercial	CPV routine commercial
Number of stratum or locations	15	15	Composite
Number of tablets	Sample 6 tablets B, M, E 3 locations between B & M and 3 locations between M & E from each output stream.	Sample 6 tablets B, M, E from each output stream	18
Number of batches	4	25 initial commercial	Routine commercial
Characterize	Within batch variation by batch and output stream.	Between batch variation and within batch	NA
Monitor	Similarity of output streams within and across batches.	Retrospective control chart where subgroup is batch.	Retrospective control chart where subgroup is batch.
Conformance to specificationTesting scheme	Each output stream Test 6 tablets B, M, E For stream A, test 6 tablets from midpoint of the B, M or midpoint M, E locations. For stream B, test 6 tablet from midpoint that was not selected for Stream A. 24 Tablets per output stream 48 Tablets total per batch	Test 2 tablets B, M, E from each output stream. 6 tablets per output stream 12 Tablet per Batch	Composite sample Tier 1: 6 tablets Tier 2: 12 tablets

Table 19.8 Acceptance criteria by stage of process validation (k values correspond to 95 % confidence of 99.7 % coverage)

PPQ	$\overline{X}_{48} - k_{48}s_{48} \geq D\%\ dissolved$	$k_{48} = 3.375$
CPV initial commercial	$\overline{X}_{12} - k_{12}s_{12} \geq D\%\ dissolved$	$k_{12} = 4.381$
CPV routine commercial	Tier 1: $\overline{X}_{6} - k_{6}s_{6} \geq D\%\ dissolved$ Tier 2: $\overline{X}_{18} - k_{18}s_{18} \geq D\%\ dissolved$	$k_{6} = 7.127$ $k_{18} = 4.130$

19.6.1 CPV for Legacy Products

For legacy products, which are currently marketed products and their manufacturing process validated prior to the 2011 FDA guidance, implementation of the recommendations in the FDA PV guidance would most likely begin in the CPV stage. The most reasonable starting place for legacy products would be to conduct retrospective process capability or process performance studies. Based on the outcome of these retrospective studies, a risk assessment approach should be used to determine whether or not data collection and data evaluation activities similar to those in PD and PPQ should be conducted. This would include consideration of within batch location of samples, the number of batches, the sample size per batch and justification of acceptance criteria for PPQ, CPV 3A and CPV 3B. This justification should be based on both well-defined and appropriately documented scientific and empirical knowledge combined with appropriate statistical practice.

19.7 Summary

The lifecycle approach to process validation consists of three stages: Process Stage 1 Process Design (PD) where the commercial process is defined based on knowledge gained through development and scale-up; Stage 2 Process Qualification (PQ) where the process is studied to demonstrate the ability to manufacture acceptable product at commercial scale; Stage 3 Continued Process Verification (CPV) consisting of an initial intensive study of the process to provide the justification for the "maturing" of the process (Stage 3A) to transition into routine manufacturing and controls (Stage 3B). The lifecycle approach stresses understanding of the total variability associated with within batch and between batch sources linked to producer and consumer risk. Routine manufacturing does not mean that the manufacturing process is fixed. Materials, equipment, production environment, personnel and manufacturing procedures can change. The impact of the changes may be evaluated by carrying out a reduced version of the original PPQ or by reverting to the level of enhanced sampling and monitoring that was performed in the initial commercial release stage. The ongoing data collection and data evaluation activities may identify process problems or opportunities for process improvement that may warrant repeating activities similar to those in PD and PPQ.

The study of the process as a lifecycle paradigm requires a perpetual learning commitment, to show on an ongoing basis that the process remains in a state of control. Statistical tools are applied throughout the validation process through modeling and risk control calculations. These would involve confidence and coverage considerations and appropriate statistical process and quality control charts to characterize and monitor the variability in the process as reflected through

Table 19.9 Comparison of objective, sampling protocol and statistical considerations in relation to stage of process validation

	PPQ	CPV initial commercial release	CPV routine commercial release
Objective	1. Estimate within batch variability. 2. Early estimate of between batch variability. 3. Preliminary risk estimates of meeting acceptance criteria. 4. Demonstrate product is of intended quality.	1. Further estimation of within batch variability. 2. Estimate between batch variability to required precision. 3. Detailed risk estimates meeting acceptance criteria, consider consumer and producer risk. 4. Statistical quality control release criteria are established. 5. Demonstrate Product is of intended quality.	1. Monitoring with appropriate SPC tools to maintain process stability and product acceptability. 2. Confirmation of between batch variability. 3. Ongoing assessment of consumer and producer risk.
Sampling protocol	1. Enhanced sampling through rational sub-grouping to characterize the within batch product variability. 2. Justify number of batches incorporating prior knowledge.	1. Less intensive rational sub-grouping within batch justified through risk calculations of PPQ information. 2. Increased number of batches to characterize between batch variability justified through prior knowledge.	1. Routine sampling and routine batch release justified by a formal variance component model of data collected from PPQ and CPV initial.
Statistical testing and confidence	1. Acceptance criteria applied to each batch separately using justified consumer risk and consumer risk quality. 2. SQC monitoring to assess within batch product stability.	1. Acceptance criteria applied to each batch separately while considering both consumer risk and producer risk. 2. SQC monitoring to assess between batch product stability.	1. Acceptance criteria applied to each batch separately while considering both consumer risk and producer risk. 2. Comprehensive statistical quality monitoring and assessments of within and between batch variability.

product CQAs. Bayesian approaches hold great promise in establishing statistical justification for number of batches necessary for each stage of PV. Table 19.9 provides a summary view of a comparison of goal, sampling protocol and statistical considerations in relation to the three stages of Process Validation.

Acknowledgements The authors thank Lynne Hare, Rick Lewis, Ron Snee and Lanju Zhang for their careful reading of this chapter and suggestions that led to many clarifications and improvements to the language. Any errors or omissions are solely the authors'.

References

American Society for Quality (ASQ) (2008) ANSI/ASQ Z1.4-2008 (Milwaukee, WI)

ASTM E2709-10 Standard Practice for Demonstrating Capability to Comply with a Lot Acceptance Procedure

Bergum JS (1990) Constructing acceptance limits for multiple stage tests. Drug Dev Ind Pharm 16(14):2153–2166

Bryder M, Etling H, Fleming J, Hu Y, Levy P (2014) Topic 1—Stage 2 process validation: determining and justifying the number of process performance qualification batches, International Society for Pharmaceutical Engineering Discussion Paper. www.ispe.org/discussion-papers/stage-2-process-validation.pdf

EMA/CHMP/CVMP/QWP/BWP/70278/2012-Rev1, Committee for Medicinal Products for Human Use (CHMP), Committee for Medicinal Products for Veterinary Use (CVMP), Guideline on process validation for finished products - information and data to be provided in regulatory submissions 27 Feb 2014

EudraLex (2015) Annex 15: Qualification and Validation; EU Guidelines for Good Manufacturing Practice for Medicinal Products for Human and Veterinary Use Ref. Ares(2015)1380025-30/03/2015. http://ec.europa.eu/health/documents/eudralex/vol-4/index_en.htm

Food and Drug Administration. Center for Drugs Evaluation Research (2014) Draft guidance for industry analytical procedures and methods validation for drugs and biologics FDA Maryland. http://www.fda.gov/downloads/drugs/guidancecomplianceregulatoryinformation/guidances/ucm386366.pdf

Food and Drug Administration. Center for Drugs Evaluation Research (2011) Guidance for industry process validation: general principles and practices. http://www.fda.gov/downloads/Drugs/Guidances/UCM070336.pdf

Hahn GJ, Meeker WQ (1991) Statistical intervals: a guide for practitioners. Wiley, New York

International Conference on Harmonization (1999) Q6A Specifications: Test Procedures and Acceptance Criteria for New Drug Substances and New Drug Products: Chemical Substances

International Conference on Harmonization (2000) Q7 Good Manufacturing Practice for Active Pharmaceutical Ingredients, guidance for industry, November 2000

International Conference on Harmonization (2009) Q8(R2) Pharmaceutical Development, guidance for industry, August 2009

International Conference on Harmonization (2005) Q9 Quality Risk Management, guidance for industry, November 2005

International Conference on Harmonization (2008) Q10 Pharmaceutical Quality System, guidance for industry, June 2008

International Conference on Harmonization (2012) Q11 Development and Manufacture of Drug Substances, May 2012

International Conference on Harmonization (2011) Quality Implementation Working Group Points to Consider (R2), December 2011

ISO 2859–1:1999, Sampling procedures for inspection by attributes—Part 1: Sampling schema indexed by acceptance quality limit (AQL) for lot-by-lot inspection

ISO 2859–2, Sampling procedures for inspection by attributes—Part 2: Sampling plans indexed by limiting quality (LQ) for isolated lot inspection

ISO 2859–10:2006, Sampling procedures for inspection by attributes—Part 10: Introduction to the ISO 2859 series of standards for sampling for inspection by attributes

ISO 3534–1, Statistics—Vocabulary and symbols—Part 1: General statistical terms and terms used in probability

ISO 3534–2, Statistics—Vocabulary and symbols—Part 2: Applied statistics

ISO 3951–1:2005, Sampling procedures for inspection by variables—Part 1: Specification for single sampling plans indexed by acceptance quality limit (AQL) for lot-by-lot inspection for a single quality characteristic and a single AQL

ISO 7879-1:2007, Control charts—General guide and introduction

ISO 16269–6:2005, Statistical interpretation of data—Part 6: Determination of statistical tolerance intervals

ISO 11462–1:2001, Guidelines for implementation of statistical process control (SPC)—Part 1: Elements of SPC

ISO 22514–1, Statistical methods in process management—Capability and performance—Part 1: General principles and concepts

ISO 22514–2:2014, Statistical methods in process management—Capability and performance—Part 2: Process capability and performance of time-dependent process models

ISO/TR 8550–1, Guidance on the selection and usage of acceptance sampling systems for inspection of discrete items in lots—Part 1: Acceptance Sampling

ISO/TR 8550–2, Guidance on the selection and usage of acceptance sampling systems for inspection of discrete items in lots—Part 2: Sampling by attributes ISO/TR 8550–3, Guidance on the selection and usage of acceptance sampling systems for inspection of discrete items in lots—Part 3: Sampling by variables

ISO/TR 18532:2009, Guidance on the application of statistical methods to quality and to industrial standardization

Iyer K (2014) "Statistics and Pharmaceutical Quality". Presentation, IFPAC conference, Arlington, VA 24 January 2014

Krishnamoorthy K, Mathew T (2009) Statistical tolerance regions: theory, applications, and computation. Wiley, New York

Little T (2001) 10 requirements for effective process control: a case study. Qual Prog 34(2):46–52 [QICID: 14351]

Marroum P (2012) Clinically relevant dissolution methods and specifications. Am Pharm Rev 15(1):36–41

Owen DB (1964) Control of percentages in both tails of the normal distribution. Technometrics 6(4):377–387

Lewis RA, Henry WE, Peterson JJ, McAllister PR (2014) Determining the Number of Process Performance Qualification Batches Presentation QSPI/FDA Summit, March 7, 2014

Novick S, Christopher D, Dey M, Lyapustina S, Golden M, Leiner S, Wyka B, Delzeit H, Novak C, Larner G (2009) A two one-sided parametric tolerance interval test for control of delivered dose uniformity—Part 1-characterization of FDA proposed test. AAPS PharmSciTech 10(3): 820–828

Sharp SS (2012) Establishing Clinically Relevant Drug Product Specifications: FDA Perspective FDA/ONDQA/Biopharmaceutics 2012 AAPS Annual Meeting and Exposition Chicago, IL, October 16, 2012 http://www.fda.gov/downloads/AboutFDA/CentersOffices/OfficeofMedicalProductsandTobacco/CDER/UCM341185.pdf

Woodall WH (2000, ASQ) ASQ Journal of Quality Technology 32(4) QICID: 14600:41–378

United States Pharmacopeia. USP 35–NF 30, General Notices 3.10. Rockville, MD: USP; 2012:4

United States Pharmacopeia. "Lifecycle management of analytical procedures: method development, procedure performance qualification, and procedure performance verification", Pharmacopeial Forum, Stimuli to the Revision 30 (5). http://www.usp.org/sites/default/files/usp_pdf/EN/USPNF/revisions/lifecycle_pdf.pdf

United States Pharmacopeia (2014) <1210>Statistical tools for validation procedure. pharmacopeial forum 40(5). http://www.usp.org/meetings-courses/courses/new-proposed-usp-general-chapter-1210-statistical-tools-procedure-validation

Yang H (2013) How many batches are needed for process validation under the new FDA guidance? PDA J Pharm Sci Technol 67:53–62

Chapter 20
Acceptance Sampling

Richard K. Burdick and Frank Ye

Abstract Acceptance sampling has a role in the overall CMC strategy for providing safe and efficacious products for patients. This chapter describes the mathematical model that supports this sampling process, describes different approaches based on the measurement scale of the data, and describes its relationship with validation testing. A recommendation concerning future work to better incorporate measurement error is also provided.

Keywords Acceptable quality level (AQL) • ANSI/ASQ Z1.4 • ANSI/ASQ Z1.9 • Attribute sampling • Average sample size (ASN) • Consumer's risk • Corrective and preventative action (CAPA) • Lot tolerance percent defective (LTPD) • MIL-STD-105E • Misclassification errors • Mixed variables-attributes plan • Operating characteristic (OC) curve • Producer's risk • Rejectable quality level (RQL) • USP < 711 > • USP < 790 > • USP < 905 > • Validation testing • Variables sampling

20.1 Introduction

This chapter concerns acceptance sampling as applied in CMC. Acceptance sampling was first introduced in the military during World War II. Early work in this area made a clear distinction between acceptance sampling as a disposition activity and acceptance quality control which focuses more on the holistic process approach of which acceptance sampling is only one element. Accordingly, acceptance sampling in the pharmaceutical industry must be incorporated into an overall strategy that employs several other tools for ensuring product quality. Some of this overall strategy includes assessment of the criticality of product quality attributes, development of manufacturing processes capable of safety and efficacy requirements, and

R.K. Burdick (✉)
Elion Labs 1450 Infinite Drive, Louisville, CO 80027, USA
e-mail: rburdick@elionlabs.com

F. Ye
Amgen Inc., Thousand Oaks, CA, USA
e-mail: fye@amgen.com

L. Zhang (ed.), *Nonclinical Statistics for Pharmaceutical and Biotechnology Industries*, Statistics for Biology and Health, DOI 10.1007/978-3-319-23558-5_20

development of analytical methods that are validated as fit for the purpose of making measurements. The next section describes several CMC applications of acceptance sampling within this larger quality system.

20.2 CMC Applications

Perhaps the most widely adopted area for acceptance sampling in CMC is incoming raw material sampling. Pharmaceutical companies procure a large number and volume of raw materials on a regular basis from numerous suppliers all over the world to support the production of their drug products. Raw materials range from simple chemicals to complicated medical delivery systems. To ensure the incoming lots meet a set of pre-defined quality requirements that ensure they are fit for use in production, manufacturers routinely employ schemes of acceptance sampling plans, often times relying on guidance from ANSI/ASQ Z1.4 (ANSI 2008a).

To demonstrate, consider a biotech company XYZ that has developed a life-extending cancer drug in a monoclonal anti-body. After completing the manufacturing steps of cell culture, recovery, and purification, the company formulates the biological product and fills it into syringes before supplying the finished product to either a clinical trial or commercial market. The primary container of the drug product is a syringe manufactured by a company-approved vendor. A lot of syringes is released to XYZ by the vendor after it passes a set of a quality tests. After receiving the lot, XYZ samples representative syringes from the lot and runs its own tests to ensure the lot quality is acceptable for its introduction into the manufacturing process. XYZ measures a set of critical and key parameters based on scientific rationale and research that reflect the quality attributes of both the syringe as a primary container and the delivery system for a liquid biological product. One tested parameter is barrel hoop strength of a glass syringe. Test forces are applied to the barrel of a syringe in a uniform manner and the force required to break the barrel is recorded and compared to a pre-defined threshold that separates "acceptable" and "unacceptable"(defective) syringes. To appropriately conduct the test, it is necessary to determine the number of syringes to be sampled from the hundreds of thousands of syringes in the lot. The company first decides on an Acceptable Quality Level (AQL) and then looks up sampling tables provided in ANSI/ASQ Z1.4 to determine sample size and an acceptance rule. In this example the company sets the AQL at 0.15% and obtains a sample size of 315 with an acceptance rule of 1 defective. That is, the lot is accepted if there are one or fewer defectives in the sample. (More detail on use of the ANSI/ASQ tables is provided in Sect. 20.4.) This means the company will randomly collect a sample of 315 representative syringes from an incoming lot, measure the breaking force of each syringe, and label each syringe as either acceptable or defective by comparing the recorded force to the specification. If the total number of defectives is one or less, the lot will be accepted by XYZ. Otherwise, the lot will be rejected.

Another example of acceptance sampling in manufacturing is the visual inspection of filled vials for visible particles. Modern day manufacturing processes of a biological product typically deploy an automated visual inspection machine to separate out filled vials containing visible particles during production. Typically, this 100 % sampling scheme is followed by another round of manual visual inspection with human eyes based on only a sample of items. Once again, an acceptance sampling plan can be adopted and used as part of this batch release procedure. Other cases of acceptance sampling in CMC come from areas of product development, process control, and stability testing.

20.3 Terminology of Acceptance Sampling

An acceptance sampling plan consists of a sampling design and a set of rules for making decisions based on the resulting sample. For situations where only a single sample is selected, the two decisions are either

a) Accept the lot, or
b) Reject the lot.

In a pre-planned multiple sample design, a third decision is to

c) Select another sample and then decide either a) or b) or continue sampling.

The fundamental tool for selecting a sampling plan is the *operating characteristic* (OC) curve. An OC curve is a bi-variate graph with probability of passing a lot on the vertical axis and some parameter of the sampling model on the horizontal axis. Figure 20.1 provides an example of an OC curve for an attribute sampling plan in which a sample of 80 items is selected at random from a lot. The parameter shown on the horizontal axis is the percentage of defective units in a lot. The lot is "accepted" if there is one or fewer defective units in the sample. The lot is "rejected" if there are two or more defective units in the sample. The terms "accepted" and "rejected" are used in a generic sense. The action that results from either conclusion depends on the particular application.

Figure 20.1 shows a vertical line drawn at the lot percent defective value of 1.5 %. The value on the vertical axis where the vertical line crosses the OC curve is 0.662=66.2 %. This means that if the true percentage of defective units in the lot is 1.5 %, there is a 66.2 % probability the lot will be accepted. One desires to accept a lot when the percentage of defective units in the lot is acceptably low, and reject the lot when the percentage of defective units in the lot is unacceptably high. Thus, definition of a sampling plan requires definition for both acceptable and unacceptable quality as well as the risks for making incorrect decisions. This process is consistent with the risk based approach advocated in ICH Q9 (ICH 2005).

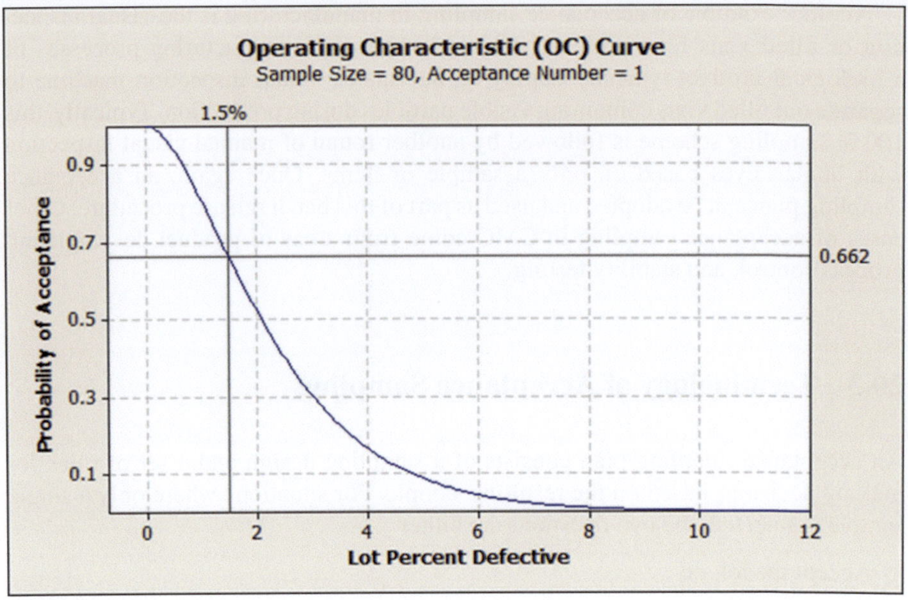

Fig. 20.1 Operating characteristic (OC) *curve*

When deciding whether to accept or reject a lot, there are two types of errors:

1. Rejecting a good lot (Type 1), and
2. Accepting a bad lot (Type 2).

The risk of committing a Type 1 error is referred to as the producer's risk and is denoted with the Greek letter α. The risk of committing a Type 2 error is called the consumer's risk and is denoted with the Greek letter β. Definitions for "good" and "bad" are typically defined in terms of the percentage of nonconforming units in the lot. Acceptable quality level (AQL) is the percentage of defective units on the horizontal axis associated with the 95 % probability of acceptance on the vertical axis of the OC curve. The lot tolerance percent defective (LTPD) is the percent defective shown on the horizontal axis associated with the 10 % probability of acceptance (see Fig. 20.2). It is useful to also define AQL and LTPD in terms of the proportions $p_1 = \text{AQL}/100\,\%$ and $p_2 = \text{LTPD}/100\,\%$.

The OC curve is based on a realistic compromise between consumer and producer under the recognition that production of perfect product is not realistic. For purposes of this chapter, AQL represents the level of producer quality associated with a probability of acceptance of 95 % and LTPD represents the level of consumer quality associated with a probability of acceptance of 10 %. (Note that LTPD is sometimes represented using other acronyms, most notably as rejectable quality level (RQL)). Using these definitions, Fig. 20.2 shows that the OC curve presented in Fig. 20.1 has an AQL value of 0.45 % and a LTPD value of 4.8 %. (Figures in this chapter were created using Minitab Version 16 Minitab Statistical Software 2015).

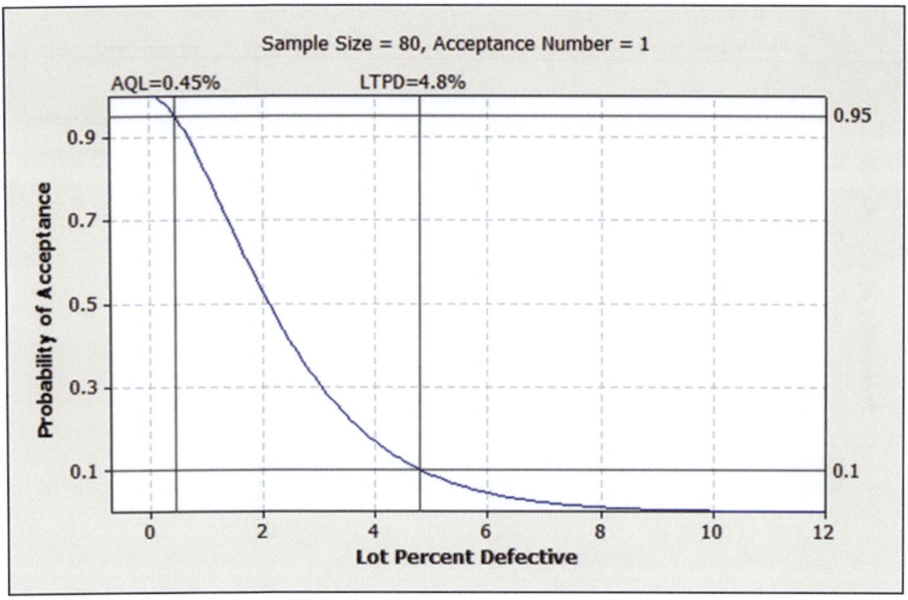

Fig. 20.2 AQL and LTPD on OC *curve*

The OC curve changes as the sample design changes. Comparison of the AQL and LTPD values from two OC curves is an appropriate manner for selecting a sampling design. Figure 20.3 shows two sampling plans. The first plan is the same one shown in Fig. 20.1 represented as $(n = 80, a = 1)$. This notation means the lot is accepted if a sample of $n = 80$ units includes only $a = 1$ or fewer defective units. Also shown in Fig. 20.3 is the plan $(n = 100, a = 2)$. For this plan, a sample of $n = 100$ units is selected, and the lot is accepted if there are $a = 2$ or fewer defective units. To determine the better sampling plan, one can compare the two OC curves as shown in Fig. 20.3. In this situation, at both the 95 % and 10 % probabilities of lot acceptance, the lot percentage defective is lesser for $(n = 80, a = 1)$.

Determination of acceptable values for AQL and LTPD require assessment of a variety of criteria including risks, costs, and consumer requirements. The first step in this process is to classify the severity of the defects that might occur. Typical classifications are Critical, Major, and Minor. Defectives of the same category would generally be expected to have the same values for AQL and LTPD.

20.4 Attribute Sampling

The example introduced in Sect. 20.3 describes an attribute sampling plan. Attribute sampling concerns units that can be classified into one of two groups: defective or non-defective. The summary measure that defines the lot quality is the percentage

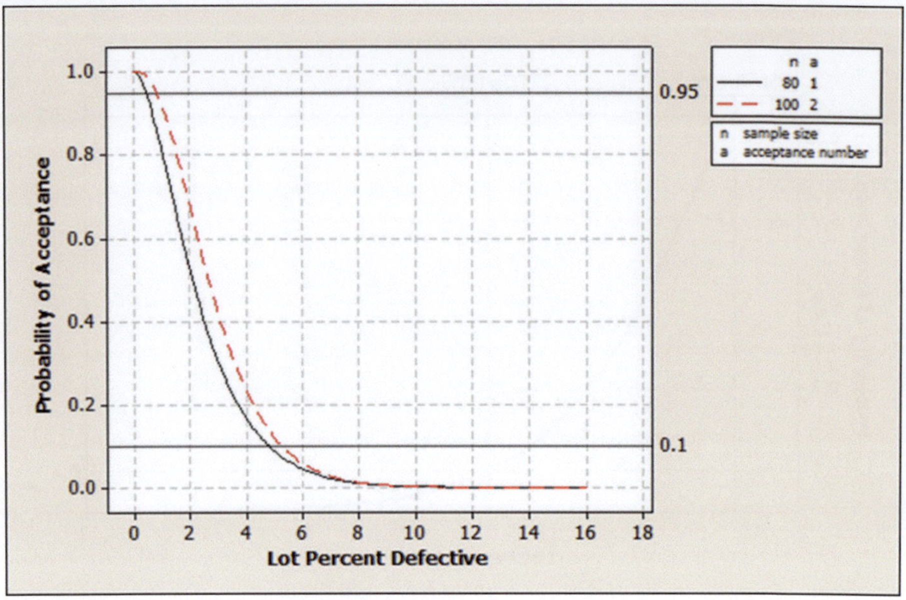

Fig. 20.3 Comparison of two sampling plans

of units in the lot that are defective. Lesser values of this percentage are associated with greater levels of quality. Calculations used to determine the probabilities shown on the vertical axis of the OC curve are based on either the hypergeometric or the binomial probability distribution. If the sample size is less than 10 % of the total lot size (as occurs in most applications), the binomial distribution is adequate for performing the calculations. The probability of acceptance, P_a, shown on the vertical axis of the OC curve is not difficult to calculate. Assuming the binomial model, the probability of accepting a lot for the plan (n, a) when the true lot percentage defective is π is

$$P_a = \sum_{i=0}^{a} \frac{n!}{i!(n-i)!} \left(\frac{\pi}{100}\right)^i \left(1 - \frac{\pi}{100}\right)^{n-i} \tag{20.1}$$

To demonstrate, the probability of accepting a lot using the plan in Fig. 20.2 ($n = 80, a = 1$) when $\pi = 0.45\,\%$ is

$$P_a = \sum_{i=0}^{1} \frac{80!}{i!(80-i)!} \left(\frac{0.45}{100}\right)^i \left(1 - \frac{0.45}{100}\right)^{80-i}$$

$$P_a = \frac{80!}{0!(80)!} \left(\frac{0.45}{100}\right)^0 \left(1 - \frac{0.45}{100}\right)^{80} + \frac{80!}{1!(79)!} \left(\frac{0.45}{100}\right)^1 \left(1 - \frac{0.45}{100}\right)^{79}$$

$$P_a = 0.697 + 0.252 = 0.95.$$

A popular set of sampling plans for attribute sampling which evolved from MIL-STD-105E are contained in ANSI/ASQ Z1.4 (ANSI 2008a). To use the tables in ANSI/ASQ Z1.4 one specifies

1. Lot size,
2. AQL,
3. Inspection level, and
4. Sampling scheme (single, double, multiple).

The sample size selected is determined based on the lot size through the use of code letters. The desired AQL and code letter then determine the values for n and a. The ANSI/ASQ process is thought to be a generally acceptable practice by most regulatory agencies. However, it is not without its drawbacks. First, strict adherence to the overall ANSI/ASQ Z1.4 sampling system requires switching rules that either increase or decrease the sample size on a given lot based on the historical performance of the output. This requires a sometimes complicated record keeping process that might not be feasible or desirable. Second, when sample size is small relative to the total number of units in a lot, lot size has no impact on the calculations needed to derive the OC curve. Thus, increasing sample size as lot size increases can lead to needless oversampling. Third, these plans consider only AQL in determining a sampling design and in that sense, give the appearance of placing lesser importance on the consumer. For these reasons, many companies prefer to select sampling plans based solely on the OC curve with well-defined and meaningful values for AQL and LTPD.

An example is useful to illustrate this point. USP 2015a recommends an ANSI/ASQ Z1.4 sampling plan for batch release purposes to monitor visible particulates in injections after 100 % inspection during manufacturing. It recommends a single sampling plan using General Inspection Level II for normal inspection with AQL $= 0.65$ %. Suppose a lot consists of 12,000 units. Using Table 20.1 of ANSI/ASQ Z1.4, the sample size code is letter M. From Table II-A, letter M requires a sample size of 315 units with an acceptable number of defective units equal to five units. This provides the plan ($n = 315, a = 5$).

The OC curve for this design is shown in Fig. 20.4. The attained value for AQL is 0.83 % and for LTPD is 2.94 % as shown in Table X-M where switching rules are not employed. If switching rules are applied, Chart XV-M reports the realized AQL is 0.781 % and LTPD is 2.12 %. Note that even when switching rules are employed, the attained AQL is greater than the desired value of 0.65 % due to the manner in which tables are constructed.

Table 20.1 Comparison of sampling designs

π	$P(A_1\|\pi)$	ASN
0.20 %	0.953	25.5
0.65 %	0.767	39.5
2.5 %	0.284	75.7
5 %	0.084	90.7

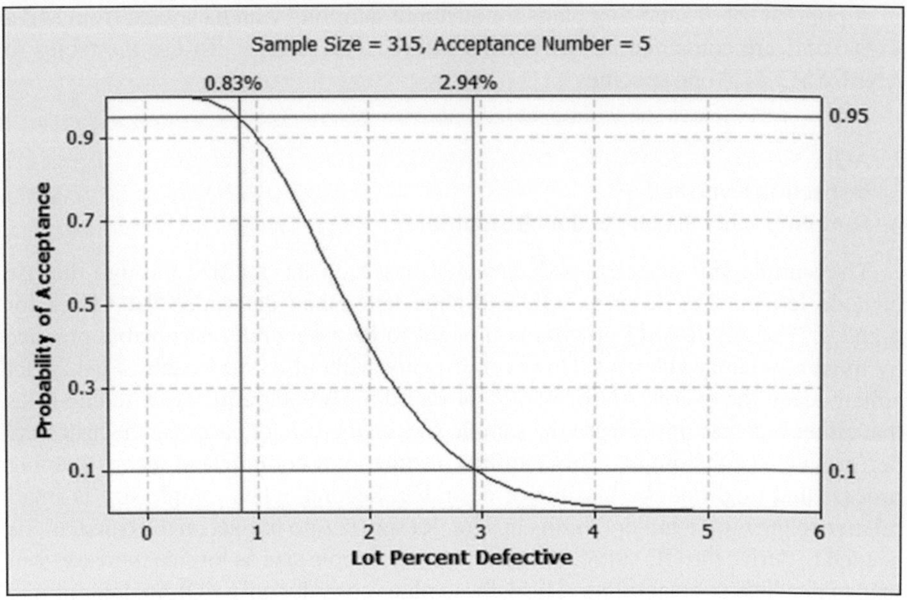

Fig. 20.4 Design for USP < 790 >

However, stand alone sampling plans that do not require switching rules can be determined directly from an OC curve for any given set of AQL, LTPD, α, and β values. In the present example, USP < 790 > states that "Alternative sampling plans (to ANSI/ASQ Z1.4) with equivalent or better protection are acceptable". Thus, suppose it is desired to construct an alternative design with AQL no greater than 0.65 % and with LTPD no greater than 2.5 % (assuming $\alpha = 0.05$ and $\beta = 0.10$). Figure 20.5 shows one possible design ($n = 155, a = 1$) compared with the OC curve shown in Fig. 20.4. The AQL is 0.23 % and the LTPD is 2.49 % for this alternative design so that USP < 790 > is satisfied. In contrast to the ANSI/ASQ Z1.4 plan, this same sampling design can be used for all lots without the record keeping required to implement the ANSI/ASQ Z1.4 switching rules.

20.5 Variables Sampling

In many situations, quality attributes are measured on a continuous scale and acceptance of a lot is based on the percentage of individual values in a lot that satisfy a numerical specification. The objective is to demonstrate process capability is sufficiently good to provide the desired AQL and LTPD values. To demonstrate, consider the process of accepting a lot of devices that release drug product from a pre-filled syringe. The upper specification limit on the injection time is 15 s.

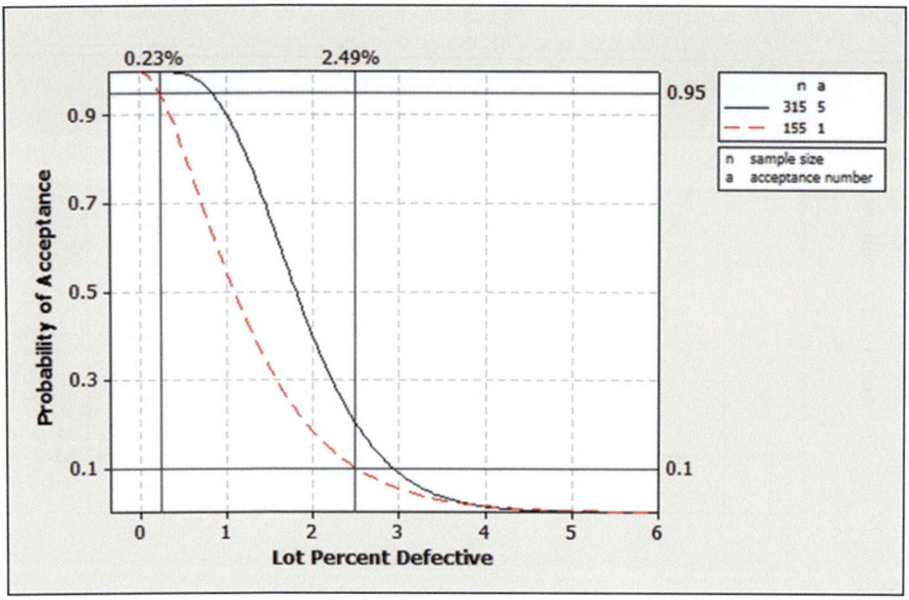

Fig. 20.5 Alternative USP < 790 > Designs

A sample of size n is randomly sampled from the lot and the sample mean is compared to the quantity

$$A = U - kS \tag{20.2}$$

where U is the upper specification limit of 15 s, S is the standard deviation for the sample of n items, and k is a constant that is a function of AQL, LTPD, and their associated risk values. The lot is accepted if the sample mean of the n items is less than or equal to A, and is rejected if the sample mean exceeds A. Schilling and Neubauer (2009, p. 186) provide the following approximate formulas for both k and n for what is commonly referred to as the k-method:

$$k = \frac{Z_{p2} Z_\alpha + Z_{p_1} Z_\beta}{Z_\alpha + Z_\beta} \tag{20.3}$$

$$n = \left(\frac{Z_\alpha + Z_\beta}{Z_{p_1} - Z_{p_2}} \right)^2 \text{ (variance known)} \tag{20.4}$$

$$n = \left(\frac{Z_\alpha + Z_\beta}{Z_{p_1} - Z_{p_2}} \right)^2 \left(1 + \frac{k^2}{2} \right) \text{ (variance unknown)} \tag{20.5}$$

where Z_δ is the standard normal percentile with area δ to the right. In the present example, Eqs. (20.3) and (20.5) are used with $p_1 = 0.0065$, $p_2 = 0.05$, $\alpha = 0.05$,

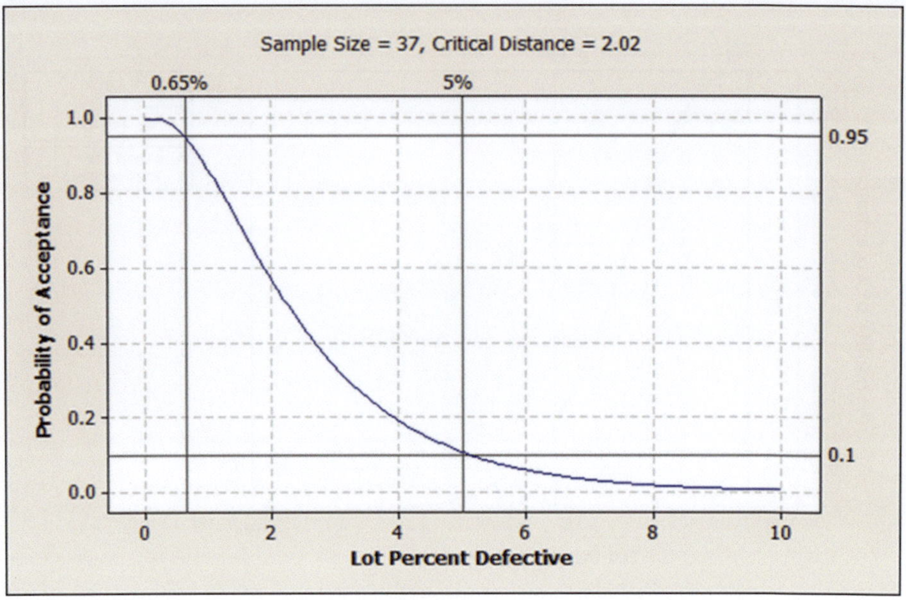

Fig. 20.6 OC *curve* for variables sampling design

$\beta = 0.10$, $Z_{p_1} = 2.484$, $Z_{p_2} = 1.645$, $Z_\alpha = 1.645$, and $Z_\beta = 1.282$ to compute $n = 37$ (rounded up) and $k = 2.02$ (rounded up). The OC curve for this example is shown in Fig. 20.6.

The probability of acceptance for the variables sampling OC curve is based on a non-central t-distribution. In particular, the probability of acceptance for the plan (n, k) when the true percentage of defective units in the lot is π is

$$P_a = Pr\left(t \geq \sqrt{n}k\right) \tag{20.6}$$

where t is a non-central t random variable with degrees of freedom $n - 1$ and non-centrality parameter $\sqrt{n}Z_{\frac{\pi}{100}}$. To demonstrate for the OC curve in Fig. 20.6, the probability of acceptance when $\pi = 5.0\,\%$ is

$$P_a = Pr\left(t \geq \sqrt{n}k\right)$$

$$P_a = Pr\left(t \geq \sqrt{37} \times 2.02\right)$$

$$P_a = Pr\left(t \geq 12.287\right) = 0.10$$

where the degrees of freedom= 36 and the non-centrality parameter is $\sqrt{37} \times Z_{0.05} = \sqrt{37} \times 1.645 = 10.006$. ANSI/ASQ Z1.9 (ANSI 2008b) provides sampling plans for inspection by variables in an AQL and batch size driven manner analogous to ANSI/ASQ Z1.4.

The primary advantage of variables sampling over attribute sampling in situations where either can be used is a reduced sample size. For example, in the above problem, suppose it was decided to design an attribute sampling plan where each unit is labeled defective if the value exceeds 15 s and non-defective if the value is less than or equal to 15 (regardless of how close the measured value is to 15 s). The minimum sample size for an attribute sampling design with AQL $= 0.65\,\%$ and LTPD $= 5\,\%$ is 45, for the sampling plan ($n = 45, a = 0$). This sample size is greater than the value of 37 required for the variables sampling plan. This advantage of a variable sampling plan requires an additional assumption. Namely, the distribution of the continuous variable must be well modeled using a normal probability distribution. Additionally, in situations where both qualitative and continuous quality characteristics are measured on each unit, it will be required to obtain the maximum sample size across all characteristics, thus possibly negating the sampling advantage of a variables scheme.

Variables sampling plans can also be constructed with a two-sided specification. In these situations, it is necessary to test whether the sample standard deviation is less than a pre-specified limit called the maximum standard deviation (MSD). Details of the two-sided procedure are provided in Chap. 10 of Schilling and Neubauer (2009).

20.6 Types of Sampling Plans

The examples discussed to this point have all considered single stage sampling plans. In a single stage plan, a single sample is selected and a decision is made after assessment of the sample. Often there are advantages for sampling with more than a single stage. As an example, consider the variable sampling plans described in the previous section.

A situation sometimes arises where the variables sampling decision rule leads to a lot rejection decision, even though all items in the sample are within specification. Such a conclusion may seem illogical to a client since all collected evidence in the sample demonstrates acceptable quality. An alternative to the single stage variables sampling plan that avoids this situation is a two-stage mixed variables-attributes plan (see, e.g., pages 275–289 of Schilling and Neubauer 2009). One such plan referred to as an independent mixed plan can be described in the following manner:

1. Obtain a sample of size n_1.
2. Test the sample against a variables-acceptance criterion that meets AQL and LTPD requirements.
    (a.) Accept the lot if the test meets the acceptance criterion.
    (b.) Collect a second sample if the test fails to meet the acceptance criterion.
3. Obtain a second sample of size n_2 if necessary per (2b).
4. Test the second sample against criterion based on an attributes design and either accept or reject the lot based on this test.

To demonstrate, suppose we desire an acceptance sampling plan for a quality attribute with an upper specification limit U. Set AQL=0.65 % ($p_1 = 0.0065$), LTPD = 5.0 % ($p_2 = 0.05$), $\alpha = 0.05$, and $\beta = 0.10$. Perform the following steps to determine the sample sizes and decision rule for the mixed design.

1. Split the probabilities of acceptance $1 - \alpha$ and β into the two stages such that $1 - \alpha \geq 1 - \alpha_1$ and $\beta \geq \beta_1$. Assume here $1 - \alpha_1 = 0.75$ ($\alpha_1 = 0.25$) and $\beta_1 = 0.08$. Then the risk probabilities for the attribute sampling design in the second stage are

$$\alpha_2 = \frac{\alpha}{\alpha_1} = \frac{0.05}{0.25} = 0.20, \tag{20.7}$$

and

$$\beta_2 = \frac{\beta - \beta_1}{1 - \beta_1} = \frac{0.10 - 0.08}{1 - 0.08} = 0.0217. \tag{20.8}$$

In this manner, $1 - \alpha = (1 - \alpha_1) + \alpha_1(1 - \alpha_2) = 0.75 + 0.25(1 - 0.20) = 0.95$ and $\beta = \beta_1 + (1 - \beta_1)\beta_2 = 0.08 + (1 - 0.08)0.0217 = 0.10$.

2. Compute the decision rule for the first stage to accept the lot if

$$\bar{X} \leq U - kS \tag{20.9}$$

where from (20.3)

$$k = \frac{Z_{p_2}Z_{\alpha_1} + Z_{p_1}Z_{\beta_1}}{Z_{\alpha_1} + Z_{\beta_1}} \tag{20.10}$$

$$k = \frac{1.645(0.674) + 2.484(1.405)}{0.674 + 1.405} = 2.22 \text{ (rounded up)}. \tag{20.11}$$

3. Compute the sample size for the first stage using (20.5) with values α_1 and β_1. Here $Z_{\alpha_1} = Z_{0.25} = 0.674$, $Z_{\beta_1} = Z_{0.08} = 1.405$, $Z_{p_1} = Z_{0.0065} = 2.484$, and $Z_{p_2} = Z_{0.05} = 1.645$ so that

$$n_1 = \left(\frac{0.674 + 1.405}{2.484 - 1.645}\right)^2 \left(1 + \frac{2.22^2}{2}\right)^2 = 21.1 \text{ (rounded up to 22)}. \tag{20.12}$$

4. Now determine an attribute design for the desired AQL and LTPD with α_2 and β_2. In this example, suppose we want the minimal sample size attribute design with AQL = 0.65 % where $\alpha_2 = 0.20$ and LTPD = 5.0 % with $\beta_2 = 0.0217$. This provides the design ($n_2 = 75, a = 0$).

To compute the OC curve for this two-stage sample plan, one can compute the probability of acceptance for a given percentage defect rate π using the equation

$$P(A|\pi) = P(A_1|\pi) + (1 - P(A_1|\pi)) \times P(A_2|\pi) \tag{20.13}$$

where $P(A_1|\pi)$ is the probability of acceptance given π in the variables plan of the first stage and $P(A_2|\pi)$ is the probability of acceptance given π in the attributes plan of the second stage.

In the example, if $\pi = 0.65\,\%$, using (20.6) $P(A_1|\pi) = 0.767$ and using (20.1) $P(A_2|\pi) = 0.613$. Using (20.13), the probability of acceptance when the true percentage defective is 0.65 % is $0.767 + (1 - 0.767)(.613) = 0.91 = 91\,\%$ which is less than the desired value of 95 % as required. If $\pi = 5.0\,\%$, $P(A_1|\pi) = 0.084$ and $P(A_2|\pi) = 0.021$. This means the probability of acceptance when the true percentage defective is 5.0 % is $0.084 + (1 - 0.084)(.021) = 0.10 = 10\,\%$ as required.

The same requirements as this design could be attained with the single sample attribute design ($n = 45, a = 0$). Table 20.1 compares the expected sample sizes required for the mixed design to this single sample design. The average sample size (ASN) for a mixed design is defined by the formula

$$\text{ASN} = n_1 + (1 - P(A_1|\pi))n_2. \tag{20.14}$$

Note that when the error rate π is relatively low, the sample size is much less for the mixed design. However, when π is relatively large, the mixed design sample size is greater than the sample size of 45 required by the single sample attribute design. Reference Schilling and Neubauer (2009) also describes a dependent mixed plan in which data from the variables sample is combined with data from the attributes sample in order to make a decision at the second stage.

Many decision procedures with multiple sampling stages, such as the (USP 2015b) standard (USP 2015a) for content uniformity and the (USP 2015c) standard (USP 2015b) for drug dissolution appear similar to acceptance sampling plans. However, as stated in the USP, these tests are simply standards that must be met whenever a sample is tested from a batch. Despite their appearance, they are not intended to be sampling plans as described in this chapter and should not be treated as such.

20.7 Describing Lot Quality in Terms of an Overall Measure

Acceptance sampling plans are closely related to validation testing commonly used for medical devices. As defined here, a validation test seeks to demonstrate that the true percentage of defectives in a population is no greater than some value with a stated level of confidence.

To demonstrate, consider the sampling plan ($n = 80, a = 1$) shown in Fig. 20.2. The computed 90 % upper bound on the true percentage of defective units (π) based on the binomial distribution when a random sample of 80 units has 1 defective unit is 4.8 %. Note this upper bound is the same value as the LTPD for the sampling plan

($n = 80, a = 1$). Thus, in contrast to what we believe is a common misconception, the single measure that best describes the overall quality of a lot is the LTPD and not the AQL. For this reason, the focus of many regulatory recommendations on AQL rather than LTPD is curious. In the language of validation testing, one states that with 90 % confidence the true percentage of defectives in the population can be no greater than 4.8 % (i.e., the LTPD).

Unlike acceptance sampling, a validation test is a one-time activity for a stable process. If validation is planned to be followed by ongoing acceptance sampling, the validation criterion might be as small as the planned AQL in order to avoid a high rate of rejected lots during the acceptance sampling process.

As an example where validation testing is performed apart from acceptance sampling, regulatory agencies expect companies to verify that changes made in response to a corrective and preventative action (CAPA) actually work to eliminate the root cause of a non-conformance failure. To do this, it is required to examine data collected after the CAPA and demonstrate that the failure rate intended to be improved by the CAPA satisfies the desired goal. Typically a protocol is drafted that states a post-change sample must satisfy some test criterion related to an upper value for the new defective rate. To demonstrate, multiple valves on a bulk drug substance (BDS) vessel were not properly maintained before use in production. A CAPA was created to improve procedural controls for BDS vessel maintenance. A protocol to test the effectiveness of the CAPA was created where a total of 150 BDS vessels used in the process are to be evaluated for any instances where the BDS vessel was used without completion of required maintenance. This CAPA will be considered successful if there are $a = 0$ failures in $n = 150$ samples selected after the CAPA was instituted. Based on the binomial probability distribution, the upper 95 % confidence bound on the true defective rate for a sample of 150 with no defects is 2 %. Thus, successful execution of the protocol ensures the true defect rate of the process is no greater than 2 % with 95 % confidence. Figure 20.7 presents the relationship between sample size and the desired upper bound on the defect rate based on the number of allowable defects (a) in a sample with 95 % confidence.

20.8 An Opportunity for Innovative Statistical Impact

An often overlooked aspect of acceptance sampling is that calculations assume one has a perfect measurement device. A perfect measurement device would correctly record every non-defective item as "non-defective", and every defective item as "defective". This is clearly not the case in many CMC applications. For example, when visual testing for visible particles, it is possible for an operator to miss seeing a visible particle, and classify a defective unit as "non-defective". It is probably less likely that the operator sees a particle when one does not exist and classifies a non-defective unit as "defective". Table 20.2 reports the probabilities of misclassification (c_1 and c_2) that can occur.

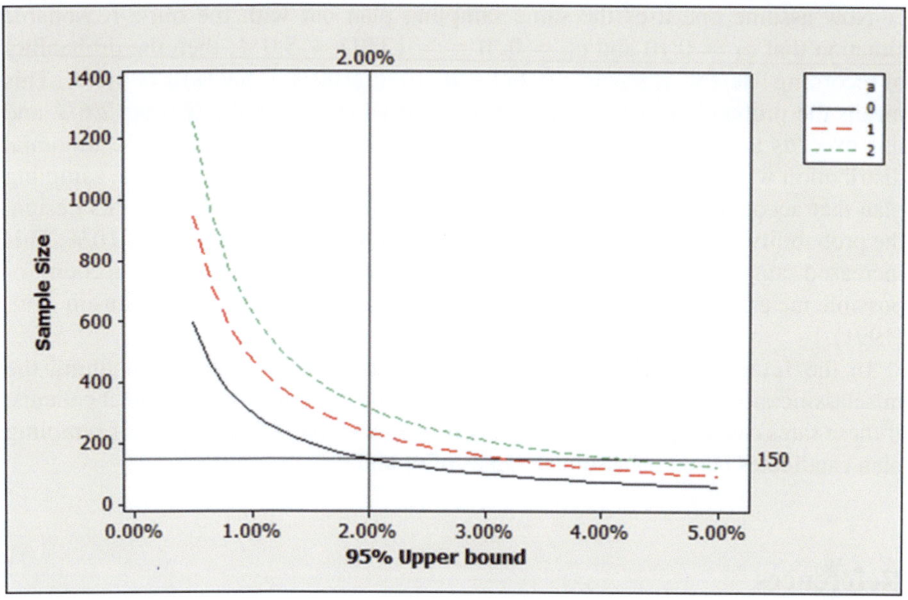

Fig. 20.7 Relationship between sample size and desired 95 % upper bound on defect rate

Table 20.2 Misclassification errors

Truth	Measured non-defective	Measured defective
Non-defective	Correct	c_1
Defective	c_2	Correct

If $c_1 = c_2 = 0$, then the measurement process is perfect. Let π represent the true percentage of defective items in the lot to be inspected (this is the value that appears on the horizontal axis of an OC curve). The decision based on the sampling plan requires a count of the number of measured defective items in the sample. The true probability of recording defect is a function of π, c_1, and c_2. In words, it is the sum of the probability of recording "defect" when a defect truly exists plus the probability of incorrectly recording "defect" when the item is non-defective. Algebraically, this is expressed as

$$\pi_O = \pi(1 - c_2) + (100\,\% - \pi)c_1. \tag{20.15}$$

To account for measurement error, calculations for selecting a sampling plan should be based on π_O as opposed to π.

To demonstrate, consider a method used for measuring the presence of visible particles in a vial. It is desired to select a single sample attribute design with $a = 0$, an AQL no greater than 0.65 %, and an LTPD no greater than 5 %. A sample design that meets these criteria assuming $c_1 = c_2 = 0$ is $(n = 45, a = 0)$.

Now assume one uses the same sampling plan but with the quite reasonable situation that $c_2 = 0.10$ and $c_1 = 0$. If $\pi = \text{LTPD} = 5.0\,\%$, then the probability of recording "defect" is $\pi_O = 5.0\,\%(1 - 0.10) + (100\,\% - 5.0\,\%)0 = 4.5\,\%$. This means the probability of passing a lot ($a = 0$ when $\pi_O = 4.5\,\%$) is 12.6 % and this exceeds the desired risk of 10 %. (This calculation is based on the binomial distribution with $n = 45$, $a = 0$, and $\pi_O = 4.5\,\%$). A more appropriate sampling plan that accounts for the measurement error is ($n = 50, a = 0$). For this design, the probability that a lot is passed with $\pi_O = 4.5\,\%$ is the desired rate of 10 %. This increased sample size is necessary to provide the desired protection and account for possible measurement error. For more information on this topic, see Johnson et al. (1991).

In the future it would seem more effort should be focused on defining the misclassification rates c_1 and c_2 and using this information to better protect patients. If these rates cannot be estimated, it may be reasonable to develop a set of sampling plan candidates based on assumed values of c_1 and c_2.

References

American National Standards Institute (2008a) American National Standard: Sampling procedures and tables for inspection by attributes, ANSI/ASQC Standard Z1.4-2008. American Society of Quality, Milwaukee

American National Standards Institute (2008b) American National Standard: Sampling procedures and tables for inspection by variables for percent nonconforming, ANSI/ASQC Standard Z1.9-2008. American Society of Quality, Milwaukee

International Conference on Harmonisation of Technical Requirements for Registration of Pharmaceuticals for Human Use, Quality Risk Management Q9 (2005)

Johnson NL, Kotz S, Wu X (1991) Inspection errors for attributes in quality control. Chapman & Hall/CRC/Taylor & Francis Group, Boca Raton

Minitab Statistical Software (2015) Version 16. www.minitab.com

Schilling RG, Neubauer DV (2009) Acceptance sampling in quality control, 2nd edn. Chapman & Hall, London

U. S. Pharmacopeia USP38-NF33 (2015a) USP <790> Visible particulates in injections

U. S. Pharmacopeia USP38-NF33 (2015b) USP <905> Uniformity of dosage units

U. S. Pharmacopeia USP38-NF33 (2015c) USP <711> Uniformity of dosage units

Chapter 21
Process Capability and Statistical Process Control

Stan Altan, Lynne Hare and Helen Strickland

Abstract Pharmaceutical drug substances and drug products are manufactured following a well-defined process in units of batches. Drug product batches consist of a large number of dosage units (tablets, capsules, patches, syringes, vials, etc.) Such processes have inherent variability in critical quality attributes (CQA) as a consequence of incipient differences in starting materials, environmental conditions, equipment control setting and other uncontrolled sources of variability at both the batch and dosage unit levels. Regulatory and business considerations intended to safeguard patient health, require careful statistical monitoring of the batch's CQAs. Adherence to specifications of every batch of drug product is a regulatory requirement for marketing. This chapter summarizes the statistical methods used to carry out the statistical characterization of the manufacturing process's ability to consistently produce acceptable product (process capability) and associated monitoring tools (statistical process control). For reporting purposes, a process capability index (e.g. C_{pk}) or process performance index (e.g. P_{pk}) is calculated based on two quantities: (1) the variability of the CQA (standard deviation, σ), and (2) the comparison of that variability with the specification, typically as the quotient of the specification width to 6σ. Statistical process control refers to graphical methods that permit detection of departures from target over time of manufacture by revealing process shifts greater than capability or inherent variability alone would allow. Details of these charts are described along with additional rules for increased vigilance in detecting assignable causes (beyond capability) of variation. Alternative charts for serially correlated data are also shown, and recommendations for the use of these tools in practical applications and for reporting quality indices are provided.

S. Altan (✉)
Janssen Pharmaceuticals R&D, Raritan, NJ 08869, USA
e-mail: saltan@its.jnj.com

L. Hare
Statistical Strategies, LLC, Plymouth, MA 02360, USA

H. Strickland
Glaxosmithkline, Raleigh, NC 27606, USA

L. Zhang (ed.), *Nonclinical Statistics for Pharmaceutical and Biotechnology Industries*, Statistics for Biology and Health, DOI 10.1007/978-3-319-23558-5_21

Keywords Quality control statistics • Shewhart charts • EWMA charts • cGMPs • ICH guidances • Quality indices

21.1 Introduction

The beginnings of Quality Control Statistics took place in 1924 with Walter A. Shewhart's description of a control chart (Shewhart 1980). Since then, the use of statistical methods for quality control and characterization of industrial manufacturing processes has rapidly expanded with important contributions made by Ott, Deming, Juran, Montgomery and others (see Grant and Leavenworth 1996; Deming 2000; Juran and DeFeo 2010; Ott et al., 2000). Deming's and Juran's methods found great success in Japan in minimizing variability and waste and in improved quality, which stimulated American industries to improve their processes as well. Many have learned that it is simply good business to focus on quality.

Unfortunately, the use of such statistical approaches by the pharmaceutical industry has generally lagged as evidenced by FDA 483 citations and warning letters for failure to meet the statistical requirements of current Good Manufacturing Practices (cGMPs). But in the face of recent regulatory documents such as FDA's guidance for industry on process validation and the International Committee for Harmonization's (ICH) guidelines Q8, Q9 and Q10 for product development and risk management, it's clear that the industry is being encouraged to embrace statistical approaches to improve product and process understanding. The large body of regulatory guidances and references that relate to concepts of quality, risk and process validation will drive more extensive statistical approaches to track and improve manufacturing processes and medicinal products destined for treating a plethora of human illnesses, keeping in mind that the common objective by producers, regulators and consumers is to enhance and protect patient health.

Given this backdrop, the objective of this chapter is to provide a concise overview of what statisticians working in the pharmaceutical industry need to know about general principles of process capability estimation and statistical process control. It is intended as a resource for gaining greater process understanding of quantifying process capability and performance and controlling processes so that they remain as consistent with their capability as possible. We also provide some useful references and cautions with practical advice based on our combined experience.

21.2 Regulatory Aspects of Pharmaceutical Manufacturing, Quality and Statistics

The pharmaceutical industry is arguably the most highly regulated industry in the United States, Japan and Europe, as well as in much of the rest of the world. Notions of quality have been embodied in various regulatory and compendial documents.

It is outside the scope of this chapter to delve very deeply into these documents but to the extent that quality has a regulatory definition, and employs statistical tools, some discussion is unavoidable. In the References section, a separate list headed *Regulatory, Compendial and Standards Guidances* lists many of the key documents that one should refer to for acquiring a more comprehensive perspective of quality and how quality is viewed by US and European regulators and the pharmaceutical industry. In the following paragraphs, we provide a brief overview of important concepts relevant to notions of quality and the role of Statistics in pharmaceutical manufacture.

While our focus is on the quality of manufactured product, readers should also understand that the principles of assessing process capability and performance and of maintaining high levels of control apply to all business phases, including procurement, marketing, sales and distribution.

An important definition of quality is given in ICH Q6A where it states that 'the quality of drug substances and drug products is determined by their design, development, in-process controls, GMP controls, and process validation, and by specifications applied to them throughout development and manufacture' and consequently it is understood as 'fitness for use'. A regulatory definition of a specification is given as follows:

> A list of tests, references to analytical procedures, and appropriate acceptance criteria that are numerical limits, ranges, or other criteria for the tests described. It establishes the set of criteria to which a drug substance or drug product should conform to be considered acceptable for its intended use. ... Specifications are critical quality standards that are proposed and justified by the manufacturer and approved by regulatory authorities.

Thus, the notion of quality is inextricably tied in with the notion of specifications, so in that sense, as long as the defined specifications are met, the quality of the product is assured. Recent statements from the FDA have extended the definition to include the reliable production of high quality drugs and are even proposing to develop metrics around pharmaceutical production quality (Woodcock 2013). They point to recent drug shortages as being connected to the quality of manufacturing and production facilities and the industry not being incentivized to invest in infrastructure changes to improve the quality of manufacturing (Woodcock and Wosinska 2013).

Specifications themselves derive in large part from experience gained during the pre-marketing approval clinical trials conducted to establish the efficacy and safety of the product. The critical quality attributes associated with efficacy and safety are then surrogates for the clinical trials, assuring patient response, and hence quality, once the product is commercialized following approval to market. We will return to the question of specifications again while discussing cGMP rules later.

The notion of what constitutes "better" quality then is somewhat debatable and other guidances and references provide alternate definitions. ICH Q10 states quality is "the degree to which a set of inherent properties of a product, system, or process fulfills requirements". ICH Q9 defines quality as "the degree to which a set of intrinsic properties of a drug product, its underpinning manufacturing process, and any supporting processes fulfils the pre-determined criteria", in other words quality is conformance to requirements. Juran (1992) defines quality as fitness for use and freedom from deficiencies and Montgomery (2000) states 'quality is inversely proportional to variability'. In general, we will understand the quality of a pharmaceutical product in relation to the extent to which it meets specifications across a series of critical tests, which in turn assures fitness for purpose, i.e. the product's suitability for its intended purpose. Regulatory oversight is necessary in this regard since product quality cannot be discerned by the patient except to a limited extent, and hence quality is not driven by end consumer demand.

cGMP rules suggest that the imposition of specifications makes quality an either-or proposition. Consequently, there has been enormous attention paid to the notion of "compliance" to specifications, that quality is synonymous with compliance. That is, if you have a compliant process, then the quality is assured. This has led to contradictions in practice, discouraging greater process understanding by rewarding minimal testing (LeBlond et al. 2005). The fear being that additional analytical testing simply meant more chances to fail a specification, rather than an opportunity to acquire greater understanding of the properties of the pharmaceutical product and the process. This paradigm is being challenged by Quality by Design (QbD) concepts (see Chap. 18 for details). This is discussed in ICH Guidances Q8, Q9, Q10 and Q11 which imply greater testing, possibly through Process Analytic Technology (PAT) tools, to gain greater understanding of the manufacturing process leading to a formal risk control approach. The goal of QbD is to build quality into a pharmaceutical product through a causal understanding of the important factors impacting product performance and their linkages to clinical outcomes.

Recent publications have advocated a multivariate Bayesian approach to manufacturing risk control (Peterson and Yahyah 2009). The multivariate region of the critical parameters that assure quality in relation to critical attributes is characterized, and a risk profile is constructed for a "Design Space" to guide manufacturing. In other words, the fear of testing quality in will disappear by designing quality into the process. Readers are encouraged to examine Chaps. 15 and 18 for a discussion on this subject. In addition, the recent Process Validation Guidance (2011) sets forth a statistical requirement to justify sampling plans in the commercialization stage based on information gathered during development and product performance qualification (PPQ) stage. This too challenges the cGMP compliance paradigm by requiring a statistical basis for greater understanding of products and processes. Commercial scale production would be coupled with a formal Quality Risk Management (QRM) program to proactively identify and control potential manufacturing issues to minimize risks to the patient from issues related to subpotent, suprapotent, adulterated, improper packaging or otherwise poorly performing product.

Several sections in the document cGMP pertain directly to the need for statistical approaches in relation to specifications, process control and other aspects of pharmaceutical manufacture. Applying statistical methods for validating the manufacturing process and control of the process are discussed in sections 211.100 and 211.110(a). Specific language given in section 211.165(d) "to assure that batches ... meet ... appropriate statistical quality control criteria ...." suggests the use of statistical control charts and process capability studies. The 2011 Process Validation guidance goes further in proposing a structured approach relying on statistical justification for developing sampling plans and process monitoring.

Setting specifications is addressed in section 211.110(b), where it states:

> Valid in-process specifications for such characteristics shall be consistent with drug product final specifications and shall be derived from previous acceptable process average and process variability estimates where possible and determined by the application of suitable statistical procedures where appropriate.

Clearly, specifications based on empirical studies and appropriate statistics are being recommended.

More recently, the role of the statistician and the application of statistical tools in quality management are strongly implied in ICH Q10 where the following requirements are discussed:

- Use effective monitoring and control systems, for assessing the capability of processes and product quality (control charts).
- Identify areas for continual improvement by identifying potential problems and preventing them before they result in rejects or recalls (regression models, root cause analyses).
- Identify sources of variation affecting process performance and product quality for potential continual improvement activity to reduce or control variation (DoEs, Variance Components modeling).

Finally, we should not overlook the requirement that qualified personnel carry out these tasks as stated in section 211.25 of the cGMPs:

> Each person engaged in the manufacture, processing, packing, or holding of a drug product shall have the education, training, and experience, or any combination thereof, to enable that person to perform the assigned functions. Training shall be in particular operations that the employee performs ....

21.2.1 More on Manufacturing Process and Quality Considerations

It is clear that both the producer and the consumer are best served when manufacturing process understanding is thorough. From the manufacturing perspective, process understanding leads to improved control, the advantages of which include:

- Greater certainty of process performance in terms of overall equipment effectiveness (OEE),
- Greater ability to guide processes to achieve the desired results, and
- Greater planning efficiencies due to the reduced uncertainty of production throughput.

Meeting regulatory requirements through analytical testing is one of the industry's biggest costs. Improved process understanding may lead to a reduced need for testing resulting in substantial savings in time, material and costs.

From the consumer's perspective, greater process understanding leads to improved process control. This leads to reduced dosage variation meaning that closer and more consistent proximity to target physical and chemical properties such as the prescribed quantity is more likely to be received, time after time, by the patient. The health implications are obvious.

Productivity and public health, therefore, are the motivators of process validation as described in the previous chapter and documented in the Guidance for Industry—Process Validation (FDA 2011).

It is important to understand that finished product is judged for conformance to specifications by some kind of sampling inspection. There are numerous resources that describe end-of-line acceptance sampling [ANSI/ASQ Z1.9-2008, Z1.4; ASTM E2709]. These documents speak to after-the-fact evaluation; that is, inference through sampling concerning conformance of finished product to specifications. And while acceptance sampling has its place in assuring quality, it is also true that when the sampling is complete, we may be left with rejected product requiring expensive disposition. It is incumbent on manufacturers, therefore, to install systems to build quality into the process; thereby assuring few, if any, surprises come sampling time. Refer to Chap. 19, Process Validation for a thorough discussion on this subject.

More recently, PAT has set the stage for real time release based on a comprehensive quality assessment from a large sample size, conceivably a large proportion of the batch. PAT considerations are outside the scope of this chapter but SPC tools could still be appropriate in the context of PAT monitored processes and products.

The notions of assessing process capability and establishing process control began as early as 1924 when Walter Shewhart, a physicist, engineer and statistician working at Western Electric Company began wondering why repeated experiments, however closely controlled, failed to produce the same results. While some of his colleagues blamed him for lack of consistency, Shewhart had a different idea. Basically, he understood that there is never a signal without noise, and in a brief memo to his boss, proposed the idea of the control chart. Mathematically, its basis is $y = f(x) + \varepsilon$, which says that the result of any measurement, y, has two components. The first, $f(x)$, is the signal or what the controlled factor or factors are, and the second, ε, is the noise. The good news is that through replication, the noise can be brought down to manageable levels so the signal can be heard over it. His early control chart capitalized on this idea by reducing the noise in order to detect process incursions, today called assignable sources of variation. It's important to point out that ε can consist of several components, for example noise related to the measurement system, another component due to variations in starting materials and yet another component due to processing conditions. In certain cases, these components may be separable but in many cases, they will remain confounded. Their composition and possible existence are important to recognize and take into account.

Shewhart's idea took hold, and over the decades we have seen many authors build on it to produce Statistical Quality Control, Statistical Process Control, and Total Quality Management. It has led to the present-day Lean Six Sigma initiative which sequences and links a variety of statistical methods and augments earlier programs by incorporating financial analyses. Resulting financial savings roll into billions of dollars (Hare 2003).

21.3 Process Capability and Performance

In business parlance, process capability is often referred to as entitlement. That is, if an owner or manager purchases a piece of equipment he or she is "entitled" to receive its most consistent results. That terminology may suffice in the board room, but it doesn't really capture the essence of the concept or aid practical application.

Admittedly, the concept of process capability is a bit nebulous and subject to various interpretations. We choose to think of it as the inherent, intrinsic variation of the process. It is the variation the process exhibits when it runs at its absolute best. There are no exterior disturbances, no haphazard shocks to the process, no tweaks, and no adjustments, illicit or otherwise; only pure random variation. For pharmaceutical drug products, this consists of both dosage unit variation as well as analytical uncertainty. Naturally, the variation due to process capability alone should be very much smaller than the specification range. For example, if the response of interest is normally distributed, its process capability standard deviation should be smaller than one-sixth of the specification range to assure consistency in producing the specified product.

Performance variation, on the other hand, is a measure of the variation experienced by the consumer or end user. It includes capability variation along with all the other sources of variation such as environmental changes, variation induced by different batches using various raw material supplies, process adjustments during time of batch manufacture, ambient conditions and so on. Ideally, we would want performance variation to be as close to capability variation as possible. But the difference between performance and capability should be a major subject of attention: if the difference is large, it is likely that the process would benefit from interventions to reduce the difference. This may be true even when all production is within specification because departure from process capability may result in production line inefficiencies.

Assessing process capability is no easy chore. Some text books teach users to wait until the process reaches equilibrium, then take roughly 30 samples and then calculate their standard deviation. We have some problems with that advice. How might we know when the process reaches a state of equilibrium? How do we know that the recommended samples are representative of the process, much less truly representative of process capability? So the measurement of process capability is more complicated than that.

For example, suppose we have a rotary tablet press that produces 30 tablets, one from each of thirty pockets per rotation, and let's say we are interested in tablet thickness. We might want to base our estimate of process capability on the standard deviation calculated from 30 consecutive tablets. Better yet, we might assure representativeness by taking those 30 consecutive tablets repeatedly over, say 8, time periods spaced evenly throughout a production run. We would pool the 8 individual standard deviations yielding a weight capability estimate based on $(8 \times (30-1)) = 232$ degrees of freedom. For greater assurance yet, we might want to include several production runs with perhaps fewer sampling times per production run. The point is that estimates of the process capability made this way would be both representative and independent of process mean changes that might take place from sampling time to sampling time.

A purist may want to drill down even further. What is the variation experienced among repeated tablet thickness from each of the 30 pockets? You could measure that by sampling 60 consecutive tablets and pair tablet 1 with 31, 2 with 32 and so on to measure the within pocket variation. Isn't the resulting variation also a component of process capability? Of course, it is. But practicality steps in at this juncture to recommend that we follow the earlier technique described above and let within-pocket variation be assigned to the capability estimate as a random component. A check to guard against blunders caused by avoiding the within-pocket detail can be made through careful examination of the control charts to be described.

The measurement of process performance is a bit more complicated. Suppose we have the same tablet press as mentioned above. To measure the thickness weight variation experienced by the consumer, we would want to assure representation of normal production. Sampling within one batch alone is not adequate. Instead, we should have upwards of 20 or 30 batches in our sample. In that case, we would carry out a variance components analysis, combining the within and between batch variance components to form the estimate of the performance variance.

21.4 Assessing Process Capability

The 2011 Process Validation guidance speaks to three stages: (1) Process Design, (2) Process Performance Qualification and (3) Continued Process Verification. During Stage 1, DoE tools are brought to bear to study important process parameters and linkages to critical quality attributes. During Stages 2 and 3, product batches are produced and evaluated for many quality related parameters including content uniformity, assay, moisture, degradation compounds, and in the case of tablet manufacture, hardness, friability and weight. Additional measurements will be taken depending on the nature of the product. Associated with each of these measurements is a target value and accompanying specification limits. One goal during these stages of new drug manufacture is to assess consistency with compendial requirements and preliminary specifications. The methodology given in ASTM 2709–12 using the specified acceptance test would be one way to achieve this goal. The methodology computes at a prescribed confidence level, a lower bound on the probability of passing the acceptance procedure. A Bayesian calculation based on the posterior predictive distribution is also an appropriate method to achieve this purpose, and has the added benefit of providing inferences about future performance. Refer to Chap. 19 for greater detail about the Process Validation requirements.

Another goal, which is entirely consistent with the first, is to assess process capability. As discussed above, process capability is the inherent, intrinsic variation of the process. It is not necessarily what the process is doing or what it has done, so much as it is what the process can do when it operates at its best. And, of course, when it operates at its best, it is unencumbered by extraneous sources of variation such as system shocks that might result from such events as shift changes, climatic changes, raw material changes and other assignable causes of variation. In this context, "assignable" is meant to contrast with "common cause" variation. Another perspective on the interpretation of capability and performance concepts and indices is given in ISO-22541 Part 1(International Standards Organization 2009).

Appropriate sampling is essential if we are to isolate the common cause or capability variation from the overall variation experienced during the initial phase of new drug manufacture. Careful consideration should be given to sampling in order to assure that the assessment of capability variation is not contaminated by assignable sources of variation. The brief rotary tablet press example described above exemplifies the capability assessment strategy. There, we spread sampling throughout the production run, but to assess capability, we bore down to very local variation assessment, pooling it across the entire production run. By sampling in that manner, we strive to be fully representative of the process, beginning to end, and we build in a check against nonhomogeneous variation across the production period. Such nonhomogeneous variation would be revealed by simple scatter plots showing replicated observations by time. If present, it might indicate erratic process behavior or it might indicate that there are additional sources of variation not accounted for by our sampling. In either event, the accumulation of process data from systematic samples provides information that would otherwise not be available. If there are sources of variation for which we have failed to account, a repeat of the sampling should be done with those taken into consideration.

As an example, suppose a process of interest produces tablets, and we are concerned about the uniformity of tablet thickness. The tableting device has 10 pockets which engineers assure us are precisely milled to the same dimensions. The sampling plan is to measure thicknesses twice per pocket at each of 36 times spaced uniformly throughout the production run. The resulting data are shown partially in Table 21.1.

A graph of the data is appropriate prior to any kind of analysis (see Fig. 21.1).

There should be no great surprises. However, we might wonder why sometimes the spread of duplicate observations is very small, while at other times it is considerably larger. Are we looking at random variation? Did the engineers get it wrong? We'll come back to those questions.

In the meantime, we might consider what the data analysis says. A fixed effects analysis of variance (ANOVA) model with terms shown in Table 21.2 shows an error

Table 21.1 Replicate tablet thickness by sample time and pocket position

Position	Replicate	Time 1	Time 2	Time 3	...	Time 36
1	1	11.30	10.60	10.57	...	11.08
2	1	10.98	10.82	11.24	...	11.09
3	1	11.11	11.35	11.14	...	11.09
4	1	11.20	10.52	11.28	...	10.45
5	1	11.25	11.11	10.37	...	10.68
6	1	10.76	10.85	10.50	...	10.39
7	1	10.42	11.30	11.13	...	10.54
8	1	10.81	10.53	10.89	...	10.41
9	1	10.70	10.82	11.26	...	10.31
10	1	10.67	11.29	10.64	...	10.42
1	2	11.13	10.98	10.87	...	10.27
2	2	10.79	10.47	10.65	...	10.64
3	2	11.21	11.18	10.88	...	11.09
4	2	10.88	10.45	11.30	...	10.85
5	2	10.51	11.07	10.40	...	11.19
6	2	10.57	10.54	10.73	...	10.46
7	2	10.72	10.87	11.23	...	10.90
8	2	10.72	11.26	11.06	...	10.34
9	2	10.67	10.98	10.61	...	10.68
10	2	11.06	10.52	10.98	...	10.39

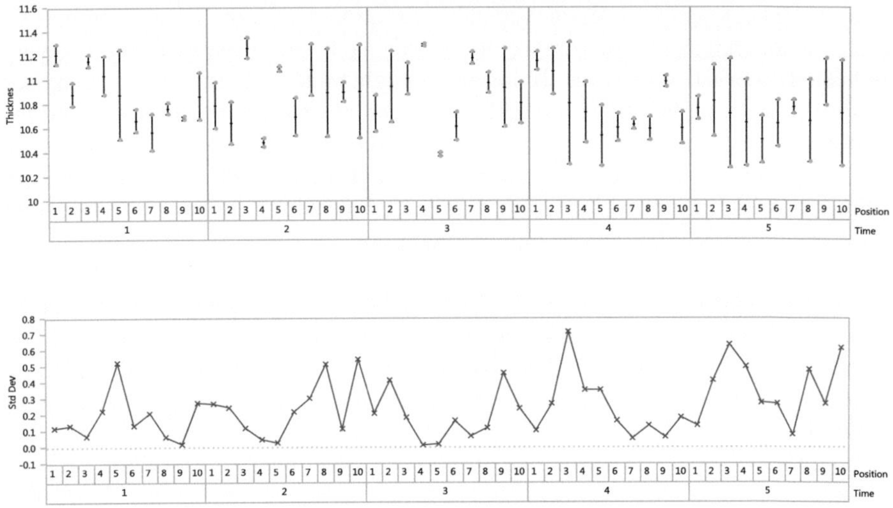

Fig. 21.1 Variability chart for thickness—first five time periods only

Table 21.2 Fixed effects analysis of variance—thickness data

Source	DF	Sum of squares	Mean squares	F ratio	Prob > F
Position	9	0.52	0.058	0.647	0.757
Time	35	6.84	0.195	2.189	< 0.001
Time*Position	315	25.61	0.081	0.911	0.804
Error	360	32.15	0.089		

Table 21.3 Variance components model—thickness data

Random effect	Variance component	Percent of total
Position	0	0
Time	0.006	6.1
Time*Position	0	0
Residual	0.085	93.9
Total	0.091	100.0

or residual mean square of 0.089. This is actually composed of pooled, single degree of freedom standard deviations from duplicate readings at each combination of time and position. As they are duplicates, quickly taken in time, there is no time for assignable cause variation to creep into their calculation. Therefore, they measure capability. The square root of this mean square, which comes to 0.299, is probably a close estimate of the true process capability standard deviation. This estimate includes measurement error which, in many cases, may be assumed to be small. However, if there is any doubt, a complete measurement systems analysis (MSA, Montgomery 2000) should be undertaken.

Notice also from the ANOVA table that the mean square for Time-by-Position is very close to the error mean square. This suggests that there is no "weaving" or changing among positions from sampling time to sampling time. Further, we can see that position differences are small, but that there are time differences in relation to the error mean square ($p < 0.001$)—the process mean is shifting. Therefore the process throughout the batch manufacture is shifting, and it is not, strictly speaking, in control.

Now, if we approach the situation from a different perspective and declare all effects random, we find variance components computed using a restricted maximum likelihood (REML) approach as shown in Table 21.3.

It shows that most of the variation is due to the process capability itself, but there is some variation due to time drift. An estimate of the performance variance is derived from the sum of the variance components. Some may point out that this is not entirely correct because there are actually fixed effects in the model. "Position" is a fixed effect, technically. So the estimate of performance variation should be taken as a useful approximation. It is 0.091, and it is a measure of the variation experienced by the end user of product from the batch under study. Its square root, the performance standard deviation, is 0.301. In those cases where the process capability is a smaller proportion of the total variance, the performance standard deviation will be large compared to the process capability standard deviation. More often than not, this will be the case in pharmaceutical manufacturing due to shifting means caused by assignable sources of variation, which may not be known.

Of course, there are other considerations. After all, this is only one batch, and other batches may show different characteristics. The process performance qualification would benefit from replication across multiple batches. If necessary, additional replication of the same sampling plan across additional batches can be carried forward into the continued process verification stage. In the case of this example, given what we've seen from the first batch, it is probably not necessary to examine duplicate samples from each position and time combination for the remainder of the batches. But the concept of duplicate or back-to-back sampling should not be dismissed from future similar studies.

21.4.1 Normality

The estimates of process capability and performance derived in the manner shown can be used for purposes of prediction provided safe assumptions can be made about the underlying data distribution. Many, but not all pharmaceutical responses can be safely assumed to follow the normal distribution. If that is the case, verified via various tests for normality or even approximately confirmed through a simple normal probability plot, then an interval of plus or minus three capability standard deviations about the mean will most likely serve as a useful test to determine if the process is capable of meeting specifications, assuming the process is stable. Similarly, an interval of plus or minus three performance standard deviations will be useful to determine if it is actually meeting those specifications. This, of course, assumes the existence of long-term, representative data involving multiple batches, 30 or more.

If the underlying data distribution is not normal, a normalizing transformation may be applied (e.g. Box-Cox or others, see Atkinson 1985). In this connection, when data sets are large, a few outliers or slight departures from normality may lead to a low p-value for the test of significance for normality. In practice, central tendency, symmetry and homogeneity are more important considerations in this context rather than passing a test for normality.

21.5 Statistical Process Control and Control Charts

The previous paragraphs have presented an overview of process capability and its context in the pharmaceutical industry. It must be understood that the assessment of process capability and the careful control of quality consistent with that capability are essential components to the key goal of producing products that consistently meet specifications throughout their expected shelf lives. To reach that overall goal, the capability and control technologies must be combined with shelf life modeling and acceptance sampling, topics discussed in Chap. 21 and Chap. 24, respectively.

We define a process as being out of statistical control if an observation falls outside the limits of their control chart. The Nelson rules (see below) are also useful for this purpose. This is not meant to suggest that manufacturing has lost control of the overall process as long as the levels of variability of the key attributes lead to product that consistently meets specifications.

21.5.1 Shewhart Control Charts

Regardless of the shape of the data distribution, the distribution of sample means is normal. For practical purposes, it is best to have at least four observations in each mean being examined, but in theory, the tendency toward normality holds true for means of any size. Shewhart's original idea for the control chart is based on this principle. The basic idea is to form a graph with time or sequential samplings on the horizontal axis and the mean measured response on the vertical axis. Add a horizontal line showing the target or the grand mean ($\overline{\overline{X}}$), as appropriate, and surround that line with limits based on the center line plus or minus three standard errors:

$$\text{Upper Control Limit (UCL)}: \ UCL = \overline{\overline{X}} + 3s_c/\sqrt{n}$$

$$\text{Lower Control Limit (LCL)}: \ LCL = \overline{\overline{X}} - 3s_c/\sqrt{n}$$

where s_c is the estimate of the capability standard deviation as described above, and n is the number of samples taken at each sampling time.

Shewhart claimed to have chosen three standard errors for economic convenience, but others later followed with probability calculations. He reasoned that if an observed mean strayed beyond the limits, there must have been an incursion of assignable cause variation in the process.

At the time control chart theory was being developed, it was recognized that the routine hand calculation of the sample standard deviation is too tedious and time-consuming to be of practical value, and control charts based on sample ranges were introduced. The control chart limits became:

$$\text{Upper Control Limit (UCL)}: \ UCL = \overline{\overline{R}} + 3\overline{X}/(d_2\sqrt{n})$$

$$\text{Lower Control Limit (LCL)}: \ LCL = \overline{\overline{X}} - 3\overline{R}/(d_2\sqrt{n})$$

where $\overline{R}$ is the average range over many, preferably more than 30, subsets of data and d_2 is a correction factor that adjusts the average range to the standard deviation. Values of d_2 corresponding to the sample size are tabulated in range charts (see Montgomery 2000, inner cover, for example). The use of d_2 leads to a biased and less efficient standard deviation estimate in general (Read 2006). The relative ease of calculating the range in the days when computers were not available made it a reasonable alternative. These days however, computers carry out the calculations. Therefore there is no advantage to using range charts except in those cases where computers and hand-held calculators are not available.

While some may find it tempting to chart extreme values such as the minimum and the maximum values of a set of observations, they should be brought to understand that the Shewhart control chart is not intended for that purpose. Extremes follow special distributions which should be used for those purposes.

Of equal importance to Shewhart control charts for the mean are control charts for the standard deviation. These should be run in parallel to mean charts to alert users to changes in process variation. The rationale for the standard deviation chart parallels that of the mean chart. Sample standard deviations are plotted across time together with a center line and upper and lower control limits:

$$\text{Upper Control Limit (UCL)}: \; UCL = c_4 s_c + 3 s_c \sqrt{1 - c_4^2}$$

$$\text{Center Line}: \; c_4 s_c$$

$$\text{Lower Control Limit (UCL)}: \; UCL = c_4 s_c - 3 s_c \sqrt{1 - c_4^2}.$$

Here $c_4 = \left(\frac{2}{n-1}\right)^{1/2} \frac{\Gamma(n/2)}{\Gamma[(n-1)/2]}$, where n is the number of samples taken during each sampling time and s_c^2 is a precise estimate of the true process capability variance σ_c^2. Also, $\Gamma(r) = \int_0^\infty x^{r-1} e^{-x} dx, \; r > 0$, easily computed in Excel with the GAMMA.DIST function.

Readers may wonder why there is a lower control limit for the standard deviation chart. For samples of size 5 or fewer, the lower control limit for the standard deviation chart is zero, but for larger sample sizes, it is positive but usually small. Still, a sample standard deviation appearing below the lower control limit suggests that the actual process capability standard deviation may be lower than that originally calculated. It may also indicate data arising from non-random or systematic sources which would affect the nominal confidence level in the control chart.

More complete tables of control chart factors are provided in quality control texts (Montgomery (2000)) For the present purposes, it is sufficient to show d_2 and c_4 corresponding to the sample size, n in Table 21.4.

Table 21.4 Constants for Shewhart control charts

Sample size, n	d_2	c_4
2	1.128	0.7979
3	1.693	0.8862
4	2.059	0.9213
5	2.326	0.9400
6	2.534	0.9515
7	2.704	0.9594
8	2.847	0.9650
9	2.970	0.9693
10	3.078	0.9727

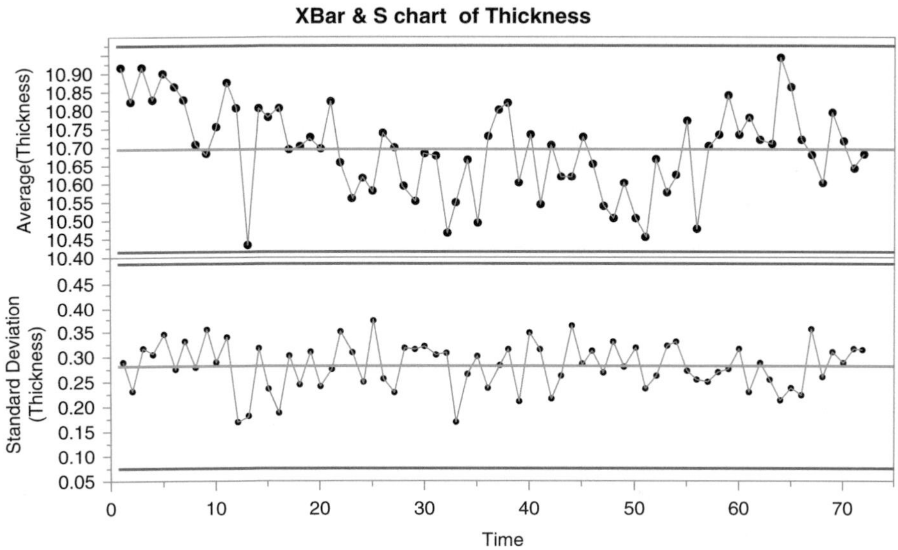

Thickness Limit Summaries

Points plotted	LCL	Avg	UCL	Limits Sigma	Sample Size
Average	10.41905	10.69925	10.97945	Standard Deviation	10
Standard Deviation	0.081502	0.287277	0.493051	Standard Deviation	10

Fig. 21.2 A Shewhart control chart of tablet thickness data

A Shewhart control chart of tablet thickness data similar to those shown in Table 21.1 above is shown in Fig. 21.2. This second data set was generated from the same process as the data shown in Table 21.1, but without benefit of back-to-back sampling from the same pocket. All the sample means, taken over the 10 pockets, are within the control limits, so it might be tempting to declare that the process is "in control" meaning that it is free of assignable cause variation. A closer look reveals some opportunities for improvement. Notice that there are five occasions where there are at least seven consecutive means on one side of the center line. Those are low probability events that should trigger curiosity regarding process interventions.

21.5.2 Nelson and Other Rules

To improve on the speed at which Shewhart control charts find assignable causes of variation, engineers at Western Electric Company developed a set of rules to be used in addition to the usual limits. These rules were rearranged and published by Lloyd Nelson (1984). They are paraphrased as follows where rules 2–8 indicate an out of trend signal and rule 1 is reserved for an out of control signal (Torbeck 2011):

1. One point beyond three standard errors.
2. Nine consecutive points on the same side of the center line within one standard error of the center line.
3. Six consecutive points increasing or decreasing.
4. Fourteen consecutive points alternating, increasing and decreasing.
5. Two of three consecutive points between two and three standard errors on either side of the center line.
6. Four of five consecutive points on either side of the center line beyond one standard error of the center line.
7. Fifteen consecutive points within one standard error of the center line.
8. Eight consecutive points on either or both sides of the center line with none within one standard error of the center line.

But the intent of their use was not so much to detect out-of-control situations and dictate adjustments as it is to help operators focus on the process to seek assignable causes of variation. For example, Rules 5 and 6 should be used if an early warning is desired, and Rules 7 and 8 are tests for stratification. If Rules 7 and/or 8 raise flags, there are two or more sources of variation in the process.

Often, practitioners will include warning limits on their control charts. These are plus or minus two standard errors distant from target. Two consecutive observations on the same side of the target in the warning zone, defined as the area between 2 and 3 standard errors from the target, would constitute cause for concern.

Modern control chart software packages usually contain these rules or variations of them. (JMP, for example, uses a variation of these rules called the Westgaard Rules.) They should be used with caution because over-use will result in false alarms. If Shewhart charts are being applied for the first time on a process, it is advisable to commence with only the 3-sigma control chart limits shown above. Use the standard chart to eliminate initial assignable causes of variation. Once this is accomplished, and more experience is gained, increasingly sophisticated rules may be applied. The rationale behind this recommendation is based on the average run length (ARL) of the chart. With the usual 3-sigma limits, a Shewhart control chart has an average run length of 370. This is the average number of sample times until a false, out-of-control signal is generated by the chart, assuming a stable process. When all eight Nelson rules are applied, the ARL is reduced to approximately 50.

Regardless of which and how many rules are used during control charting, it is important to understand that when a process adjustment is made, the counters for all rules should be reset to zero.

21.5.3 More on Control Chart Limits

In some applications, control charts may be considered too restrictive because they call for action more frequently than is practical. Two situations are addressed below.

21.5.3.1 Charts Permitting Within-Batch Variation

Especially during early applications, before the process can be brought under strict control, it is important to incorporate estimates of variation due to response differences among sampling times within batches into the calculation of control chart limits. These limits would become:

$$\text{Upper Control Limit (UCL)}: \ \text{UCL} = \overline{X}' + 3\left(s_b^2 + \frac{s_c^2}{n}\right)^{1/2}$$

$$\text{Lower Control Limit (LCL)}: \ \text{LCL} = \overline{X}' - 3\left(s_b^2 + \frac{s_c^2}{n}\right)^{1/2}$$

where s_b^2 is the estimate of the between sample time variation within batches and s_c^2 is as defined above. Both estimates of variation can be obtained from a simple variance components analysis. Limits for the standard deviation chart should remain based on the capability standard deviation, as above.

Chart interpretation under these circumstances can be challenging: the chart for the mean will be less sensitive to process shifts due to assignable cause variation. If the between sample time variation is very much larger than the capability standard deviation, the chart will be powerless to detect differences that may be important to process and product integrity. It is incumbent on practitioners, therefore, to work out of the large between-time variation environment as quickly as possible by engaging in efforts to eliminate assignable cause variation. This is accomplished one step at a time, and following each step of assignable cause variation elimination, new limits should be calculated.

21.5.4 Acceptance Control Charts

While process engineers, quality control experts and statisticians focus on keeping a process in a stable state of statistical control, business managers focus on allocation of resources and the assurance that products are within specification limits. The former would argue that if the process is initially shown capable of meeting specifications and, if it is in control, then specification violations should be nil, whereas the latter might argue that if the process capability variation is tiny by comparison to the width of the specification range, the urgency of finding assignable causes of variation revealed by an out-of-control mean could be relaxed to a certain degree, with vigilance on the specification limits.

In that case, acceptance control limits (Freund 1957) might be used. These are:

$$UCL_A = USL - \left(Z_\gamma + \frac{Z_\beta}{\sqrt{n}}\right)\sigma \text{ and } LCL_A = LSL + \left(Z_\gamma + \frac{Z_\beta}{\sqrt{n}}\right)\sigma,$$

where USL, LSL are upper and lower specification limits, respectively γ is the process fraction nonconformance to be rejected with probability $1 - \beta$ and Z_α is the standard normal deviate corresponding to the probability of the area to the right of Z_α , for example, $Z_{0.05} = 1.65$.

These limits can be used in conjunction with the usual Shewhart limits to show conformance to both capability and specifications (Fig. 21.3). In any event, it is important that a control chart for the standard deviation (or range) be maintained in conjunction with acceptance control charts because the acceptance control chart is no longer valid if the standard deviation goes out of control.

One should exercise a little care in placing specification limits on this chart. The plotted points represent means of samples of size n. The scale of the chart could be confused if specification limits apply to individual reported values or dosage units, rather than means.

One pitfall to the use of acceptance control charts is that they might encourage complacency with regard to seeking assignable sources of variation when an out-of-control, but not out-of-specification situation is encountered.

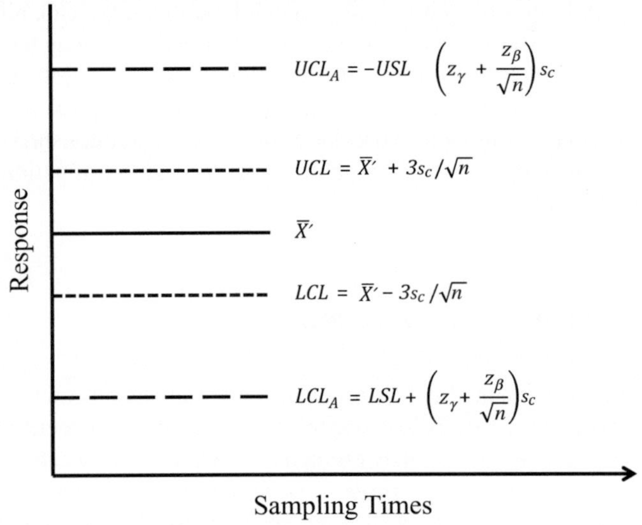

Fig. 21.3 Acceptance control chart limits

21.5.5 Rational Subgroup Sampling

The question of how many samples to take at each sampling time usually arises during Phase 1 of Process Validation. Clearly, 2 samples are probably too few whereas more than 10 may become unwieldy. During the early applications of control chart technology, favorite sample sizes were 5 and 10 for ease of calculation. With computers doing the work, this no longer applies although vestiges remain. A better approach than relying on tradition is to examine the underlying rational subgroups of the process.

In the tableting example above, there were 10 positions on the press. Rational subgroup sampling would dictate samples of size 10 after initial assurances that the position differences do not contribute to the overall variation. Sampling in sets of 10 would provide that advantage of vigilance against position misalignment which might be detected by the standard deviation chart and might also be reflected in the chart for the mean.

If there were 30 positions, instead, and if practicality would allow, we might still sample in sets of 10, but with positions 1–10 in the first sampling, 11–20 in the second and 21–30 in the third, continuing in this rotation. Such sampling is not ideal, but it may be expedient.

In principle, the purpose of rational subgroup sampling is to take data over some structural (internal, assignable cause) source of variation in order to provide a balance over all elements of that source. Alternatives are, in the case of more samples than rational subgroup elements, to over-sample some elements at the expense of others or, in the case of fewer samples than rational subgroup elements, to deny representation of some element. In either case, the effectiveness of the control chart will be diminished.

21.5.6 Exponentially Weighted Moving Average Charts

Shewhart's basic assumption is one of process stability; the process mean is unchanging unless it is subject to random shocks from the outside. To be sure, Shewhart control charts have stood the test of time and served well. But there is a different view, one that suggests that there is no such thing as stability. All processes wander.

Wandering is easy to see, especially when consecutive observations are taken. A time series plot will often show a sinusoidal or similar pattern. The spacing of sampling times by ½ hour or more is often sufficient to allow these patterns to damp in most processes. Therefore, these patterns are not seen, and Shewhart charts serve their intended purposes. But there are processes whose wandering or drifting patterns persist longer due to carryover effects of adjustments, changes in raw materials, changes in shifts and so on. In those situations, charts that take process memory into account are useful.

One such chart is the chart for exponentially weighted moving averages (EWMA), introduced by Roberts (1959). Its underlying model is:

$$z_i = \lambda x_i + (1 - \lambda)\, z_{i-1}$$

where,
z_i is the current exponentially weighted moving average,
z_{i-1} is its predecessor,
λ is the smoothing constant ($0 < \lambda \leq 1$),
x_i is the observed average at the ith sequential location.

The model is recursive in that z_i for any value of i is dependent on its predecessor, all the way back to the beginning of the process where the initial value, z_0, is set at the intended process mean or target, μ_0.

In practice, the EWMA is plotted on a control chart with the center line, $CL = \mu_0$,

$$UCL = \mu_0 + L\sigma \sqrt{\tfrac{\lambda}{2-\lambda}\left[1 - (1 - \lambda)^{2i}\right]}$$

$$LCL = \mu_0 - L\sigma \sqrt{\tfrac{\lambda}{2-\lambda}\left[1 - (1 - \lambda)^{2i}\right]}$$

where σ is the residual variation among sample means after the EWMA model is fitted to the data.

As i increases, the expression $[1-(1 - \lambda)^{2i}]$ approaches one, so the control limits reduce to

$$UCL = \mu_0 + L\sigma \sqrt{\tfrac{\lambda}{2-\lambda}}$$

$$LCL = \mu_0 - L\sigma \sqrt{\tfrac{\lambda}{2-\lambda}}.$$

Values of L and λ determine the average run length (ARL) of the control scheme. They can be used to fine tune the chart to a desired level of responsiveness. The ARL is the average number of sampling times before a process which is in control will signal as being out of control. More information on the impact of L and λ on the ARL can be found in Lucas and Saccucci (1990). For the EWMA chart to have behavior close to the Shewhart chart with run rules, Hunter (1989) recommended $L = 3.054$ and $\lambda = 0.40$.

Another way to arrive at a useful value for λ is to obtain process representative data, fit an integrated moving average model of the form $x_t - x_{t-1} = a_t - \theta a_{t-1}$ and calculate $\lambda = 1 - \theta$, see Montgomery (2000) for details. Fortunately, for most practical purposes, the EWMA is not greatly influenced by the choices of L and λ, and initial default values of 3 and 0.20, respectively, will be effective.

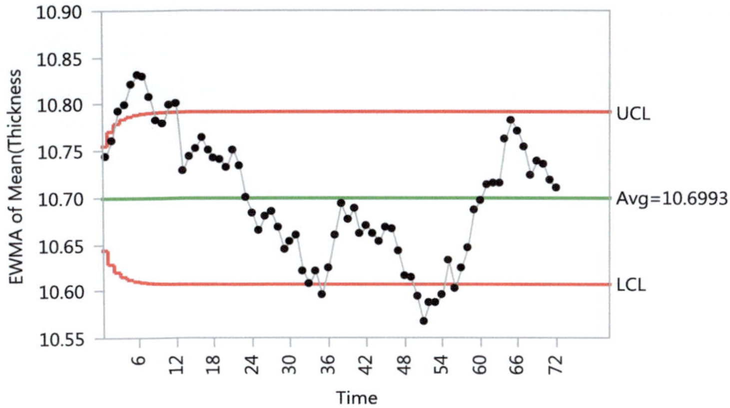

Fig. 21.4 EWMA chart of thickness data corresponding to the data of Fig. 21.2

Figure 21.4 shows the EWMA chart with $L = 3$ and $\lambda = 0.20$ for the data shown in the Shewhart chart of Fig. 21.2. The EWMA chart triggers an out-of-control situation at the third observation whereas the Shewhart chart hints at it with eight consecutive observations on one side of the center line.

Box and Luceño (2009) make an important distinction between process monitoring and process regulation. They point out that "waiting for the deviation from the target to be statistically significant before making a change would usually be a very poor strategy if the objective was to keep the adjusted process as close as possible to target." Recall that the Shewhart strategy is to seek and eliminate assignable sources of variation. It is not intended as an aid to process steering. By contrast, Box and Luceño provide extensions of EWMA monitoring schemes for steering or regulation with adjustments to help minimize deviations from target.

A final word on the use of control charts of either Shewhart or time series type is in order. Quite often, the measurement process lags the charting process by a substantial margin. This happens, for example, when analytical time is needed for precise results. Of course, control charts are meant to be real-time tools, so it would seem that a delay would render them only partially effective. While this is true to a certain extent, control charts, being very informative graphical displays, also serve as excellent communications devices, revealing improvement opportunities that might otherwise go unnoticed. Their use should not be abandoned due to delays in the measurement process.

In the preceding sections, we have discussed only charts for continuous variables because those are most prevalent in the manufacturing setting. However, for organizational success, quality improvement efforts must extend beyond manufacturing and into all phases of corporate endeavors. Practitioners are encouraged to examine control chart methods for attributes and to avail themselves of other quality tools as covered in texts on quality control such as Montgomery (2000).

21.6 Quality Indices

While process capability and process performance are defined above in terms of the corresponding standard deviations, some practitioners find it convenient to index these estimates relative to process targets and specifications. Picture a production summary report for top executives. Brevity is of the essence.

Originally proposed by Juran (2010), the derived index provides a measure of process standing along a continuum, with an index of 1 representing conformance with specifications and higher numbers indicating greater degrees of conformance with specifications. In its simplest form the process capability index, Cp, sometimes referred to as the process capability ratio (PCR) is:

$$C_p = \frac{USL - LSL}{6\sigma}$$

but sometimes the process under examination is not centered between the lower and upper specification limits, LSL and USL.

A more popular index that accounts for the lack of centering is:

$$C_{pk} = min\left\{\frac{\overline{X} - LSL}{3\sigma}, \frac{USL - \overline{X}}{3\sigma}\right\}$$

Essentially, C_{pk} measures the distance from the process mean to the nearer specification limit, relative to the process variation.

Variations on these themes include C_{pL} and C_{pU} which are the left and right terms, respectively of C_{pk}, and then all the same indices with the Cs replaced by Ps: P_c, P_{pk}, P_{pL} and P_{pU}. The difference between the Cs and the Ps lies with the value in the denominator, specifically σ. If σ is an estimate of the process capability as defined above, then the index is a capability index; if σ is an estimate of the process performance, then the index is a performance index. As discussed previously, the estimate of σ should come from a variance components analysis.

Extreme care should be taken in the use of these indices:

- They should not be thought to take the place of the capability and performance standard deviations. Doing so deprives the scientists and engineers of the information essential to quality improvement in that it obscures, rather than clarifies, the opportunity.
- They should be based on customer driven, functional specifications. Otherwise there will be strong temptations to widen specifications to make the indices look better.
- The use of the capability indices assumes the process is stable and that the data follow a normal distribution.
- The sample sizes on which the indices are based should exceed 100 and perhaps be higher than 200 (Hare 2007). As a case in point, the first five sets of 10 observations from the data in Fig. 21.2 were used to calculate Cpk based on

specifications of (10–13) and a target of 10.7. The resulting value of Cpk is 0.991. Users might believe that is sufficiently close to 1.0 for acceptability, assuming a true value of Cpk = 1.0 is adequate. But the confidence interval is very wide, stretching from 0.774 to 1.207. When all the data of Fig. 21.2 are included, the calculated value is 0.774 with lower and upper bounds of 0.728 and 0.821, respectively. The confidence interval is narrower and the conclusion is very different.

21.7 Key Chapter Points to Remember

- Increased regulatory scrutiny, taken together with strong business incentives for improved productivity and quality, incentivize the pharmaceutical industry to improve process understanding and control. This includes the quantification of capability and performance variation and the use of the statistical body of knowledge to guide these activities.
- Process capability is the inherent, intrinsic variation of the process. Process performance is a measure of the variation experienced by the consumer.
- Shewhart control charts are based on process capability and are aimed at detecting assignable causes of variation in processes. Charts for the mean should always be accompanied by charts for the standard deviation.
- When control charts are first applied to a process, it is likely that several sources of large variation will be discovered. These are the low hanging fruit. A finely tuned control chart is not needed, and using only 3-sigma limits will suffice. As the process performance improves, it is increasingly advantageous to apply Nelson or other rules to increase the sensitivity of the chart to ever more subtle assignable sources of variation.
- Special adaptations of Shewhart control charts include the acceptance control chart which can be applied to situations where the capability variation is very much smaller than the specification range, where control is generally very good and where only vigilance against out-of-specification situations is needed.
- Control chart sampling should be carried out to include structural sources of variation at each sampling time whenever possible. If that is not possible, systematic sampling should be applied in order to assure that each structural element appears in the sample as often as possible.
- When process response values observed on a temporal scale influence subsequent observations, the process data are said to be autocorrelated. Exponentially Weighted Moving Average (EWMA) charts and other charts that take autocorrelation into account are useful for finding assignable sources of variation in such cases.

- Quality indices are useful for reporting purposes, especially in executive reports where the health of many processes must be reduced to as few words and numbers as possible. They should not be used to replace the capability and performance standard deviations, and care should be taken to report them only when the process data are normally distributed, when the process is stable and when there are sufficient data for reporting with a high level of certainty.

References

Atkinson AC (1985) Plots, transformation and regression. Clarendon Press, Oxford

Box GEP, Luceño A (2000) Statistical control by monitoring and feedback adjustment. Wiley, New York

Box GEP, Paniagua-Quiñones Luceño M (2009) Statistical control by monitoring and feedback adjustment, 2nd edn. Wiley, New York

Deming WE (2000) Out of the crisis. MIT Press, Cambridge, MA

Freund RA (1957) Acceptance control charts. Ind Qual Control 14(4):13–23

Grant EL, Leavenworth RS (1996) Statistical quality control, 7th edn. McGraw-Hill, New York

Hare LB (2003) From chaos to wiping the floor. Qual Prog 53–63 http://asq.org/quality-progress/2003/07/statistical-process-control-spc/chaos-wiping-the-floor.html

Hare LB (2007) The Ubiquitous C_{pk}. Qual Prog 72–73 http://asq.org/quality-progress/2007/01/statistics-roundtable/the-ubiquitous-cpk.html

Hare LB (2013a) Follow the rules. Qual Prog 56–57 http://asq.org/quality-progress/2013/01/statistics-roundtable/follow-the-rules.html

Hare LB (2013b) Wanderlust and memory. Qual Prog 52–53 http://asq.org/quality-progress/2013/07/statistics-roundtable/wanderlust-and-memory.html

Hunter JS (1989) One point plot equivalent to the Shewhart chart with western electric rules. Qual Eng 2:13–19

JMP®, Version 11. SAS Institute Inc., Cary, NC, 1989–2007

Juran JM (1992) Juran on Quality by Design. The Free Press, New York, NY

Juran JM, DeFeo JA (2010) Quality control handbook, 6th edn. McGraw-Hill, New York

LeBlond D, Schofield T, Altan S (2005) Revisiting the notion of singlet testing requirements. Pharm Technol 29:85–86

Lucas J, Saccucci M (1990) Exponentially weighted moving average control schemes: properties and enhancements. Technometrics 32:1–12

Montgomery DC (2000) Introduction to statistical quality control, 7th edn. Wiley, New York, NY

Nelson L (1984) The Shewhart control chart – tests for special causes. J Qual Technol 16:238–239

Ott ER, Neubauer DV, Schilling EG (2000) Process control: troubleshooting and interpretation of data. McGraw Hill, New York, NY

Peterson JJ, Yahyah M (2009) A Bayesian design space approach to robustness and system suitability for pharmaceutical assays and other processes. Stat Biopharm Res 1(4):441–449

Read C (2006) Ranges. In: Balakrishnan N et al (eds) Encyclopedia of statistical sciences. Wiley, Hoboken, New Jersey, pp 6899–6902

Roberts SW (1959) Control chart tests based on geometric moving averages. Technometrics 1:239–250

Shewhart WA (1980) Economic control of quality of manufactured product. ASQ Quality Press, Milwaukee (republished). Van Nostrand, New York, 1931

Torbeck L (2011) Statistics in the service of quality. Pharm Technol 35(6):34

Woodcock J (2013) Interview with Dr. Janet Woodcock. Angle: US Life Science Industry, June 2013. http://www.nnepharmaplan.com/insights/angle-magazine/us-life-science-industry/articles/interview-with-dr-janet-woodcock-director-cder-fda/

Woodcock J, Wosinska M (2013) Economic and technological drivers of generic sterile injectable drug shortages. Clin Pharm Ther 93(2):170–176, http://www.nature.com/clpt/journal/v93/n2/pdf/clpt2012220a.pdf

Regulatory, Compendial and Standards Guidances

FDA Documents 21 *CFR* 210 and 211 cGMP in Manufacturing, Processing, Packing, or Holding of Drugs and Finished Pharmaceuticals

- 21 *CFR* 600 Biological Products: General
- 21 *CFR* 820 Quality Systems Regulations
- *Guidance for Industry: Investigating Out-of-Specification Test Results for Pharmaceutical Production* (FDA, Oct. 2006)
- *Guidance for Industry: Process Validation: General Principles and Practices* (FDA, Jan. 2011)

ICH harmonized quality guidelines available online at www.ich.org:

- Q2(R1) *Validation of Analytical Procedures: Text and Methodology, 1997*
- Q6A *Specifications: Test Procedures and Acceptance Criteria for New Drug Substances and New Drug Products: Chemical Substances*, 2000
- Q6B *Specifications: Test Procedures and Acceptance Criteria for Biotechnology/Biological Products, 1999*
- Q7 *Good Manufacturing Practice Guide for Active Pharmaceutical Ingredients, 2001*
- Q8(R2) *Pharmaceutical Development*
- Q9 *Quality Risk Management.*
- Q10 *Pharmaceutical Quality System*
- Q11 *Development and Manufacture of Drug Substances.*

Compendial references

- *USP*, Guide to General Chapters:
- *<905> Uniformity of Dosage Units*
- *<1010> Analytical Data–Interpretation and Treatment*

Standards references

- ANSI/ASQ Z1.9–2003 (R2013), "Sampling Procedures and Tables for Inspection by Variables for Percent Nonconforming", Milwaukee, WI http://asq.org/quality-press/display-item/index.html?item=T965
- ASTM Standard E2709, 2012, "Standard Practice for Demonstrating Capability to Comply with an Acceptance Procedure", ASTM International, West Conshohocken, PA, 2003, DOI 10.1520/E2709-12, www.astm.org.
- *ISO 7870 Control charts – Part 1 (General guidelines, 2014), Part 2 (Shewhart charts. 2013), Part 3 (Acceptance control charts, 2012).* International Standard, Geneva. https://www.iso.org/obp/ui/#iso:std:iso:7870:-1:ed-2:v1:en
- *ISO 22541 Statistical Methods in process management – Capability and Performance. Part 1 (General Principles and Concepts, 2009), Part 2 (Process capability and performance of time-dependent process models. 2013), Part 4 (Process capability estimates and performance measures, 2007).* International Standard, Geneva http://www.iso.org/iso/home/store/catalogue_tc/catalogue_detail.htm?csnumber=64135

Chapter 22
Statistical Considerations for Stability and the Estimation of Shelf Life

Walter W. Stroup and Michelle Quinlan

Abstract Stability testing is required to demonstrate that a pharmaceutical product meets its acceptance criteria throughout its shelf life and to gain regulatory approval. Stability testing involves many disciplines. Huynh-Ba (Handbook of stability testing in pharmaceutical development: regulations, methodologies, and best practices. Springer, New York, 2009) presents a comprehensive survey of these various aspects. This chapter will focus on aspects that involve statistical science: specifically, the probability distribution, modeling, and estimation of shelf life. Capen et al. (On the shelf life of pharmaceutical products. AAPS PharmSciTech 13:3, 2012) and Quinlan et al. (On the distribution of batch shelf lives. J Biopharm Stat 23(4): 897–920, 2013a) gave a framework for characterizing shelf life distribution which is reviewed in the chapter. Quinlan et al. (Evaluating the performance of the ICH guidelines for shelf life estimation. J Biopharm Stat 23(4), 881–896, 2013b) investigated the consequences of currently mandated shelf life estimation procedures and proposed alternatives. These are reviewed and expanded upon in this chapter.

Keywords Shelf life • Product and shelf life distribution • ICH Q1E guidance • Linear mixed model • Best linear unbiased prediction • Random coefficient linear regression

22.1 Introduction

The *ICH Harmonised Tripartite Guideline*: *Stability Testing of New Drug Substances and Products Q1A* (*R2*) (2003a) defines shelf life as follows:

W.W. Stroup (✉)
Department of Statistics, University of Nebraska, Lincoln, NE, USA
e-mail: wstroup1@unl.edu

M. Quinlan
Novartis Oncology, East Hanover, NJ, USA

L. Zhang (ed.), *Nonclinical Statistics for Pharmaceutical and Biotechnology Industries*, Statistics for Biology and Health, DOI 10.1007/978-3-319-23558-5_22

> **Shelf life** (also referred to as expiration dating period): The time period during which a drug product is expected to remain within the approved shelf life specification, provided that it is stored under the conditions defined on the container label.

Shelf life is estimated using data obtained from stability testing. The *ICH* document *Q1A* outlines procedures recommended for conducting stability trials. Readers interested in a comprehensive treatment of the multidisciplinary aspects of stability trials are referred to Huynh-Ba (2009). Our focus in this chapter is on statistical methods to estimate shelf life. Document *Q1E*, whose full title is *ICH Harmonised Tripartite Guideline*: *Evaluation For Stability Data Q1E* (2003b), specifies guidelines for statistical analysis of stability data. Under Section 2.1, General Principles, document *Q1E* states:

> The purpose of a stability study is to establish, based on testing a minimum of three batches of the drug substance or product, a retest period or shelf life and label storage instructions applicable to all future batches manufactured and packaged under similar circumstances.

As we shall see, the phrase that begins, "applicable to all future batches," is crucial. In this chapter, we present the essential statistical aspects of data obtained from stability studies. We review the *ICH* recommended statistical analysis, as given in *Q1E*. We describe both the conceptual difficulties and the observed behavior of the *ICH* recommendation. We then survey alternatives to the *ICH* recommendations. The organization of this chapter is as follows:

- Section 22.2 describes the model setting for data from stability studies and how this model setting can be used to visualize, and give tangible meaning to, the definition and goal of shelf life estimation.
- Section 22.3 reviews the *ICH* recommended procedure, as specified in *Q1E*, and describes its consequences.
- Section 22.4 outlines the standard linear mixed model (LMM) alternative to the *Q1E* procedure. The standard mixed model is referred to in this chapter as the "naive mixed model." In this section we explain why.
- Section 22.5 presents alternatives to the *Q1E* and naive mixed model procedures, and explores the conceptual justification and observed behavior of each alternative.
- Section 22.6 considers the effect on the *Q1E* and alternative procedures from Sect. 22.5 when batch-to-batch variation in slope is zero.
- Section 22.7 provides summary, conclusions, discussion and suggestions for future work.

22.2 Model Setting

Following Stroup (2013) a statistical model has two aspects: how the data arose and how the data are to be analyzed. As Stroup notes, practitioners often focus on the latter and tend to neglect the former. Indeed, the statistical procedure outlined by the *Q1E* guidelines could be criticized on this basis—a criticism that will be elaborated upon later in this chapter. Stroup and Quinlan (2010) and Quinlan et al. (2013a) address the question, "How do we describe the *process* by which the data arose, and how do we construct a sensible model consistent with that process?" The section begins with a review of their work.

We obtain stability data from a sample of batches. In theory, the batches are drawn from a population that includes currently available batches that could have been drawn, and "all future batches manufactured and packaged under similar circumstances." While the observed batches ideally form a random sample of batches available at the time of the stability study, in order to address the intent of the *ICH Q1A* and *Q1E* guidelines (hereafter referred to as "the *ICH* guidance"), we must also assume that the batches we observe are representative of "*all future batches*." The main features of the process giving rise to stability data are

1. batch responses change over time
2. all batches are not equal, i.e. there is random variation among batch responses over time

Depending on the stability-limiting characteristic, the response variable of interest may increase or decrease over time. Typically (but not always) measures of efficacy or potency decrease over time, whereas measures of a degradant increase over time. The *Q1E* guidelines state that changes over time may be linear, or they may follow a quadratic or more complex polynomial regression, or they may be nonlinear, e.g. an exponential or logistic growth or decay curve. In the interest of clarity to develop a conceptual framework, this discussion focuses on the simple linear case. As the *Q1E* document notes, concepts developed using linear regression extend in a straightforward manner to more complex regressions.

Using the linear regression framework, we can characterize changes in the response variable over time for a given batch as

$$E\left(Y_{ij}\middle|B_{0i},B_{1i}\right)=B_{0i}+B_{1i}T_j$$

where T_j denotes the jth time (typically given in months of storage), Y_{ij} denotes the stability limiting response at the jth time for the ith batch, B_{0i} denotes the intercept for the ith batch, and B_{1i} denotes the slope for the ith batch. Variation among batches, as exhibited by the variation among batch intercepts and slopes, is accounted for by regarding B_{0i} and B_{1i} as random variables. In principle, B_{0i} and B_{1i} can have any plausible distribution. Typically, and for the purposes of this discussion, we assume normality. Specifically,

$$\begin{bmatrix}B_{0i}\\B_{1i}\end{bmatrix}\sim N\left(\begin{bmatrix}\beta_0\\\beta_1\end{bmatrix},\begin{bmatrix}\sigma_0^2&\sigma_{01}\\\sigma_{01}&\sigma_1^2\end{bmatrix}\right).$$

Often, B_{0i} and B_{1i} are further assumed to be uncorrelated, that is, $\sigma_{01}=0$. Equivalently, we can characterize the process giving rise to the data as

$$E\left(Y_{ij}\middle|B_{0i},B_{1i}\right)=\beta_0+b_{0i}+\left(\beta_1+b_{1i}\right)T_j,$$

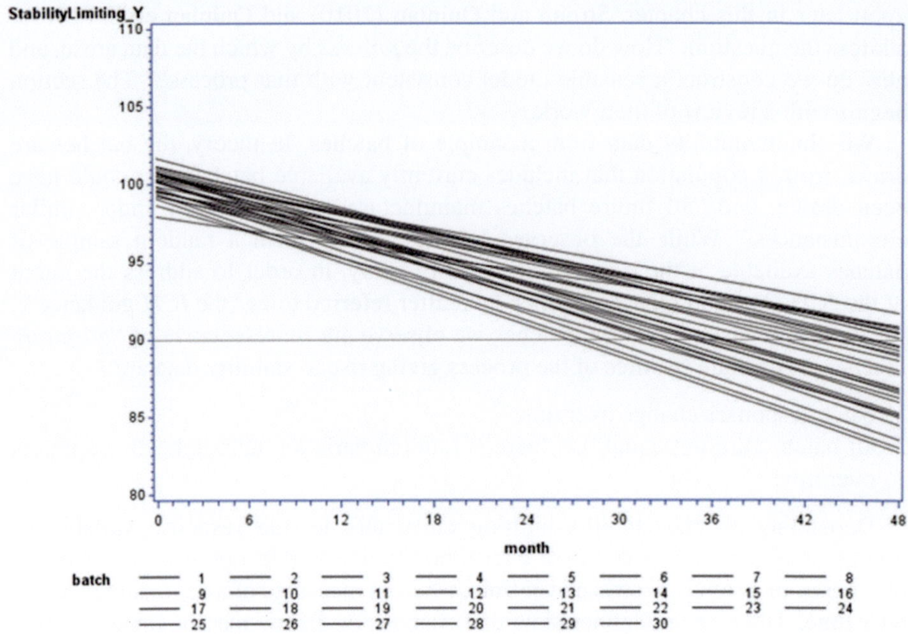

Fig. 22.1 Visualization of variation among batch response over time

where $\begin{bmatrix} b_{0i} \\ b_{1i} \end{bmatrix} \sim N\left(\begin{bmatrix} 0 \\ 0 \end{bmatrix}, \begin{bmatrix} \sigma_0^2 & \sigma_{01} \\ \sigma_{01} & \sigma_1^2 \end{bmatrix}\right)$. Note that the right-hand side of this description of the process is identical to the linear predictor of a random coefficient linear regression mixed model. Figure 22.1 shows the distribution of expected batch responses under this framework.

To illustrate how this framework extends to more complex models, suppose that the response over time is nonlinear. For example, consider a logistic growth curve $Y_{ij} = B_{0i}/\left\{1 + \exp\left[-\left(B_{1i} + B_{2i}T_j\right)\right]\right\}$. In this case, B_{0i}, B_{1i} and B_{2i} are batch-specific coefficients. As with the linear case, we account for the inherent variability among batches by regarding B_{0i}, B_{1i} and B_{2i} as random variables. Regardless of the response-over-time model used to describe the process giving rise to the data, the discussion below can be adapted to fit the model's particulars.

Assuming that the regression coefficients are random variables can be controversial, at least in certain cases. The controversy stems, in part, from a misunderstanding of what it means for a process to be in a state of statistical control. The slope of a linear regression illustrates the issue. The process being in a state of statistical control, goes the argument, means that degradation rates do not vary among batches, i.e. no batch-to-batch variability among slopes. However, "in a state of statistical control" does *not* equate to zero variance. While we hope that the variance is small, strictly speaking it is never zero.

There are, however, situations in which variation among certain batch characteristics is too small to be clinically relevant. Altan et al. (2013) gives examples of products for which variation in the initial response (the intercept) makes sense, but variation in degradation rate (the slope) does not. In such cases, the variance among slopes is indistinguishable from, even if in theory not exactly equal to, zero. Both cases—non-negligible and negligible variance among batch slopes—will be considered as this chapter proceeds.

Continuing with the linear regression, define A as the acceptance criterion. If the stability limiting response increases over time, A will be the value of Y_{ij} above which the product is no longer acceptable; if response decreases over time, A will be the value of Y_{ij} below which the product is no longer acceptable. It follows that the batch-specific shelf life for the ith batch can be determined by solving for the time at which the response, $Y_{ij} = A$, i.e. set $A = B_{0i} + B_{1i}T_{Si}$, where T_{Si} denotes the ith batch-specific shelf life. This yields $T_{Si} = (A - B_{0i})/B_{1i}$. Notice that T_{Si} is a random variable whose distribution depends on the distributions of B_{0i} and B_{1i}. Figure 22.2 shows the distributions in question: changes in response over time and the resulting shelf life for the batches in the population.

Stroup and Quinlan (2010) noted a lack of standardized terminology, stating, "... conversations about shelf life are prone to going astray unless participants take care at the beginning of the discussion to make sure that they agree on a common understanding of all key terminology." Capen et al. (2012) pursued this point, and provided precise definitions of key shelf life vocabulary. Specifically,

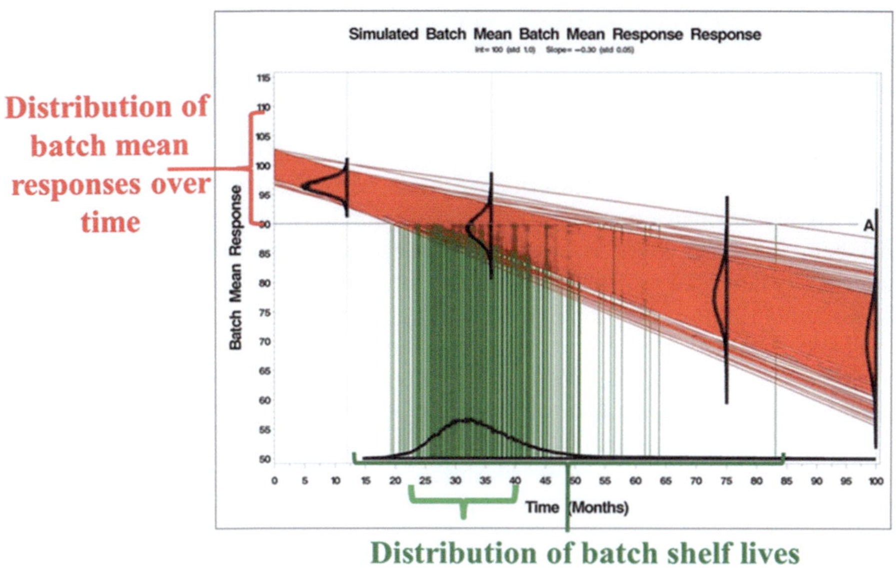

Fig. 22.2 Visualization of Relationship between Batch Mean and Shelf Life Distributions

with regard to Figs. 22.1 and 22.2, following Capen et al. we use the following terms:

- **Product distribution**: this is the distribution of the observations, Y_{ij}. In the linear regression case, we assume $Y_{ij}\big|B_{0i}, B_{1i} \sim NI\left(B_{0i} + B_{1i}X_j, \sigma^2\right)$ where NI denotes "normally and independently." Hence, if B_{0i} and B_{1i} are uncorrelated, $Y_{ij} \sim NI\left(\beta_0 + \beta_1 X_j, \sigma_0^2 + \sigma_1^2 X_j^2 + \sigma^2\right)$. In Figs. 22.1 and 22.2, visualize this as the distribution on the Y-axis.
- **Shelf life distribution**: also called "true shelf life distribution." This is the distribution of the T_{S_i}. Visualize this distribution on the X-axis of Fig. 22.2. Also note that the shelf life distribution is always a consequence of the product distribution
- **Product shelf life**: this is a value on the X-axis that is the target of shelf life estimation. This is the principal controversy of shelf life estimation: what target does one set that addresses the goal stated in the *ICH* guidance, is realistic, and can be implemented using available statistical methodology? In the next paragraph, we further clarify how statistical science understands and attempts to work with the product shelf life estimation problem.

Figure 22.2 allows us to visualize the goal at the heart of the *ICH* guidance. Denote T_S as the true batch shelf life and $\widehat{T}_S$ as the estimate of the product shelf life. Strictly speaking, because the distribution of T_S is continuous for all $T_S \geq 0$, setting a shelf life such that *all* future batches meet the acceptance criterion—i.e. Y_{ij} stays on the correct side of A with probability *one* for *all* times $\leq$ the stated product shelf life—would require setting $\widehat{T}_S = 0$. This is both impractical and unrealistic. Realistically, a suitable estimate of the product's shelf life should address two considerations. First, it should provide assurance that *most* of the distribution of Y_{ij} lies on the acceptance side of A for all times $\leq \widehat{T}_S$, where "most of the distribution" is defined in some agreed upon manner. Second, it would seem reasonable to avoid estimates of shelf life *below* the effective minimum of the distribution of T_S.

Shelf life estimates below the effective minimum of the shelf life distribution risk being undesirable for several possible reasons. For example, if $\widehat{T}_S$ is below a target—say below 12 months on Fig. 22.2—further development of the product may be discontinued, thus preventing a potentially beneficial product from being made available. Alternatively, an unrealistically short labelled shelf life will cause users to discard and replace the product prematurely. There may be other undesirable contingencies depending on the nature of the product and its intended use.

Shelf life estimates above an agreed upon target informed by the shelf life distribution also present problems because they would fail to address the primary goal of the *ICH* guidance, i.e. to provide a time before which the product should be expected to meet acceptance criteria. For example, if the estimated shelf life $\widehat{T}_S$ exceeds the 5th percentile of the distribution of T_S, this in effect means that the probability of a future batch failing to meet the acceptance criterion for the stated life of the product is 5 %. If $\widehat{T}_S$ exceeds the 10th percentile of the distribution of T_S, the probability of a future batch failing to meet acceptance criteria for the stated life of the product is 10 %. The greater the estimate $\widehat{T}_S$ relative to the distribution of T_S, the greater the likelihood of a product failing to meet acceptance criteria within its stated shelf life.

While it is not the purpose of this chapter to prescribe acceptable—or unacceptable—targets for $\widehat{T}_S$, the upper and lower thresholds of the distribution of T_S described above necessarily play a central role in setting these targets, and, as we shall see below, in evaluating the relative merits and shortcomings of competing methods of obtaining $\widehat{T}_S$.

This concludes the first aspect of statistical modeling: how did the data arise? In the following sections, we consider the second aspect of modeling—what template do we use to define and estimate relevant parameters and how do we use these estimates to address our objectives?

22.3 Existing Methodology

In this section we describe the methodology *Q1E* guidelines prescribe for estimating shelf life. Following the *ICH Q1E* document, we focus on the linear regression case, noting that this approach can be extended to polynomial or nonlinear regression, but the linear regression case is sufficient to illustrate the essential principles involved.

22.3.1 *The Q1E Estimation Procedure*

As noted in Sect. 22.1, *Q1E* prescribes linear regression over time based on data from at least three batches. Batches are typically observed at 0, 3, 6, 9 and 12 months initially. Subsequent observations are often taken at 18 and 24 months. Occasionally, there may be additional observations after 24 months, e.g. at 30 or 36 months or even later. Data are then analyzed using the following modeling process:

Analysis begins with an unequal slopes analysis of covariance model:

$$y_{ijk} = \beta_{0i} + \beta_{1i}X_j + e_{ijk}$$

where y_{ijk} denotes the kth observation on the ith batch at the jth time, β_{0i} and β_{1i} denote the intercept and slope parameters for the ith batch, X_j denotes the time (usually in months) when observation at the jth time is taken, and e_{ijk} denotes random variation among observations within a batch, assumed i.i.d. $N(0, \sigma^2)$. Note that to be technically complete, we should refer to this as an unequal slopes *and intercepts* model. In the interest of readability, and because this is the only model with unequal slopes allowed under the *Q1E* guidance, we refer to this model hereafter as the "unequal slopes" analysis of covariance model.

The next step of the *Q1E* prescribed analysis is the so-called "poolability" step. First, one tests the equal slopes hypothesis, $H_0 : \beta_{1i} = \beta_1$ for all i. That is, are regression slopes equal for all batches? In most cases, the suggested criterion for rejection is $\alpha = 0.25$. The idea here is to use the unequal slopes model unless there is strong evidence for doing otherwise. If we fail to reject the hypothesis of equal slopes, then we test the equal intercept hypothesis, that is, $H_0 : \beta_{0i} = \beta_0$ for all i.

Depending on the result of these tests, one of the following three models will be selected:

I. Unequal Slopes: as given above
II. Common slope, unequal intercepts: $y_{ijk} = \beta_{0i} + \beta_1 X_j + e_{ijk}$
III. Common slope and intercept: $y_{ijk} = \beta_0 + \beta_1 X_j + e_{ijk}$

Subsequent estimation depends on the model selected.

If model I, unequal slopes, is selected, batch-specific 95 % confidence bands are computed for the estimates $\widehat{\beta}_{0i} + \widehat{\beta}_{1i}X$. If the response decreases over time, the band whose lower one-sided confidence bound first intersects the acceptance criterion determines the shelf life: the shelf life is the time X at which this intersection occurs.

If model II, common slope, unequal intercepts, is selected, confidence bands for $\widehat{\beta}_{0i} + \widehat{\beta}_1 X$ of the worst batch, that is the batch that intersects the acceptance criterion first, are computed. As before, the shelf life is the time X at which this intersection occurs.

If model III, common slope and intercept, is selected, a confidence band for $\widehat{\beta}_0 + \widehat{\beta}_1 X$ is computed. The shelf life is the time X at which the relevant confidence bound intersects the acceptance criterion, A.

22.3.2 Criticism and Behavior of the *Q1E* Procedure

We can understand the performance of the *Q1E* procedure, and why improvements may be needed, by considering two aspects: the procedure's theoretical shortcomings and its small sample behavior. We examine both in this section.

First, we examine the theoretical. The analysis of covariance model used as the basis of the *Q1E* procedure is a fixed-batch intercept and slope model. By definition the model does not account for random variation among batch intercepts and slopes. Inference on a fixed-effects model is technically limited to *only* those levels of the effect actually observed—in this case, inference is technically limited to the three batches in the stability study. It does not apply to other batches in the population—specifically, it does *not* apply to "all future batches." Thus, it does not, by definition, address the objectives of shelf life estimation articulated in *ICH*.

As a practical matter, we know that a mismatch between a model that plausibly describes the process giving rise to the observed data and the model used to analyze the data can lead to inaccurate analysis. In this case, the analysis of covariance model implicitly assumes that all future batches are identical to those observed—that the only source of uncertainty in predicting the behavior of future batches derives from within-batch variability, and no uncertainty whatever comes from between-batch variation. We know that in practice this is an implausible assumption. Variation among batch intercepts and slopes may be small—we assume that it is when the product is in a state of statistical control—but it is never zero.

This raises the second issue: do *Q1E*'s theoretical limitations translate into problems with its small sample performance that need to be addressed? Quinlan et al. (2013b) and Schwenke (2010) explored the small sample behavior of the *Q1E* procedure using both simulated data and data provided by industry participants in PQRI's Stability Shelf Life Working Group. They found that with three batches, the *Q1E* procedure yielded erratic results, with ample opportunity for both excessively low and excessively high shelf life estimates. Despite the perception that increasing the number of batches increases the likelihood of a lower estimated shelf life, thereby creating a conservative estimate of shelf life, but also a disincentive to collect more data, neither Quinlan nor Schwenke found evidence that this actually happens in practice.

To illustrate their findings, we present below the results from two simulations using parameters that represent a synthesis of intercept and slope characteristics suggested by Altan et al. (2013) and those used by Quinlan et al. and Schwenke in their studies mentioned above. Altan et al. obtained representative intercept and slope characteristics from "33 recent stability trials." These new simulations not only provide verification of Quinlan et al. and Schwenke's results, but will allow us to compare the *Q1E* procedure's behavior to alternatives proposed in subsequent sections of this chapter.

In the first simulation, one thousand data sets were generated from the process described in Sect. 22.2, with random intercept $B_{0i} \sim NI\,(100, 0.5)$, where *NI* denotes normally and independently distributed, and random slopes $B_{1i} \sim NI\,(-0.25, 0.000625)$. These characteristics describe a stability limiting characteristic with an average of 100 at time zero, and a 3 % annual rate of decrease—equivalent to the 0.25 unit decrease per month used as the slope coefficient. Following Quinlan et al. (2013b) and Schwenke (2010) we assume that B_{0i} and B_{1i} are uncorrelated. For each simulated data set, observations were generated for each batch at times 0, 3, 6, 9, 12, 18 and 24 months. Thus, each observation was generated as $B_{0i} + B_{1i}M + w_{ij}$, where $M = 0, 3, 6, 9, 12, 18, 24$ denotes time in months, and $w_{ij} \sim NI\,(0, 0.5)$ denotes random variability among observations resulting from analytical variation, measurement error, etc. One set of 1000 simulated experiments was generated with 3 batches per trial, and another set of 1000 was generated with 6 batches per trial. In this simulation, the negative slope indicates that the stability limiting characteristic decreases over time. The acceptance criterion was set at $A = 90$, i.e. the product is considered "acceptable" as long as the response is 90 or above.

As mentioned in Sect. 22.2, with some products, while random variation is expected among the B_{0i} (intercepts), batch-to-batch variation among the B_{1i} (slopes) may be too small to be clinically relevant, and therefore effectively zero. To assess the behavior of the *Q1E* procedure and selected alternatives deemed promising in

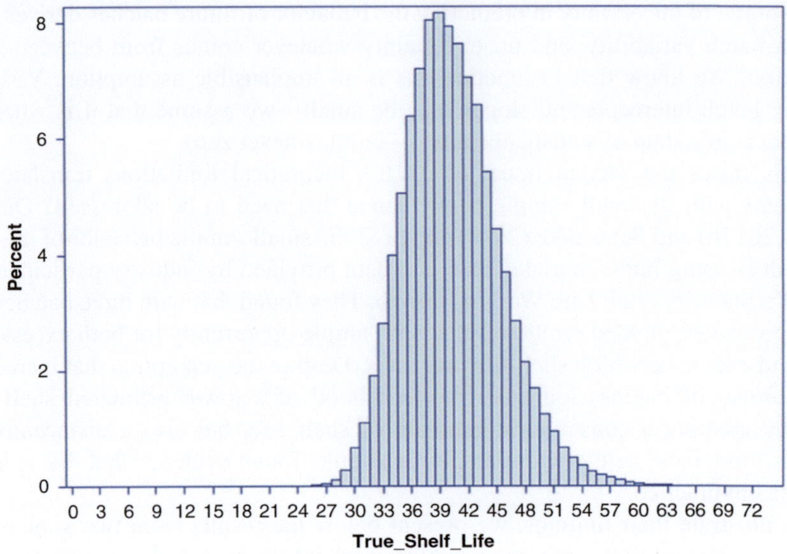

Fig. 22.3 Empirical distribution of true batch shelf with nonzero variation among batch slopes

the first simulation, we generated an additional 1000 simulated stability trials with $B_{1i} = -0.25$ for all trials (hence no random slope term) and 3 batches per trial.

Results for the simulated data with non-negligible variance among slopes are discussed from this point through Sect. 22.5. Section 22.6 gives results for the simulated data generated with no random variation among slopes.

Note that the true shelf life of each simulated batch can be determined as $T_{Si} = (90 - B_{0i}) / B_{1i}$. Figure 22.3 shows the empirical distribution of true shelf lives for the first simulation, with non-zero variance among slopes. Figure 22.4 shows the empirical distribution for the second scenario, with no random slope effect. Each empirical distribution was created by generating 1,000,000 samples using the parameters given above.

Summary statistics important when evaluating the performance of the estimating procedures shown below are as follows. For the first scenario, Fig. 22.3, of the 1,000,000 simulated shelf lives, the lowest was 23.9 months. The 1st percentile of the empirical distribution was 30.3; 5th percentile, 32.9; 10th percentile, 34.3; 25th percentile, 36.9; median, 40.0. The mean of the empirical distribution was 40.4. These percentile values are given in the first column of Table 22.1. Information for the second scenario appears in Table 22.2 and Fig. 22.4. These will be discussed in Sect. 22.6.

We can interpret these numbers approximately as follows. First, consider the empirical distribution shown in Fig. 22.3. Suppose that this is the process by which data arise in stability trials under our consideration. If we obtain a shelf life estimate 24 months, we would be able to give 24 months as the stated shelf life—i.e. the product's label states that it will last at least 24 months. The empirical distribution

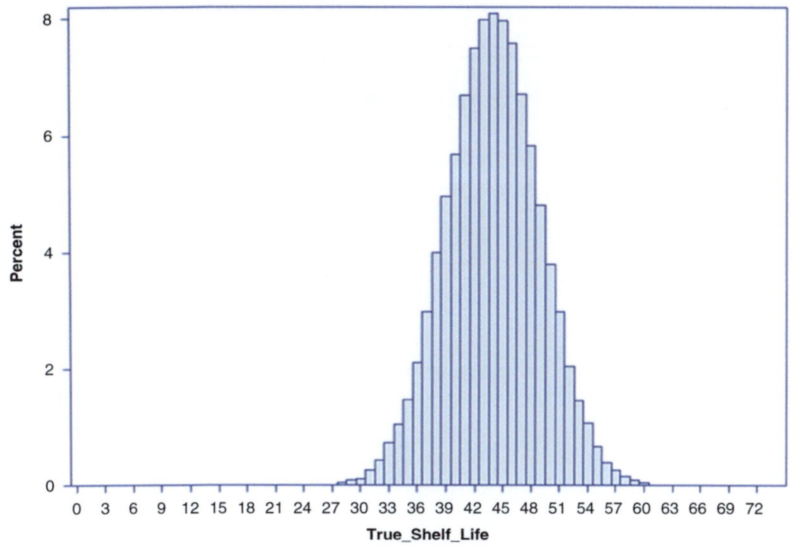

Fig. 22.4 Empirical distribution of true batch shelf with no variation among batch slopes

tells us that we would expect future batches to meet the acceptance criterion for the stated shelf life of at least 24 months with probability essentially equal to one. If the estimated shelf life is 30 months, future batches would be expected to meet acceptance criterion for at least 30 months with probability $\geq$ 0.99. For an estimated shelf life of 32 months, future batches would be expected to meet acceptance criterion with probability $\geq$ 0.95. For an estimated shelf life of 34, the probability is $\geq$ 0.90, for an estimated shelf life of 36, the probability is $\geq$ 0.75, and so forth. An estimated shelf life of 40 months (the population median and just below the population mean) would imply that future batches have an expected probability $\cong$ 0.5 of meeting the acceptance criterion for the entire stated life of the product.

With these numbers in mind, we have a basis for distinguishing between reasonable and unreasonable shelf life estimates. For example, again using Fig. 22.3, in evaluating simulation results, we regard shelf life estimates of less than 24 as unreasonably low, because the probability of a batch actually having a shelf life this short is essentially zero. On the other hand, depending on the level of risk deemed acceptable, we regard estimates above the corresponding percentile of the shelf life distribution as unreasonably high. For example, if the agreed upon target mandates that future batches meet acceptance criteria for their entire stated shelf life with 95 % probability, then any estimate greater than 32.9 months—the 5th percentile of the shelf life distribution—would be considered unreasonably high.

Christopher (2010) and Capen et al. (2012) stressed the importance of a quality statement. Among other things, a quality statement could guide decision criteria when evaluating simulation study results. A quality statement that includes a target probability that future batches meet acceptance criteria for the stated life of the

Table 22.1 Descriptive statistics for empirical true shelf life distribution and sampling distribution of shelf life estimators

Quantile	Empirical true shelf life	Q1E		Naive_LMM		LCL_RegCoeff		Adj_BLUP		LB_Reg_BLUP		Direct	
		3 batches	6 batches	3 batches	6 batches	3 batches	6 batches	3 batches	6 batches	3 batches	6 batches	3 batches	6 batches
Effective Max[a]	**70**[a]	61	66	44	43	42.0	40.4	42.9	41.0	42.0	40.0	44	41
95 %	**49.3**	42.5	44	37	38	35.4	36.0	36.3	36.0	35.0	34.6	38	35
90 %	**46.9**	40.5	42	35	37	34.1	35.1	35.2	34.7	33.8	33.8	37	34
75 % Q3	**43.5**	38	38	32	35	31.9	33.5	32.9	32.8	31.6	31.4	34	32
50 % Q2	**40.0**	35	35	28	34	29.3	31.6	30.4	30.0	29.0	28.8	31	30
25 % Q1	**36.9**	32	32	25	32	26.8	30.1	27.7	26.6	26.4	25.4	28	28
10 %	**34.3**	29	30	22	30	24.0	28.7	24.8	24.5	23.6	23.2	25	26
5 %	**32.9**	28	28	20	29	22.4	27.8	23.4	23.5	22.3	21.7	23	25
1 %	**30.3**	25.5	25	13	27	18.9	26.1	20.4	21.4	18.1	15.6	21	23
Effective Min	**24**[b]	22	23	2	22	11.3	25.3	16.8	19.2	6.0	6.2	17	22
Mean	**40.4**	**34.9**	**35.6**	**28.4**	**33.7**	**29.2**	**31.8**	**30.2**	**29.7**	**28.8**	**28.4**	**30.7**	**30.1**
SD	**5.1**	**4.8**	**5.0**	**5.5**	**2.8**	**4.0**	**2.5**	**4.0**	**4.0**	**4.1**	**4.3**	**4.4**	**3.0**
Pct Estimates between[c]		**coverage**											
Eff. min and Q5		**30**	**25**	**58**	**32**	**72**	**68**	**69**	**69**	**73**	**72**	**60**	**75**
Eff. min and Q10		**49**	**42**	**67**	**64**	**81**	**84**	**78**	**80**	**81**	**80**	**73**	**89**

Simulation scenario 1: process generating data includes random variation among slopes

Note: Q1E, Naive LMM and Direct shelf life estimates were determined to integer value; BLUP and LCL Reg Coeff estimation used computations that yielded fractional values

[a]Effective Max: quantile of empirical distribution such proportion of true shelf lives exceeding value < 1 in 100,000

[b]Effective Min: quantile of empirical distribution such that proportion of true shelf live less than value < 1 in 100,000

[c]Percentage of shelf life estimates between effective minimum and specified quantile; Q5 denotes 5th percentile; Q10 denotes 10th percentile; also referred to as coverage

Table 22.2 Descriptive statistics for empirical true shelf life distribution and sampling distribution of shelf life estimators

Quantile	Empirical true shelf life	Q1E	LCL_RegCoeff	Adj_BLUP	LB_BLUP	Direct
Effective Max[a]	**50**[a]	49	40.5	41.3	40.5	44
95%	**44.6**	41	35.3	36.2	35.0	38
90%	**43.6**	39	34.2	35.2	33.8	36
75% Q3	**41.9**	37	31.9	33.2	31.8	34
50% Q2	**40.0**	35	29.7	31.1	29.7	32
25% Q1	**38.1**	33	27.8	29.4	28.0	29
10%	**36.4**	31	25.3	27.3	26.0	27
5%	**35.4**	29	23.7	25.6	24.2	26
1%	**33.5**	27	21.4	22.7	19.8	24
Effective Min[b]	**27.5**[b]	25	18.6	20.4	1.5	22
Mean	**40.0**	**34.9**	**29.7**	**31.1**	**29.7**	**31.6**
SD	**2.8**	**3.5**	**3.4**	**3.1**	**3.5**	**3.5**
Pct Estimates between[c]		**Coverage**				
Eff. min and Q5		**58**	**72**	**80**	**76**	**75**
Eff. min and Q10	**68**	**74**	**84**	**79**	**80**	

Simulation scenario 2: process generating data has constant slope across batches; 3 batches per simulated trial

Note: Q1E and Direct shelf life estimates were determined to integer value; BLUP and LCL Reg Coeff estimation used computations that yielded fractional values

[a]Effective Max: quantile of empirical distribution such proportion of true shelf lives exceeding value < 1 in 100,000

[b]Effective Min: quantile of empirical distribution such that proportion of true shelf live less than value < 1 in 100,000

[c]Percentage of shelf life estimates between effective minimum and specified quantile; Q5 denotes 5th percentile; Q10 denotes 10th percentile; also referred to as coverage

product allow us to determine the likelihood of a given estimation procedure yielding an estimate that meets this target. It also provides a basis for comparing competing estimation procedures.

To illustrate, in assessing simulation results, we could regard the effective minimum of the empirical shelf life distribution as a lower "red line" and the percentile of the empirical shelf life distribution corresponding to the mandated probability that future batches meet the acceptance criterion as the upper "red line." For instance, using Fig. 22.3, if the mandated probability is 95 %, the lower and upper red lines are 24 and 32.9, respectively. If the mandated probability is 90 %, the lower and upper red lines are 24 and 34.3. We can judge the quality of an estimation procedure by the observed proportion of estimates that fall between the lower and upper red lines. In subsequent discussion of the various estimators, unless stated otherwise, "between the lower and upper red lines as defined above," refers to the proportion of shelf life estimates between the effective minimum and the empirical 5th percentile of the true shelf life distribution (Q5 in Table 22.1 or Table 22.2,

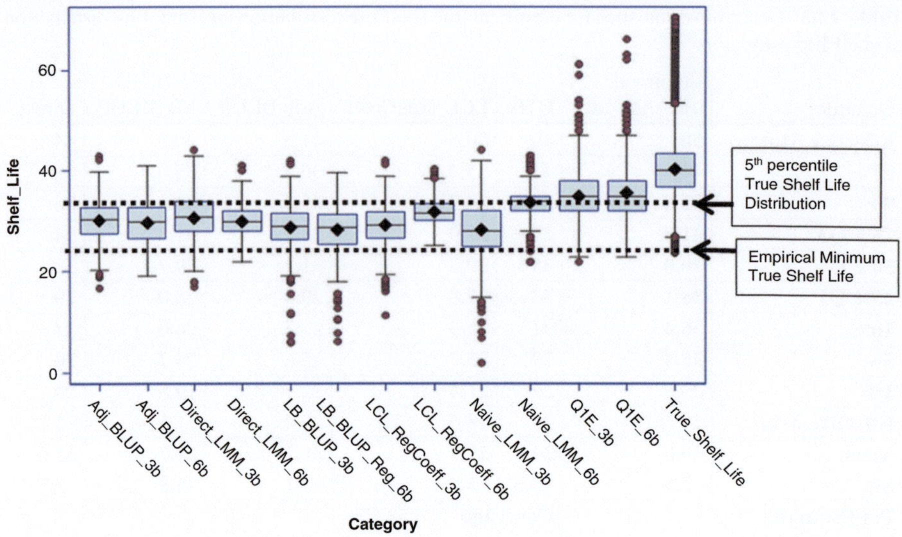

Fig. 22.5 Sampling distributions of estimated shelf life relative to the empirical minimum and 5th percentile of true shelf life distribution. True shelf life distribution includes random variation among slopes

depending on the scenario under discussion). At various points in the discussion, we will also refer to this proportion as "percent coverage."

Figures 22.5 and 22.6 show the performance of the *Q1E* procedure and all alternative estimation procedures discussed in subsequent sections of this chapter. Both figures show box-and-whisker plots of the distribution of shelf life estimates for simulated data assuming a stability study with 3 batches, and from the simulated data using 6 batches. The upper dotted line on Fig. 22.5 locates the 5th percentile of the empirical true shelf life distribution. The upper dotted line on Fig. 22.6 locates the 10th percentile. Shelf life estimates above the upper dotted line would imply a stated shelf life that more than 5 % (on Fig. 22.5) or 10 % (on Fig. 22.6) of future batches would fail to meet. Estimates based on the *Q1E* procedure are labelled *Q1E_3b* and *Q1E_6b* for estimates obtained from 3-batch or 6-batch simulated data, respectively. In addition to Figs. 22.5 and 22.6, Table 22.1 provides detailed information on the sampling distribution for each estimation procedure.

For the 3 batch case, the *Q1E* procedure yields a mean estimated shelf life of 34.9 months with a standard deviation of 4.8. The quantiles of the estimates are: lowest estimate, 22 months; 1st percentile, 25.5 months; 10th percentile, 29 months; 25th percentile, 32 months; median, 35 months. This means fewer than 1 % of the 1000 simulated shelf life estimates technically satisfies *ICH*'s stated objective: an estimate of 24 months or less effectively guarantees that all future batches will meet the acceptance criterion for the stated shelf life. The median *Q1E*-based shelf life estimate would result in an expected probability between 10 and 25 % of future batches failing to meet acceptance criteria during the stated life of the product.

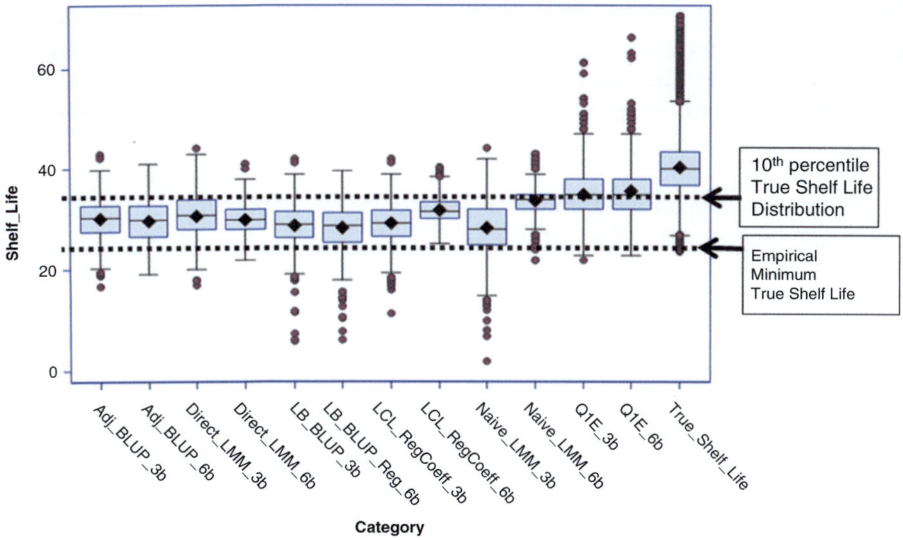

Fig. 22.6 Sampling distributions of estimated shelf life relative to the empirical minimum and 10th percentile of true shelf life distribution. True shelf life distribution includes random variation among slopes

Put another way, if we set the mandated probability for future batches meeting acceptance criterion at 90 %, i.e. setting the upper red line at the 10th percentile, the likelihood of *Q1E* producing an estimate between the lower and upper red is less than 50 %; setting the mandated probability at 95 %, drops *Q1E*'s likelihood of an estimate between the lower and upper red line to 30 %. One would be hard pressed to call this performance acceptable. Clearly, these results raise hard questions about the *Q1E* procedure.

We might speculate that a possible reason for the *Q1E* procedure's poor performance may be inadequate sample size. Perhaps 3 batches are not enough. Increasing to 6 batches, the mean estimated shelf life is 35.6 months with a standard deviation of 5.0. Interestingly, the standard deviation actually increases. This occurs because using 6 batches increases the likelihood of rejecting the equal slopes hypothesis and using the unequal slopes model: in this simulation, with 3 batches, 391 shelf life estimates were determined using the unequal slopes model and 242 were based on the common slope and intercept model; with 6 batches, 498 estimates used the unequal slopes model and only 100 used the common slope and intercept model. Basing shelf life estimates on the unequal slopes model increases variation among estimates because they are driven by the worst batch. This has the effect of driving down the estimated shelf life. With 6 batches, this is offset to some extent, but not completely, by the increased precision (and hence narrower confidence bounds) associated with increasing the number of batches. The net result, observed in this simulation as well as by Quinlan et al. (2013b), is close to a wash. The quality of decisions is not much improved by increasing the number of batches using the

Q1E procedure. The lowest shelf life estimate using 6 batches is 23 and the 1st percentile is 25, meaning the likelihood of an estimate technically satisfying the *ICH* objective is less than 0.01. The 25th percentile is 32. The median estimate is 35. These percentiles are essentially identical to those obtained using 3 batches. Quinlan et al. (2013b) and Schwenke (2010) found that increasing from 3 to 6 batches slightly decreased mean *Q1E*-based estimated shelf life but found similar results to this simulation in terms of impact on the standard deviation of the estimates and the overall quality of decision making.

These results dispel two common misconceptions about the *Q1E* procedure. First is the claim that *Q1E* is an inherently conservative procedure. The evidence clearly does not support this. Second, it is widely believed that increasing the number of batches moves the *Q1E*-based $\widehat{T}_S$ downward, making the procedure *too* conservative, and thereby creating a disincentive to increase sample size in cases where it might otherwise make sense. Again, the facts do not support this perception. There is little evidence from this simulation, or from results reported by Quinlan et al. or Schwenke, that increasing batches from 3 to 6 moves *Q1E*-based shelf life estimates consequentially in either direction. Moreover, there is no evidence that increasing the number of batches while using the *Q1E* estimation procedure yields any advantage in terms of accurate decision making.

A well-conceived estimating procedure ought to provide some advantage—e.g., yield better estimates or enable better decisions—as sample size increases. Clearly, the *Q1E* procedure fails this test.

22.4 The Linear Mixed Model

This section could easily have been called "Naive Mixed Model based Estimation." In this section, we examine the rationale for, and behavior of, the random coefficient mixed model defined by the process giving rise to the data given in Sect. 22.2.

As noted in Sect. 22.3, the *Q1E*-prescribed procedure's unimpressive performance stems from a mismatch between the process believed to plausibly explain the way the data arise and the model used to analyze the data and estimate shelf life. Specifically, the *Q1E* procedure uses a fixed-effects unequal slopes model, whereas the random coefficient mixed model more plausibly describes the process giving rise to the data. It seems reasonable, therefore, to base shelf life estimation on a model more closely aligned with the process by which stability data arises.

A random coefficient linear mixed model (LMM) for data from a multi-batch stability study, assuming linear change over time, is given by

$$y_{ijk} = \beta_0 + b_{0i} + (\beta_1 + b_{1i})X_j + e_{ijk}$$

where y_{ijk} and e_{ijk} denote the observation and random residual variation, respectively, as defined for the model used by *Q1E*. The intercept and slope parameters, β_0 and β_1, and the intercept and slope random batch effects, b_{0i} and b_{1i}, are as defined in Sect. 22.2. As with previous models, X_j denotes the jth time of observation. As described in Sect. 22.2, the standard assumption for the random intercept and batch effects is

$$\begin{bmatrix} b_{0i} \\ b_{1i} \end{bmatrix} \sim N\left(\begin{bmatrix} 0 \\ 0 \end{bmatrix}, \begin{bmatrix} \sigma_0^2 & \sigma_{01} \\ \sigma_{01} & \sigma_1^2 \end{bmatrix}\right)$$

Often, the covariance term σ_{01} is assumed to be zero. Alternatively, we could test $H_0 : \sigma_{01} = 0$, and set $\sigma_{01} = 0$ if we fail to reject H_0, or we could fit the model with and without σ_{01}, compare information criteria, setting $\sigma_{01} = 0$ if the information criteria so indicate. In some instances, it may be reasonable to test the null hypothesis that a variance component is equal to zero. One such case, noted earlier, occurs when the degradation rates are suspected to be consistent among batches. One could test, for example, $H_0 : \sigma_1^2 = 0$, e.g. using the COVTEST statement in SAS® PROC GLIMMIX. For some of the procedures presented below, we explore the consequences of doing so.

Adapting the strategy of the *Q1E* procedure, we compute 95 % confidence bands for the population-averaged linear regression estimate $\widehat{\beta}_0 + \widehat{\beta}_1 X$. As with the *Q1E* procedure, shelf life is then determined by the time at which the confidence band intersects the acceptance criterion. Ordinarily, variation among intercepts and slopes is assumed, with variation characterized by the random intercept and slope variance components, σ_0^2 and σ_1^2. Unlike the fixed batch effect model, mixed model based regression estimates yield standard errors that include contributions from variance among batches as well as variance among observations within batches. It is well-known that doing so improves confidence interval coverage.

On Figs. 22.5 and 22.6, estimates of shelf life yielded by the random coefficient mixed model are labelled *Naive_LMM_3b* for estimates from the 3 batch simulation and *Naive_LMM_6b* for estimates from the 6 batch simulation. Table 22.1 provides percentiles discussed below.

For the 3-batch case, the mean estimated shelf life is 28.4 with a standard deviation of 5.5. The quantiles of the estimates are: lowest estimate, 2 months; 10th percentile, 22 months; 25th percentile, 25 months; median, 28 months; 95th percentile 37 months. Recall that the *Q1E* procedure yielded a mean and standard deviation of 40.4 and 5.1, respectively, and a lowest estimate and median of 22 and 35 months, respectively. The random effect model's lower median estimate translates to a greater likelihood of future batches meeting the acceptance criteria for the estimated shelf life of the product. For a labelled shelf life based on the median estimate of 28, future batches would have a probability >0.99 of meeting acceptance criteria for their entire stated shelf life, whereas using the fixed batch median estimate of 35, this probability drops just under 0.75. Noting that the median true shelf life is 40.0, the mixed model procedure has a less than 5 % chance of producing estimates greater than or equal to the true shelf life median, whereas the *Q1E* approach will produce such estimates with probability between 0.10 and 0.25.

With regard to the lower and upper red lines defined in the previous section, notice that the LMM yields estimated shelf lives below the lower red line with probability between 0.10 and 0.25. Probability of estimates above the upper red line are between 0.10 and 0.25 if the mandated shelf life quantile is 95 %, approximately 0.10 if the mandated shelf life quantile is 90 %. Put another way, referring to Table 22.1, if the mandated shelf life quantile is 95 %, the LMM would be expected to yield estimates between the lower and upper red lines with probability 0.58, whereas *Q1E* yields similar estimates with probability 0.49.

If the performance of the mixed model procedure with 3 batches seems to be an improvement relative to *Q1E*, the story changes when we increase the number of batches. With the 6-batch stability studies, the mean estimated shelf life is 23.7 with a standard deviation of 2.8. The quantiles of the estimates are: lowest estimate, 22 months; 10th percentile, 30 months; 25th percentile, 32 months; median, 34 months. Although the precision with 6 batches increases substantially, the distribution of shelf life estimates also moves dramatically to the right.

Quinlan et al. (2013a) showed that as the number of batches increases, the LMM-based shelf life estimates tend toward a distribution centered at the shelf life distribution mean minus the expected confidence interval width. That is $\beta_0 + \beta_1 X_A - 1.96 \times \sqrt{\left(\sigma_0^2 + \sigma_1^2 X_A^2\right)/b}$, assuming random intercept and slope are uncorrelated and X_A denotes the time at which this bound intersects the acceptance criterion. In a standard estimation problem, this would be a good thing, but in the context of shelf life estimates, and given the objectives stated in the *ICH* guidance, movement to the right translates to increased likelihood that future batches will fail to stay within acceptance criterion for the stated life of the product. For labelled shelf lives at the median estimate of 34 when 6 batches are observed, future batches would have a probability approximately 0.90 of meeting acceptance criteria for their entire stated life, compared with >0.99 when 3 batches are observed. If the mandated shelf life quantile is 95 %, the LMM with 6 batches would be expected to yield estimates between the lower and upper red lines with probability 0.32—versus 0.58 for LMM with 3 batches. As the number of batches increases, the probability of obtaining a useable shelf life estimate decreases.

Using confidence bounds based on the population averaged regression estimates from the random coefficient mixed model, as the number of batches increases, confidence bounds narrow, and the distribution of shelf life estimates moves toward the mean of the true shelf life distribution. This is consistent with what Quinlan et al. (2013a) found in their simulation with the LMM. Increasing the number of batches moves a narrower range of shelf life estimates to the right of the region between the lower and upper red lines, decreasing the likelihood that future batches will meet acceptance criteria for the stated life of the product. This explains the term "Naive Mixed Model Alternative." The mixed model offers a clear advantage in precision but, as implemented in this section, does not address the *ICH* mandated objective. Something better is needed.

22.5 Alternative Estimation Using Linear Mixed Models

The previous section established that estimating shelf life requires both aligning the model used for estimation with the process we think describes how the data arose, and paying attention to the objectives given in the *ICH* guidance. The *Q1E* procedure fails the former requirement; the naive LMM procedure fails the latter.

Quinlan et al. (2013a, 2014) considered LMM-based quantile regression and LMM-based tolerance interval estimation as possible ways to address LMM issues illustrated in the previous section. While these methods showed promise, and are the subject of continuing development, they do not at this point provide a viable, ready-to-use replacement for current *Q1E* guideline. The problem concerns the relationship between the quantile regression or tolerance interval targets, which one must express in terms to the distribution of y_{ijk} and the probability that future batches meet acceptance criteria within the stated shelf life, which is defined in terms of the distribution of T_S. Refer to Fig. 22.2 to visualize this. The relationship is not one-to-one: targeting the 5th percentile of the distribution of y_{ijk}, for example, does not target the 5th percentile of the distribution of batch shelf lives. If the random batch intercept and slope variance components are known, the relationship between quantiles of these two distributions is straightforward to calculate. Quinlan et al. (2013a) showed how to do this. On the other hand, if the variance components are not known, then estimated variance components must be used to determine the quantile regression or tolerance interval targets for a given objective percentile of the shelf life distribution. Quinlan et al. found that this introduces so much additional variability that the sampling distribution of the resulting estimators is actually considerably worse than those of either the *Q1E* or naive LMM estimates.

Kroenker (2005) presents a comprehensive introduction to quantile regression theory and methods. The methods in Kroenker's book can be implemented in statistical software such as SAS PROC QUANTREG. Indeed, the PQRI Stability Shelf Life Working Group explored shelf life estimation using PROC QUANTREG. However, quantile regression as presented by Kroenker and implemented by generally available software is limited to fixed-effects-only models. Kroenker does have one chapter devoted to mixed models, but it is speculative in nature. Quinlan's work extending quantile regression to shelf life estimation with mixed models has so far been limited by the problems noted in the previous paragraph. However, quantile regression approaches which account for within batch correlation, such as that described in Tang and Leng (2011) may prove to be a more viable approach.

In the rest of this section, we present four LMM-based approaches that appear to have potential. All share the goal of taking advantage of the LMM's merits—superior precision and its alignment with the process giving rise to the data—and at the same time explicitly addressing the objectives of the *ICH* guidance.

Four Mixed Model Based Alternatives to the Naive LMM

The "naive" LMM estimates use confidence bands from the population-averaged regression equation estimate, $\widehat{\beta}_0 + \widehat{\beta}_1 X$. This makes sense when the objective is to estimate the mean. However, *ICH* clearly states that the objective is to determine how long *all future batches* can be expected to meet acceptance criteria. In other words, how long do we expect batches that meet the acceptance criterion the shortest length of time to last?

With this in mind, we explore four mixed model based approaches. This is not an exhaustive list, but these approaches are simple to implement and appear to have potential. They are

1. Use the confidence bound of the intercept and slope estimates rather than the estimates per se. That is, if the response decreases over time, base the shelf life estimate on $\widehat{\beta}_{0,L} + \widehat{\beta}_{1,L}X$, where $\widehat{\beta}_{0,L}$ and $\widehat{\beta}_{1,L}$ are the lower confidence bounds of the intercept and slope estimates, respectively. If the response increases over time, use the upper bounds.
2. Base the shelf life estimate on the BLUP of the batch-specific regression equation $\widehat{\beta}_0 + \widehat{b}_{0i} + \left(\widehat{\beta}_1 + \widehat{b}_{1i}\right) X$. Specifically, identify the batch-specific regression with the shortest shelf life and use it as a basis for the shelf life estimate.
3. Base the shelf life estimate on the confidence bounds of the batch-specific intercept and slope BLUPs of the batch with the shortest shelf life. This is similar to (1) but use $\left(\widehat{\beta}_0 + \widehat{b}_{0i}\right)_L$ and $\left(\widehat{\beta}_1 + \widehat{b}_{1i}\right)_L$ instead of $\widehat{\beta}_{0,L}$ and $\widehat{\beta}_{1,L}$.
4. Use the product distribution directly. For example, assuming $\sigma_{01} = 0$, the product distribution is $y_{ij} \sim N\left(\beta_0 + \beta_1 X_j, \sigma_0^2 + \sigma_1^2 X_j^2 + \sigma^2\right)$. The lower and upper α quantiles of this distribution are $\beta_0 + \beta_1 X \pm Z_\alpha \sqrt{\sigma_0^2 + \sigma_1^2 X^2}$ where Z_α is the α quantile value from the standard normal distribution. Depending on whether the stability limiting characteristic increases or decreases over time, one can determine shelf life by equating the upper or lower α quantile to A and solving for shelf life.

In the following sections, all simulation results were obtained implementing the above procedures with SAS PROC GLIMMIX.

22.5.1 Estimates Based on the Confidence Bounds of the Regression Coefficients

In this subsection we explore an LMM-based alternative that, like the naive LMM, uses population averaged intercept and slope estimates, but explicitly takes the *ICH* objective into account. Specifically, if the response decreases over time, use the lower confidence bounds of the intercept and slope fixed effect estimates, $\widehat{\beta}_{0,L}$ and $\widehat{\beta}_{1,L}$. If the response increases over time, use the upper bounds.

The approach is as follows:

Begin by using the random effects coefficient model defined in Sect. 22.4. Obtain the confidence bounds for $\widehat{\beta}_0$ and $\widehat{\beta}_1$ using standard mixed model estimation. For our simulation study, whose response decreases over time, we use the lower bounds, which we denote $\widehat{\beta}_{0,L}$ and $\widehat{\beta}_{1,L}$. Then the resulting shelf life estimate is $\widehat{T}_S = \left(A - \widehat{\beta}_{0,L}\right) / \widehat{\beta}_{1,L}$. In this simulation, a first step was added to test the random batch effect, i.e. $H_0 : \sigma_1^2 = 0$. The test was implemented using the likelihood ratio statistic computed by the SAS PROC GLIMMIX COVTEST statement. The random batch effect was retained if H_0 was rejected at $\alpha = 0.20$. Otherwise, the random batch effect was dropped. If one does not introduce this step, all of the LMM-based procedure described from this point forward in this chapter show excessively conservative behavior. The use of the $\alpha = 0.20$ criterion was the result of trial and error, determined by which cutoff yielded the greatest coverage as defined by the upper and lower red lines described in previous sections.

The box plots of the estimates from the 3-batch and 6-batch simulations are labelled *LCL_RegCoeff_3b* and *LCL_RegCoeff_6b* respectively on Figs. 22.5 and 22.6. With 3 batches, the mean shelf life estimate is 29.2 with a standard deviation of 4.0; the lowest shelf life estimate is 11.3; the 10th percentile is 24.0; the 25th percentile is 26.8; the median is 29.3. Using this approach, one would expect approximately 10 % of the shelf life estimates to be below the shortest true shelf life in the population. The median estimate is just below the 1st percentile of the empirical shelf life distribution. Coverage, the likelihood that the shelf life estimate is between the empirical minimum and 5th percentile of the true shelf life distribution, is 72 %. If one uses the empirical 10th percentile, coverage increases to just over 80 %. This is a dramatic improvement relative to the *Q1E* and naive LMM estimators.

With 6 batches, the mean estimate is 31.8 with a standard deviation of 2.5. The lowest shelf life estimate is 25.3. Selected percentiles are: 10th percentile, 28.7; 25th percentile, 30.1; median, 31.6; 75th percentile, 33.5. As with the naive LMM procedure, the regression coefficient lower bound method shows a tendency for the observed distribution of estimates to narrow and move to the right as the number of batches increases. However, unlike the naive LMM, increasing to six does not adversely affect coverage: using the 5th percentile as the upper red line, coverage is 68 %; using the 10th percentile, it is 84 %.

These simulation results make clear the potential value of procedures that simultaneously take advantage of the LMM's increased precision with increasing sample size and its alignment with plausible processes that model how stability data arise, and at the same time realistically take into account the objective of shelf life estimation as stated in the *ICH* guidance. However, while the outer confidence bound regression coefficient moves us in the right direction, it is not the only mixed model base procedure the merits consideration. We turn now to alternatives that use best linear unbiased prediction.

22.5.2 Estimation Based on the BLUP of the Worst Batch

In this subsection, we focus on estimation using the batch-specific BLUPs. We present two variations. Both use batch-specific estimates of the intercept and slope, but in different ways.

The batch-specific BLUP uses the same LMM as the "naive" mixed model shown in Sect. 22.4. The intercept for the ith batch is $B_{0i} = \beta_0 + b_{0i}$, for which we can obtain the BLUP $\widehat{B}_{0i} = \widehat{\beta}_0 + \widehat{b}_{0i}$. Similarly, the BLUP for the ith batch slope is $\widehat{B}_{1i} = \widehat{\beta}_1 + \widehat{b}_{1i}$. It follows that the BLUP for the ith batch shelf life is $\widehat{T}_{Si} = \left(A - \widehat{B}_{0i}\right)/\widehat{B}_{1i}$. With respect to shelf life estimate, one could think of these BLUPs as regression estimates for cumulative segments of the shelf life distribution, where the number of segments is determined by the number of batches. For example, in a stability study with 3 batches, the batch-specific BLUP-based regressions represent, in some sense, the $16 - 2/3^{rd}$, 50th, and $83 - 1/3^{rd}$ percentiles, i.e. the centers of the lower, middle and upper third of the distribution. Admittedly, this is an ad hoc and approximate way of regarding these BLUPs, and this alone does not completely address the *ICH* objective. However, a BLUP-based confidence bound reflecting an agreed upon target based on this line of reasoning could provide a useful estimate.

One could obtain confidence bands in a matter similar to the unequal slopes model used in the *Q1E* procedure. That is, track the confidence bound for the shortest-lived batch and define shelf life as the time at which the confidence bound intersects the acceptance criterion. However, confidence bounds for the batch-specific regression, $\widehat{\beta}_0 + \widehat{b}_{0i} + \left(\widehat{\beta}_1 + \widehat{b}_{1i}\right)X$, whether obtained via fixed-effects estimation used in *Q1E* or via BLUP, only use within-batch variance and hence underestimate the confidence interval width needed to accurately apply inference to the entire population. Used in this way, the advantage BLUP has over fixed-effect estimation is that BLUP takes among-batch variability into account, whereas *Q1E* does not, thereby providing more accurate batch-specific estimates. However, to fully realize BLUP's advantage, the confidence bands applicable to a population-wide rather than batch-specific inference space must be used.

Instead of using batch-specific confidence bands, one could obtain a confidence interval for the estimated shelf life, $\widehat{T}_{Si} = \left(A - \widehat{B}_{0i}\right)/\widehat{B}_{1i}$, by noting that $\widehat{T}_{Si}$ is a ratio of two BLUPs, each of whose variances can be estimated. Bounds can be approximated using Fieller's Theorem. The resulting interval is $\widehat{T}_{Si} \pm \left(t_{(1-\alpha),\nu}/\widehat{B}_{1i}\right)\sqrt{\left[s.e.\left(\widehat{B}_{0i}\right)\right]^2 + \widehat{T}_{Si}^2\left[s.e.\left(\widehat{B}_{1i}\right)\right]^2}$. For a response that decreases over time, use the lower bound, $\widehat{T}_{Si} - \left(t_{(1-\alpha),\nu}/\widehat{B}_{1i}\right)\sqrt{\left[s.e.\left(\widehat{B}_{0i}\right)\right]^2 + \widehat{T}_{Si}^2\left[s.e.\left(\widehat{B}_{1i}\right)\right]^2}$, as the estimate of shelf life.

Estimates obtained using this procedure are labelled *Adj_BLUP_3b* and *Adj_BLUP_6b*, respectively, for the 3-batch and 6-batch simulation on Figs. 22.5 and 22.6. For the 3-batch case, the mean shelf life estimate is 30.2 months with a standard deviation of 4.0, compared with 34.9 ± 4.8 for *Q1E* and 28.4 ± 5.5 for the naive LMM. The lowest shelf life estimate is 16.8 months. Selected percentiles are: 10th percentile, 24.8 months; 25th percentile, 27.7 months; median, 30.4 months; 75th percentile, 32.9 months. Using the median expected estimate, the probability of future batches meeting the *ICH* criterion is greater than 0.95. Even if a given stability study yields an estimate as high as the 75th percentile, future batches will still meet acceptance criteria for their stated lifetime with probability greater than 0.90. Coverage, i.e. the proportion of simulated trials with estimates between the upper and lower red lines was 69 % using the 5th percentile of the empirical shelf life distribution as the upper red line, and 78 % using the empirical distribution's 10th percentile. On the negative side, more than 10 % of the estimates were below the effective minimum of the empirical shelf life distribution.

Note that these estimates use $\alpha = 0.05$ to obtain the t-value in the Fieller computation. Quinlan et al. (2013a) showed that setting the confidence criterion, α, with respect to the product distribution (i.e. distribution of batch means over time) does not correspond to the quantile α with respect to the shelf life distribution. In this case, α with respect to the product distribution corresponds to a shelf life quantile less than α. Quinlan et al. found that this resulted in excessively conservative shelf life estimates. If one regards the batch-specific BLUP as an estimate of the $16{-}1/3^{rd}$ percentile of the shelf life distribution, then the lower confidence bound determined by $\alpha = 0.05$ may actually represent something closer to a $0.05 \times 0.167 = 0.0083$ confidence bound. This would certainly account for the estimates' conservative tendency.

With 6 batches, the mean shelf life estimate is 29.7 with a standard deviation of 4.0. The lowest shelf life estimate is 19.2. Selected percentiles are: 10th percentile, 24.5 months; 25th percentile, 26.6 months; median, 30.0 months; 75th percentile, 32.8. Increasing the number of batches moved the observed distribution of BLUP-based estimates to the left, but the change in precision was negligible. 80 % of the shelf life estimates were between the lower and upper red lines using the empirical distribution 10th percentile as the upper red line, compared with 78 % with 3 batches. The proportion of estimates below the minimum of the empirical true shelf life distribution dropped from 10 % with 3 batches to 5 % with 6 batches.

Whereas increasing the number of batches with QIE showed no evidence of enabling improved decision making, and increasing batches with the naive LMM actually leads to *fewer* appropriate decisions, increasing batches using the BLUP procedure leads to fewer unrealistically low estimates without increasing the proportion of inappropriately high estimates. Even with a shelf life at the 75th percentile of the sampling distribution, 32.7 months, the probability of a future batch meeting acceptance criteria for the stated lifetime would be greater than 0.95.

A variation on BLUP-based shelf life estimation uses the confidence bounds of the batch-specific intercept and slope instead of the BLUPs themselves for the batch that reaches the acceptance criterion first. If the stability-limiting characteristic increases over time, use upper confidence bounds. If, as is the case in this simulation, the response decreases over time, use the lower bounds. Denote the lower bounds of the batch specific BLUPs by $\left(\widehat{\beta}_0 + \widehat{b}_{0i}\right)_L$ and $\left(\widehat{\beta}_1 + \widehat{b}_{1i}\right)_L$ for the intercept and slope, respectively. The resulting shelf life estimate is $\widehat{T}_S = \left[A - \left(\hat{\beta}_0 + b_{0i}\right)_L\right] / \left(\hat{\beta}_1 + b_{1i}\right)_L$.

Estimates obtained using this procedure are labelled *LB_Reg_BLUP_3b* and *LB_Reg_BLUP_6b*, respectively, for the 3-batch and 6-batch simulation on Figs. 22.5 and 22.6. With 3 batches, the mean shelf life estimate is 28.8 months with a standard deviation of 4.1, compared with 30.1 ± 4.0 for the Fieller-adjusted BLUP estimates. The lowest shelf life estimate was 6.0. Selected percentiles are: 10th percentile, 23.6 months; 25th percentile, 26.4 months; median, 29.0 months; 75th percentile, 31.6 months. Using the median expected estimate, the probability of future batches meeting the *ICH* criterion is greater than 0.99. Even if a given stability study yields an estimate as high as the 75th percentile, future batches will still meet acceptance criteria for their stated lifetime with probability greater than 0.95. Coverage was 73 % using the 5th percentile of the empirical shelf life distribution as the upper red line, and 81 % using the empirical distribution's 10th percentile. These are improvements on the 69 and 78 % coverages with the Fieller-adjusted BLUP procedure. On the downside, the likelihood of an estimate below the empirical lower bound of the true shelf life distribution is over 10 %, and is somewhat higher that the corresponding proportion using the Fieller-adjusted estimate.

With 6 batches, the mean estimate is 28.3 with a standard deviation of 4.3. Selected percentiles are: 10th percentile, 23.2 months; 25th percentile, 25.4 months; median, 28.8 months; 75th percentile, 31.5 months. Coverage, using the empirical distribution's 10th percentile as the upper red line, was 80 %, compared to 81 % percent for 3 batches. This reduction in coverage may be attributable to the fact that the discrepancy between the nominal and the actual confidence bounds referred to earlier is exacerbated as the number of batches increases. That is, with 6 batches, the effective confidence bound being computed may be closer to $(1/2) \times (1/6) \times 0.05 = 0.00416$ than the nominal $\alpha = 0.05$.

These results suggest that BLUP-based procedures are promising approaches to shelf life estimation. They are relatively easy to implement, they are based on models consistent with *ICH*'s intent and a process that plausibly explains how the data arose, and their small sample behavior is superior to alternatives thus far considered. The main issue to be addressed is how to adjust the confidence coefficient α used for estimation so that it accurately targets the desired effective level of confidence and accounts for changing sample size.

22.5.3 Estimation Based on the Distribution of the Observations

In Sects. 22.5.1 and 22.5.2 we noted that increasing the number of batches reduces standard errors and hence confidence interval widths on which shelf life estimates are based, while the actual distribution of the stability limiting response variable does not depend on sample size. Both the lowest BLUP and regression coefficient lower bound method appear to require some tradeoff between the confidence coefficient α and the number of batches, although how to do this is unclear. In this section, we explore what this tradeoff entails. This suggests another possible approach to estimating shelf life.

From previous sections, we model the observed response as

$$y_{ijk} = \beta_0 + b_{0i} + (\beta_1 + b_{1i}) X_j + e_{ijk}$$

Following Stroup (2013) we can express the model in probability distribution form as

$$y_{ijk} \Big| b_{0i}, b_{1i} \sim N\left(\beta_0 + b_{0i} + (\beta_1 + b_{1i}) X_j, \sigma^2\right).$$

It follows that the marginal distribution of the observations is

$$y_{ijk} \sim N\left(\beta_0 + \beta_1 X_j, \sigma_0^2 + \sigma_1^2 X_j^2 + \sigma^2\right),$$

assuming b_{0i} and b_{1i} are uncorrelated. Hence, at any time X_j, $(1-\alpha)\,100\%$ of the distribution of the stability limiting response lies between $\beta_0 + \beta_1 X_j \pm Z_\alpha\sqrt{\sigma_0^2 + \sigma_1^2 X_j^2 + \sigma^2}$, where Z_α denotes the appropriate value from the standard normal distribution. Depending on whether the response increases or decreases over time, the time X_j at which the implied bound intersects the acceptance criterion A defines the shelf life. For example, if the response decreases over time, as in the simulation example, for a target quantile of α, we can estimate shelf life $X_j = T_S$ such that $A = \beta_0 + \beta_1 X_j - Z_\alpha\sqrt{\sigma_0^2 + \sigma_1^2 X_j^2 + \sigma^2}$.

The reason adjustments are needed with all of the shelf life estimation methods shown above is that $Z_\alpha\sqrt{\sigma_0^2 + \sigma_1^2 X_j^2 + \sigma^2}$, the term that determines where the response distribution intersects the acceptance criteria does not change with sample size, whereas the interval-based methods using shortest-lived batch BLUP or lower bound regression coefficient estimates do depend on sample size. What we need is an interval-based method that accurately reflects the width of the product distribution and does not depend on the number of batches.

In theory, once we have estimates of the required regression coefficients and variance components, we could use them in the equation $A = \widehat{\beta}_0 + \widehat{\beta}_1 X - Z_\alpha\sqrt{\widehat{\sigma}_0^2 + \widehat{\sigma}_1^2 X^2 + \widehat{\sigma}^2}$ and solve for X. This is similar to the TI method presented by Quinlan et al. (2013a), except that the standard error expression used in Quinlan et al. follows from the variance of a generic $\mathbf{X}\widehat{\boldsymbol{\beta}}$ whereas the standard error expression here follows directly from the variance of the product distribution. We refer to this estimation method here as the *direct* method. It would seem reasonable to replace Z_α by $t_{\alpha,\nu}$ where we determine the degrees of freedom based on, say, the number of batches in the stability trial.

The results for this method are labelled *Direct_LMM_3b* and *Direct_LMM_6b* for the 3-batch and 6-batch cases, respectively, in Figs. 22.5 and 22.6. For 3 batches, the mean estimated shelf life is 30.7 with a standard deviation of 4.4. The lowest shelf life estimate was 17. The percentiles are: 10th, 25; 25th, 28; median, 31; 75th,

34. Even using a relatively high, 75th percentile, estimate, the likelihood of future batches meeting acceptance criterion for the stated product lifetime would exceed 0.9. Just over 5 % of the estimates yielded by this method are less than the lower red line and, referring to Table 22.1, coverage was 73 % using the 10th percentile of the empirical shelf life distribution as the upper red line. This is a noticeable improvement relative to *Q1E* and the naive LMM, but lower than coverage obtained by the BLUP-based procedures..

The direct procedure is crude, but shows promise. It does have one important advantage: the quantile of the product distribution that defines shelf life is stated explicitly. Unlike all other procedures described in this chapter, the relationship between the α level used for estimation and the target percentile of the product distribution that defines shelf life is completely transparent. The main disadvantage stems from the fact that with 3 batches, σ_0^2 and σ_1^2 are in effect estimated with only 2 degrees of freedom. This has a greater impact when the variance component estimates must be used directly, rather that indirectly as they are in the regression confidence limit and BLUP-based procedures.

The summary overview of this section is that, while literal application of the linear mixed model—what was called the "naive LMM" procedure in Sect. 22.4—is not suitable for shelf life estimation, several adaptations of LMM methods that address the *ICH* guidance regarding "all future batches" provide promising alternatives to current practice. Each addresses the *ICH* guidance differently, but all produce estimates with distinctly better coverage, as defined in this chapter, than the *Q1E* procedure currently in use. While not shown here, simulations varying the slope's expected value and variance have been done to see if the results shown here might be an artifact of the parameters used in the simulation. The results are not artifacts.

One important remaining question the simulations discussed in this section do not address is, "what happens when the batch-to-batch variation among slopes truly is negligible, essentially equal to zero?" The next section addresses that question.

22.6 Simulation Scenario 2: No Variation Among Batch Slopes Giving Rise to Data

Reviewers of this chapter expressed a concern that the results in the previous sections might not apply when the degradation rate is essentially identical across batches, i.e. when there is no random slope effect in the process giving rise to the data. To address this question, a second simulation study using the same parameters as the first simulation, except the random slope was removed from the process generating the data.

This scenario used one thousand data sets generated as $B_{0i} + B_{1i}M + w_{ij}$, where M denotes time in months as in scenario 1. As in scenario 1, random intercepts for each batch were generated as $B_{0i} \sim NI\,(100, 0.5)$. The slope, B_{1i}, was -0.25

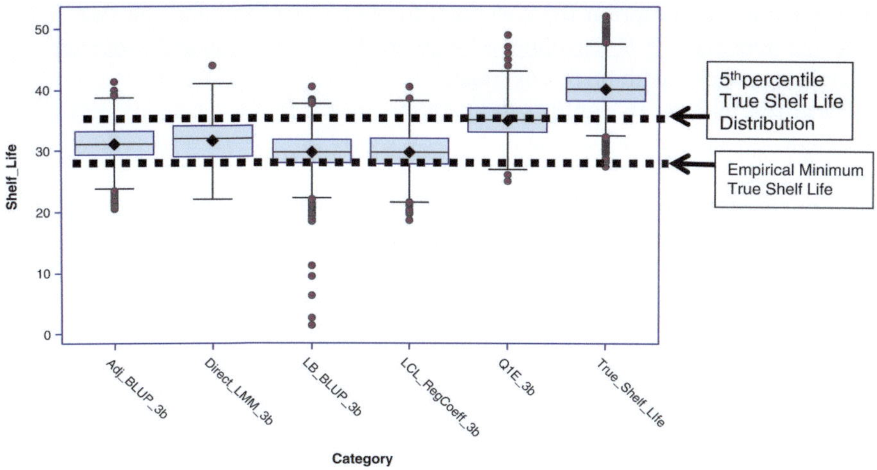

Fig. 22.7 Sampling distributions of estimated shelf life relative to the empirical minimum and 10th percentile of true shelf life distribution. True shelf life distribution with slopes identical

for all batches. As before, the distribution of random variability among observations resulting from analytical variation, measurement error, etc., was assumed to be $w_{ij} \sim NI(0, 0.5)$. This scenario addressed simulated stability studies with 3 batches per trial, only. Given the results from the first scenario, the naive LMM was eliminated from further consideration, and the results for trials with 6 batches were judged to be of lesser interest.

Figure 22.4 shows the empirical true shelf life distribution for this scenario. Figure 22.7 shows side-by-side box plots of the empirical true shelf life distribution and the sampling distributions of the estimates from the *Q1E* procedure and the LMM-based alternatives. Table 22.2 gives these results in numerical form.

Assuming no variation in batch slopes did not change the overall picture. The *Q1E* procedure shows a large proportion of shelf life estimates above the 5th and 10th percentiles of the empirical shelf life distribution—roughly 50 and 30 % respectively. Coverage was 58 % if the 5th percentile of the empirical true shelf life distribution defines the upper red line, and 68 % if the 10th percentile is used. All of the LMM-based alternatives showed higher coverage. Basing labelled shelf life on the *Q1E* procedure would result in a relatively high proportion of future batches failing to meet the acceptance criterion before the stated shelf life of the product.

All of the LMM-based procedures showed a reduced proportion of shelf life estimates above the red line. Just over 10 % of the Direct method shelf life estimates were greater than the 5th percentile of the empirical true shelf life distribution. The LCL Regression and Lower Bound BLUP methods yielded fewer than 5 % estimates above the empirical distribution's 5th percentile. As for the lower red line, roughly 10 % of the shelf life estimates from each of the LMM-based methods were below the effective minimum of the empirical true shelf life distribution, with the Lower Bound BLUP procedure having the greatest proportion of low estimates. Using the

5th percentile as a criterion for coverage, the Direct method and the two BLUP-based methods had 75 % coverage or greater; the Fieller-adjusted procedure was the highest of these, with 80 % coverage.

These results tend to confirm the patterns shown in the previous sections. If one characterized the various methods, they would be classified as follows. *Q1E*: excessively liberal. Fieller-adjusted BLUP: essentially neutral. Lower bound BLUP and Direct: somewhat conservative. Lower confidence limit regression: conservative with 3 batches but tends to be more liberal as number of batches increases.

One caveat: all of the LMM-based methods were implemented using $\alpha = 0.05$ as the confidence coefficient criterion. As noted in previous sections, this does not accurately reflect the relationship between the estimate and the shelf life distribution percentile. All of the LMM-based methods could thus be improved by addressing this issue. Nonetheless, even in their crude form, and even with 3 batches per stability trial, the LMM-based estimators perform quite well.

22.7 Summary and Conclusions

Christopher (2010) and Capen et al. (2012) state that a well-defined quality statement is a prerequisite for determining shelf life. The *ICH Q1A* and *Q1E* documents provide a basis for such a quality statement. A well-conceived estimator for shelf life should, as directed by *ICH*, provide reasonable assurance that future batches meet acceptance criteria for the stated shelf life based on that estimate. In this chapter, we attempted to make such a quality statement tangible in the context of how one might evaluate competing estimation methods in a simulation study. An estimator should minimize the likelihood of estimates below shelf lives that never—or rarely—occur. When there is an upper bound on shelf life corresponding to some percentile of the shelf life distribution related to acceptable risk, the estimator should also minimize the likelihood of estimated shelf lives that exceed that bound.

The estimation procedure prescribed by the *ICH Q1E* guidance and several LMM-based alternatives were reviewed in this chapter. Tables 22.1 and 22.2 summarize their performance via mean, standard deviation, maximum, minimum, and several percentiles. Figures 22.5, 22.6 and 22.7 portray their performance relative to one another in side-by-side box-and-whisker plots. The figures also superimpose ranges of desirable estimates determined by possible quality statements. Figures 22.5 and 22.7 show a range if the quality statement targets estimates between the effective minimum and the 5th percentile of the true shelf life distribution. Figure 22.5 applies to scenario 1, in which the process giving rise to the data includes non-negligible variation among batch slopes. Figure 22.7 applies to scenario 2. Figure 22.6 shows a range between the effective minimum and the 10th percentile.

Several conclusions are evident from discussion in this chapter, reinforced by inspection of Figs. 22.5 through 22.7. The *Q1E* procedure does not perform well with regard to any imaginable defensible quality statement. The likelihood of obtaining a shelf life estimate from *Q1E* that is consistent with the *ICH* "all future batches" objective is quite low. This is consistent with other work, e.g. Schwenke (2010) and Quinlan et al. (2013b). The *Q1E* assumes a model that does not align with plausible probability processes we think lead to the data observed in a stability study, and its inference space is inconsistent with the stated intent of *ICH*. Therefore, its poor performance should not be surprising. The myth that the *Q1E* procedure is conservative appears to be exactly that: a myth. Finally, there is no evidence that increasing sample size improves the *Q1E* procedure's performance.

The LMM is clearly better aligned with the process giving rise to the data, but the "Naive LMM" estimation procedure does not perform well. In fact, increasing sample size actually worsens the performance of the naive LMM procedure, increasing the likelihood of unusable estimates. Failure to align the naive LMM with the quality statement implicit in *ICH* accounts for its performance.

The LMM-based procedures, lower confidence limit (LCL) regression coefficient, batch-specific BLUP, and direct lower quantile methods, all show promise. Because all are LMM-based, they are aligned with a plausible process giving rise to the data. All are explicitly defined with reference to the *ICH* objective. The main problems with these procedures stem from issues determining suitable interval estimate widths. The BLUP and LCL regression coefficient method standard errors depend on sample size in a way that is not consistent with the quantile of the distribution they are attempting to estimate. All use variance component estimates based on few degrees of freedom. The relationship between quantiles of the product distribution and shelf life distribution cannot be determined without knowing the variance components, making it difficult to set α levels for determining appropriate confidence bounds.

Despite all of these difficulties, the BLUP, LCL regression coefficient, and direct methods performed remarkably well in this simulation, even with 3 batch stability studies. Their performance was unambiguously superior to that of the *Q1E* procedure. While these are obviously only two simulation studies, and not a comprehensive and definitive comparison, the results are consistent with work by other researchers, and they provide a template by which more exhaustive comparisons can be conducted. Future work could include refined methods of determining α, comparing relative performance with the distributions giving rise to the data that have different mean and variance parameters or are non-normal.

Finally, all of these methods, as described in this chapter, use standard mixed model computing algorithms. For example, all results shown in this chapter were obtained using SAS PROC GLIMMIX. In principle, all are amenable to estimation via Bayesian methodology. Because stability studies are usually conducted at the end of a lengthy development process, it is likely that there is ancillary information available that could be used for modestly but defensibly informative priors. Standard LMM estimation is essentially equivalent to Bayesian estimation with non-informative priors. Defensible informative priors could greatly improve the precision of variance estimates thereby greatly improving the accuracy of shelf life estimates and their alignment with *ICH* objectives. The LMM-based procedures reviewed in this chapter form a basis for future work along these lines.

Acknowledgements The authors gratefully acknowledge the Product Quality Research Institute and the members of the PQRI Stability Shelf Life Working Group (SSL WG). The authors had the privilege of serving as active members of the PQRI SSL WG from its inception through 2010, and have continued their association with the working group from 2010 to the present. Citations by Capen, Christopher, Quinlan, Stroup and Schwenke all represent PQRI SSL WG efforts. The working group continues its efforts to understand current shelf life estimation practice and investigate alternative methodology with the potential to improve existing practice. Without our time as members of the working group, this chapter would not be possible. We also acknowledge Stan Altan. His review, suggestions and overall support are much appreciated.

References

Altan S, Manola A, Shoung J-M, Shen Y (2013) Perspectives on pooling as described in the ICH Q1E guidance. Session Nonclinical Statistics and Computing. Proceedings of the biopharmaceutical section. Joint Statistical Meetings

David Christopher J (2010) The philosophy and intent of stability shelf life. Topic Contributed Session Pharmaceutical Stability Shelf Life: Philosophy, Intent and Estimation. Joint Statistical Meetings

Huynh-Ba K (ed) (2009) Handbook of stability testing in pharmaceutical development: regulations, methodologies, and best practices. Springer, New York

International Conference on Harmonization (2003) ICH Harmonised Tripartite Guideline: Stability Testing of New Drug Substances and Products Q1A (R2)

International Conference on Harmonization (2003) ICH Harmonised Tripartite Guideline: Evaluation For Stability Data Q1E (R2)

Kroenker R (2005) Quantile regression. Cambridge Press, Cambridge

Quinlan M, Stroup W, Christopher JD, Schwenke J (2013a) On the distribution of batch shelf lives. J Biopharm Stat 23(4):897–920

Quinlan M, Stroup W, Schwenke J, Christopher JD (2013b) Evaluating the performance of the *ICH* guidelines for shelf life estimation. J Biopharm Stat 23(4):881–896

Quinlan M, Stroup W, Christopher JD (2014) Alternative procedures for shelf life estimation utilizing mixed models. Contributed Session Joint Statistical Meetings

Capen R, Christopher JD, Forenzo P, Ireland C, Liu O, Lyapustina S, O'Neill J, Patterson N, Quinlan M, Sandell D, Schwenke J, Stroup W, Tougas T (2012) On the shelf life of pharmaceutical products. AAPS PharmSciTech 13:3

Schwenke J (2010) Current practices in shelf life estimation. Topic Contributed Session Pharmaceutical Stability Shelf Life: Philosophy, Intent and Estimation. Joint Statistical Meetings

Stroup W, Quinlan M (2010) Alternative shelf life estimation methodologies. Proceedings of the biopharmaceutical section. Joint Statistical Meetings

Stroup W (2013) Generalized linear mixed models: modern concepts, methods and applications. CRC Press, Boca Raton

Tang C, Leng C (2011) Empirical likelihood and quantile regression in longitudinal data analysis. Biometrika 98(4):1001–1006

Chapter 23
In Vitro Dissolution Testing: Statistical Approaches and Issues

David LeBlond

Abstract Dissolution (or in vitro release) studies play an important role during pharmaceutical drug development. They are useful as a quality control tool, establishing an IVIVC, justifying a biowaiver and supporting equivalence between generics and innovator products. This chapter provides an overview of regulatory guidances related to dissolution testing. The important question of dissolution profile comparisons and the challenges of defining similarity are discussed. The limitations of the f_2 statistic, a regulatory driven test, are described, with emphasis on the lack of a statistical basis for the test. The strengths and weaknesses of alternative approaches are described. A Bayesian test procedure is given that mitigates to a large extent the weaknesses of other approaches by providing a coherent connection between a parameter defining "fit for use" to a statistical probability statement of similarity.

Keywords In vitro dissolution • In vitro release • Dissolution profile similarity • f_2 statistic • Bayesian model

23.1 Introduction

Many pharmacologically active molecules are formulated as solid dosage form drug products. Following oral administration, the diffusion of an active molecule from the gastrointestinal tract into systemic distribution requires the disintegration of the dosage form followed by the dissolution of the molecule in the stomach lumen. Its dissolution properties may have a direct impact on its bioavailability and subsequent therapeutic effect. Consequently, dissolution (or in vitro release) testing has been the subject of intense scientific and regulatory interest over the past several decades. In vitro dissolution testing as an analytical methodology measures drug release in liquid media. The method requires specialized laboratory

---

D. LeBlond (✉)
Consultant in CMC Statistical Studies, Wadsworth, IL 60083, USA
e-mail: David.leblond@sbcglobal.net

L. Zhang (ed.), *Nonclinical Statistics for Pharmaceutical and Biotechnology Industries*, Statistics for Biology and Health, DOI 10.1007/978-3-319-23558-5_23

equipment, following a well-defined protocol such as described in the United States Pharmacopeial convention reference chapter <711> (2011).

Dissolution (or in vitro drug release) studies of a solid oral dosage form are extensively carried out during pharmaceutical drug development. The dissolution profile is comprised of a concentration curve of dissolved active pharmaceutical ingredient (API), expressed as a % of label claim, versus time. These studies are important for a number of reasons.

- A quality control tool, implemented through a specification. The specification typically incorporates a sum of variance components related to analytical method and manufacturing variability. The innovator company proposes specifications as a way to maintain lot quality and consistency between development batches and post-approval commercial batches.
- Investigation of in vitro in vivo correlations (IVIVC). In vitro dissolution testing is frequently used as a surrogate measure of bioavailability. This avoids the risk and expense of human trials, and facilitates the implementation of improvements in processes and products and serves to guide development of new formulations.
- Provide a scientific basis to justify a biowaiver for post-approval changes (FDA 1997a, b) which requires establishing equivalence between the new and old product (i.e., changes in formulation, the manufacturing process, the site of manufacture, and the scale-up of the manufacturing process).
- Authorization to market a generic compound. Dissolution profile comparisons is required to establish the claim of equivalence between generic and innovator products.

Consequently, in vitro dissolution testing has played an increasingly important role in drug development.

23.2 Dissolution Testing

23.2.1 USP Standards and Informational Guidance

The USP chapters <711> and <724> provide dissolution standards. USP chapter <711> describes the equipment, media, testing protocols, acceptance limits for dissolution testing of immediate release (IR), extended or modified release (ER or MR), and delayed release (DR) dosage forms. Chapter <724> extends the standards to transdermal delivery (TD) dosage forms. Units tested from any batch at any time during its shelf life are expected to meet the following standards.

In Table 23.1, Y_i's are dissolution values of individual units expressed as a % of label claim, $\overline{Y}_n$ are averages of n individual unit dissolution values tested up to that stage, $\#_A$ gives the number of dissolution values of individual units tested up to that stage that fall within range defined by A. Q, L and U are values having units of % label claim that must be justified by the product sponsor.

Table 23.1 Dissolution testing and acceptance criteria

		Acceptance criteria				
		IR			DR	
Stage/level	Number tested	Individual samples	Pooled sample	ER, MR, or TD	Acid medium	Buffer medium
1	6	$Y_i \geq Q + 5$	$\overline{Y}_6 \geq Q + 10$	$L \leq Y_i \leq U$	$Y_i \leq 10$	$Y_i \geq Q + 5$
2	6	$\overline{Y}_{12} \geq Q$ $Y_i \geq Q + 15$	$\overline{Y}_{12} \geq Q + 5$	$L \leq \overline{Y}_{12} \leq U$ $L - 10 \leq Y_i \leq U + 10$	$\overline{Y}_{12} \leq 10$ $Y_i \leq 25$	$\overline{Y}_{12} \geq Q$ $Y_i \geq Q - 15$
3	12	$\overline{Y}_{24} \geq Q$ $\#_{<Q-15} \leq 2$ $Y_i \geq Q + 25$	$\overline{Y}_{24} \geq Q$	$L \leq \overline{Y}_{24} \leq U$ $\#_{<L-10 \cup >U+10} \leq 2$ $L - 20 \leq Y_i \leq U + 20$	$\overline{Y}_{24} \leq 10$ $Y_i \leq 25$	$\overline{Y}_{24} \geq Q$ $\#_{<Q-15} \leq 2$ $Y_i \geq Q - 25$

For IR products, a single dissolution time is also selected (generally less than 60 min). Similarly for DR products, dissolution periods in acid and buffer media respectively are chosen based on considerations given in USP chapter <711>. For ER, MR, or TD products, multiple time points are chosen and the requirements in Table 23.1 must be satisfied for all time points. Early time points are chosen as a check on dose-dumping and later times are chosen to assure complete dissolution/release. A different L and U will generally be justified for each time point. The acceptance criteria must be met for each time point tested.

USP standards are subject to change as technology and quality expectations evolve. Because of its potential relationship to safety and efficacy, dissolution specifications are the subject of debate among stake-holders. The current USP standards do not purport to control population mean (parametric) behavior of a product. Recent reports suggests that dissolution limits should be based on limiting the percentages of slow and fast dissolving units (Tsong et al. 2004).

An important aspect of USP standards that is often overlooked is that they are not intended as batch release tests. Instead they represent a standard of quality that should be met by any product sample tested at any time during the product shelf life. The practice of employing a USP standard as a batch release test only assures that the standard will be met about 50 % of the time. In practice, many manufacturers use batch release tests more stringent than USP standards to ensure that the USP standard is met with high confidence.

The USP also provides informational guidance on dissolution. Chapter <1088>, In vitro and in vivo evaluation of dosage forms, discusses method development for IR and MR dosage forms, some basic aspects of IVIVC, and establishment of dissolution specification ranges. Chapter <1090>, Assessment of drug product performance—bioavailability, bioequivalence, and dissolution, discusses the relationship of in vitro dissolution to in vivo performance. It gives specific guidance on in vitro dissolution comparisons, statistical methods, and the justification of biowaivers. It provides a good overview of other regulatory guidance from FDA, ICH, and WHO. Finally, USP chapter <1092>, The dissolution procedure: development and validation, discusses factors that impact dissolution testing such as choice of medium, apparatus, equipment operation, sampling, choice of time points, validation and dissolution acceptance criteria.

23.2.2 Setting Specifications

The justification of the values of Q, L, and U as well as choice of dissolution testing times should be based on their relationship (or lack thereof) to in vivo performance. Often the choices for Q, L, and U are based on the observed mean dissolution profile of units from clinical study batches. A further objective is to ensure batch to batch consistency over the product life-cycle. These limits may also be useful in demonstrating equivalency during scale-up or in justification of certain post-approval changes. A suggested starting range for acceptable release is ±10 % LC from the mean dissolution profile. Wider ranges may be accepted if it can be demonstrated or argued that clinical lots with the same variability are bioequivalent. An IVIVC (see below) may sometimes be used to justify wider acceptance ranges.

USP <1092> recommends, with respect to IR products, a Q value in the range of 75–80 % LC. Acceptance criteria, including test times, are usually established on the basis of an evaluation of the dissolution profile and should be consistent with historical batches. There is an expectation that batches consistent with the in vivo performance, composition, manufacturing procedure of clinical batches will have results that generally fall within the acceptance criteria. However, some failure at the first stage of testing may be expected by regulators to assure an adequate consumer risk level. In negotiation of acceptance criteria with regulators, Hofer and Gray (2003) point out that there is a large discriminatory ability difference between stage 1 and subsequent testing stages and argue against rounding Q values to 5 % LC intervals (i.e., 70, 75, 80, 85, ...). They suggest choosing a Q value based on computer simulations, or calculated from estimated variance components, that predict the future rate of stage 2 testing (e.g., such as 20 % probability of advancing to stage 2). Their approach should also be extended to ER, MR, and DR products.

23.2.3 Stability

An important aspect of USP standards is that they are expected to be met for any sample taken from any batch of the product at any point during its shelf life. When dosage form characteristics such as enteric coating integrity, moisture content, or other excipient properties drift over time, dissolution can change on product storage. Thus it is critical that dissolution stability be taken into account in the design and justification of dissolution testing procedures and acceptance limits.

When dissolution testing is conducted over multiple dissolution time points (e.g., for MR, DR, TD, or ER products) the stability indicating response should be treated as multivariate due to correlations among results at different dissolution time points. Various statistical procedures such as principle component analysis or a dissolution profile model are available to reduce the dimensionality, but a multivariate stability model may still be required.

23.3 Dissolution Profile Comparisons

23.3.1 Regulatory Guidance

The FDA guidance (1997b) suggests that for certain drugs, in vitro dissolution results might be sufficient to gain regulatory approval for post-marketing changes and waiver of bioequivalence requirements for lower strength dosage forms (Moore and Flanner 1996a, b). A formal similarity evaluation is a regulatory (FDA 1995, 1997a, b, c, 2000a, b, 2003; EMA 2008) for this purpose. Thus the question of demonstrating similarity or equivalence between the reference and test drug dissolution curves is of both scientific and regulatory importance.

Various methods comparing drug dissolution profiles have been proposed. In general, they can be classified into three categories: (1) model-independent approaches based on a similarity factor, (2) model-independent methods using multivariate statistical distance (MSD) test, and (3) model-dependent methods using parametric curves to describe dissolution profiles.

The profile comparison methods recommended in regulatory guidance lacks detail. Perhaps this is intentional since the scope of products regulated is wide and regulatory bodies often expect sponsors to justify approaches scientifically appropriate for a given product. Below are some details that sponsors should take into consideration when justifying a profile comparison method.

- **Define what is meant by "similar":** Dissolution profiles can differ in many ways. When possible, it is good to identify those aspects of the reference profile that, for safety and/or efficacy reasons, be maintained in the test profile. Depending on the product, dose dumping, incomplete dissolution, slow dissolution, unusual deviations at critical time points, dissolution stability, or unit to unit variability may be most critical. Popular similarity approaches such as f_2 and MSD may not always be the best choices as they are aggregate summary measures and do not consider deviations at specific time points. Availability of an IVIVC may place limits on how much the test profile can deviate from the reference profile.

  Similarity can be expressed either empirically or parametrically. Empirical similarity is expressed by placing limits on observed data or statistics calculated from observed data. The f_2 statistic is an example of empirical similarity. When defined empirically, similarity depends on aspects of the study design. For instance, changing the number and location of time points introduces bias into the f_2 statistic.

  Parametrical similarity requires specification of a statistical model of the dissolution profile (ideally based on mechanistic understanding) and is expressed by placing limits on underlying model parameters. Parametric similarity is not described in dissolution regulatory guidance, but is a central aspect of bioequivalence regulatory guidance. When defined parametrically, similarity is not dependent on study design.
- **Clarification of inference space**: The inference space is the population to which the inference (i.e. decision about profile equivalence) applies. Depending on the data available and the statistical model used for the comparison, the inference space can be a) a pair of batches (i.e., 1 test batch and 1 reference batch), b) 2 groups of batches (i.e., multiple batches each of test and reference), or c) to the populations (test and reference) of batches from which tested batches are drawn. Ideally the last of these is the only one that provides any assurance about the behavior of future batches made by the new process. Clarifying the inference space will help determine some aspects of the study design—for instance how many batches of each type should be tested.

- **Clarification of desired confidence level and power for inference**: For clinical studies, it is generally expected that a study plan will include a specification for type I error (=1 − confidence level) and type II error (=1 − power). Similarly it is good practice to prospectively state the operating characteristics of a dissolution similarity comparison study. By specifying the confidence level and power in advance, it is possible to determine the required number of units that must be tested.
- **Clarify whether a difference or equivalence test is appropriate**: A difference test would provide a statistical test of the null hypothesis of similarity. In other words, similarity would be concluded unless contradicted by the data. A difference test would be appropriate in cases where the risk that a change in the process produced a change in dissolution profile was a priori low. The burden of proof is on demonstrating a lack of similarity.

  An equivalence test on the other hand is a test of the null hypothesis of non-similarity. This reverses the usual roles of type I and type II errors. The burden of proof is on showing acceptable degree of similarity. In an equivalence test, a "region" of similarity is defined (ideally parametrically). If the estimated similarity metric is well within the similarity region, similarity is concluded. An equivalence test would be called for if there is a substantial risk a priori that a change to a process would produce a change to the dissolution profile. Equivalence tests may require larger sample sizes.

23.3.2 Reasons for Comparisons

There are a number of situations in which a comparison of in vitro dissolution profiles may be required.

- Development of an IVIVC (see below).
- Development and/or validation of a dissolution test procedure.
- Demonstration of equivalency during manufacturing process scale-up.
- Demonstration of equivalency after manufacturing process, site, or excipient changes.

23.3.3 In Vitro–In Vivo (IVIVC) Correlations

In vitro drug release profiles are known to be affected by formulation factors such as API particle size, excipients, presence of surfactants or polymers. Besides these formulation factors, the in vivo absorption is also complicated by interactions with physiologic parameters such as pH, transit times, presence or absence of food, and gastrointestinal physiology (Kesisoglou and Wu 2008). Characterizing the link between in vitro release and clinical outcome has been a challenging task for

Table 23.2 BCS classification system

		Aqueous solubility	
		High	Low
Intestinal permeability	High	I	II
	Low	III	IV

immediate release formulations and remains an area of intensive scientific interest. Recent papers have reported the use of dissolution media simulating physiologic conditions as an improved way to develop an IVIVC exploiting mechanistic models (Jamei et al. 2009; Sugano 2009; Sjögren et al. 2013) relating physical and chemical properties of the API with oral absorption and dose.

23.3.3.1 Bio-Pharmaceutical Classification System (BCS)

The BCS system (Amidon et al. 1995) categorizes drug substances into four groups depending on aqueous solubility and intestinal permeability as shown in Table 23.2.

The BCS system also takes into account the in vitro dissolution characteristics of a drug product. Together, the aqueous solubility, intestinal permeability and in vitro dissolution rate are often considered key processes that govern the rate and extent of oral absorption for an IR solid oral dosage form. When in vitro dissolution is conducted under conditions that mimic those in the gastro-intestinal tract, it may be supposed that the observed dissolution behavior reflects that which occurs in vivo. If this observed in vitro dissolution is rapid relative to other pharmacokinetic processes such as membrane absorption, transport to the site of action, metabolism, and elimination, moderate differences in dissolution rate between different dosage forms may not impact bioavailability characteristics, and thus efficacy and safety, of the product. In fact, it may be difficult or impossible to demonstrate an IVIVC for drug substances in IR dosage forms classified as BCS I or III. For drug substances classified as BCS I, depending on stability in the gastro-intestinal tract, effect of excipients, therapeutic index, site of absorption, a biowaiver for bioavailability and bioequivalence studies, and wider dissolution testing limits may be justified.

23.3.4 Dissolution Profile Models

Much interest has focused on models describing in vitro release profiles over a time scale. A good review of mechanistic profile models is given by Costa and Lobo (2001). These and some empirical release models are considered by Costa et al. (2003). Some of these models are given in Table 23.3.

The models do not consider a lag time or background. They assume that the drug release at $t = 0$ is zero and that drug release commences immediately. However, many of the models in the table above can be modified to accommodate lag

Table 23.3 Some dissolution profile models

Model	Interpretation	Expression for fraction dissolved
Higuchi	Fickian diffusion	$k\sqrt{t}$
Korsmeyer-Peppas	Diffusion based	$k_2 \cdot t^{k_1}$
Quadratic	Empirical, parabolic curve	$k_1 t + k_2 t^2$
Hixson-Crowell	Erosion release	$k_2\left(1-(1-k_1 t)^3\right)$
Exponential	First order kinetics	$(1 - exp(-k_1 t))$
Probit	Normal cumulative distribution function	$\Phi(k_1 + k_2 \cdot \log(t))$
Logistic	Empirical sigmoid shape	$\frac{exp(k_1 + k_2 \cdot \log(t))}{1 + exp(k_1 + k_2 \cdot \log(t))}$
Weibull	Life-time distribution function	$1 - exp\left(-\left(\frac{t}{k_1}\right)^{k_2}\right)$
Gompertz	Growth/distribution function curve	$1 - exp(-k_1(exp(k_2 \cdot t) - 1))$

times and/or background levels of dissolved drug. Additional parameters such as mechanistic constants can be accommodated in some models. The maximum dissolution can also be accommodated in the exponential, probit, logistic, Weibull, and Gompertz models. Of the models presented in the table, the Weibull model, which is has a mechanistic interpretation, often provides the most satisfactory fits to dissolution data.

Each parameter in these models can be treated as fixed and estimated using nonlinear least squares methods. In some cases the models can be linearized by various transformations and the parameters estimated by linear least squares. Random variation can also be incorporated into the models either as measurement variance or as variance associated with one or more of the model parameters (i.e. as a nonlinear mixed model).

Hierarchical models (see Gelman and Hill 2007) can be considered which account for random variation at various levels, such as batch to batch, run to run, or vessel to vessel. The appropriate model can be identified using available data by the usual modeling approaches (e.g., minimum AIC). However, model selection should not be based on statistics, but should consider known underlying mechanistic processes.

23.3.5 Challenges in Defining Similarity

In order to make a judgment about the similarity of two dissolution profiles, a clear, prospective definition of similarity must be established. It is convenient to view similarity geometrically as a region in some space. If the projection of two profiles onto this space falls within a defined region, similarity can be concluded. Similarity regions can either be defined within the parameter (population) space or the observed data space. This concept is illustrated in Fig. 23.1.

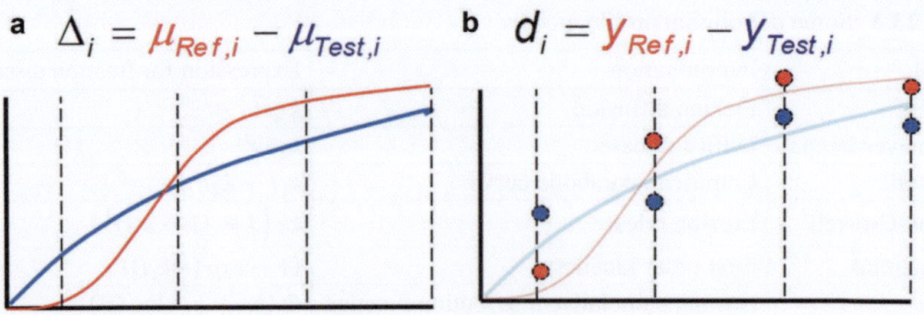

Fig. 23.1 (**a**) Parametric similarity region. (**b**) Data space similarity region

Consider a comparison of a Test and Reference profile. The true (population) profiles are shown in Fig. 23.1a in blue and red, respectively. These true profiles are unknown, not based on observed data, and are therefore conceptual. A region of similarity might be defined in the p-space of the true (population) differences (Δ_i's where i = 1, 2, . . . ,p and p = number of time points) between the profiles. Restrictions may be placed on the magnitudes of the Δ_i, or some function of them, based on bioavailability and bioequivalence considerations, to define a similarity region. The resulting region of similarity is thus defined in the space of population parameters. This population based similarity definition is useful because a decision rule for similarity can then be based on statistical equivalence testing of the true population parameters (in this case the Δ_i) once observed data are available. This is the approach that was taken in defining bioequivalence (FDA 2001). This was also the approach also advocated by Eaton et al. (2003) for similarity comparisons. Statistical modeling presumes some understanding of the underlying dissolution process and thus represents a basis for risk based decision making and continuous knowledge building. It is important to recognize that safety and efficacy risk is associated with an underlying true population state and that similarity must be defined parametrically. Parametric definitions often permit us to leverage known theory and understanding of underlying mechanisms within a statistical model.

Figure 23.1b illustrates how similarity may be defined in the p-space of observed profile differences (d_i's based on the difference between the red and blue symbols). In this case the true profiles (faded blue and red lines) are not considered explicitly. The decision rule is based solely on whether the observed d_i, or some function of them, fall within some pre-established similarity region. In this case, no statistical equivalence test is possible because the underlying parameters are not defined and cannot be statistically estimated. The decision rule is essentially based on point estimates, but the underlying estimands are unclear. This definition of similarity is therefore not very useful. No statistical test of hypothesis or equivalence is possible and therefore the operating characteristics of the decision rule cannot be understood. Unfortunately, it is current practice to define dissolution similarity in the observed data space. The f_2 based similarity decision rule, discussed below, is usually employed without regard to any underlying statistical model. This represents a poor basis for risk based decision making and knowledge building.

Unfortunately, the inference space in dissolution comparisons is rarely explicitly defined. The use of hierarchical models (mentioned in the previous section) is important if the inference is to be made to the populations of batches made by two processes, rather than merely comparing the dissolution of samples from two batches on hand. In most cases, the desired inference space goes beyond a simple comparison of two batches. For instance in demonstrating that a manufacturing process change does not impact bioavailability, our inference must extend to all future batches made by the new process. A proper definition of inference space for dissolution comparisons, and a corresponding proper study design and statistical modeling remain significant challenges in current practice and regulatory guidance.

23.3.6 Model Independent Approaches

23.3.6.1 The f_2 Similarity Factor

Moore and Flanner (1996a, b) defined the f_2 "fit factor" in terms of observed data as given in the following formula:

$$f_2 = 50\log_{10}\left\{100\left(1+\frac{1}{k}\sum_{i=1}^{k} w_i(R_i - T_i)^2\right)^{-1/2}\right\}$$

where R_i and T_i are reference and test results at time t_i, $i = 1, \ldots, k$, respectively and the w_i values are appropriately chosen weights to reflect possible heterogeneity. Typically the weights are set to 1. This became the recommended similarity test in regulatory guidances for dissolution profile comparison (FDA 1995, 1997b, c, 2000a, b) with the similarity criterion rule of $f_2 > 50$. The f_2 statistic has been adopted by regulatory bodies world-wide, although implementation details and limits in specific guidances may differ somewhat.

The f_2 similarity statistic is very simple for scientists to calculate and use. However, there has been little discussion on a clear scientific rationale on the choice of this decision criterion. Despite its widespread use, it has serious technical shortcomings. LeBlond et al. (2016) have commented extensively on the many statistical deficiencies of the f_2 statistic. These are briefly given below.

1. Current regulatory guidances (EMA 2008; FDA 1997a, b, c, 2000a, b; Japan 2012; WHO 2006) fail to define the population characteristic that f_2 estimates. The f_2 statistic implies a similarity region in the data space, not in the parameter space. This precludes statistical modeling and leaves the operation characteristics undefineable.
2. It does not leverage any model of profile shape. Test and reference must use exactly the same time points, a requirement that is not always met in practice.

3. If the median sampling distribution of f_2, for a particular comparison, equals 50, the Type I error of the determination is 50 % (equivalent to a decision based on a fair coin flip).
4. The unweighted f_2 statistic does not account for differences in variance among the time points. Yet it is commonly observed that the variance is larger near 50 % dissolution than near 0 or 100 % dissolution.
5. The f_2 statistic is a univariate summary statistic. It ignores any large differences that may be present at individual time points.
6. As the number of time points included in the comparison is increased, the f_2 criterion becomes more liberal in that larger deviations can be accommodated.
7. The f_2 statistic is more of a Euclidean distance metric than a profile shape metric. It would not distinguish between non-similarity of the test units due to dose dumping from that due to failure to deliver the labeled dose at longer times.

In addition to the above deficiencies, current regulatory guidance is often insufficient or contradictory about the use of f_2. This can lead to conflicting approaches and potentially different conclusions about similarity. Practicing statisticians must be aware of the contradictory nature and inadequacy of regulatory guidance to properly advise their clients about risk.

1. EMA (2008) mentions briefly the use of a Weibull model and FDA (1997b) permits the use of a model dependent approach. Other guidances provide no explicit provision for use of a profile model.
2. Current regulatory guidances provide insufficient direction regarding experimental design of the comparison trial.
3. Guidance typically requires 12 dosage units of both test and reference. No justification for sample size is given and no suggestion about how to proceed if more units are available.
4. It is not clear whether the units should come from single or multiple batches. FDA (1997b) states that the reference either be the most recent manufactured lot, prechange or the last two or more consecutive lots, prechange. FDA (1997c) states that the reference should be three consecutive recent lots, prechange. The EMA (2008), WHO (2006) and Japanese (2012) guidances give no information on the selection of test or reference batches.
5. The issue of batch to batch variation is ignored. Pooling tests or procedures are not given and multiple comparison issues are not discussed. Thus the intended inference space is unclear.
6. EMA (2008), FDA (1997b) and WHO (2006) state that no more than a single time point past 85 % dissolution should be used for either test or reference batches. However, it is not clear whether this requirement applies to the observed dissolution mean or to that of individual dose units.
7. FDA (2000a, b) states the number of time points to be used should be a "sufficient number of points". EMA (2008), FDA (1997b) and WHO (2006) recommend at least three. The Japanese guideline specifies 4 according to a formula.
8. Four of the guidances (EMA 2008; FDA 1997b, 2000a, b; WHO 2006) disallow the use of the f_2 statistic in cases of excess variation in the determination of

percent dissolution. Two of the guidances (EMA 2008; WHO 2006) state an RSD strictly less than 20 % for the first time point and strictly less than 10 % at later times, whereas FDA (1997a, 2000a, b) allow up to and including 20 and 10 % respectively for the corresponding time points. FDA (1997c) also imposes an additional constraint that no difference between reference and test means across any of the time points can exceed 15 %. The guidances do not make clear whether the RSD requirement is based on intra-batch RSD, inter-batch RSD, or some other (e.g., total RSD) measure.

Some of the deficiencies noted above about f_2 may be corrected by defining f_2 parametrically. For this we capitalize the F to indicate that it is a function of population parameters (Δ_i's).

$$F_2 = 50 \times \log\left\{\left[1 + \frac{1}{p}\sum_{i=1}^{p}\Delta_i^2\right]^{-1/2} \times 100\right\} > 50$$

or equivalently,

$$\sum_{i=1}^{p}\Delta_i^2 < 99p$$

The F_2 similarity region is based on a hypersphere having a dimension equal to the number of time points. The resulting similarity region can be visualized for $p = 2$ as shown in Fig. 23.2.

When $p = 2$, the similarity region is a circle with its center at the origin (perfect overlap of test and reference profiles). In higher dimensions the similarity region will be a spheroid with radius equal to $p\sqrt{99}$. The radius of the sphere increases with the number of time points, being 29 % larger for five time points than for three.

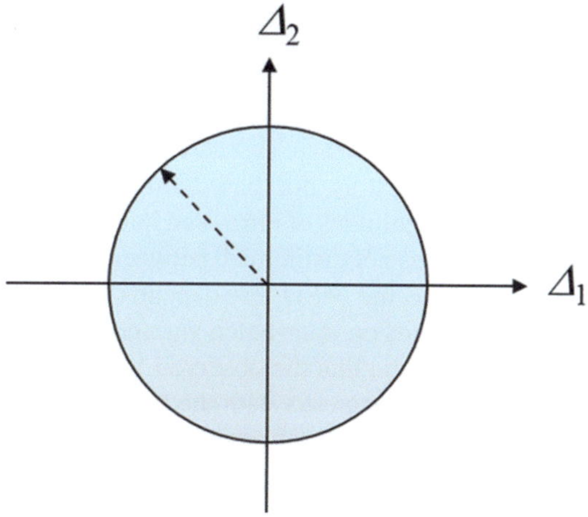

Fig. 23.2 F_2 similarity region

This dependence radius on p shows that as the number of time points increases, larger differences at individual time points can be accommodated. This leads to an opportunity to increase the likelihood of claiming similarity by increasing the number of time points used to calculate the f_2 statistic, or by choosing an excess of early or late time points where test and reference percent dissolution may be physically constrained to be similar Thus when p is large, f_2 may exceed 50, even though large differences exist between dissolution profiles at one time point. For these reasons, choice of p may be critical in defining that actual meaning of similarity.

23.3.6.2 Multivariate Squared Distance (MSD) Test

The switch to use of the MSD test represents a fundamental change in the definition of similarity. Whereas the f_2 statistic is based on a transformed Euclidean distance dependent only on the mean differences, the MSD is based on a Mahalanobis distance metric whose value depends on the pooled variances at each time point and the correlations among time points. While the f_2 test of similarity is based on an observed f_2, the MSD test of similarity is based on the outcome of a formal multivariate statistical equivalence test with a specified maximum Type I error level (0.05). Despite an apparent advantage in relation to known statistical properties, the MSD also fails to provide a satisfactory approach to a test for similarity. The reasons are summarized as follows:

1. The decision whether to use f_2 or MSD, and thus on the fundamental definition of similarity, depends on observed data (i.e., sample variances or % CVs). Further, the tolerance limit, *TL*, is a function of the observed pooled sample variance, S_p.
2. The MSD similarity limit is unclear. Whereas the decision limit for the f_2 statistic ($f_2 > 50$) is provided in the guidances, the MSD limit is not provided, but must be determined on a case-by-case basis and is dependent on differences among reference batches. No direction is provided on the amount of historical data required or the degree of conservatism to be used in setting this limit. A greater inter-batch variance among reference batches implies a wider acceptance limit for MSD. Thus there is no clear understanding of the meaning of similarity across products.
3. The MSD test is not based on a hierarchical model that accounts for the relative magnitudes of intra- and inter-batch variance components. Whereas the f_2 statistic does not distinguish between intra- and inter-batch variance, the decision to switch to the MSD test depends on intra-batch variance and the MSD limit itself depends on inter-batch variance (of the reference formulation). Yet there is no requirement that the observed MSD take into account the relative magnitudes of these variances. As with the f_2 statistic, it is not clear whether the conclusion regarding similarity is meant to apply to the test and reference populations of batches or merely to the batches used to conduct the MSD equivalence test.

4. The MSD metric is not a measure of profile shape similarity. The MSD metric is a unitless pivotal quantity. It is essentially a ratio of squared differences divided by their pooled measurement variances. Processes that exhibit larger variance can accommodate greater mean differences for a given fixed value of the MSD metric. For this reason, the decision criterion for the MSD test requires either an intersection-union/ likelihood ratio test or an interval estimation involving a multivariate confidence region. The complexity of the MSD test requires greater statistical expertise and specialized statistical software. The need to use the MSD approach thus leads to a considerable "culture shock" for the dissolution scientists who may not have anticipated the need for statistical support prior to collection of the data.
5. Current regulatory guidances do not mention the use of an intersection-union or likelihood ratio test based on the MSD. Berger and Hsu (1996) have shown that this test has a nominal size, given the usual assumption of normality. However, literature examples have used an inversion of Hotelling's T^2 to obtain a test based on an estimated confidence region. Such a test can be very conservative and increasingly so as the number of test points (or dimensions) increases (Eaton et al. 2003).
6. Multivariate confidence region estimates are not unique. While the minimum coverage of the MSD based multivariate confidence region (0.90) is specified (FDA 1997b), other aspects of the region are not defined. Confidence regions for a given coverage can take on arbitrary shape (hyper-elliptical, hyper-rectangular, 1- or 2 sided in some or all dimensions, etc.) and, in general, will not conform to the shape of the defined similarity region. The location and extent of the confidence region extrema thus depend on these shape choices and the way in which risk is allocated. The MSD equivalence test requires that the entire confidence region (including extrema) be contained within whatever similarity region is determined. Thus the MSD equivalence test can be made more or less conservative, depending on the shape choices made by those conducting and reporting the equivalence test results.
7. Literature implementations of the MSD test have been controversial. For instance the implementation reported by Tsong et al. (1996a, b) and Sathe et al. (1996) illustrate the subjective impact of similarity and confidence region shape choices. They illustrate rectangular similarity neighborhoods, whereas the f_2 statistic utilizes a spherical similarity neighborhood. Their stated measure of similarity (the pivotal Mahalanobis distance statistic) is independent of model parameters. The implementation reported by Saranadasa and Krishnamoorthy (2005) assumes that the curve shapes of test and reference are parallel which is likely never exactly true, and therefore cannot be a consistently tenable approximation.

The multivariate statistical distance (MSD) metric has been proposed in cases where the dissolution measurements are highly variable (Tsong et al. 1996a). One important feature of the MSD metric is that it is parametrically defined and arises from a statistical model whereas the f_2 is based on observed statistics. The MSD metric also takes into account the covariance structure across the dissolution time points which the f_2 does not. But no decision criterion for MSD has been established as there is with f_2 mainly because the notion of similarity is redefined with each different set of data. It is not possible to show a correspondence to the $f_2 > 50$ criterion for similarity although some researchers have proposed decision criteria for MSD or MSD-like statistical procedures. There currently is no industry standard and users of MSD must justify their own criteria for similarity. One approach (Tsong et al. 1996b) defines an MSD_{max} limit as follows:

$$\Delta^T \Sigma^{-1} \Delta \leq MSD_{\max}$$

where Δ is a vector of maximum allowable differences and Σ is a known covariance matrix. The above relation describes the outer limits of the similarity region as the ellipse centered on the origin. Figure 23.3 illustrates this similarity region for $p = 2$ dimensions. In higher dimensions this similarity region would be a hyper ellipsoid.

A 90 % confidence region for the observed difference is given by

$$(\Delta - d)^T S_{pooled}^{-1} (\Delta - d) \leq \frac{4p(n-1)}{n(2n-p-1)} F(p, 2n-p-1, 0.9)$$

where d is the vector of observed mean differences, S^{-1}_{pooled} is the pooled, observed covariance matrix, and F(a,b,c) is the c = 90th percentile of the F distribution with a and b degrees of freedom. The confidence region, illustrated for $p = 2$, is the smaller ellipsoid in Fig. 23.3. The requirement for similarity is that the 90 % confidence region be completely contained within the similarity region. Figure 23.3 illustrates the situation where this is not the case. A test for similarity can be devised based on checking whether the most extreme point in the confidence region, illustrated by a dot, is contained in the similarity region. The method of Lagrange multipliers can be used to locate this point and perform the test.

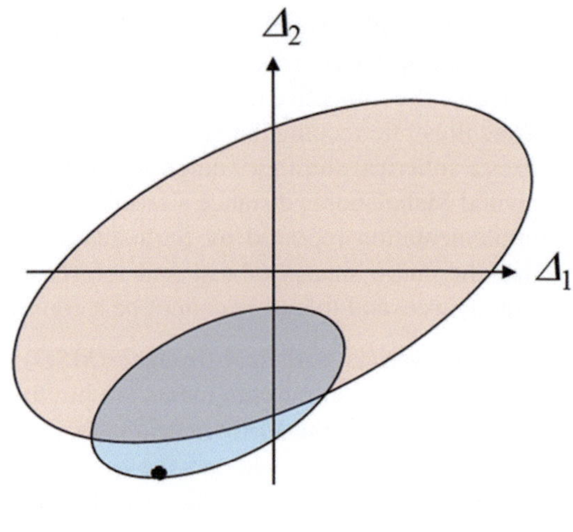

Fig. 23.3 MSD p = 2 elliptical similarity region (centered at the origin) and corresponding estimated MSD elliptical confidence region. The point on the edge of the confidence region indicates the most extreme point identified by the method of Lagrange multipliers

As the above description illustrates, some culture shock is encountered when the f_2 test is not applicable and the complexities of the MSD approach are encountered. Statistical support is called for in this situation. However, there are some other troubling aspects of employing the MSD approach as a back-up to the f_2. First switching from f_2 to MSD changes the actual definition of similarity from a simple univariate point estimate to a multivariate confidence ellipsoid. Since no regulatory limit is given, there is no guaranteed correspondence between the two definitions. It would be desirable to define profile similarity in relation to safety and efficacy aspects, where possible, rather than analytical variability. It should also be pointed out that confidence regions are not unique. The ellipsoid in Fig. 23.3 is only 1 possible 90 % confidence region. An infinite number of other valid 90 % confidence regions exist depending on the chosen shape and how confidence is allocated among the time points.

Saranadasa (2001) suggested a model-independent method under the assumption that the reference and test dissolution curves are parallel—one overlaps with the other after an upward or downward shift of d. A maximum shift d_M is determined as the solution to the equation

$$\left[\left(\overline{R}-\overline{T}\right)-dJ\right]' S_p^{-1}\left[\left(\overline{R}-\overline{T}\right)-dJ\right]=\frac{n_1+n_2}{n_1 n_2}\frac{nk}{n-k+1}F_{k,n-k+1;\alpha},$$

where $n = n_1 + n_2 - 2$. The global similarity is achieved if $d_M \leq \delta$. As the reference and test dissolution curves are seldom parallel, the method is of limited utility in practice.

23.3.6.3 Similarity Region Shape Considerations

Figure 23.2 illustrates the F_2 circular similarity region for p = 2. In higher dimensions this region is a hyper-spheroid. Table 23.4 below lists the radius of this hyper-sphere as a function of p. Notice that a point on the surface of the hyper-sphere (i.e. indicating borderline similarity) is a constant distance from the center regardless of direction. However as Table 23.4 indicates, this radius increases with p.

Table 23.4 Dimensions of spheroid or cuboid similarity regions

p	F_2 spheroid surface	δ(p) cuboid face	δ(p) cuboid vertices
2	14.0	15	21.2
3	17.2	15	26.0
4	19.9	15	30.0
5	22.3	15	33.5
6	24.4	15	36.7
7	26.3	15	39.7

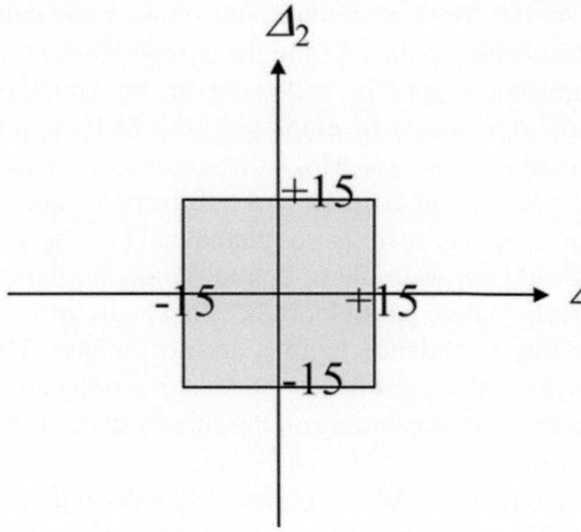

Fig. 23.4 Similarity region based on cuboid face for p = 2

The FDA SUPAC-MR guidance (FDA 1997a, b, c) suggests the following similarity region

$$\delta\,(p) = \max_{t=1,\ldots,p} |\Delta_i| < 15$$

The corresponding similarity region is illustrated in Fig. 23.4 for p = 2. For p = 2, the region is a square but in higher dimensions it is hyper-cuboid. The dimensions of this similarity region that are of interest are a) the distance from the center (exact similarity) to the edge of each side (or face) and b) the distance from the center to each vertex or corner where p sides (or faces) meet. These distances are indicated as "δ(p) cuboid face" and "δ(p) cuboid vertices" as a function of p in Table 23.4 respectively. Notice that the distance of the face from the center is a constant 15 % LC regardless of p, however a point on the hyper-cube (representing borderline similarity) varies in distance from the center depending on the direction of the deviation, being minimum at a face center and maximum at a vertex.

The F_2 similarity region has the virtue of being directionally constant and the cuboid similarity region has the virtue of guaranteeing a conservative fixed allowable deviation (at least at the face) that is independent of p. It would seem desirable to have a similarity region which combines the virtues of both. That can be achieved by defining a similarity region that requires both the F_2 and cuboid requirements be met. Such a similarity region is illustrated in Fig. 23.5 for p = 3. In this approach, the similarity region includes only the spherical region, except for the "bowls" that extend beyond the cuboidal faces. Notably, this definition of similarity excludes the cuboidal vertices that, as indicated in Table 23.4, can extend so far from the center of similarity in higher dimensions. This "combined" similarity region can give better control over deviations of individual time points, while at the same time

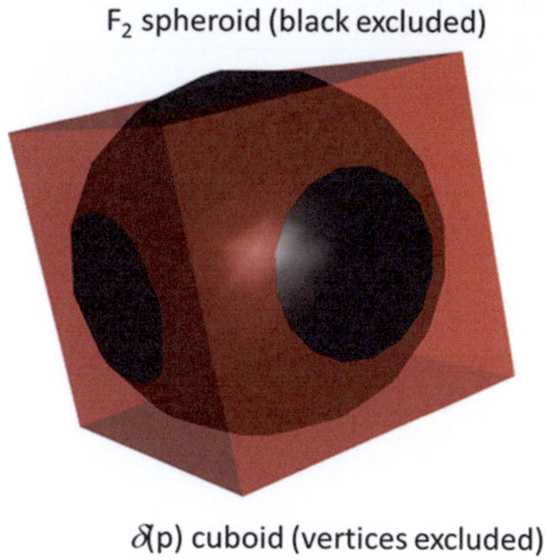

Fig. 23.5 A similarity region combining the virtues of both the F_2 and cuboidal similarity requirements

avoiding large systematic deviations at multiple time points (i.e., at the vertices) allowed by the cuboidal similarity region alone. This definition of similarity has been advocated by Novick et al. (2015).

23.3.7 Model Dependent Approaches

23.3.7.1 Profile Models

Dissolution of a solid oral dosage form involves a number of physical/chemical processes (e.g., disaggregation, disintegration, mass-transport, solvation, convection, laminar flow, diffusion) which are well-studied and for which theoretical models are available.

It is often desirable that an in vitro dissolution test maintain "sink condition". This term may be confusing to a statistician so we provide some insight here. One of the requirements to conduct an appropriate drug dissolution test is to use a sufficient volume of dissolution medium, which should be able to dissolve the expected amount of drug released from a product. This ability of the medium to dissolve the expected amount of drug is known as a "sink condition".

23.3.8 Bayesian Approaches

If a similarity region is defined—as it should be—parametrically, then an assessment of similarity can be based on the sampling distribution of an estimate of the true

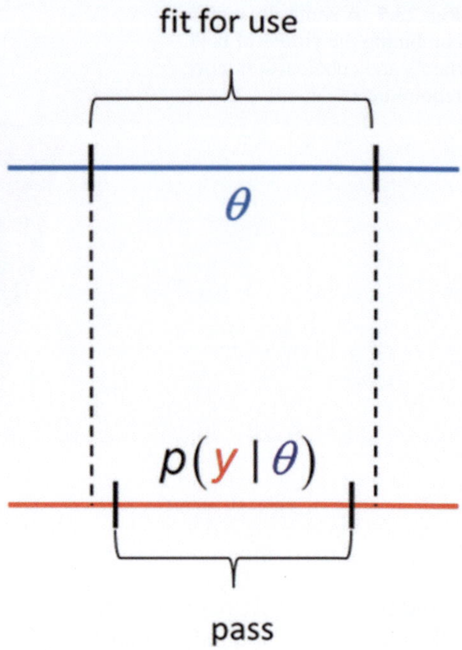

Fig. 23.6 Illustration of a univariate traditional approach to dissolution profile similarity

model parameter (or function of parameters). Figure 23.6 illustrates this approach. Fitness for use (i.e. similarity) is defined as a range of values for some relevant hypothetical model parameter known to govern critical aspects of safety and efficacy related to dissolution θ. A corresponding acceptance range for an estimate, y, of θ is defined based on the similarity range for θ. The sampling distribution of y is used to estimate a (say 90 %) confidence interval for θ. If the confidence interval for θ is within the acceptance range for y, similarity is concluded. This is the traditional statistical approach to equivalence testing. The difficulty with this approach is that the confidence level applies to the properties of the repeated sampling concept associated with the confidence interval method (which treats θ as fixed and the confidence interval as random). There is no inference on θ, the parameter of interest. We rely indirectly on the confidence we have in the statistical methodology to "capture" the true fixed θ with stated confidence on repeated sampling and execution of the methodology. For a given study, we do not learn the probability that θ is truly within the similarity region.

A Bayesian approach, on the other hand, allows us to estimate directly the probability of similarity (i.e., that likely values of θ is located within the similarity region. This implies an uncertainty in θ which can be represented by a probability distribution. A direct probability statement can be a component in risk assessment. Uncertainty in θ , is introduced into the Bayesian paradigm through prior distributions (subjective distributions based on prior knowledge). After data are observed, knowledge about a parameter (or function) is captured in its posterior distribution as illustrated in Fig. 23.7.

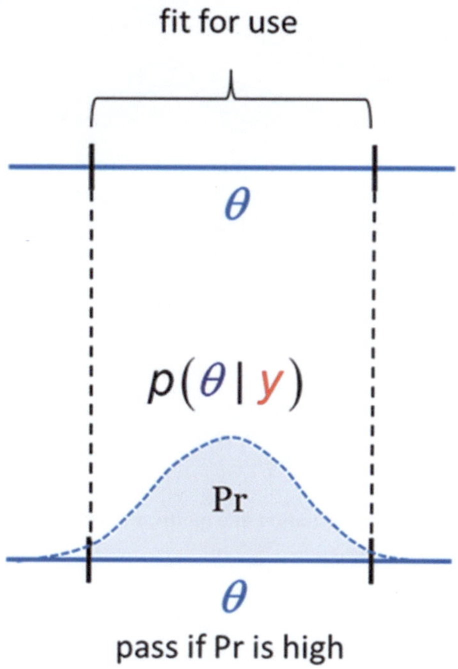

Fig. 23.7 Illustration of a univariate Bayesian approach to dissolution profile similarity

The univariate approach illustrated in Fig. 23.7 is easily extended to a multivariate situation in which q is multidimensional. Figure 23.8 illustrates a bivariate situation in which the similarity region is defined on a two dimensional parametric space. In this case the required probability of equivalence (or similarity) is simply the integrated volume of the posterior density above the two dimensional similarity region.

The above multivariate Bayesian approach was used to re-analyze a data set that appeared in Sathe et al. 1996. In this example, the region of similarity was defined as a rectangle in the space of differences (test-reference) of two parameters, ln(alpha) and ln(beta), of the nonlinear Weibull model used to describe the dissolution profile. The 2D similarity region is shown as a black square in Fig. 23.9. The similarity in two test products ("minor change" indicated by pink circles and "major change" indicated by blue circles in Fig. 23.9) were compared to a reference product. The plotted circles represent draws from the marginal bivariate posteriors of alpha and beta. For the minor change product, 94.9 % of the posterior density was included within the similarity region, but essentially 0 % the posterior density of the major change product was included. These probabilities were estimated simply by counting the number of MCMC posterior draws within the rectangular similarity region. The posterior densities illustrated in Fig. 23.9 are virtually identical to that in Fig. 4 of Shah et al. (1996b) who used a traditional confidence region approach. However these authors were not able to provide a direct probability estimate similar to that given by the Bayesian approach.

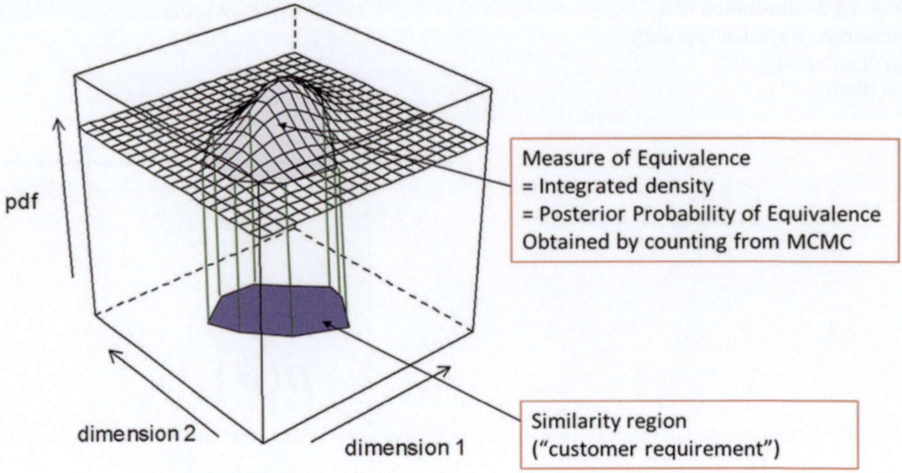

Fig. 23.8 Illustration of a multivariate Bayesian approach to dissolution profile similarity

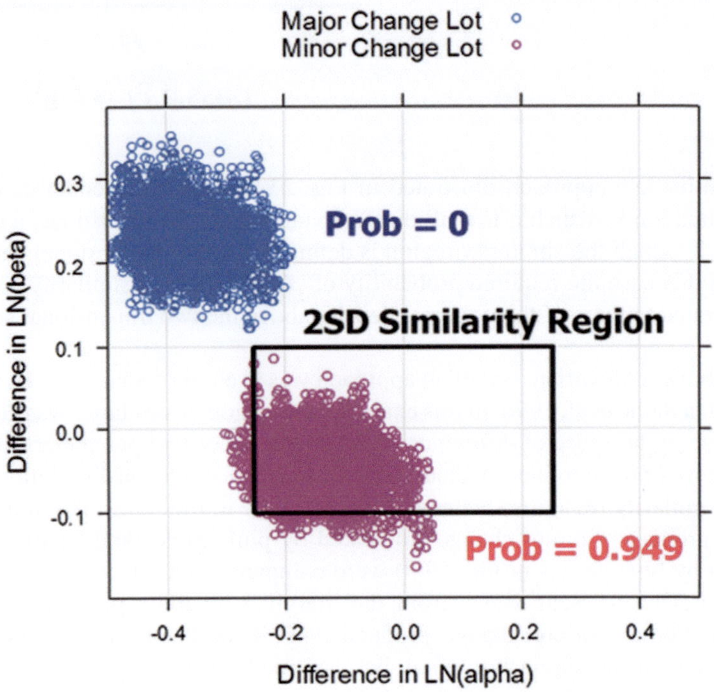

Fig. 23.9 Application of a multivariate Bayesian approach to dissolution profile similarity using data that appeared in Sathe et al. (1996), Table III

While prior knowledge is always employed in traditional statistical methods through choice of likelihood, only Bayesian methodology provides a seamless integration of prior knowledge and data. Usually prior distributions are diffuse or non-informative representing a lack (or unwillingness to use) prior information. However, when informative prior distributions are justified, the similarity test can in principle be conducted with less data (i.e., smaller study). Use of such prior information represents knowledge building expected as a result of modern regulatory initiatives. Other advantages of a Bayesian approach include the ease with which they handle even complex hierarchical models (e.g., variation both within and between batches of the same type), the insightful modeling environment provided by modern MCMC statistical software (such as BUGS, JAGS, STAN, and SAS Proc MCMC), and the ability to handle complex nonlinear modeling situations in an exact manner without approximations or tedious analytical derivations.

The Bayesian approach to similarity testing has a number of attractive features. Unlike confidence region methods, the integrated posterior corresponds exactly to the region of similarity. There is no conservatism. A Bayesian approach makes possible an intuitive statement such as "The probability of similarity is P". Such a probability statement takes into account the uncertainty of the unknown model parameters. In contrast, the results of a frequentist approach to inference through a confidence interval or hypothesis test must be interpreted indirectly through a repeated sampling paradigm. Thus frequentist inferences are about the reliability of the statistical methodology, not about the behavior of the process under study. Frequentist inference may also require the use of large-sample asymptotic assumptions which further complicates the interpretation of the resulting inferential statements.

Using MCMC sampling approaches, it is possible to handle in an exact way (i.e., estimation accuracy limited only by the size of the posterior sample drawn) complex situations which require conservative, asymptotic, or large sample approximations using traditional approaches. This includes the following types of models which can be usefully employed in profile comparisons:

- Multivariate,
- nonlinear, including theoretical models of dissolution profile shape such as Weibull,
- hierarchical models that include both intra- and inter-batch variance,
- "mixed" models in which some parameters are considered fixed and others random,
- predictive models in which we are concerned with behavior of future batches,
- missing data.

Because MCMC technology obtains the joint posterior distribution of model parameters as a sample, it is straightforward to calculate the posterior distribution of any fixed or random function of these parameters (such as a given similarity metric). This circumvents the need to derive a sampling distribution for such functions. The availability of the posterior as an MCMC sample also makes it possible to calculate various probabilities by simple counting. This circumvents the need to analytically or numerically integrate the distributions.

Bayesian and MCMC computations can be efficiently carried out by a number of freely available software such as R, WinBUGS, Stan and JAGS. WinBUGS in particular represents over 20 years of experience by thousands of users world-wide and is a very mature and stable computing environment. A plethora of computing examples are available with this software and in many text books, including the kind of models appropriate for dissolution similarity comparisons (see for instance Lunn et al. 2000; Cowles 2004; Gelman et al. 2004; Gelman and Hill 2007; Lunn et al. 2013).

The Bayesian approach permits prior knowledge to be incorporated into the decision process in a quantitative, objective way. When there is no relevant prior knowledge available, or where the data must "stand on its own", non-informative or vague priors are well known and their impact on the decision can be determined. The incorporation of prior knowledge can be of great utility in community decision making.

As noted above, the sampling distribution of f_2 is intractable which inhibits development of a statistical equivalence test based on f_2. However, it is trivial to obtain the posterior distribution of f_2 (given an appropriate definition of f_2 in terms of model parameters). Thus the Bayesian approach can provide a link to the established metric.

We see two principle barriers to taking a Bayesian approach. First, while the results of Bayesian analyses will be intuitive and understandable to scientists, statisticians, and regulators alike, the use of Bayesian technology is probably unfamiliar to many. Consequently, statistical support is recommended. Certain aspects such as model parameterization, prior elicitation and choice, and convergence verification require care. However, our experiences with Bayesian methodology have been positive. We feel this barrier can be overcome with training. Second, mindful of the quality perspective, there is a need to quantify the risks associated with a decision rule, particularly when an analytical power calculation is not possible. This can be computationally intensive, but in principle, it is straightforward to accomplish using a Bayesian simulation approach. Such methodology is becoming more common place and we see this as part of best statistical practice. For these reasons, we believe Bayesian methodology has the potential of overcoming many of the issues with f_2 and MSD, while maintaining a link to established criteria.

References

Amidon GL, Lennernas H, Shah VP, Crison JR (1995) A theoretical basis for a biopharmaceutical drug classification: the correlation of in vitro drug product dissolution and in vivo bioavailability. Pharm Res 12:413–420

Berger RL, Hsu JC (1996) Bioequivalence trials, intersection-union tests and equivalence confidence sets. Stat Sci 11(4):283–319

Costa P, Lobo JMS (2001) Modeling and comparison of dissolution profiles. Eur J Pharm Sci 13:123–133

Costa FO, Sousa JJS, Pais AACC, Formosinho SJ (2003) Comparison of dissolution profiles of Ibuprofen pellets. J Controlled Release 89:199–212

Cowles MC (2004) Review of WinBUGS 1.4. Am Stat 58(4):330–336

Eaton ML, Muirhead RJ, Steeno GS (2003) Aspects of the dissolution profile testing problem. Biopharm Rep 11(2):2–7

EMA (2008) Guideline on the investigation of bioequivalence. Committee for medicinal products for human use. European Medicines Agency (Doc. Ref. CPMP/EWP/QWP/1401/98 Rev. http://www.ema.europa.eu/docs/en_GB/document_library/Scientific_guideline/2009/09/WC500003011.pdf

FDA (1995) Guidance for industry: immediate release solid oral dosage forms. scale-up and postapproval changes: chemistry, manufacturing and controls, in vitro dissolution testing and in vivo bioequivalence documentation. U.S. Department of Health and Human Services, FDA. Center of Drug Evaluation and Research (CDER), Rockville, MD. http://www.fda.gov/downloads/drugs/guidancecomplianceregulatoryinformation/guidances/ucm070636.pdf

FDA (1997a) Guidance for industry: extended release oral dosage forms: development, evaluation, and application of in vitro/in vivo correlations. U.S. Department of Health and Human Services, FDA. Center of Drug Evaluation and Research (CDER), Rockville, MD. http://www.fda.gov/downloads/drugs/guidancecomplianceregulatoryinformation/guidances/ucm070239.pdf

FDA (1997b) Guidance for industry: dissolution testing of immediate release solid oral dosage forms. U.S. Department of Health and Human Services, FDA. Center of Drug Evaluation and Research (CDER), Rockville, MD. http://www.fda.gov/downloads/drugs/guidancecomplianceregulatoryinformation/guidances/ucm070237.pdf

FDA (1997c) Guidance for industry: SUPAC-MR: modified release solid oral dosage forms. scale-up and postapproval changes: chemistry, manufacturing and controls; in vitro dissolution testing and in vivo bioequivalence documentation. U.S. Department of Health and Human Services, FDA. Center of Drug Evaluation and Research (CDER), Rockville, MD. http://www.fda.gov/downloads/drugs/guidancecomplianceregulatoryinformation/guidances/ucm070640.pdf

FDA (2000) Guidance for industry: bioavailability and bioequivalence studies for orally administered drug products – general considerations

FDA (2000) Guidance for industry: waiver of in vivo bioavailability and bioequivalence studies for immediate- release solid oral dosage forms based on biopharmaceutics classification system. U.S. Department of Health and Human Services, FDA. Center of Drug Evaluation and Research (CDER), Rockville, MD. http://www.fda.gov/downloads/drugs/guidancecomplianceregulatoryinformation/guidances/ucm070246.pdf

FDA (2001) Guidance for industry: statistical approaches to bioequivalence

FDA (2003) Guidance for industry: bioavailability and bioequivalence studies for orally administered drug products – general considerations. U.S. Department of Health and Human Services, FDA. Center of Drug Evaluation and Research (CDER), Rockville, MD. http://www.fda.gov/downloads/drugs/guidancecomplianceregulatoryinformation/guidances/ucm070124.pdf

Gelman A, Hill J (2007) Data analysis using regression and multilevel/hierarchical models. Cambridge University Press, New York, pp 286–287, 377

Gelman A, Carlin J, Stern H, Rubin D (2004) Bayesian data analysis, 2nd edn. Chapman & Hall, New York

Hofer JD, Gray VA (2003) Examination of selection of immediate release dissolution acceptance criteria. Dissolution Technol 10:16–20

Jamei M, Turner D, Yang J, Neuhoff S, Polak S, Rostami-Hodjegan A, Tucker G (2009) Population-based mechanistic prediction of oral drug absorption. AAPS J 11(2):225–237

Japanese National Institute of Health Sciences (2012) Pharmaceutical and Food Safety Bureau. English translation of attachment 3 of division-notification 0229 No. 10. Appendix 1. Similarity factor and time points. Tokyo, Japan. http://www.nihs.go.jp/drug/be-guide%28e%29/Generic/GL-E_120229_BE.pdf

Kesisoglou F, Wu Y (2008) Understanding the effect of API properties on bioavailability through absorption modeling. AAPS J 10(4):516–525. doi:10.1208/s12248-008-9061-4

LeBlond D, Altan S, Novick S, Peterson J, Shen Y, Yang H (2016) In vitro dissolution curve comparisons: a critique of current practice. Dissolution Technologies (accepted for publication)

Lunn DJ, Thomas A, Best N, Spiegelhalter D (2000) WinBUGS – a Bayesian modelling framework: concepts, structure, and extensibility. Stat Comp 10(4):325–337

Lunn D, Jackson C, Best N, Thomas A, Spiegelhalter D (2013) The BUGS book: a practical introduction to Bayesian analysis. CRC, Boca Raton

Moore JW, Flanner HH (1996a) Mathematical comparison of dissolution profiles. Pharm Technol 24:46–54

Moore JW, Flanner HH (1996b) Mathematical comparison of curves with an emphasis on in vitro dissolution profiles. Pharm Tech 20(6):64–74

Novick S, Shen Y, Yang H, Peterson J, LeBlond D, Altan S (2015) Dissolution curve comparisons through the F2 parameter, a Bayesian extension of the f2 statistic. J Biopharm Stat 25(2): 351–371. doi:10.1080/10543406.2014.971175

Plummer M (2003) JAGS: a program for analysis of Bayesian graphical models using Gibbs sampling. In: Proceedings of the 3rd international workshop on distributed statistical computing (DSC 2003), March 20–22, Vienna, Austria. ISSN 1609-395X. http://mcmc-jags.sourceforge.net/

Saranadasa H (2001) Defining similarity of dissolution profiles through Hotelling's T2 statistic. Pharma Technol 2001:46–54

Saranadasa H, Krishnamoorthy K (2005) A multivariate test for similarity of two dissolution profiles. J Biopharm Stat 15:265–278

Sathe PM, Tsong Y, Shah VP (1996) In vitro dissolution profile comparisons: statistics and analysis, model dependent approach. Pharm Res 13(12):1799–1803

Sjögren E, Westergren J, Grant I, Hanisch G, Lindfors L, Lennernäs H, Abrahamsson B, Tonnergren C (2013) In silico predictions of gastrointestinal drug absorption in pharmaceutical product development: application of the mechanistic absorption model GI-Sim. Eur J Pharm Sci 49(4):679–698

Sugano K (2009) Introduction to computational oral absorption simulation. Expert Opin Drug Metab Toxicol 5(3):259–293

Tsong Y, Sathe P, Shah VP (1996a) Comparing 2 dissolution data sets for similarity. In: ASA proceedings of the biopharmaceutical section, pp 129–134

Tsong Y, Hammerstrom T, Sathe P, Shah VP (1996b) Statistical assessment of mean differences between two dissolution data sets. Drug Inf J 30(4):1105–1112

Tsong Y, Shen M, Shah VP (2004) Three-stage sequential statistical dissolution testing rules. J Biopharm Stat 14(3):757–779

United States Pharmacopeial Convention (2011). General chapters: Dissolution <711>, Drug release <724>, Intrinsic dissolution <1087>, In vitro and in vivo evaluation <1088>, Assessment of product performance <1090>, The dissolution procedure <1092>. Washington D.C.

World Health Association (2006). WHO expert committee on specifications for pharmaceutical preparations. WHO technical report series fortieth report. Dissolution profile comparison, Geneva, p 382 (http://whqlibdoc.who.int/trs/WHO_TRS_937_eng.pdf)

Chapter 24
Assessing Content Uniformity

Buffy Hudson-Curtis and Steven Novick

Abstract Content Uniformity tests are used to establish that the dosage units of a drug product consistently contain the specified amount of drug (active pharmaceutical ingredient). The term uniformity may refer to uniformity within a batch, or within-product uniformity when evaluating multi-dose units such as inhaled and topical products. Various guidance documents regarding the establishment of content uniformity of drug product exist and are discussed in this chapter. In particular, we consider USP standards <905>, <3>, as well as the parametric tolerance interval test (PTIT). While the current version of USP <601> does not offer criteria to establish content uniformity, we discuss the method suggested in the 2011 proposed revision to the USP to illustrate the &zero tolerance' approach to establishing content uniformity. While the current version of USP <601> does not offer criteria to establish content uniformity, we discuss the method suggested in the 2011 proposed revision to the USP to illustrate the 'zero tolerance' approach to establishing content uniformity. It is worth noting that there is much discussion within the industry and amongst regulatory agencies regarding the appropriateness of using the current USP standards to evaluate content uniformity. Consideration is being given to revision of these standards, and implementing alternatives such as the parametric tolerance interval test (PTIT) approach.

Keywords USP • Parametric tolerance interval test • CuDAL method

24.1 U.S. Pharmacopeial Convention (USP) Standards for Content Uniformity

USP standards are often referenced for establishing content uniformity of a product. According to their website, the U.S. Pharmacopeial Convention (USP) "*. . . is a scientific nonprofit organization that sets standards for the identity, strength, quality,*

B. Hudson-Curtis (✉)
GSK, Research Triangle Park, NC, USA

S. Novick
MedImmune, Gaithersburg, MD, USA

L. Zhang (ed.), *Nonclinical Statistics for Pharmaceutical and Biotechnology Industries*, Statistics for Biology and Health, DOI 10.1007/978-3-319-23558-5_24

and purity of medicines, food ingredients, and dietary supplements manufactured, distributed and consumed worldwide. USP's drug standards are enforceable in the United States by the Food and Drug Administration, and these standards are used in more than 140 countries."

This chapter will focus specifically on three USP standards: <905> Uniformity of Dosage Units, <3> Topical and Transdermal Drug Products—Product Quality Tests, and the zero tolerance approach in the 2011 suggested revisions to USP <601> Aerosols, Nasal Sprays, Metered-Dose Inhalers, and Dry Powder Inhalers. In general, standards <905>, the 2011 suggested revisions to <601> and <3> are two tiered procedures for testing content uniformity of a batch of drug product. They are implemented by selecting a set $n = n_1 + n_2$ dosage units for evaluation. For tier 1, the acceptance criteria are applied to n_1 of the dosage units. If the acceptance criteria are met, content uniformity is established under that USP standard. If not, the second set of n_2 dosage units may be evaluated. The second tier testing combines the results of the n_2 dosage units with those from tier 1 and then evaluates the $n_1 + n_2$ total dose units against a second set of acceptance criteria. The following three sections of this chapter will discuss the details and statistical properties of these USP standards.

It is worth noting that these tests are revised and updated over time; the latest USP chapter should be referenced to ensure the most up to date standard is utilized. The USP methods may not be harmonized with the European Pharmacopoeia and the Japanese Pharmacopoeia, and so those resources should be consulted when applicable. USP and other regulatory guidance documents may specify the method used to obtain content uniformity results (e.g. content uniformity, weight variation), and is beyond the scope of this chapter.

24.2 Process for Establishing Content Uniformity Using USP <905>

USP <905>, Uniformity of Dosage Units, applies to single dosage units (e.g., tablets), and is not intended to apply to suspensions, emulsions, or gels in unit-dose containers intended for external, cutaneous administration. The USP <905> content uniformity testing calls for a two-tiered testing approach where thirty dosage units are selected. Ten of these dosage units are assayed individually for tier 1 using the appropriate analytical method. If the ten tablets meet the acceptance criteria, then content uniformity has been established. If the tablets in this sample do not meet the acceptance criteria, the remaining 20 dosage units are assayed and evaluated in combination with the individual assay results from tier 1, using the criteria for tier 2. Note that the sample sizes in tier 1 and tier 2 are inflexibly fixed. The specific criteria are as follows:

1. Calculate the mean ($\overline{X}$) and standard deviation (s) of the assay results where $n = 10$ for tier 1, and $n = 30$ for tier 2,

$$\overline{X} = \frac{1}{n}\sum_{i=1}^{n} X_i, \text{ and } s = \left[\frac{\sum_{i=1}^{n}(X_i-\overline{X})^2}{n-1}\right]^{\frac{1}{2}}.$$

2. Establish the Reference Value (M). This value depends on the target content (T) per dosage unit at the time of manufacture and the observed mean of the sample ($\overline{X}$). The value T is expressed as a percent of the label claim, and is typically 100 %.

   a. Case 1: $T \leq 101.5$

      i. If $98.5\% \leq \overline{X} \leq 101.5$ then $M = \overline{X}$
      ii. If $\overline{X} < 98.5$, then $M = 98.5$
      iii. If $\overline{X} > 101.5$, then $M = 101.5$

   b. Case 2: $T > 101.5$

      i. If $98.5\% \leq \overline{X} \leq T$ then $M = \overline{X}$
      ii. If $\overline{X} < 98.5$, then $M = 98.5$
      iii. If $\overline{X} > T$, then $M = T$

3. Calculate the Acceptance Value (AV) using the acceptability constant k, where $k = 2.4$ for tier 1 ($n = 10$), and $k = 2.0$ for tier 2 ($n = 30$).

$$AV = \left|M - \overline{X}\right| + ks$$

4. The requirements for dosage uniformity are met if:

   a. Tier 1: The AV of the first ten dosage units is less than $L_1 = 15.0$ %. tier 2: If the AV for the first ten dosage units is greater than $L_1 = 15.0$ %, the criteria are met if (tier 2) the final acceptance value of the 30 dosage units is less than $L_1 = 15.0$ % and if all individual dosage units (X_i, $i = 1, \ldots, 30$) meet the requirement $[1 - (0.01) * L_2] * M \leq X_i \leq [1 + (0.01) * L_2] * M$, where $L_2 = 25.0$ %.

Example
Ten capsules are assayed from a batch, and the following ten results are observed: {96.4, 104.9, 104, 103.5, 97.5, 92.4, 96.2, 107.8, 91.2, 100.2}. The target value (T) for this product is 100. The mean and standard deviation of these results are $\overline{X} = 99.4$ and $s = 5.6$, respectively. Since $98.5\% \leq \overline{X} \leq 101.5$, $M = \overline{X}$, and the AV is calculated as $k*s = 2.4*5.6 = 13.44$. Since this value is less than 15.0 (L_1), the criteria for establishing content uniformity using USP <905> are met.

24.2.1 Discussion USP <905>

Operating characteristic curves were generated to illustrate the probability to pass USP <905> for $T = 96$, 98, 100, 102 and 104, assuming the true batch mean is μ, and the true batch standard deviation is σ. These curves, shown in Fig. 24.1, were created by computer-generating 10,000 tier 1 data sets from a normal distribution

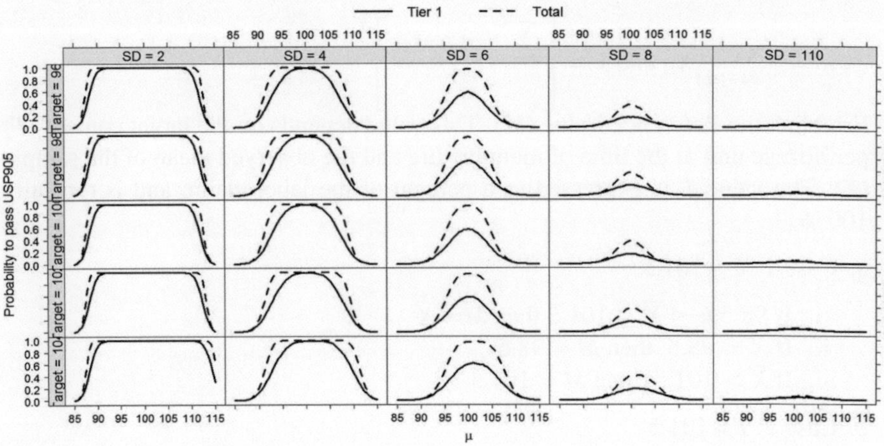

Fig. 24.1 OC curves for USP <905>, Rows display T (Target) values between 96 and 104

with mean μ and standard deviation σ. The probability to pass tier 1 testing was estimated by taking the proportion of tier 1 data sets that passed the tier 1 criteria. For computer-generated data sets that failed tier 1 testing, additional tier 2 data were generated from the same normal distribution. The combined tier-1/tier-2 data were then evaluated through the tier 2 criteria. The total passing rate probability was estimated by calculating the proportion of data sets that passed either tier 1 or tier 2.

As seen in Fig. 24.1, for batches with smaller variability ($\sigma = 2, 4$), there is little increase in the probability that a batch will pass the criteria in USP <905> at tier 2 if it has failed to pass tier 1. For batches with larger standard deviation (e.g., $\sigma = 6$), tier 2 testing increases the probability of passing when the batch is on target. When the standard deviation is larger (e.g., $\sigma = 8$), there is little probability of passing USP <905> criteria even at tier 2. The graphs in Fig. 24.1 also demonstrate how use of the reference value (M) creates a zone of indifference to the mean. The zone of indifference, discussed in Lostritto (2012) establishes the same requirements on the size of the standard deviation whether the average of the dosage units is at target (100 %), or slightly off target (as low as 98.5 %, and as high as 101.5 %). Lostritto (2012) also discusses the potential use of an alternative test based on a parametric tolerance interval approach, discussed later in this chapter.

24.3 Process for Establishing Content Uniformity Using USP <601>: Aerosols, Nasal Sprays, Metered-Dose Inhalers, and Dry Powder Inhalers

USP <601> applies to aerosols, nasal sprays, metered dose-inhalers and dry powder inhalers. Specific tests for uniformity are provided for inhalation and nasal products that are not packaged in single unit dosage forms; these products have multiple doses within one unit. The USP test places requirements on doses collected at the beginning of unit life (sometimes called beginning of use) and end of unit life (or end of use). While the current version of USP <601> does not offer criteria to establish content uniformity, we discuss the method suggested in the 2011 proposed revision to the USP to illustrate the 'zero tolerance' approach to establishing content uniformity. As with USP <905>, this is a two tiered test. The test described in the suggested revisions to USP <601> is as follows.

Tier 1
Sample one beginning-of-use delivered dose, and one end-of-use delivered dose from each of ten containers for a total of twenty results, $X_{1B}, X_{2B}, \ldots X_{nB}, n = 10$ (beginning of use results) and $X_{1E}, X_{2E}, \ldots X_{nE}, n = 10$ (end of use results). These 20 results must meet the following criteria (tier 1).

1. No more than 2 of the 20 results are outside of the range 80–120 % of label claim.
2. No results are outside of the range of 75–125 % of label claim.
3. The mean at each of beginning of use falls within the range of 85–115 % of label claim. In other words,

$$85 \le \frac{1}{10}\sum_{i=1}^{10} X_{iB} \le 115 \text{ and } 85 \le \frac{1}{10}\sum_{i=1}^{10} X_{iE} \le 115.$$

If tier 1 criteria are not met, tier 2 criteria may be applied if:

1. No more than 6 doses of the 20 doses collected are outside of 80–120 % of the label claim,
2. None of the 20 results is outside of 75–125 % of label claim,
3. The means for each of the beginning and end delivered doses fall within 85–115 % of label claim.

Tier 2 Criteria
Sample one beginning-of-use delivered dose, and one end-of-use delivered dose from each of an additional twenty containers and combine these results with those from tier 1 for a total of sixty results: $X_{1B}, X_{2B}, \ldots X_{nB}, n = 30$ (beginning of use results) and $X_{1E}, X_{2E}, \ldots X_{nE}, n = 30$ (end of use results). These results must meet the following criteria (tier 2):

1. No more than 6 of the 60 results are outside of the range 80–120 % of label claim,
2. No results are outside of the range of 75–125 % of label claim,

Table 24.1 Example 2, metered dose inhaler

Canister	BOU	EOU
1	109.6	93.7
2	78.8	110.1
3	75.6	116.8
4	103.7	116.5
5	102.6	112.1
6	86.1	104.5
7	102.6	108.6
8	101.3	92.4
9	102.8	105.7
10	104.7	95.5
Mean	*96.78*	*105.59*

3. The mean at each of beginning of use falls within the range of 85–115 % of label claim. In other words,
$$85 \leq \frac{1}{30}\sum_{i=1}^{30} X_{iB} \leq 115 \text{ and } 85 \leq \frac{1}{30}\sum_{i=1}^{30} X_{iE} \leq 115.$$

Example
Ten metered dose inhalers are tested at beginning of use (BOU) and end of use (EOU); the results from canisters 1–10 are shown in Table 24.1. Applying tier 1 criteria, only two results are outside of the 80–120 % label claim, and none is outside of 75–125 %. The mean results for BOU and EOU are both within 85–115 %, so the requirements for establishing content uniformity per the suggested revisions to USP <601> are met.

24.3.1 Discussion of Zero Tolerance Approach Described in the 2011 Suggested Revision USP <601>

Operating characteristic curves were generated for the suggested revisions to USP <601>, shown in Fig. 24.2. These curves were constructed by creating and subsequently testing 10,000 normally-distributed tier 1 data sets with mean μ and standard deviation σ in a fashion similar to that given in the USP <905> section, but using the USP <601> criteria. Additional tier 2 data were generated for failed tier 1 data sets that met criteria to move to tier 2. The tier 1 and total passing rate probability were estimated by calculating the proportion of data sets that respectively passed tier 1 testing and either tier 1 or tier 2. The OC curves demonstrate that tier 2 testing provides little or no benefit over tier 1 testing for virtually any batch mean and standard deviation. Because of this testing characteristic, there has been significant discussion around the use and properties of this test. An alternative test based on a parametric tolerance interval approach was proposed by the FDA (Nasr 2005) and will be discussed later in this chapter.

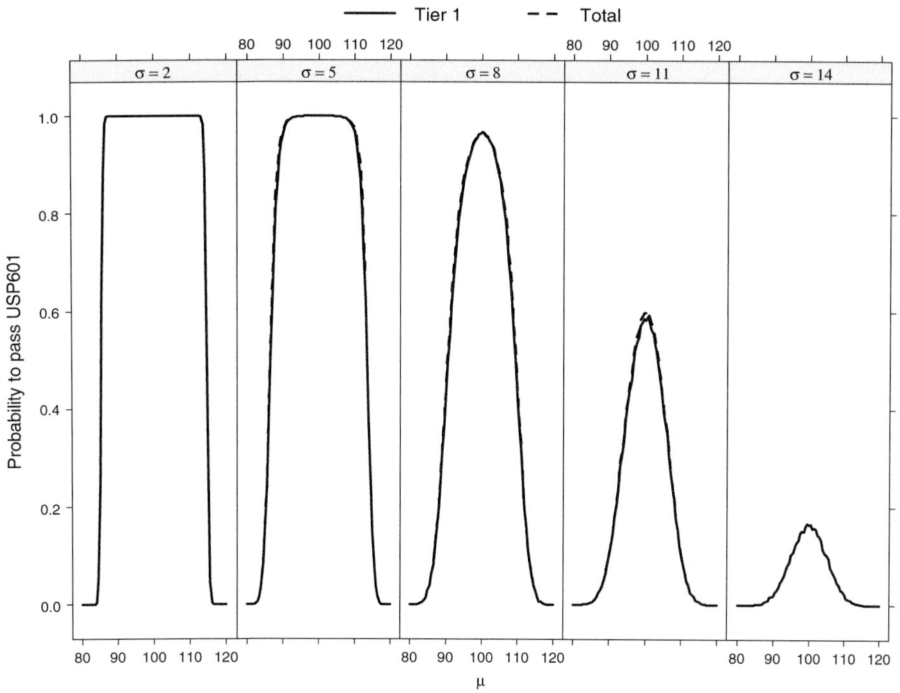

Fig. 24.2 OC curves for Revision Suggestions (2011) USP <601>

24.4 Process for Establishing Content Uniformity Using USP <3> Topical and Transdermal Drug Products

USP <3> may be used to assess uniformity in containers of multi-dose topical products. In contrast, USP <905> may be used to assess transdermal drug products and for dosage forms packed in single-unit containers. Topically applied drug products include, but are not restricted to creams, gels, ointments, pastes, suspensions, lotions, and foams. The uniformity of dosage units test described in USP <905> is appropriate for dosage forms packed in single-unit containers; however, as topically applied semi-solid drug products may experience physical separation during manufacturing processes and during their shelf life, the within-tube content uniformity should also be evaluated.

The procedure for evaluating within-tube content uniformity described in USP <3> depends on the size of the container. For multiple dose products that contain 5 g or more, there are two procedures that may be followed.

Procedure 1 (products 5 g or more)

1. From a single tube, assay the active ingredient from an appropriate amount of product taken from the top, middle, and bottom portions of the tube for a total of three assay results: X_1, X_2, and X_3..
2. Acceptance Criteria *A* are met if all assay results are within the product assay range and the relative standard deviation (RSD) is no more than (NMT) 6 % or as specified in the product specification or compendial monograph. The RSD is calculated as $100\% * s/\overline{X}$, where $n = 3$, and

$$\overline{X} = \frac{1}{n}\sum_{i=1}^{n} X_i, \quad s = \left[\frac{\sum_{i=1}^{n}\left(X_i - \overline{X}\right)^2}{n-1}\right]^{\frac{1}{2}}.$$

3. If the product fails Acceptance Criteria *A*, test three additional tubes from the same batch following step 1 described above and apply Acceptance Criteria *B*.
4. Acceptance Criteria *B* are met if all of the results are within the product assay range and the RSD of the 12 assay results is NMT 6 % or as specified in the product specification or compendial monograph. In determining the RSD from multiple tubes, first determine the variance from the three measurements for each tube and average across the tubes. The RSD is calculated using this average variance so that RSD $= 100\% * s/\overline{X}$ where

$$\overline{X} = \tfrac{1}{n}\sum_{i=1}^{n} X_i, \; n = 12, \; s = \left[\tfrac{1}{4}\sum_{j=1}^{4} Var_{Tube\ j}\right]^{\frac{1}{2}},$$

$$Var_{Tube\ j} = \frac{\sum_{i=1}^{n}\left(X_{i,\ Tube\ j} - \overline{X}_{Tube_j}\right)^2}{n-1}, \; n = 3, j = 1, 2, \ldots 4.$$

Procedure 2 (products 5 g or more)

1. Using two tubes, assay the active ingredient from an appropriate amount of product taken from the top, middle, and bottom portions of the tube for a total of six assay results: $X_1, X_2, \ldots X_6$..
2. Acceptance Criteria *A* are met if all assay within the product assay range and the RSD is NMT 6 %. In determining the RSD from multiple tubes, first determine the variance from the three measurements for each tube and average across the tubes. The RSD is calculated using this average variance so that RSD $= 100\% * s/\overline{X}$ where

$$\overline{X} = \tfrac{1}{n}\sum_{i=1}^{n} X_i, \; n = 6, \; s = \left[\tfrac{1}{2}\sum_{j=1}^{2} Var_{Tube\ j}\right]^{\frac{1}{2}},$$

$$Var_{Tube\ j} = \frac{\sum_{i=1}^{n}\left(X_{i,\ Tube\ j} - \overline{X}_{Tube_j}\right)^2}{n-1}, \; n = 3, j = 1, 2.$$

3. If the product fails Acceptance Criteria *A*, test two additional tubes from the same batch following step 1 described above and apply Acceptance criteria *B*.
4. Acceptance Criteria *B* are met if all assay results are within the product assay range and the RSD of the 12 assay results is NMT 6 %. In determining the RSD

from multiple tubes, first determine the variance from the three measurements for each tube and average across the tubes. The RSD is calculated using this average variance so that RSD $= 100\% * s/\overline{X}$ where

$$\overline{X} = \frac{1}{n}\sum_{i=1}^{n} X_i,\ n = 12,\ s = \left[\frac{1}{4}\sum_{j=1}^{4} Var_{Tube\ j}\right]^{\frac{1}{2}},$$

$$Var_{Tube\ j} = \frac{\sum_{i=1}^{n}\left(X_{i,\ Tube\ j} - \overline{X}_{Tube j}\right)^2}{n-1},\ n = 3, j = 1, 2, \ldots 4.$$

For Multiple-Dose Products That Contain Less Than 5 g of Product: The procedure to assess products with less than 5 g of product uses the criteria for products with more than 5 g, but is modified such that samples are taken only from the top and bottom portions of the tube, for a total of $n = 2$ samples from each tube. There is some lack of clarity around the text for this procedure, as Acceptance Criteria *A* uses $n = 3$ results for procedure 1, and only $n = 2$ results will be available for product with less than 5 g, and should be clarified before implementing this procedure.

The guidance states:

> Test the top and bottom portions of two tubes using Procedure 1 or Procedure 2 as described above. Evaluate the test results using Acceptance Criteria A.
>
> If the product fails Acceptance Criteria A, test 2 additional tubes from the same batch following step 1 described above, and evaluate all 8 test results using Acceptance Criteria B.

Example
A 10 g product is evaluated for content uniformity using procedure 1, assuming a product assay range of 90–110 % of label claim. Samples are taken from the top, middle, and bottom of a tube with the resulting data values of 105.2, 91.1, and 96.7 % of label claim. Since the RSD of this dataset is 7.3 %, the standard for content uniformity as described by Criteria *A* in USP <3> is not met. Because all of the results are within 90–110 % of label claim, Criteria *B* can be applied by sampling an additional three tubes. The results are shown in Table 24.2. The average variance was calculated by summing the variances from each tube, then dividing by the number of tubes. This, combined with the overall average of the twelve tubes, was used to calculate the RSD for evaluation using Criteria *B*. Since the new RSD was 5.9 % and none of the values were outside the range 90–110 %, content uniformity is established according to USP <3>.

24.4.1 Discussion USP <3>

Operating characteristic curves were generated for USP <3> assessing container uniformity for products greater than 5 g, and the results are shown in Figs 24.3 and 24.4. The OC curves for USP <3> procedure 1 were constructed in a manner similar to that given in the USP <905> section, with 10,000 tier 1 (and tier 2, if necessary) computer-generated data sets drawn from a normal distribution with mean μ and

Table 24.2 Example topical product USP <3>

	Tube 1	Tube 2	Tube 3	Tube 4
Top	105.2	99.1	96.2	104.7
Middle	91.1	103.1	100.5	97.5
Bottom	96.7	95.2	109.6	107.6
Mean	97.7	99.1	102.1	103.3
Stdev	7.1	4.0	6.8	5.2
RSD	7.3	4.0	6.7	5.0
Variance	50.4	15.6	46.8	27.0
Overall mean	100.5			
Average variance	35.0			
Stdev (Tier 2)	5.9			
RSD(Tier 2)	5.9			

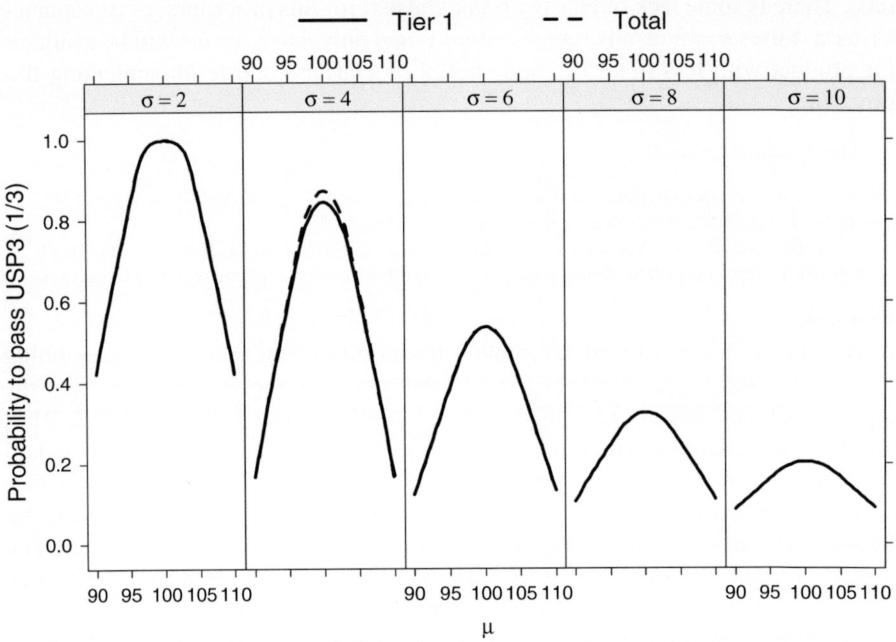

Fig. 24.3 OC curves for USP <3> testing with $n = 1$ unit in tier 1 and $n = 3$ additional units in tier 2. Loess smoothing was performed on USP<3> OC curves to remove rough edges

standard deviation σ. Each data set was tested via the USP <3> procedure 1 criteria with the tier 1 batch-passing probability estimated by the proportion of data sets that passed tier 1 testing and the total batch-passing probability from the proportion of data sets that passed either tier 1 or tier 2 testing. The entire process was repeated to create the OC curves for USP <3> Procedure 2. Both tests used a product assay range of 90–110 %.

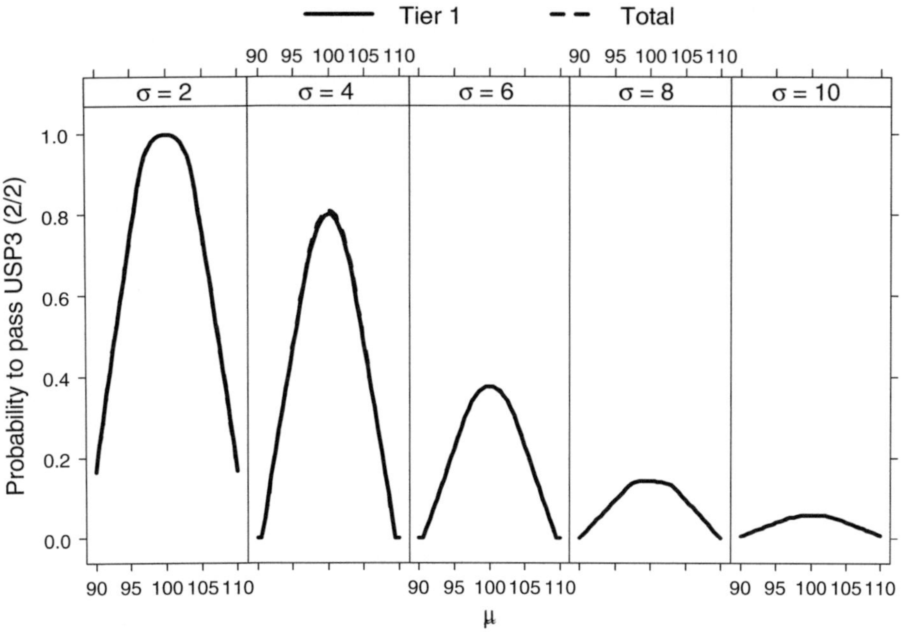

Fig. 24.4 OC curves for USP <3> testing with $n = 2$ units in tier 1 and $n = 2$ additional units in tier 2. Loess smoothing was performed on USP <3> OC curves to remove rough edges

24.5 Parametric Tolerance Interval Test as an Alternative to USP <905> and USP <601>

A parametric tolerance interval test (PTIT) has been proposed as an alternative to both USP <601> (October 2005 Advisory Committee meeting) and USP <905> (Lostritto 2012). This section discusses some basic properties of the test, and how it can be applied to single dosage unit forms and orally inhaled and nasal drug products (OINDP's). A far more detailed discussion of the properties of the test can be found in Novick et al. (2009), and further thoughts on the PTIT may be seen in Larner et al. (2011).

The PTIT discussed in this chapter is based on the parametric tolerance interval for a univariate normal distribution (Odeh and Owen 1980) and possesses well-known statistical properties. The test is constructed to offer a high level of confidence that batches with unacceptable content uniformity properties have a low probability of passing, regardless of the sample size tested.

In general terms, a tolerance interval is constructed from observed data for a particular confidence level, $100(1-\alpha)\%$, and coverage value $100p\%$, to yield a statistical interval with endpoints T_L and T_U. These endpoints can be used to assess whether (or not) $100p\%$ of units within a batch fall within a pre-specified interval [L, U] with $100(1-\alpha)\%$ confidence. For example, after sampling n units from a batch, with confidence $100(1-\alpha)\%$ (or, alternatively stated, with type I error level α), one declares that 100p% of units in the batch fall within [L, U] if $T_L > L$ and $T_U < U$. If either $T_L \leq L$ or $T_U \geq U$, then one cannot make such a declaration and typically the batch testing fails. This operation forms the basis of the PTIT.

This concept was expanded to propose a two-tiered procedure to establish content uniformity (Novick et al. 2009; Lostritto 2005; Hauck et al. 2012), called the PTIT, or parametric tolerance interval test. A unique feature of this test compared to the methods described in the USP is that it can be used for a range of sample sizes. The acceptance criteria are calculated based on the sample size, ensuring that the Type I error level α is fixed. The exact details of such a test are described in Novick et al. (2009). The level of confidence, coverage, and the target interval [L, U] along with the sample size determine the acceptance criteria used for the PTIT. For instance, a test may be described as being performed at the 95 % confidence level ($\alpha = 0.05$) with 87.5 % coverage of the target interval = 80–120 % (L and U) of the label claim (LC). A test described as such provides 95 % confidence that no more than (100–87.5 %)/2 = 6.25 % of units fall below the target limit of 80 % and no more than 6.25 % of units fall above the target limit of 120 %.

Specifically, a PTIT can be executed by sampling a total of $n_1 + n_2$ dosage units. For tier 1, n_1 of these dosage units are assayed individually using the appropriate analytical method. If these units meet the acceptance criteria, then content uniformity has been established. If the units in this sample do not meet the acceptance criteria, the remaining n_2 dosage units are assayed, and evaluated in combination with the individual assay results from tier 1, using the criteria for tier 2.

An alternative to USP <905> approach for establishing content uniformity is to implement a PTIT for single dosage units (Lostritto 2012) as shown below.

Tier 1

1. The content (or weight, where appropriate) values of n_1 individual dosage units (tier 1) are collected.
2. Calculate the mean ($\overline{X}_{n1}$) and standard deviation (s_{n1}) of the assay results

$$\overline{X}_{n1} = \frac{1}{n_1}\sum_{i=1}^{n_1} X_i \text{ and } s_{n1} = \left[\frac{\sum_{i=1}^{n_1}\left(X_i - \overline{X}_{n1}\right)^2}{n_1 - 1}\right]^{\frac{1}{2}}.$$

3. The batch passes tier 1 of the PTIT test if:
   $T_{L1} = \overline{X}_{n1} - k_1 s_{n1} \geq L$ and $T_{U1} = \overline{X}_{n1} + k_1 s_{n1} \leq U$.
   If the batch does not meet the acceptance criteria at tier 1, proceed to tier 2.

Tier 2

1. The content (or weight, where appropriate) of an additional n_2 dosage units are collected for a total of $n_1 + n_2$ results.
2. Calculate the mean ($\overline{X}_{n1+n2}$) and standard deviation (s_{n1+n2}) of the assay results

$$\overline{X}_{n1+n2} = \frac{1}{n1 + n2}\sum_{i=1}^{n1+n2} X_i,\ s_{n1+n2} = \left[\frac{\sum_{i=1}^{n1+n2}\left(X_i - \overline{X}_{n1+n2}\right)^2}{n_1 + n_2 - 1}\right]^{\frac{1}{2}}.$$

Table 24.3 The k coefficients for the two-tiered PTIT for a variety of coverage values and sample sizes

Coverage	# of tier 1 samples (n_1)	# of additional tier 2 samples (n_2)	Total number of samples	k_1	k_2
85 %	10	20	30	2.957	2.037
	20	40	60	2.317	1.830
87.5 %	10	20	30	3.119	2.155
	20	40	60	2.448	1.940
90 %	10	20	30	3.310	2.294
	20	40	60	2.601	2.068
95 %	10	20	30	3.859	2.692
	20	40	60	3.043	2.435

For all cases, $\alpha_1 = 0.026$ and $\alpha_2 = 0.0340$. For inhaled products with BOU and EOU sampling, number of units = (number of samples/2)

Table 24.4 Example data to illustrate the PTIT for inhaled products

	BOU	EOU	Statistics
Tier 1	85.5, 89, 90.1, 93.2, 93.2, 93.4, 95.9, 97.3, 100.9, 103	87.9, 99.2, 99.8, 102.2, 103.9, 104, 104.6, 110.1, 110.5, 114.9	$N_1 = 20$ $\overline{X}_{BOU} = 94.2$ $\overline{X}_{EOU} = 103.7$ $\overline{X}_{n1} = 98.9$ $s_{n1} = 8{,}0$
Tier 2	83.4, 86.9, 99.4, 88.2, 104.9, 96.9, 101.8, 103.1, 101.8, 107.2, 90.4, 90.4, 90.6, 94.6, 93.1, 88.1, 104.7, 101.9, 105.6, 101.7	89.5, 99.5, 95.6, 98.3, 105.8, 111.9, 103.7, 110.7, 107.3, 116, 97.2, 97.6, 102.2, 108.3, 107.1, 108.4, 109.8, 105.7, 103.7, 118.3	$N_2 = 40$ (total = 60) $\overline{X}_{BOU} = 95.9$ $\overline{X}_{EOU} = 104.5$ $\overline{X}_{n1+n2} = 100.2$ $s_{n1+n2} = 8.1$

3. The batch passes tier 2 of the PTIT if:
$$T_{L2} = \overline{X}_{n1+n2} - k_2 s_{n1+n2} \geq L \text{ and } T_{U2} = \overline{X}_{n1+n2} + k_2 s_{n1+n2} \leq U.$$

The values used establishing acceptance criteria k_1 and k_2 are calculated as described in Novick et al. (2009), and are based on the coverage, confidence level, and sample size. These values as well as the values of L and U should be selected such that they are appropriate for the product. Values that have been discussed specifically for OINDP's are 87.5 % coverage, 95 % confidence, and a target interval [80, 120]. These values may be considered for single unit dosage forms as well. The values of k for a range of coverage values and sample sizes, assuming 95 % overall confidence, are provided in Table 24.3. Novick et al. (2009) discuss in detail how the k_1 and k_2 values are calculated.

An alternative content uniformity test for OINDPs (Lostritto 2005) modifies the criterion above to reflect the multi-dose nature of these products. For inhaled products, $n_1/2$ units are selected for tier 1, and beginning of use (BOU) and end of use (EOU) results are collected for each unit, for a total of n_1 results. The same procedure is used to collect results for tier 2. The BOU and EOU data are combined and evaluated using the tier 1 and tier 2 criteria, with the added requirement that

$85 \le \overline{X}_{BOU} \le 115$ and $85 \le \overline{X}_{EOU} \le 115$. The PTIT test described in Novick et al. (2009) was built under the assumption that BOU and EOU results within a unit are uncorrelated and stem from the same normal distribution. Lewis and Novick (2010) demonstrate that the PTIT for OINDPs may be anti-conservative when there are differences in BOU and EOU means or standard deviations or if within-unit correlation is present. The authors offer a modified PTIT to accommodate these scenarios.

Example

A metered dose inhaler (e.g., albuterol), manufactured to contain 200 sprays of 120 mcg of product is evaluated for content uniformity using a PTIT approach. Samples of BOU and EOU collected on each of 10 inhaled units in tier 1 and an additional 20 inhaled units in tier 2 are shown in Table 24.4. Testing was performed with the PTIT using $n_1 = 20$, $n_2 = 40$, $\alpha = 0.05$, $p = 0.875$, $L = 80$, and $U = 120$. In tier 1 testing, the BOU and EOU means are 94.2 and 103.7, respectively, and so both $\overline{X}_{BOU}$ and $\overline{X}_{EOU}$ are between 85 – 115. The tier 1 mean and standard deviation are $\overline{X}_{n1} = 98.9$ and $s_{n1} = 8.0$. Using the tier 2 k value from Table 24.3, the tolerance interval is $T_{L1} = 98.9 - 2.448\ (8.0) = 79.3$ and $T_{U1} = 98.9 + 2.448\ (8.0) = 118.5$. Because $T_{L1} < 80$, tier 1 testing failed and so tier 2 testing must be initiated. After an additional 20 units are sampled, in tier 2 testing, the BOU and EOU means are 95.9 and 104.5, respectively, and so both $\overline{X}_{BOU}$ and $\overline{X}_{EOU}$ are between 85 and 115. The

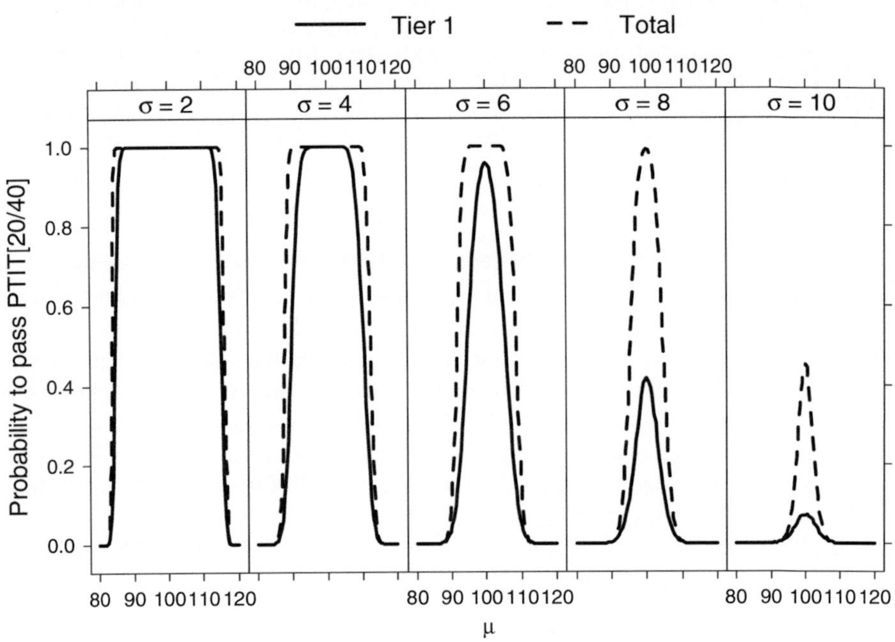

Fig. 24.5 OC curves for PTIT (without the means test for BOU and EOU) with $n_1 = 20$ samples in tier 1 and $n_2 = 40$ additional samples in tier 2

overall mean $\overline{X}_{n1+n2} = 100.2$ and the overall standard deviation $s_{n1+n2} = 8.1$. Using the tier 2 k value from Table 24.3, the tolerance interval is $T_{L2} = 100.2 - 1.940$ $(8.1) = 84.5$ and $T_{U2} = 100.2 + 1.940\ (8.1) = 115.9$. Since T_{L2} and T_{U2} lie within 80–120, the batch passes the PTIT in the tier 2.

24.5.1 Discussion PTIT

Operating characteristic curves were generated for the PTIT assessing content uniformity for inhaled products and results are shown in Fig. 24.5. The OC curves for the PTIT in Fig. 24.5 were constructed to reflect testing for an OINDP. OC curve probabilities were calculated from 10,000 computer-generated data sets of $n_1 = 20$ tier 1 samples and, if necessary, $n_2 = 40$ additional tier 2 samples, drawn from a normal distribution with mean μ and standard deviation σ. Each data set was tested via the PTIT criteria without the means test for BOU and EOU. The tier 1 batch-passing probability was estimated by the proportion of data sets that passed tier 1 testing and the total batch-passing probability was estimated from the proportion of data sets that passed either tier 1 or tier 2 testing. It is clear from the sets of OC curves that tier 2 testing does generally increase the probability for a batch to meet the PTIT criteria, which may alleviate one of the major concerns of the USPs.

Operating characteristic curves were also generated for the PTIT in a manner similar to that given in the USP <905> section, with 10,000 computer-generated data sets of $n_1 = 10$ tier 1 samples and, if necessary, $n_2 = 20$ additional tier 2 samples drawn from a normal distribution with mean μ and standard deviation σ. The PTIT testing limits were modified to match with USP <905> so that $L = 85$ and $U = 115$. The tier 1 batch-passing probability was estimated by the proportion of data sets that passed tier 1 testing and the total batch-passing probability was estimated from the proportion of data sets that passed either tier 1 or tier 2 testing. Figure 24.6 shows the tier 1 and total probability to pass the modified PTIT along with the total probability to pass USP<905> when Target = 100. Comparing this figure to Fig. 24.1, the PTIT offers the advantage of a higher conditional pass rate when units fail to pass in tier 1 testing. Under the set of PTIT parameters, the PTIT total passing probability is universally lower when compared with the total passing probability of USP <905>.

24.6 Establishing Content Uniformity Criterion for Batch Release

The following statement from the USP General Notices discusses a very important feature of these USP tests:

> At times, compendial standards take on the character of statistical procedures, with multiple units involved and perhaps a sequential procedural design to allow the user to determine that

the tested article meets or does not meet the standard. The similarity to statistical procedures may seem to suggest an intent to make inference to some larger group of units, but in all cases, statements about whether the compendial standard is met apply only to the units tested.

In recent times there have been numerous discussions about alternatives to USP standards for evaluating content uniformity of a batch, and whether they are appropriate for determining content uniformity for release of a batch of product. Multiple approaches, including the CuDAL method (Bergum and Li 2007) and PTIT (Lostritto 2005, 2012), have been proposed as techniques to address these questions. The CuDAL efforts are reflected in the ASTM standards E2810-11 (2011b) and E2709-11 (2011a). Given the particular focus on content uniformity in the industry, a thorough exploration of current and relevant guidances and recommendations is advisable before implementing any particular approach.

24.6.1 CuDAL Approach

The USP tests and the PTIT in this chapter can be used to determine whether the observed sample mean $\bar{x}$ and observed sample standard deviation s from a batch meet a specific criterion. If $\bar{x}$ and s from the particular sample meet the criteria, the batch passes the test. Would a different sample from the same batch also meet the criteria? In 2007, Bergum and Li published a novel method called CuDAL (content uniformity and dissolution acceptance limits) to estimate, with a given level of confidence, a lower bound on the probability that a random sample from a batch would pass the USP <905> standard for content uniformity based on a sample of size n. We generalize the Bergum and Li approach here. Assume that the n samples

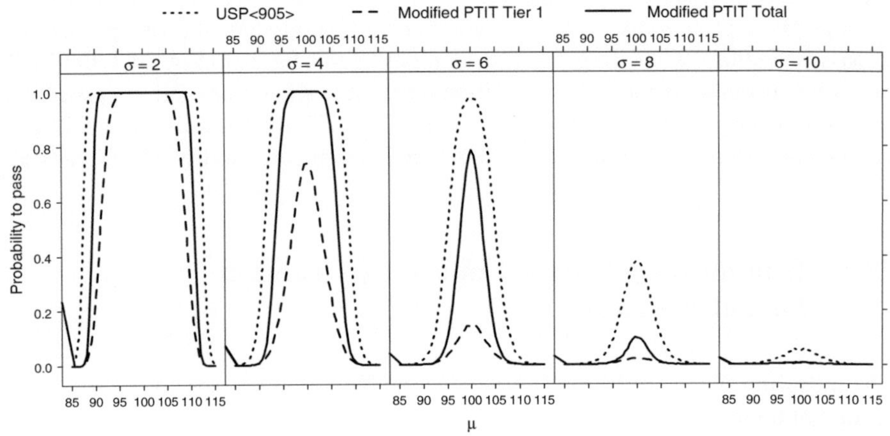

Fig. 24.6 OC curves for modified PTIT with $n_1 = 10$ samples in tier 1, $n_2 = 20$ additional samples in tier 2, L = 85, and U = 115

of dosage units drawn from a batch stem from a normal distribution with mean μ and standard deviation σ, resulting in the sample mean $\bar{x}$ and sample standard deviation s. Create a simultaneous $100(1-\alpha)\%$ confidence region for (μ, σ), based on $(\bar{x}, s)$. The construction of the confidence region depends on the statistical methodology, the sampling plan used to collect the samples and may also reflect multiple sources of variation, such as location within the batch. For a given USP test, the operating characteristic probability (which may be computed by Monte Carlo simulation) is calculated for every value (μ, σ) in the confidence region. If the operating characteristic probabilities are all larger than p_0 for the set of every (μ, σ) in the confidence region, then with $100(1-\alpha)\%$ confidence, the acceptance probability for a random sample from the tested manufactured batch is at least p_0. Bergum and Li (2007) apply this method to USP <905>, using statistical theory and numerical approximations to cut down on the number of computational steps necessary to compute this probability. Various sample sizes may be used to make such an assessment; OC curves for various sample sizes may be found in Bergum (2011).

24.6.2 PTIT

The PTIT approach addresses a slightly different question than that of the CuDAL approach. Rather than assessing the probability that samples from a given batch will pass a particular test, it assesses the probability that the dosage units within a batch will fall within certain limits with $(1-\alpha)$ confidence. The details of this test and its implementation have been described in detail in a previous section of this chapter.

24.6.3 Discussion and Future Research

The CuDAL method addresses the concept of coverage through assessment of the probability that a random sample from a batch collected according to a particular test (in this case, USP <905>) will pass the test to establish content uniformity. Conversely, the PTIT test assesses coverage by evaluating the percent of units within a batch that would fall within a specified range, say $80-120\ \%$ of label claim. In other words, if one is looking to assess with confidence that the individual content values of a particular batch fall within a specified range, the PTIT test might be the approach of choice. If the intent is to insure with high probability that a specified portion of samples from a batch would pass a particular test, the CuDAL approach might be a better. The different approaches are complementary, not contradictory in nature, but as with all statistical tests, the appropriate test should be selected to address the question at hand.

24.7 Evaluating Risk

Little literature is available to evaluate the manufacturer's process capability to successfully evaluate a future batch and pass one of the USP test methods. In this chapter, we created operating characteristic (OC) curves, which represent the probability to pass a USP test for a batch with characteristics given by a true mean μ and true standard deviation σ. A very natural question that follows an OC curve calculation is "will our manufacturing process lend itself to a high probability to pass the batch testing?" Put another way, "what mean and standard deviation should be expected from our manufacturing process?" For a given USP test, the CuDAL approach provides a statistical method to evaluate the risk of taking another random sample from within the same batch, illustrating the robustness of the within-batch sampling process against a USP test; but, it does not provide insights into the nature of future batches.

It is unlikely that the true values for the mean and standard deviation for all future batches are known and so interpretation of the OC curve for a particular batch manufacturing process requires development work to estimate these values. To evaluate the process capability, at a minimum, one must estimate the mean and two components of variation, batch-to-batch standard deviation and within-batch standard deviation. Other sources of variability might also be of interest (and hence should be measured and estimated), such as manufacturing site, analyst, or machine line.

As an example, suppose a manufacturing process must be evaluated against USP <905> and the data shown in Table 24.5 and Fig. 24.7 were collected.

By fitting a linear mixed effects model to the data, we estimated that the mean is 97.0, the batch-to-batch standard deviation is 5.2, and the within-batch standard deviation is 3.2. If these model parameters were estimated perfectly, then to calculate the probability that a new batch from this process will be accepted, the following procedure might be considered.

Table 24.5 Example development data: content uniformity from six batches with 10 units per batch

	Batch					
Unit	1	2	3	4	5	6
1	100.3	107.3	97.0	90.1	93.1	98.9
2	95.6	103.1	98.9	87.5	90.7	103.6
3	93.9	108.0	95.6	89.9	92.7	94.0
4	96.2	101.0	103.5	94.9	94.5	100.1
5	98.7	98.6	98.0	92.9	90.9	105.7
6	99.1	103.3	98.4	91.6	88.9	95.7
7	98.8	100.4	105.6	89.6	95.8	101.9
8	97.5	101.3	93.8	91.3	87.8	98.7
9	96.8	108.8	103.1	88.8	82.9	98.8
10	97.9	103.9	101.9	89.7	88.5	104.5

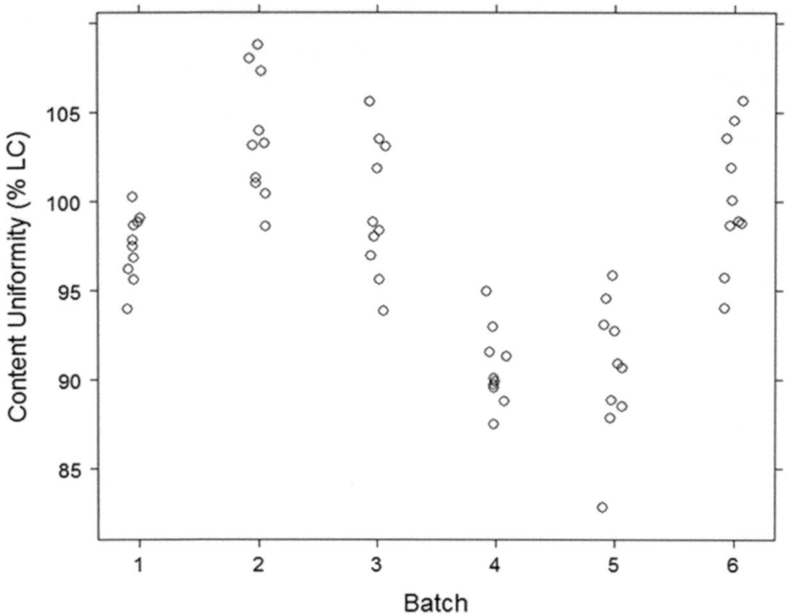

Fig. 24.7 Example development data: six batches with 10 units per batch

1. Simulate a new batch mean μ_B from a univariate normal distribution with mean 97.0 and standard deviation 5.2.
2. Simulate a new data set with $n = 10$ from a univariate normal distribution with mean μ_B and standard deviation 3.2.
3. Evaluate the data set under USP <905> tier 1 rules.
4. If failure results in step 3, create 20 additional units from the same distribution (in step 2) and evaluate the full data set under USP <905> tier 2 rules.
5. Repeat steps 1–4 many times (~10,000 times).
6. The tier 1 probability = the proportion of times the data set passes the tier 1 testing. The total probability = the proportion of times the data sets passes either tier 1 or tier 2 testing.

By following this method, the tier 1 probability is approximately 83 % and the total probability is approximately 91 %. This procedure does not, however, take into account our uncertainty in the estimated parameters, resulting in overly-high probability estimates. LeBlond and Mockus (2014) detail a more modern statistical method that uses Bayesian posterior probabilities. Their technique is described as follows:

1. Fit a linear mixed effects model to generate the posterior distribution of values for each of parameters: the mean μ, the batch-to-batch standard deviation σ_B, and the within-batch standard deviation σ_e. The posterior distribution is a Bayesian paradigm for estimating parameter values with uncertainty, a method

that significantly differs from classical statistical methods. The uninitiated reader is instructed to see Gelman et al. (2004) for a primer on Bayesian analyses.

2. Draw one posterior sample for the set of parameters in step 1. Simulate a new batch mean μ_B from a univariate normal distribution with mean μ and standard deviation σ_B.
3. Simulate a new data set with $n = 10$ from a univariate normal distribution with mean μ_B and standard deviation σ_e.
4. Evaluate the data set under USP <905> tier 1 rules.
5. If failure results in tier 1, create 20 additional units from the same distribution (as step 3) and evaluate the full data set under USP <905> tier 2 rules.
6. Repeat steps 1–5 many times (~10,000 times).
7. The tier 1 probability = the proportion of times the data set passes the tier 1 testing. The total probability = the proportion of times the data sets passes either tier 1 or tier 2 testing.

This technique does take the model parameter estimation uncertainty into account and yields a more-realistic tier 1 probability of approximately 78 % and a total probability of approximately 85 %. The LeBlond and Mockus proposed method better reflects our knowledge of the mean and reproducibility of the batch process.

24.8 Summary

Approaches for assessing content uniformity vary depending on dosage form, and the regulatory and technical landscape regarding approaches is constantly evolving. This chapter summarizes many of those approaches, particularly those described in the USP, but appropriate guidances should always be referenced in practice.

Acknowledgements The authors would like to thank Stan Altan for his insightful comments and suggestions during the revision process.

References

Odeh RE, Owen DB (1980) Tables for normal tolerance limits, sampling plans, and screening. Marcel Dekker, New York

Bergum J, Li, H (2007) Acceptance limits for the new ICH USP 29 content uniformity test. Pharm Technol 31(10)

Bergum J (2011) PQRI Sample Size Workshop, "Demonstrating capability to comply with a test procedure – the content uniformity and dissolution acceptance limits (CuDAL) Approach"

Gelman A, Carlin J, Stern H, Rubin D (2004) Bayesian data analysis, 2nd edn. Chapman & Hall/CRC, Boca Raton

Hauck W, DeStefano A, Tyle P, Williams R (2012) Methods for Measuring Uniformity in USP. United States Pharmacopeia, Stimuli to the Revision Process, 38(6), November, 2012

Lewis RA, Novick SJ (2010) A generalized pivotal quantity approach to the parametric tolerance interval test for dose-content uniformity batch testing. Stat Biopharm Res 4(1):28–36

Lostritto R (2012) Content uniformity (CU) testing for the 21st century: CDER perspective, October 17, 2012, AAPS Annual Meeting. http://www.fda.gov/downloads/AboutFDA/CentersOffices/OfficeofMedicalProductsandTobacco/CDER/UCM341168.pdf. Accessed 14 July 2014

Lostritto R (2005) Advisory Committee for Pharmaceutical Science Meeting (in transcripts), vol, 25 October 2005, p 361. http://www.fda.gov/ohrms/dockets/ac/05/transcripts/2005-4187T1.pdf. Accessed 11 Sept 2008

Larner G, Cooper A, Lyapustina S, Leiner S, Christopher D, Strickland H, Golden M, Delzeit H, Friedman E (2011) Challenges and opportunities in implementing the FDA default parametric tolerance interval two one-sided test for delivered dose uniformity of orally inhaled products. AAPS PharmSciTech 12(4): 1144–1156. http://www.ncbi.nlm.nih.gov/pmc/articles/PMC3225528/citedby/

LeBlond D, Mockus L (2014) "The posterior probability of passing a compendial standard, part 1: uniformity of dosage units", accepted May 2014 to Statistics in Biopharmaceutical Research.

Nasr MM (2005) Parametric tolerance interval test for delivered dose uniformity. Presentation to the FDA Advisory Committee, 2005. http://www.fda.gov/ohrms/dockets/ac/05/slides/2005-4187S1_13_Nasr.ppt. Accessed 8 May 2010

Novick S, Christopher D, Dey M, Lyapustina S, Golden M, Leiner S, Wyka B, Delzeit HJ, Novak C, Larner G (2009) A two one-sided parametric tolerance interval test for control of delivered dose uniformity. Part 1—characterization of FDA proposed test. AAPS PharmSciTech 10(3):820–828

ASTM E2709-11 (2011a) Standard practice for demonstrating capability to comply with an acceptance procedure, ASTM E2709-11, 2011

ASTM E2810-11 (2011b) Standard practice for demonstrating capability to comply with the test for uniformity of dosage units, ASTM E2810-11, 2011

USP (2012) USP 35-NF 30, General Notices 3.10. USP, Rockville, p 4

The U.S. Pharmacopeial Convention (USP) (2014). http://www.usp.org/about-usp. Accessed 06 June 2014

USP (2015) USP <3>, Topical and Transdermal Drug Products - Product Quality Tests, USP 38-NF 33 S1. http://www.uspnf.com/uspnf/pub/index?usp=38nf=33s=1officialOn=August1, 2015. Accessed October 2015

USP (2015) USP <601>, Inhalation and Nasal Drug Products, USP 38 -NF 33 S1. http://www.uspnf.com/uspnf/pub/index?usp=38nf=33s=1officialOn=August1, 2015. Accessed October 2015

USP (2015) USP <905>, Uniformity of Dosage Units, USP 38 -NF 33 S1. http://www.uspnf.com/uspnf/pub/index?usp=38nf=33s=1officialOn=August1, 2015. Accessed October 2015

USP (2011) Proposed New and Revised USP Chapters <601>, <5> [PF 2011; 37(4)]

Chapter 25
Chemometrics and Predictive Modelling

Wen Wu and Athula Herath

Abstract The focus of this chapter is Chemometrics and Predictive Modelling. Chemometrics is a multivariate statistical methodology that has a parallel and independent path of development that grew out of the need to statistically analyse chemical measurements with moderate to large numbers of variables, especially in cases when there are more variables than samples (or objects). In recent years, more and more of the Chemometrics methods have been fused into the mainstream of multivariate and high dimensional statistics. In this chapter, we explore the methodological foundations of Chemometrics and supplement it with a motivating example for the reader to appreciate the methodology. We also provide a comprehensive reference list for the readers who may want to read more about Chemometric methods.

Keywords Chemometrics • Predictive modelling • High dimensional statistics • Data mining • Principal components analysis (PCA) • Partial least-squares (PLS) • Dimension reduction • Genetic algorithms • Model validation • Predicting efficacy • Predicting toxicity

25.1 Chemometrics

Chemometrics is a discipline that uses mathematical, statistical, and other methods to select optimal measurement procedures and to provide maximum relevant chemical information by analyzing chemical data (Massart et al. 1988, 1997, 1998). Chemometrics was born in the 1970s and the term 'Chemometrics' was first invented by Svante Wold in a grant application in 1971 (Wold 1995). The International Chemometrics Society was formed in 1974 (Geladi and Esbensen 2005; Esbensen and Geladi 2005; Kvalheim 1996). Chemometrics is concerned with

W. Wu (✉)
Manufacturing Science and Technology, AstraZeneca, Liverpool L24 9JW, UK
e-mail: wenwu2009@googlemail.com

A. Herath
Statistical Sciences, MedImmune, Cambridge CB21 6GH, UK

L. Zhang (ed.), *Nonclinical Statistics for Pharmaceutical and Biotechnology Industries*, Statistics for Biology and Health, DOI 10.1007/978-3-319-23558-5_25

applications of statistics, optimization, signal processing, resolution, calibration, parameter estimation, structure-activity relationships, pattern recognition, library searching, artificial intelligence and data mining (Rathore et al. 2011, 2014; Brown et al. 1994; Brereton 1990). Some large chemometric application areas have gone on to form or impact new domains, such as molecular modeling Quantitative Structure–Activity Relationship (QSAR), cheminformatics, Quality by Design (QbD) and Process Analytical Technology (PAT). Chemometrics has widely been applied to analyze large-scale experiments of genomic, proteomic and metabolomic data within the pharmaceutical industry (Daszykowski et al. 2007; Czekaj et al. 2008; Cutler et al. 2008; Wu et al. 2003a, 2005; Connor et al. 2004; Cordingley et al. 2003).

Many early applications of chemometrics (in the late 1970s and early 1980s) involved multivariate classification. The data from infrared and UV/visible spectroscopy experiments are of large scale and often yield thousands of measurements per sample. Mass spectrometry, nuclear magnetic resonance, atomic emission/absorption and chromatography experiments are also by nature highly multivariate and yield tens of thousands of measurements per sample. The structure of these data indicates a need for dimension reduction using techniques such as principal components analysis (PCA) and partial least-squares (PLS). This is primarily because variables in such high dimensional datasets are often correlated. PLS in particular was used ubiquitously in chemometric applications for many years before it began to find regular use in other fields.

The professional journals on Chemometrics first appeared in the 1980s: *Journal of Chemometrics*, and *Chemometrics and Intelligent Laboratory Systems* remain to be the prominent two journals to date. Most fundamental and methodological research work in chemometrics has been published in these journals. The routine applications of existing chemometric methods are usually published in application-oriented journals, such as Analytica Chimica Acta, Applied Spectroscopy and Analytical Chemistry.

25.2 High Dimensional Problems

Obtaining high dimensional spectroscopic, chromatographic and nuclear magnetic resonance (NMR) intensities on individual samples in a short time is not only possible but also done routinely. In genomics and proteomics, each microarray chip (sample) can contain the quantitative expression data of 50,000 probe sets (variables). In the Next-Generation Sequencing measurements several million of short sequences (variables) are obtained (Menaa 2014).

Usually, measurements are presented as a data matrix where each row corresponds to an object (say a patient) and each column a variable (say a gene/protein). For high dimensional data, the ratio of the number of variables to the number of objects is very high, that is often the number of variables measured is far higher than the number of objects. This results in a singular matrix problem. Those variables are often highly correlated (i.e. collinear) and contain a lot of redundant information to describe a system. Therefore, finding meaningful hidden patterns within these datasets becomes a challenge. The collinearity makes models unstable and thereby difficult to use as robust predictive models for new samples. Many traditional methods such as multivariate linear regression (MLR) and linear discriminant analysis will not handle this type of data due to the singularity and collinearity problems. Chemometric methods have been developed to analyze such high dimensional data (Wu et al. 1995, 1996a, b, c, d, 1997a, b, 1998, 1999, 2001, 2002, 2003b; Walczak and Wu 2005; Guo et al. 1999, 2000, 2001, 2002; Leardi et al. 1992; Leardi 2000; Kalivas et al. 1989; Baldovin et al. 1996, 1997; Wu and Manne 2000; Czekaj et al. 2005; Wu and Massart 1996, 1997; Candolfi et al. 1998).

25.3 Data Analysis Procedures

The key steps in applying chemometric tools can be broken down further into the following:

1) Data exploration,
2) Data pretreatment,
3) Training set selection,
4) Dimension reduction,
5) Model optimization,
6) Model evaluation and prediction.

25.3.1 Data Exploration

Before analyzing the data set, the quality of the data should be investigated. One simple way to do this is to inspect the scatter plot of the data with an eye to identify strange and/or unexpected patterns. The most commonly used multivariate procedure is Principal Component Analysis (PCA, see Sect. 25.3.4.1) *due to its ability* to uncover multivariate outliers or clusters in the data. PCA is also applied to visualize the structure of the data and to check if the training set and the test set can represent each other (see Sect. 25.3.3). PCA score plot is limited and only able to visualize up to 3-dimension at one time. Alternative methods such as star plots (Wu et al. 1998) and parallel coordinate plots (Wegman 1990; Young et al. 2006) can visualize high dimensional PCA data.

25.3.2 Data Pretreatment

In many cases, data for analysis by chemometric methods should be transformed before the modeling step. This step is called data pretreatment. The purpose of pretreatment is to remove some irrelevant mechanical process variations and possible artifacts of measurement of the data. For instance, in spectral data, pretreatment can eliminate the effects due to particle size, scatter light and other effects. There are different techniques to implement pretreatment depending on the type of data (Walczak and Wu 2005; Wu et al. 1995, 1999, 2001; Guo et al. 1999). For example, in Fig. 25.1, one cannot recognize the number of peaks in the spectra because of unwanted local peak shifts caused by matrix and instrument variability. After pretreatment of the data by fuzzy warping method (Walczak and Wu 2005), the shift is easily eliminated and the peaks are clearly aligned. As a result the downstream analysis will be greatly improved after the pretreatment.

In order to eliminate the effects caused by the differing magnitudes of different variables, auto-scaling (or center and scale) can be used. This is implemented as

$$\mathbf{z} = [\mathbf{x} - \text{mean}\,(\mathbf{x})]\,/\,\text{std}\,(\mathbf{x})$$

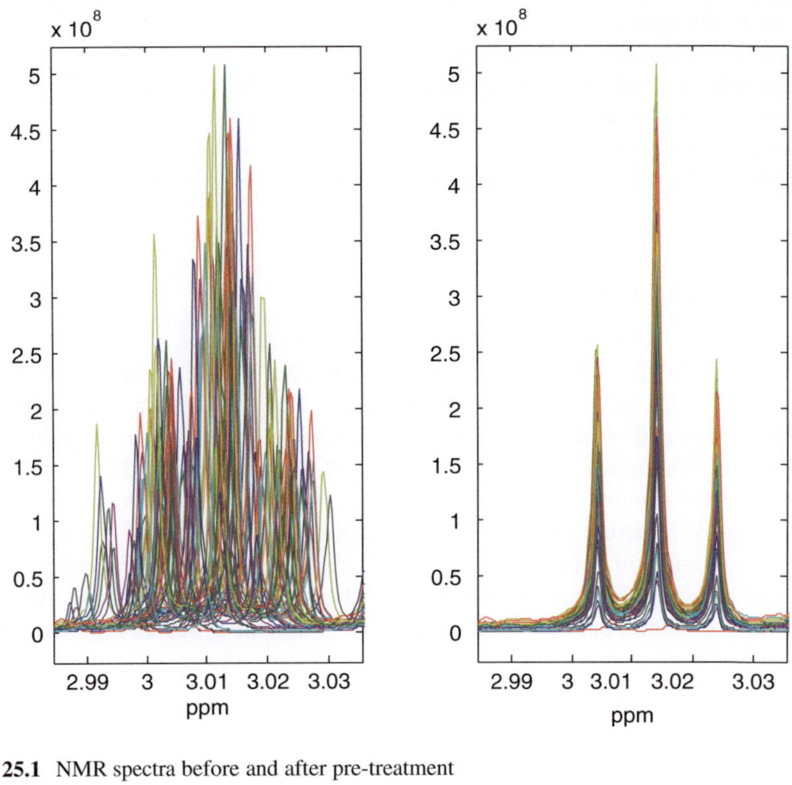

Fig. 25.1 NMR spectra before and after pre-treatment

where **x** is a vector of measurements for a particular variable, mean(**x**) is the mean of **x**, std(**x**) is the standard deviation of **x**, and **z** denotes the transformed variable. After auto scaling, the data dimension remains the same as before and all variables have the same mean and standard deviation.

25.3.3 Training Set Selection

In order to build a predictive model, *a training set* and an independent *test set* are utilized. The training set is used to construct or train a model. When the model is a regression or a calibration model, it is also called a calibration set. The model is then validated using the test set. If the validation is successful, the model can then be applied to new unknown samples for prediction. Before building a proper model, it is very important to correctly split the data into a training set and a test set. The training set should be as representative of the whole data as possible. It should cover all of the spread in the population including the new unknown objects. Therefore, the measurements should always be carefully designed by the expert in order to control the span of data variance. In an article by Wu et al. (1996d), four different data splitting methods were compared. These were: D-optimal design, Kennard-Stone design, Kohonen self-organizing mapping, and random selection. It was found that the Kennard-Stone design out-performed all other methods by selecting objects that were evenly distributed in the X-space.

25.3.4 Dimension Reduction

There are three main reasons to reduce the dimensions in data analysis. The first is to reduce and eliminate the variables that are irrelevant to the study being undertaken. The second is to preserve only the variables that carry information within the data. The third is known as parsimony, which is to simplify the model with the necessary and sufficient number of variables as a model. A simple model with low number of parameters is more stable and easily interpretable than a complex model.

In general there are two ways to reduce the dimensionality of data. The first method is known as *feature selection* (Leardi et al. 1992; Leardi 2000; Kalivas et al. 1989; Guo et al. 2001, 2002; Wu et al. 1996a, 2003b; Baldovin et al. 1996), the process of finding the most adequate subset available of input variables for a prediction or modeling task. The second method is *feature reduction* (Wu et al. 1996b, 1997a, b, 2002; Wu and Manne 2000; Guo et al. 2000), such as PCA and Partial Least Squares (PLS), which extract a small number of orthogonal latent variables to replace the original variables. The latent variables are linear or non-linear combinations of the original variables and the number of orthogonal latent variables is usually lower than the number of objects.

25.3.4.1 Feature Reduction

PCA and PLS are the most popular feature reduction methods to reduce the dimensionality of the data. In PCA, the latent variables (also called factors) are linear combinations of the original variables (i.e. the weighted sum of the original variables). The number of factors is much lower than the number of original variables. The first factor is obtained in such a way as to maximize the total variance (information) in the data. The second factor is selected to be orthogonal to the first and has the maximal remaining variance, and so on. PCA is also often used as a visualization method (see Sect. 25.3.1) and can be effective when the first two factors explain most of the information i.e. variance of the data. Similar to PCA, Partial Least Squares (PLS) extracts factors by maximizing the covariance between **X** and **Y** instead of the total variance of **X**.

When the latent variables of principal components (PC) are used instead of the original variables to build a linear regression model, this is called Principal Component Regression (PCR). When the latent variables of PLS factors are used to build a linear regression model, this is called Partial Least Squares Regression (PLSR). After the feature reduction by PCA or PLS, the number of factors is much lower than the number of objects, and the factors are orthogonal to each other. Therefore, the two high dimensional problems of matrix singularity and collinearity are solved, and one can have enough degrees of freedom to build a regression model.

25.3.4.2 Feature Selection

In feature selection, the only possible way to be sure that "the best" set of variables are picked up is the "all-possible-models" technique, where all possible model combinations are tested. With k variables, the number of possible combinations is 2^k-1; therefore this approach cannot be used due to the computational complexity unless the number of variables is low. So a compromise is used in most situations.

The simplest (but less effective) way of performing a feature selection is to operate on a "univariate" basis, by retaining those variables having the greatest correlation with the response. Here, each variable is taken into account by itself, without considering how its information "integrates" with the information brought by the other (selected or unselected) variables. As a result, if several highly correlated variables are "good", they are all selected independent of their correlation, and the information may be highly redundant. On the other hand, those variables that are not taken into account become very important when their information is integrated with other variables. To improve the method, several multivariate methods such as the stepwise variable selection, genetic algorithms (Leardi et al. 1992; Leardi 2000) and simulated annealing (Kalivas et al. 1989) may be used.

Genetic Algorithms (GAs) are a general optimization technique that has found use in many fields (Guo et al. 2001, 2002; Wu et al. 2003b; Niazi and Leardi 2012). GAs are especially useful when the problem becomes so complex that it cannot be solved by standard techniques. In Chemometrics, GAs have been found useful in feature selection (Niazi and Leardi 2012). GAs are inspired by the theory of evolution: in a living environment, the best individuals have a greater chance of survival and a greater probability to spread their genes by reproduction. The mating of two "good" individuals causes the optimization of the offspring due to the mixing of their genomes, which may result in a "better" offspring. The terms "good", "better", and the "best" denote the degree of adaptation/fitness of the individuals to their environment. The application of a GA to a problem requests the implementation of five basic steps: (1) coding of variables, (2) initiation of population, (3) evaluation of the response, (4) reproduction and (5) mutation. Steps 3–5 alternate until a termination criterion is reached; the criterion can be based on a lack of improvement in the response, simply on a maximum number of generations or on the total time allowed for the process.

In the case study (see Sect. 25.4), four methods were applied for dimension reduction. The GA and Stepwise variable selection methods were applied as feature selection methods; PCA and PLS were applied as feature reduction methods to reduce the dimension of data. All these methods were integrated with an MLR method, namely GA_MLR, stepwise MLR (STP_MLR), PCR and PLSR. The results after applying these methods on data were then compared.

25.3.5 Model Optimization

There are many different modeling techniques that can be classified as linear and non-linear methods. Some classification and regression methods are discussed and their performances are compared in refs (Czekaj et al. 2005; Wu and Massart 1996, 1997; Wu et al. 1996c; Candolfi et al. 1998; Baldovin et al. 1997). In chemometrics, the most commonly used models for regression are the PCR and PLSR. In PCR and PLSR, one should optimize the number of latent variables or factors included in the model in order to reduce model complexity. Cross-validation is one of the most popular techniques to optimize the model complexity. The data are randomly divided into a given number of segments. Each segment is comprised of primary units referred to as "objects". The cross validation method is implemented in a way that each time one segment of objects is left out and the remaining objects are used to build models with different number of factors. The responses (**Y**) corresponding to the left-out objects are predicted by the models. Then another segment of objects is left out and subjected to the same procedures. This is repeated until all objects have been left out once. After completion of the procedure, all objects will have been predicted once, and the entire predicted **Y** values are recorded. The root mean square error of cross-validation prediction (RMSECV) can be obtained (by comparing the predicted **Y** and observed **Y** values) for each model with different factors. The optimal model is denoted as the model giving the lowest RMSECV.

25.3.6 Model Evaluation and Prediction

Model over-fitting is a frequent problem and as a result the predictions may not be accurate when applied to new objects. Therefore, it is necessary to evaluate the prediction power of the model before it is applied to predict new objects. The prediction power of a model can be estimated by predicting for an independent test set of objects with known responses (**Y**).

When no dimension reduction is involved in the modeling, no optimization step is required as all variables are used in modeling. In such situation, cross-validation can be applied for model validation, as the predicted responses (**Y**) of the left-out objects are independent of the model. However, when dimension reduction is used, cross-validation is not a good validation method as all objects within the model building data set have been used in constructing the model (for optimizing for the number of factors). It is therefore no longer an independent evaluation. Therefore, an independent test set has to be acquired to validate the model.

In the case study (Sects. 25.5 and 25.6), an independent test set was used to validate the model obtained after dimension reduction.

If the prediction results of the test set are satisfactory, validation is successful. The model is fit and applied to predict the response of a new (unknown response) object. Otherwise the model has to be rejected/revised and cannot be used for prediction. In order to establish the prediction power of the optimized model, the scatter plot of the predicted response value against the observed response value of the objects in the independent test set can be looked at. If the predicted values are close to the observed values, the scatter plot will fall near the 45° line and the Pearson correlation coefficient (r) will be near 1, which indicates a model with good prediction power. When r is near 0, it indicates a poor prediction model. However, when r is in the middle such as 0.5, it is difficult to decide whether to accept or reject a model just based on r itself. A randomization test (Wu et al. 2002; Eugene 1964) can be applied in such a circumstance. The rationale of the randomization test is to examine whether the correlation coefficient obtained in the independent test set outperforms the correlation coefficient obtained in any random compositions of test sets, i.e. $H_0: r = 0$; $H_1: r > 0$ (Eugene 1964; Edgington 1995). This procedure can be implemented by following the steps below:

1) Apply the model to objects in the independent test set ($\mathbf{X}_{test}$) to predict the response **Y** values, and calculate the correlation coefficient referred to as r_{orig} between the predicted and observed **Y** values;
2) Randomly permute each column of $\mathbf{X}_{test}$ to obtain a permuted $\mathbf{X}_{test}$;
3) Apply the model to the permuted $\mathbf{X}_{test}$ to predict the responses (**Y**), and calculate correlation coefficient referred to as r_{perm} between the predicted and observed **Y** values;
4) Repeat steps 2–3 large enough number of times (say 1000) to obtain 1000 r_{perm}'s;

5) Calculate p-value (probability of prediction of random test set) by the ratio of number of r_{perm}'s having values greater than r_{orig} divided by 1000, or calculate the 95 % percentile of r_{perm}'s as Upper Limit;
6) If the p-value is less than 0.05 or r_{orig} is higher than the Upper Limit, we can accept the model; otherwise, we reject the model.

25.4 Case Study

The example data were from a clinical phase III study (McInnes et al. 2010, 2011). There were 69 patients in the drug treatment group and 63 patients in the placebo group. The goal was to predict both the efficacy and toxicity variables after 12 weeks drug treatment using baseline variables. In the study, 66 baseline variables were measured and collected as independent variables. To eliminate the effects caused by magnitude of the measures, the auto-scale was applied to pretreat the data. In order to assess the model, the patients presented in the drug treatment group were divided into training and test sets. The training set, containing 35 subjects, was used to build and optimize a model. The test set, containing 34 subjects, was used to validate the optimal model. The other 63 subjects in the placebo group were used as a new data set to verify if the models obtained from the drug treatment group were applicable to the placebo group.

All the multivariate Chemometric methods were programmed in Matlab (MATLAB 6.1 2000).

The aim here is to build predictive models of efficacy and toxicity measures at week 12 from the baseline parameters for the drug treatment patients.

Before modelling, PCA was applied to visualize the structure of the **X**-data. Figure 25.2 was the score plot obtained after PCA of the patients in the DRUG TREATMENT group. It shows that the patents selected in both the training and test sets were evenly distributed and covered the whole **X**-space of the data. Therefore, the datasets are representative.

25.4.1 Predictive Model for TOXICITY MEASUREMENT at Week 12

PCR, PLSR, Stepwise MLR (STW_MLR) and GA_MLR were applied to build the predictive models for TOXICITY MEASUREMENT at week 12 for the DRUG TREATMENT group. To compare the results of different methods, the Root Mean Square Error of the prediction (RMSEP) of the independent test set objects in the DRUG TREATMENT group was used. The best model should give the smallest RMSEP. In Table 25.1, the results of the prediction of the test set are listed for comparison. The results show that GA_MLR gave the smallest RMSEP (359.3) and highest correlation ($r = 0.67$) between the predicted and observed TOXICITY MEASUREMENT values. Therefore, GA_MLR with 10 baseline variables outperformed all the other studied methods and gave the best prediction model.

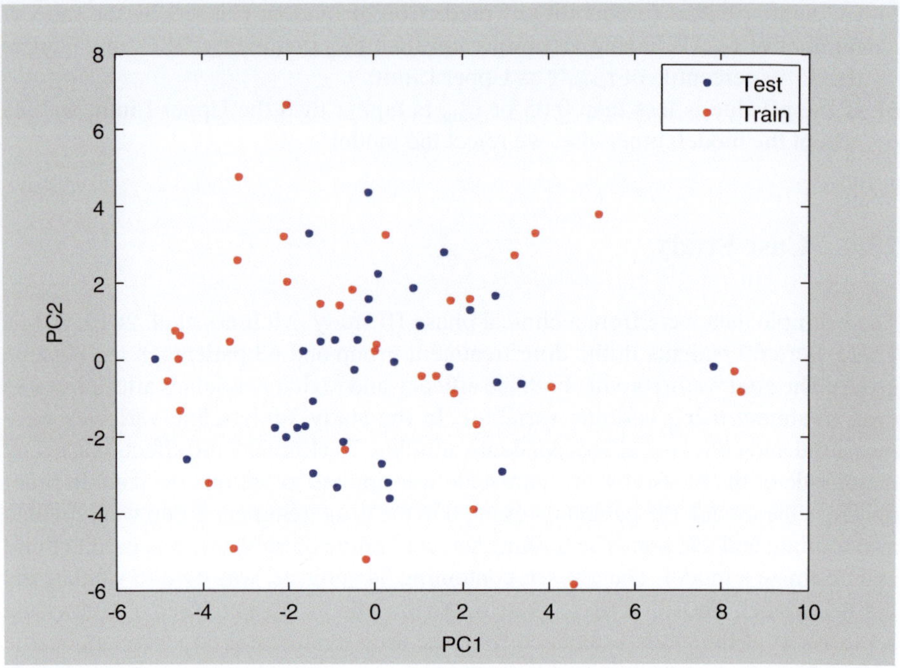

Fig. 25.2 PCA score plot of all the patients in the treatment group

Table 25.1 Model comparison by the prediction of the test set for TOXICITY MEASUREMENT at week 12

Method	RMSEP	r	Features
PCR	438.6	0.50	19
PLSR	417.1	0.53	2
STW_MLR	426.5	0.52	4
GA_MLR	359.3	0.67	10

Figure 25.3 was the scatter plot of calculated **Y** (Yfit) vs observed **Y** (TOXICITY MEASUREMENT) when the best model obtained from GA_MLR was applied to the training set objects. When a model is applied to the objects in the training set, the calculated **Y** is a measure of fitness of the model as opposed to "predicted Y" as the data have been used to build the model. Figure 25.3 shows that the best model obtained from GA_MLR fitted data in the training set very well, and the Pearson correlation coefficient between the calculated and observed TOXICITY MEASUREMENT (r) was 0.94.

Figure 25.4 demonstrates that the prediction of the model for the test set, and the correlation coefficient (r) between the predicted and observed TOXICITY MEASUREMENT values was 0.67. Figure 25.5 shows the distribution of r obtained from the 1000 randomly permuted test sets. The red line was the 95th percentile or so-called Upper Limit of the r. Only if the r of the test set is higher than the upper limit, the model is significant ($p < 0.05$). The results show that the r of the test set

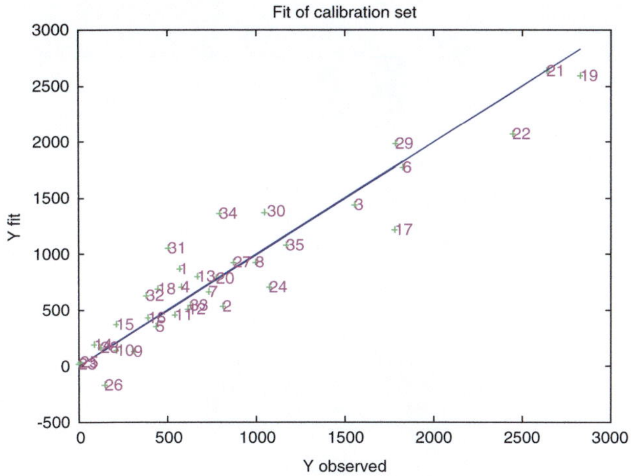

Fig. 25.3 Fit of the training set or calibration set for the TOXICITY MEASUREMENT model at week 12

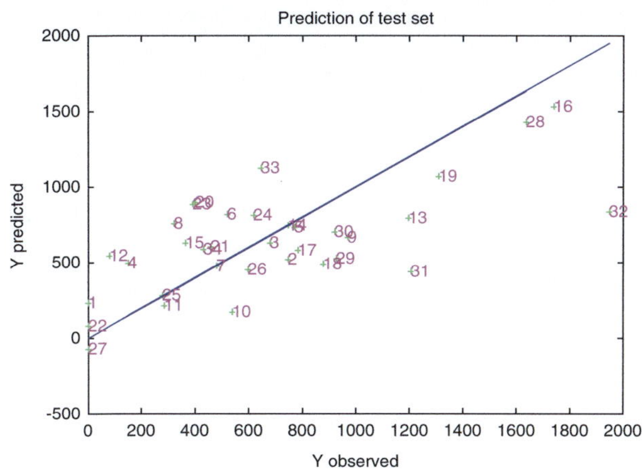

Fig. 25.4 Prediction of the test set for the TOXICITY MEASUREMENT model at week 12

($r = 0.67$) was significantly higher than any of the r's of the randomised test set ($p < 0.001$). Therefore, the best model passes the validation by the test set, and can be used for prediction for new subjects.

Figure 25.6 shows the prediction of the placebo patients for the TOXICITY MEASUREMENT at week 12. The correlation between the predicted and observed TOXICITY MEASUREMENT values ($r = 0.63$) in Fig. 25.6 was similar to that ($r = 0.67$) in Fig. 25.4. This indicates that the best model obtained from the DRUG TREATMENT group could also be used to predict the TOXICITY MEASUREMENT at week 12 for the placebo group.

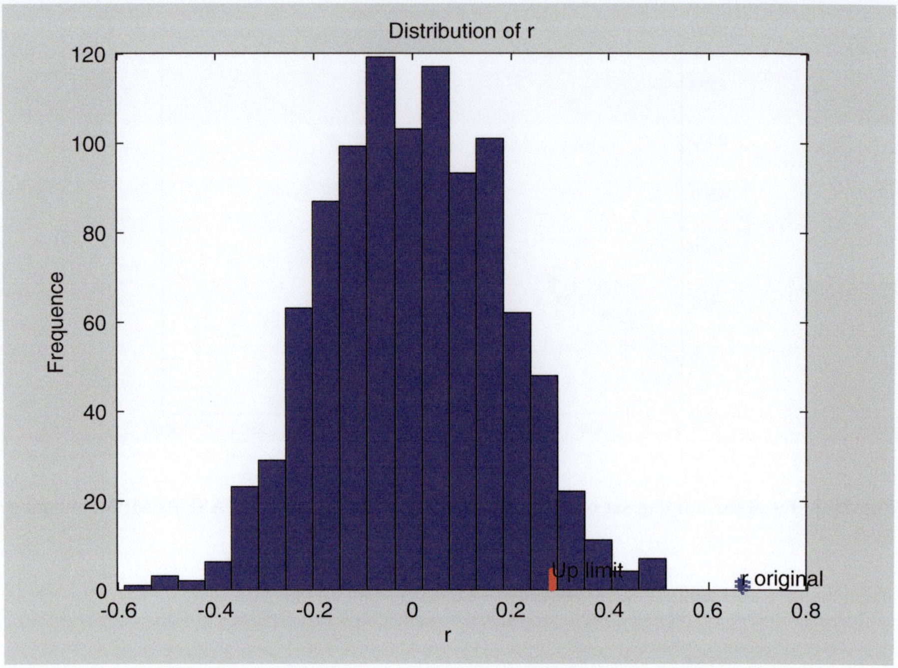

Fig. 25.5 Randomisation test for the TOXICITY MEASUREMENT model at week 12

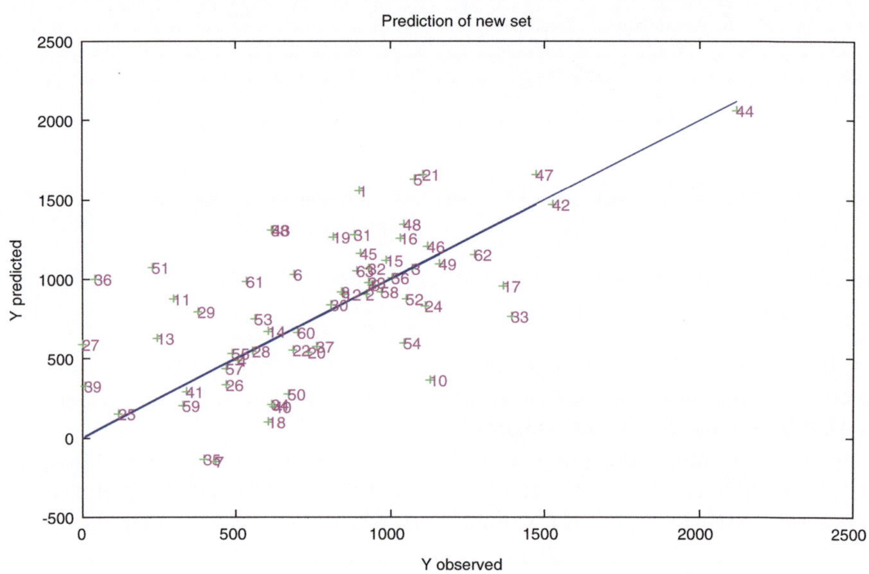

Fig. 25.6 Prediction of the placebo patients for the TOXICITY MEASUREMENT model at week 12

Table 25.2 Prediction of low CV risk using the optimal predictive model for TOXICITY MEASUREMENT at week 12

	Training set	Test set	Placebo set
Model accuracy	0.77	0.68	0.75
Sensitivity	0.69	0.65	0.63
Specificity	0.84	0.71	0.80
Sensitivity/(1-specificity)	4.4	2.2	3.1

This was consistent with the univariate t-test result: no significant change of TOXICITY MEASUREMENT has been observed at week 12 between the DRUG TREATMENT and placebo groups.

After the model validation, the best model was applied to predict TOXICITY MEASUREMENT at week 12 from the 10 baseline variables. TOXICITY MEASUREMENT was used to measure the cardiovascular (CV) risk where low TOXICITY MEASUREMENT indicates low CV risk. If the predicted TOXICITY MEASUREMENT of a patient was lower than 600, the patient was predicted as low CV risk (LCVR); otherwise the patient was predicted as non-LCVR. Table 25.2 shows the summary of the prediction results. The model accuracy is the number of individuals correctly predicted to be LCVR or non-LCVR divided by the total number of individuals. The closer the value is to one, the better the accuracy. Sensitivity is the proportion of individuals with LCVR who are correctly predicted by the model. Specificity is the proportion of individuals with non-LCVR who are correctly predicted by the model. The ratio of Sensitivity/(1-Specificity), namely likelihood ratio (LR), is equivalent to the probability that a person with the disease (LCVR) tested positive for the disease (true positive) divided by the probability that a person without the disease tested positive for the disease (false positive). LR indicates how useful the model is, i.e. a model with LR >1 indicating that the model is useful.

For the training set, the model correctly predicted 77 % patients; the sensitivity and specificity were 69 % and 84 % respectively. The LRs of 4.4, 2.2 and 3.1 for the training, test and placebo sets indicate that the model was useful. The total model accuracies of the test set (68 %) and the placebo set (75 %) also further confirmed that the optimised model could be a valuable tool to predict low CV risk patients.

25.4.2 Predictive Model for EFFICACY MEASUREMENT at Week 12

PCR, PLSR, Stepwise MLR (STW_MLR) and GA_MLR were applied to build the predictive models for EFFICACY MEASUREMENT at week 12. Table 25.3 lists the results of the prediction for the test set. The results show that GA_MLR gave the smallest RMSEP (1.22) and highest correlation (0.52) between the predicted

Table 25.3 Model comparison by the prediction of the test set for EFFICACY MEASUREMENT at week 12

Method	RMSEP	r	Features
PCR	1.25	0.05	4
PLSR	1.35	0.29	4
STW_MLR	1.58	0.31	6
GA_MLR	1.22	0.52	6

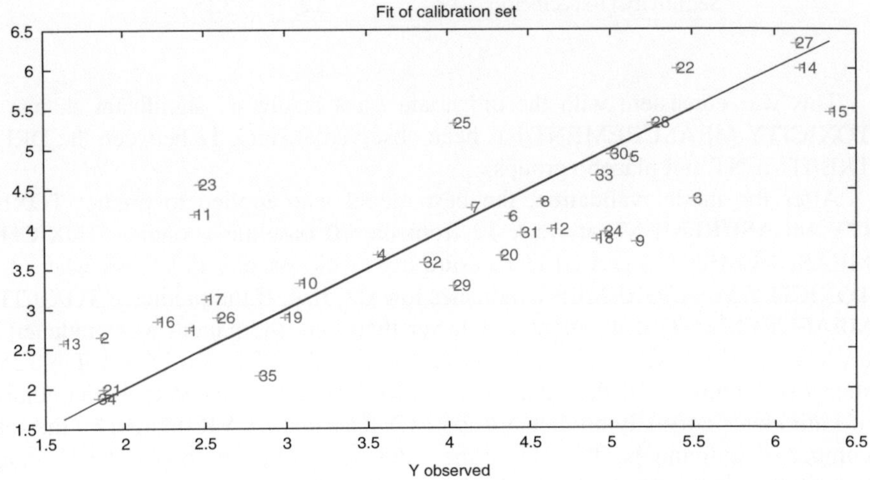

Fig. 25.7 Fit of the training set (calibration set) for the EFFICACY MEASUREMENT model at week 12

and observed EFFICACY MEASUREMENT values in the test set. Therefore, GA_MLR outperformed all the other studied methods and gave the best prediction model.

Figure 25.7 is the scatter plot of calculated **Y** (Yfit) vs observed **Y** (EFFICACY MEASUREMENT) when the best model built via GA_MLR was applied to the training set objects. The plot shows that the best model fitted the data for the training set reasonably well. The Pearson correlation coefficient—r (between the calculated and observed EFFICACY MEASUREMENT) was 0.83. Figure 25.8 demonstrates the prediction of the model of the test set with $r = 0.52$. Figure 25.9 shows the randomisation test results, and the r of the test set ($r = 0.52$) was higher than the upper limit ($p = 0.002$). Therefore, the best model passed the validation by means of the test set, and can be used for prediction of new objects.

Figure 25.10 shows the prediction of the placebo patients applying the EFFICACY MEASUREMENT model at week 12. Most of the predicted EFFICACY MEASUREMENT values were lower than their corresponding observed values. The correlation between the predicted and observed EFFICACY MEASUREMENT values was 0.31, which was 0.21 lower than that ($r = 0.52$) of the DRUG TREATMENT patients in the test set. This demonstrates that the optimal model obtained from the DRUG TREATMENT group could not be used to predict the EFFICACY

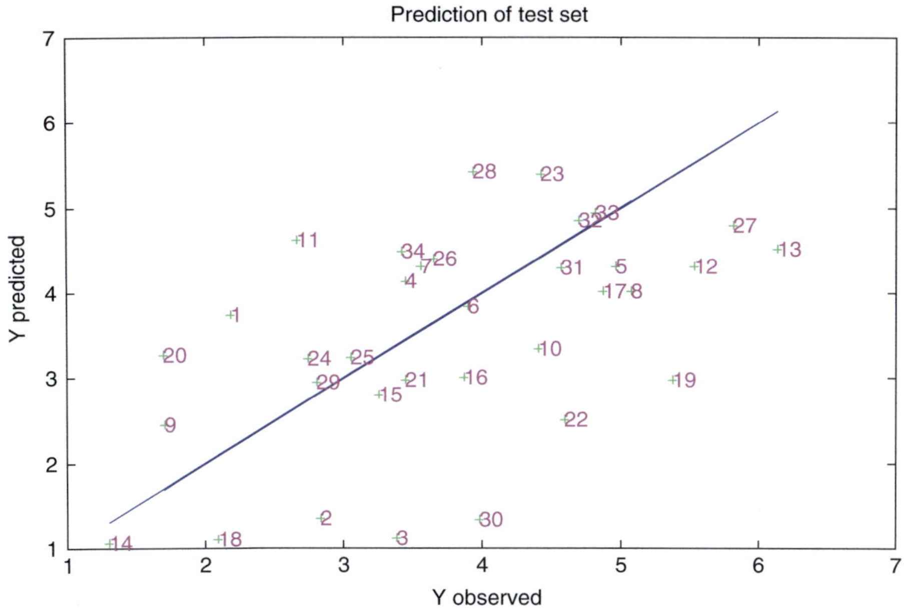

Fig. 25.8 Prediction of the test set for the EFFICACY MEASUREMENT model at week 12

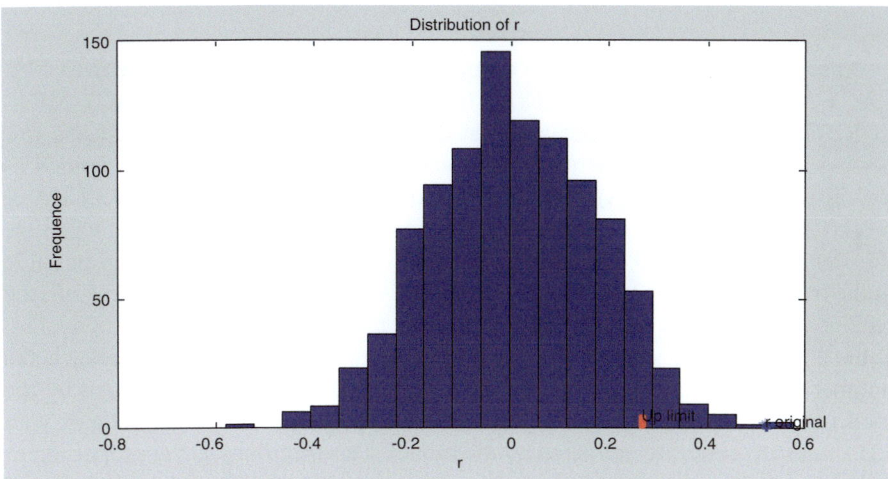

Fig. 25.9 Randomisation test for the EFFICACY MEASUREMENT model at week 12

MEASUREMENT for the placebo group. In fact, the model was purposely used to predict EFFICACY MEASUREMENT for the placebo patients as if they had been treated by DRUG. The reason why most of the predicted EFFICACY MEASUREMENT values were lower than their corresponding observed values was that the DRUG significantly decreased EFFICACY MEASUREMENT in the treatment group compared to the placebo group.

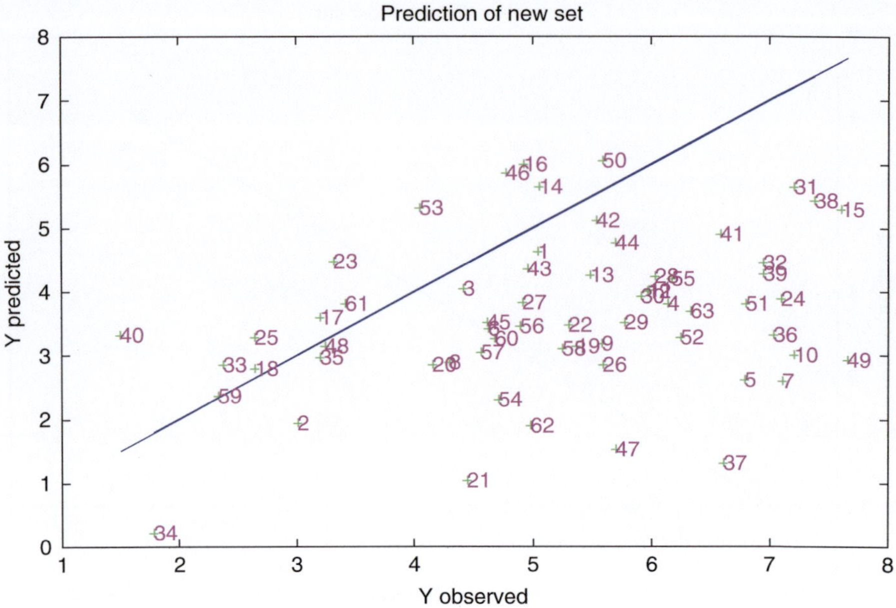

Fig. 25.10 Prediction of placebo patients for the EFFICACY MEASUREMENT model at week 12

After the model validation, the optimal model can be applied to predict EFFICACY MEASUREMENT at week 12 from baseline variables. EFFICACY MEASUREMENT was used to measure the disease activity (a low score means low disease activity.) After patients with Efficacy measurement <3.2 were defined as low disease activity (LDA), the optimal model then was used to predict LDA at week 12 from the 6 baseline variables. The cut-off of 3.66 was identified when the threshold was optimised to bring the sensitivity and specificity as close as possible in the training set (Fig. 25.11). If the predicted score was lower than the cut-off, the patient was predicted as LDA; otherwise, the patient was predicted as non-LDA. Table 25.4 shows the summary of the prediction results. The model accuracy is the number of individuals correctly predicted to be LDA or non-LDA divided by the total number of individuals. The sensitivity is the proportion of individuals with LDA who are correctly predicted by the model. The specificity is the proportion of individuals with non-LDA who are correctly predicted by the model. The LR, i.e. Sensitivity/(1-Specificity), was also recorded.

For the training set, the model correctly predicted 86 % patients; the sensitivity and specificity were 85 % and 86 % respectively. The LR of 6.2 for the training set and 2.4 for the test set indicate that the model was useful. The reasonable accuracy (71 %), sensitivity (80 %) and specificity (67 %) of the test set confirmed that the optimised model could be a valuable tool to predict LDA.

The Sensitivity and Specificity depend on the cut-off value. All possible cut-off points give a unique pair of values for Sensitivity and Specificity. Figure 25.11 shows that as the threshold was increased from 1 to 7, the sensitivity was increased

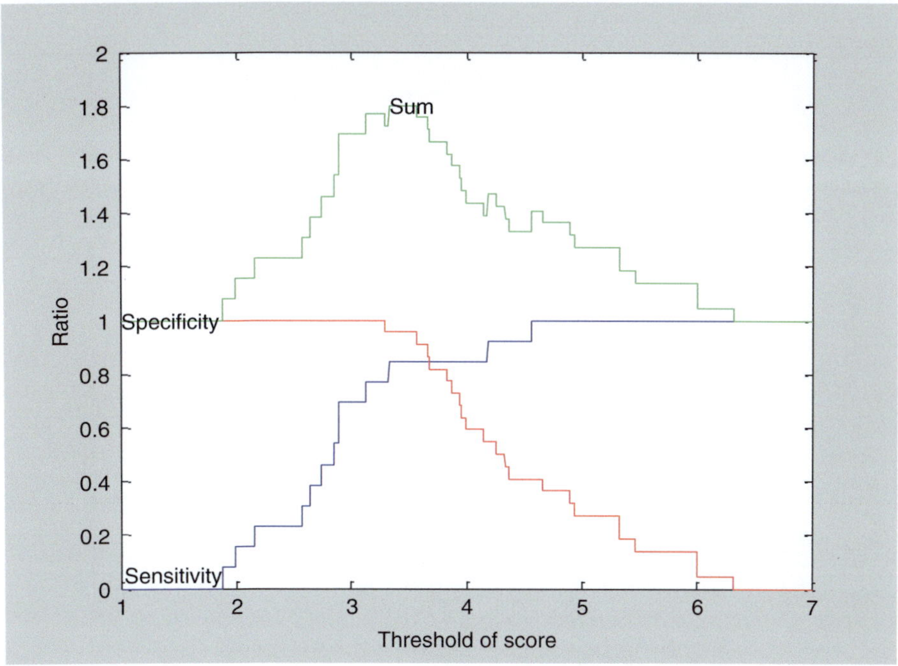

Fig. 25.11 Sensitivity and specificity vs threshold of the training set

Table 25.4 Prediction of LDA using the optimal predictive model for EFFICACY MEASUREMENT at week 12

	Training set	Test set
Model accuracy	0.86	0.71
Sensitivity	0.85	0.80
Specificity	0.86	0.67
Sensitivity/(1-specificity)	6.2	2.4

from 0 to 1 while the specificity correspondingly decreased from 1 to 0. Theoretically, when the sensitivity is equal to the specificity, the sum of the sensitivity and specificity will reach the maximum. The figure shows that when the cut-off of 3.66 was selected, the sensitivity was almost equal to the specificity.

Another way to illustrate the usefulness of the predicted score is to use the receiver operating characteristic (ROC) curve. The ROC curve is obtained by plotting the Sensitivity against (1-Specificity) for every possible cut-off value, and connecting these points by lines. The curve for a score that has a good discriminative ability lies in the upper left-hand quadrant of the plot with the area under the curve (AUC) > 0.5 and that for a score that is no better than a chance at discriminating will lie along the 45° diagonal (AUC = 0.5). Figures 25.12 and 25.13 were ROC curves of the training and test sets respectively. These figures show that the predicted score from the optimal model had discriminative abilities for both training and test sets, although the performance was better in the training set than that in the test set.

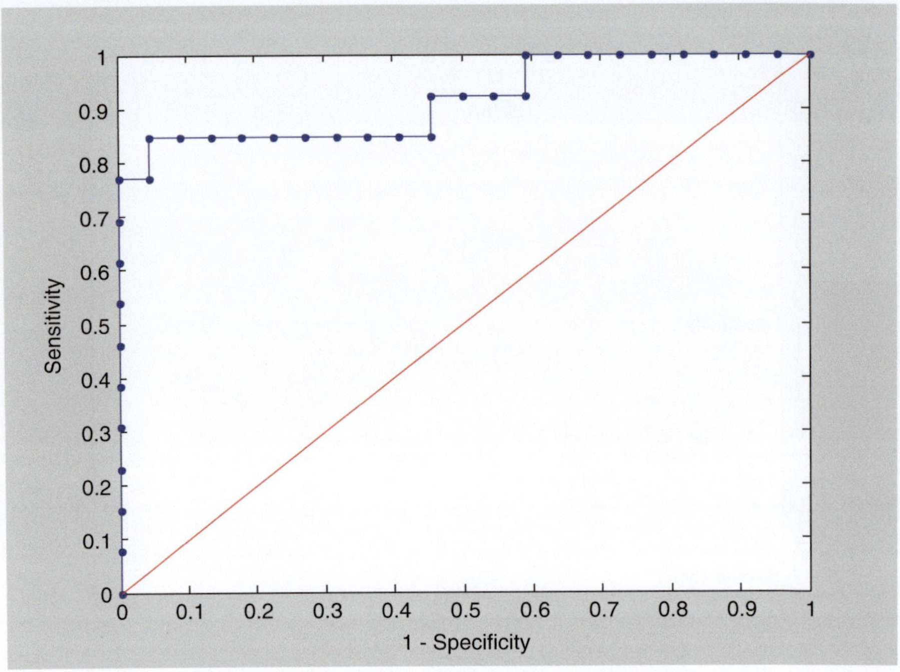

Fig. 25.12 Receiver operating characteristic (ROC) curve of the training set

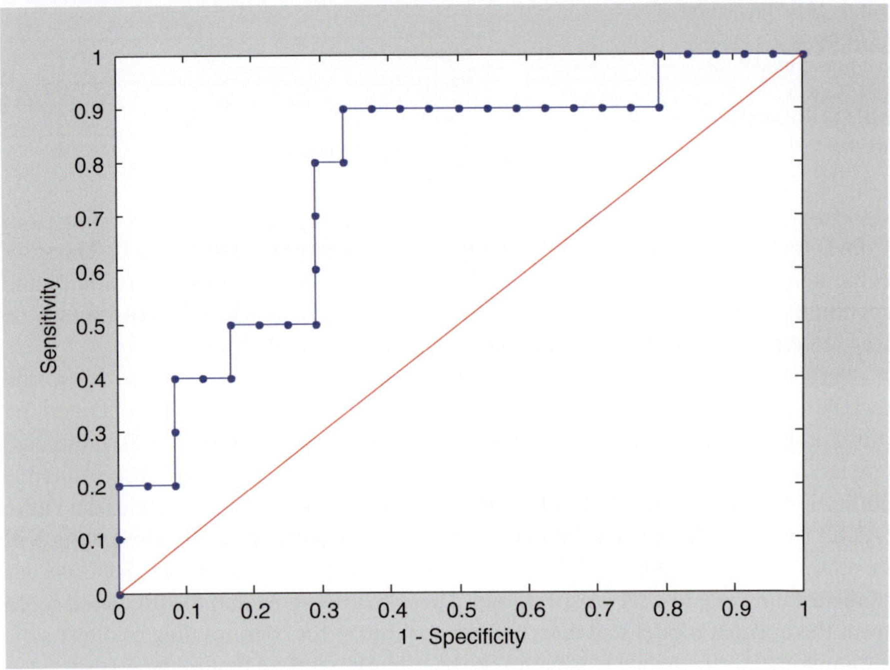

Fig. 25.13 Receiver operating characteristic (ROC) curve of the test set

25.5 Conclusions

Two optimal models were built to predict TOXICITY MEASUREMENT and EFFICACY MEASUREMENT at week 12 from baseline variables. The results showed that GA_MLR gave the best models in all the four of the tested methods. The prediction of the test set of patients treated by DRUG TREATMENT showed that the models passed the validation.

The two optimal models were further applied to predict low CV risk and LDA, and reasonable accuracies, sensitivities and specificities were obtained from both training and test sets. These results indicate that Chemometrics is a powerful methodology to develop predictive models from noisy high dimensional data. The optimal models could be used to predict the TOXICITY MEASUREMENT and EFFICACY MEASUREMENT at week 12 for a patient from baseline variables before the treatment of DRUG. It is important to note that only 69 DRUG treated patients were available to build and to validate the models, and further validation of the models is recommended.

References

Baldovin A, Wu W, Centner V, Jouan-Rimbaud D, Massart DL, Favretto L, Turello A (1996) Feature selection for the discrimination between pollution types with partial least squares modelling. Analyst 121:1603–1608

Baldovin A, Wu W, Massart DL, Turello A (1997) Regularized discriminant analysis (RDA)-modelling for the binary discrimination between pollution types. Chemometr Intell Lab Syst 38:25–37

Brereton RG (1990) Chemometrics, applications of mathematics and statistics to laboratory systems. Ellis Horwood Limited, Chichester

Brown SD, Blank TB, Sum ST, Weyer LG (1994) Chemometrics. Anal Chem 66:315R–359R

Candolfi A, Wu W, Centner V, Massart DL (1998) Comparison of classification approaches applied to NIR-spectra of clinical study lots. J Pharm Biomed Anal 16:1329–1347

Connor SC, Wu W, Sweatman BC, Manini J, Haselden JN, Crowther DJ, Waterfield CJ (2004) The effects of feeding and body weight loss on the 1H NMR-based urine metabolic profiles of male Wistar Han rats: implications for biomarker discovery. Biomarkers 9:156–179

Cordingley HC, Rpberts SLL, Tooke P, Armitage JR, Lane PW, Wu W, Wildsmith WE (2003) Multifactorial screening design and analysis of SELDI-TOF ProteinChip array optimisation experiments. Biotechniques 34:364–373

Cutler P, Akuffo EL, Bodnar WM, Briggs DM, Davis JB, Debouck CM, Fox SM, Gibson RA, Gormley DA, Holbrook JD, Jacqueline Hunter A, Kinsey EE, Prinjha R, Richardson JC, Roses AD, Smith MA, Tsokanas N, Willé DR, Wu W, Yates JW, Gloger IS (2008) Proteomic identification and early validation of complement 1 inhibitor and pigment epithelium-derived factor: two novel biomarkers of Alzheimer's disease in human plasma. Proteomics Clin Appl 2:467–477

Czekaj T, Wu W, Walczak B (2005) About kernel latent variable approaches and SVM. J Chemometr 19:341–354

Czekaj T, Wu W, Walczak B (2008) Classification of genomic data: some aspects of feature selection. Talanta 76:564–574

Daszykowski M, Wu W, Nicholls AW, Ball RJ, Walczak B (2007) Identifying potential biomarkers in LC-MS data. J Chemometr 21:292–302

Edgington ES (1995) Randomization tests, 3rd edn. Wiley, New York

Esbensen K, Geladi P (2005) The start and early history of chemometrics: selected interviews. Part 2. J Chemometr 4:389–412

Edgington ES (1964) Randomization tests. J Psychol 57(2):445–449

Geladi P, Esbensen K (2005) The start and early history of chemometrics: selected interviews. Part 1. J Chemometr 4:337–354

Guo Q, Wu W, Massart DL (1999) The robust normal variate transform for pattern recognition with near-infrared data. Anal Chim Acta 382:87–103

Guo Q, Wu W, Questier F, Massart DL, Boucon C, de Jong S (2000) Sequential projection pursuit using genetic algorithms for data mining. Anal Chem 72:2846–2855

Guo Q, Wu W, Massart DL, Boucon C, de Jong S (2001) Feature selection in sequential projection pursuit. Anal Chem Acta 446:85–96

Guo Q, Wu W, Massart DL, Boucon C, de Jong S (2002) Feature selection in principal component analysis of analytical data. Chemometr Intell Lab Syst 61:123–132

Kalivas JH, Roberts N, Sutter JM (1989) Global optimization by simulated annealing with wavelength selection for ultraviolet-visible spectrophotometry. Anal Chem 61:2024–2030

Kvalheim OM (1996) Chemometrics, quality, information and the third waves. Chemometr Intell Lab Syst 33:1–2

Leardi R (2000) Application of genetic algorithm-PLS for feature selection in spectral data sets. J Chemometr 14:643–655

Leardi R, Boggia R, Terrile M (1992) Genetic algorithms as a strategy for feature selection. J Chemometr 6:267–281

Massart DL, Vandeginste BGM, Deming SN, Michotte Y, Kaufman L (1988) Chemometrics: a textbook. Elsevier Science Publishers B. V, Amsterdam

Massart DL, Vandeginste BGM, Buydens LMC, De Jong S, Lewi PJ, Smeyers-Verbeke J (1997) Handbook of chemometrics and qualimetrics. Part A. In: Data Handling in Science and Technology, vol 20A. Elsevier, Amsterdam

Massart DL, Vandeginste BGM, Buydens LMC, De Jong S, Lewi PJ, Smeyers-Verbeke J (1998) Handbook of chemometrics and qualimetrics. Part B. In: Data Handling in Science and Technology, vol 20A. Elsevier, Amsterdam

MATLAB 6.1, The MathWorks Inc., Natick, MA, 2000

McInnes IB, Lee JS, Wu W, Giles JT, Bathon J, Salmon J, Beaulieu A, Codding C, Delles C, Sattar N (2010) Lipid and inflammation parameters: a translational, randomized placebo-controlled study to evaluate effects of tocilizumab: the MEASURE study. Oral presentation, 2010 ACR/ARHP annual scientific meeting, Atlanta, GA, 6–11 November 2010

McInnes IB, Lee JS, Wu W, Giles JT, Bathon JM, Salmon JE, Beaulieu AD, Codding CE, Delles C, Sattar N (2011) MEASURE: A translational, randomized, placebo (PBO)-controlled study to evaluate the effects of tocilizumab (TCZ) on parameters of lipids and inflammation. Oral presentation, EULAR 2011, European League Against Rheumatism, London, 25–28 May 2011

Menaa F (2014) Next-generation sequencing or the dilemma of large-scale data analysis: opportunities, insights, and challenges to translational, preventive and personalized medicine. J Investig Genomics 1(1):00005

Niazi A, Leardi R (2012) Genetic algorithms in chemometrics. J Chemometr 26:345–351

Rathore AS, Bhushan N, Hadpe S (2011) Chemometrics applications in biotech processes: a review. Biotechnol Prog 27:307–315

Rathore AS, Mittal S, Pathak M, Mahalingam V (2014) Chemometrics application in biotech processes: assessing comparability across processes and scales. J Chem Technol Biotechnol 89:7

Walczak B, Wu W (2005) Fuzzy warping of chromatograms. Chemometr Intell Lab Syst 77: 173–180

Wegman EJ (1990) Hyperdimensional data analysis using parallel coordinates. J Am Stat Assoc 85:664–675

Wold S (1995) Chemometrics; what do we mean with it, and what do we want from it? Chemometr Intell Lab Syst 30:109–115
Wu W, Manne R (2000) Fast regression methods in a Lanczos (or PLS-1) basis. Theory and applications. Chemometr Intell Lab Syst 51:145–161
Wu W, Massart DL (1996) Artificial neural networks in classification of NIR spectral data: selection of the input. Chemometr Intell Lab Syst 35:127–135
Wu W, Massart DL (1997) Regularised nearest neighbour classification method in pattern recognition of near infrared spectra. Anal Chim Acta 349:253–261
Wu W, Walczak B, Massart DL, Prebble KA, Last IR (1995) Spectral transformation and wavelength selection in NIR spectra classification. Anal Chim Acta 315:243–255
Wu W, Rutan SC, Baldovin A, Massart DL (1996a) Feature selection using the Kalman filter for classification of multivariate data. Anal Chim Acta 335:11–22
Wu W, Walczak B, Penninckx W, Massart DL (1996b) Feature reduction by Fourier transform in pattern recognition of NIR data. Anal Chim Acta 331:75–83
Wu W, Mallet Y, Walczak B, Penninckx W, Massart DL, Heuerding S, Erni F (1996c) Comparison of regularized discriminant analysis, linear discriminant analysis and quadratic discriminant analysis, applied to NIR data. Anal Chim Acta 329:257–265
Wu W, Walczak B, Massart DL, Heuerding S, Erni F, Last IR, Prebble KA (1996d) Artificial neural networks in classification of NIR spectral data: design of the training set. Chemometr Intell Lab Syst 33:35–46
Wu W, Massart DL, de Jong S (1997a) The kernel PCA algorithms for wide data, Part I: theory and algorithms. Chemometr Intell Lab Syst 36:165–172
Wu W, Massart DL, de Jong S (1997b) Kernel PCA algorithms for wide data, Part II: Fast cross-validation and application in classification of NIR data. Chemometr Intell Lab Syst 37:271–280
Wu W, Guo Q, de Aguiar PF, Massart DL (1998) The star plot: an alternative display method for multivariate data in the analysis of food and drugs. J Pharm Biomed Anal 17:1001–1013
Wu W, Guo Q, Jouan-Rimbaud D, Massart DL (1999) Using contrasts as data pretreatment method in pattern recognition of multivariate data. Chemometr Intell Lab Syst 45:39–53
Wu W, Wildsmith SE, Winkley AJ, Yallop RM, Elcock F, Bugelski PJ (2001) Chemometric strategies for normalisation of gene expression data obtained from cDNA microarrays. Anal Chem Acta 446:451–466
Wu W, Guo Q, de Jong S, Massart DL (2002) Randomisation test for the number of dimensions of the group average space in generalised Procrustes analysis. Food Qual Prefer 13:191–200
Wu W, Roberts SLL, Cordingley HC, Armitage JR, Tooke P, Wildsmith SE (2003a) Validation of consensus between proteomic and clinical chemical data by applying a new randomisation F-test for generalised Procrustes analysis. Anal Chim Acta 490:365–378
Wu W, Guo Q, Massart DL, Boucon C, de Jong S (2003b) Structure preserving feature selection in PARAFAC using a genetic algorithm and Procrustes analysis. Chemometr Intell Lab Syst 65:83–95
Wu W, Shaw P, Ruan J, Elcock FJ, Wildsmith SE (2005) Optimisation of image analysis process for cDNA microarrays by experimental designs. Chemometr Intell Lab Syst 76:175–184
Young FW, Valero-Mora PM, Friendly M (2006) Visual statistics – seeing data with dynamic interactive graphics. Wiley, Hoboken

Chapter 26
Statistical Methods for Comparability Studies

Jason J.Z. Liao

Abstract Biological products are complex mixtures of molecular species. Their individual entities are difficult to characterize. During development and post-approval, improvements are made in production methods, process and control test methods for product characterization, and equipment or facilities. These could result in fundamental changes to the biological product itself, perhaps requiring additional clinical studies to demonstrate that the product's safety, identity, purity and potency have not been impacted. Consequently, regulations require the sponsor to demonstrate product comparability between the post-change and pre-change products. A stepwise approach to extensively characterize the post-change product and the pre-change product with state-of-the-art technology is the first step in assessing the potential impact on safety and efficacy. These comparability studies should have direct side-by-side comparisons of the pre-change "reference" product and post-change "test" product. In this chapter, statistical methods are reviewed for establishing comparability in relation to critical quality attributes and animal pharmacokinetics (PK) to systematically evaluate the data.

Keywords Biologics • Comparability • In vivo pharmacokinetics • Quality attributes • Stepwise approach

26.1 Introduction

A biological product is defined as "a virus, therapeutic serum, toxin, antitoxin, vaccine, blood, blood component or derivative, allergenic product, or analogous product, or arsphenamine, or derivative of arsphenamine or any other trivalent organic arsenic compound, applicable to the prevention, treatment or cure of a disease or condition of human beings" (Public Health Services Act 42 U.S.C. § 262(i)). One of the earliest biological products, a blood protein called Factor VIII was first introduced to the U.S. marketplace in 1966. The earliest FDA approved

J.J.Z. Liao (✉)
Novartis Pharmaceuticals, One Health Plaza, East Hanover, NJ 07936, USA
e-mail: Jason.liao@novartis.com

L. Zhang (ed.), *Nonclinical Statistics for Pharmaceutical and Biotechnology Industries*, Statistics for Biology and Health, DOI 10.1007/978-3-319-23558-5_26

modern biotech product designed for human therapeutic use was human insulin in 1982. Approval was given in 1985 to a human growth hormone (HGH) for the treatment of dwarfism. Since then, biological products have played an increasing role in treating unmet medical needs. Biologics were 7 of the top 10 bestselling drugs in 2012 and 2013 including 6 mAbs (Genetic Engineering & Biotechnology News (GEN) 2013, 2014). It is projected that 10 of the top 20 selling drugs in 2016 will be biologics including 7 mAbs (McCamish and Woollett 2011).

Biological products, like other drugs, are used for the treatment, prevention or cure of disease in humans. In contrast to chemically synthesized small molecular weight drugs, which have a well-defined structure and can be thoroughly characterized, biological products are generally derived from living material; human, animal, or microorganism. They are complex in structure, and thus are usually not fully characterized as individual entities. Biologics typically have significant heterogeneity, seen through subtle differences associated with the primary amino acid sequence, post-translational modifications and three dimensional structures. In addition to structural complexity, the source materials and manufacturing processes for proteins and antibodies may lead to a variety of product variants and impurities that can impact product efficacy and safety.

Manufacturing biologics is a complicated process. It can take highly experienced companies many years to devise the technologies necessary to cultivate and genetically engineer the living cells that produce the proteins and antibodies that are filtered and refined into successful biological drugs. A biologics manufacturing process depends on many steps of production and purification which can influence the biological and clinical properties. These sophisticated steps are monitored by sensitive analytical methods, which need to be adapted to specific biological products. Because of the limited ability to characterize the identity and structure and measure the activity of the therapeutically-active component(s), it is the manufacturing process that defines a biological product.

Improvements are commonly applied to biological products in production methods, process and control test methods for product characterization, or changes in equipment or facilities both during development and after approval. A biological product manufacturer may seek to make changes in the manufacturing process used to make a particular product for a variety of reasons. These include improvement of product quality, yield, and manufacturing efficiency involving increase in manufacturing scale, improving product stability, and complying with changes in regulatory requirements. Innovation and improvements in manufacturing processes and test methods have been encouraged because they bring important and improved products to market more efficiently and rapidly. International regulatory agencies acknowledge that product and process changes are necessary for the biotech industry to evolve. Changes in the manufacturing process may in some cases affect the biological product itself in fundamental ways. In such cases, additional clinical studies would be required to demonstrate that the product's safety, identity, purity and potency have not been impacted. It is a regulatory requirement for the sponsor to demonstrate product comparability between the post-change and pre-change products.

As specified in regulatory guidances (FDA 1996, 2003, 2005; ICH 2005), determination of comparability can be based on a combination of analytical testing, biological assays, and, in some cases, nonclinical and clinical data. The demonstration of comparability does not necessarily mean that the quality attributes and other key parameters of interest of the pre-change and post-change product are identical, but that they are highly similar and that the existing knowledge is sufficiently predictive to ensure that any differences in quality attributes and the key parameters of interest have no adverse impact upon safety or efficacy of the drug product. To assess the impact of the change, a careful evaluation of all foreseeable consequences for the product should be performed using a totality of the evidence in a stepwise approach with fingerprint-like techniques to extensively characterize the post-change product with appropriate state-of-the-art technology. The extent of the studies necessary to demonstrate comparability will depend on many factors including the manufacturing operations for which changes were made, careful assessment of the potential impact of the changes, the availability of suitable analytical techniques to detect potential product modifications and the results of these studies. Determination of comparability can be based solely on quality considerations if the manufacturer can provide sufficient assurance of comparability through analytical studies related to quality attributes. Additional evidence from nonclinical or clinical studies is considered appropriate when quality data are insufficient to establish comparability. The extent and nature of nonclinical and clinical studies will be determined on a case-by-case basis in consideration of various factors, including quality findings, the nature and the level of knowledge of the product, and existing nonclinical and clinical data relevant to the product. An extensive characterization may be conducted using multiple lots of representative product to determine the impact on lot-to-lot variability and provide a comprehensive understanding of physico-chemical and biological characteristics. This is performed with an in-depth side by side comparison of the post-change and the pre-change product in relation to chemical, physical, and bioactivity properties and possibly other specific suitable attributes. As outlined in the guidances, the information submitted in the comparability study may include the structural and functional characterization, nonclinical evaluation, pharmacokinetics (PK) studies, pharmacodynamics (PD) studies, PK/PD studies, clinical efficacy studies, specific safety studies, immunogenicity studies and pharmacovigilance studies.

Recently, developing biosimilars have been a hot discussion topic both from the industry and health authorities (FDA 2012a, b, c, 2014; WHO 2009; EMA 2005). The regulatory science on which comparability is based has subsequently been applied to the review and approval of biosimilars (Weise et al. 2012; EMA 2005; FDA 2012a, b, c, 2014). A comparability study is defined as a quantitative assessment of the extent of similarity of drug substances and drug products made before and after a bioprocess change or between two products (FDA 2005; ICH 2005). The purpose of comparability studies is to ensure that the test product does not have any significant or clinically meaningful difference or changes that impact the material/product/process quality, safety, purity, and potency or efficacy compared to the reference product (FDA 2005, 2012a, b, c; ICH 2005).

In this chapter, we focus on statistical approaches to demonstrate comparability between the post-change product and the pre-change product in terms of the critical quality attributes including examples for purity and potency, and pharmacokinetics in animal studies.

26.2 Statistical Methods for Comparability Studies

26.2.1 Critical Quality Attributes (CQA)

CQAs are attributes that may affect safety and efficacy, and are defined in relation to the subject population and dosing regimen. The contribution of different sources of variability in defining comparability range is illustrated in Fig. 26.1.

The first step in comparability determination is defining the target CQAs (T). To assess manufacturing capability (ΔM), one must have two inputs, analytical variability (ΔA) and process variability (ΔP). Based on the knowledge of structure activity relationships, existing clinical and pre-clinical experience, and manufacturing capability, the comparability acceptance criteria (ΔC) should be proposed for each CQA.

To ensure consistency in comparability study practice, a statistics-based systematic approach is recommended. The comparison of the results to the acceptance criteria allows for an objective assessment of whether or not the two products are comparable. This section outlines the general statistical methods to be used in evaluating comparability. Assessing comparability is NOT just about meeting the

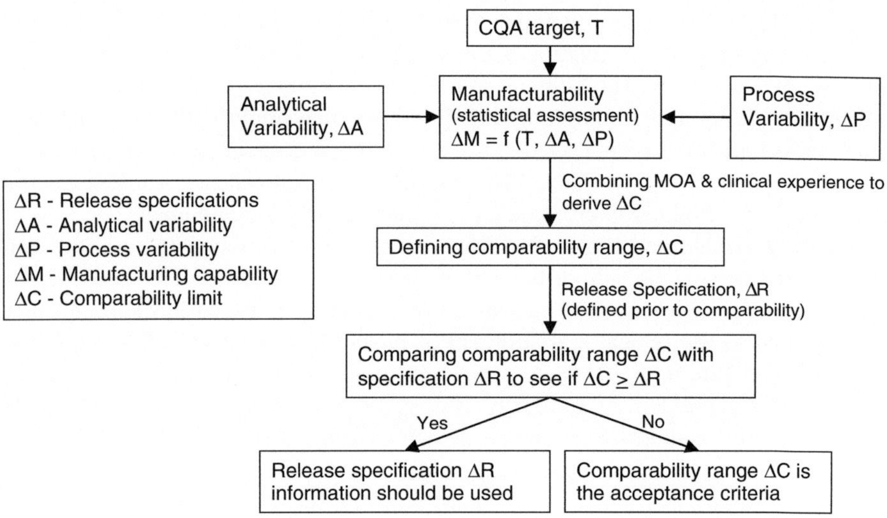

Fig. 26.1 Decision tree—defining critical quality attributes comparability criteria

predefined specification/criterion where the specification/criterion of the proposed post-change product should not be wider than the variability of pre-change product. Developing an analytical method in differentiating meaningful difference is critical and important.

With those considerations, three statistical models are proposed below: (1) a non-paired head-to-head comparison where the two substances/products can be tested within a single assay run, such as sequential measurements by HPLC, (2) a paired head-to-head analytical comparability where pre-change and proposed post-change product materials can be tested simultaneously, such as the measurements in a potency assay, and (3) one-sided testing of proposed post-change product materials against a pre-existing database for the pre-change product. It is recommended that the actual study design and data analysis should be carried out with the close collaboration of a statistician.

26.2.1.1 Unpaired Quality Attributes

Consider the most commonly seen data structure, for example, HPLC generated data where non-paired observations are produced from both the proposed post-change and the pre-change product. The data structure is

$$\begin{array}{ll} x_{11},\cdots,x_{1n_x} & y_{11},\cdots,y_{1n_y} \\ x_{21},\cdots,x_{2n_x} & y_{21},\cdots,y_{2n_y} \\ \cdots\cdots\cdots\cdots & \cdots\cdots\cdots\cdots \\ x_{m_x1},\cdots,x_{m_xn_x} & y_{m_y1},\cdots,y_{m_yn_y} \end{array}$$

where $x_{ij} = x_i^0 + \delta_{ij}$, $y_{kj} = y_k^0 + \varepsilon_{kj}$, $\delta_{ij} \sim N\left(0, \sigma_\delta^2\right)$, $\varepsilon_{kj} \sim N\left(0, \sigma_\varepsilon^2\right)$, $x_i^0 \sim N\left(\mu_x, \sigma_x^2\right)$, $y_k^0 \sim N\left(\mu_y, \sigma_y^2\right)$. Let x_{ij} be the observed j^{th} critical quality attribute measurement for the pre-change (reference) product for lot (batch, or run) i, where $i = 1, \cdots, m_x$, $j = 1, \cdots, n_x$, n_x is the number of measurements for each reference lot, and m_x is the number of reference lots used. Let y_{kj} be the observed j^{th} critical quality attribute measurement for the proposed post-change product (test) for lot k, where $k = 1, \cdots, m_y$, $j = 1, \cdots, n_y$, n_y is the number of measurements for each test product lot, and m_y is the number of test product lots used. The σ_x^2 and σ_δ^2 are the process variability and the assay variability for the reference product, respectively. Similarly, the σ_y^2 and σ_ε^2 are the process variability and the assay variability for the test product, respectively. The goal is to compare the two products (test vs reference) and see how similar they are.

For this type of data, formal statistical approaches are traditionally used such as the two one-sided test (TOST) to assess equivalence of means with the pre-specified acceptance criteria (Chatfield et al. 2011). However, the assessment of how similar the test and the reference product are to each other is not necessarily straightforward since how similar is similar enough is not well defined and scientific/clinical judgment may not be easily available. Thus, Liao and Darken (2013) proposed a method using a tolerance interval (TI) and a plausibility interval (PI) to define the comparability criteria. The basic idea is described here. Consider a hypothesized study where the test is also the reference. Since the reference product is an established or approved product, therefore the test (here it is the reference also in this hypothesized study) should be almost always comparable to the reference (itself). Any observed difference is due to chance and the difference is clinically negligible. Thus if any difference is within the reference variability, it is reasonable to conclude comparability between the test and the reference. Toward this end, an interval called the Plausibility Interval (PI) was proposed to quantify this reference variability. The assay + process PI for the difference between the reference and itself is defined as follows:

$$\left(-k\sqrt{2\left(\sigma_x^2+\sigma_\delta^2\right)},+k\sqrt{2\left(\sigma_x^2+\sigma_\delta^2\right)}\right) \tag{26.1}$$

where the critical value k is a factor to control the sponsor's tolerance, the σ_x^2 and σ_δ^2 are the process variability and the assay variability for the reference product, respectively, which can be estimated from the commonly used variance decomposition method. Note that the interval in Eq. (26.1) is defined for the difference of test and reference. The PI defines the acceptable range for the quality attribute difference between the test and the reference to fall within. Any difference within this PI should be considered practically acceptable, and thus it can serve as a goalpost for judging comparability. The concept of "goal posts," whereby the attributes of the post-change product fall within the demonstrated variation of the reference over its lifetime, is becoming a widely accepted approach for creation of a "highly similar" comparable product (McCamish and Woollett 2013).

An at least p-content with confidence $1-\alpha$ tolerance interval for the difference between the test and the reference is constructed so that

$$\Pr\left[\Pr\left(L<Y-X<U\right)\geq p\right]\geq 1-\alpha \tag{26.2}$$

where L and U are statistics calculated from the data, X(Y) is the quality attribute of the reference (test). In most practical applications, $1-\alpha$ and p are typically chosen from the set of values {0.90, 0.95, 0.99} (Krishnamoorthy and Mathew 2009). The choice of the combination can be used to control the risk for both consumer and the sponsor. The interval (L, U) is usually termed as the approximate $[100\times(1-\alpha)\,\%]\,/\,[100\times p\%]$ TI. In order to construct the approximate $[100\times(1-\alpha)\,\%]\,/\,[100\times p\%]$ TI, the method based on the Satterthwaite approximation recommended by Krishnamoorthy et al. (2011) is used, where the lower bound L and the upper bound U are defined as follows.

$$L = \widehat{\mu}_y - \widehat{\mu}_x - z_{(1+p)/2}\sqrt{\frac{\widehat{f}\left(a_x s_x^2 + a_y s_y^2\right)}{\chi^2_{\widehat{f},1-\alpha}}}, \text{ and}$$

$$U = \widehat{\mu}_y - \widehat{\mu}_x + z_{(1+p)/2}\sqrt{\frac{\widehat{f}\left(a_x s_x^2 + a_y s_y^2\right)}{\chi^2_{\widehat{f},1-\alpha}}}$$

where $z_{(1+p)/2}$ is the $100 \times (1+p)/2^{th}$ percentile of a normal distribution, $\chi^2_{\widehat{f},1-\alpha}$ is the $100 \times (1-\alpha)^{th}$ percentile of a chi-square distribution with df $= f$, s_x^2 and s_y^2 are the estimates for the total variance of the reference and the test product, respectively and can be estimated from the commonly used variance decomposition method, $n_1 = m_x \times n_x$, $n_2 = m_y \times n_y$, $a_x = 1 + \frac{1}{n_1}$, $a_y = 1 + \frac{1}{n_2}$ and $\widehat{f} = \frac{\left(a_x s_x^2 + a_y s_y^2\right)^2}{a_x^2 s_x^4/(n_1-1) + a_y^2 s_y^4/(n_2-1)}$, $\widehat{\mu}_x$ and $\widehat{\mu}_y$ can be estimated using the weighted average.

The test and the reference are claimed comparable if: (1) the approximate $[100 \times (1-\alpha)\,\%] / [100 \times p\%]$ tolerance interval for the difference between the test and the reference defined in Eq. (26.2) is within the plausibility interval defined in Eqs. (26.1); and (2) if the estimated mean ratio is within a specified boundary, for example, [0.8, 1.25].

Note that the tolerance interval is used here instead of the commonly used confidence interval. A confidence interval's width is due entirely to sampling error but in contrast, the width of a tolerance interval is due to both sampling error and variance in the population. As the sample size approaches the entire population, the width of the confidence interval approaches zero but in contrast, the width of the confidence interval gives the estimated percentiles approaching the true population percentiles. When comparing to the reference variability scaled range, the tolerance interval is a more appropriate choice.

The first condition in the comparability acceptance criteria is to control false failing comparability claims and the second condition is to control false passing comparability claims. It is a capability based approach which could falsely pass a large mean difference for the test product due to a large reference product variability. The additional point-estimate constraint in the second condition eliminates the potential that a test product with a large mean difference would enter the market (Haidar et al. 2007, 2008).

In contrast to the traditional equivalence approach for just testing the equivalence of means, one of the advantages of the new approach is that it also considers the variability instead of just the mean difference. The performance and feasibility of this approach was demonstrated through simulation studies and an example, and details can be found in Liao and Darken (2013). The simulation results showed that the between batch variability of the reference product plays a very important role in passing comparability. The number of batches that should be used in the comparability study heavily depends on the between batch variability. Larger number of batches is needed when the between batch variability is higher. At least two different batches should always be used in the head-to-head comparison. Using simulation and real data, Liao and Darken (2013) recommended using $k = 2.5$ or 3 in constructing the PI in Eq. (26.1). However, the actual k value can be chosen case-by-case depending on the nature of the reference product and still be consistent with health authorities' requirements.

26.2.1.2 Paired Quality Attributes

Consider the paired data structure, a comparison of the relative potency of the proposed post-change product (test) with this potency of reference standard. As recommended in ICH Q6B, an in-house reference standard(s) should always be qualified and used for control of the manufacturing process and product. Thus, the relative potency data for both the pre-change product (reference) and the post-change product (test) are a relative potency of the reference product to the in-house standard and the relative potency of the test product to the in-house standard. The data structure is

$$\begin{array}{l} x_{11}, \cdots, x_{1n1} \leftrightarrow y_{11}, \cdots, y_{1n1} \\ x_{21}, \cdots, x_{2n2} \leftrightarrow y_{21}, \cdots, y_{2n2} \\ \cdots\cdots\cdots\cdots\cdots\cdots\cdots\cdots\cdots\cdots \\ x_{m1}, \cdots, x_{mnm} \leftrightarrow y_{m1}, \cdots, y_{mnm} \end{array}$$

where $x_{ij} = x_i^0 + \delta_{ij}$, $y_{ij} = y_i^0 + \varepsilon_{ij}$, $\delta_{ij} \sim N\left(0, \sigma_\delta^2\right)$, $\varepsilon_{ij} \sim N\left(0, \sigma_\varepsilon^2\right)$, $x_i^0 \sim N\left(\mu_x, \sigma_x^2\right)$, $y_i^0 \sim N\left(\mu_y, \sigma_y^2\right)$, x_{ij} is the observed j^{th} critical quality attribute measurement for the reference product for lot (batch, or run) i, $i = 1, \cdots, m, j = 1, \cdots, n_i$, m is the number of reference lots, and n_i is the number of measurements for reference lot i, y_{ij} is the observed j^{th} critical quality attribute measurement for the test product for lot i, $i = 1, \cdots, m, j = 1, \cdots, n_i$, m is the number of test product lot, and n_i is the number of measurements for the test product lot i. The σ_x^2 and σ_δ^2 are the process variability and the assay variability for the reference product, respectively, and σ_y^2 and σ_ε^2 are the process variability and the assay variability for the test product, respectively. Thus, the observed data are (i, x_{ij}, y_{ij}), where $i = 1, \cdots, m, j = 1, \cdots, n_i$. For lot i, it is reasonable to assume that there exists a linear relationship $y_i^0 = \alpha_i + \beta_i x_i^0$, $i = 1, \cdots, m$. The goal is to compare the two products (test vs reference) and see how similar they are.

Given $N = \sum_{i=1}^{m} n_i$ paired observations (i, x_{ij}, y_{ij}), where x_{ij} is the observation of independent variable from the reference product and y_{ij} is the observation of the response from the test product, $i = 1, \ldots m, j = 1, \ldots, n_i$, and m is the total number of lots. In the current case with the paired observations, consider a linear structural measurement error model (Fuller 1987) for lot (batch, or run) i as follows.

$$y_{ij} = y_i^0 + \varepsilon_{ij} = \alpha_i + \beta_i x_i^0 + \varepsilon_{ij} \tag{26.3}$$

$$x_{ij} = x_i^0 + \delta_{ij} \tag{26.4}$$

where ε_{ij} and δ_{ij} are independent with normal distributions $N(0, \sigma_\varepsilon^2)$ and $N(0, \sigma_\delta^2)$, respectively. Note that both x_i^0 and y_i^0 are still a random variable in a structural measurement error model (Fuller 1987). In order to avoid the unidentifiability problem, the reliability ratio $\lambda = \frac{\sigma_\delta^2}{\sigma_\varepsilon^2}$ is assumed fixed and known. In the comparability study setting, it is reasonable to assume the reliability ratio $\lambda = 1$. The parameters in each source are a sample from a bivariate normal distribution such as

$$(\alpha_i, \beta_i)^T = (\alpha, \beta)^T + \Delta_i, \tag{26.5}$$

where Δ_i is a bivariate normal distribution with mean 0 and covariance matrix $\sum = \begin{pmatrix} \sigma_1^2 & \rho\sigma_1\sigma_2 \\ \rho\sigma_1\sigma_2 & \sigma_2^2 \end{pmatrix}$. Combining Eqs. (26.3), (26.4) and (26.5) results in a linear mixed-effects structural measurement error model.

Following the same ideas laid out for analyzing the unpaired critical quality attributes, a $[100 \times (1 - \alpha)\,\%] / [100 \times p\%]$ tolerance interval for the difference of the test and the reference and a plausibility interval will be constructed. Thus, the assay + process PI for the difference between the reference against the reference itself is defined as follows:

$$\left(-k\sqrt{2\left(\sigma_x^2 + \sigma_\delta^2\right)}, +k\sqrt{2\left(\sigma_x^2 + \sigma_\delta^2\right)}\right) \tag{26.6}$$

where the critical value k is a factor to control the sponsor's tolerance, the σ_x^2 and σ_δ^2 are the process variability and the assay variability for the reference product, respectively. The PI defines the acceptable range for the difference between the test and the reference to fall within and any difference within this PI should be considered practically acceptable. Following the same recommendation for the non-paired case, the $k = 2.5$ or 3 is recommended for constructing the PI in Eq. (26.6). However, the actual k value can be chosen case-by-case depending on the nature of the reference product and still be in alignment with health authorities' requirements.

Again, an at least p-content with confidence $1 - \alpha$ tolerance interval for the difference between the test and the reference will be constructed so that

$$\Pr\left[\Pr\left(L < Y - X < U\right) \geq p\right] \geq 1 - \alpha \tag{26.7}$$

where L and U are statistics calculated from the data, X(Y) is the quality attribute of the reference (test). The interval (L, U) obtained from above is usually termed as the approximate $[100 \times (1 - \alpha)\,\%] / [100 \times p\%]$ TI, where $L = \widehat{\mu}_y - \widehat{\mu}_x - k_2 s$, $U = \widehat{\mu}_y - \widehat{\mu}_x + k_2 s$, $\widehat{\mu}_y = \widehat{\alpha} + \widehat{\beta} \times \widehat{\mu}_x$, s is the estimate for the standard deviation of the $\widehat{\mu}_y - \widehat{\mu}_x$ and can be estimated using the following formula

$$\widehat{\mu}_y - \widehat{\mu}_x = \widehat{\alpha} + \left(\widehat{\beta} - 1\right)\widehat{\mu}_x,$$
$$\mathrm{var}\left(\widehat{\mu}_y - \widehat{\mu}_x\right) = E\left(\mathrm{var}\left(\widehat{\mu}_y - \widehat{\mu}_x \middle| \widehat{\mu}_x\right)\right) + Var\left(E\left(\widehat{\mu}_y - \widehat{\mu}_x \middle| \widehat{\mu}_x\right)\right).$$

Assuming a normally distributed population, an approximate value for the k_2 factor as a function of p and $1-\alpha$ for a two-sided tolerance interval (Howe 1969) is

$$k_2 = \sqrt{\frac{v\left(1+\frac{1}{N}\right)z^2_{(1-p)/2}}{\chi^2_{v,1-\alpha}}}$$

where $\chi^2_{v,1-\gamma}$ is the critical value of the chi-square distribution with degrees of freedom ν that is exceeded with probability γ, and $Z_{(1-p)/2}$ is the critical value of the normal distribution associated with cumulative probability $(1-p)/2$. The quantity ν represents the degrees of freedom used to estimate the standard deviation. Most of the time the same sample will be used to estimate both the mean and standard deviation so that $\nu = N-1$, but the formula allows for other possible values of ν.

Similar to the non-paired case, the test and the reference are claimed comparable if (1) the approximate $[100 \times (1-\alpha)\,\%] / [100 \times p\%]$ tolerance interval for the difference between the test and the reference defined in Eq. (26.7) is within the Plausibility interval for the difference between the reference against the reference itself defined in Eq. (26.6); and (2) if the estimated mean ratio is within a specified boundary, for example, [0.8, 1.25].

Note that the assay + process variability interval in the paired head-to-head study and the assay + process variability interval for the non-paired head-to-head study have the same form. However, one of the two quantities in the paired head-to-head setting is much smaller than the value from the non-paired head-to-head setting. This is why the paired head-to-head setting is more powerful than the non-paired head-to-head setting and the paired head-to-head setting is more preferred if possible.

There are different ways to estimate the parameters in the linear mixed-effects structural measurement error model. Fan and Zhang (2000) developed a two-step approach. They estimate the parameters $(\alpha_i, \beta_i)^T$ in each source and then smooth these estimates from each source and derive the final estimate, called the two-step method. In other words, estimates $\widehat{\alpha}_i$ and $\widehat{\beta}_i$ can be obtained through Eqs. (26.3) and (26.4) for each lot as a linear measurement error model. Therefore, $\widehat{\alpha}_i$ and $\widehat{\beta}_i$ have a normal distribution $N(\alpha, w^2_{1i})$ and $N(\beta, w^2_{2i})$, respectively. Then a smoothed estimator, here a weighted estimator is given as

$$\widehat{\alpha}_{TS} = \frac{\sum_{i=1}^m \widehat{\alpha}_i / w^2_{1i}}{\sum_{i=1}^m 1/w^2_{1i}},$$

$$\widehat{\beta}_{TS} = \frac{\sum_{i=1}^m \widehat{\beta}_i / w^2_{2i}}{\sum_{i=1}^m 1/w^2_{2i}},$$

where $\widehat{\alpha}_{TS}$ and $\widehat{\beta}_{TS}$ are the estimates for the intercept and the slope, respectively, from the two-step method. An alternative approach (Tosteson et al. 1998; Buonaccorsi et al. 2000) is to use an estimate of $\widehat{x}_{ij}^0$ and then using observations $\left(i, \widehat{x}_{ij}^0, y_{ij}\right)$ to fit a regular linear mixed-effects model, called the regression calibration. The popular choice of $\widehat{x}_{ij}^0$ can be obtained from Eqs. (26.3) and (26.4) using a linear measurement error model for each lot by the following formula (see Fuller 1987; Casella and Berger 1990).

$$\widehat{x}_{ij}^0 = \frac{\lambda x_{ij} + \widehat{\beta}_i \left(y_{ij} - \widehat{\alpha}_i\right)}{\lambda + \widehat{\beta}_i^2}.$$

Note that the variance obtained from the regression calibration approach usually underestimates the true variance. The variance from the regression calibration approach is a conditional variance given the estimate $\widehat{x}_{ij}^0$. Thus, the true variance should be calculated based on the following formula.

$$Var\left(\widehat{\beta}_{RC}\right) = E\left(Var\left(\widehat{\beta}_{RC} \middle| \widehat{x}_{ij}^0\right)\right) + Var\left(E\left(\widehat{\beta}_{RC} \middle| \widehat{x}_{ij}^0\right)\right),$$

$$Var\left(\widehat{\alpha}_{RC}\right) = E\left(Var\left(\widehat{\alpha}_{RC} \middle| \widehat{x}_{ij}^0\right)\right) + Var\left(E\left(\widehat{\alpha}_{RC} \middle| \widehat{x}_{ij}^0\right)\right),$$

where $\widehat{\alpha}_{RC}$ and $\widehat{\beta}_{RC}$ are the estimates for the intercept and the slope, respectively, from the regression calibration approach. The second part should be added in the variance calculation. The second part of the variance in the regression calibration approach is $Var\left(E\left(\widehat{\beta}_{RC} \middle| \xi_{ij}\right)\right) = \frac{1}{m-1}\sum_{i-1}^{m}\left(\widehat{\beta}_i - \overline{\widehat{\beta}}\right)^2$, which is the sample variance of $\left(\widehat{\beta}_{1,} \ldots, \widehat{\beta}_m\right)$ obtained from each source and it measures the variation among different sources, where $\overline{\widehat{\beta}} = \frac{1}{m}\sum_{i=1}^{m}\widehat{\beta}_i$. Similarly, one can estimate $Var\left(E\left(\widehat{\alpha}_{RC} \middle| \widehat{x}_{ij}^0\right)\right) = \frac{1}{m-1}\sum_{i=1}^{m}\left(\widehat{\alpha}_i - \overline{\widehat{\alpha}}\right)^2$, where $\overline{\widehat{\alpha}} = \frac{1}{m}\sum_{i=1}^{m}\widehat{\alpha}_i$. Liao et al. (2005) conducted a simulation comparison of the regression calibration approach and the two-step approach and found that the regression calibration approach and the two-step approach have a similar mean squared error (MSE) performance but the regression calibration approach has a smaller bias than that of two-step approach. Therefore, the regression calibration approach is preferred.

Zhong et al. (2002) presented a unified method based on the corrected score function for the linear mixed-effects model with measurement error only in the fixed effect. Cui et al. (2004) presented the method of moments approach for the linear mixed-effects model with measurement error in both the fixed and the random effects.

26.2.1.3 Only Post-Change Analytical Comparability

Sometimes, the manufacturer may just generate data for the proposed post-change product (test) in an early development stage and then compare the generated data of the proposed test product to the historical data of the pre-change product (reference). In this case, the data structure is

$$\begin{array}{l} y_{11}, \cdots, y_{1n1} \\ y_{21}, \cdots, y_{2n2} \\ \cdots\cdots\cdots\cdots \\ y_{m1}, \cdots, y_{mnm} \end{array}$$

where $y_{kj} = y_k^0 + \varepsilon_{kj}$, $\varepsilon_{kj} \sim N\left(0, \sigma_\varepsilon^2\right)$, $y_k^0 \sim N\left(\mu_y, \sigma_y^2\right)$, $j = 1, \cdots, n_k$, $k = 1, \cdots, m$. There are two types of variabilities involved in the observed data set. They are the assay variability and the process variability. The process variability is defined by σ_y^2 for the proposed post-change product. The assay variability is defined by σ_ε^2 for the proposed test product.

To evaluate this type of comparability study, a random-effects statistical model is used as follows.

$$y_{kj} = y_k^0 + \varepsilon_{kj},\ \ \varepsilon_{kj} \sim N\left(0, \sigma_\varepsilon^2\right),\ j = 1, \cdots, n_k$$

$$y_k^0 \sim N\left(\mu_y, \sigma_y^2\right),\ k = 1, \cdots, m.$$

Comparability is claimed if (1) the approximate $[100 \times (1-\alpha)\ \%]\ /\ [100 \times p\%]$ for the test product is within the specification for the reference product; and (2) the proposed test has the same trend as the reference product.

Depending on whether the specification for the reference product is one-sided or two-sided, a corresponding one-sided or two-sided TI can be constructed accordingly (Hoffman 2010; Hoffman and Kringle 2005). The statistical process control chart can be used to assess the trend. Note that this type of comparability study has less power compared to the head-to-head studies and thus is least preferable.

26.2.2 *In Vivo Pharmacokinetics (PK) Aspects*

Under certain circumstances, a single-dose study in animals comparing the proposed post-change product (test) and pre-change product (reference) using PK and PD measurements may contribute to the totality of evidence that supports a demonstration of comparability (ICH 2005). Specifically, the product manufacturer can use results from animal studies to support the degree of comparability based on PK and PD profiles of the proposed test product and the reference product. Where appropriate, PK and PD measurements also can be incorporated into a single animal toxicity study. Animal PK and PD assessment will not negate the need for human PK and PD studies.

If a preclinical comparability study is required after evaluating the analytical data, a pharmacologically relevant rodent species should be chosen for such a study. However, if the rodent species is not pharmacologically relevant, a higher species may need to be used. If no pharmacologically relevant species are available, transgenic rodents expressing functional human targets may be appropriate.

To conduct the animal study, a parallel design is preferred over a cross-over design to demonstrate the comparability of therapeutic proteins because a carryover effect of immune responses may develop towards the previously administered protein. In addition, a significantly longer half-life ($t_{1/2}$) of protein therapeutics requires a long wash-out period. Duration of the PK comparability study is based on the known half-life ($t_{1/2}$) of the molecule. In general, about 4–5 half-lives are required for the wash out period to adequately determine the PK profiles free of carryover effects.

The number of animals to be used in the comparability exercise is usually small and should be determined based on the variability of the key PK parameters. In general, PK results with intravenous (IV) administration tend to be less variable than those of subcutaneous (SC) administration; therefore fewer animals for IV than for SC route of administration are required to demonstrate comparability statistically. The intended route of administration should be used for the comparability study. In general, if the variability (%CV) of key PK parameters such as area under the concentration curve (AUC), maximum or peak concentration (C_{max}), or clearance (CL) is less than 20 % (usually the case for IV administration), about 10 animals per each product are adequate to demonstrate comparability. If the %CV is >40 % (usually the case for SC administration), about 15 animals are recommended.

The method for analyzing clinical pharmacokinetic data is usually borrowed for analyzing animal pharmacokinetic data. Typically, to demonstrate comparability of PK measurements of the proposed test product and the reference product, a bioequivalence (BE) approach will be used. To demonstrate bioequivalence, the 90 % confidence interval (CI) for the estimated ratio of geometric means (GMR) for both AUC and C_{max} of two products must fall between the predefined limits 0.80 and 1.25 (FDA 2001, 2014). However, due to the usually small sample size in an animal study and the high variability for biological products because of the complexity and the possible pleiotropic activities of biologics, the CI of the estimated GMR will be wide. Consequently, there is a high probability of falling out of the predetermined limits (0.8, 1.25). To reduce this high probability of producer risk, many different approaches have been proposed for high variability drug products with a reasonable sample size requirement. The type of acceptance criteria (reference-scaled method) proposed by the FDA working group rather than the traditional bioequivalent criteria is recommended for the demonstration of PK comparability in this situation. The acceptance criteria are based on the variability of reference material and generally reduce the producer's risk without increasing the consumer risk. It is important to note that sometimes, a smaller CI, for example, 80 % CI may be used for animal studies.

Let μ_T and μ_R be the average logarithmic pharmacokinetic response (i.e., AUC or C_{max}) of the test and the reference products, respectively. The classical paradigm for the determination of similarity of a chemical generic product to the original product is based on the 90 % confidence interval of the difference $\mu_T - \mu_R$ falling within the predetermined acceptable interval (log(0.8), log(1.25)), i.e., (−0.223, +0.223). That is

$$|\mu_T - \mu_R| \leq 0.223 \tag{26.8}$$

This approach does not depend on the underlying variability, or the ratio of geometric means of the PK parameters.

To reduce the producer risk for the case with small sample size and high variability, Haidar et al. (2007, 2008) compared several approaches and proposed the following FDA ACPS reference-scaled acceptance criterion

$$(\mu_T - \mu_R)^2/\sigma_{WR}^2 - \theta_F \leq 0 \tag{26.9}$$

where $\theta_F = \frac{(\log(1.25))^2}{\sigma_{W0}^2} = \frac{(\log(1.25))^2}{0.25^2} \approx 0.8$ using a "switching" variability $\sigma_{W0} = 0.25$, σ_{WR}^2 is the within-subject variance for the reference product in a cross-over design but the total variance for the reference product in a parallel design. Comparability is claimed if the upper 95 % CI for the statistic on the left side of Eq. (26.9) is negative or zero, and the point estimate of GMR (test/reference geometric mean ratio) falls within (0.8, 1.25). The authors claimed that this FDA ACPS reference-scaled approach effectively decreases the sample size needed for a demonstration of comparability. The additional point-estimate constraint eliminates the potential that a test product with a large mean difference would enter the market.

There are two kinds of risks in evaluating the PK comparability: One is the consumer risk which requires a larger switching variability, and the other is the sponsor risk which requires a smaller switching variability. However, the FDA ACPS approach uses only one switching variability. When there is no mean difference in the reference and the test compounds, the ideal test will have a high probability of accepting comparability. So if the reference compound is compared to the reference itself (i.e., GMR = 1) then it should always be comparable regardless of the variability. However, for this case, the FDA ACPS method only gives 74 % chance of accepting comparability, and leads to a large producer's risk. In other words, the "switching" variability σ_{W0} is too large in this case and a smaller "switching" variability should be used. In the other direction, a "switching" variability that is too small will lead to a large consumer's risk. For a more reasonable balance and control of the two risks, Liao and Heyse (2011) proposed using two "switching" variabilities: one for controlling the producer's risk when GMR = 1 and a high accepting comparability rate is desired, and a second one for controlling the consumer's risk when GMR $\geq$ 1.25 and a low accepting comparability rate is desired. For this purpose, Liao and Heyse (2011) proposed a new reference-scaled approach using the acceptance criterion

$$(\mu_T - \mu_R)^2/\sigma_{WR}^2 - 2\theta_L(GMR) \leq 0 \tag{26.10}$$

where $\theta_L(GMR) = \left[\log(1.25)/\left(\sqrt{2} \times (3.763 \times GMR - 2.763) \times 0.152\right)\right]^2$, σ_{WR}^2 is the within-subject variance for the reference product in a cross-over design but the total variance for the reference product in a parallel design. Note that GMR should always be greater than or equal to 1 and the inverse should be used in (26.10) if GMR < 1. Comparability is claimed if the upper 95 % CI for the statistic in Eq. (26.10) is negative or zero.

For the criteria in the form of Eqs. (26.9) and (26.10), an approximate linearized regulatory model (Hyslop et al. 2000) $\eta = (\mu_T - \mu_R)^2 - \theta\sigma_R^2 < 0$ is suggested. However, Tothfalusi and Endrenyi (2003) used simulations and demonstrated that the approximate linearized regulatory model and the non-central t-distribution or the approximate normal distribution for a larger number of subjects gave very good agreement in terms of the study acceptance passing comparability. Thus, a normal approximation can be used to calculate the confidence limits rather than the more complicated linearized regulatory model.

The performance of the two reference scaled approaches were compared using the simulation and an example. There are several advantages for using the new reference-scaled approach in Eq. (26.10) (Liao and Heyse 2011). First, when there is no subject-by-product interaction and no mean shift, this new criterion using a "switching" variability $\sigma_{W0} = 0.152$ leads to an 85 % accepting comparability percentage and thereby, it reduces the producer's risk compared to the FDA ACPS. Second, when there is a mean shift at the GMR = 1.25, the "switching" variability is set at $\sigma_{W0} = 0.292$ for controlling the consumer's risk. Third, this new test increases the sensitivity to deviations between the within-subject variance. Fourth, the new approach assigns a reasonable penalty for a test product exhibiting a larger variance than the reference. Finally, the approach (26.10) gives a narrower boundary for a larger GMR, which is very reasonable since the probability of claiming comparability should be getting smaller when GMR is deviating from 1. The two "switching" variabilities in this new reference scaled method in (26.10) can be optimized based on a specified consumer's risk and a producer's risk, or based on the mutually agreed risks between the producer and the regulatory agencies for a specified biologic. For two given "switching" variabilities σ_{w1} and σ_{w2}, the $\theta_L(GMR)$ in Eq. (26.10) should be

$$\theta_L(GMR) = \left[\log(1.25)/\left(\sqrt{2} \times ((a+1) \times GMR - a) \times \sigma_{w1}\right)\right]^2,$$

where $a = 4 \times \sigma_{w2}/\sigma_{w1} - 5$, σ_{w1} is the smaller "switching" variability controlling the producer's risk and σ_{w2} is the larger "switching" variability controlling the consumer's risk.

26.3 An Illustration

Consider a hypothetical comparability study of the post-change (test) product against the pre-change (reference) cancer product. The paired relative potency values of the reference product and the test product against a house standard were measured from each of three reference product lots. The raw data are plotted in Fig. 26.2 and summarized in Table 26.1. Figure 26.2 indicates that there is some lot variability among the three lots and Table 26.1 indicates both the test and the reference products have similar relative potency value but the reference has slightly less variability.

Following the method mentioned in Sect. 26.2.1.2 for paired quality attributes, the estimated intercept and slope using the regression calibration approach from the linear mixed-effects structural measurement error model are summarized in Table 26.2 which confirms the lot variability. Based on the data, the between lot variability is $\widehat{\sigma}_x = 0.296$ and the within lot variability is $\widehat{\sigma}_\delta = 0.064$.

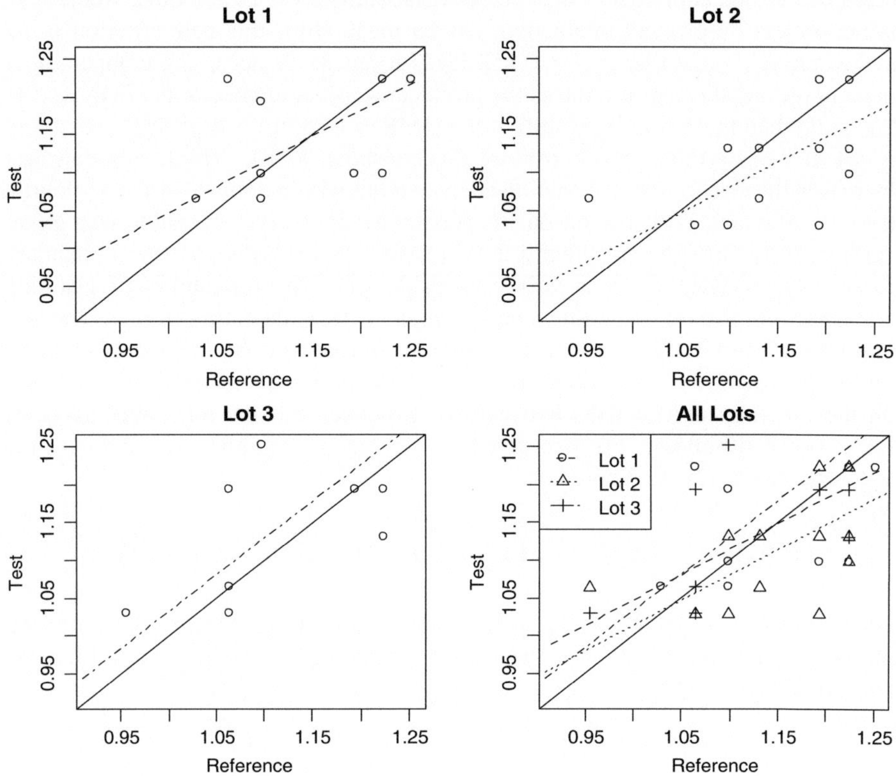

Fig. 26.2 Raw data for the relative potency for three lots. The *solid line* is the identical line and the *non-solid line* is the regression line

Table 26.1 Summary statistics for the relative potency data

	Mean (SD)			
	Lot 1 (n = 9)	Lot 2 (n = 14)	Lot 3 (n = 10)	Combined (n = 33)
Test	1.14 (0.081)	1.15 (0.077)	1.13 (0.091)	1.14 (0.080)
Reference	1.14 (0.071)	1.11 (0.063)	1.16 (0.089)	1.13 (0.074)

Table 26.2 Estimated parameters for the relative potency data

	Lot 1	Lot 2	Lot 3	Combined
Intercept	0.370	0.344	0.059	0.257
Slope	0.677	0.670	0.973	0.773

The Plausibility Interval (PI) for the difference between the reference against the reference itself is (−0.856, +0.856) for $k = 2$ and (−1.283, +1.283) for $k = 3$, respectively. Based on the data, $\widehat{\mu}_x = 1.139$ and $\widehat{\mu}_y = 1.137$. The estimated standard deviation of the $\widehat{\mu}_y - \widehat{\mu}_x$ is $\widehat{s} = 0.097$. The critical value for the 90%/95% TI is $k_2 = 1.725$ with the degrees of freedom $\nu = 32$. The 90%/95% Tolerance Interval (TI) for the difference between the test and the reference (test-reference) is (−0.170, +0.166). According to the acceptance criteria, the TI for the difference between the test and the reference is within the PI for the difference of the reference against the reference itself, and the estimated mean ratio 0.998 is within the boundary [0.8, 1.25]. Thus, the test product is comparable in terms of the relative potency compared to the reference product.

26.4 Summary and Discussion

Manufacturers of biological products frequently make changes to manufacturing processes both during development and after approval. It is a regulatory requirement for the sponsor to demonstrate product comparability between the post-change and pre-change products and to ensure the modifications have not adversely impacted the safety and efficacy of the drug product. Establishing comparability is a stepwise approach and depends on the totality of evidence with the use of fingerprint-like techniques for extensive characterization. A direct head to head comparison between the post-change and the pre-change is crucial and it begins with the in vivo and in vitro critical quality attributes, and ends with the clinical efficacy comparison, if needed.

When planning a comparability study, a careful evaluation of all foreseeable impacts and consequences of the potential differences of the proposed post-change product to the pre-change product should be performed. To achieve this, adequately sensitive bioanalytical methods to detect small but clinically meaningful difference are critical. General aspects to consider when designing a comparability exercise are:

- Potential impact to product quality, safety, and efficacy
- Nature of the potential difference in the manufacturing process comparing to the pre-change reference product
- Stage of development
- Complexity of product
- Understanding of mechanism of action
- Suitability and capability of analytical methods
- Acceptance criteria
- Preclinical experience
- Clinical characteristics
- Regulatory authority guidance

The Quality by Design (QbD) and associated concept of Design Space (DS) as outlined in ICH Q8, and robust process techniques will have a positive impact on the scope of comparability assessments. A QbD submission has the potential to reduce the intensity and frequency of incidents where comparability studies are necessary, since some process changes or modifications or deviations are allowed through the establishment of knowledge management and data supported design space. The design of a comparability study is based on the knowledge of the reference product and the process, and an understanding of how specific process parameters impacts product quality. The documentation generated at each stage of the process development, validation, and post-approval assessment lifecycle of the product is essential for effective internal and external communication in complex, lengthy, and multidisciplinary projects.

The criteria for a comparability study should be established at the planning stage. Likewise, defining critical quality attributes (CQAs) and performing a risk assessment will be required to plan the comparability study appropriately. To ensure consistency in comparability study practice, a statistically based systematic approach is recommended. The comparison of the results to the predefined acceptance criteria allows an objective assessment of whether or not the two products are comparable. The statistical approaches defined in this chapter provide tools for this purpose. Different aspects of the post-change product and the available statistical approaches rely on the pre-change product (reference) variability. The reference against reference boundary was used as a goalpost in deriving the acceptance criteria. Simulations and examples have been used for the demonstration of the feasibility of these proposed statistical methods. These statistical methods can be useful to comparability studies.

Disclaimers The thoughts and opinions presented in this chapter only represent the author's positions.

Acknowledgements The author thanks Drs. Stan Altan, Rick Burdick, Lisa Hendricks and Lanju Zhang for their valuable comments that dramatically improved the presentation of this chapter.

References

Buonaccorsi J, Demidenko E, Tosteson T (2000) Estimation in longitudinal random effects models with measurement error. Statistica Sinica 10:885–903

Casella G, Berger RL (1990) Statistical inference. Duxbury Press, Belmont, CA

Chatfield MJ, Borman PJ, Damjanov I (2011) Evaluating change during pharmaceutical product development and manufacture-comparability and equivalence. Qual Reliability Eng Int 17: 629–640

Cui H, Ng KW, Zhu L (2004) Estimation in mixed effects model with errors in variables. J Multivar Anal 91:53–73

EMA Committee for Medicinal Products for Human Use (CHMP) (2005) Guideline on similar biological medicinal products. London

Fan J, Zhang JT (2000) Two-step estimation of functional linear models with applications to longitudinal data. J R Stat Soc B 62:303–322

Food and Drug Administration (FDA) of U.S.A. (1996) Demonstration of comparability of human biological products, including therapeutic biotechnology-derived products. Washington, D.C.

Food and Drug Administration (FDA) of U.S.A. (2001) Guidance for industry: statistical approaches to establishing bioequivalence. Washington, D.C.

Food and Drug Administration (FDA) of U.S.A. (2003), Guidance for industry: comparability protocols—chemistry, manufacturing, and controls information. Washington, D.C.

Food and Drug Administration (FDA) of U.S.A. (2005) Guidance for industry: Q5E comparability of biotechnological/biological products subject to changes in their manufacturing process. Washington, D.C.

Food and Drug Administration (FDA) of U.S.A. (2012a) Guidance for industry: scientific considerations in demonstrating biosimilarity to a reference product. Washington, D.C. February 2012

Food and Drug Administration (FDA) of U.S.A. (2012b) Guidance for industry: quality considerations in demonstrating biosimilarity to a reference product. Washington, D.C. February 2012

Food and Drug Administration (FDA) of U.S.A. (2012c) Guidance for industry: biosimilars: questions and answers regarding implementation of the biologics price competition and innovation act of 2009. Washington, D.C. February 2012

Food and Drug Administration (FDA) of U.S.A. (2014) Guidance for industry: clinical pharmacology data to support a demonstration of biosimilarity to a reference product. Washington, D.C. May 2014

Fuller WA (1987) Measurement error models. Wiley, New York

Genetic Engineering & Biotechnology News (GEN, 2013) Top 20 best-selling drugs of 2012. March 5, 2013. Website: www.genengnews.com/insight-and-intelligence/top-20-best-selling-drugs-of2012/77899775/?page=2

Genetic Engineering & Biotechnology News (GEN, 2014) The top 25 best-selling drugs of 2013. September 16, 2014. Website: http://www.genengnews.com/insight-and-intelligence/the-top-25-best-selling-drugs-of-2013/77900053/

Haidar SH et al (2007) Bioequivalence approaches for highly viable drugs and drug products. Pharm Res 25:237–241

Haidar SH et al (2008) Evaluation of scaling approach for the bioequivalence of highly variable drugs. AAPS J 10:450–454

Hoffman D (2010) One-sided tolerance limits for balanced and unbalanced random effects models. Technometrics 52(3):303–312

Hoffman D, Kringle R (2005) Two-sided tolerance intervals for balanced and unbalanced random effects models. J Biopharm Stat 15:283–293

Howe WG (1969) Two-sided tolerance limits for normal populations – some improvements. J Am Stat Assoc 64:610–620

Hyslop T, Hsuan F, Holder DJ (2000) A small sample confidence interval approach to assess individual bioequivalence. Stat Med 19:2885–2897

ICH guidance (2005): Q5E comparability of biotechnological/biological products subject to changes in their manufacturing process. June 2005

Krishnamoorthy K, Mathew T (2009) Statistical tolerance regions: theory, applications, and computation. Wiley, Hoboken

Krishnamoorthy K, Lian X, Mondal S (2011) Tolerance intervals for the distribution of the difference between two independent normal random variables. Commun Stat Theory Methods 40(1):117–129

Liao JJZ, Darken PF (2013) Comparability of critical quality attributes for establishing biosimilarity. Stat Med 32:462–469

Liao JJZ, Heyse JF (2011) Biosimilarity for follow-on biologics. Stat Biopharm Res 3(3):445–455

Liao JJZ, Schofield TL, Bennett PS (2005) Analyzing highly variable potency data using a linear mixed-effects measurement error model. J Agric Biol Environ Stat 4:388–397

McCamish M, Woollett G (2011) Worldwide experience with biosimilar development. mAbs 3(2):209–217

McCamish M, Woollett G (2013) The continuum of comparability extends to biosimilarity: how much is enough and what clinical data are necessary? Clin Pharmacol Ther 93(4):315–317

Tosteson T, Buonaccorsi J, Demidenko E (1998) Covariate measurement error and the estimation of random effect parameters in a mixed model for longitudinal data. Stat Med 17:1959–1971

Tothfalusi L, Endrenyi L (2003) Limits for the scaled average bioequivalence of highly variable drugs and drug products. Pharm Res 20:382–389

Weise M et al (2012) Biosimilars: what clinicians should know. Blood 120:5111–5117

World Health Organization (WHO) (2009) Guidelines on evaluation of similar biotherapeutic products (SBPs). World Health Organization, Geneva

Zhong XP, Fung WK, Wei BC (2002) Estimation in linear models with random effects and errors-in-variables. Ann Inst Stat Mat 54(3):595–606

Index

L. Zhang (ed.), *Nonclinical Statistics for Pharmaceutical and Biotechnology Industries*, Statistics for Biology and Health, DOI 10.1007/978-3-319-23558-5

MIX
Papier aus verantwortungsvollen Quellen
Paper from responsible sources
FSC® C105338

If you have any concerns about our products,
you can contact us on
ProductSafety@springernature.com

In case Publisher is established outside the EU,
the EU authorized representative is:
Springer Nature Customer Service Center GmbH
Europaplatz 3, 69115 Heidelberg, Germany

Printed by Libri Plureos GmbH
in Hamburg, Germany